Student's Solutions Manual
Single Variable

Mark Woodard
Furman University

Calculus
Early Transcendentals
Third Edition

William Briggs
University of Colorado at Denver

Lyle Cochran
Whitworth University

Bernhard Gillett
University of Colorado, Boulder

Eric Schulz
Walla Walla Community College

The author and publisher of this book have used their best efforts in preparing this book. These efforts include the development, research, and testing of the theories and programs to determine their effectiveness. The author and publisher make no warranty of any kind, expressed or implied, with regard to these programs or the documentation contained in this book. The author and publisher shall not be liable in any event for incidental or consequential damages in connection with, or arising out of, the furnishing, performance, or use of these programs.

Reproduced by Pearson from electronic files supplied by the author.

Copyright © 2019, 2015, 2011 Pearson Education, Inc.
Publishing as Pearson, 330 Hudson Street, NY NY 10013

All rights reserved. No part of this publication may be reproduced, stored in a retrieval system, or transmitted, in any form or by any means, electronic, mechanical, photocopying, recording, or otherwise, without the prior written permission of the publisher. Printed in the United States of America.

1 18

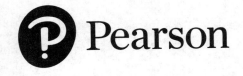

ISBN-13: 978-0-13-477048-2
ISBN-10: 0-13-477048-X

Contents

1 Functions — 3
  1.1 Review of Functions — 3
  1.2 Representing Functions — 7
  1.3 Inverse, Exponential and Logarithmic Functions — 16
  1.4 Trigonometric Functions and Their Inverses — 21
  Chapter One Review — 28

2 Limits — 37
  2.1 The Idea of Limits — 37
  2.2 Definition of a Limit — 40
  2.3 Techniques for Computing Limits — 48
  2.4 Infinite Limits — 53
  2.5 Limits at Infinity — 58
  2.6 Continuity — 65
  2.7 Precise Definitions of Limits — 72
  Chapter Two Review — 77

3 Derivatives — 83
  3.1 Introducing the Derivative — 83
  3.2 The Derivative as a Function — 87
  3.3 Rules of Differentiation — 97
  3.4 The Product and Quotient Rules — 102
  3.5 Derivatives of Trigonometric Functions — 107
  3.6 Derivatives as Rates of Change — 112
  3.7 The Chain Rule — 120
  3.8 Implicit Differentiation — 126
  3.9 Derivatives of Logarithmic and Exponential Functions — 134
  3.10 Derivatives of Inverse Trigonometric Functions — 140
  3.11 Related Rates — 145
  Chapter Three Review — 152

4 Applications of the Derivative — 159
  4.1 Maxima and Minima — 159
  4.2 Mean Value Theorem — 166
  4.3 What Derivatives Tell Us — 170
  4.4 Graphing Functions — 179
  4.5 Optimization Problems — 197
  4.6 Linear Approximation and Differentials — 207
  4.7 L'Hôpital's Rule — 212
  4.8 Newton's Method — 219
  4.9 Antiderivatives — 226
  Chapter Four Review — 231

5 Integration — 243
- 5.1 Approximating Areas under Curves — 243
- 5.2 Definite Integrals — 254
- 5.3 Fundamental Theorem of Calculus — 264
- 5.4 Working with Integrals — 273
- 5.5 Substitution Rule — 278
- Chapter Five Review — 284

6 Applications of Integration — 293
- 6.1 Velocity and Net Change — 293
- 6.2 Regions Between Curves — 300
- 6.3 Volume by Slicing — 307
- 6.4 Volume by Shells — 311
- 6.5 Length of Curves — 316
- 6.6 Surface Area — 319
- 6.7 Physical Applications — 323
- Chapter Six Review — 328

7 Logarithmic, Exponential, and Hyperbolic Functions — 337
- 7.1 Logarithmic and Exponential Functions Revisited — 337
- 7.2 Exponential Models — 341
- 7.3 Hyperbolic Functions — 344
- Chapter Seven Review — 350

8 Integration Techniques — 353
- 8.1 Basic Approaches — 353
- 8.2 Integration by Parts — 358
- 8.3 Trigonometric Integrals — 367
- 8.4 Trigonometric Substitutions — 372
- 8.5 Partial Fractions — 380
- 8.6 Integration Strategies — 387
- 8.7 Other Methods of Integration — 404
- 8.8 Numerical Integration — 409
- 8.9 Improper Integrals — 415
- Chapter Eight Review — 422

9 Differential Equations — 433
- 9.1 Basic Ideas — 433
- 9.2 Direction Fields and Euler's Method — 436
- 9.3 Separable Differential Equations — 442
- 9.4 Special First-Order Linear Differential Equations — 447
- 9.5 Modeling with Differential Equations — 451
- Chapter Nine Review — 455

10 Sequences and Infinite Series — 461
- 10.1 An Overview — 461
- 10.2 Sequences — 465
- 10.3 Infinite Series — 472
- 10.4 The Divergence and Integral Tests — 478
- 10.5 Comparison Tests — 483
- 10.6 Alternating Series — 487
- 10.7 The Ratio and Root Tests — 490
- 10.8 Choosing a Convergence Test — 493
- Chapter Ten Review — 502

11 Power Series — **509**
- 11.1 Approximating Functions With Polynomials 509
- 11.2 Properties of Power Series 516
- 11.3 Taylor Series 520
- 11.4 Working with Taylor Series 528
- Chapter Eleven Review 536

12 Parametric and Polar Curves — **541**
- 12.1 Parametric Equations 541
- 12.2 Polar Coordinates 549
- 12.3 Calculus in Polar Coordinates 559
- 12.4 Conic Sections 567
- Chapter Twelve Review 578

Copyright © 2019 Pearson Education, Inc.

Chapter 1

Functions

1.1 Review of Functions

1.1.1 A function is a rule that assigns each to each value of the independent variable in the domain a unique value of the dependent variable in the range.

1.1.3 Graph A does not represent a function, while graph B does. Note that graph A fails the vertical line test, while graph B passes it.

1.1.5 Item i. is true while item ii. isn't necessarily true. In the definition of function, item i. is stipulated. However, item ii. need not be true – for example, the function $f(x) = x^2$ has two different domain values associated with the one range value 4, because $f(2) = f(-2) = 4$.

1.1.7 The domain of this function is the set of a real numbers. The range is $[-10, \infty)$.

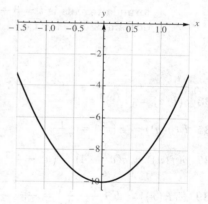

1.1.9 The independent variable h is the height of the water in the tank and the dependent variable V is the volume of water in the tank. The domain in context is $[0, 50]$

1.1.11 $f(g(1/2)) = f(-2) = -3$; $g(f(4)) = g(9) = \frac{1}{8}$; $g(f(x)) = g(2x+1) = \dfrac{1}{(2x+1)-1} = \dfrac{1}{2x}$.

1.1.13 The domain of $f \circ g$ consists of all x in the domain of g such that $g(x)$ is in the domain of f.

1.1.15

a. $(f \circ g)(2) = f(g(2)) = f(2) = 4$.

b. $g(f(2)) = g(4) = 1$.

c. $f(g(4)) = f(1) = 3$.

d. $g(f(5)) = g(6) = 3$.

e. $f(f(8)) = f(8) = 8$.

f. $g(f(g(5))) = g(f(2)) = g(4) = 1$.

5

1.1.17 $\dfrac{f(5) - f(0)}{5 - 0} = \dfrac{83 - 6}{5} = 15.4$; the radiosonde rises at an average rate of 15.4 ft/s during the first 5 seconds after it is released.

1.1.19 $f(-2) = f(2) = 2$; $g(-2) = -g(2) = -(-2) = 2$; $f(g(2)) = f(-2) = f(2) = 2$; $g(f(-2)) = g(f(2)) = g(2) = -2$.

1.1.21 Function A is symmetric about the y-axis, so is even. Function B is symmetric about the origin so is odd. Function C is symmetric about the y-axis, so is even.

1.1.23 $f(x) = \dfrac{x^2 - 5x + 6}{x - 2} = \dfrac{(x-2)(x-3)}{x-2} = x - 3$, $x \ne 2$. The domain of f is $\{x : x \ne 2\}$. The range is $\{y : y \ne -1\}$.

1.1.25 The domain of the function is the set of numbers x which satisfy $7 - x^2 \ge 0$. This is the interval $[-\sqrt{7}, \sqrt{7}]$. Note that $f(\sqrt{7}) = 0$ and $f(0) = \sqrt{7}$. The range is $[0, \sqrt{7}]$.

1.1.27 Because the cube root function is defined for all real numbrs, the domain is $\mathbb{R}$, the set of all real numbers.

1.1.29 The domain consists of the set of numbers x for which $9 - x^2 \ge 0$, so the interval $[-3, 3]$.

1.1.31 a. The formula for the height of the rocket is valid from $t = 0$ until the rocket hits the ground, which is the positive solution to $-16t^2 + 96t + 80 = 0$, which the quadratic formula reveals is $t = 3 + \sqrt{14}$. Thus, the domain is $[0, 3 + \sqrt{14}]$.

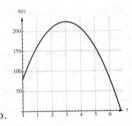

b. The maximum appears to occur at $t = 3$. The height at that time would be 224.

1.1.33 $g(1/z) = (1/z)^3 = \dfrac{1}{z^3}$

1.1.35 $F(g(y)) = F(y^3) = \dfrac{1}{y^3 - 3}$

1.1.37 $g(f(u)) = g(u^2 - 4) = (u^2 - 4)^3$

1.1.39 $F(F(x)) = F\left(\dfrac{1}{x-3}\right) = \dfrac{1}{\frac{1}{x-3} - 3} = \dfrac{1}{\frac{1}{x-3} - \frac{3(x-3)}{x-3}} = \dfrac{1}{\frac{10-3x}{x-3}} = \dfrac{x-3}{10-3x}$

1.1.41 $f(\sqrt{x+4}) = (\sqrt{x+4})^2 - 4 = x + 4 - 4 = x$.

1.1.43 $g(x) = x^3 - 5$ and $f(x) = x^{10}$.

1.1.45 $g(x) = x^4 + 2$ and $f(x) = \sqrt{x}$.

1.1.47 $(f \circ g)(x) = f(g(x)) = f(x^2 - 4) = |x^2 - 4|$. The domain of this function is the set of all real numbers.

1.1.49 $(f \circ G)(x) = f(G(x)) = f\left(\dfrac{1}{x-2}\right) = \left|\dfrac{1}{x-2}\right| = \dfrac{1}{|x-2|}$. The domain of this function is the set of all real numbers except for the number 2.

1.1.51 $(G \circ g \circ f)(x) = G(g(f(x))) = G(g(|x|)) = G(x^2 - 4) = \dfrac{1}{x^2 - 4 - 2} = \dfrac{1}{x^2 - 6}$. The domain of this function is the set of all real numbers except for the numbers $\pm\sqrt{6}$.

1.1. Review of Functions

1.1.53 $(g \circ g)(x) = g(g(x)) = g(x^2 - 4) = (x^2 - 4)^2 - 4 = x^4 - 8x^2 + 16 - 4 = x^4 - 8x^2 + 12$. The domain is the set of all real numbers.

1.1.55 Because $(x^2 + 3) - 3 = x^2$, we may choose $f(x) = x - 3$.

1.1.57 Because $(x^2 + 3)^2 = x^4 + 6x^2 + 9$, we may choose $f(x) = x^2$.

1.1.59 Because $(x^2)^2 + 3 = x^4 + 3$, this expression results from squaring x^2 and adding 3 to it. Thus we may choose $f(x) = x^2$.

1.1.61

a. True. A real number z corresponds to the domain element $z/2 + 19$, because $f(z/2 + 19) = 2(z/2 + 19) - 38 = z + 38 - 38 = z$.

b. False. The definition of function does not require that each range element comes from a unique domain element, rather that each domain element is paired with a unique range element.

c. True. $f(1/x) = \frac{1}{1/x} = x$, and $\frac{1}{f(x)} = \frac{1}{1/x} = x$.

d. False. For example, suppose that f is the straight line through the origin with slope 1, so that $f(x) = x$. Then $f(f(x)) = f(x) = x$, while $(f(x))^2 = x^2$.

e. False. For example, let $f(x) = x+2$ and $g(x) = 2x-1$. Then $f(g(x)) = f(2x-1) = 2x-1+2 = 2x+1$, while $g(f(x)) = g(x+2) = 2(x+2) - 1 = 2x + 3$.

f. True. This is the definition of $f \circ g$.

g. True. If f is even, then $f(-z) = f(z)$ for all z, so this is true in particular for $z = ax$. So if $g(x) = cf(ax)$, then $g(-x) = cf(-ax) = cf(ax) = g(x)$, so g is even.

h. False. For example, $f(x) = x$ is an odd function, but $h(x) = x + 1$ isn't, because $h(2) = 3$, while $h(-2) = -1$ which isn't $-h(2)$.

i. True. If $f(-x) = -f(x) = f(x)$, then in particular $-f(x) = f(x)$, so $0 = 2f(x)$, so $f(x) = 0$ for all x.

1.1.63 $\dfrac{f(x+h) - f(x)}{h} = \dfrac{3(x+h) - 3x}{h} = \dfrac{3x + 3h - 3x}{h} = \dfrac{3h}{h} = 3$.

1.1.65 $\dfrac{f(x+h) - f(x)}{h} = \dfrac{(x+h)^2 - x^2}{h} = \dfrac{(x^2 + 2hx + h^2) - x^2}{h} = \dfrac{h(2x+h)}{h} = 2x + h$.

1.1.67 $\dfrac{f(x+h) - f(x)}{h} = \dfrac{\frac{2}{x+h} - \frac{2}{x}}{h} = \dfrac{\frac{2x - 2(x+h)}{x(x+h)}}{h} = \dfrac{2x - 2x - 2h}{hx(x+h)} = -\dfrac{2h}{hx(x+h)} = -\dfrac{2}{x(x+h)}$.

1.1.69 $\dfrac{f(x) - f(a)}{x - a} = \dfrac{x^2 + x - (a^2 + a)}{x - a} = \dfrac{(x^2 - a^2) + (x - a)}{x - a} = \dfrac{(x-a)(x+a) + (x-a)}{x-a} = \dfrac{(x-a)(x+a+1)}{x-a} = x + a + 1$.

1.1.71 $\dfrac{f(x) - f(a)}{x - a} = \dfrac{x^3 - 2x - (a^3 - 2a)}{x - a} = \dfrac{(x^3 - a^3) - 2(x - a)}{x - a} = \dfrac{(x-a)(x^2 + ax + a^2) - 2(x-a)}{x-a} = \dfrac{(x-a)(x^2 + ax + a^2 - 2)}{x-a} = x^2 + ax + a^2 - 2$.

1.1.73 $\dfrac{f(x) - f(a)}{x - a} = \dfrac{\frac{-4}{x^2} - \frac{-4}{a^2}}{x - a} = \dfrac{\frac{-4a^2 + 4x^2}{a^2 x^2}}{x - a} = \dfrac{4(x^2 - a^2)}{(x-a)a^2 x^2} = \dfrac{4(x-a)(x+a)}{(x-a)a^2 x^2} = \dfrac{4(x+a)}{a^2 x^2}$.

Copyright © 2019 Pearson Education, Inc.

1.1.75

a. The slope is $\frac{12227-10499}{3-1} = 864$ ft/h. The hiker's elevation increases at an average rate of 874 feet per hour.

b. The slope is $\frac{12144-12631}{5-4} = -487$ ft/h. The hiker's elevation decreases at an average rate of 487 feet per hour.

c. The hiker might have stopped to rest during this interval of time or the trail is level on this portion of the hike.

1.1.77

a.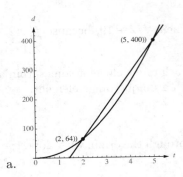

b. The slope of the secant line is given by $\frac{400-64}{5-2} = \frac{336}{3} = 112$ feet per second. The object falls at an average rate of 112 feet per second over the interval $2 \leq t \leq 5$.

1.1.79 This function is symmetric about the y-axis, because $f(-x) = (-x)^4 + 5(-x)^2 - 12 = x^4 + 5x^2 - 12 = f(x)$.

1.1.81 This function has none of the indicated symmetries. For example, note that $f(-2) = -26$, while $f(2) = 22$, so f is not symmetric about either the origin or about the y-axis, and is not symmetric about the x-axis because it is a function.

1.1.83 This curve (which is not a function) is symmetric about the x-axis, the y-axis, and the origin. Note that replacing either x by $-x$ or y by $-y$ (or both) yields the same equation. This is due to the fact that $(-x)^{2/3} = ((-x)^2)^{1/3} = (x^2)^{1/3} = x^{2/3}$, and a similar fact holds for the term involving y.

1.1.85 This function is symmetric about the origin. Note that $f(-x) = (-x)|(-x)| = -x|x| = -f(x)$.

1.1.87

a. $f(g(-2)) = f(-g(2)) = f(-2) = 4$

b. $g(f(-2)) = g(f(2)) = g(4) = 1$

c. $f(g(-4)) = f(-g(4)) = f(-1) = 3$

d. $g(f(5) - 8) = g(-2) = -g(2) = -2$

e. $g(g(-7)) = g(-g(7)) = g(-4) = -1$

f. $f(1 - f(8)) = f(-7) = 7$

1.1.89 We will make heavy use of the fact that $|x|$ is x if $x > 0$, and is $-x$ if $x < 0$. In the first quadrant where x and y are both positive, this equation becomes $x - y = 1$ which is a straight line with slope 1 and y-intercept -1. In the second quadrant where x is negative and y is positive, this equation becomes $-x - y = 1$, which is a straight line with slope -1 and y-intercept -1. In the third quadrant where both x and y are negative, we obtain the equation $-x - (-y) = 1$, or $y = x + 1$, and in the fourth quadrant, we obtain $x + y = 1$. Graphing these lines and restricting them to the appropriate quadrants yields the following curve:

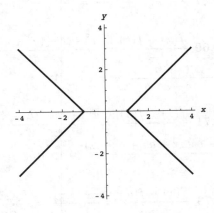

1.1.91 We have $y = 2 - \sqrt{-x^2 + 6x + 16}$, so by subtracting 2 from both sides and squaring we have $(y-2)^2 = -x^2 + 6x + 16$, which can be written as

$$x^2 - 6x + (y-2)^2 = 16.$$

To complete the square in x, we add 9 to both sides, yielding

$$x^2 - 6x + 9 + (y-2)^2 = 16 + 9,$$

or

$$(x-3)^2 + (y-2)^2 = 25.$$

This is the equation of a circle of radius 5 centered at $(3,2)$. Because $y \leq 2$, we see that the graph of f is the lower half of this circle. The domain of the function is $[-2, 8]$ and the range is $[-3, 2]$.

1.1.93 Because the composition of f with itself has first degree, f has first degree as well, so let $f(x) = ax+b$. Then $(f \circ f)(x) = f(ax+b) = a(ax+b) + b = a^2x + (ab+b)$. Equating coefficients, we see that $a^2 = 9$ and $ab + b = -8$. If $a = 3$, we get that $b = -2$, while if $a = -3$ we have $b = 4$. So the two possible answers are $f(x) = 3x - 2$ and $f(x) = -3x + 4$.

1.1.95 Let $f(x) = ax^2 + bx + c$. Then $(f \circ f)(x) = f(ax^2 + bx + c) = a(ax^2 + bx + c)^2 + b(ax^2 + bx + c) + c$. Expanding this expression yields $a^3x^4 + 2a^2bx^3 + 2a^2cx^2 + ab^2x^2 + 2abcx + ac^2 + abx^2 + b^2x + bc + c$, which simplifies to $a^3x^4 + 2a^2bx^3 + (2a^2c + ab^2 + ab)x^2 + (2abc + b^2)x + (ac^2 + bc + c)$. Equating coefficients yields $a^3 = 1$, so $a = 1$. Then $2a^2b = 0$, so $b = 0$. It then follows that $c = -6$, so the original function was $f(x) = x^2 - 6$.

1.1.97 $\dfrac{f(x+h) - f(x)}{h} = \dfrac{\sqrt{x+h} - \sqrt{x}}{h} = \dfrac{\sqrt{x+h} - \sqrt{x}}{h} \cdot \dfrac{\sqrt{x+h} + \sqrt{x}}{\sqrt{x+h} + \sqrt{x}} = \dfrac{(x+h) - x}{h(\sqrt{x+h} + \sqrt{x})} = \dfrac{1}{\sqrt{x+h} + \sqrt{x}}.$

$\dfrac{f(x) - f(a)}{x - a} = \dfrac{\sqrt{x} - \sqrt{a}}{x - a} = \dfrac{\sqrt{x} - \sqrt{a}}{x - a} \cdot \dfrac{\sqrt{x} + \sqrt{a}}{\sqrt{x} + \sqrt{a}} = \dfrac{x - a}{(x-a)(\sqrt{x} + \sqrt{a})} = \dfrac{1}{\sqrt{x} + \sqrt{a}}.$

1.1.99 $\dfrac{f(x+h) - f(x)}{h} = \dfrac{\frac{-3}{\sqrt{x+h}} - \frac{-3}{\sqrt{x}}}{h} = \dfrac{-3(\sqrt{x} - \sqrt{x+h})}{h\sqrt{x}\sqrt{x+h}} = \dfrac{-3(\sqrt{x} - \sqrt{x+h})}{h\sqrt{x}\sqrt{x+h}} \cdot \dfrac{\sqrt{x} + \sqrt{x+h}}{\sqrt{x} + \sqrt{x+h}} = \dfrac{-3(x - (x+h))}{h\sqrt{x}\sqrt{x+h}(\sqrt{x} + \sqrt{x+h})} = \dfrac{3}{\sqrt{x}\sqrt{x+h}(\sqrt{x} + \sqrt{x+h})}.$

$\dfrac{f(x) - f(a)}{x - a} = \dfrac{\frac{-3}{\sqrt{x}} - \frac{-3}{\sqrt{a}}}{x - a} = \dfrac{-3\left(\frac{\sqrt{a} - \sqrt{x}}{\sqrt{a}\sqrt{x}}\right)}{x - a} = \dfrac{(-3)(\sqrt{a} - \sqrt{x})}{(x-a)\sqrt{a}\sqrt{x}} \cdot \dfrac{\sqrt{a} + \sqrt{x}}{\sqrt{a} + \sqrt{x}} = \dfrac{(3)(x - a)}{(x-a)(\sqrt{a}\sqrt{x})(\sqrt{a} + \sqrt{x})} = \dfrac{3}{\sqrt{ax}(\sqrt{a} + \sqrt{x})}.$

1.1.101 This would not necessarily have either kind of symmetry. For example, $f(x) = x^2$ is an even function and $g(x) = x^3$ is odd, but the sum of these two is neither even nor odd.

1.1.103 This would be an even function, so it would be symmetric about the y-axis. Suppose f is even and g is odd. Then $g(f(-x)) = g(f(x))$, because $f(-x) = f(x)$.

1.2 Representing Functions

1.2.1 Functions can be defined and represented by a formula, through a graph, via a table, and by using words.

1.2.3 The slope of the line shown is $m = \frac{-3 - (-1)}{3 - 0} = -2/3$. The y-intercept is $b = -1$. Thus the function is given by $f(x) = -\frac{2}{3}x - 1$.

1.2.5 The domain of a rational function $\frac{p(x)}{q(x)}$ is the set of all real numbers for which $q(x) \neq 0$.

1.2.7 For $x < 0$, the graph is a line with slope 1 and y-intercept 3, while for $x > 0$, it is a line with slope $-1/2$ and y-intercept 3. Note that both of these lines contain the point $(0, 3)$. The function shown can thus be written

$$f(x) = \begin{cases} x + 3 & \text{if } x < 0; \\ -\frac{1}{2}x + 3 & \text{if } x \geq 0. \end{cases}$$

1.2.9 Compared to the graph of $f(x)$, the graph of $f(x + 2)$ will be shifted 2 units to the left.

1.2.11 Compared to the graph of $f(x)$, the graph of $f(3x)$ will be compressed horizontally by a factor of $\frac{1}{3}$.

1.2.13 $f(x) = |x - 2| + 3$, because the graph of f is obtained from that of $|x|$ by shifting 2 units to the right and 3 units up.

$g(x) = -|x + 2| - 1$, because the graph of g is obtained from the graph of $|x|$ by shifting 2 units to the left, then reflecting about the x-axis, and then shifting 1 unit down.

1.2.15

The slope is given by $\frac{5-3}{2-1} = 2$, so the equation of the line is $y - 3 = 2(x - 1)$, which can be written as $f(x) = 2x - 2 + 3$, or $f(x) = 2x + 1$.

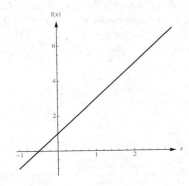

1.2.17 We are looking for the line with slope 3 that goes through the point $(3, 2)$. Using the point-slope form of the equation of a line, we have $y - 2 = 3(x - 3)$, which can be written as $y = 2 + 3x - 9$, or $y = 3x - 7$.

1.2.19 We have $571 = C_s(100)$, so $C_s = 5.71$. Therefore $N(150) = 5.71(150) = 856.5$ million.

1.2.21 Using price as the independent variable p and the average number of units sold per day as the dependent variable d, we have the ordered pairs $(250, 12)$ and $(200, 15)$. The slope of the line determined by these points is $m = \frac{15-12}{200-250} = \frac{3}{-50}$. Thus the demand function has the form $d(p) = (-3/50)p + b$ for some constant b. Using the point $(200, 15)$, we find that $15 = (-3/50) \cdot 200 + b$, so $b = 27$. Thus the demand function is $d = (-3p/50) + 27$. While the domain of this linear function is the set of all real numbers, the formula is only likely to be valid for some subset of the interval $(0, 450)$, because outside of that interval either $p \leq 0$ or $d \leq 0$.

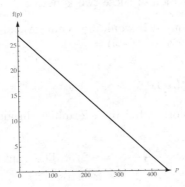

1.2. Representing Functions

1.2.23

a. Using the points $(1986, 1875)$ and $(2000, 6471)$ we see that the slope is about 328.3. At $t = 0$, the value of p is 1875. Therefore a line which reasonably approximates the data is $p(t) = 328.3t + 1875$.

b. Using this line, we have that $p(9) = 4830$ breeding pairs.

1.2.25 For $x \leq 3$, we have the constant function 3. For $x \geq 3$, we have a straight line with slope 2 that contains the point $(3, 3)$. So its equation is $y - 3 = 2(x - 3)$, or $y = 2x - 3$. So the function can be written as $f(x) = \begin{cases} 3 & \text{if } x \leq 3; \\ 2x - 3 & \text{if } x > 3 \end{cases}$

1.2.27

The cost is given by

$$c(t) = \begin{cases} 0.05t & \text{for } 0 \leq t \leq 60 \\ 1.2 + 0.03t & \text{for } 60 < t \leq 120 \end{cases}.$$

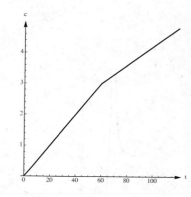

1.2.29

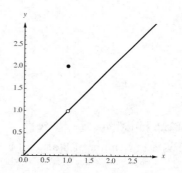

1.2.31

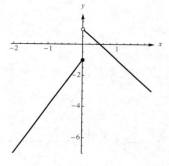

1.2.33

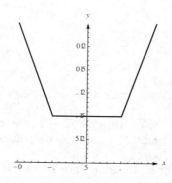

1.2.35

a.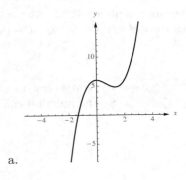

b. The function is a polynomial, so its domain is the set of all real numbers.

c. It has one peak near its y-intercept of $(0,6)$ and one valley between $x = 1$ and $x = 2$. Its x-intercept is near $x = -4/3$.

1.2.37

a.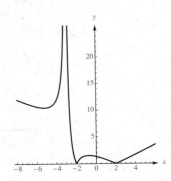

b. The domain of the function is the set of all real numbers except -3.

c. There is a valley near $x = -5.2$ and a peak near $x = -0.8$. The x-intercepts are at -2 and 2, where the curve does not appear to be smooth. There is a vertical asymptote at $x = -3$. The function is never below the x-axis. The y-intercept is $(0, 4/3)$.

1.2.39

a.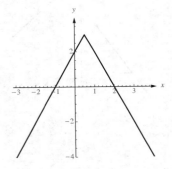

b. The domain of the function is $(-\infty, \infty)$

c. The function has a maximum of 3 at $x = 1/2$, and a y-intercept of 2.

1.2.41

a. The zeros of f are the points where the graph crosses the x-axis, so these are points A, D, F, and I.

b. The only high point, or peak, of f occurs at point E, because it appears that the graph has larger and larger y values as x increases past point I and decreases past point A.

c. The only low points, or valleys, of f are at points B and H, again assuming that the graph of f continues its apparent behavior for larger values of x.

d. Past point H, the graph is rising, and is rising faster and faster as x increases. It is also rising between points B and E, but not as quickly as it is past point H. So the marked point at which it is rising most rapidly is I.

e. Before point B, the graph is falling, and falls more and more rapidly as x becomes more and more negative. It is also falling between points E and H, but not as rapidly as it is before point B. So the marked point at which it is falling most rapidly is A.

1.2.43

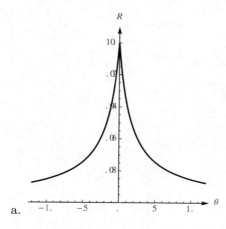

a.

b. This appears to have a maximum when $\theta = 0$. Our vision is sharpest when we look straight ahead.

c. For $|\theta| \leq .19°$. We have an extremely narrow range where our eyesight is sharp.

1.2.45 The slope of this line is constantly 2, so the slope function is $S(x) = 2$.

1.2.47 The slope function is given by $S(x) = \begin{cases} 1 & \text{if } x < 0; \\ -1/2 & \text{if } x > 0. \end{cases}$

1.2.49

a. Because the area under consideration is that of a rectangle with base 2 and height 6, $A(2) = 12$.

b. Because the area under consideration is that of a rectangle with base 6 and height 6, $A(6) = 36$.

c. Because the area under consideration is that of a rectangle with base x and height 6, $A(x) = 6x$.

1.2.51

a. Because the area under consideration is that of a trapezoid with base 2 and heights 8 and 4, we have $A(2) = 2 \cdot \frac{8+4}{2} = 12$.

b. Note that $A(3)$ represents the area of a trapezoid with base 3 and heights 8 and 2, so $A(3) = 3 \cdot \frac{8+2}{2} = 15$. So $A(6) = 15 + (A(6) - A(3))$, and $A(6) - A(3)$ represents the area of a triangle with base 3 and height 2. Thus $A(6) = 15 + 6 = 21$.

c. For x between 0 and 3, $A(x)$ represents the area of a trapezoid with base x, and heights 8 and $8 - 2x$. Thus the area is $x \cdot \frac{8+8-2x}{2} = 8x - x^2$. For $x > 3$, $A(x) = A(3) + A(x) - A(3) = 15 + 2(x-3) = 2x + 9$. Thus
$$A(x) = \begin{cases} 8x - x^2 & \text{if } 0 \leq x \leq 3; \\ 2x + 9 & \text{if } x > 3. \end{cases}$$

1.2.53

a. True. A polynomial $p(x)$ can be written as the ratio of polynomials $\frac{p(x)}{1}$, so it is a rational function. However, a rational function like $\frac{1}{x}$ is not a polynomial.

b. False. For example, if $f(x) = 2x$, then $(f \circ f)(x) = f(f(x)) = f(2x) = 4x$ is linear, not quadratic.

c. True. In fact, if f is degree m and g is degree n, then the degree of the composition of f and g is $m \cdot n$, regardless of the order they are composed.

d. False. The graph would be shifted two units to the left.

1.2.55

a.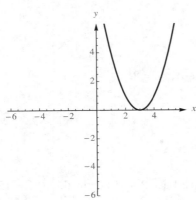
Shift 3 units to the right.

b.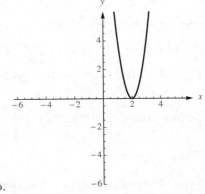
Horizontal compression by a factor of $\frac{1}{2}$, then shift 2 units to the right.

c.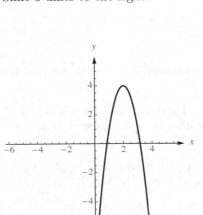
Shift to the right 2 units, vertically stretch by a factor of 3, reflect across the x-axis, and shift up 4 units.

d.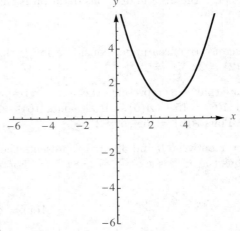
Horizontal stretch by a factor of 3, horizontal shift right 2 units, vertical stretch by a factor of 6, and vertical shift up 1 unit.

1.2. Representing Functions

1.2.57 The graph is obtained by shifting the graph of x^2 two units to the right and one unit up.

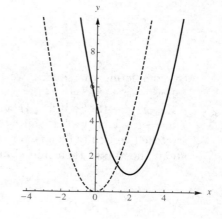

1.2.59 Stretch the graph of $y = x^2$ vertically by a factor of 3 and then reflect across the x-axis.

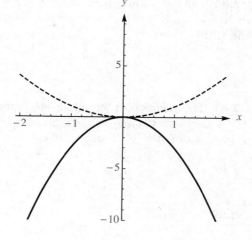

1.2.61 Shift the graph of $y = x^2$ left 3 units and stretch vertically by a factor of 2.

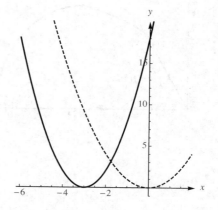

1.2.63 By completing the square, we have that $h(x) = -4(x^2 + x - 3) = -4\left(x^2 + x + \frac{1}{4} - \frac{1}{4} - 3\right) = -4(x + \frac{1}{2})^2 + 13$. So it is $-4f(x + (\frac{1}{2})) + 13$ where $f(x) = x^2$. The graph is shifted $\frac{1}{2}$ unit to the left, stretched vertically by a factor of 4, then reflected about the x-axis, then shifted up 13 units.

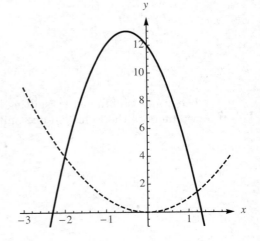

1.2.65 The curves intersect where $4\sqrt{2x} = 2x^2$. If we square both sides, we have $32x = 4x^4$, which can be written as $4x(8 - x^3) = 0$, which has solutions at $x = 0$ and $x = 2$. So the points of intersection are $(0, 0)$ and $(2, 8)$.

1.2.67 The points of intersection are found by solving $x^2 = -x^2 + 8x$. This yields the quadratic equation $2x^2 - 8x = 0$ or $(2x)(x - 4) = 0$. So the x-values of the points of intersection are 0 and 4. The actual points of intersection are $(0, 0)$ and $(4, 16)$.

1.2.69

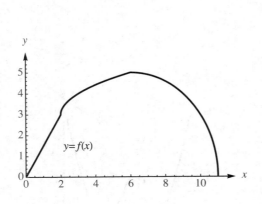

1.2.71

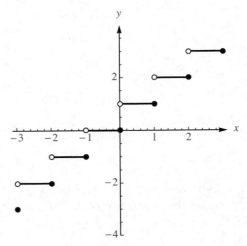

1.2. Representing Functions

1.2.73

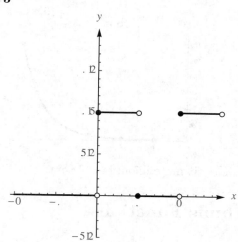

1.2.75

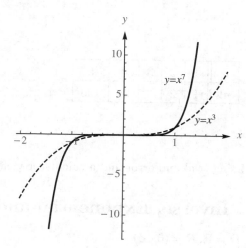

1.2.77

a. $f(0.75) = \frac{.75^2}{1-2(.75)(.25)} = .9$. There is a 90% chance that the server will win from deuce if they win 75% of their service points.

b. $f(0.25) = \frac{.25^2}{1-2(.25)(.75)} = .1$. There is a 10% chance that the server will win from deuce if they win 25% of their service points.

1.2.79

a. Because you are paying $350 per month, the amount paid after m months is $y = 350m + 1200$.

b. After 4 years (48 months) you have paid $350 \cdot 48 + 1200 = 18000$ dollars. If you then buy the car for $10,000, you will have paid a total of $28,000 for the car instead of $25,000. So you should buy the car instead of leasing it.

1.2.81

a. The volume of the box is $x^2 h$, but because the box has volume 125 cubic feet, we have that $x^2 h = 125$, so $h = \frac{125}{x^2}$. The surface area of the box is given by x^2 (the area of the base) plus $4 \cdot hx$, because each side has area hx. Thus $S = x^2 + 4hx = x^2 + \frac{4 \cdot 125 \cdot x}{x^2} = x^2 + \frac{500}{x}$.

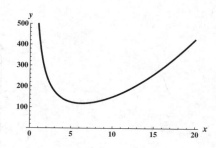

b. By inspection, it looks like the value of x which minimizes the surface area is about 6.3.

1.2.83 Suppose that the parabola f crosses the x-axis at a and b, with $a < b$. Then a and b are roots of the polynomial, so $(x-a)$ and $(x-b)$ are factors. Thus the polynomial must be $f(x) = c(x-a)(x-b)$ for some non-zero real number c. So $f(x) = cx^2 - c(a+b)x + abc$. Because the vertex always occurs at the x value which is $\frac{-\text{coefficient on x}}{2 \cdot \text{coefficient on } x^2}$ we have that the vertex occurs at $\frac{c(a+b)}{2c} = \frac{a+b}{2}$, which is halfway between a and b.

1.2.85

a.

n	1	2	3	4	5
$n!$	1	2	6	24	120

b.
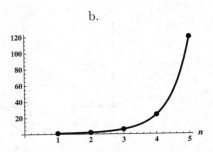

c. Using trial and error and a calculator yields that 10! is more than a million, but 9! isn't.

1.3 Inverse, Exponential and Logarithmic Functions

1.3.1 $D = \mathbb{R}, R = (0, \infty)$.

1.3.3
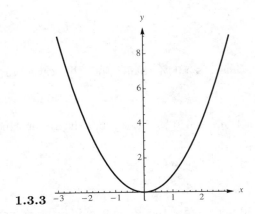

1.3.5 f is one-to-one on $(-\infty, -1]$, on $[-1, 1]$, and on $[1, \infty)$.

1.3.7 If a function f is not one-to-one, then there are domain values $x_1 \neq x_2$ with $f(x_1) = f(x_2)$. If f^{-1} were to exist, then $f^{-1}(f(x_1)) = f^{-1}(f(x_2))$ which would imply that $x_1 = x_2$, a contradiction.

1.3.9 Suppose $x = 2y$, then $y = \frac{1}{2}x$, so the inverse of f is $f^{-1}(x) = \frac{1}{2}x$. Then $f(f^{-1}(x)) = f(\frac{1}{2}x) = 2 \cdot \frac{1}{2}x = x$. Also, $f^{-1}(f(x)) = f^{-1}(2x) = \frac{1}{2} \cdot 2x = x$.

1.3.11
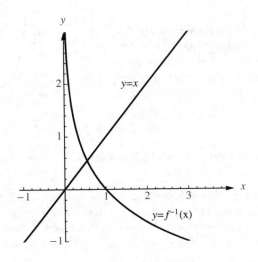

1.3. Inverse, Exponential and Logarithmic Functions

1.3.13 $g_1(x)$ is the right side of the standard parabola shifted up one unit. So $g_1(x) = x^2 + 1$, $x \geq 0$. The domain for g_1 is $[0, \infty)$ and the range is $[1, \infty)$. The inverse of g_1 is therefore the square root function shifted one unit to the right. So $g_1^{-1}(x) = \sqrt{x - 1}$, and its domain is $[1, \infty)$ and its range is $[0, \infty)$.

1.3.15 $\log_b x$ represents the power to which b must be raised in order to obtain x. So, $b^{\log_b x} = x$.

1.3.17 Because the domain of b^x is $\mathbb{R}$ and the range of b^x is $(0, \infty)$, and because $\log_b x$ is the inverse of b^x, the domain of $\log_b x$ is $(0, \infty)$ and the range is $\mathbb{R}$.

1.3.19

a. Because $10^3 = 1000$, $\log_{10} 1000 = 3$.

b. Because $2^4 = 16$, $\log_2 16 = 4$.

c. Because $10^{-2} = \frac{1}{100} = 0.01$, $\log_{10} 0.01 = -2$.

d. Because e^x and $\ln x$ are inverses, $\ln e^3 = 3$.

e. Because e^x and $\ln x$ are inverses, $\ln \sqrt{e} = \ln e^{1/2} = \frac{1}{2}$.

1.3.21 f is one-to-one on $(-\infty, \infty)$, so it has an inverse on $(-\infty, \infty)$.

1.3.23 f is one-to-one on its domain, which is $(-\infty, 5) \cup (5, \infty)$, so it has an inverse on that set.

1.3.25 f is one-to-one on the interval $(0, \infty)$, so it has an inverse on that interval. (Alternatively, it is one-to-one on the interval $(-\infty, 0)$, so that interval could be used as well.)

1.3.27 Switching x and y gives $x = 8 - 4y$. Solving this for y yields $y = f^{-1}(x) = \frac{8-x}{4}$.

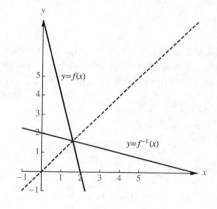

1.3.29 Switching x and y, we have $x = \sqrt{y + 2}$. Solving for y in terms of x we have $y = f^{-1}(x) = x^2 - 2$. Note that because the range of f is $[0, \infty)$, that is also the domain of f^{-1}.

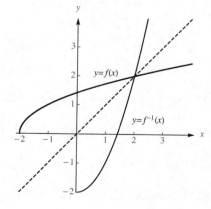

1.3.31 Switching x and y gives $x = (y-2)^2 - 1$. Then $x + 1 = (y-2)^2$, so $\sqrt{x+1} = |y-2|$, but because in the original function the variable is greater than or equal to 2, we choose the the positive portion of the graph of $|y-2|$. So we have $y = 2 + \sqrt{x+1}$.

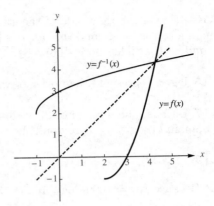

1.3.33 Switching x and y, we have $x = \frac{2}{y^2+1}$. Then $y^2 + 1 = \frac{2}{x}$, so $y^2 = \frac{2}{x} - 1$. So $|y| = \sqrt{\frac{2}{x} - 1}$. We choose the positive portion, so that $y = \sqrt{\frac{2}{x} - 1}$. Note that the domain of f is $[0, \infty)$ while the range of f is $(0, 2]$. So the domain of f^{-1} is $(0, 2]$ and the range is $[0, \infty)$.

1.3.35 Switching x and y, we have $x = e^{2y+6}$. Then $\ln x = 2y + 6$, so $2y = \ln x - 6$, and $y = f^{-1}(x) = \frac{1}{2}\ln x - 3$.

1.3.37 Switching x and y, we have $x = \ln(3y+1)$. Then $e^x = 3y+1$, so $3y = e^x - 1$ and $y = f^{-1}(x) = \frac{e^x - 1}{3}$.

1.3.39 Switching x and y, we have $x = 10^{-2y}$. Then $\log_{10} x = -2y$, so $y = f^{-1}(x) = -\frac{1}{2}\log_{10} x$.

1.3.41 Switching x and y, we have $x = \frac{e^y}{e^y+2}$. Taking the reciprocal of both sides, we have $\frac{1}{x} = \frac{e^y+2}{e^y} = 1 + 2e^{-y}$. Then $2e^{-y} = \frac{1}{x} - 1$, and $e^{-y} = \frac{1}{2x} - \frac{1}{2}$. So $-y = \ln\left(\frac{1}{2x} - \frac{1}{2}\right)$, and $y = f^{-1}(x) = -\ln\left(\frac{1}{2x} - \frac{1}{2}\right) = -\ln\left(\frac{1-x}{2x}\right) = \ln\left(\frac{2x}{1-x}\right)$.

1.3.43 First note that because the expression is symmetric, switching x and y doesn't change the expression. Solving for y gives $|y| = \sqrt{1-x^2}$. To get the four one-to-one functions, we restrict the domain and choose either the upper part or lower part of the circle as follows:

a. $f_1(x) = \sqrt{1-x^2}$, $0 \leq x \leq 1$
$f_2(x) = \sqrt{1-x^2}$, $-1 \leq x \leq 0$
$f_3(x) = -\sqrt{1-x^2}$, $-1 \leq x \leq 0$
$f_4(x) = -\sqrt{1-x^2}$, $0 \leq x \leq 1$

b. Reflecting these functions across the line $y = x$ yields the following:
$f_1^{-1}(x) = \sqrt{1-x^2}$, $0 \leq x \leq 1$
$f_2^{-1}(x) = -\sqrt{1-x^2}$, $0 \leq x \leq 1$
$f_3^{-1}(x) = -\sqrt{1-x^2}$, $-1 \leq x \leq 0$
$f_4^{-1}(x) = \sqrt{1-x^2}$, $-1 \leq x \leq 0$

1.3.45 $\log_b\left(\frac{x}{y}\right) = \log_b x - \log_b y = 0.36 - .056 = -0.2$.

1.3.47 $\log_b xz = \log_b x + \log_b z = 0.36 + 0.83 = 1.19$.

1.3.49 $\log_b \frac{\sqrt{x}}{\sqrt[3]{z}} = \log_b x^{1/2} - \log_b z^{1/3} = (1/2)\log_b x - (1/3)\log_b z = (0.36)/2 - (0.83)/3 = -.09\overline{6}$.

1.3.51 If $\log_{10} x = 3$, then $10^3 = x$, so $x = 1000$.

1.3. Inverse, Exponential and Logarithmic Functions

1.3.53 If $\log_8 x = 1/3$, then $x = 8^{1/3} = 2$.

1.3.55 $\ln x = -1$, then $e^{-1} = x$, so $x = \frac{1}{e}$.

1.3.57 Since $7^x = 21$, we have that $\ln 7^x = \ln 21$, so $x \ln 7 = \ln 21$, and $x = \frac{\ln 21}{\ln 7}$.

1.3.59 Since $3^{3x-4} = 15$, we have that $\ln 3^{3x-4} = \ln 15$, so $(3x - 4)\ln 3 = \ln 15$. Thus, $3x - 4 = \frac{\ln 15}{\ln 3}$, so $x = \frac{(\ln 15)/(\ln 3) + 4}{3} = \frac{\ln 15 + 4\ln 3}{3\ln 3} = \frac{\ln 5 + \ln 3 + 4\ln 3}{3\ln 3} = \frac{\ln 5}{3\ln 3} + \frac{5}{3}$.

1.3.61 We are seeking t so that $50 = 100e^{-t/650}$. This occurs when $e^{-t/650} = \frac{1}{2}$, which is when $-\frac{t}{650} = \ln(1/2)$, so $t = 650 \ln 2 \approx 451$ years.

1.3.63 We need to solve $1100 = 1000\left(1 + \frac{0.01}{12}\right)^{12t}$ for t. We have $1.1 = \left(1 + \frac{0.01}{12}\right)^{12t}$, so $\ln 1.1 = 12t \ln\left(1 + \frac{0.01}{12}\right)$. Then $t = \frac{\ln 1.1}{12 \ln\left(1 + \frac{0.01}{12}\right)} \approx 9.53$ years.

1.3.65

a. No. The function takes on the values from 0 to 64 as t varies from 0 to 2, and then takes on the values from 64 to 0 as t varies from 2 to 4, so h is not one-to-one.

b. Solving for h in terms of t we have $h = 64t - 16t^2$, so (completing the square) we have $h - 64 = -16(t^2 - 4t + 4)$. Thus, $h - 64 = -16(t-2)^2$, and $(t-2)^2 = \frac{64-h}{16}$. Therefore $|t - 2| = \frac{\sqrt{64-h}}{4}$. When the ball is on the way up we know that $t < 2$, so the inverse of f is $f^{-1}(h) = 2 - \frac{\sqrt{64-h}}{4}$.

c. Using the work from the previous part of this problem, we have that when the ball is on the way down (when $t > 2$) we have that the inverse of f is $f^{-1}(h) = 2 + \frac{\sqrt{64-h}}{4}$.

d. On the way up, the ball is at a height of 30 ft at $2 - \frac{\sqrt{64-30}}{4} \approx 0.542$ seconds.

e. On the way down, the ball is at a height of 10 ft at $2 + \frac{\sqrt{64-10}}{4} \approx 3.837$ seconds.

1.3.67 $\log_2 15 = \frac{\ln 15}{\ln 2} \approx 3.9069$.

1.3.69 $\log_4 40 = \frac{\ln 40}{\ln 4} \approx 2.6610$.

1.3.71 Let $2^x = z$. Then $\ln 2^x = \ln z$, so $x \ln 2 = \ln z$. Taking the exponential function of both sides gives $z = e^{x \ln 2}$.

1.3.73 Let $z = \ln |x|$. Then $e^z = |x|$. Taking logarithms with base 5 of both sides gives $\log_5 e^z = \log_5 |x|$, so $z \cdot \log_5 e = \log_5 |x|$, and thus $z = \frac{\log_5 |x|}{\log_5 e}$.

1.3.75 Let $z = a^{1/\ln a}$. Then $\ln z = \ln\left(a^{1/\ln a}\right) = \frac{1}{\ln a} \cdot \ln a = 1$. Thus $z = e$.

1.3.77

a. False. For example, $3 = 3^1$, but $1 \neq \sqrt[3]{3}$.

b. False. For example, suppose $x = y = b = 2$. Then the left-hand side of the equation is equal to 1, but the right-hand side is 0.

c. False. $\log_5 4^6 = 6 \log_5 4 > 4 \log_5 6$.

d. True. This follows because 10^x and $\log_{10}$ are inverses of each other.

e. False. $\ln 2^e = e \ln 2 < 2$.

f. False. For example $f(0) = 1$, but the alleged inverse function evaluated at 1 is not 0 (rather, it has value 1/2.)

g. True. f is its own inverse because $f(f(x)) = f(1/x) = \frac{1}{1/x} = x$.

1.3.79 A is $\log_2 x$, B is $\log_4 x$, C is $\log_{10} x$.

1.3.81

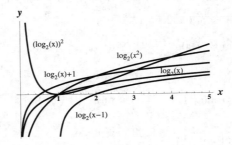

Note: need better pic from back of book

1.3.83

a. The relevant graph is:

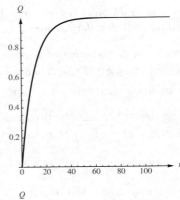

b. Varying a while holding c constant scales the curve vertically. It appears that the steady-state charge is equal to a.

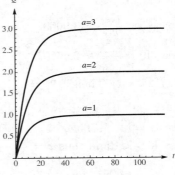

c. Varying c while holding a constant scales the curve horizontally. It appears that the steady-state charge does not vary with c.

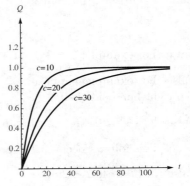

1.4. Trigonometric Functions and Their Inverses

d. As t grows large, the term $ae^{-t/c}$ approaches zero for any fixed c and a. So the steady-state charge for $a - ae^{-t/c}$ is a.

1.3.85 Begin by completing the square: $f(x) = x^2 - 2x + 6 = (x^2 - 2x + 1) + 5 = (x-1)^2 + 5$. Switching x and y yields $x = (y-1)^2 + 5$. Solving for y gives $|y-1| = \sqrt{x-5}$. Choosing the principal square root (because the original given interval has x positive) gives $y = f^{-1}(x) = \sqrt{x-5} + 1$, $x \geq 5$.

1.3.87 Note that f is one-to-one, so there is only one inverse. Switching x and y gives $x = (y+1)^3$. Then $\sqrt[3]{x} = y + 1$, so $y = f^{-1}(x) = \sqrt[3]{x} - 1$. The domain of f^{-1} is $\mathbb{R}$.

1.3.89 Note that to get a one-to-one function, we should restrict the domain to either $[0, \infty)$ or $(-\infty, 0]$. Switching x and y yields $x = \frac{2}{y^2+2}$, so $y^2 + 2 = (2/x)$. So $y = \pm\sqrt{(2/x) - 2}$. So the inverse of f when the domain of f is restricted to $[0, \infty)$ is $f^{-1}(x) = \sqrt{(2/x) - 2}$, while if the domain of f is restricted to $(-\infty, 0]$ the inverse is $f^{-1}(x) = -\sqrt{(2/x) - 2}$. In either case, the domain of f^{-1} is $(0, 1]$.

1.3.91 Using the change of base formula, we have $\log_{1/b} x = \frac{\ln x}{\ln 1/b} = \frac{\ln x}{\ln 1 - \ln b} = \frac{\ln x}{-\ln b} = -\frac{\ln x}{\ln b} = -\log_b x$.

1.3.93 Let $x = b^p$ and $y = b^q$. Then $\frac{x}{y} = \frac{b^p}{b^q} = b^{p-q}$. Thus $\log_b \frac{x}{y} = \log_b b^{p-q} = p - q = \log_b x - \log_b y$.

1.3.95

a. f is one-to-one on $(-\infty, -\sqrt{2}/2]$, on $[-\sqrt{2}/2, 0]$, on $[0, \sqrt{2}/2]$, and on $[\sqrt{2}/2, \infty)$.

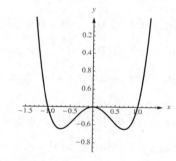

b. If $u = x^2$, then our function becomes $y = u^2 - u$. Completing the square gives $y + (1/4) = u^2 - u + (1/4) = (u - (1/2))^2$. Thus $|u - (1/2)| = \sqrt{y + (1/4)}$, so $u = (1/2) \pm \sqrt{y + (1/4)}$, with the "+" applying for $u = x^2 > (1/2)$ and the "−" applying when $u = x^2 < (1/2)$. Now letting $u = x^2$, we have $x^2 = (1/2) \pm \sqrt{y + (1/4)}$, so $x = \pm\sqrt{(1/2) \pm \sqrt{y + (1/4)}}$. Now switching the x and y gives the following inverses:

Domain of f	$(-\infty, -\sqrt{2}/2]$	$[-\sqrt{2}/2, 0]$	$[0, \sqrt{2}/2]$	$[\sqrt{2}/2, \infty)$
Range of f	$[-1/4, \infty)$	$[-1/4, 0]$	$[-1/4, 0]$	$[-1/4, \infty)$
Inverse of f	$-\sqrt{(1/2) + \sqrt{x + (1/4)}}$	$-\sqrt{(1/2) - \sqrt{x + (1/4)}}$	$\sqrt{(1/2) - \sqrt{x + (1/4)}}$	$\sqrt{(1/2) + \sqrt{x + (1/4)}}$

1.3.97 Using the change of base formulas $\log_b c = \frac{\ln c}{\ln b}$ and $\log_c b = \frac{\ln b}{\ln c}$ we have

$$(\log_b c) \cdot (\log_c b) = \frac{\ln c}{\ln b} \cdot \frac{\ln b}{\ln c} = 1.$$

1.4 Trigonometric Functions and Their Inverses

1.4.1 Let O be the length of the side opposite the angle x, let A be length of the side adjacent to the angle x, and let H be the length of the hypotenuse. Then $\sin x = \frac{O}{H}$, $\cos x = \frac{A}{H}$, $\tan x = \frac{O}{A}$, $\csc x = \frac{H}{O}$, $\sec x = \frac{H}{A}$, and $\cot x = \frac{A}{O}$.

1.4.3 We have $t = \dfrac{v\sin\theta}{16} = \dfrac{96\sin\frac{\pi}{6}}{16} = \dfrac{96/2}{16} = \dfrac{48}{16} = 3$ seconds.

1.4.5 The radian measure of an angle θ is the length of the arc s on the unit circle associated with θ.

1.4.7 $\sin^2 x + \cos^2 x = 1$, $1 + \cot^2 x = \csc^2 x$, and $\tan^2 x + 1 = \sec^2 x$.

1.4.9 The only point on the unit circle whose second coordinate is -1 is the point $(0, -1)$, which is the point associated with $\theta = \frac{3\pi}{2}$. So that is the only solution for $0 \leq \theta < 2\pi$.

1.4.11 The tangent function is undefined where $\cos x = 0$, which is at all real numbers of the form $\dfrac{\pi}{2} + k\pi$, k an integer.

1.4.13 The sine function is not one-to-one over its whole domain, so in order to define an inverse, it must be restricted to an interval on which it is one-to-one.

1.4.15 $\cos\dfrac{5\pi}{4} = -\dfrac{\sqrt{2}}{2}$, so $\cos^{-1}\left(\cos\dfrac{5\pi}{4}\right) = \cos^{-1}\left(-\dfrac{\sqrt{2}}{2}\right) = \frac{3\pi}{4}$.

1.4.17 The numbers $\pm\pi/2$ are not in the range of $\tan^{-1} x$. The range is $(-\pi/2, \pi/2)$. However, it is true that as x increases without bound, the values of $\tan^{-1} x$ get close to $\pi/2$, and as x decreases without bound, the values of $\tan^{-1} x$ get close to $-\pi/2$.

1.4.19 The point on the unit circle associated with $2\pi/3$ is $(-1/2, \sqrt{3}/2)$, so $\cos(2\pi/3) = -1/2$.

1.4.21 The point on the unit circle associated with $-3\pi/4$ is $(-\sqrt{2}/2, -\sqrt{2}/2)$, so $\tan(-3\pi/4) = 1$.

1.4.23 The point on the unit circle associated with $-13\pi/3$ is $(1/2, -\sqrt{3}/2)$, so $\cot(-13\pi/3) = -1/\sqrt{3} = -\sqrt{3}/3$.

1.4.25 The point on the unit circle associated with $-17\pi/3$ is $(1/2, \sqrt{3}/2)$, so $\cot(-17\pi/3) = 1/\sqrt{3} = \sqrt{3}/3$.

1.4.27 Because the point on the unit circle associated with $\theta = 0$ is the point $(1, 0)$, we have $\cos 0 = 1$.

1.4.29 Because $-\pi$ corresponds to a half circle clockwise revolution, the point on the unit circle associated with $-\pi$ is the point $(-1, 0)$. Thus $\cos(-\pi) = -1$.

1.4.31 Because $5\pi/2$ corresponds to one and a quarter counterclockwise revolutions, the point on the unit circle associated with $5\pi/2$ is the same as the point associated with $\pi/2$, which is $(0, 1)$. Thus $\sec 5\pi/2$ is undefined.

1.4.33 Using the fact that $\dfrac{\pi}{12} = \dfrac{\pi/6}{2}$ and the half-angle identity for cosine:

$$\cos^2(\pi/12) = \dfrac{1 + \cos(\pi/6)}{2} = \dfrac{1 + \sqrt{3}/2}{2} = \dfrac{2 + \sqrt{3}}{4}.$$

Thus, $\cos(\pi/12) = \sqrt{\dfrac{2 + \sqrt{3}}{4}}$.

1.4.35 First note that $\tan x = 1$ when $\sin x = \cos x$. Using our knowledge of the values of the standard angles between 0 and 2π, we recognize that the sine function and the cosine function are equal at $\pi/4$. Then, because we recall that the period of the tangent function is π, we know that $\tan(\pi/4 + k\pi) = \tan(\pi/4) = 1$ for every integer value of k. Thus the solution set is $\{\pi/4 + k\pi, \text{where } k \text{ is an integer}\}$.

1.4.37 Given that $\sin^2\theta = \dfrac{1}{4}$, we have $|\sin\theta| = \dfrac{1}{2}$, so $\sin\theta = \dfrac{1}{2}$ or $\sin\theta = -\dfrac{1}{2}$. It follows that $\theta = \pi/6, 5\pi/6, 7\pi/6, 11\pi/6$.

1.4. Trigonometric Functions and Their Inverses

1.4.39 The equation $\sqrt{2}\sin(x) - 1 = 0$ can be written as $\sin x = \dfrac{1}{\sqrt{2}} = \dfrac{\sqrt{2}}{2}$. Standard solutions to this equation occur at $x = \pi/4$ and $x = 3\pi/4$. Because the sine function has period 2π the set of all solutions can be written as:

$$\{\pi/4 + 2k\pi, \text{where } k \text{ is an integer}\} \cup \{3\pi/4 + 2l\pi, \text{where } l \text{ is an integer}\}.$$

1.4.41 If $\sin\theta\cos\theta = 0$, then either $\sin\theta = 0$ or $\cos\theta = 0$. This occurs for $\theta = 0, \pi/2, \pi, 3\pi/2$.

1.4.43 Let $u = 3x$. Then we are interested in the solutions to $\cos u = \sin u$, for $0 \leq u < 6\pi$. This would occur for $u = 3x = \pi/4, 5\pi/4, 9\pi/4, 13\pi/4, 17\pi/4,$ and $21\pi/4$. Thus there are solutions for the original equation at

$$x = \pi/12, 5\pi/12, 3\pi/4, 13\pi/12, 17\pi/12, \text{ and } 7\pi/4.$$

1.4.45 Using a computer algebra system or graphing calculator, we find that the roots are approximately 0.1007 and 1.4701.

1.4.47 We are seeking solutions to the equation $400 = \dfrac{150^2}{32}\sin 2\theta$, or $\sin 2\theta = 0.56\overline{8}$. Using a computer algebra system or graphing calculator, we find that the solutions are about 0.30257 radians or about 17.3 degrees, and about 1.2682 radians which is about 72.7 degrees.

1.4.49 Let $z = \sin^{-1} 1$. Then $\sin z = 1$, and because $\sin \pi/2 = 1$, and $\pi/2$ is in the desired interval, $z = \pi/2$.

1.4.51 Let $z = \sin^{-1}(-1/2)$. Then $\sin z = 1/2$, and because $\sin(-\pi/6) = -1/2$, and $-\pi/6$ is in the desired interval, $z = -\pi/6$.

1.4.53 $\sin^{-1}(\sqrt{3}/2) = \pi/3$, because $\sin(\pi/3) = \sqrt{3}/2$.

1.4.55 $\cos^{-1}(-1/2) = 2\pi/3$, because $\cos(2\pi/3) = -1/2$.

1.4.57 $\cos(\cos^{-1}(-1)) = \cos(\pi) = -1$.

1.4.59 Because $\theta = \cos^{-1}(5/13)$, we know that $\cos\theta = 5/13$. The triangle in question has a leg of length 5 and a hypotenuse of length 13, so we can deduce using the Pythagorean theorem that the other leg as length 12. So $\sin\theta = 12/13$. Then $\tan\theta = \dfrac{\sin\theta}{\cos\theta} = \dfrac{12/13}{5/13} = \dfrac{12}{5}$.

1.4.61

$$\cos(\sin^{-1}(x)) = \dfrac{\text{side adjacent to } \sin^{-1}(x)}{\text{hypotenuse}} = \dfrac{\sqrt{1-x^2}}{1} = \sqrt{1-x^2}.$$

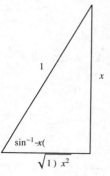

1.4.63

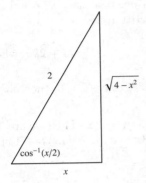

$$\sin(\cos^{-1}(x/2)) = \frac{\text{side opposite of } \cos^{-1}(x/2)}{\text{hypotenuse}} = \frac{\sqrt{4-x^2}}{2}.$$

1.4.65

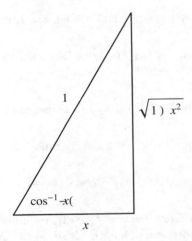

Using the identity given, we have $\sin(2\cos^{-1}(x)) = 2\sin(\cos^{-1}(x))\cos(\cos^{-1}(x)) = 2x\sin(\cos^{-1}(x)) = 2x\sqrt{1-x^2}$.

1.4.67 From our definitions of the trigonometric functions via a point $P(x,y)$ on a circle of radius $r = \sqrt{x^2+y^2}$, we have $\sec\theta = \frac{r}{x} = \frac{1}{x/r} = \frac{1}{\cos\theta}$.

1.4.69 We have already established that $\sin^2\theta + \cos^2\theta = 1$. Dividing both sides by $\cos^2\theta$ gives $\tan^2\theta + 1 = \sec^2\theta$.

1.4. Trigonometric Functions and Their Inverses

1.4.71

Using the triangle pictured, we see that
$$\sec(\pi/2 - \theta) = \frac{c}{a} = \csc\theta.$$

This also follows from the sum identity $\cos(a+b) = \cos a \cos b - \sin a \sin b$ as follows: $\sec(\pi/2 - \theta) = \dfrac{1}{\cos(\pi/2 + (-\theta))} = \dfrac{1}{\cos(\pi/2)\cos(-\theta) - \sin(\pi/2)\sin(-\theta)} = \dfrac{1}{0 - (-\sin(\theta))} = \csc(\theta).$

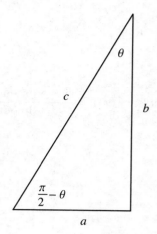

1.4.73

Let $\theta = \cos^{-1}(x)$, and note from the diagram that it then follows that $\cos^{-1}(-x) = \pi - \theta$. So $\cos^{-1}(x) + \cos^{-1}(-x) = \theta + \pi - \theta = \pi.$

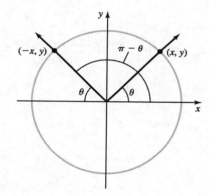

1.4.75 $\tan^{-1}(\sqrt{3}) = \tan^{-1}\left(\dfrac{\sqrt{3}/2}{1/2}\right) = \pi/3$, because $\sin(\pi/3) = \sqrt{3}/2$ and $\cos(\pi/3) = 1/2$.

1.4.77 $\sec^{-1}(2) = \sec^{-1}\left(\dfrac{1}{1/2}\right) = \pi/3$, because $\sec(\pi/3) = \dfrac{1}{\cos(\pi/3)} = \dfrac{1}{1/2} = 2.$

1.4.79 $\tan^{-1}(\tan(\pi/4)) = \tan^{-1}(1) = \pi/4.$

1.4.81 Let $\csc^{-1}(\sec 2) = z$. Then $\csc z = \sec 2$, so $\sin z = \cos 2$. Now by applying the result of problem 64, we see that $z = \sin^{-1}(\cos 2) = \pi/2 - 2 = \dfrac{\pi - 4}{2}.$

Copyright © 2019 Pearson Education, Inc.

1.4.83 $\cos(\tan^{-1}(x)) = \dfrac{\text{side adjacent to } \tan^{-1}(x)}{\text{hypotenuse}} = \dfrac{1}{\sqrt{1+x^2}}.$

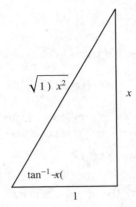

1.4.85

$\cos(\sec^{-1}(x)) = \dfrac{\text{side adjacent to } \sec^{-1} x}{\text{hypotenuse}} = \dfrac{1}{x}.$

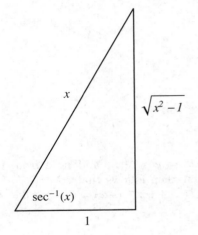

1.4.87

Assume $x > 0$. Then $\sin\left(\sec^{-1}\left(\dfrac{\sqrt{x^2+16}}{4}\right)\right) =$

$\dfrac{\text{side opposite of } \sec^{-1}\left(\dfrac{\sqrt{x^2+16}}{4}\right)}{\text{hypotenuse}} =$

$\dfrac{x}{\sqrt{x^2+16}}.$

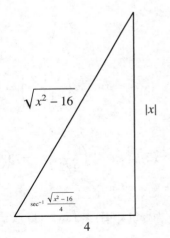

1.4.89 Because $\sin\theta = \dfrac{x}{6}$, $\theta = \sin^{-1}(x/6)$. Also, $\theta = \tan^{-1}\left(\dfrac{x}{\sqrt{36-x^2}}\right) = \sec^{-1}\left(\dfrac{6}{\sqrt{36-x^2}}\right).$

1.4. Trigonometric Functions and Their Inverses

1.4.91

a. False. For example, $\sin(\pi/2 + \pi/2) = \sin(\pi) = 0 \neq \sin(\pi/2) + \sin(\pi/2) = 1 + 1 = 2$.

b. False. That equation has zero solutions, because the range of the cosine function is $[-1, 1]$.

c. False. It has infinitely many solutions of the form $\pi/6 + 2k\pi$, where k is an integer (among others.)

d. False. It has period $\dfrac{2\pi}{\pi/12} = 24$.

e. True. The others have a range of either $[-1, 1]$ or $(-\infty, -1] \cup [1, \infty)$.

f. False. For example, suppose $x = .5$. Then $\sin^{-1}(x) = \pi/6$ and $\cos^{-1}(x) = \pi/3$, so that $\dfrac{\sin^{-1}(x)}{\cos^{-1}(x)} = \dfrac{\pi/6}{\pi/3} = .5$. However, note that $\tan^{-1}(.5) \neq .5$,

g. True. Note that the range of the inverse cosine function is $[0, \pi]$.

h. False. For example, if $x = .5$, we would have $\sin^{-1}(.5) = \pi/6 \neq 1/\sin(.5)$.

1.4.93 If $\cos\theta = 5/13$, then the Pythagorean identity gives $|\sin\theta| = 12/13$. But if $0 < \theta < \pi/2$, then the sine of θ is positive, so $\sin\theta = 12/13$. Thus $\tan\theta = 12/5$, $\cot\theta = 5/12$, $\sec\theta = 13/5$, and $\csc\theta = 13/12$.

1.4.95 If $\csc\theta = 13/12$, then $\sin\theta = 12/13$, and the Pythagorean identity gives $|\cos\theta| = 5/13$. But if $0 < \theta < \pi/2$, then the cosine of θ is positive, so $\cos\theta = 5/13$. Thus $\tan\theta = 12/5$, $\cot\theta = 5/12$, and $\sec\theta = 13/5$.

1.4.97 The amplitude is 3, and the period is $\dfrac{2\pi}{1/3} = 6\pi$.

1.4.99 The amplitude is 3.6, and the period is $\dfrac{2\pi}{\pi/24} = 48$.

1.4.101 Note that $\sin A = \dfrac{h}{c}$ and $\sin C = \dfrac{h}{a}$, so $h = c \sin A = a \sin C$. Thus

$$\frac{\sin A}{a} = \frac{\sin C}{c}.$$

Now drop a perpendicular from the vertex A to the line determined by $\overline{BC}$, and let h_2 be the length of this perpendicular. Then $\sin C = \dfrac{h_2}{b}$ and $\sin B = \dfrac{h_2}{c}$, so $h_2 = b \sin C = c \sin B$. Thus

$$\frac{\sin C}{c} = \frac{\sin B}{b}.$$

Putting the two displayed equations together gives

$$\frac{\sin A}{a} = \frac{\sin B}{b} = \frac{\sin C}{c}.$$

1.4.103 The area of the entire circle is πr^2. The ratio $\dfrac{\theta}{2\pi}$ represents the proportion of the area swept out by a central angle θ. Thus the area of a sector of a circle is this same proportion of the entire area, so it is $\dfrac{\theta}{2\pi} \cdot \pi r^2 = \dfrac{r^2 \theta}{2}$.

1.4.105

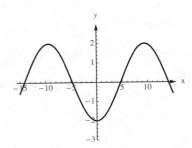

1.4.107

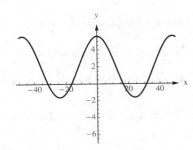

1.4.109 It is helpful to imagine first shifting the function horizontally so that the x intercept is where it should be, then stretching the function horizontally to obtain the correct period, and then stretching the function vertically to obtain the correct amplitude, and then shifting the whole graph up. Because the old x-intercept is at $x = 0$ and the new one should be at $x = 9$ (halfway between where the maximum and the minimum occur), we need to shift the function 9 units to the right. Then to get the right period, we need to multiply (before applying the sine function) by $\pi/12$ so that the new period is $\dfrac{2\pi}{\pi/12} = 24$. Finally, to get the right amplitude and to get the max and min at the right spots, we need to multiply on the outside by 3, and then shift the whole thing up 13 units. Thus, the desired function is:

$$f(x) = 3\sin((\pi/12)(x - 9)) + 13 = 3\sin((\pi/12)x - 3\pi/4) + 13.$$

1.4.111 Let C be the circumference of the earth. Then the first rope has radius $r_1 = \dfrac{C}{2\pi}$. The circle generated by the longer rope has circumference $C + 38$, so its radius is $r_2 = \dfrac{C + 38}{2\pi} = \dfrac{C}{2\pi} + \dfrac{38}{2\pi} \approx r_1 + 6$, so the radius of the bigger circle is about 6 feet more than the smaller circle.

1.4.113 We are seeking a function with amplitude 10 and period 1.5, and value 10 at time 0, so it should have the form $10\cos(kt)$, where $\dfrac{2\pi}{k} = 1.5$. Solving for k yields $k = \dfrac{4\pi}{3}$, so the desired function is $d(t) = 10\cos(4\pi t/3)$.

1.4.107 The area of the entire circle is πr^2. The ratio $\dfrac{\theta}{2\pi}$ represents the proportion of the area swept out by a central angle θ. Thus the area of a sector of a circle is this same proportion of the entire area, so it is $\dfrac{\theta}{2\pi} \cdot \pi r^2 = \dfrac{r^2 \theta}{2}$.

1.4.116

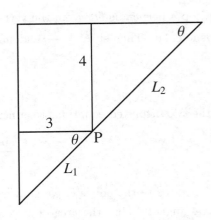

Let the corner point P divide the pole into two pieces, L_1 (which spans the 3-ft hallway) and L_2 (which spans the 4-ft hallway.) Then $L = L_1 + L_2$.
Now $L_2 = \dfrac{4}{\sin\theta}$, and $\dfrac{3}{L_1} = \cos\theta$ (see diagram.)
Thus $L = L_1 + L_2 = \dfrac{3}{\cos\theta} + \dfrac{4}{\sin\theta}$. When $L = 10$, $\theta \approx .9273$.

Chapter One Review

1

a. True. For example, $f(x) = x^2$ is such a function.

b. False. For example, $\cos(\pi/2 + \pi/2) = \cos(\pi) = -1 \neq \cos(\pi/2) + \cos(\pi/2) = 0 + 0 = 0$.

c. False. Consider $f(1+1) = f(2) = 2m + b \neq f(1) + f(1) = (m+b) + (m+b) = 2m + 2b$. (At least these aren't equal when $b \neq 0$.)

d. True. $f(f(x)) = f(1-x) = 1 - (1-x) = x$.

e. False. This set is the union of the disjoint intervals $(-\infty, -7)$ and $(1, \infty)$.

f. False. For example, if $x = y = 10$, then $\log_{10} xy = \log_{10} 100 = 2$, but $\log_{10} 10 \cdot \log_{10} 10 = 1 \cdot 1 = 1$.

g. True. $\sin^{-1}(\sin(2\pi)) = \sin^{-1}(0) = 0$.

3 f is a one-to-one function, but g isn't (it fails the horizontal line test.)

5 The denominator must not be zero, so we must have $w \neq 2$. The domain is $\{w : w \neq 2\}$. Note that when $w \neq 2$, the function becomes $\dfrac{(w-2)(2w+1)}{w-2} = 2w + 1$. So the graph of f is a line of slope 2 with the point $(2, 5)$ missing, so the range is $\{y : y \neq 5\}$.

7 Because h can be written $h(z) = \sqrt{(z-3)(z+1)}$, we see that the domain is $(-\infty, -1] \cup [3, \infty)$. The range is $[0, \infty)$. (Note that as z gets large, $h(z)$ gets large as well.)

9 Yes, $\tan(\tan^{-1} x) = x$ because the range of the inverse tangent function is a subset of the domain of the tangent function, and the functions are inverses. However, $\tan^{-1}(\tan x)$ does not always equal x, for example: $\tan^{-1}(\tan \pi) = \tan^{-1} 0 = 0$.

11 $g(f(4)) = g(5) = 8$.

13 $g^{-1}(5) = 7$ (Because $g(7) = 5$.)

15 $g^{-1}(f(3)) = g^{-1}(4) = 8$.

17 $f^{-1}(1 + f(-3)) = f^{-1}(1 - f(3)) = f^{-1}(1 - 4) = f^{-1}(-3) = -f(3) = -2$.

19

a. $h(g(\pi/2)) = h(1) = 1$

b. $h(f(x)) = h(x^3) = x^{3/2}$.

c. $f(g(h(x))) = f(g(\sqrt{x})) = f(\sin(\sqrt{x})) = (\sin(\sqrt{x}))^3$.

d. The domain of $g(f(x))$ is $\mathbb{R}$, because the domain of both functions is the set of all real numbers.

e. The range of $f(g(x))$ is $[-1, 1]$. This is because the range of g is $[-1, 1]$, and on the restricted domain $[-1, 1]$, the range of f is also $[-1, 1]$.

21

$$\frac{f(x+h) - f(x)}{h} = \frac{(x+h)^2 - 2(x+h) - (x^2 - 2x)}{h} = \frac{x^2 + 2hx + h^2 - 2x - 2h - x^2 + 2x}{h}$$

$$= \frac{2hx + h^2 - 2h}{h} = 2x + h - 2.$$

$$\frac{f(x) - f(a)}{x - a} = \frac{x^2 - 2x - (a^2 - 2a)}{x - a} = \frac{(x^2 - a^2) - 2(x - a)}{x - a} = \frac{(x-a)(x+a) - 2(x-a)}{x - a} = x + a - 2.$$

23

$$\frac{f(x+h)-f(x)}{h} = \frac{(x+h)^3+2-(x^3+2)}{h} = \frac{x^3+3x^2h+3xh^2+h^3+2-x^3-2}{h}$$
$$= \frac{h(3x^2+3xh+h^2)}{h} = 3x^2+3xh+h^2.$$

$$\frac{f(x)-f(a)}{x-a} = \frac{x^3+2-(a^3+2)}{x-a} = \frac{x^3-a^3}{x-a} = \frac{(x-a)(x^2+ax+a^2)}{x-a} = x^2+ax+a^2.$$

25

a. This line has slope $\frac{2-(-3)}{4-2} = \frac{5}{2}$. Therefore the equation of the line is $y - 2 = \frac{5}{2}(x-4)$, so $y = \frac{5}{2}x - 8$.

b. This line has the form $y = \frac{3}{4}x + b$, and because $(-4, 0)$ is on the line, $0 = (3/4)(-4) + b$, so $b = 3$. Thus the equation of the line is given by $y = \frac{3}{4}x + 3$.

c. This line has slope $\frac{0-(-2)}{4-0} = \frac{1}{2}$, and the y-intercept is given to be -2, so the equation of this line is $y = \frac{1}{2}x - 2$.

27 We are looking for the line between the points $(0, 212)$ and $(6000, 200)$. The slope is $\frac{212-200}{0-6000} = -\frac{12}{6000} = -\frac{1}{500}$. Because the intercept is given, we deduce that the line is $B = f(a) = -\frac{1}{500}a + 212$.

29

a. This is a straight line with slope $2/3$ and y-intercept $10/3$.

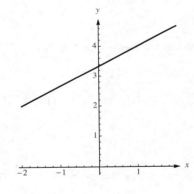

b. Completing the square gives $y = (x^2+2x+1) - 4$, or $y = (x+1)^2 - 4$, so this is the standard parabola shifted one unit to the left and down 4 units.

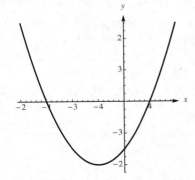

Chapter One Review

c. Completing the square, we have $x^2+2x+1+y^2+4y+4 = -1+1+4$, so we have $(x+1)^2+(y+2)^2 = 4$, a circle of radius 2 centered at $(-1,-2)$.

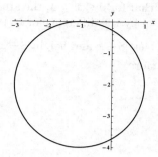

d. Completing the square, we have $x^2-2x+1+y^2-8y+16 = -5+1+16$, or $(x-1)^2+(y-4)^2 = 12$, which is a circle of radius $\sqrt{12}$ centered at $(1,4)$.

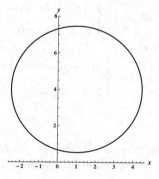

31

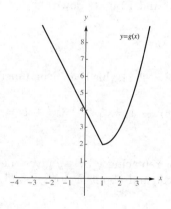

33

a.

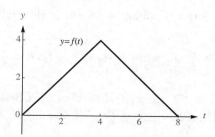

b. $A(2) = \dfrac{1}{2} \cdot 2 \cdot 2 = 2$ and $A(6) = 16 - \dfrac{1}{2} \cdot 2 \cdot 2 = 14$.

c. Note that for $0 \le x \le 4$, the area is that of a triangle with base x and height x. For $4 \le x \le 8$, the area is given by the difference of 16 (the total area under the curve from 0 to 8) minus the area of a triangle with base $8-x$ and height $8-x$. So the area for x in that range is $16 - \dfrac{(8-x)^2}{2} = 16 - 32 + 8x - \dfrac{x^2}{2}$. Therefore,

$$A(x) = \begin{cases} \frac{x^2}{2} & \text{if } 0 \le x \le 4 \\ -\frac{x^2}{2} + 8x - 16 & \text{if } 4 \le x \le 8. \end{cases}$$

35

Because $|x| = \begin{cases} -x & \text{if } x < 0; \\ x & \text{if } x \ge 0, \end{cases}$

we have

$2(x - |x|) = \begin{cases} 2(x - (-x)) = 4x & \text{if } x < 0; \\ 2(x - x) = 0 & \text{if } x \ge 0. \end{cases}$

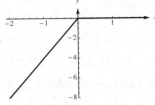

37 The domain of $x^{1/7}$ is the set of all real numbers, as is its range. The domain of $x^{1/4}$ is the set of non-negative real numbers, as is its range.

39 Completing the square, we can write $x^2 + 6x - 3 = x^2 + 6x + 9 - 3 - 9 = (x+3)^2 - 12$, so the graph is obtained by shifting $y = x^2$ 3 units right and 12 units down.

41

a. Because $f(-x) = \cos -3x = \cos 3x = f(x)$, this is an even function, and is symmetric about the y-axis.

b. Because $f(-x) = 3(-x)^4 - 3(-x)^2 + 1 = 3x^4 - 3x^2 + 1 = f(x)$, this is an even function, and is symmetric about the y-axis.

c. Because replacing x by $-x$ and/or replacing y by $-y$ gives the same equation, this represents a curve which is symmetric about the y-axis and about the origin and about the x-axis.

43 If $\log x^2 + 3 \log x = \log 32$, then $\log(x^2 \cdot x^3) = \log(32)$, so $x^5 = 32$ and $x = 2$. The answer does not depend on the base of the log.

45 If $3\ln(5t+4) = 12$, then $\ln(5t+4) = 4$, and $e^4 = 5t + 4$. It then follows that $5t = e^4 - 4$ and so $t = \dfrac{e^4 - 4}{5}$.

47 If $1 - 2\sin^2 \theta = 0$, then $\sin^2 \theta = \frac{1}{2}$, so $|\sin \theta| = \frac{1}{\sqrt{2}} = \frac{\sqrt{2}}{2}$. So $\theta = \frac{\pi}{4}, \frac{3\pi}{4}, \frac{5\pi}{4}, \frac{7\pi}{4}$.

49 First note that if θ is between $-\pi/2$ and $\pi/2$, that 2θ is then between $-\pi$ and π. If $4\cos^2 2\theta = 3$, then $\cos^2 2\theta = \frac{3}{4}$, and $|\cos 2\theta| = \frac{\sqrt{3}}{2}$. Thus $2\theta = \pm\frac{\pi}{6}, \pm\frac{5\pi}{6}$, and $\theta = \pm\frac{\pi}{12}, \pm\frac{5\pi}{12}$.

51 In 2010 (when $t = 0$), the population is $P(0) = 100$. So we are seeking t so that $200 = 100e^{t/50}$, or $e^{t/50} = 2$. Taking the natural logarithm of both sides yields $\frac{t}{50} = \ln 2$, or $t = 50 \ln 2 \approx 35$ years.

53

By graphing, it is clear that this function is not one-to-one on its whole domain, but it is one-to-one on the interval $(-\infty, 0]$, on the interval $[0, 2]$, and on the interval $[2, \infty)$, so it would have an inverse if we restricted it to any of these particular intervals.

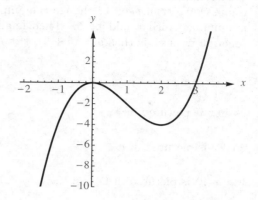

55 Switching x and y gives $x = 6 - 4y$. Then $x - 6 = -4y$, so
$$y = \frac{x-6}{-4} = \frac{6-x}{4} = -\frac{1}{4}x + \frac{3}{2}.$$

57 Completing the square gives $f(x) = x^2 - 4x + 4 + 1 = (x-2)^2 + 1$. Switching the x and y and solving for y yields $(y-2)^2 = x - 1$, so $|y - 2| = \sqrt{x-1}$, and thus $y = f^{-1}(x) = 2 + \sqrt{x-1}$ (we choose the "+" rather than the "−" because the domain of f is $x \geq 2$, so the range of f^{-1} must also consist of numbers greater than or equal to 2).

59 Switching x and y gives $x = 3y^2 + 1$, so $3y^2 = x - 1$, and $y^2 = \frac{x-1}{3}$. Then $|y| = \sqrt{\frac{x-1}{3}}$, and $y = -\sqrt{\frac{x-1}{3}}$ (we choose the "−" rather than the "+" because the domain of f is $x \leq 0$, so the range of f^{-1} must also consist of numbers less than or equal to 0).

61 Switching x and y gives $x = e^{y^2+1}$. Then $\ln x = y^2 + 1$, so $y^2 = \ln x - 1$, and $|y| = \sqrt{\ln x - 1}$, so $y = \sqrt{\ln x - 1}$ (we choose the "+" rather than the "−" because the domain of f is $x \geq 0$, so the range of f^{-1} must also consist of numbers greater than or equal to 0).

63

Switching x and y gives $x = \frac{6\sqrt{y}}{\sqrt{y}+2}$. Then $x\sqrt{y} + 2x = 6\sqrt{y}$, and so $x\sqrt{y} - 6\sqrt{y} = -2x$. Then $\sqrt{y}(x-6) = -2x$, so $\sqrt{y} = \frac{2x}{6-x}$, and $y = \left(\frac{2x}{6-x}\right)^2 = \frac{4x^2}{(6-x)^2}$. Because the range of f must match the domain of f^{-1}, we must restrict our inverse function to $[0, 6]$.

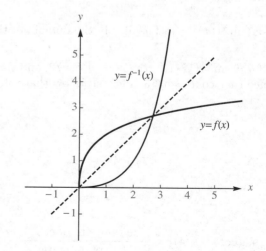

65

a. We need to scale the ordinary cosine function so that its period is 6, and then shift it 3 units to the right, and multiply it by 2. So the function we seek is $y = 2\cos((\pi/3)(t-3)) = -2\cos(\pi t/3)$.

b. We need to scale the ordinary cosine function so that its period is 24, and then shift it to the right 6 units. We then need to change the amplitude to be half the difference between the maximum and minimum, which would be 5. Then finally we need to shift the whole thing up by 15 units. The function we seek is thus $y = 15 + 5\cos((\pi/12)(t-6)) = 15 + 5\sin(\pi t/12)$.

67

a. $-\sin x$ is pictured in F.

b. $\cos 2x$ is pictured in E.

c. $\tan(x/2)$ is pictured in D.

d. $-\sec x$ is pictured in B.

e. $\cot 2x$ is pictured in C.

f. $\sin^2 x$ is pictured in A.

69 $\sin x = -\frac{1}{2}$ for $x = 7\pi/6$ and for $x = 11\pi/6$, so the intersection points are $(7\pi/6, -1/2)$ and $(11\pi/6, -1/2)$.

71 Note that $\frac{7\pi}{8} = \frac{7\pi/4}{2}$. Using the half-angle identity,

$$\cos\left(\frac{7\pi/4}{2}\right) = -\sqrt{\frac{1+\cos 7\pi/4}{2}} = -\sqrt{\frac{1+\sqrt{2}/2}{2}} = -\sqrt{\frac{2+\sqrt{2}}{4}} = -\frac{\sqrt{2+\sqrt{2}}}{2}.$$

73 Because $\cos(\pi/6) = \sqrt{3}/2$, $\cos^{-1}(\sqrt{3}/2) = \pi/6$.

75 Because $\sin(-\pi/2) = -1$, $\sin^{-1}(-1) = -\pi/2$.

77 $\sin(\sin^{-1}(x)) = x$, for all x in the domain of the inverse sine function.

79 If $\theta = \sin^{-1}(12/13)$, then $0 < \theta < \pi/2$, and $\sin\theta = 12/13$. Then (using the Pythagorean identity) we can deduce that $\cos\theta = 5/13$. It must follow that $\tan\theta = 12/5$, $\cot\theta = 5/12$, $\sec\theta = 13/5$, and $\csc\theta = 13/12$.

81

$$\sin(\cos^{-1}(x/4)) = \frac{\text{side opposite of } \cos^{-1}(x/4)}{\text{hypotenuse}} = \frac{\sqrt{16-x^2}}{4}.$$

83

Note that
$$\tan\theta = \frac{a}{b} = \cot(\pi/2 - \theta).$$

Thus,
$$\cot^{-1}(\tan\theta) = \cot^{-1}(\cot(\pi/2 - \theta)) = \pi/2 - \theta.$$

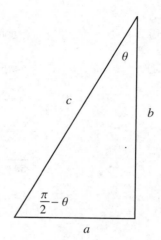

85 Let $\theta = \sin^{-1}(x)$. Then $\sin\theta = x$ and note that then $\sin(-\theta) = -\sin\theta = -x$, so $-\theta = \sin^{-1}(-x)$. Then $\sin^{-1}(x) + \sin^{-1}(-x) = \theta + -\theta = 0$.

87 Using the definition of the tangent function in terms of sine and cosine, we have:
$$\tan 2\theta = \frac{\sin 2\theta}{\cos 2\theta} = \frac{2\sin\theta\cos\theta}{\cos^2\theta - \sin^2\theta}.$$

If we divide both the numerator and denominator of this last expression by $\cos^2\theta$, we obtain
$$\frac{2\tan\theta}{1 - \tan^2\theta}.$$

89

a.

n	1	2	3	4	5	6	7	8	9	10
$T(n)$	1	5	14	30	55	91	140	204	285	385

b. The domain of this function consists of the positive integers.

c. Using trial and error and a calculator yields that $T(n) > 1000$ for the first time for $n = 14$.

91 To find $s(t)$ note that we are seeking a periodic function with period 365, and with amplitude 87.5 (which is half of the number of minutes between 7:25 and 4:30). We need to shift the function 4 days plus one fourth of 365, which is about 95 days so that the max and min occur at $t = 4$ days and at half a year later. Also, to get the right value for the maximum and minimum, we need to multiply by negative one and add 117.5 (which represents 30 minutes plus half the amplitude, because $s = 0$ corresponds to 4:00 AM.) Thus we have
$$s(t) = 117.5 - 87.5\sin\left(\frac{\pi}{182.5}(t - 95)\right).$$

A similar analysis leads to the formula
$$S(t) = 844.5 + 87.5\sin\left(\frac{\pi}{182.5}(t - 67)\right).$$

The graph pictured shows $D(t) = S(t) - s(t)$, the length of day function, which has its max at the summer solstice which is about the 172nd day of the year, and its min at the winter solstice.

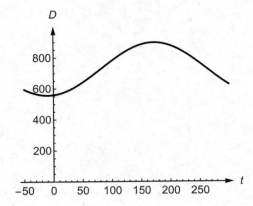

Chapter 2

Limits

2.1 The Idea of Limits

2.1.1 The average velocity of the object between time $t = a$ and $t = b$ is the change in position divided by the elapsed time: $v_{av} = \dfrac{s(b) - s(a)}{b - a}$.

2.1.3 The average velocity is $\dfrac{s(3) - s(2)}{3 - 2} = 156 - 136 = 20$.

2.1.5

a. $\dfrac{s(2) - s(0)}{2 - 0} = \dfrac{72 - 0}{2} = 36$.

b. $\dfrac{s(1.5) - s(0)}{1.5 - 0} = \dfrac{66 - 0}{1.5} = 44$.

c. $\dfrac{s(1) - s(0)}{1 - 0} = \dfrac{52 - 0}{1} = 52$.

d. $\dfrac{s(.5) - s(0)}{.5 - 0} = \dfrac{30 - 0}{.5} = 60$.

2.1.7 $\dfrac{s(1.01) - s(1)}{.01} = 47.84$, while $\dfrac{s(1.001) - s(1)}{.001} = 47.984$ and $\dfrac{s(1.0001) - s(1)}{.0001} = 47.9984$. It appears that the instantaneous velocity at $t = 1$ is approximately 48.

2.1.9 The slope of the secant line between points $(a, f(a))$ and $(b, f(b))$ is the ratio of the differences $f(b) - f(a)$ and $b - a$. Thus $m_{\text{sec}} = \dfrac{f(b) - f(a)}{b - a}$.

2.1.11 Both problems involve the same mathematics, namely finding the limit as $t \to a$ of a quotient of differences of the form $\dfrac{g(t) - g(a)}{t - a}$ for some function g.

2.1.13

a. Over $[1, 4]$, we have $v_{av} = \dfrac{s(4) - s(1)}{4 - 1} = \dfrac{256 - 112}{3} = 48$.

b. Over $[1, 3]$, we have $v_{av} = \dfrac{s(3) - s(1)}{3 - 1} = \dfrac{240 - 112}{2} = 64$.

c. Over $[1, 2]$, we have $v_{av} = \dfrac{s(2) - s(1)}{2 - 1} = \dfrac{192 - 112}{1} = 80$.

d. Over $[1, 1+h]$, we have

$$v_{\text{av}} = \frac{s(1+h) - s(1)}{1+h-1} = \frac{-16(1+h)^2 + 128(1+h) - (112)}{h} = \frac{-16h^2 - 32h + 128h}{h}$$

$$= \frac{h(-16h + 96)}{h} = 96 - 16h = 16(6-h).$$

2.1.15

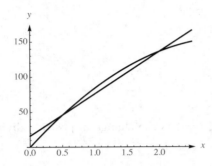

The slope of the secant line is given by $\frac{s(2) - s(0.5)}{2 - 0.5} = \frac{136 - 46}{1.5} = 60$. This represents the average velocity of the object over the time interval $[0.5, 2]$.

2.1.17

Time Interval	$[1, 2]$	$[1, 1.5]$	$[1, 1.1]$	$[1, 1.01]$	$[1, 1.001]$
Average Velocity	80	88	94.4	95.84	95.984

The instantaneous velocity appears to be 96 ft/s.

2.1.19

Time Interval	$[2, 3]$	$[2.9, 3]$	$[2.99, 3]$	$[2.999, 3]$	$[2.9999, 3]$	$[2.99999, 3]$
Average Velocity	20	5.6	4.16	4.016	4.002	4.0002

The instantaneous velocity appears to be 4 ft/s.

2.1.21

Time Interval	$[3, 3.1]$	$[3, 3.01]$	$[3, 3.001]$	$[3, 3.0001]$
Average Velocity	-17.6	-16.16	-16.016	-16.002

The instantaneous velocity appears to be -16 ft/s.

2.1.23

Time Interval	$[0, 0.1]$	$[0, 0.01]$	$[0, 0.001]$	$[0, 0.0001]$
Average Velocity	79.468	79.995	80.000	80.0000

The instantaneous velocity appears to be 80 ft/s.

2.1.25

x Interval	$[2, 2.1]$	$[2, 2.01]$	$[2, 2.001]$	$[2, 2.0001]$
Slope of Secant Line	8.2	8.02	8.002	8.0002

The slope of the tangent line appears to be 8.

2.1.27

x Interval	$[0, 0.1]$	$[0, 0.01]$	$[0, 0.001]$	$[0, 0.0001]$
Slope of the Secant Line	1.05171	1.00502	1.0005	1.00005

The slope of the tangent line appears to be 1.

2.1. The Idea of Limits

2.1.29

a. Note that the graph is a parabola with vertex $(2, -1)$.

b. At $(2, -1)$ the function has tangent line with slope 0.

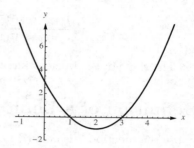

c.

x Interval	$[2, 2.1]$	$[2, 2.01]$	$[2, 2.001]$	$[2, 2.0001]$
Slope of the Secant Line	0.1	0.01	0.001	0.0001

The slope of the tangent line at $(2, -1)$ appears to be 0.

2.1.31

a. Note that the graph is a parabola with vertex $(4, 448)$.

b. At $(4, 448)$ the function has tangent line with slope 0, so $a = 4$.

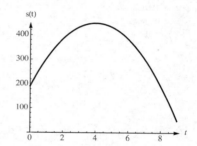

c.

x Interval	$[4, 4.1]$	$[4, 4.01]$	$[4, 4.001]$	$[4, 4.0001]$
Slope of the Secant Line	-1.6	$-.16$	$-.016$	$-.0016$

The slopes of the secant lines appear to be approaching zero.

d. On the interval $[0, 4)$ the instantaneous velocity of the projectile is positive.

e. On the interval $(4, 9]$ the instantaneous velocity of the projectile is negative.

2.1.33 For line AD, we have

$$m_{AD} = \frac{y_D - y_A}{x_D - x_A} = \frac{f(\pi) - f(\pi/2)}{\pi - (\pi/2)} = \frac{1}{\pi/2} \approx 0.63662.$$

For line AC, we have

$$m_{AC} = \frac{y_C - y_A}{x_C - x_A} = \frac{f(\pi/2 + .5) - f(\pi/2)}{(\pi/2 + .5) - (\pi/2)} = -\frac{\cos(\pi/2 + .5)}{.5} \approx 0.958851.$$

For line AB, we have

$$m_{AB} = \frac{y_B - y_A}{x_B - x_A} = \frac{f(\pi/2 + .05) - f(\pi/2)}{(\pi/2 + .05) - (\pi/2)} = -\frac{\cos(\pi/2 + .05)}{.05} \approx 0.999583.$$

Computing one more slope of a secant line:

$$m_{\text{sec}} = \frac{f(\pi/2 + .01) - f(\pi/2)}{(\pi/2 + .01) - (\pi/2)} = -\frac{\cos(\pi/2 + .01)}{.01} \approx 0.999983.$$

Conjecture: The slope of the tangent line to the graph of f at $x = \pi/2$ is 1.

2.2 Definition of a Limit

2.2.1 Suppose the function f is defined for all x near a except possibly at a. If $f(x)$ is arbitrarily close to a number L whenever x is sufficiently close to (but not equal to) a, then we write $\lim_{x \to a} f(x) = L$.

2.2.3

a. $h(2) = 5$.

b. $\lim_{x \to 2} h(x) = 3$.

c. $h(4)$ does not exist.

d. $\lim_{x \to 4} f(x) = 1$.

e. $\lim_{x \to 5} h(x) = 2$.

2.2.5

a. $f(1) = -1$.

b. $\lim_{x \to 1} f(x) = 1$.

c. $f(0) = 2$.

d. $\lim_{x \to 0} f(x) = 2$.

2.2.7

a.
x	1.9	1.99	1.999	1.9999	2.1	2.01	2.001	2.0001
$f(x) = \dfrac{x^2 - 4}{x - 2}$	3.9	3.99	3.999	3.9999	4.1	4.01	4.001	4.0001

b. $\lim_{x \to 2} f(x) = 4$.

2.2.9

a.
t	8.9	8.99	8.999	9.1	9.01	9.001
$g(t) = \dfrac{t - 9}{\sqrt{t} - 3}$	5.98329	5.99833	5.99983	6.01662	6.00167	6.00017

b. $\lim_{t \to 9} \dfrac{t - 9}{\sqrt{t} - 3} = 6$.

2.2.11 Suppose the function f is defined for all x near a but greater than a. If $f(x)$ is arbitrarily close to L for x sufficiently close to (but strictly greater than) a, then we write $\lim_{x \to a^+} f(x) = L$.

2.2.13 It must be true that $L = M$.

2.2. Definition of a Limit

2.2.15

a. $f(1) = 0$.
b. $\lim_{x \to 1^-} f(x) = 1$.
c. $\lim_{x \to 1^+} f(x) = 0$.
d. $\lim_{x \to 1} f(x)$ does not exist, since the two one-sided limits aren't equal.

2.2.17

a. $f(1) = 3$.
b. $\lim_{x \to 1^-} f(x) = 2$.
c. $\lim_{x \to 1^+} f(x) = 2$.
d. $\lim_{x \to 1} f(x) = 2$.
e. $f(3) = 2$.
f. $\lim_{x \to 3^-} f(x) = 4$.
g. $\lim_{x \to 3^+} f(x) = 1$.
h. $\lim_{x \to 3} f(x)$ does not exist.
i. $f(2) = 3$.
j. $\lim_{x \to 2^-} f(x) = 3$.
k. $\lim_{x \to 2^+} f(x) = 3$.
l. $\lim_{x \to 2} f(x) = 3$.

2.2.19

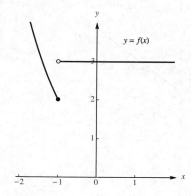

$f(-1) = 2$, $\lim_{x \to -1^-} f(x) = 2$, $\lim_{x \to -1^+} f(x) = 3$, $\lim_{x \to -1} f(x)$ does not exist.

2.2.21

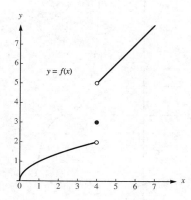

$f(4) = 3$, $\lim_{x \to 4^-} f(x) = 2$, $\lim_{x \to 4^+} f(x) = 5$, $\lim_{x \to 4} f(x)$ does not exist.

2.2.23

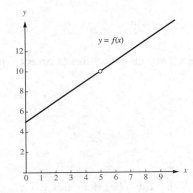

$f(5)$ does not exist. $\lim_{x \to 5^-} f(x) = 10$, $\lim_{x \to 5^+} f(x) = 10$, $\lim_{x \to 5} f(x) = 10$.

2.2.25

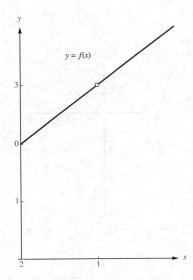

$f(1)$ does not exist. $\lim_{x \to 1^-} f(x) = 3$, $\lim_{x \to 1^+} f(x) = 3$, $\lim_{x \to 1} f(x) = 3$.

2.2.27

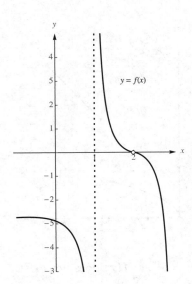

2.2. Definition of a Limit

x	1.99	1.999	1.9999
$f(x) = \frac{x-2}{\ln|x-2|}$	0.0021715	0.00014476	0.000010857

x	2.0001	2.001	2.01
$f(x) = \frac{x-2}{\ln|x-2|}$	-0.000010857	-0.00014476	-0.0021715

From the graph and the table, the limit appears to be 0.

2.2.29

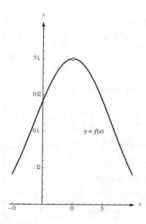

x	0.9	0.99	0.999	1.001	1.01	1.1
$f(x)$	1.993342	1.999933	1.999999	1.999999	1.999933	1.993342

From both the graph and the table, the limit appears to be 2.

2.2.31

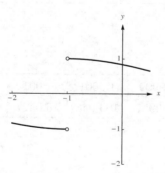

x	-1.1	-1.01	-1.001	-0.999	-0.99	-0.9
$f(x)$	-0.9983342	-0.9999833	-0.9999998	0.9999998	0.9999833	0.9983342

From both the graph and the table, it appears that the limit does not exist.

2.2.33

a. False. In fact $\lim\limits_{x \to 3} \dfrac{x^2 - 9}{x - 3} = \lim\limits_{x \to 3}(x+3) = 6$.

b. False. For example, if $f(x) = \begin{cases} x^2 & \text{if } x \neq 0; \\ 5 & \text{if } x = 0 \end{cases}$ and if $a = 0$ then $f(a) = 5$ but $\lim\limits_{x \to a} f(x) = 0$.

c. False. For example, the limit in part a of this problem exists, even though the corresponding function is undefined at $a = 3$.

d. False. It is true that the limit of $\sqrt{x}$ as x approaches zero from the right is zero, but because the domain of $\sqrt{x}$ does not include any numbers to the left of zero, the two-sided limit doesn't exist.

e. True. Note that $\lim_{x \to \pi/2} \cos x = 0$ and $\lim_{x \to \pi/2} \sin x = 1$, so $\lim_{x \to \pi/2} \dfrac{\cos x}{\sin x} = \dfrac{0}{1} = 0$.

2.2.35

a. Note that the function is piecewise constant.

b. $\lim_{w \to 2.3} f(w) = .89$.

c. $\lim_{w \to 3^+} f(w) = 1.1$ corresponds to the fact that for any piece of mail that weighs slightly over 3 ounces, the postage will cost $1.1 cents. $\lim_{w \to 3^-} f(w) = \$0.89$ corresponds to the fact that for any piece of mail that weighs slightly less than 3 ounces, the postage will cost 89 cents. Because the two one-sided limits are not equal, $\lim_{w \to 3} f(w)$ does not exist.

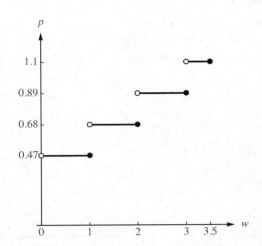

2.2.37

x	1.37	1.47	1.57	1.58	1.68	1.78
$\dfrac{\cot 3x}{\cos x}$	3.44773	3.10016	3.00001	3.0008	3.11834	3.49316

$\lim_{x \to \pi/2} \dfrac{\cot 3x}{\cos x} = 3$.

2.2.39

x	0.99	0.999	0.9999	1.0001	1.001	1.01
$\dfrac{18(\sqrt[3]{x}-1)}{x^3-1}$	15.5803	15.9574	15.9957	16.0043	16.0427	16.4339

$\lim_{x \to 1} \dfrac{9\sqrt{2x-x^4} - \sqrt[3]{x}}{1 - x^{3/4}} = 16$.

2.2.41

h	0.01	0.001	0.0001	-0.0001	-0.001	-0.01
$\dfrac{\ln(1+h)}{h}$	0.995033	0.9995	0.99995	1.00005	1.0005	1.00503

$\lim_{h \to 0} \dfrac{\ln(1+h)}{h} = 1$.

2.2.43

a.

x	$\dfrac{2}{\pi}$	$\dfrac{2}{3\pi}$	$\dfrac{2}{5\pi}$	$\dfrac{2}{7\pi}$	$\dfrac{2}{9\pi}$	$\dfrac{2}{11\pi}$
$f(x) = \sin(1/x)$	1	-1	1	-1	1	-1

If $x_n = \dfrac{2}{(2n+1)\pi}$, then $f(x_n) = (-1)^n$ where n is a non-negative integer.

2.2. Definition of a Limit

b. As $x \to 0$, $1/x \to \infty$. So the values of $f(x)$ oscillate dramatically between -1 and 1.

c. $\lim_{x \to 0} \sin(1/x)$ does not exist.

2.2.45

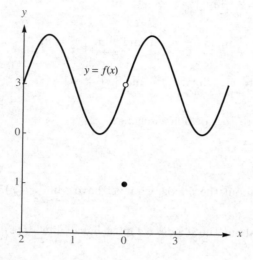

2.2.47

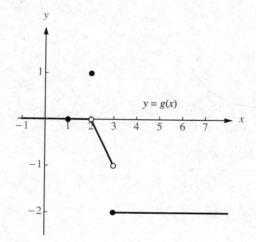

2.2.49

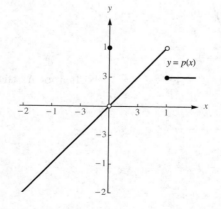

2.2.51

a. $\lim_{x\to -1^-}\lfloor x\rfloor = -2$, $\lim_{x\to -1^+}\lfloor x\rfloor = -1$, $\lim_{x\to 2^-}\lfloor x\rfloor = 1$, $\lim_{x\to 2^+}\lfloor x\rfloor = 2$.

b. $\lim_{x\to 2.3^-}\lfloor x\rfloor = 2$, $\lim_{x\to 2.3^+}\lfloor x\rfloor = 2$, $\lim_{x\to 2.3}\lfloor x\rfloor = 2$.

c. In general, for an integer a, $\lim_{x\to a^-}\lfloor x\rfloor = a-1$ and $\lim_{x\to a^+}\lfloor x\rfloor = a$.

d. In general, if a is not an integer, $\lim_{x\to a^-}\lfloor x\rfloor = \lim_{x\to a^+}\lfloor x\rfloor = \lfloor a\rfloor$.

e. $\lim_{x\to a}\lfloor x\rfloor$ exists and is equal to $\lfloor a\rfloor$ for non-integers a.

2.2.53

a. Because of the symmetry about the y axis, we must have $\lim_{x\to -2^+} f(x) = 8$.

b. Because of the symmetry about the y axis, we must have $\lim_{x\to -2^-} f(x) = 5$.

2.2.55

a.
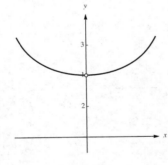
$$\lim_{x\to 0}\frac{\tan 2x}{\sin x} = 2.$$

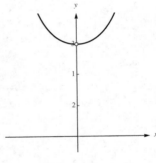
$$\lim_{x\to 0}\frac{\tan 3x}{\sin x} = 3.$$

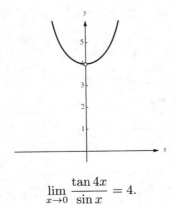
$$\lim_{x\to 0}\frac{\tan 4x}{\sin x} = 4.$$

b. It appears that $\lim_{x\to 0}\dfrac{\tan(px)}{\sin x} = p$.

2.2.57

For $p = 8$ and $q = 2$, it appears that the limit is 4.

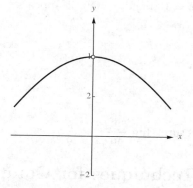

For $p = 12$ and $q = 3$, it appears that the limit is 4.

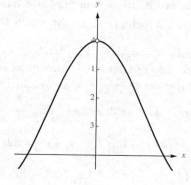

For $p = 4$ and $q = 16$, it appears that the limit is $1/4$.

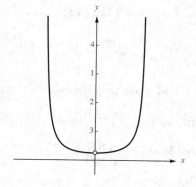

For $p = 100$ and $q = 50$, it appears that the limit is 2.

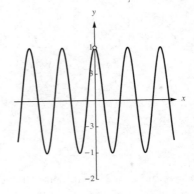

Conjecture: $\lim\limits_{x \to 0} \dfrac{\sin px}{\sin qx} = \dfrac{p}{q}$.

2.3 Techniques for Computing Limits

2.3.1 If $p(x) = a_n x^n + a_{n-1} x^{n-1} + \cdots + a_1 x + a_0$, then $\lim\limits_{x \to a} p(x) = \lim\limits_{x \to a} (a_n x^n + a_{n-1} x^{n-1} + \cdots + a_1 x + a_0)$
$= a_n (\lim\limits_{x \to a} x)^n + a_{n-1} (\lim\limits_{x \to a} x)^{n-1} + \cdots + a_1 \lim\limits_{x \to a} x + \lim\limits_{x \to a} a_0$
$= a_n a^n + a_{n-1} a^{n-1} + \cdots + a_1 a + a_0. = p(a)$.

2.3.3 For a rational function $r(x)$, we have $\lim\limits_{x \to a} r(x) = r(a)$ exactly for those numbers a which are in the domain of r. (Which are those for which the denominator isn't zero.)

2.3.5 Because $\dfrac{x^2 - 7x + 12}{x - 3} = \dfrac{(x-3)(x-4)}{x-3} = x - 4$ (for $x \neq 3$), we can see that the graphs of these two functions are the same except that one is undefined at $x = 3$ and the other is a straight line that is defined everywhere. Thus the function $\dfrac{x^2 - 7x + 12}{x - 3}$ is a straight line except that it has a "hole" at $(3, -1)$. The two functions have the same limit as $x \to 3$, namely $\lim\limits_{x \to 3} \dfrac{x^2 - 7x + 12}{x - 3} = \lim\limits_{x \to 3} (x - 4) = -1$.

2.3.7 $\lim\limits_{x \to 1} 4f(x) = 4 \lim\limits_{x \to 1} f(x) = 4 \cdot 8 = 32$. This follows from the Constant Multiple Law.

2.3.9 $\lim\limits_{x \to 1} (f(x) - g(x)) = \lim\limits_{x \to 1} f(x) - \lim\limits_{x \to 1} g(x) = 8 - 3 = 5$. This follows from the Difference Law.

2.3.11 $\lim\limits_{x \to 1} \dfrac{f(x)}{g(x) - h(x)} = \dfrac{\lim\limits_{x \to 1} f(x)}{\lim\limits_{x \to 1} [g(x) - h(x)]} = \dfrac{\lim\limits_{x \to 1} f(x)}{\lim\limits_{x \to 1} g(x) - \lim\limits_{x \to 1} h(x)} = \dfrac{8}{3 - 2} = 8$. This follows from the Quotient and Difference Laws.

2.3.13 $\lim\limits_{x \to 1} f(x)^{2/3} = \left(\lim\limits_{x \to 1} f(x) \right)^{2/3} = 8^{2/3} = 2^2 = 4$. This follows from the Root and Power Laws.

2.3.15 $\lim\limits_{x \to 0} g(x) = \lim\limits_{x \to 0} (2x + 1) = 1$, while $g(0) = 5$.

2.3.17 If p and q are polynomials then $\lim\limits_{x \to 0} \dfrac{p(x)}{q(x)} = \dfrac{\lim\limits_{x \to 0} p(x)}{\lim\limits_{x \to 0} q(x)} = \dfrac{p(0)}{q(0)}$. Because this quantity is given to be equal to 10, we have $\dfrac{p(0)}{2} = 10$, so $p(0) = 20$.

2.3.19 $\lim\limits_{x \to 4} (3x - 7) = 3 \lim\limits_{x \to 4} x - 7 = 3 \cdot 4 - 7 = 5$.

2.3.21 $\lim\limits_{x \to -9} (5x) = 5 \lim\limits_{x \to -9} x = 5 \cdot -9 = -45$.

2.3. Techniques for Computing Limits

2.3.23 $\lim\limits_{x\to 1}(2x^3-3x^2+4x+5) = \lim\limits_{x\to 1}2x^3 - \lim\limits_{x\to 1}3x^2 + \lim\limits_{x\to 1}4x + \lim\limits_{x\to 1}5 = 2(\lim\limits_{x\to 1}x)^3 - 3(\lim\limits_{x\to 1}x)^2 + 4(\lim\limits_{x\to 1}x) + 5 = 2(1)^3 - 3(1)^2 + 4\cdot 1 + 5 = 8.$

2.3.25 $\lim\limits_{x\to 1}\dfrac{5x^2+6x+1}{8x-4} = \dfrac{\lim\limits_{x\to 1}(5x^2+6x+1)}{\lim\limits_{x\to 1}(8x-4)} = \dfrac{5(\lim\limits_{x\to 1}x)^2 + 6\lim\limits_{x\to 1}x + \lim\limits_{x\to 1}1}{8\lim\limits_{x\to 1}x - \lim\limits_{x\to 1}4} = \dfrac{5(1)^2+6\cdot 1+1}{8\cdot 1-4} = 3.$

2.3.27 $\lim\limits_{p\to 2}\dfrac{3p}{\sqrt{4p+1}-1} = \dfrac{\lim\limits_{p\to 2}3p}{\lim\limits_{p\to 2}(\sqrt{4p+1}-1)} = \dfrac{3\lim\limits_{p\to 2}p}{\lim\limits_{p\to 2}\sqrt{4p+1}-\lim\limits_{p\to 2}1} = \dfrac{3\cdot 2}{\sqrt{\lim\limits_{p\to 2}(4p+1)}-1} = \dfrac{6}{3-1} = 3.$

2.3.29 $\lim\limits_{x\to 3}-\dfrac{5x}{\sqrt{4x-3}} = \dfrac{\lim\limits_{x\to 3}-5x}{\lim\limits_{x\to 3}\sqrt{4x-3}} = \dfrac{-5\lim\limits_{x\to 3}x}{\sqrt{\lim\limits_{x\to 3}(4x-3)}} = -\dfrac{5\cdot 3}{\sqrt{4\lim\limits_{x\to 3}x-\lim\limits_{x\to 3}3}} = -\dfrac{15}{\sqrt{4\cdot 3-3}} = -5.$

2.3.31 $\lim\limits_{x\to 2}(5x-6)^{3/2} = (5\cdot 2-6)^{3/2} = 4^{3/2} = 2^3 = 8.$

2.3.33 $\lim\limits_{x\to 1}\dfrac{x^2-1}{x-1} = \lim\limits_{x\to 1}\dfrac{(x+1)(x-1)}{x-1} = \lim\limits_{x\to 1}(x+1) = 2.$

2.3.35 $\lim\limits_{x\to 4}\dfrac{x^2-16}{4-x} = \lim\limits_{x\to 4}\dfrac{(x+4)(x-4)}{-(x-4)} = \lim\limits_{x\to 4}[-(x+4)] = -8.$

2.3.37 $\lim\limits_{x\to b}\dfrac{(x-b)^{50}-x+b}{x-b} = \lim\limits_{x\to b}\dfrac{(x-b)^{50}-(x-b)}{x-b} = \lim\limits_{x\to b}\dfrac{(x-b)((x-b)^{49}-1)}{x-b} = \lim\limits_{x\to b}[(x-b)^{49}-1] = -1.$

2.3.39

$$\lim_{x\to -1}\dfrac{(2x-1)^2-9}{x+1} = \lim_{x\to -1}\dfrac{(2x-1-3)(2x-1+3)}{x+1}$$

$$= \lim_{x\to -1}\dfrac{2(x-2)\cdot 2(x+1)}{x+1} = \lim_{x\to -1}4(x-2) = 4\cdot(-3) = -12.$$

2.3.41 $\lim\limits_{x\to 9}\dfrac{\sqrt{x}-3}{x-9} = \lim\limits_{x\to 9}\dfrac{(\sqrt{x}-3)(\sqrt{x}+3)}{(x-9)(\sqrt{x}+3)} = \lim\limits_{x\to 9}\dfrac{x-9}{(x-9)(\sqrt{x}+3)} = \lim\limits_{x\to 9}\dfrac{1}{\sqrt{x}+3} = \dfrac{1}{6}.$

2.3.43

$$\lim_{t\to 5}\left(\dfrac{1}{t^2-4t-5}-\dfrac{1}{6(t-5)}\right) = \lim_{t\to 5}\left(\dfrac{1}{(t-5)(t+1)}-\dfrac{1}{6(t-5)}\right)$$

$$= \lim_{t\to 5}\left(\dfrac{6}{6(t-5)(t+1)}-\dfrac{t+1}{6(t-5)(t+1)}\right)$$

$$= \lim_{t\to 5}\dfrac{5-t}{6(t-5)(t+1)} = -\lim_{t\to 5}\dfrac{1}{6(t+1)} = -\dfrac{1}{36}.$$

2.3.45 $\lim\limits_{x\to a}\dfrac{x-a}{\sqrt{x}-\sqrt{a}} = \lim\limits_{x\to a}\dfrac{x-a}{\sqrt{x}-\sqrt{a}}\cdot\dfrac{\sqrt{x}+\sqrt{a}}{\sqrt{x}+\sqrt{a}} = \lim\limits_{x\to a}\dfrac{(x-a)(\sqrt{x}+\sqrt{a})}{x-a} = \lim\limits_{x\to a}(\sqrt{x}+\sqrt{a}) = 2\sqrt{a}.$

2.3.47

$$\lim_{h\to 0}\dfrac{\sqrt{16+h}-4}{h} = \lim_{h\to 0}\dfrac{(\sqrt{16+h}-4)(\sqrt{16+h}+4)}{h(\sqrt{16+h}+4)} = \lim_{h\to 0}\dfrac{(16+h)-16}{h(\sqrt{16+h}+4)}$$

$$= \lim_{h\to 0}\dfrac{h}{h(\sqrt{16+h}+4)} = \lim_{h\to 0}\dfrac{1}{(\sqrt{16+h}+4)} = \dfrac{1}{8}.$$

2.3.49 $\lim_{x\to 4} \dfrac{\frac{1}{x}-\frac{1}{4}}{x-4} = \lim_{x\to 4} \dfrac{\frac{4-x}{4x}}{x-4} = \lim_{x\to 4} \dfrac{4-x}{4x(x-4)} = -\lim_{x\to 4} \dfrac{1}{4x} = -\dfrac{1}{16}.$

2.3.51
$$\lim_{x\to 1} \dfrac{\sqrt{10x-9}-1}{x-1} = \lim_{x\to 1} \dfrac{(\sqrt{10x-9}-1)(\sqrt{10x-9}+1)}{(x-1)(\sqrt{10x-9}+1)} = \lim_{x\to 1} \dfrac{(10x-9)-1}{(x-1)(\sqrt{10x-9}+1)}$$
$$= \lim_{x\to 1} \dfrac{10(x-1)}{(x-1)(\sqrt{10x-9}+1)} = \lim_{x\to 1} \dfrac{10}{(\sqrt{10x-9}+1)} = \dfrac{10}{2} = 5.$$

2.3.53 $\lim_{h\to 0} \dfrac{(5+h)^2-25}{h} = \lim_{h\to 0} \dfrac{25+10h+h^2-25}{h} = \lim_{h\to 0} \dfrac{h(10+h)}{h} = \lim_{h\to 0}(10+h) = 10.$

2.3.55 $\lim_{x\to 1} \dfrac{x-1}{\sqrt{x}-1} = \lim_{x\to 1} \dfrac{(x-1)(\sqrt{x}+1)}{(\sqrt{x}-1)(\sqrt{x}+1)} = \lim_{x\to 1} \dfrac{(x-1)(\sqrt{x}+1)}{x-1} = \lim_{x\to 1}(\sqrt{x}+1) = 2.$

2.3.57
$$\lim_{x\to 4} \dfrac{3(x-4)\sqrt{x+5}}{3-\sqrt{x+5}} = \lim_{x\to 4} \dfrac{3(x-4)(\sqrt{x+5})(3+\sqrt{x+5})}{(3-\sqrt{x+5})(3+\sqrt{x+5})} = \lim_{x\to 4} \dfrac{3(x-4)(\sqrt{x+5})(3+\sqrt{x+5})}{9-(x+5)}$$
$$= \lim_{x\to 4} \dfrac{3(x-4)(\sqrt{x+5})(3+\sqrt{x+5})}{-(x-4)}$$
$$= \lim_{x\to 4} [-3(\sqrt{x+5})(3+\sqrt{x+5})] = (-3)(3)(3+3) = -54$$

2.3.59 $\lim_{x\to 0} x\cos x = 0\cdot 1 = 0.$

2.3.61 $\lim_{x\to 0} \dfrac{1-\cos x}{\cos^2 x - 3\cos x + 2} = \lim_{x\to 0} \dfrac{1-\cos x}{(\cos x - 2)(\cos x - 1)} = -\lim_{x\to 0} \dfrac{1}{\cos x - 2} = -\dfrac{1}{1-2} = 1.$

2.3.63 $\lim_{x\to 0^-} \dfrac{x^2-x}{|x|} = \lim_{x\to 0^-} \dfrac{x(x-1)}{-x} = -\lim_{x\to 0^-}(x-1) = 1.$

2.3.65 $\lim_{t\to 2^+} \dfrac{|2t-4|}{t^2-4} = \lim_{t\to 2^+} \dfrac{2(t-2)}{(t-2)(t+2)} = \dfrac{2}{4} = \dfrac{1}{2}.$

2.3.67 $\lim_{x\to 3^+} \dfrac{x-3}{|x-3|} = \lim_{x\to 3^+} \dfrac{x-3}{x-3} = \lim_{x\to 3^+} 1 = 1.$ On the other hand, $\lim_{x\to 3^-} \dfrac{x-3}{|x-3|} = \lim_{x\to 3^-} \dfrac{x-3}{3-x} = \lim_{x\to 3^-}(-1) = -1.$ Therefore, $\lim_{x\to 3} \dfrac{x-3}{|x-3|}$ does not exist.

2.3.69 Because the domain of $f(x) = \dfrac{x^3+1}{\sqrt{x-1}}$ is the interval $(1,\infty)$, the limit doesn't exist.

2.3.71

a. False. For example, if $f(x) = \begin{cases} x & \text{if } x \neq 1; \\ 4 & \text{if } x = 1, \end{cases}$ then $\lim_{x\to 1} f(x) = 1$ but $f(1) = 4$.

b. False. For example, if $f(x) = \begin{cases} x+1 & \text{if } x \leq 1; \\ x-6 & \text{if } x > 1, \end{cases}$ then $\lim_{x\to 1^-} f(x) = 2$ but $\lim_{x\to 1^+} f(x) = -5$.

c. False. For example, if $f(x) = \begin{cases} x & \text{if } x \neq 1; \\ 4 & \text{if } x = 1, \end{cases}$ and $g(x) = 1$, then f and g both have limit 1 as $x\to 1$, but $f(1) = 4 \neq g(1)$.

d. **False.** For example $\lim_{x\to 2} \frac{x^2-4}{x-2}$ exists and is equal to 4.

e. **False.** For example, it would be possible for the domain of f to be $[1,\infty)$, so that the one-sided limit exists but the two-sided limit doesn't even make sense. This would be true, for example, if $f(x) = x-1$.

2.3.73

a. $\lim_{x\to -1^-} f(x) = \lim_{x\to -1^-} (x^2+1) = (-1)^2 + 1 = 2$.

b. $\lim_{x\to -1^+} f(x) = \lim_{x\to -1^+} \sqrt{x+1} = \sqrt{-1+1} = 0$.

c. Because the two one-sided limits differ, $\lim_{x\to -1} f(x)$ does not exist.

2.3.75

a. $\lim_{x\to 2^+} \sqrt{x-2} = \sqrt{2-2} = 0$.

b. The domain of $f(x) = \sqrt{x-2}$ is $[2,\infty)$. Thus, any question about this function that involves numbers less than 2 doesn't make any sense, because those numbers aren't in the domain of f.

2.3.77 $\lim_{x\to 10} E(x) = \lim_{x\to 10} \frac{4.35}{x\sqrt{x^2+0.01}} = \frac{4.35}{10\sqrt{100.01}} \approx 0.0435$ N/C.

2.3.79 $\lim_{S\to 0^+} r(S) = \lim_{S\to 0^+} (1/2)\left(\sqrt{100 + \frac{2S}{\pi}} - 10\right) = 0$.

The radius of the circular cylinder approaches zero as the surface area approaches zero.

2.3.81

a. The statement we are trying to prove can be stated in cases as follows: For $x > 0$, $-x \le x\sin(1/x) \le x$, and for $x < 0$, $x \le x\sin(1/x) \le -x$.

Now for all $x \ne 0$, note that $-1 \le \sin(1/x) \le 1$ (because the range of the sine function is $[-1,1]$). We will consider the two cases $x > 0$ and $x < 0$ separately, but in each case, we will multiply this inequality through by x, switching the inequalities for the $x < 0$ case.

For $x > 0$ we have $-x \le x\sin(1/x) \le x$, and for $x < 0$ we have $-x \ge x\sin(1/x) \ge x$, which are exactly the statements we are trying to prove.

b.

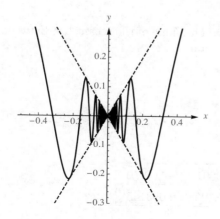

c. Because $\lim_{x\to 0} -|x| = \lim_{x\to 0} |x| = 0$, and because $-|x| \le x\sin(1/x) \le |x|$, the Squeeze Theorem assures us that $\lim_{x\to 0} [x\sin(1/x)] = 0$ as well.

2.3.83

a.

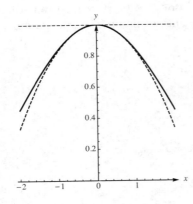

b. Note that $\lim_{x \to 0}\left[1 - \frac{x^2}{6}\right] = 1 = \lim_{x \to 0} 1$. So because $1 - \frac{x^2}{6} \leq \frac{\sin x}{x} \leq 1$, the squeeze theorem assures us that $\lim_{x \to 0} \frac{\sin x}{x} = 1$ as well.

2.3.85 Using the definition of $|x|$ given, we have $\lim_{x \to 0^-} |x| = \lim_{x \to 0^-} (-x) = -0 = 0$. Also, $\lim_{x \to 0^+} |x| = \lim_{x \to 0^+} x = 0$. Because the two one-sided limits are both 0, we also have $\lim_{x \to 0} |x| = 0$.

2.3.87 $\lim_{x \to 3} f(x) = \lim_{x \to 3} \frac{x^2 - 5x + 6}{x - 3} = \lim_{x \to 3} \frac{(x-3)(x-2)}{x-3} = \lim_{x \to 3}(x-2) = 1$. So $a = 1$.

2.3.89 In order for $\lim_{x \to -1} g(x)$ to exist, we need the two one-sided limits to exist and be equal. We have $\lim_{x \to -1^-} g(x) = \lim_{x \to -1^-} (x^2 - 5x) = 6$, and $\lim_{x \to -1^+} g(x) = \lim_{x \to -1^+} (ax^3 - 7) = -a - 7$. So we need $-a - 7 = 6$, so we require that $a = -13$. Then $\lim_{x \to -1} f(x) = 6$.

2.3.91 $\lim_{x \to 1} \frac{x^6 - 1}{x - 1} = \lim_{x \to 1} \frac{(x-1)(x^5 + x^4 + x^3 + x^2 + x + 1)}{x - 1} = \lim_{x \to 1}(x^5 + x^4 + x^3 + x^2 + x + 1) = 6$.

2.3.93 $\lim_{x \to a} \frac{x^5 - a^5}{x - a} = \lim_{x \to a} \frac{(x-a)(x^4 + ax^3 + a^2x^2 + a^3x + a^4)}{x - a} = \lim_{x \to a}(x^4 + ax^3 + a^2x^2 + a^3x + a^4) = 5a^4$.

2.3.95

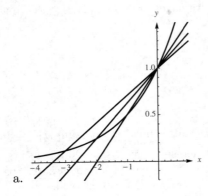

a.

b. The slope of the secant line between $(0, 1)$ and $(x, 2^x)$ is $\frac{2^x - 1}{x}$.

c.

x	-1	-0.1	-0.01	-0.001	-0.0001	-0.00001
$\frac{2^x - 1}{x}$	0.5	0.66967	0.69075	0.692907	0.693123	0.693145

It appears that $\lim_{x \to 0^-} \frac{2^x - 1}{x} \approx 0.693$.

2.3.97 $\lim_{x\to -1^-} f(x) = \lim_{x\to 1^+} f(x) = 6$. $\lim_{x\to -1^+} f(x) = \lim_{x\to 1^-} f(x) = 5$.

2.3.99 $\lim_{x\to 1} \dfrac{\sqrt[3]{x}-1}{x-1} = \lim_{x\to 1} \dfrac{\sqrt[3]{x}-1}{(\sqrt[3]{x}-1)(\sqrt[3]{x^2}+\sqrt[3]{x}+1)} = \lim_{x\to 1} \dfrac{1}{\sqrt[3]{x^2}+\sqrt[3]{x}+1} = \dfrac{1}{3}$.

2.3.101 Let $f(x) = x-1$ and $g(x) = \dfrac{5}{x-1}$. Then $\lim_{x\to 1} f(x) = 0$, $\lim_{x\to 1} f(x)g(x) = \lim_{x\to 1} \dfrac{5(x-1)}{x-1} = \lim_{x\to 1} 5 = 5$.

2.3.103 Let $p(x) = x^2 + 2x - 8$. Then $\lim_{x\to 2} \dfrac{p(x)}{x-2} = \lim_{x\to 2} \dfrac{(x-2)(x+4)}{x-2} = \lim_{x\to 2}(x+4) = 6$.

The constants are unique. We know that 2 must be a root of p (otherwise the given limit couldn't exist), so it must have the form $p(x) = (x-2)q(x)$, and q must be a degree 1 polynomial with leading coefficient 1 (otherwise p wouldn't have leading coefficient 1.) So we have $p(x) = (x-2)(x+d)$, but because $\lim_{x\to 2} \dfrac{p(x)}{x-2} = \lim_{x\to 2}(x+d) = 2+d = 6$, we are forced to realize that $d = 4$. Therefore, we have deduced that the only possibility for p is $p(x) = (x-2)(x+4) = x^2+2x-8$.

2.3.105 As $x \to 0^+$, $(1-x) \to 1^-$. So $\lim_{x\to 0^+} g(x) = \lim_{(1-x)\to 1^-} f(1-x) = \lim_{z\to 1^-} f(z) = 6$. (Where $z = 1-x$.)
As $x \to 0^-$, $(1-x) \to 1^+$. So $\lim_{x\to 0^-} g(x) = \lim_{(1-x)\to 1^+} f(1-x) = \lim_{z\to 1^+} f(z) = 4$. (Where $z = 1-x$.)

2.3.107

$$\lim_{x\to a} p(x) = \lim_{x\to a}(a_n x^n + a_{n-1} x^{n-1} + \cdots + a_1 x + a_0)$$
$$= \lim_{x\to a}(a_n x^n) + \lim_{x\to a}(a_{n-1} x^{n-1}) + \cdots + \lim_{x\to a}(a_1 x) + \lim_{x\to a} a_0$$
$$= a_n \lim_{x\to a} x^n + a_{n-1} \lim_{x\to a} x^{n-1} + \cdots + a_1 \lim_{x\to a} x + a_0$$
$$= a_n(\lim_{x\to a} x)^n + a_{n-1}(\lim_{x\to a} x)^{n-1} + \cdots + a_1(\lim_{x\to a} x) + a_0$$
$$= a_n a^n + a_{n-1} a^{n-1} + \cdots + a_1 a + a_0 = p(a).$$

2.4 Infinite Limits

2.4.1 As x approaches a from the right, the values of $f(x)$ are negative and become arbitrarily large in magnitude.

2.4.3 A vertical asymptote for a function f is a vertical line $x = a$ so that one or more of the following are true: $\lim_{x\to a^-} f(x) = \pm\infty$, $\lim_{x\to a^+} f(x) = \pm\infty$.

2.4.5

x	$\dfrac{x+1}{(x-1)^2}$	x	$\dfrac{x+1}{(x-1)^2}$
1.1	210	.9	190
1.01	20,100	.99	19,900
1.001	2,001,000	.999	1,999,000
1.0001	200,010,000	.9999	199,990,000

From the data given, it appears that $\lim_{x\to 1} f(x) = \infty$.

2.4.7

a. $\lim_{x\to 1^-} f(x) = \infty$.

b. $\lim_{x\to 1^+} f(x) = \infty$.

c. $\lim_{x\to 1} f(x) = \infty$.

d. $\lim_{x\to 2^-} f(x) = \infty$.

e. $\lim_{x\to 2^+} f(x) = -\infty$.

f. $\lim_{x\to 2} f(x)$ does not exist.

2.4.9

a. $\lim_{x \to -2^-} h(x) = -\infty$.

b. $\lim_{x \to -2^+} h(x) = -\infty$.

c. $\lim_{x \to -2} h(x) = -\infty$.

d. $\lim_{x \to 3^-} h(x) = \infty$.

e. $\lim_{x \to 3^+} h(x) = -\infty$.

f. $\lim_{x \to 3} h(x)$ does not exist.

2.4.11

a. $\lim_{x \to 0^-} \dfrac{1}{x^2 - x} = \infty$.

b. $\lim_{x \to 0^+} \dfrac{1}{x^2 - x} = -\infty$.

c. $\lim_{x \to 1^-} \dfrac{1}{x^2 - x} = -\infty$.

d. $\lim_{x \to 1^+} \dfrac{1}{x^2 - x} = \infty$.

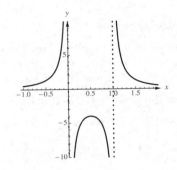

2.4.13 Because the numerator is approaching a non-zero constant while the denominator is approaching zero, the quotient of these numbers is getting big – at least the absolute value of the quotient is getting big. The quotient is actually always negative, because a number near 100 divided by a negative number is always negative. Thus $\lim_{x \to 2} \dfrac{f(x)}{g(x)} = -\infty$.

2.4.15 Note that $\lim_{x \to 1} \dfrac{x^2 - 4x + 3}{x^2 - 3x + 2} = \lim_{x \to 1} \dfrac{(x-3)(x-1)}{(x-2)(x-1)} = \lim_{x \to 1} \dfrac{x - 3}{x - 2} = \dfrac{-2}{-1} = 2$. So there is *not* a vertical asymptote at $x = 1$. On the other hand, $\lim_{x \to 2^+} \dfrac{x^2 - 4x + 3}{x^2 - 3x + 2} = \lim_{x \to 2^+} \dfrac{(x-3)(x-1)}{(x-2)(x-1)} = \lim_{x \to 2^+} \dfrac{x - 3}{x - 2} = -\infty$, so there is a vertical asymptote at $x = 2$.

2.4.17

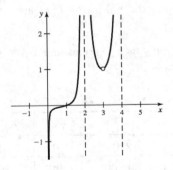

2.4.19 Both *a* and *b* are true statements.

2.4.21

a. $\lim_{x \to 2^+} \dfrac{1}{x - 2} = \infty$.

b. $\lim_{x \to 2^-} \dfrac{1}{x - 2} = -\infty$.

c. $\lim_{x \to 2} \dfrac{1}{x - 2}$ does not exist.

2.4. Infinite Limits

2.4.23

a. $\lim\limits_{x \to 4^+} \dfrac{x-5}{(x-4)^2} = -\infty$.

b. $\lim\limits_{x \to 4^-} \dfrac{x-5}{(x-4)^2} = -\infty$.

c. $\lim\limits_{x \to 4} \dfrac{x-5}{(x-4)^2} = -\infty$.

2.4.25

a. $\lim\limits_{x \to 3^+} \dfrac{(x-1)(x-2)}{(x-3)} = \infty$.

b. $\lim\limits_{x \to 3^-} \dfrac{(x-1)(x-2)}{(x-3)} = -\infty$.

c. $\lim\limits_{x \to 3} \dfrac{(x-1)(x-2)}{(x-3)}$ does not exist.

2.4.27

a. $\lim\limits_{x \to 2^+} \dfrac{x^2 - 4x + 3}{(x-2)^2} = -\infty$.

b. $\lim\limits_{x \to 2^-} \dfrac{x^2 - 4x + 3}{(x-2)^2} = -\infty$.

c. $\lim\limits_{x \to 2} \dfrac{x^2 - 4x + 3}{(x-2)^2} = -\infty$.

2.4.29

a. $\lim\limits_{x \to 2^+} \dfrac{1}{\sqrt{x(x-2)}} = \infty$.

b. $\lim\limits_{x \to 2^-} \dfrac{1}{\sqrt{x(x-2)}}$ does not exist. Note that the domain of the function is $(-\infty, 0) \cup (2, \infty)$.

c. $\lim\limits_{x \to 2} \dfrac{1}{\sqrt{x(x-2)}}$ does not exist.

2.4.31

a. $\lim\limits_{x \to 0} \dfrac{x-3}{x^4 - 9x^2} = \lim\limits_{x \to 0} \dfrac{x-3}{x^2(x-3)(x+3)} = \lim\limits_{x \to 0} \dfrac{1}{x^2(x+3)} = \infty$.

b. $\lim\limits_{x \to 3} \dfrac{x-3}{x^4 - 9x^2} = \lim\limits_{x \to 3} \dfrac{x-3}{x^2(x-3)(x+3)} = \lim\limits_{x \to 3} \dfrac{1}{x^2(x+3)} = \dfrac{1}{54}$.

c. $\lim\limits_{x \to -3} \dfrac{x-3}{x^4 - 9x^2} = \lim\limits_{x \to -3} \dfrac{x-3}{x^2(x-3)(x+3)} = \lim\limits_{x \to -3} \dfrac{1}{x^2(x+3)}$, which does not exist.

2.4.33 $\lim\limits_{x \to 0} \dfrac{x^3 - 5x^2}{x^2} = \lim\limits_{x \to 0} \dfrac{x^2(x-5)}{x^2} = \lim\limits_{x \to 0}(x-5) = -5$.

2.4.35 $\lim\limits_{x \to 1^+} \dfrac{x^2 - 5x + 6}{x-1} = \lim\limits_{x \to 1^+} \dfrac{(x-2)(x-3)}{x-1} = \infty$. (Note that as $x \to 1^+$, the numerator is near 2, while the denominator is near zero, but is positive. So the quotient is positive and large.)

2.4.37 $\lim_{x \to 6^+} \frac{x-7}{\sqrt{x-6}} = -\infty$. (Note that as $x \to 6^+$ the numerator is near -1 and the denominator is near zero but is positive. So the quotient is negative with large absolute value.)

2.4.39 $\lim_{\theta \to 0^+} \csc \theta = \lim_{\theta \to 0^+} \frac{1}{\sin \theta} = \infty$.

2.4.41 $\lim_{x \to 0^+} -10 \cot x = \lim_{x \to 0^+} \frac{-10 \cos x}{\sin x} = -\infty$. (Note that as $x \to 0^+$, the numerator is near -10 and the denominator is near zero, but is positive. Thus the quotient is a negative number whose absolute value is large.)

2.4.43 $\lim_{\theta \to 0} \frac{2 + \sin \theta}{1 - \cos^2 \theta} = \infty$. (Note that as $\theta \to 0$, the numerator is near 2 and the denominator is near 0, but is positive. Thus the quotient is a positive number whose absolute value is large.)

2.4.45

a. $\lim_{x \to 5} \frac{x-5}{x^2 - 25} = \lim_{x \to 5} \frac{1}{x+5} = \frac{1}{10}$, so there isn't a vertical asymptote at $x = 5$.

b. $\lim_{x \to -5^-} \frac{x-5}{x^2 - 25} = \lim_{x \to -5^-} \frac{1}{x+5} = -\infty$, so there is a vertical asymptote at $x = -5$.

c. $\lim_{x \to -5^+} \frac{x-5}{x^2 - 25} = \lim_{x \to -5^+} \frac{1}{x+5} = \infty$. This also implies that $x = -5$ is a vertical asymptote, as we already noted in part b.

2.4.47 $f(x) = \frac{x^2 - 9x + 14}{x^2 - 5x + 6} = \frac{(x-2)(x-7)}{(x-2)(x-3)}$. Note that $x = 3$ is a vertical asymptote, while $x = 2$ appears to be a candidate but isn't one. We have $\lim_{x \to 3^+} f(x) = \lim_{x \to 3^+} \frac{x-7}{x-3} = -\infty$ and $\lim_{x \to 3^-} f(x) = \lim_{x \to 3^-} \frac{x-7}{x-3} = \infty$, and thus $\lim_{x \to 3} f(x)$ doesn't exist. Note that $\lim_{x \to 2} f(x) = 5$.

2.4.49 $f(x) = \frac{x+1}{x^3 - 4x^2 + 4x} = \frac{x+1}{x(x-2)^2}$. There are vertical asymptotes at $x = 0$ and $x = 2$. We have $\lim_{x \to 0^-} f(x) = \lim_{x \to 0^-} \frac{x+1}{x(x-2)^2} = -\infty$, while $\lim_{x \to 0^+} f(x) = \lim_{x \to 0^+} \frac{x+1}{x(x-2)^2} = \infty$, and thus $\lim_{x \to 0} f(x)$ doesn't exist.

Also we have $\lim_{x \to 2^-} f(x) = \lim_{x \to 2^-} \frac{x+1}{x(x-2)^2} = \infty$, while $\lim_{x \to 2^+} f(x) = \lim_{x \to 2^+} \frac{x+1}{x(x-2)^2} = \infty$, and thus $\lim_{x \to 2} f(x) = \infty$ as well.

2.4.51

a. $\lim_{x \to (\pi/2)^+} \tan x = -\infty$.

b. $\lim_{x \to (\pi/2)^-} \tan x = \infty$.

c. $\lim_{x \to (-\pi/2)^+} \tan x = -\infty$.

d. $\lim_{x \to (-\pi/2)^-} \tan x = \infty$.

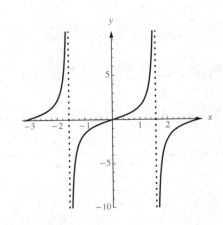

2.4. Infinite Limits

2.4.53

a. False. $\lim_{x \to 1^-} f(x) = \lim_{x \to 1^+} f(x) = \lim_{x \to 1} f(x) = \lim_{x \to 1} \frac{(x-1)(x-6)}{(x-1)(x+1)} = -\frac{5}{2}$.

b. True. For example, $\lim_{x \to -1^+} f(x) = \lim_{x \to -1^+} \frac{(x-1)(x-6)}{(x-1)(x+1)} = -\infty$.

c. False. For example $g(x) = \frac{1}{x-1}$ has $\lim_{x \to 1^+} g(x) = \infty$, but $\lim_{x \to 1^-} g(x) = -\infty$.

2.4.55 We are seeking a function with a factor of $x - 1$ in the denominator, but there should be more factors of $x - 1$ in the numerator, and there should be a factor of $(x - 2)^2$ in the denominator. This will accomplish the desired results. So

$$r(x) = \frac{(x-1)^2}{(x-1)(x-2)^2}.$$

2.4.57 One example is $f(x) = \frac{1}{x-6}$.

2.4.59 $f(x) = \frac{x^2 - 3x + 2}{x^{10} - x^9} = \frac{(x-2)(x-1)}{x^9(x-1)}$. f has a vertical asymptote at $x = 0$, because $\lim_{x \to 0^+} f(x) = -\infty$ (and $\lim_{x \to 0^-} f(x) = \infty$.) Note that $\lim_{x \to 1} f(x) = -1$, so there isn't a vertical asymptote at $x = 1$.

2.4.61 $h(x) = \frac{e^x}{(x+1)^3}$ has a vertical asymptote at $x = -1$, because

$$\lim_{x \to -1^+} \frac{e^x}{(x+1)^3} = \infty \quad \text{and} \quad \lim_{x \to -1^-} h(x) = -\infty.$$

2.4.63 $g(\theta) = \tan(\pi\theta/10) = \frac{\sin(\pi\theta/10)}{\cos(\pi\theta/10)}$ has a vertical asymptote at each $\theta = 10n + 5$ where n is an integer. This is due to the fact that $\cos(\pi\theta/10) = 0$ when $\pi\theta/10 = \pi/2 + n\pi$ where n is an integer, which is the same as $\{\theta : \theta = 10n + 5, n \text{ an integer}\}$. Note that at all of these numbers which make the denominator zero, the numerator isn't zero.

2.4.65 $f(x) = \frac{1}{\sqrt{x} \sec x} = \frac{\cos x}{\sqrt{x}}$ has a vertical asymptote at $x = 0$.

2.4.67

a. Note that the numerator of the given expression factors as $(x - 3)(x - 4)$. So if $a = 3$ or if $a = 4$ the limit would be a finite number. In fact, $\lim_{x \to 3} \frac{(x-3)(x-4)}{x-3} = -1$ and $\lim_{x \to 4} \frac{(x-3)(x-4)}{x-4} = 1$.

b. For any number other than 3 or 4, the limit would be either $\pm\infty$. Because $x - a$ is always positive as $x \to a^+$, the limit would be $+\infty$ exactly when the numerator is positive, which is for a in the set $(-\infty, 3) \cup (4, \infty)$.

c. The limit would be $-\infty$ for a in the set $(3, 4)$.

2.4.69

a. The slope of the secant line is $\frac{f(h) - f(0)}{h} = \frac{h^{2/3}}{h} = h^{-1/3}$.

b. $\lim_{h \to 0^+} \frac{1}{h^{1/3}} = \infty$, and $\lim_{h \to 0^-} \frac{1}{h^{1/3}} = -\infty$. The tangent line is infinitely steep at the origin (i.e., it is a vertical line.)

2.5 Limits at Infinity

2.5.1 As $x < 0$ becomes arbitrarily large in absolute value, the corresponding values of f approach 10.

2.5.3 $\lim_{x \to \infty} x^{12} = \infty$. Note that x^{12} is positive when $x > 0$.

2.5.5 $\lim_{x \to \infty} x^{-6} = \lim_{x \to \infty} \dfrac{1}{x^6} = 0$.

2.5.7 $\lim_{t \to \infty} (-12t^{-5}) = \lim_{t \to \infty} -\dfrac{12}{t^5} = 0$.

2.5.9 $\lim_{x \to \infty} (3 + 10/x^2) = 3 + \lim_{x \to \infty} (10/x^2) = 3 + 0 = 3$.

2.5.11 If $f(x) \to 100,000$ as $x \to \infty$ and $g(x) \to \infty$ as $x \to \infty$, then the ratio $\dfrac{f(x)}{g(x)} \to 0$ as $x \to \infty$. (Because *eventually* the values of f are small compared to the values of g.)

2.5.13 $\lim_{t \to \infty} e^t = \infty$, $\lim_{t \to -\infty} e^t = 0$, and $\lim_{t \to \infty} e^{-t} = 0$.

2.5.15 Because $\lim_{x \to \infty} 3 - \dfrac{1}{x^2} = 3$ and $\lim_{x \to \infty} 3 + \dfrac{1}{x^2} = 3$, by the Squeeze Theorem we must have $\lim_{x \to \infty} g(x) = 3$. Similarly, because $\lim_{x \to -\infty} 3 - \dfrac{1}{x^2} = 3$ and $\lim_{x \to -\infty} 3 + \dfrac{1}{x^2} = 3$, by the Squeeze Theorem we must have $\lim_{x \to -\infty} g(x) = 3$.

2.5.17 $\lim_{\theta \to \infty} \dfrac{\cos \theta}{\theta^2} = 0$. Note that $-1 \leq \cos \theta \leq 1$, so $-\dfrac{1}{\theta^2} \leq \dfrac{\cos \theta}{\theta^2} \leq \dfrac{1}{\theta^2}$. The result now follows from the Squeeze Theorem.

2.5.19 $\lim_{x \to \infty} \dfrac{\cos x^5}{\sqrt{x}} = 0$. Note that $-1 \leq \cos x^5 \leq 1$, so $\dfrac{-1}{\sqrt{x}} \leq \dfrac{\cos x^5}{\sqrt{x}} \leq \dfrac{1}{\sqrt{x}}$. Because $\lim_{x \to \infty} \dfrac{1}{\sqrt{x}} = \lim_{x \to \infty} \dfrac{-1}{\sqrt{x}} = 0$, we have $\lim_{x \to \infty} \dfrac{\cos x^5}{\sqrt{x}} = 0$ by the Squeeze Theorem.

2.5.21 $\lim_{x \to \infty} (3x^{12} - 9x^7) = \infty$.

2.5.23 $\lim_{x \to -\infty} (-3x^{16} + 2) = -\infty$.

2.5.25 $\lim_{x \to \infty} \dfrac{(14x^3 + 3x^2 - 2x)}{(21x^3 + x^2 + 2x + 1)} \cdot \dfrac{1/x^3}{1/x^3} = \lim_{x \to \infty} \dfrac{14 + (3/x) - (2/x^2)}{21 + (1/x) + (2/x^2) + (1/x^3)} = \dfrac{14}{21} = \dfrac{2}{3}$.

2.5.27 $\lim_{x \to -\infty} \dfrac{(3x^2 + 3x)}{(x+1)} \cdot \dfrac{1/x}{1/x} = \lim_{x \to -\infty} \dfrac{3x + 3}{1 + (1/x)} = -\infty$.

2.5.29 Note that for $w > 0$, $w^2 = \sqrt{w^4}$. We have
$$\lim_{w \to \infty} \dfrac{(15w^2 + 3w + 1)}{\sqrt{9w^4 + w^3}} \cdot \dfrac{1/w^2}{1/\sqrt{w^4}} = \lim_{w \to \infty} \dfrac{15 + (3/w) + (1/w^2)}{\sqrt{9 + (1/w)}} = \dfrac{15}{\sqrt{9}} = 5.$$

2.5.31 Note that for $x < 0$, $\sqrt{x^2} = -x$. We have
$$\lim_{x \to -\infty} \dfrac{\sqrt{16x^2 + x}}{x} \cdot \dfrac{\sqrt{1/x^2}}{-1/x} = \lim_{x \to -\infty} \dfrac{\sqrt{16 + (1/x)}}{-1} = -\sqrt{16} = -4.$$

2.5. Limits at Infinity

2.5.33 $\lim\limits_{x \to \infty} \dfrac{(x^2 - \sqrt{x^4 + 3x^2})}{1} \cdot \dfrac{(x^2 + \sqrt{x^4 + 3x^2})}{(x^2 + \sqrt{x^4 + 3x^2})} = \lim\limits_{x \to \infty} \dfrac{x^4 - (x^4 + 3x^2)}{x^2 + \sqrt{x^4 + 3x^2}} = \lim\limits_{x \to \infty} \dfrac{-3x^2}{x^2 + \sqrt{x^4 + 3x^2}}$. Now divide the numerator and denominator by x^2 to give

$$\lim_{x \to \infty} \frac{-3x^2}{x^2 + \sqrt{x^4 + 3x^2}} \cdot \frac{1/x^2}{1/x^2} = \lim_{x \to \infty} \frac{-3}{1 + \sqrt{1 + (3/x^2)}} = -\frac{3}{2}.$$

2.5.35 Note that because $-1 \leq \sin x \leq 1$, we have $-\dfrac{1}{e^x} \leq \dfrac{\sin x}{e^x} \leq \dfrac{1}{e^x}$. Then because $\lim\limits_{x \to \infty} \dfrac{\pm 1}{e^x} = 0$, the Squeeze Theorem tells us that $\lim\limits_{x \to \infty} \dfrac{\sin x}{e^x} = 0$.

2.5.37 $\lim\limits_{x \to \infty} \dfrac{4x}{20x + 1} = \lim\limits_{x \to \infty} \dfrac{4x}{20x + 1} \cdot \dfrac{1/x}{1/x} = \lim\limits_{x \to \infty} \dfrac{4}{20 + 1/x} = \dfrac{4}{20} = \dfrac{1}{5}$. Thus, the line $y = \dfrac{1}{5}$ is a horizontal asymptote.

$\lim\limits_{x \to -\infty} \dfrac{4x}{20x + 1} = \lim\limits_{x \to -\infty} \dfrac{4x}{20x + 1} \cdot \dfrac{1/x}{1/x} = \lim\limits_{x \to -\infty} \dfrac{4}{20 + 1/x} = \dfrac{4}{20} = \dfrac{1}{5}$. This shows that the curve is also asymptotic to the asymptote in the negative direction.

2.5.39 $\lim\limits_{x \to \infty} \dfrac{(6x^2 - 9x + 8)}{(3x^2 + 2)} \cdot \dfrac{1/x^2}{1/x^2} = \lim\limits_{x \to \infty} \dfrac{6 - 9/x + 8/x^2}{3 + 2/x^2} = \dfrac{6 - 0 + 0}{3 + 0} = 2$. Similarly $\lim\limits_{x \to -\infty} f(x) = 2$. The line $y = 2$ is a horizontal asymptote.

2.5.41 $\lim\limits_{x \to \infty} \dfrac{3x^3 - 7}{x^4 + 5x^2} = \lim\limits_{x \to \infty} \dfrac{3x^3 - 7}{x^4 + 5x^2} \cdot \dfrac{3/x^4}{1/x^4} = \lim\limits_{x \to \infty} \dfrac{1/x - (7/x^4)}{1 + (5/x^2)} = \dfrac{0 - 0}{1 + 0} = 0$. Thus, the line $y = 0$ (the x-axis) is a horizontal asymptote.

$\lim\limits_{x \to -\infty} \dfrac{3x^3 - 7}{x^4 + 5x^2} = \lim\limits_{x \to -\infty} \dfrac{3x^3 - 7}{x^4 + 5x^2} \cdot \dfrac{3/x^4}{1/x^4} = \lim\limits_{x \to -\infty} \dfrac{1/x - (7/x^4)}{1 + (5/x^2)} = \dfrac{0 - 0}{1 + 0} = 0$. Thus, the curve is asymptotic to the x-axis in the negative direction as well.

2.5.43 $\lim\limits_{x \to \infty} \dfrac{(40x^5 + x^2)}{(16x^4 - 2x)} \cdot \dfrac{1/x^4}{1/x^4} = \lim\limits_{x \to \infty} \dfrac{40x + 1/x^2}{16 - 2/x^3} = \infty$. Similarly $\lim\limits_{x \to -\infty} f(x) = -\infty$. There are no horizontal asymptotes.

2.5.45 Note that for all x, $\sqrt{x^8} = x^4$. Then $\lim\limits_{x \to \pm\infty} \dfrac{1}{(2x^4 - \sqrt{4x^8 - 9x^4})} \cdot \dfrac{(2x^4 + \sqrt{4x^8 - 9x^4})}{(2x^4 + \sqrt{4x^8 - 9x^4})}$

$= \lim\limits_{x \to \pm\infty} \dfrac{(2x^4 + \sqrt{4x^8 - 9x^4})}{(4x^8 - (4x^8 - 9x^4))} \cdot \dfrac{1/x^4}{1/x^4} = \lim\limits_{x \to \pm\infty} \dfrac{2 + \sqrt{4 - (9/x^4)}}{9} = \dfrac{4}{9}$.

So $y = \dfrac{4}{9}$ is the only horizontal asymptote.

2.5.47 First note that $\sqrt{x^6} = x^3$ if $x > 0$, but $\sqrt{x^6} = -x^3$ if $x < 0$. We have $\lim\limits_{x \to \infty} \dfrac{4x^3 + 1}{(2x^3 + \sqrt{16x^6 + 1})} \cdot \dfrac{1/x^3}{1/x^3} =$

$\lim\limits_{x \to \infty} \dfrac{4 + 1/x^3}{2 + \sqrt{16 + 1/x^6}} = \dfrac{4 + 0}{2 + \sqrt{16 + 0}} = \dfrac{2}{3}$.

However, $\lim\limits_{x \to -\infty} \dfrac{4x^3 + 1}{(2x^3 + \sqrt{16x^6 + 1})} \cdot \dfrac{1/x^3}{1/x^3} = \lim\limits_{x \to -\infty} \dfrac{4 + 1/x^3}{2 - \sqrt{16 + 1/x^6}} = \dfrac{4 + 0}{2 - \sqrt{16 + 0}} = \dfrac{4}{-2} = -2$.

So $y = \dfrac{2}{3}$ is a horizontal asymptote (as $x \to \infty$) and $y = -2$ is a horizontal asymptote (as $x \to -\infty$).

2.5.49 First note that $\sqrt[3]{x^6} = x^2$ and $\sqrt{x^4} = x^2$ for all x (even when $x < 0$.) We have $\lim\limits_{x \to \infty} \dfrac{\sqrt[3]{x^6 + 8}}{(4x^2 + \sqrt{3x^4 + 1})} \cdot$

$\dfrac{1/x^2}{1/x^2} = \lim\limits_{x \to \infty} \dfrac{\sqrt[3]{1 + 8/x^6}}{4 + \sqrt{3 + 1/x^4}} = \dfrac{1}{4 + \sqrt{3 + 0}} = \dfrac{1}{4\sqrt{3}}$.

The calculation as $x \to -\infty$ is similar. So $y = \dfrac{1}{4\sqrt{3}}$ is a horizontal asymptote.

2.5.51

a. $f(x) = \dfrac{x^2 - 3}{x + 6} = x - 6 + \dfrac{33}{x + 6}$. The oblique asymptote of f is $y = x - 6$.

b. Because $\lim\limits_{x \to -6^+} f(x) = \infty$, there is a vertical asymptote at $x = -6$. Note also that $\lim\limits_{x \to -6^-} f(x) = -\infty$.

c.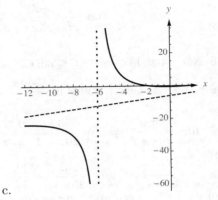

2.5.53

a. $f(x) = \dfrac{x^2 - 2x + 5}{3x - 2} = (1/3)x - 4/9 + \dfrac{37}{9(3x - 2)}$. The oblique asymptote of f is $y = (1/3)x - 4/9$.

b. Because $\lim\limits_{x \to (2/3)^+} f(x) = \infty$, there is a vertical asymptote at $x = 2/3$. Note also that $\lim\limits_{x \to (2/3)^-} f(x) = -\infty$.

c.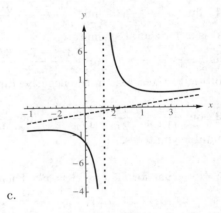

2.5.55

a. $f(x) = \dfrac{4x^3 + 4x^2 + 7x + 4}{1 + x^2} = 4x + 4 + \dfrac{3x}{1 + x^2}$. The oblique asymptote of f is $y = 4x + 4$.

b. There are no vertical asymptotes.

c.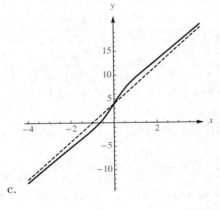

2.5. Limits at Infinity

2.5.57

$\lim\limits_{x \to \infty} (-3e^{-x}) = -3 \cdot 0 = 0.$ $\lim\limits_{x \to -\infty} (-3e^{-x}) = -\infty.$

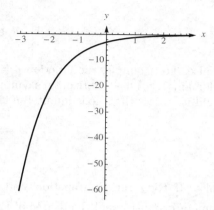

2.5.59

$\lim\limits_{x \to \infty} (1 - \ln x) = -\infty.$ $\lim\limits_{x \to 0^+} (1 - \ln x) = \infty.$

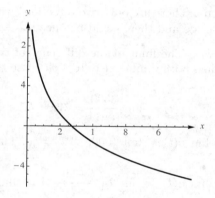

2.5.61

$y = \sin x$ has no asymptotes. $\lim\limits_{x \to \infty} \sin x$ and $\lim\limits_{x \to -\infty} \sin x$ do not exist.

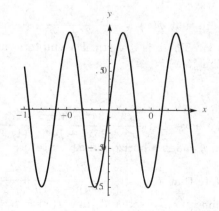

2.5.63

a. False. For example, the function $y = \frac{\sin x}{x}$ on the domain $[1, \infty)$ has a horizontal asymptote of $y = 0$, and it crosses the x-axis infinitely many times.

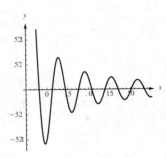

b. False. If f is a rational function, and if $\lim_{x \to \infty} f(x) = L \neq 0$, then the degree of the polynomial in the numerator must equal the degree of the polynomial in the denominator. In this case, both $\lim_{x \to \infty} f(x)$ and $\lim_{x \to -\infty} f(x) = \frac{a_n}{b_n}$ where a_n is the leading coefficient of the polynomial in the numerator and b_n is the leading coefficient of the polynomial in the denominator. In the case where $\lim_{x \to \infty} f(x) = 0$, then the degree of the numerator is strictly less than the degree of the denominator. This case holds for $\lim_{x \to -\infty} f(x) = 0$ as well.

c. True. There are only two directions which might lead to horizontal asymptotes: there could be one as $x \to \infty$ and there could be one as $x \to -\infty$, and those are the only possibilities.

d. False. The limit of the difference of two functions can be written as the difference of the limits only when both limits exist. It is the case that $\lim_{x \to \infty} (x^3 - x) = \infty$.

2.5.65 $\lim_{t \to \infty} p(t) = \lim_{t \to \infty} \frac{3500t}{t+1} = 3500$. The steady state exists. The steady state value is 3500.

2.5.67 $\lim_{t \to \infty} v(t) = \lim_{t \to \infty} 1000 e^{0.065t} = \infty$. The steady state does not exist.

2.5.69 $\lim_{t \to \infty} a(t) = \lim_{t \to \infty} 2 \left(\frac{t + \sin t}{t} \right) = \lim_{t \to \infty} 2 \left(1 + \frac{\sin t}{t} \right) = 2$. The steady state exists. The steady state value is 2.

2.5.71

a. $\lim_{x \to \infty} \frac{2x^3 + 10x^2 + 12x}{x^3 + 2x^2} \cdot \frac{(1/x^3)}{(1/x^3)} = \lim_{x \to \infty} \frac{2 + 10/x + 12/x^2}{1 + 2/x} = 2$. Similarly, $\lim_{x \to -\infty} f(x) = 2$. Thus, $y = 2$ is a horizontal asymptote.

b. Note that $f(x) = \frac{2x(x+2)(x+3)}{x^2(x+2)}$. So $\lim_{x \to 0^+} f(x) = \lim_{x \to 0^+} \frac{2(x+3)}{x} = \infty$, and similarly, $\lim_{x \to 0^-} f(x) = -\infty$. There is a vertical asymptote at $x = 0$. Note that there is no asymptote at $x = -2$ because $\lim_{x \to -2} f(x) = -1$.

2.5.73

a. We have $\lim_{x \to \infty} \frac{3x^4 + 3x^3 - 36x^2}{x^4 - 25x^2 + 144} \cdot \frac{(1/x^4)}{(1/x^4)} = \lim_{x \to \infty} \frac{3 + 3/x - 36/x^2}{1 - 25/x^2 + 144/x^4} = 3$. Similarly, $\lim_{x \to -\infty} f(x) = 3$. So $y = 3$ is a horizontal asymptote.

b. Note that $f(x) = \frac{3x^2(x+4)(x-3)}{(x+4)(x-4)(x+3)(x-3)}$. Thus, $\lim_{x \to -3^+} f(x) = -\infty$ and $\lim_{x \to -3^-} f(x) = \infty$. Also, $\lim_{x \to 4^-} f(x) = -\infty$ and $\lim_{x \to 4^+} f(x) = \infty$. Thus there are vertical asymptotes at $x = -3$ and $x = 4$.

2.5. Limits at Infinity

2.5.75

a. $\lim\limits_{x \to \infty} \dfrac{x^2 - 9}{x^2 - 3x} \cdot \dfrac{(1/x^2)}{(1/x^2)} = \lim\limits_{x \to \infty} \dfrac{1 - 9/x^2}{1 - 3/x} = 1$. A similar result holds as $x \to -\infty$. So $y = 1$ is a horizontal asymptote.

b. Because $\lim\limits_{x \to 0^+} f(x) = \lim\limits_{x \to 0^+} \dfrac{x + 3}{x} = \infty$ and $\lim\limits_{x \to 0^-} f(x) = -\infty$, there is a vertical asymptote at $x = 0$.

2.5.77

a. First note that $f(x)$ can be written s

$$\dfrac{\sqrt{x^2 + 2x + 6} - 3}{x - 1} \cdot \dfrac{\sqrt{x^2 + 2x + 6} + 3}{\sqrt{x^2 + 2x + 6} + 3} = \dfrac{x^2 + 2x + 6 - 9}{(x - 1)(\sqrt{x^2 + 2x + 6} + 3)} = \dfrac{(x - 1)(x + 3)}{(x - 1)(\sqrt{x^2 + 2x + 6} + 3)}.$$

Thus

$$\lim\limits_{x \to \infty} f(x) = \lim\limits_{x \to \infty} \dfrac{x + 3}{\sqrt{x^2 + 2x + 6} + 3} \cdot \dfrac{1/x}{1/x} = \lim\limits_{x \to \infty} \dfrac{1 + 3/x}{\sqrt{1 + 2/x + 6/x^2} + 3/x} = 1.$$

Using the fact that $\sqrt{x^2} = -x$ for $x < 0$, we have $\lim\limits_{x \to -\infty} f(x) = -1$. Thus the lines $y = 1$ and $y = -1$ are horizontal asymptotes.

b. f has no vertical asymptotes.

2.5.79

a. Note that when $x > 1$, we have $|x| = x$ and $|x - 1| = x - 1$. Thus

$$f(x) = (\sqrt{x} - \sqrt{x - 1}) \cdot \dfrac{\sqrt{x} + \sqrt{x - 1}}{\sqrt{x} + \sqrt{x - 1}} = \dfrac{1}{\sqrt{x} + \sqrt{x - 1}}.$$

Thus $\lim\limits_{x \to \infty} f(x) = 0$.

When $x < 0$, we have $|x| = -x$ and $|x - 1| = 1 - x$. Thus

$$f(x) = (\sqrt{-x} - \sqrt{1 - x}) \cdot \dfrac{\sqrt{-x} + \sqrt{1 - x}}{\sqrt{-x} + \sqrt{1 - x}} = -\dfrac{1}{\sqrt{-x} + \sqrt{1 - x}}.$$

Thus, $\lim\limits_{x \to -\infty} f(x) = 0$. There is a horizontal asymptote at $y = 0$.

b. f has no vertical asymptotes.

2.5.81

a. $\lim\limits_{x \to \infty} \dfrac{\cos x + 2\sqrt{x}}{\sqrt{x}} = \lim\limits_{x \to \infty} \left(2 + \dfrac{\cos x}{\sqrt{x}}\right) = 2$. $y = 2$ is a horizontal asymptote.

b. $\lim\limits_{x \to 0^+} \dfrac{\cos x + 2\sqrt{x}}{\sqrt{x}} = \infty$. and $\lim\limits_{x \to 0^-} \dfrac{\cos x + 2\sqrt{x}}{\sqrt{x}}$ does not exist. $x = 0$ is a vertical asymptote.

2.5.83

a. $\lim\limits_{x \to \infty} \sec^{-1} x = \pi/2$.

b. $\lim\limits_{x \to -\infty} \sec^{-1} x = \pi/2$.

2.5.85

a. $\lim_{x\to\infty} \dfrac{e^x - e^{-x}}{2} = \infty$.

$\lim_{x\to-\infty} \dfrac{e^x - e^{-x}}{2} = -\infty$.

b. $\sinh(0) = \dfrac{e^0 - e^0}{2} = \dfrac{1-1}{2} = 0$.

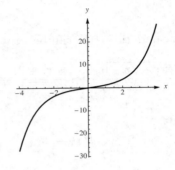

2.5.87

One possible such graph is:

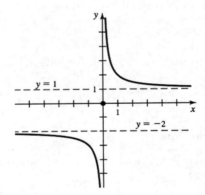

2.5.89 $\lim\limits_{n\to\infty} f(n) = \lim\limits_{n\to\infty} \dfrac{n-1}{n} = \lim\limits_{n\to\infty}[1 - (1/n)] = 1$.

2.5.91 $\lim\limits_{n\to\infty} f(n) = \lim\limits_{n\to\infty} \dfrac{n+1}{n^2} = \lim\limits_{n\to\infty}[1/n + 1/n^2] = 0$.

2.5.93

a. No. If $m = n$, there will be a horizontal asymptote, and if $m = n+1$, there will be a slant asymptote.

b. Yes. For example, $f(x) = \dfrac{x^4}{\sqrt{x^6 + 1}}$ has slant asymptote $y = x$ as $x \to \infty$ and slant asymptote $y = -x$ as $x \to -\infty$.

2.5.95 $\lim\limits_{x\to\infty} \dfrac{2e^x + 3}{e^x + 1} = \lim\limits_{x\to\infty} \dfrac{(2e^x + 3)}{(e^x + 1)} \cdot \dfrac{1/e^x}{1/e^x} = \lim\limits_{x\to\infty} \dfrac{2 + 3/e^x}{1 + 1/e^x} = \dfrac{2 + 0}{1 + 0} = 2$. Thus the line $y = 2$ is a horizontal asymptote. Also $\lim\limits_{x\to-\infty} \dfrac{2e^x + 3}{e^x + 1} = \dfrac{0 + 3}{0 + 1} = 3$, so $y = 3$ is a horizontal asymptote.

2.5.97 Using the rules of logarithms, $f(x) = \dfrac{6 \ln x}{3 \ln x - 1}$. The domain of f is $(0, \sqrt[3]{e}) \cup (\sqrt[3]{e}, \infty)$. We first examine the end behavior of the function. Observe that $\lim\limits_{x \to \infty} \dfrac{6 \ln x}{3 \ln x - 1} = \lim\limits_{x \to \infty} \dfrac{6}{3 - (1/\ln x)} = \dfrac{6}{3} = 2$ and $\lim\limits_{x \to 0^+} \dfrac{6 \ln x}{3 \ln x - 1} = \lim\limits_{x \to 0^+} \dfrac{6}{3 - (1/\ln x)} = \dfrac{6}{3} = 2$. So the function has a horizontal asymptote of $y = 2$ and it is undefined at $x = 0$ but has limit 2 as x approaches 0 from the right. Notice also that as $x \to \sqrt[3]{e}^+$, $6 \ln x \to 2$ and $3 \ln x - 1$ is positive and approaches 0. Therefore, $\lim\limits_{x \to \sqrt[3]{e}^+} \dfrac{6 \ln x}{3 \ln x - 1} = \infty$ and by a similar argument, $\lim\limits_{x \to \sqrt[3]{e}^-} \dfrac{6 \ln x}{3 \ln x - 1} = -\infty$.

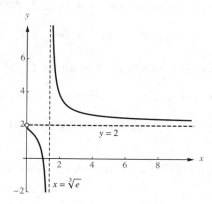

2.6 Continuity

2.6.1

a. $a(t)$ is a continuous function during the time period from when she jumps from the plane and when she touches down on the ground, because her position is changing continuously with time.

b. $n(t)$ is not a continuous function of time. The function "jumps" at the times when a quarter must be added.

c. $T(t)$ is a continuous function, because temperature varies continuously with time.

d. $p(t)$ is not continuous – it jumps by whole numbers when a player scores a point.

2.6.3 A function f is continuous on an interval I if it is continuous at all points in the interior of I, and it must be continuous from the right at the left endpoint (if the left endpoint is included in I) and it must be continuous from the left at the right endpoint (if the right endpoint is included in I.)

2.6.5 f is discontinuous at $x = 1$, at $x = 2$, and at $x = 3$. At $x = 1$, $f(1)$ is not defined (so the first condition is violated). At $x = 2$, $f(2)$ is defined and $\lim\limits_{x \to 2} f(x)$ exists, but $\lim\limits_{x \to 2} f(x) \neq f(2)$ (so condition 3 is violated). At $x = 3$, $\lim\limits_{x \to 3} f(x)$ does not exist (so condition 2 is violated).

2.6.7 f is discontinuous at $x = 1$, at $x = 2$, and at $x = 3$. At $x = 1$, $\lim\limits_{x \to 1} f(x)$ does not exist, and $f(1)$ is not defined (so conditions 1 and 2 are violated). At $x = 2$, $\lim\limits_{x \to 2} f(x)$ does not exist (so condition 2 is violated). At $x = 3$, $f(3)$ is not defined (so condition 1 is violated).

2.6.9

a. A function f is continuous from the left at $x = a$ if a is in the domain of f, and $\lim\limits_{x \to a^-} f(x) = f(a)$.

b. A function f is continuous from the right at $x = a$ if a is in the domain of f, and $\lim_{x \to a^+} f(x) = f(a)$.

2.6.11 f is continuous on $(0,1)$, on $(1,2)$, on $(2,3]$, and on $(3,4)$. It is continuous from the left at 3.

2.6.13 f is continuous on $[0,1)$, on $(1,2)$, on $[2,3)$, and on $(3,5)$. It is continuous from the right at 2.

2.6.15 The domain of $f(x) = \dfrac{e^x}{x}$ is $(-\infty, 0) \cup (0, \infty)$, and f is continuous everywhere on this domain.

2.6.17 The number -5 is not in the domain of f, because the denominator is equal to 0 when $x = -5$. Thus, the function is not continuous at -5.

2.6.19 f is discontinuous at 1, because 1 is not in the domain of f; $f(1)$ is not defined.

2.6.21 f is discontinuous at 1, because $\lim_{x \to 1} f(x) \neq f(1)$. In fact, $f(1) = 3$, but $\lim_{x \to 1} f(x) = 2$.

2.6.23 f is discontinuous at 4, because 4 is not in the domain of f; $f(4)$ is not defined.

2.6.25 Because p is a polynomial, it is continuous on all of $\mathbb{R} = (-\infty, \infty)$.

2.6.27 Because f is a rational function, it is continuous on its domain. Its domain is $(-\infty, -3) \cup (-3, 3) \cup (3, \infty)$.

2.6.29 Because f is a rational function, it is continuous on its domain. Its domain is $(-\infty, -2) \cup (-2, 2) \cup (2, \infty)$.

2.6.31 Because $f(x) = (x^8 - 3x^6 - 1)^{40}$ is a polynomial, it is continuous everywhere, including at 0. Thus $\lim_{x \to 0} f(x) = f(0) = (-1)^{40} = 1$.

2.6.33 Because $x^3 - 2x^2 - 8x = x(x^2 - 2x - 8) = x(x-4)(x+2)$, we have (as long as $x \neq 4$)

$$\sqrt{\dfrac{x^3 - 2x^2 - 8x}{x - 4}} = \sqrt{x(x+2)}.$$

Thus, $\lim_{x \to 4} \sqrt{\dfrac{x^3 - 2x^2 - 8x}{x - 4}} = \lim_{x \to 4} \sqrt{x(x+2)} = \sqrt{24}$, using Theorem 2.12 and the fact that the square root is a continuous function.

2.6.35 Because $f(x) = \left(\dfrac{x+5}{x+2}\right)^4$ is a rational function, it is continuous at all points in its domain, including at $x = 1$. Thus $\lim_{x \to 1} f(x) = f(1) = 16$.

2.6.37 Note that

$$\lim_{x \to 5} \dfrac{6(\sqrt{x^2 - 16} - 3)}{5(x - 5)} = \lim_{x \to 5} \dfrac{6(\sqrt{x^2 - 16} - 3)}{5(x - 5)} \cdot \dfrac{(\sqrt{x^2 - 16} + 3)}{(\sqrt{x^2 - 16} + 3)} = \lim_{x \to 5} \dfrac{6(x^2 - 25)}{5(x - 5)(\sqrt{x^2 - 16} + 3)}$$

$$= \lim_{x \to 5} \dfrac{6(x + 5)}{5(\sqrt{x^2 - 16} + 3)} = \dfrac{60}{30} = 2.$$

2.6.39

a. f is defined at 1. We have $f(1) = 1^2 + (3)(1) = 4$. To see whether or not $\lim_{x \to 1} f(x)$ exists, we investigate the two one-sided limits. $\lim_{x \to 1^-} f(x) = \lim_{x \to 1^-} 2x = 2$, and $\lim_{x \to 1^+} f(x) = \lim_{x \to 1^+} (x^2 + 3x) = 4$, so $\lim_{x \to 1} f(x)$ does not exist. Thus f is discontinuous at $x = 1$.

b. f is continuous from the right, because $\lim_{x \to 1^+} f(x) = 4 = f(1)$.

2.6. Continuity

c. f is continuous on $(-\infty, 1)$ and on $[1, \infty)$.

2.6.41 f is defined and is continuous on $(-\infty, 5]$. It is continuous from the left at 5.

2.6.43 f is continuous on $(-\infty, -\sqrt{8}]$ and on $[\sqrt{8}, \infty)$. It is continuous from the left at $-\sqrt{8}$ and from the right at $\sqrt{8}$.

2.6.45 Because f is the composition of two functions which are continuous on $(-\infty, \infty)$, it is continuous on $(-\infty, \infty)$.

2.6.47 Because f is the composition of two functions which are continuous on $(-\infty, \infty)$, it is continuous on $(-\infty, \infty)$.

2.6.49 $\lim\limits_{x \to 2} \sqrt{\dfrac{4x+10}{2x-2}} = \sqrt{\dfrac{18}{2}} = 3.$

2.6.51 $\lim\limits_{x \to \pi} \dfrac{\cos^2 x + 3\cos x + 2}{\cos x + 1} = \lim\limits_{x \to \pi} \dfrac{(\cos x + 1)(\cos x + 2)}{\cos x + 1} = \lim\limits_{x \to \pi} (\cos x + 2) = 1.$

2.6.53 $\lim\limits_{x \to 3} \sqrt{x^2 + 7} = \sqrt{9 + 7} = 4.$

2.6.55 $\lim\limits_{x \to \pi/2} \dfrac{\sin x - 1}{\sqrt{\sin x} - 1} = \lim\limits_{x \to \pi/2} (\sqrt{\sin x} + 1) = 2.$

2.6.57 $\lim\limits_{x \to 0} \dfrac{\cos x - 1}{\sin^2 x} = \lim\limits_{x \to 0} \dfrac{\cos x - 1}{1 - \cos^2 x} = \lim\limits_{x \to 0} \dfrac{\cos x - 1}{(1 - \cos x)(1 + \cos x)} = \lim\limits_{x \to 0} -\dfrac{1}{1 + \cos x} = -\dfrac{1}{2}.$

2.6.59 $\lim\limits_{x \to 0} \dfrac{e^{4x} - 1}{e^x - 1} = \lim\limits_{x \to 0} \dfrac{(e^{2x} + 1)(e^{2x} - 1)}{e^x - 1} = \lim\limits_{x \to 0} \dfrac{(e^{2x} + 1)(e^x - 1)(e^x + 1)}{e^x - 1} = \lim\limits_{x \to 0} (e^{2x} + 1)(e^x + 1) = 2 \cdot 2 = 4.$

2.6.61 $f(x) = \csc x$ isn't defined at $x = k\pi$ where k is an integer, so it isn't continuous at those points. So it is continuous on intervals of the form $(k\pi, (k+1)\pi)$ where k is an integer. $\lim\limits_{x \to \pi/4} \csc x = \sqrt{2}.$ $\lim\limits_{x \to 2\pi^-} \csc x = -\infty.$

2.6.63 f isn't defined for any number of the form $\pi/2 + k\pi$ where k is an integer, so it isn't continuous there. It is continuous on intervals of the form $(\pi/2 + k\pi, \pi/2 + (k+1)\pi)$, where k is an integer.
$\lim\limits_{x \to \pi/2^-} f(x) = \infty.$ $\lim\limits_{x \to 4\pi/3} f(x) = \dfrac{1 - \sqrt{3}/2}{-1/2} = \sqrt{3} - 2.$

2.6.65 This function is continuous on its domain, which is $(-\infty, 0) \cup (0, \infty)$.
$\lim\limits_{x \to 0^-} f(x) = \lim\limits_{x \to 0^-} \dfrac{e^x}{1 - e^x} = \infty$, while $\lim\limits_{x \to 0^+} f(x) = \lim\limits_{x \to 0^+} \dfrac{e^x}{1 - e^x} = -\infty.$

2.6.67

a. Note that $f(x) = 2x^3 + x - 2$ is continuous everywhere, so in particular it is continuous on $[-1, 1]$. Note that $f(-1) = -5 < 0$ and $f(1) = 1 > 0$. Because 0 is an intermediate value between $f(-1)$ and $f(1)$, the Intermediate Value Theorem guarantees a number c between -1 and 1 where $f(c) = 0$.

b. Using a graphing calculator and a computer algebra system, we see that the root of f is about 0.835.

c.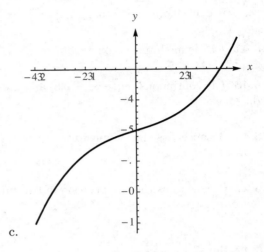

2.6.69

a. Note that $f(x) = x^3 - 5x^2 + 2x$ is continuous everywhere, so in particular it is continuous on $[-1, 5]$. Note that $f(-1) = -8 < -1$ and $f(5) = 10 > -1$. Because -1 is an intermediate value between $f(-1)$ and $f(5)$, the Intermediate Value Theorem guarantees a number c between -1 and 5 where $f(c) = -1$.

b. Using a graphing calculator and a computer algebra system, we see that there are actually three different values of c between -1 and 5 for which $f(c) = -1$. They are $c \approx -0.285$, $c \approx 0.778$, and $c \approx 4.507$.

c.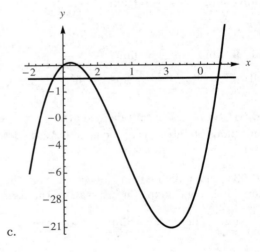

2.6.71

a. Note that $f(x) = e^x + x$ is continuous on its domain, so in particular it is continuous on $[-1, 0]$. Note that $f(-1) = \frac{1}{e} - 1 < 0$ and $f(0) = 1 > 0$. Because 0 is an intermediate value between $f(-1)$ and $f(0)$, the Intermediate Value Theorem guarantees a number c between -1 and 0 where $f(c) = 0$.

2.6. Continuity

b. Using a graphing calculator and a computer algebra system, we see that the value of c guaranteed by the theorem is about -0.567.

c.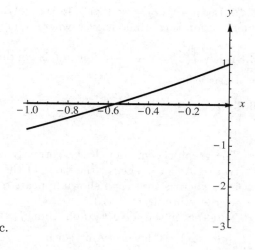

2.6.73

a. True. If f is right continuous at a, then $f(a)$ exists and the limit from the right at a exists and is equal to $f(a)$. Because it is left continuous, the limit from the left exists — so we now know that the limit as $x \to a$ of $f(x)$ exists, because the two one-sided limits are both equal to $f(a)$.

b. True. If $\lim_{x \to a} f(x) = f(a)$, then $\lim_{x \to a^+} f(x) = f(a)$ and $\lim_{x \to a^-} f(x) = f(a)$.

c. False. The statement would be true if f were continuous. However, if f isn't continuous, then the statement doesn't hold. For example, suppose that $f(x) = \begin{cases} 0 & \text{if } 0 \leq x < 1; \\ 1 & \text{if } 1 \leq x \leq 2, \end{cases}$ Note that $f(0) = 0$ and $f(2) = 1$, but there is no number c between 0 and 2 where $f(c) = 1/2$.

d. False. Consider $f(x) = x^2$ and $a = -1$ and $b = 1$. Then f is continuous on $[a, b]$, but $\frac{f(1)+f(-1)}{2} = 1$, and there is no c on (a, b) with $f(c) = 1$.

2.6.75

a. Because A is a continuous function of r on $[0, 0.08]$, and because $A(0) = 5000$ and $A(0.08) \approx 11098.2$, (and 7000 is an intermediate value between these two numbers) the Intermediate Value Theorem guarantees a value of r between 0 and 0.08 where $A(r) = 7000$.

b. Solving $5000(1 + (r/12))^{120} = 7000$ for r, we see that $(1 + (r/12))^{120} = 7/5$, so $1 + r/12 = \sqrt[120]{7/5}$, so $r = 12(\sqrt[120]{7/5} - 1) \approx 0.034$.

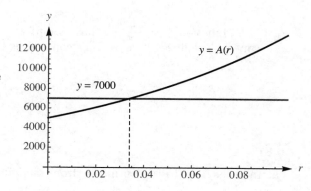

2.6.77 Consider the function $f(x) = \cos x - 2x$ on the interval $[0, 1]$. Note that $f(0) = 1$ and $f(\pi/2) = -\pi < 0$. So by the Intermediate Value Theorem, there must be a root of f on the interval $[0, \pi/2]$. Using a computer algebra system, we find a root of approximately 0.45.

2.6.79 Because $f(x) = x^3 + 3x - 18$ is a polynomial, it is continuous on $(-\infty, \infty)$, and because the absolute value function is continuous everywhere, $|f(x)|$ is continuous everywhere.

2.6.81 Let $f(x) = \dfrac{1}{\sqrt{x}-4}$. Then f is continuous on $[0, 16) \cup (16, \infty)$. So $h(x) = |f(x)|$ is continuous on this set as well.

2.6.83

The graph shown isn't drawn correctly at the integers. At an integer a, the value of the function is 0, whereas the graph shown appears to take on all the values from 0 to 1.
Note that in the correct graph, $\lim\limits_{x \to a^-} f(x) = 1$ and $\lim\limits_{x \to a^+} f(x) = 0$ for every integer a.

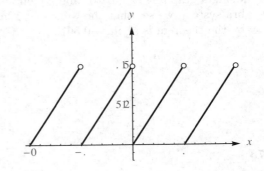

2.6.85 With slight modifications, we can use the examples from the previous two problems.

a. The function $y = x - \lfloor x \rfloor$ is defined at $x = 1$ but isn't continuous there.

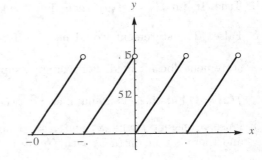

b. The function $y = \dfrac{\sin(x-1)}{x-1}$ has a limit at $x = 1$, but isn't defined there, so isn't continuous there.

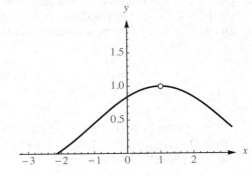

2.6.87

a. In order for g to be continuous from the left at $x = 1$, we must have $\lim\limits_{x \to 1^-} g(x) = g(1) = a$. We have $\lim\limits_{x \to 1^-} g(x) = \lim\limits_{x \to 1^-} (x^2 + x) = 2$. So we must have $a = 2$.

b. In order for g to be continuous from the right at $x = 1$, we must have $\lim\limits_{x \to 1^+} g(x) = g(1) = a$. We have $\lim\limits_{x \to 1^+} g(x) = \lim\limits_{x \to 1^+} (3x + 5) = 8$. So we must have $a = 8$.

2.6. Continuity

c. Because the limit from the left and the limit from the right at $x = 1$ don't agree, there is no value of a which will make the function continuous at $x = 1$.

2.6.89 $\lim_{x \to 0} \dfrac{2e^x + 10e^{-x}}{e^x + e^{-x}} = \dfrac{12}{2} = 6$.

$\lim_{x \to -\infty} \dfrac{2e^x + 10e^{-x}}{e^x + e^{-x}} = \lim_{x \to -\infty} \dfrac{2e^x + 10e^{-x}}{e^x + e^{-x}} \cdot \dfrac{e^x}{e^x} = \lim_{x \to -\infty} \dfrac{2e^{2x} + 10}{e^{2x} + 1} = \dfrac{10}{1} = 10$.

$\lim_{x \to \infty} \dfrac{2e^x + 10e^{-x}}{e^x + e^{-x}} = \lim_{x \to \infty} \dfrac{2e^x + 10e^{-x}}{e^x + e^{-x}} \cdot \dfrac{e^{-x}}{e^{-x}} = \lim_{x \to \infty} \dfrac{2 + 10e^{-2x}}{1 + e^{-2x}} = \dfrac{2}{1} = 2$.

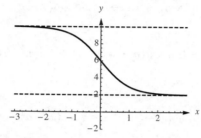

There are no vertical asymptotes. The lines $y = 2$ and $y = 10$ are horizontal asymptotes.

2.6.91 Let $f(x) = 70x^3 - 87x^2 + 32x - 3$. Note that $f(0) < 0$, $f(0.2) > 0$, $f(0.55) < 0$, and $f(1) > 0$. Because the given polynomial is continuous everywhere, the Intermediate Value Theorem guarantees us a root on $(0, 0.2)$, at least one on $(0.2, 0.55)$, and at least one on $(0.55, 1)$. Because there can be at most 3 roots and there are at least 3 roots, there must be exactly 3 roots. The roots are $x_1 = 1/7$, $x_2 = 1/2$ and $x_3 = 3/5$.

2.6.93 We can argue essentially like the previous problem, or we can imagine an identical twin to the original monk, who takes an identical version of the original monk's journey up the winding path while the monk is taking the return journey down. Because they must pass somewhere on the path, that point is the one we are looking for.

2.6.95 The discontinuity is not removable, because $\lim_{x \to a} f(x)$ does not exist. The discontinuity pictured is a jump discontinuity.

2.6.97 Note that $\lim_{x \to 2} \dfrac{x^2 - 7x + 10}{x - 2} = \lim_{x \to 2} \dfrac{(x-2)(x-5)}{x-2} = \lim_{x \to 2}(x - 5) = -3$. Because this limit exists, the discontinuity is removable.

2.6.99 Note that $h(x) = \dfrac{x^3 - 4x^2 + 4x}{x(x-1)} = \dfrac{x(x-2)^2}{x(x-1)}$. Thus $\lim_{x \to 0} h(x) = -4$, and the discontinuity at $x = 0$ is removable. However, $\lim_{x \to 1} h(x)$ does not exist, and the discontinuity at $x = 1$ is not removable (it is infinite.)

2.6.101

a. Note that $-1 \leq \sin(1/x) \leq 1$ for all $x \neq 0$, so $-x \leq x \sin(1/x) \leq x$ (for $x > 0$. For $x < 0$ we would have $x \leq x \sin(1/x) \leq -x$.) Because both $x \to 0$ and $-x \to 0$ as $x \to 0$, the Squeeze Theorem tells us that $\lim_{x \to 0} x \sin(1/x) = 0$ as well. Because this limit exists, the discontinuity is removable.

b. Note that as $x \to 0^+$, $1/x \to \infty$, and thus $\lim_{x \to 0^+} \sin(1/x)$ does not exist. So the discontinuity is not removable.

2.6.103

a. Consider $g(x) = x + 1$ and $f(x) = \dfrac{|x-1|}{x-1}$. Note that both g and f are continuous at $x = 0$. However $f(g(x)) = f(x+1) = \dfrac{|x|}{x}$ is not continuous at 0.

b. The previous theorem says that the composition of f and g is continuous at a if g is continuous at a and f is continuous at $g(a)$. It does not say that if g and f are both continuous at a that the composition is continuous at a.

2.6.105

a. Using the hint, we have
$$\sin x = \sin(a + (x-a)) = \sin a \cos(x-a) + \sin(x-a) \cos a.$$

Note that as $x \to a$, we have that $\cos(x-a) \to 1$ and $\sin(x-a) \to 0$.

So,
$$\lim_{x \to a} \sin x = \lim_{x \to a} \sin(a + (x-a)) = \lim_{x \to a} (\sin a \cos(x-a) + \sin(x-a) \cos a) = (\sin a) \cdot 1 + 0 \cdot \cos a = \sin a.$$

b. Using the hint, we have
$$\cos x = \cos(a + (x-a)) = \cos a \cos(x-a) - \sin a \sin(x-a).$$

So,
$$\lim_{x \to a} \cos x = \lim_{x \to a} \cos(a + (x-a)) = \lim_{x \to a} ((\cos a) \cos(x-a) - (\sin a) \sin(x-a))$$
$$= (\cos a) \cdot 1 - (\sin a) \cdot 0 = \cos a.$$

2.7 Precise Definitions of Limits

2.7.1 Note that all the numbers in the interval $(1,3)$ are within 1 unit of the number 2. So $|x-2| < 1$ is true for all numbers in that interval. In fact, $\{x \colon 0 < |x-2| < 1\}$ is exactly the set $(1,3)$ with $x \ne 2$.

2.7.3

a. This is symmetric about $x = 5$, because $\dfrac{1+9}{2} = 5$.

b. This is symmetric about $x = 5$, because $\dfrac{4+6}{2} = 5$.

c. This is not symmetric about $x = 5$, because $\dfrac{3+8}{2} \ne 5$.

d. This is symmetric about $x = 5$, because $\dfrac{4.5+5.5}{2} = 5$.

2.7.5 $\lim_{x \to a} f(x) = L$ if for any arbitrarily small positive number ε, there exists a number δ, so that $f(x)$ is within ε units of L for any number x within δ units of a (but not including a itself).

2.7.7 We are given that $|f(x) - 5| < 0.1$ for values of x in the interval $(0,5)$, so we need to ensure that the set of x values we are allowing fall in this interval.

Note that the number 0 is two units away from the number 2 and the number 5 is three units away from the number 2. In order to be sure that we are talking about numbers in the interval $(0,5)$ when we write $|x-2| < \delta$, we would need to have $\delta = 2$ (or a number less than 2). In fact, the set of numbers for which $|x-2| < 2$ is the interval $(0,4)$ which is a subset of $(0,5)$.

If we were to allow δ to be any number greater than 2, then the set of all x so that $|x-2| < \delta$ would include numbers less than 0, and those numbers aren't on the interval $(0,5)$.

2.7.9

a. In order for f to be within 2 units of 5, it appears that we need x to be within 1 unit of 2. So $\delta = 1$.

b. In order for f to be within 1 unit of 5, it appears that we would need x to be within 1/2 unit of 2. So $\delta = 0.5$.

2.7.11

a. In order for f to be within 3 units of 6, it appears that we would need x to be within 2 units of 3. So $\delta = 2$.

b. In order for f to be within 1 unit of 6, it appears that we would need x to be within 1/2 unit of 3. So $\delta = 1/2$.

2.7.13

a. If $\varepsilon = 1$, we need $|x^3 + 3 - 3| < 1$. So we need $|x| < \sqrt[3]{1} = 1$ in order for this to happen. Thus $\delta = 1$ will suffice.

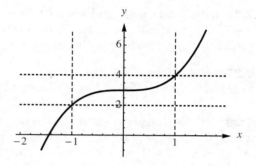

b. If $\varepsilon = 0.5$, we need $|x^3 + 3 - 3| < 0.5$. So we need $|x| < \sqrt[3]{0.5}$ in order for this to happen. Thus $\delta = \sqrt[3]{0.5} \approx 0.79$ will suffice.

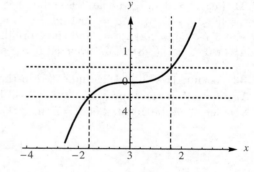

2.7.15

a. For $\varepsilon = 1$, the required value of δ would also be 1. A larger value of δ would work to the right of 2, but this is the largest one that would work to the left of 2.

b. For $\varepsilon = 1/2$, the required value of δ would also be 1/2.

c. It appears that for a given value of ε, it would be wise to take $\delta = \min(\varepsilon, 2)$. This assures that the desired inequality is met on both sides of 2.

2.7.17

a. For $\varepsilon = 2$, it appears that a value of $\delta = 1$ (or smaller) would work.

b. For $\varepsilon = 1$, it appears that a value of $\delta = 1/2$ (or smaller) would work.

c. For an arbitrary ε, a value of $\delta = \varepsilon/2$ or smaller appears to suffice.

2.7.19 For any $\varepsilon > 0$, let $\delta = \varepsilon/8$. Then if $0 < |x - 1| < \delta$, we would have $|x - 1| < \varepsilon/8$. Then $|8x - 8| < \varepsilon$, so $|(8x + 5) - 13| < \varepsilon$. This last inequality has the form $|f(x) - L| < \varepsilon$, which is what we were attempting to show. Thus, $\lim_{x \to 1}(8x + 5) = 13$.

2.7.21 First note that if $x \neq 4$, $f(x) = \frac{x^2 - 16}{x - 4} = x + 4$.

Now if $\varepsilon > 0$ is given, let $\delta = \varepsilon$. Now suppose $0 < |x - 4| < \delta$. Then $x \neq 4$, so the function $f(x)$ can be described by $x + 4$. Also, because $|x - 4| < \delta$, we have $|x - 4| < \varepsilon$. Thus $|(x + 4) - 8| < \varepsilon$. This last inequality has the form $|f(x) - L| < \varepsilon$, which is what we were attempting to show. Thus, $\lim_{x \to 4} \frac{x^2 - 16}{x - 4} = 8$.

2.7.23 Let $\varepsilon > 0$ be given and assume that $0 < |x - 0| < \delta$ where $\delta = \varepsilon$. It follows that $||x| - 0| = |x| = |x - 0| < \delta = \varepsilon$. We have shown that for any $\varepsilon > 0$, $||x| - 0| < \varepsilon$ whenever $0 < |x - 0| < \delta$, provided $0 < \delta \leq \varepsilon$.

2.7.25 Let $\varepsilon > 0$ be given and assume that $0 < |x - 7| < \delta$ where $\delta = \varepsilon/3$. If $x < 7$, $|f(x) - 9| = |3x - 12 - 9| = 3|x - 7| < 3\delta = 3(\varepsilon/3) = \varepsilon$; if $x > 7$, then $|f(x) - 9| = |x + 2 - 9| = |x - 7| < \delta = \varepsilon/3 < \varepsilon$. We've shown that for any $\varepsilon > 0$, $|f(x) - 9| < \varepsilon$ whenever $0 < |x - 7| < \delta$, provided $0 < \delta \leq \varepsilon/3$.

2.7.27 Let $\varepsilon > 0$ be given. Let $\delta = \sqrt{\varepsilon}$. Then if $0 < |x - 0| < \delta$, we would have $|x| < \sqrt{\varepsilon}$. But then $|x^2| < \varepsilon$, which has the form $|f(x) - L| < \varepsilon$. Thus, $\lim_{x \to 0} f(x) = 0$.

2.7.29 Let $\varepsilon > 0$ be given and assume that $0 < |x - 2| < \delta$ where $\delta = \min\{1, \varepsilon/8\}$. By factoring $x^2 + 3x - 10$, we find that $|x^2 + 3x - 10| = |x - 2||x + 5|$. Because $|x - 2| < \delta$ and $\delta \leq 1$, we have $|x - 2| < 1$, which implies that $-1 < x - 2 < 1$, or $1 < x < 3$. It follows that $|x + 5| = x + 5 \leq 8$. We also know that $|x - 2| < \varepsilon/8$ because $0 < |x - 2| < \delta$ and $\delta \leq \varepsilon/8$. Therefore $|x^2 + 3x - 10| = |x - 2||x + 5| < (\varepsilon/8) \cdot 8 = \varepsilon$. We have shown that for any $\varepsilon > 0$, $|x^2 + 3x - 10| < \varepsilon$ whenever $0 < |x - 2| < \delta$, provided $0 < \delta \leq \min\{1, \varepsilon/8\}$.

2.7.31 Let $\varepsilon > 0$ be given and assume that $0 < |x - (-3)| < \delta$ where $\delta = \varepsilon/2$. Using the inequality $||a| - |b|| \leq |a - b|$ with $a = 2x$ and $b = -6$, it follows that $||2x| - 6| = ||2x| - |-6|| \leq |2x - (-6)| = 2|x - (-3)| < 2\delta = 2(\varepsilon/2) = \varepsilon$ and therefore $||2x| - 6| < \varepsilon$. We have shown that for any $\varepsilon > 0$, $||2x| - 6| < \varepsilon$ whenever $0 < |x - (-3)| < \delta$, provided $0 < \delta \leq \varepsilon/2$.

2.7.33 Assume $|x - 3| < 1$, as indicated in the hint. Then $2 < x < 4$, so $\frac{1}{4} < \frac{1}{x} < \frac{1}{2}$, and thus $\left|\frac{1}{x}\right| < \frac{1}{2}$.

Also note that the expression $\left|\frac{1}{x} - \frac{1}{3}\right|$ can be written as $\left|\frac{x-3}{3x}\right|$.

Now let $\varepsilon > 0$ be given. Let $\delta = \min(6\varepsilon, 1)$. Now assume that $0 < |x - 3| < \delta$. Then

$$|f(x) - L| = \left|\frac{x-3}{3x}\right| < \left|\frac{x-3}{6}\right| < \frac{6\varepsilon}{6} = \varepsilon.$$

Thus we have established that $\left|\frac{1}{x} - \frac{1}{3}\right| < \varepsilon$ whenever $0 < |x - 3| < \delta$.

2.7.35 Assume $|x - (1/10)| < (1/20)$, as indicated in the hint. Then $1/20 < x < 3/20$, so $\frac{20}{3} < \frac{1}{x} < \frac{20}{1}$, and thus $\left|\frac{1}{x}\right| < 20$.

Also note that the expression $\left|\frac{1}{x} - 10\right|$ can be written as $\left|\frac{10x-1}{x}\right|$.

Let $\varepsilon > 0$ be given. Let $\delta = \min(\varepsilon/200, 1/20)$. Now assume that $0 < |x - (1/10)| < \delta$. Then

$$|f(x) - L| = \left|\frac{10x - 1}{x}\right| < |(10x - 1) \cdot 20|$$

$$\leq |x - (1/10)| \cdot 200 < \frac{\varepsilon}{200} \cdot 200 = \varepsilon.$$

Thus we have established that $\left|\frac{1}{x} - 10\right| < \varepsilon$ whenever $0 < |x - (1/10)| < \delta$.

2.7.37 Let $\varepsilon > 0$ be given and assume that $0 < |x - 0| < \delta$ where $\delta = \min\{1, \sqrt{\varepsilon}/\sqrt{2}\}$. Because $|x - 0| < \delta$, we have $|x| < 1$ and $|x| < \sqrt{\varepsilon}/\sqrt{2}$, which implies that $x^2 < 1$ and $x^2 < \varepsilon/2$. It follows that $|x^2 + x^4 - 0| = x^2 + x^4 = x^2(1 + x^2) \leq \frac{\varepsilon}{2} \cdot 2 = \varepsilon$. We have shown that for any $\varepsilon > 0$, $|x^2 + x^4 - 0| < \varepsilon$ whenever $0 < |x - 0| < \delta$, provided $0 < \delta \leq \min\{1, \sqrt{\varepsilon}/\sqrt{2}\}$.

2.7. Precise Definitions of Limits

2.7.39 Let $m = 0$, then the proof is as follows: Let $\varepsilon > 0$ be given and assume that $0 < |x - a| < \delta$ where $\delta = 1$ (or any other positive number). Then $|f(x) - b| = |b - b| = 0 < \varepsilon$. We have shown that for any $\varepsilon > 0$, $|b - b| < \varepsilon$ whenever $0 < |x - a| < \delta$, provided δ equals any positive number.

Now assume that $m \neq 0$. Let $\varepsilon > 0$ be given and assume that $0 < |x - a| < \delta$ where $\delta = \varepsilon/|m|$. Then

$$|(mx + b) - (ma + b)| = |m||x - a| < |m|\delta = |m|(\varepsilon/|m|) = \varepsilon.$$

Therefore it has been shown that for any $\varepsilon > 0$, $|(mx+b) - (ma+b)| < \varepsilon$ whenever $0 < |x - a| < \delta$, provided $\delta = \varepsilon/|m|$.

2.7.41 Let $\varepsilon > 0$ be given and assume that $0 < |x - 1| < \delta$ where $\delta = \min\left\{\dfrac{1}{2}, \dfrac{8\varepsilon}{65}\right\}$. Observe that $|x^4 - 1| = |(x^2 - 1)(x^2 + 1)| = |x - 1||x + 1||x^2 + 1|$. Because $|x - 1| < \delta$ and $\delta \leq \dfrac{1}{2}$, we have $|x - 1| < \dfrac{1}{2}$, which implies that $-\dfrac{1}{2} < x - 1 < \dfrac{1}{2}$, or $\dfrac{1}{2} < x < \dfrac{3}{2}$. It follows that $|x + 1| = x + 1 \leq \dfrac{5}{2}$. Also $x^2 < \dfrac{9}{4}$, so $|x^2 + 1| = x^2 + 1 \leq \dfrac{13}{4}$. We also know that $|x - 1| < \dfrac{8\varepsilon}{65}$ because $|x - 1| < \delta$ and $\delta \leq \dfrac{8\varepsilon}{65}$. Therefore

$$|x^4 - 1| = |x - 1||x + 1||x^2 + 1| \leq \dfrac{8\varepsilon}{65} \cdot \dfrac{5}{2} \cdot \dfrac{13}{4} = \varepsilon.$$

We have shown that for any $\varepsilon > 0$, $|x^4 - 1| < \varepsilon$ whenever $0 < |x - 1| < \delta$, provided $0 < \delta = \min\left\{\dfrac{1}{2}, \dfrac{8\varepsilon}{65}\right\}$.

2.7.43 Let $\varepsilon > 0$ be given.

Because $\lim\limits_{x \to a} f(x) = L$, we know that there exists a $\delta_1 > 0$ so that $|f(x) - L| < \varepsilon/2$ when $0 < |x - a| < \delta_1$. Also, because $\lim\limits_{x \to a} g(x) = M$, there exists a $\delta_2 > 0$ so that $|g(x) - M| < \varepsilon/2$ when $0 < |x - a| < \delta_2$.

Now let $\delta = \min(\delta_1, \delta_2)$.

Then if $0 < |x - a| < \delta$, we would have $|f(x) - g(x) - (L - M)| = |(f(x) - L) + (M - g(x))| \leq |f(x) - L| + |M - g(x)| = |f(x) - L| + |g(x) - M| \leq \varepsilon/2 + \varepsilon/2 = \varepsilon$. Note that the key inequality in this sentence follows from the triangle inequality.

2.7.45 Let $N > 0$ be given. Let $\delta = 1/\sqrt{N}$. Then if $0 < |x - 4| < \delta$, we have $|x - 4| < 1/\sqrt{N}$. Taking the reciprocal of both sides, we have $\dfrac{1}{|x - 4|} > \sqrt{N}$, and squaring both sides of this inequality yields $\dfrac{1}{(x - 4)^2} > N$. Thus $\lim\limits_{x \to 4} f(x) = \infty$.

2.7.47 Let $N > 1$ be given. Let $\delta = 1/\sqrt{N - 1}$. Suppose that $0 < |x - 0| < \delta$. Then $|x| < 1/\sqrt{N - 1}$, and taking the reciprocal of both sides, we see that $1/|x| > \sqrt{N - 1}$. Then squaring both sides yields $dsfrac1x^2 > N - 1$, so $\dfrac{1}{x^2} + 1 > N$. Thus $\lim\limits_{x \to 0} f(x) = \infty$.

2.7.49

a. False. In fact, if the statement is true for a specific value of δ_1, then it would be true for any value of $\delta < \delta_1$. This is because if $0 < |x - a| < \delta$, it would automatically follow that $0 < |x - a| < \delta_1$.

b. False. This statement is not equivalent to the definition – note that it says "for an arbitrary δ there exists an ε" rather than "for an arbitrary ε there exists a δ."

c. True. This is the definition of $\lim\limits_{x \to a} f(x) = L$.

d. True. Both inequalities describe the set of x's which are within δ units of a.

2.7.51 Because we are approaching a from the right, we are only considering values of x which are close to, but a little larger than a. The numbers x to the right of a which are within δ units of a satisfy $0 < x - a < \delta$.

2.7.53

a. Let $\varepsilon > 0$ be given. let $\delta = \varepsilon/2$. Suppose that $0 < x < \delta$. Then $0 < x < \varepsilon/2$ and
$$|f(x) - L| = |2x - 4 - (-4)| = |2x| = 2|x|$$
$$= 2x < \varepsilon.$$

b. Let $\varepsilon > 0$ be given. let $\delta = \varepsilon/3$. Suppose that $0 < 0 - x < \delta$. Then $-\delta < x < 0$ and $-\varepsilon/3 < x < 0$, so $\varepsilon > -3x$. We have
$$|f(x) - L| = |3x - 4 - (-4)| = |3x| = 3|x|$$
$$= -3x < \varepsilon.$$

c. Let $\varepsilon > 0$ be given. Let $\delta = \varepsilon/3$. Because $\varepsilon/3 < \varepsilon/2$, we can argue that $|f(x) - L| < \varepsilon$ whenever $0 < |x| < \delta$ exactly as in the previous two parts of this problem.

2.7.55 Let $\varepsilon > 0$ be given, and let $\delta = \varepsilon^2$. Suppose that $0 < x < \delta$, which means that $x < \varepsilon^2$, so that $\sqrt{x} < \varepsilon$. Then we have
$$|f(x) - L| = |\sqrt{x} - 0| = \sqrt{x} < \varepsilon.$$
as desired.

2.7.57

a. We say that $\lim_{x \to a^+} f(x) = \infty$ if for each positive number N, there exists $\delta > 0$ such that
$$f(x) > N \quad \text{whenever} \quad a < x < a + \delta.$$

b. We say that $\lim_{x \to a^-} f(x) = -\infty$ if for each negative number N, there exists $\delta > 0$ such that
$$f(x) < N \quad \text{whenever} \quad a - \delta < x < a.$$

c. We say that $\lim_{x \to a^-} f(x) = \infty$ if for each positive number N, there exists $\delta > 0$ such that
$$f(x) > N \quad \text{whenever} \quad a - \delta < x < a.$$

2.7.59 Let $N > 0$ be given. Let $\delta = 1/N$, and suppose that $1 - \delta < x < 1$. Then $\frac{N-1}{N} < x < 1$, so $\frac{1-N}{N} > -x > -1$, and therefore $1 + \frac{1-N}{N} > 1 - x > 0$, which can be written as $\frac{1}{N} > 1 - x > 0$. Taking reciprocals yields the inequality $N < \frac{1}{1-x}$, as desired.

2.7.61 Let $M < 0$ be given. Let $\delta = \sqrt[4]{-10/M}$. Suppose that $0 < |x+2| < \delta$. Then $(x+2)^4 < -10/M$, so $\frac{1}{(x+2)^4} > \frac{M}{-10}$, and $-\frac{10}{(x+2)^4} < M$, as desired.

2.7.63 Let $N > 0$ be given. Because $\lim_{x \to a} f(x) = \infty$, there exists $\delta_1 > 0$ such that $f(x) > \frac{N}{2}$ whenever $0 < |x-a| < \delta_1$. Similarly, because $\lim_{x \to a} g(x) = \infty$, there exists $\delta_2 > 0$ such that $g(x) > \frac{N}{2}$ whenever $0 < |x-a| < \delta_2$. Let $\delta = \min\{\delta_1, \delta_2\}$ and assume that $0 < |x-a| < \delta$. Because $\delta = \min\{\delta_1, \delta_2\}$, $\delta \leq \delta_1$ and $\delta \leq \delta_2$. It follows that $0 < |x-a| < \delta_1$ and $0 < |x-a| < \delta_2$ and therefore $f(x) + g(x) > \frac{N}{2} + \frac{N}{2} = N$. So for any $N > 0$, there exists $\delta > 0$ such that $f(x) + g(x) > N$ whenever $0 < |x-a| < \delta$.

2.7.65 Let $\varepsilon > 0$ be given. Let $N = 1/\varepsilon$. Suppose that $x > N$. Then $\frac{1}{x} < \varepsilon$, and so
$$|f(x) - L| = |2 + \frac{1}{x} - 2| < \varepsilon.$$

2.7.67 Let $M > 0$ be given. Let $N = M - 1$. Suppose that $x > N$. Then $x > M - 1$, so $x + 1 > M$, and thus $\dfrac{x^2 + x}{x} > M$, as desired.

2.7.69 Let $\varepsilon > 0$ be given. Let $N = \lfloor (1/\varepsilon) \rfloor + 1$. By assumption, there exists an integer $M > 0$ so that $|f(x) - L| < 1/N$ whenever $|x - a| < 1/M$. Let $\delta = 1/M$.

Now assume $0 < |x - a| < \delta$. Then $|x - a| < 1/M$, and thus $|f(x) - L| < 1/N$. But then

$$|f(x) - L| < \frac{1}{\lfloor (1/\varepsilon) \rfloor + 1} < \varepsilon,$$

as desired.

2.7.71 Let $f(x) = \dfrac{|x|}{x}$ and suppose $\lim\limits_{x \to 0} f(x)$ does exist and is equal to L. Let $\varepsilon = 1/2$. There must be a value of δ so that when $0 < |x| < \delta$, $|f(x) - L| < 1/2$. Now consider the numbers $\delta/3$ and $-\delta/3$, both of which are within δ of 0. We have $f(\delta/3) = 1$ and $f(-\delta/3) = -1$. However, it is impossible for both $|1 - L| < 1/2$ and $|-1 - L| < 1/2$, because the former implies that $1/2 < L < 3/2$ and the latter implies that $-3/2 < L < -1/2$. Thus $\lim\limits_{x \to 0} f(x)$ does not exist.

2.7.73 Because f is continuous at a, we know that $\lim\limits_{x \to a} f(x)$ exists and is equal to $f(a) > 0$. Let $\varepsilon = f(a)/3$. Then there is a number $\delta > 0$ so that $|f(x) - f(a)| < f(a)/3$ whenever $|x - a| < \delta$. Then whenever x lies in the interval $(a - \delta, a + \delta)$ we have $-f(a)/3 \leq f(x) - f(a) \leq f(a)/3$, so $2f(a)/3 \leq f(x) \leq 4f(a)/3$, so f is positive in this interval.

Chapter Two Review

1

a. False. Because $\lim\limits_{x \to 1} \dfrac{x-1}{x^2-1} = \lim\limits_{x \to 1} \dfrac{1}{x+1} = \dfrac{1}{2}$, f doesn't have a vertical asymptote at $x = 1$.

b. False. In general, these methods are too imprecise to produce accurate results.

c. False. For example, the function $f(x) = \begin{cases} 2x & \text{if } x < 0; \\ 1 & \text{if } x = 0; \\ 4x & \text{if } x > 0 \end{cases}$ has a limit of 0 as $x \to 0$, but $f(0) = 1$.

d. True. When we say that a limit exists, we are saying that there is a real number L that the function is approaching. If the limit of the function is ∞, it is still the case that there is no real number that the function is approaching. (There is no real number called "infinity.")

e. False. It could be the case that $\lim\limits_{x \to a^-} f(x) = 1$ and $\lim\limits_{x \to a^+} f(x) = 2$.

f. False.

g. False. For example, the function $f(x) = \begin{cases} 2 & \text{if } 0 < x < 1; \\ 3 & \text{if } 1 \leq x < 2, \end{cases}$ is continuous on $(0, 1)$, and on $[1, 2)$, but isn't continuous on $(0, 2)$.

h. True. $\lim\limits_{x \to a} f(x) = f(a)$ if and only if f is continuous at a.

3 For various values of b, we calculate $v_{\text{avg}} = \dfrac{s(b) - s(1.5)}{b - 1.5}$.

b	1.6	1.51	1.501	1.5001	1.50001
v_{avg}	10.4	11.84	11.984	11.9984	11.9998

We estimate that the instantaneous velocity is 12.

5 This function is discontinuous at $x = -1$, at $x = 1$, and at $x = 3$. At $x = -1$ it is discontinuous because $\lim_{x \to -1} f(x)$ does not exist. At $x = 1$, it is discontinuous because $\lim_{x \to 1} f(x) \neq f(1)$. At $x = 3$, it is discontinuous because $f(3)$ does not exist, and because $\lim_{x \to 3} f(x)$ does not exist.

7

a.

x	$0.9\pi/4$	$0.99\pi/4$	$0.999\pi/4$	$0.9999\pi/4$
$f(x)$	1.4098	1.4142	1.4142	1.4142

x	$1.1\pi/4$	$1.01\pi/4$	$1.001\pi/4$	$1.0001\pi/4$
$f(x)$	1.4098	1.4142	1.4142	1.4142

The limit appears to be approximately 1.4142.

b. $\lim_{x \to \pi/4} \dfrac{\cos 2x}{\cos x - \sin x} = \lim_{x \to \pi/4} \dfrac{\cos^2 x - \sin^2 x}{\cos x - \sin x} = \lim_{x \to \pi/4} (\cos x + \sin x) = \sqrt{2}.$

9

There are infinitely many different correct functions which you could draw. One of them is:

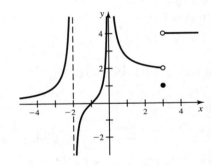

11 $\lim_{x \to 1} \sqrt{5x + 6} = \sqrt{11}.$

13

$$\lim_{h \to 0} \dfrac{(h+6)^2 + (h+6) - 42}{h} = \lim_{h \to 0} \dfrac{h^2 + 12h + 36 + h + 6 - 42}{h} = \lim_{h \to 0} \dfrac{h^2 + 13h}{h} = \lim_{h \to 0} (h + 13) = 13.$$

15 $\lim_{x \to 1} \dfrac{x^3 - 7x^2 + 12x}{4 - x} = \dfrac{1 - 7 + 12}{4 - 1} = \dfrac{6}{3} = 2.$

17 $\lim_{x \to 1} \dfrac{1 - x^2}{x^2 - 8x + 7} = \lim_{x \to 1} \dfrac{(1-x)(1+x)}{(x-7)(x-1)} = \lim_{x \to 1} \dfrac{-(x+1)}{x - 7} = \dfrac{1}{3}.$

19

$$\lim_{x \to 3} \dfrac{1}{x-3}\left(\dfrac{1}{\sqrt{x+1}} - \dfrac{1}{2}\right) = \lim_{x \to 3} \dfrac{2 - \sqrt{x+1}}{2(x-3)\sqrt{x+1}} \cdot \dfrac{(2 + \sqrt{x+1})}{(2 + \sqrt{x+1})}$$

$$= \lim_{x \to 3} \dfrac{4 - (x+1)}{2(x-3)(\sqrt{x+1})(2 + \sqrt{x+1})}$$

$$= \lim_{x \to 3} \dfrac{-(x-3)}{2(x-3)(\sqrt{x+1})(2 + \sqrt{x+1})}$$

$$= \lim_{x \to 3} -\dfrac{1}{2\sqrt{x+1}(2 + \sqrt{x+1})} = -\dfrac{1}{16}.$$

Chapter Two Review

21 $\lim\limits_{x\to 3}\dfrac{x^4-81}{x-3}=\lim\limits_{x\to 3}\dfrac{(x-3)(x+3)(x^2+9)}{x-3}=\lim\limits_{x\to 3}(x+3)(x^2+9)=108.$

23 $\lim\limits_{x\to 81}\dfrac{\sqrt[4]{x}-3}{x-81}=\lim\limits_{x\to 81}\dfrac{\sqrt[4]{x}-3}{(\sqrt{x}+9)(\sqrt[4]{x}+3)(\sqrt[4]{x}-3)}=\lim\limits_{x\to 81}\dfrac{1}{(\sqrt{x}+9)(\sqrt[4]{x}+3)}=\dfrac{1}{108}.$

25 $\lim\limits_{x\to\pi/2}\dfrac{\frac{1}{\sqrt{\sin x}}-1}{x+\pi/2}=\dfrac{0}{\pi}=0.$

27 $\lim\limits_{x\to 5}\dfrac{x-7}{x(x-5)^2}=-\infty.$

29 $\lim\limits_{x\to 3^-}\dfrac{x-4}{x^2-3x}=\lim\limits_{x\to 3^-}\dfrac{x-4}{x(x-3)}=\infty.$

31 $\lim\limits_{x\to 1^+}\dfrac{4x^3-4x^2}{|x-1|}=\lim\limits_{x\to 1^+}\dfrac{4x^2(x-1)}{x-1}=\lim\limits_{x\to 1^+}4x^2=4.$

33 $\lim\limits_{x\to 0^-}\dfrac{2}{\tan x}=-\infty.$

35 $\lim\limits_{x\to\infty}\dfrac{2x-3}{4x+10}=\lim\limits_{x\to\infty}\dfrac{2-(3/x)}{4+(10/x)}=\dfrac{2}{4}=\dfrac{1}{2}.$

37 Note that for $x<0$, $x=-\sqrt{x^2}$. Then we have
$$\lim\limits_{x\to -\infty}\dfrac{(3x+1)}{\sqrt{ax^2+2}}\cdot\dfrac{1/x}{-1/\sqrt{x^2}}=-\lim\limits_{x\to -\infty}\dfrac{3+(1/x)}{\sqrt{a+(2/x^2)}}=-\dfrac{3}{\sqrt{a}}.$$

39 We multiply the numerator and denominator by the conjugate of the denominator (i.e., the expression $\sqrt{x^2-ax}+\sqrt{x^2-x}$). This gives
$$\lim\limits_{x\to\infty}\dfrac{\sqrt{x^2-ax}+\sqrt{x^2-x}}{(x^2-ax)-(x^2-x)}=\lim\limits_{x\to\infty}\dfrac{\sqrt{x^2-ax}+\sqrt{x^2-x}}{-ax+x}.$$

We now multiply by $\dfrac{1/\sqrt{x^2}}{1/x}$ to obtain
$$\lim\limits_{x\to\infty}\dfrac{\sqrt{1-(a/x)}+\sqrt{1-(1/x)}}{-a+1}=\dfrac{2}{1-a}.$$

41 $\lim\limits_{x\to\infty}(3\tan^{-1}x+2)=\dfrac{3\pi}{2}+2.$

43 For $x<1$, $|x-1|+x=-(x-1)+x=1$. Therefore $\lim\limits_{x\to -\infty}(|x-1|+x)=\lim\limits_{x\to -\infty}1=1$.
For $x>1$, $|x-1|+x=x-1+x=2x-1$. Therefore $\lim\limits_{x\to\infty}(|x-1|+x)=\lim\limits_{x\to\infty}2x-1=\infty$.

45 $\lim\limits_{w\to\infty}\dfrac{\ln w^2}{\ln w^3+1}=\lim\limits_{w\to\infty}\dfrac{2\ln w}{(3\ln w+1)}\cdot\dfrac{1/\ln w}{1/\ln w}=\lim\limits_{w\to\infty}\dfrac{2}{3+(1/\ln w)}=\dfrac{2}{3}.$

47 $\lim\limits_{r\to\infty}\dfrac{(2e^{4r}+3e^{5r})}{(7e^{4r}-9e^{5r})}\cdot\dfrac{1/e^{5r}}{1/e^{5r}}=\lim\limits_{r\to\infty}\dfrac{(2/e^r)+3}{(7/e^r)-9}=\dfrac{3}{-9}=-\dfrac{1}{3}.$
$\lim\limits_{r\to -\infty}\dfrac{(2e^{4r}+3e^{5r})}{(7e^{4r}-9e^{5r})}\cdot\dfrac{1/e^{4r}}{1/e^{4r}}=\lim\limits_{r\to -\infty}\dfrac{2+3e^r}{7-9e^r}=\dfrac{2}{7}.$

49 We know that $0 \leq \cos^4 x \leq 1$. Dividing each part of this inequality by $x^2 + x + 1$ and then adding 5, we have $5 \leq 5 + \dfrac{\cos^4 x}{x^2 + x + 1} \leq 5 + \dfrac{1}{x^2 + x + 1}$. Note that $\lim\limits_{x \to \infty} 5 = 5$ and $\lim\limits_{x \to \infty} \left(5 + \dfrac{1}{x^2 + x + 1}\right) = 5$, so by the Squeeze Theorem we can conclude that $\lim\limits_{x \to \infty} \left(5 + \dfrac{\cos^4 x}{x^2 + x + 1}\right) = 5$.

51 $\lim\limits_{x \to 1^-} \dfrac{x}{\ln x} = -\infty$.

53

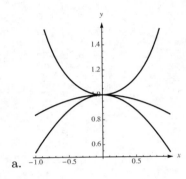

a.

b. Because $\lim\limits_{x \to 0} \cos x = \lim\limits_{x \to 0} \dfrac{1}{\cos x} = 1$, the Squeeze Theorem assures us that $\lim\limits_{x \to 0} \dfrac{\sin x}{x} = 1$ as well.

55 $\lim\limits_{x \to \infty} \dfrac{4x^3 + 1}{1 - x^3} = \lim\limits_{x \to \infty} \dfrac{4 + (1/x^3)}{(1/x^3) - 1} = \dfrac{4 + 0}{0 - 1} = -4$. A similar result holds as $x \to -\infty$. Thus, $y = -4$ is a horizontal asymptote as $x \to \infty$ and as $x \to -\infty$.

57 $\lim\limits_{x \to \infty} (1 - e^{-2x}) = 1$, while $\lim\limits_{x \to -\infty} (1 - e^{-2x}) = -\infty$.
$y = 1$ is a horizontal asymptote as $x \to \infty$.

59 $\lim\limits_{x \to \infty} \dfrac{(6e^x + 20)}{(3e^x + 4)} \cdot \dfrac{1/e^x}{1/e^x} = \lim\limits_{x \to \infty} \dfrac{6 + (20/e^x)}{3 + (4/e^x)} = \dfrac{6}{3} = 2$.
$\lim\limits_{x \to -\infty} \dfrac{6e^x + 20}{3e^x + 4} = \dfrac{0 + 20}{0 + 4} = 5$.

61

a. $\lim\limits_{x \to \infty} \dfrac{(3x^2 + 5x + 7)}{(x + 1)} \cdot \dfrac{1/x}{1/x} = \lim\limits_{x \to \infty} \dfrac{3x + 5 + (7/x)}{1 + (1/x)} = \infty$.

$\lim\limits_{x \to -\infty} \dfrac{(3x^2 + 5x + 7)}{(x + 1)} \cdot \dfrac{1/x}{1/x} = \lim\limits_{x \to -\infty} \dfrac{3x + 5 + (7/x)}{1 + (1/x)} = -\infty$.

b. By long division, we see that $\dfrac{3x^2 + 5x + 7}{x + 1} = 3x + 2 + \dfrac{5}{x+1}$, so the line $y = 3x + 2$ is a slant asymptote.

63

a. $\lim\limits_{x \to \infty} \dfrac{1 + x - 2x^2 - x^3}{x^2 + 1} = \lim\limits_{x \to \infty} \dfrac{1 + x - 2x^2 - x^3}{x^2 + 1} \cdot \dfrac{1/x^2}{1/x^2} = \lim\limits_{x \to \infty} \dfrac{1/x^2 + 1/x - 2 - x}{1 + 1/x^2} = -\infty$.

$\lim\limits_{x \to -\infty} \dfrac{1 + x - 2x^2 - x^3}{x^2 + 1} = \lim\limits_{x \to -\infty} \dfrac{1 + x - 2x^2 - x^3}{x^2 + 1} \cdot \dfrac{1/x^2}{1/x^2} = \lim\limits_{x \to -\infty} \dfrac{1/x^2 + 1/x - 2 - x}{1 + 1/x^2} = \infty$.

b. By long division, we can write $f(x)$ as $f(x) = -x - 2 + \dfrac{2x+3}{x^2+1}$, so the line $y = -x - 2$ is the slant asymptote.

65

a. $\displaystyle\lim_{x\to\infty}\frac{(4x^3+x^2+7)}{(x^2-x+1)}\cdot\frac{1/x^2}{1/x^2}=\lim_{x\to\infty}\frac{4x+1+(7/x)}{1-(1/x)+(1/x^2)}=\infty.$

$\displaystyle\lim_{x\to-\infty}\frac{(4x^3+x^2+7)}{(x^2-x+1)}\cdot\frac{1/x^2}{1/x^2}=\lim_{x\to-\infty}\frac{4x+1+(7/x)}{1-(1/x)+(1/x^2)}=-\infty.$

b. By long division, we can write $\dfrac{4x^3+x^2+7}{x^2-x+1}=4x+5+\dfrac{x+2}{x^2-x+1}$. Therefore $y=4x+5$ is a slant asymptote.

67 Recall that $\tan^{-1}x=0$ only for $x=0$. The only vertical asymptote is $x=0$.
$\displaystyle\lim_{x\to\infty}\frac{1}{\tan^{-1}x}=\frac{1}{\pi/2}=\frac{2}{\pi}.$
$\displaystyle\lim_{x\to-\infty}\frac{1}{\tan^{-1}x}=\frac{1}{-\pi/2}=-\frac{2}{\pi}.$ So $y=\frac{2}{\pi}$ is a horizontal asymptote as $x\to\infty$ and $y=-\frac{2}{\pi}$ is a horizontal asymptote as $x\to-\infty$.

69 Observe that
$$f(x)=\frac{x+xe^x+10e^x}{2(e^x+1)}=\frac{x(1+e^x)+10e^x}{2(e^x+1)}=\frac{1}{2}x+\frac{5e^x}{e^x+1}.$$
Because $\displaystyle\lim_{x\to\infty}\frac{5e^x}{e^x+1}=\lim_{x\to\infty}\frac{5}{1+(1/e^x)}=5$, the graph of f and the line $y=\frac{1}{2}x+5$ approach each other as $x\to\infty$. Similarly, $\displaystyle\lim_{x\to-\infty}\frac{5e^x}{e^x+1}=\frac{0}{0+1}=0$ and therefore the graph of f and the line $y=\frac{1}{2}x$ approach each other as $x\to-\infty$.

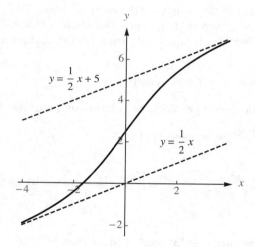

71 The function f is not continuous at 5 because $f(5)$ is not defined.

73 Observe that $h(5)=-2(5)+14=4$. Because $\displaystyle\lim_{x\to 5^-}h(x)=\lim_{x\to 5^-}(-2x+14)=4$ and $\displaystyle\lim_{x\to 5^+}h(x)=\lim_{x\to 5^+}\sqrt{x^2-9}=\sqrt{25-9}=4$, we have $\displaystyle\lim_{x\to 5}h(x)=4$. Thus f is continuous at $x=5$.

75 The domain of f is $(-\infty,-\sqrt{5}]$ and $[\sqrt{5},\infty)$, and f is continuous on that domain. It is left continuous at $-\sqrt{5}$ and right continuous at $\sqrt{5}$.

77 The domain of h is $(-\infty,-5)$, $(-5,0)$, $(0,5)$, $(5,\infty)$, and like all rational functions, it is continuous on its domain.

79 In order for g to be left continuous at 1, it is necessary that $\displaystyle\lim_{x\to 1^-}g(x)=g(1)$, which means that $a=3$. In order for g to be right continuous at 1, it is necessary that $\displaystyle\lim_{x\to 1^+}g(x)=g(1)$, which means that $a+b=3+b=3$, so $b=0$.

81

One such possible graph is pictured to the right.

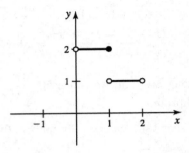

83

a. Rewrite The equation as $x - \cos x = 0$ and let $f(x) = x - \cos x$. Because x and $\cos x$ are continuous on the given interval, so is f. Because $f(0) = -1 < 0$ and $f(\pi/2) = \pi/2 > 0$, it follows from the Intermediate Value Theorem that the equation has a solution on $(0, \pi/2)$.

b. Using a computer algebra system, one can find that $c \approx 0.739085$ is a root.

85

a. Note that $m(0) = 0$ and $m(5) \approx 38.34$ and $m(15) \approx 21.2$. Thus, 30 is an intermediate value between both $m(0)$ and $m(5)$, and $m(5)$ and $m(15)$. Note also that m is a continuous function. By the IVT, there must be a number c_1 between 0 and 5 with $m(c_1) = 30$, and a number c_2 between 5 and 15 with $m(c_2) = 30$.

b. A little trial and error leads $c_1 \approx 2.4$ and $c_2 \approx 10.8$.

c. No. The graph of the function on a graphing calculator suggests that it peaks at about 38.5

87 Let $\varepsilon > 0$ be given. Let $\delta = \varepsilon$. Now suppose that $0 < |x - 5| < \delta$.
Then
$$|f(x) - L| = \left|\frac{x^2 - 25}{x - 5} - 10\right| = \left|\frac{(x-5)(x+5)}{x-5} - 10\right| = |x + 5 - 10|$$
$$= |x - 5| < \varepsilon.$$

89 Let $\varepsilon > 0$ be given. Let $\delta = \min\{1, \varepsilon/15\}$ and assume that $0 < |x - 2| < \delta$. Then $|3x^2 - 4 - 8| = 3|x^2 - 4| = 3|x - 2||x + 2|$. Because $0 < |x - 2| < \delta$ and $\delta \leq 1$, $-1 < x - 2 < 1$ and so $1 < x < 3$. It follows that $x + 2 < 5$. Therefore $|3x^2 - 4 - 8| = 3|x - 2||x + 2| < 3 \cdot \frac{\varepsilon}{15} \cdot 5 = \varepsilon$. So we've shown that for any $\varepsilon > 0$, $|3x^2 - 4 - 8| < \varepsilon$ whenever $0 < |x - 2| < \delta$, provided $0 < \delta \leq \min\{1, \varepsilon/15\}$.

91 Let $N > 0$ be given. Let $\delta = 1/\sqrt[4]{N}$. Suppose that $0 < |x - 2| < \delta$. Then $|x - 2| < \frac{1}{\sqrt[4]{N}}$, so $\frac{1}{|x-2|} > \sqrt[4]{N}$, and $\frac{1}{(x-2)^4} > N$, as desired.

Chapter 3

Derivatives

3.1 Introducing the Derivative

3.1.1 The secant line through the points $(a, f(a))$ and $(x, f(x))$ for x near a, of the graph of f, is given by $m_{\text{sec}} = \dfrac{f(x) - f(a)}{x - a}$. As x approaches a, we obtain the limit $m_{\tan} = \lim\limits_{x \to a} \dfrac{f(x) - f(a)}{x - a} = \lim\limits_{x \to a} m_{\text{sec}}$.

3.1.3 The average rate of change of f over $[a, x]$ is the slope of the secant line $m_{\text{sec}} = \dfrac{f(x) - f(a)}{x - a}$. As x approaches a, the length of the interval $x - a$ goes to zero, and in the limit we obtain the instantaneous rate of change of f at a given by $m_{\tan} = \lim\limits_{x \to a} \dfrac{f(x) - f(a)}{x - a}$.

3.1.5 $f'(a)$ is the value of the derivative of f at a. Also, $f'(a)$ is the slope of the tangent line to the graph of f at $(a, f(a))$. Furthermore, $f'(a)$ is the instantaneous rate of change of f at a.

3.1.7 $f(2) = 4(2) - 1 = 7$. $f'(2)$ is the slope of the tangent line, so $f'(2) = 4$.

3.1.9 Using the point-slope form of the equation of a line, we have $y - 2 = 3(x - 1)$, or $y = 3x - 1$.

3.1.11 $m_{\tan} = \lim\limits_{x \to 1} \dfrac{-5x + 1 + 4}{x - 1} = \lim\limits_{x \to 1} \dfrac{-5x + 5}{x - 1} = \lim\limits_{x \to 1} \left(-5\left(\dfrac{x-1}{x-1}\right)\right) = -5$.

3.1.13

$$s'(1) = \lim_{t \to 1} \dfrac{-16t^2 + 100t - 84}{t - 1} = -4 \lim_{t \to 1} \dfrac{4t^2 - 25t + 21}{t - 1} = -4 \lim_{t \to 1} \dfrac{(4t - 21)(t - 1)}{t - 1}$$
$$= -4 \lim_{t \to 1}(4t - 21) = 68 \, \text{ft/s}.$$

3.1.15

a. $m_{\tan} = \lim\limits_{x \to 3} \dfrac{x^2 - 5 - 4}{x - 3} = \lim\limits_{x \to 3} \dfrac{x^2 - 9}{x - 3} = \lim\limits_{x \to 3} \dfrac{(x - 3)(x + 3)}{x - 3} = \lim\limits_{x \to 3}(x + 3) = 6$.

b. Using the point-slope form of the equation of a line, we obtain $y - 4 = 6(x - 3)$, or $y = 6x - 14$.

c.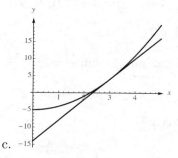

3.1.17

a. $m_{\tan} = \lim_{x \to -1} \dfrac{\frac{1}{x}+1}{x+1} = \lim_{x \to -1} \dfrac{\frac{1+x}{x}}{x+1} = \lim_{x \to -1} \dfrac{1}{x} = -1.$

b. $y - (-1) = -1(x+1)$, or $y = -x - 2$.

c.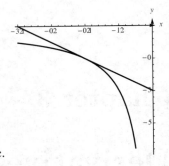

3.1.19

a. $m_{\tan} = \lim_{x \to 2} \dfrac{\sqrt{3x+3} - 3}{(x-2)} \cdot \dfrac{\sqrt{3x+3}+3}{\sqrt{3x+3}+3} =$
$\lim_{x \to 2} \dfrac{3(x-2)}{(x-2)(\sqrt{3x+3}+3)} = \lim_{x \to 2} \dfrac{3}{\sqrt{3x+3}+3} = \dfrac{1}{2}.$

b. $y - 3 = \tfrac{1}{2}(x-2)$, or $y = \tfrac{1}{2}x + 2$.

c.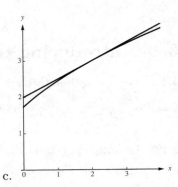

3.1.21

a. $m_{\tan} = \lim_{h \to 0} \dfrac{2(0+h)+1-1}{h} = \lim_{h \to 0} \dfrac{2h}{h} = 2.$

b. $y - 1 = 2x$, or $y = 2x + 1$.

3.1.23

a. $m_{\tan} = \lim_{h \to 0} \dfrac{3(1+h)^2 - 4(1+h)+1}{h} = \lim_{h \to 0} \dfrac{3+6h+3h^2-4-4h+1}{h} = \lim_{h \to 0} \dfrac{3h^2+2h}{h} = \lim_{h \to 0}(3h+2) = 2.$

b. $y + 1 = 2(x-1)$, or $y = 2x - 3$.

3.1.25

a. $m_{\tan} = \lim_{h \to 0} \dfrac{(2+h)^2 - 4 - 0}{h} = \lim_{h \to 0} \dfrac{4+4h+h^2-4}{h} = \lim_{h \to 0}(4+h) = 4.$

b. $y - 0 = 4(x-2)$, or $y = 4x - 8$.

3.1.27

a. $m_{\tan} = \lim_{h \to 0} \dfrac{(1+h)^3 - 1}{h} = \lim_{h \to 0} \dfrac{1+3h+3h^2+h^3-1}{h} = \lim_{h \to 0} \dfrac{h(3+3h+h^2)}{h} = \lim_{h \to 0}(3+3h+h^2) = 3.$

b. $y - 1 = 3(x-1)$, or $y = 3x - 2$.

3.1.29

a. $m_{\tan} = \lim_{h \to 0} \dfrac{\frac{1}{3-2(h-1)} - \frac{1}{5}}{h} = \lim_{h \to 0} \dfrac{\frac{5-(3-2h+2)}{15-10(h-1)}}{h} = \lim_{h \to 0} \dfrac{2}{15-10(h-1)} = \dfrac{2}{25}.$

3.1. Introducing the Derivative

b. $y - \frac{1}{5} = \frac{2}{25}(x+1)$, or $y = \frac{2}{25}x + \frac{7}{25}$.

3.1.31

a. $m_{\tan} = \lim_{h \to 0} \frac{\sqrt{1+h+3} - 2}{h} = \lim_{h \to 0} \frac{\sqrt{4+h} - 2}{h} \cdot \frac{\sqrt{4+h} + 2}{\sqrt{4+h} + 2} = \lim_{h \to 0} \frac{4+h-4}{h(\sqrt{4+h} + 2)} = \lim_{h \to 0} \frac{1}{\sqrt{4+h} + 2} = \frac{1}{4}$.

b. $y - 2 = \frac{1}{4}(x-1)$, or $y = \frac{1}{4}x + \frac{7}{4}$.

3.1.33

a. $f'(-3) = \lim_{h \to 0} \frac{8(-3+h) + 24}{h} = \lim_{h \to 0} \frac{8h}{h} = 8$.

b. $y - (-24) = 8(x+3)$, or $y = 8x$.

3.1.35

a. $f'(-2) = \lim_{h \to 0} \frac{4(-2+h)^2 + 2(-2+h) - 12}{h} = \lim_{h \to 0} \frac{16 - 16h + 4h^2 - 4 + 2h - 12}{h} = \lim_{h \to 0} \frac{-14h + 4h^2}{h} = -14$.

b. $y - 12 = -14(x+2)$, or $y = -14x - 16$.

3.1.37

a. $f'\left(\frac{1}{4}\right) = \lim_{h \to 0} \frac{\frac{1}{\sqrt{\frac{1}{4}+h}} - 2}{h} = \lim_{h \to 0} \frac{1 - 2\sqrt{\frac{1}{4}+h}}{h\sqrt{\frac{1}{4}+h}} = \lim_{h \to 0} \frac{\left(1 - 2\sqrt{\frac{1}{4}+h}\right)\left(1 + 2\sqrt{\frac{1}{4}+h}\right)}{h\sqrt{\frac{1}{4}+h}\left(1 + 2\sqrt{\frac{1}{4}+h}\right)} = \lim_{h \to 0} \frac{1 - 4\left(\frac{1}{4}+h\right)}{h\sqrt{\frac{1}{4}+h}\left(1 + 2\sqrt{\frac{1}{4}+h}\right)} = \lim_{h \to 0} -\frac{4}{\sqrt{\frac{1}{4}+h}\left(1 + 2\sqrt{\frac{1}{4}+h}\right)} = -4$.

b. $y - 2 = -4\left(x - \frac{1}{4}\right)$, or $y = -4x + 3$.

3.1.39

a. $f'(4) = \lim_{h \to 0} \frac{\sqrt{2(4+h)+1} - 3}{h} = \lim_{h \to 0} \frac{\sqrt{9+2h} - 3}{h} \cdot \frac{\sqrt{9+2h} + 3}{\sqrt{9+2h} + 3} = \lim_{h \to 0} \frac{9+2h-9}{h(\sqrt{9+2h}+3)} = \lim_{h \to 0} \frac{2}{\sqrt{9+2h}+3} = \frac{1}{3}$.

b. $y - 3 = \frac{1}{3}(x-4)$, or $y = \frac{1}{3}x + \frac{5}{3}$.

3.1.41

a. $f'(5) = \lim_{h \to 0} \frac{\frac{1}{5+h+5} - \frac{1}{10}}{h} = \lim_{h \to 0} \frac{10 - (10+h)}{10h(10+h)} = \lim_{h \to 0} \frac{-1}{10(10+h)} = -\frac{1}{100}$.

b. $y - \frac{1}{10} = -\frac{1}{100}(x-5)$, or $y = -\frac{1}{100}x + \frac{3}{20}$.

3.1.43 $m_{\tan} = \lim_{h \to 0} \frac{\frac{1}{1+h+1} - \frac{1}{2}}{h} = \lim_{h \to 0} \frac{2 - (2+h)}{h(2+h)2} = \lim_{h \to 0} \frac{-1}{(2+h)2} = -\frac{1}{4}$.

3.1.45 $m_{\tan} = \lim_{h \to 0} \frac{2\sqrt{25+h} - 1 - (2\sqrt{25} - 1)}{h} = \lim_{h \to 0} \frac{2(\sqrt{25+h} - \sqrt{25})}{h} = \lim_{h \to 0} \frac{2(\sqrt{25+h} - \sqrt{25})(\sqrt{25+h} + \sqrt{25})}{h(\sqrt{25+h} + \sqrt{25})} = \lim_{h \to 0} \frac{2(25+h-25)}{h(\sqrt{25+h} + \sqrt{25})} = \lim_{h \to 0} \frac{2}{\sqrt{25+h} + \sqrt{25}} = \frac{1}{5}$.

3.1.47

a. True. Because the graph is a line, any secant line has the same graph as the function and thus the same slope.

b. False. For example, take $f(x) = x^2$, $P = (0,0)$ and $Q = (1,1)$. Then the secant line has slope $m_{\text{sec}} = \frac{1-0}{1-0} = 1$, but the the graph has a horizontal tangent at P so $m_{\text{tan}} = 0$ and $m_{\text{sec}} > m_{\text{tan}}$.

c. True. $m_{\text{sec}} = \frac{(x+h)^2 - x^2}{h} = \frac{2xh + h^2}{h} = 2x + h$, while $m_{\text{tan}} = \lim_{h \to 0}(2x+h) = 2x$. Because we assume that $h > 0$, we have $m_{\text{sec}} = 2x + h > 2x = m_{\text{tan}}$.

3.1.49 $d'(4) = \lim_{t \to 4} \frac{16t^2 - 256}{t - 4} = 16 \lim_{t \to 4} \frac{(t-4)(t+4)}{t-4} = 16 \lim_{t \to 4}(t+4) = 16 \cdot 8 = 128\,\text{ft/s}$. The object is falling with an instantaneous speed of 128 ft/s 4 seconds after being dropped.

3.1.51 $v'(3) = \lim_{t \to 3} \frac{(t-5)^2 - 4}{t-3} = \lim_{t \to 3} \frac{(t-5-2)(t-5+2)}{t-3} = \lim_{t \to 3}(t-7) = -4\,\text{m/s}$ per second. The instantaneous rate of change in the car's speed is $-4\,\text{m/s}^2$ at time 3 seconds.

3.1.53

a. $L'(15) \approx \frac{6.75 - 3.5}{2 - 1.25} \approx 4.3\,\text{mm/week}$. At 1.5 weeks, the talon is growing at a rate of about 4.3 mm/week.

b. $L'(a) \approx 0$ for $a \geq 4$. At 4 weeks the talons have stopped growing.

3.1.55 $D'(60) \approx 0.05\,\text{hr/day}$. 60 days after January 1, the daylight hours are increasing at about 0.05 hours per day (3 minutes per day). $D'(170) \approx 0\,\text{hr/day}$. The number of daylight hours is neither increasing nor decreasing 170 days after January 1.

3.1.57 Consider $a = 2$ and $f(x) = 5x^2$.

Then $f'(2) = \lim_{x \to 2} \frac{5x^2 - 20}{x - 2}$ as desired.

We have $f'(2) = \lim_{x \to 2} \frac{5x^2 - 20}{x - 2} = \lim_{x \to 2} \frac{5(x-2)(x+2)}{x-2} = \lim_{x \to 2} 5(x+2) = 20$.

3.1.59 Consider $a = 2$ and $f(x) = x^4$.

Then $f'(2) = \lim_{h \to 0} \frac{(2+h)^4 - 16}{h}$ as desired.

We have

$$f'(2) = \lim_{h \to 0} \frac{(2+h)^4 - 16}{h} = \lim_{h \to 0} \frac{16 + 32h + 24h^2 + 8h^3 + h^4 - 16}{h}$$
$$= \lim_{h \to 0} \frac{h(32 + 24h + 8h^2 + h^3)}{h} = \lim_{h \to 0}(32 + 24h + 8h^2 + h^3) = 32.$$

3.1.61 Consider $a = -1$ and $f(x) = |x|$.

Then $f'(-1) = \lim_{h \to 0} \frac{|-1+h| - 1}{h}$ as desired.

We have $f'(-1) = \lim_{h \to 0} \frac{|-1+h| - 1}{h} = \lim_{h \to 0} \frac{1 - h - 1}{h} = \lim_{h \to 0} -1 = -1$.

3.1.63

a. Note that the slope generated by the centered difference quotient is $\frac{f(4.5) - f(3.5)}{2(0.5)} = \sqrt{4.5} - \sqrt{3.5} \approx 0.250492$. The centered difference quotient line is very close to the tangent line, which closely approximates the function near the point of tangency.

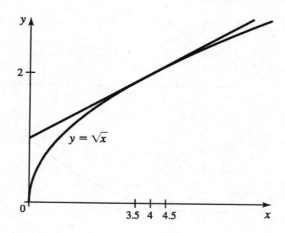

b.

h	Approximation	Error
0.1	0.25002	2.0×10^{-5}
0.01	≈ 0.25000	2.0×10^{-7}
0.001	≈ 0.25000	2.0×10^{-9}

c. The centered difference quotient is symmetric about zero, so using the corresponding negative values yields the same results.

d. The centered difference quotient appears to be more accurate than the approximation in the previous problem.

3.1.65

a. Forward:
$$\frac{\text{erf}(1.05) - \text{erf}(1)}{0.05} = \frac{0.862436 - 0.842701}{0.05} = 0.3947.$$

Centered:
$$\frac{\text{erf}(1.05) - \text{erf}(0.95)}{2(0.05)} = \frac{0.862436 - 0.820891}{0.1} = 0.41545.$$

b. Forward:
$$\left| 0.3947 - \frac{2}{e\sqrt{\pi}} \right| \approx 0.02.$$

Centered:
$$\left| 0.41545 - \frac{2}{e\sqrt{\pi}} \right| \approx 0.003$$

3.2 The Derivative as a Function

3.2.1 It represents the slope function of f.

3.2.3 $\frac{dy}{dx}$ is the limit of $\frac{\Delta y}{\Delta x}$ and is the rate of change of y with respect to x.

3.2.5

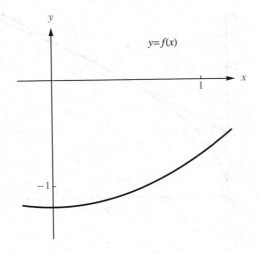

3.2.7 Yes, differentiable functions are continuous by Theorem 3.1.

3.2.9 The graph is a line with y-intercept of 1 and a slope of 3.

3.2.11 $f'(x) = \lim\limits_{h\to 0} \dfrac{f(x+h)-f(x)}{h} = \lim\limits_{h\to 0} \dfrac{7(x+h)-7x}{h} = \lim\limits_{h\to 0} \dfrac{7h}{h} = \lim\limits_{h\to 0} 7 = 7.$

3.2.13 $\dfrac{dy}{dx} = \lim\limits_{h\to 0} \dfrac{(x+h)^2 - x^2}{h} = \lim\limits_{h\to 0} \dfrac{x^2 + 2hx + h^2 - x^2}{h} = \lim\limits_{h\to 0} \dfrac{h(2x+h)}{h} = \lim\limits_{h\to 0} 2x + h = 2x.$ Therefore, $\dfrac{dy}{dx}\bigg|_{x=3} = 6$ and $\dfrac{dy}{dx}\bigg|_{x=-2} = -4.$

3.2.15 (c) is the only line with negative slope, so it corresponds to derivative (A). Since (d) contains the points $(2,0)$ and $(0,1)$, it has slope $\frac{1}{2}$, so it corresponds to derivative (B). Finally, lines (a) and (b) are parallel; since (b) contains the points $(0,1)$ and $(-1,0)$, it has slope 1, so that (a) has slope 1 as well. They both correspond to derivative (C).

3.2.17

The function f is not differentiable at $x = -2, 0, 2$, so f' is not defined at those points. Elsewhere, the slope is constant.

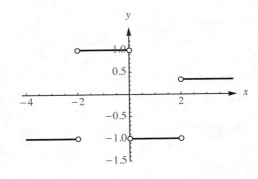

3.2.19

a. f is not continuous at $x = 1$.

b. f is not differentiable at $x = 1$ and at $x = 0$.

3.2. The Derivative as a Function

3.2.21

a. $f'(x) = \lim_{h \to 0} \dfrac{5(x+h)+2-(5x+2)}{h} = \lim_{h \to 0} \dfrac{5h}{h} = \lim_{h \to 0} 5 = 5.$

b. $f'(1) = 5$ and $f'(2) = 5.$

3.2.23

a. $f'(x) = \lim_{h \to 0} \dfrac{4(x+h)^2+1-(4x^2+1)}{h} = \lim_{h \to 0} \dfrac{4x^2+8hx+4h^2+1-4x^2-1}{h} = \lim_{h \to 0} \dfrac{h(8x+4h)}{h} = \lim_{h \to 0} 8x+4h = 8x.$

b. $f'(2) = 16$ and $f'(4) = 32.$

3.2.25

a. $f'(x) = \lim_{h \to 0} \dfrac{\frac{1}{x+h+1}-\frac{1}{x+1}}{h} = \lim_{h \to 0} \dfrac{(x+1)-(x+h+1)}{h(x+1)(x+h+1)} = \lim_{h \to 0} \dfrac{-h}{h(x+1)(x+h+1)} = -\lim_{h \to 0} \dfrac{1}{(x+1)(x+h+1)} = -\dfrac{1}{(x+1)^2}.$

b. $f'(-1/2) = -4$ and $f'(5) = -\dfrac{1}{36}.$

3.2.27

a. $f'(t) = \lim_{h \to 0} \dfrac{\frac{1}{\sqrt{t+h}}-\frac{1}{\sqrt{t}}}{h} = \lim_{h \to 0} \dfrac{\sqrt{t}-\sqrt{t+h}}{h\sqrt{t}\sqrt{t+h}} = \lim_{h \to 0} \dfrac{(\sqrt{t}-\sqrt{t+h})}{h\sqrt{t}\sqrt{t+h}} \cdot \dfrac{(\sqrt{t}+\sqrt{t+h})}{(\sqrt{t}+\sqrt{t+h})} = \lim_{h \to 0} \dfrac{t-(t+h)}{h\sqrt{t}\sqrt{t+h}(\sqrt{t}+\sqrt{t+h})} = -\lim_{h \to 0} \dfrac{1}{\sqrt{t}\sqrt{t+h}(\sqrt{t}+\sqrt{t+h})} = -\dfrac{1}{2t^{3/2}}.$

b. $f'(9) = -\dfrac{1}{54}$ and $f'(1/4) = -4.$

3.2.29

a.
$$f'(s) = \lim_{h \to 0} \dfrac{4(s+h)^3+3(s+h)-(4s^3+3s)}{h}$$
$$= \lim_{h \to 0} \dfrac{4s^3+12s^2h+12sh^2+4h^3+3s+3h-4s^3-3s}{h}$$
$$= \lim_{h \to 0} \dfrac{h(12s^2+12sh+4h^2+3)}{h} = \lim_{h \to 0}(12s^2+12sh+4h^2+3) = 12s^2+3.$$

b. $f'(-3) = 111$ and $f'(-1) = 15.$

3.2.31

a. $v(t) = s'(t) = \lim_{h \to 0} \dfrac{-16(t+h)^2+100(t+h)-(-16t^2+100t)}{h} = \lim_{h \to 0} \dfrac{-16t^2-32th-16h^2+100t+100h+16t^2-100t}{h} = \lim_{h \to 0} \dfrac{h(-32t-16h+100)}{h} = -32t+100.$

b. $v(1) = 68$ ft/s and $v(2) = 36$ ft/s.

3.2.33

$$\frac{dy}{dx} = \lim_{h \to 0} \frac{\frac{x+h+1}{x+h+2} - \frac{x+1}{x+2}}{h}$$

$$= \lim_{h \to 0} \frac{(x+2)(x+h+1) - (x+1)(x+h+2)}{h(x+h+2)(x+2)}$$

$$= \lim_{h \to 0} \frac{x^2 + xh + x + 2x + 2h + 2 - (x^2 + xh + 2x + x + h + 2)}{h(x+h+2)(x+2)}$$

$$= \lim_{h \to 0} \frac{h}{h(x+h+2)(x+2)} = \frac{1}{(x+2)^2}$$

Therefore $\left.\frac{dy}{dx}\right|_2 = \frac{1}{16}$.

3.2.35

a.
$$f'(x) = \lim_{h \to 0} \frac{3(x+h)^2 + 2(x+h) - 10 - (3x^2 + 2x - 10)}{h}$$

$$= \lim_{h \to 0} \frac{3x^2 + 6xh + 3h^2 + 2x + 2h - 10 - 3x^2 - 2x + 10}{h}$$

$$= \lim_{h \to 0} \frac{6xh + 2h + 3h^2}{h} = \lim_{h \to 0} (6x + 2 + 3h) = 6x + 2.$$

b. We have $f'(1) = 8$, and the tangent line is given by $y + 5 = 8(x - 1)$, or $y = 8x - 13$.

3.2.37

a.
$$f'(x) = \lim_{h \to 0} \frac{\sqrt{3(x+h)+1} - \sqrt{3x+1}}{h}$$

$$= \lim_{h \to 0} \frac{\sqrt{3(x+h)+1} - \sqrt{3x+1}}{h} \cdot \frac{\sqrt{3x+3h+1} + \sqrt{3x+1}}{\sqrt{3x+3h+1} + \sqrt{3x+1}}$$

$$= \lim_{h \to 0} \frac{3x + 3h + 1 - 3x - 1}{h(\sqrt{3x+3h+1} + \sqrt{3x+1})} = \lim_{h \to 0} \frac{3}{\sqrt{3x+3h+1} + \sqrt{3x+1}} = \frac{3}{2\sqrt{3x+1}}.$$

b. We have $f'(8) = \frac{3}{10}$. Using the point-slope form, we get that the tangent line has equation $y - 5 = \frac{3}{10}(x - 8)$, which can be written as $y = \frac{3}{10}x + \frac{13}{5}$.

3.2.39

a. $f'(x) = \lim_{h \to 0} \frac{\frac{2}{3(x+h)+1} - \frac{2}{3x+1}}{h} = \lim_{h \to 0} \frac{6x + 2 - (6x + 6h + 2)}{h(3x+1)(3x+3h+1)} = \lim_{h \to 0} \frac{-6h}{h(3x+1)(3x+3h+1)} =$
$-\frac{6}{(3x+1)^2}$.

b. We have $f'(-1) = -\frac{3}{2}$. Using the point-slope form, we get that the tangent line has equation $y + 1 = -\frac{3}{2}(x+1)$, which can be written as $y = -\frac{3}{2}x - \frac{5}{2}$.

3.2. The Derivative as a Function

3.2.41

a. From the graph we approximate the derivative by the slope of a secant line: For example we see that $E(6) = 250\,\text{kWh}$ and $E(18) = 350\,\text{kWh}$, so the power after 10 hours is approximately the slope of the secant line through these points, so $P(10) \approx m_{sec} = \dfrac{E(18) - E(6)}{18 - 6} = \dfrac{350\,\text{kWh} - 250\,\text{kWh}}{12\text{h}} \approx 8.3\,\text{kW}$. Similarly, after 20 hours, using 18 hours and 25 hours, that $P(20) \approx m_{sec} = \dfrac{E(22) - E(18)}{22 - 18} = \dfrac{325\,\text{kWh} - 350\,\text{kWh}}{4\text{h}} \approx -6.25\text{kW}$.

b. The power is zero where the graph of $E(t)$ has a horizontal tangent line, which happens approximately at $t = 6$ hours and $t = 18$ hours.

c. The power has a maximum where the graph of $E(t)$ has the steepest increase, which is approximately at $t = 12$ hours.

3.2.43

a. $\dfrac{d}{dx}\left(ax^2 + bx + c\right) =$
$\lim\limits_{h \to 0} \dfrac{a(x+h)^2 + b(x+h) + c - \left(ax^2 + bx + c\right)}{h} = \lim\limits_{h \to 0} \dfrac{ax^2 + 2axh + ah^2 + bx + bh + c - ax^2 - bx - c}{h}$
$= \lim\limits_{h \to 0} \dfrac{2axh + ah^2 + bh}{h} = \lim\limits_{h \to 0}(2ax + ah + b) = 2ax + b.$

b. With $a = 4, b = -3, c = 10$ we have $\frac{d}{dx}(4x^2 - 3x + 10) = 2 \cdot 4 \cdot x + (-3) = 8x - 3$.

c. From part (b), $f'(1) = 8 \cdot 1 - 3 = 5$.

3.2.45

a. At C and D, the slope of the tangent line (and thus of the curve) is negative.

b. At A, B, and E, the slope of the curve is positive.

c. The graph is in its steepest ascent at A followed by B. At E it barely increases, at D it slightly decreases and at C it is decreasing the most, so the points in decreasing order of slope are A, B, E, D, C.

3.2.47

a. The function has non-negative slope everywhere, and as there is a horizontal tangent at $x = 0$, so the derivative has to be zero at zero. The graph of the derivative has to be above the x-axis and touching it at $x = 0$, so (D) is the graph of the derivative.

b. The graph of this function has three horizontal tangent lines, at $x = -1, 0, 1$, and the matching graph of the derivative with three zeros is (C).

c. The function has negative slope on $(-1, 0)$, and positive slope on $(0, 1)$ and has a horizontal tangent at $x = 0$, so the derivative has to be negative on $(-1, 0)$, positive on $(0, 1)$ and zero at $x = 0$; the graph is (B).

d. The function has negative slope everywhere so the graph of the derivative has to be negative everywhere, which is graph (A).

3.2.49

The function always has non-negative slope, so the derivative is never below the x axis.
However, it does have slope zero at about $x = 2$.

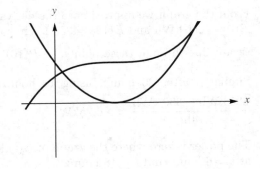

3.2.51 Note that f is undefined at $x = -2$ and $x = 1$, but is differentiable elsewhere. It is decreasing, and increasingly rapidly, as x increases towards $x = -2$. It decreases, but increasingly slowly, and towards a zero slope, as x increases from 1. Finally, between $x = -2$ and $x = 1$, the function increases, but increasingly slowly, until $x = 0$ and then decreases, but increasingly rapidly, as x approaches 1. A graph of the derivative is

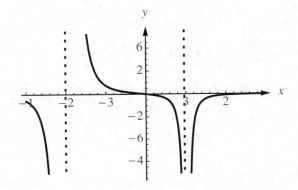

3.2.53

a. The function f is not continuous at $x = 1$, because the graph has a jump there.

b. The function f is not differentiable at $x = 1$ because it is not continuous at that point (Theorem 3.1 Alternate Version), and it is also not differentiable at $x = 2$ because the graph has a corner there.

c.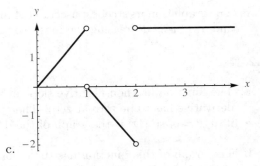

3.2.55

a. The tangent line for Q looks to be the steepest at $t = 0$.

b. All tangent lines for Q have positive slope, so Q' is positive for $t \geq 0$.

c. The tangent lines appear to be getting less steep as t increases, so Q' is decreasing.

d.

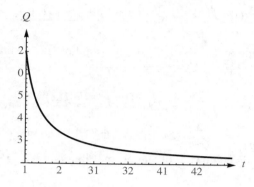

3.2.57

a. True. Differentiability implies continuity, by Theorem 3.1.

b. True. Because the absolute value function is continuous, and $y = x+1$ is continuous, and the composition of continuous functions is continuous, we know that this function is continuous. Note that it is not differentiable at $x = -1$ because the absolute value function is not differentiable at $x = 0$.

c. False. In order for f to be differentiable on $[a,b]$, f would need to be defined at a and at b. Because the domain of f doesn't include these endpoints, this situation is not possible.

3.2.59 In order for f to be differentiable at $x = 1$, it would need to be continuous there. Thus, $\lim_{x \to 1^-} f(x) = \lim_{x \to 1^-} 2x^2 = 2 = \lim_{x \to 1^+} f(x) = \lim_{x \to 1^+} ax - 2 = a - 2$, so the only possible value for a is 4. Now checking the differentiability at 1, we have (from the left)

$$\lim_{x \to 1^-} \frac{f(x) - f(1)}{x - 1} = \lim_{x \to 1^-} \frac{2x^2 - 2}{x - 1} = \lim_{x \to 1^-} 2(x+1) = 4.$$

Also, from the right we have

$$\lim_{x \to 1^+} \frac{f(x) - f(1)}{x - 1} = \lim_{x \to 1^+} \frac{4x - 2 - 2}{x - 1} = \lim_{x \to 1^+} 4 = 4,$$

so f is differentiable at 1 for $a = 4$.

3.2.61

Because $f'(x) = x$ is negative for $x < 0$ and positive for $x > 0$, we have that the graph of f has to have negative slope on $(-\infty, 0)$ and positive slope on $(0, \infty)$ and has to have a horizontal tangent at $x = 0$. Because f' only gives us the slope of the tangent line and not the actual value of f, there are infinitely many graphs possible, they all have the same shape, but are shifted along the y-axis.

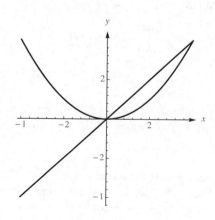

3.2.63 With $f(x) = 3x - 4$, we have
$$f'(x) = \lim_{h \to 0} \frac{f(x+h) - f(x)}{h} = \lim_{h \to 0} \frac{3(x+h) - 4 - (3x - 4)}{h} = \lim_{h \to 0} \frac{3h}{h} = \lim_{h \to 0} 3 = 3.$$

The slope of the tangent line at $(1, -1)$ is 3, so the slope of the normal line is $-\frac{1}{3}$. The equation of the normal line is thus $y - (-1) = -\frac{1}{3}(x - 1)$, or $y = -\frac{1}{3}x - \frac{2}{3}$.

3.2.65 With $f(x) = \frac{2}{x}$, we have $f'(x) = \lim_{h \to 0} \frac{f(x+h) - f(x)}{h} = \lim_{h \to 0} \frac{\frac{2}{x+h} - \frac{2}{x}}{h} = \lim_{h \to 0} \frac{\frac{2x - (2(x+h))}{x(x+h)}}{h} = \lim_{h \to 0} \frac{-2h}{hx(x+h)} = \lim_{h \to 0} \frac{-2}{x(x+h)} = -\frac{2}{x^2}$. At the point $(1, 2)$ the slope of the tangent line is -2, so the slope of the normal line is $\frac{1}{2}$. The equation of the normal line is $y - 2 = \frac{1}{2}(x - 1)$ or $y = \frac{x}{2} + \frac{3}{2}$.

3.2.67 With $f(x) = x^2 + 1$, we have $f'(x) = \lim_{h \to 0} \frac{f(x+h) - f(x)}{h} = \lim_{h \to 0} \frac{(x+h)^2 + 1 - (x^2 + 1)}{h} = \lim_{h \to 0} \frac{x^2 + 2hx + h^2 + 1 - x^2 - 1}{h} = \lim_{h \to 0} \frac{2hx + h^2}{h} = \lim_{h \to 0} \frac{h(2x + h)}{h} = 2x$. We are looking for points (x, x^2+1) where the slope of the line between this point and $Q(3, 6)$ is equal to $2x$. So we seek solutions to the equation
$$\frac{6 - (x^2 + 1)}{3 - x} = 2x,$$
which can be written as $5 - x^2 = 2x(3 - x)$, or $x^2 - 6x + 5 = 0$. Factoring, we obtain $(x - 5)(x - 1) = 0$, so the solutions are $x = 5$ and $x = 1$. Note that at the point $(5, 26)$ on the curve the tangent line is $y - 26 = 10(x - 5)$ which does contain the point $Q(3, 6)$ and at the point $(1, 2)$ the equation of the tangent line is $y - 2 = 2(x - 1)$, which also contains the point $Q(3, 6)$.

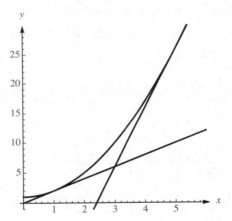

3.2.69 With $f(x) = \frac{1}{x}$, we have
$$f'(x) = \lim_{h \to 0} \frac{f(x+h) - f(x)}{h} = \lim_{h \to 0} \frac{\frac{1}{x+h} - \frac{1}{x}}{h} = \lim_{h \to 0} \frac{\frac{x - (x+h)}{x(x+h)}}{h}$$
$$= \lim_{h \to 0} \frac{-h}{hx(x+h)} = \lim_{h \to 0} \frac{-1}{x(x+h)} = -\frac{1}{x^2}.$$

We are looking for points $(x, 1/x)$ where the slope of the line between this point and $Q(-2, 4)$ is equal to $\frac{-1}{x^2}$. So we seek solutions to the equation
$$\frac{1/x - 4}{x + 2} = -\frac{1}{x^2},$$
which can be written as $x - 4x^2 = -x - 2$, or $4x^2 - 2x - 2 = 0$, or $2x^2 - x - 1 = 0$. This factors as $(2x + 1)(x - 1) = 0$, so the solutions are $x = 1$ and $x = -1/2$. Note that at $x = 1$ the equation of the tangent line is $y - 1 = -1(x - 1)$ which does contain the point $Q(-2, 4)$, and at $x = -1/2$ the equation of the tangent line is $y + 2 = -4\left(x + \frac{1}{2}\right)$ or $y = -4x - 4$ which also contains the point $Q(-2, 4)$.

3.2. The Derivative as a Function

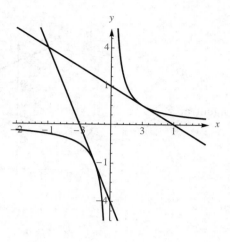

3.2.71

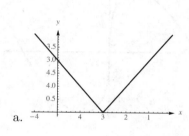

a.

b. $f'_+(2) = \lim\limits_{h \to 0^+} \dfrac{|2+h-2|-0}{h} = \lim\limits_{h \to 0^+} \dfrac{h}{h} = 1$, because for $h > 0$, we have $|h| = h$. Similarly, $f'_-(2) = \lim\limits_{h \to 0^-} \dfrac{|2+h-2|-0}{h} = \lim\limits_{h \to 0^-} -\dfrac{h}{h} = -1$, because for $h < 0$, we have $|h| = -h$.

c. Because f is defined at $a = 2$ and the graph of f does not jump, f is continuous at $a = 2$. Because the left-hand and right-hand derivatives are not equal, f is not differentiable at $a = 2$.

3.2.73

a. The graph has a vertical tangent at $x = 2$.

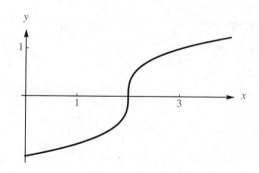

b. The graph has a vertical tangent at $x = -1$.

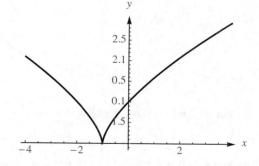

c. The graph has a vertical tangent at $x = 4$.

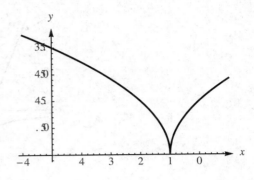

d. The graph has a vertical tangent at $x = 0$.

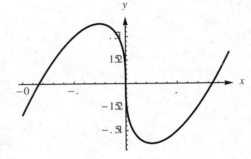

3.2.75 $f'(0) = \lim_{h \to 0} \dfrac{h^{1/3}}{h} = \lim_{h \to 0} \dfrac{1}{h^{2/3}} = +\infty$. Thus the graph of f has a vertical tangent at $x = 0$.

3.2.77

a.

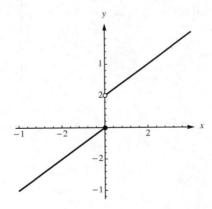

b. For $x < 0$, we have $f'(x) = \lim_{h \to 0} \dfrac{x + h - x}{h} = \lim_{h \to 0} \dfrac{h}{h} = \lim_{h \to 0} 1 = 1$.

c. For $x > 0$, we have $f'(x) = \lim_{h \to 0} \dfrac{x + h + 1 - (x + 1)}{h} = \lim_{h \to 0} \dfrac{h}{h} = \lim_{h \to 0} 1 = 1$.

d.

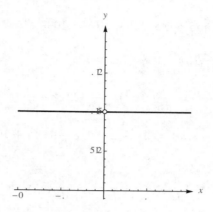

e. f is not differentiable at $x = 0$ as it is not continuous there. Also, if we were to compute the derivative of f from the right at 0 we would have

$$\lim_{h \to 0^+} \frac{f(0+h) - f(0)}{h} = \lim_{h \to 0^+} \frac{h + 1 - 0}{h} = \lim_{h \to 0^+} \frac{h+1}{h},$$

which does not exist.

3.3 Rules of Differentiation

3.3.1 Often the limit definition of f' is difficult to compute, especially for functions which are reasonably complicated. The rules for differentiation allow us to easily compute the derivatives of complex functions.

3.3.3 The function $f(x) = e^x$ is an example of a function with this property.

3.3.5 By the constant multiple rule, the derivative of the function cf where c is a constant and f is a function is cf'. That is, the derivative of a constant times a function is that same constant times the derivative of the function.

3.3.7 $(f + g)'(3) = f'(3) + g'(3) = 6 + (-2) = 4$.

3.3.9 $F'(2) = f'(2) + g'(2) = -1 + \frac{1}{2} = -\frac{1}{2}$.

3.3.11 $H'(2) = 3f'(2) + 2g'(2) = 3(-1) + 2(\frac{1}{2}) = -2$.

3.3.13 $\frac{d}{dx}[1.5f(x)]_{x=2} = 1.5f'(2) = 1.5 \cdot 5 = 7.5$.

3.3.15 $f'(t) = 10t^9$, $f''(t) = 90t^8$, and $f'''(t) = 720t^7$.

3.3.17 $\frac{d}{dx} 4f(x))|_{x=5} = 4f'(5) = \frac{4}{10} = \frac{2}{5}$.

3.3.19 By the power rule, $y' = 5x^{5-1} = 5x^4$.

3.3.21 By the constant rule, $f'(x) = 0$.

3.3.23 By the constant multiple rule and the power rule, $f'(x) = 5 \cdot \frac{d}{dx} x^3 = 5 \cdot 3x^2 = 15x^2$.

3.3.25 By the power rule, the sum rule, the constant multiplier rule, and the constant rule, $h'(x) = \frac{2t}{2} + 0 = t$.

3.3.27 By the constant multiple and power rules, $p'(x) = 8 \cdot \frac{d}{dx} x = 8 \cdot 1 = 8$.

3.3.29 By the constant multiple and power rules, $g'(t) = 100 \frac{d}{dt} t^2 = 100 \cdot 2t = 200t$.

3.3.31 $f'(x) = \frac{d}{dx}(3x^4 + 7x) = \frac{d}{dx}(3x^4) + \frac{d}{dx}(7x) = 12x^3 + 7$.

3.3.33 $f'(x) = \frac{d}{dx}(10x^4 - 32x + e^2) = \frac{d}{dx}(10x^4) - \frac{d}{dx}(32x) + \frac{d}{dx}(e^2) = 40x^3 - 32 - 0 = 40x^3 - 32$.

3.3.35 $g'(w) = \frac{d}{dw}(2w^3 + 3w^2 + 10w) = 2\frac{d}{dw}(w^3) + 3\frac{d}{dw}(w^2) + 10\frac{d}{dw}(w) = 2(3w^2) + 3(2w) + 10(1) = 6w^2 + 6w + 10$.

3.3.37 $f'(x) = \frac{d}{dx}(3e^x + 5x + 5) = 3\frac{d}{dx}(e^x) + 5\frac{d}{dx}(x) + \frac{d}{dx}(5) = 3e^x + 5$.

3.3.39 $f'(x) = \begin{cases} 2x & \text{if } x < 0 \\ 4x + 1 & \text{if } x > 0. \end{cases}$

3.3.41

a. $d'(t) = 32t$ is the velocity of the stone after t seconds, measured in feet per second.

b. The stone travels $d(6) = 16 \cdot 6^2 = 576$ feet and strikes the ground with a velocity of $32 \cdot 6 = 192$ feet per second. Converting to miles per hour, we have $192 \cdot \frac{3600}{5280} \approx 130.9$ miles per hour.

3.3.43

a. $A'(t) = -\frac{1}{25}t + 2$ square miles per year.

b. First we must find when $A(t) = 38$. This occurs when $-\frac{t^2}{50} + 2t + 28 = 38$, which can be written as $t^2 - 100t + 900 = 0$. This factors as $(t - 90)(t - 10) = 0$, so the only solution on the given domain is $t = 10$. At this time, we have $A'(10) = -0.4 + 2 = 1.6$ square miles per year.

c. Note that $A'(20) = -0.8 + 2 = 1.2$ square miles per year. In order to maintain a density of 1000 people per square mile, we must multiply the density of people per square mile times the number of square miles per year in order to obtain the rate of people per year required to maintain that density. Thus the population growth rate must be $1000 \cdot 1.2 = 1200$ people per year.

3.3.45 $w'(x) = \begin{cases} 0.4 & \text{if } 19 < x < 21 \\ 0.8 & \text{if } 21 < x < 32 \\ 1.5 & \text{if } x > 32. \end{cases}$ The derivative measures the rate of change (in lb/inch) of Atlantic salmon. Longer salmon put on more weight per inch than shorter salmon.

3.3.47 Expanding the product yields $f(x) = 6x^3 + 3x^2 + 4x + 2$. So

$$f'(x) = \frac{d}{dx}(6x^3 + 3x^2 + 4x + 2) = \frac{d}{dx}(6x^3) + \frac{d}{dx}(3x^2) + \frac{d}{dx}(4x) + \frac{d}{dx}(2)$$
$$= 18x^2 + 6x + 4.$$

3.3.49 f simplifies as $f(w) = w^2 - 1$, so $f'(w) = 2w$ for $w \neq 0$.

3.3.51 Expanding the product yields $h(x) = x^4 + 2x^2 + 1$. So

$$h'(x) = \frac{d}{dx}(x^4 + 2x^2 + 1) = \frac{d}{dx}(x^4) + \frac{d}{dx}(2x^2) + \frac{d}{dx}(1)$$
$$= 4x^3 + 4x.$$

Copyright © 2019 Pearson Education, Inc.

3.3. Rules of Differentiation

3.3.53 g simplifies as $g(x) = \dfrac{(x-1)(x+1)}{x-1} = x+1$. Thus $g'(x) = 1$ for $x \neq 1$.

3.3.55 y simplifies as $y = \dfrac{(\sqrt{x}-\sqrt{a})(\sqrt{x}+\sqrt{a})}{\sqrt{x}-\sqrt{a}} = \sqrt{x}+\sqrt{a}$. Thus $\dfrac{dy}{dx} = \dfrac{1}{2\sqrt{x}}$ for $x \neq a$.

3.3.57 g simplifies as $g(w) = \dfrac{e^{2w}}{e^w} + \dfrac{e^w}{e^w} = e^w + 1$. So $g'(w) = e^w$.

3.3.59

a. $y' = -6x$, so the slope of the tangent line at $a = 1$ is -6. Thus, the tangent line at the point $(1, -1)$ is $y + 1 = -6(x-1)$, or $y = -6x + 5$.

b.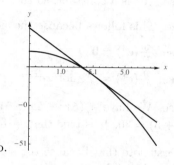

3.3.61

a. $y' = e^x$, so the slope of the tangent line at $a = \ln 3$ is 3. Thus, the tangent line at the point $(\ln 3, 3)$ is $y - 3 = 3(x - \ln 3)$, or $y = 3x + 3 - 3\ln 3$.

b.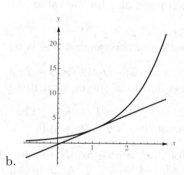

3.3.63

a. $f'(x) = 2x - 6$, so the slope is zero when $2x - 6 = 0$, which is at $x = 3$.

b. The slope is 2 when $2x - 6 = 2$ which is at $x = 4$.

3.3.65

a. The slope of the tangent line is given by $f'(x) = 6x^2 - 6x - 12$, and this quantity is zero when $x^2 - x - 2 = 0$, or $(x-2)(x+1) = 0$. The two solutions are thus $x = -1$ and $x = 2$, so the points on the graph are $(-1, 11)$ and $(2, -16)$.

b. The slope of the tangent line is 60 when $6x^2 - 6x - 12 = 60$, which is when $6x^2 - 6x - 72 = 0$. Simplifying this quadratic expression yields the equation $x^2 - x - 12 = 0$, which has solutions $x = -3$ and $x = 4$, so the points on the graph are $(-3, -41)$, and $(4, 36)$.

3.3.67

a. The slope of the tangent line is given by $\dfrac{2}{\sqrt{x}} - 1$. This is equal to zero when $\sqrt{x} = 2$, or $x = 4$. The point on the graph is $(4, 4)$.

b. The slope of the tangent line is $-\frac{1}{2}$ when $\frac{2}{\sqrt{x}} - 1 = -\frac{1}{2}$. Solving for x gives $x = 16$. The point on the graph is $(16, 0)$.

3.3.69 $f'(x) = 20x^3 + 30x^2 + 3$, $f''(x) = 60x^2 + 60x$, and $f^{(3)}(x) = 120x + 60$.

3.3.71 f simplifies as $f(x) = \frac{(x-8)(x+1)}{x+1} = x - 8$. So for $x \ne -1$, $f'(x) = 1$, $f''(x) = 0$, and $f^{(3)}(x) = 0$.

3.3.73

a. False. 10^5 is a constant, so the constant rule assures us that $\frac{d}{dx}(10^5) = 0$.

b. True. This follows because the slope is given by $f'(x) = e^x > 0$ for all x.

c. False. $\frac{d}{dx}(e^3) = 0$.

d. False. $\frac{d}{dx}(e^x) = e^x$, not xe^{x-1}.

e. False. We have $\frac{d}{dx}(5x^3 + 2x + 5) = 15x^2 + 2$. Thus we have $\frac{d^2}{dx^2}(5x^3 + 2x + 5) = 30x$, and $\frac{d^3}{dx^3}(5x^3 + 2x + 5) = 30$. It is true that $\frac{d^n}{dx^n}(5x^3 + 2x + 5) = 0$ for $n \geq 4$.

3.3.75 First note that because the slope of $4x + 1$ is 4, it must be the case that $f'(2) = 4$. Also, at $x = 2$, we have $y = 4 \cdot 2 + 1 = 9$, so $f(2) = 9$. Because the line tangent to the graph of g at 2 has slope 3, we know that $g'(2) = 3$. The tangent line to g at $x = 2$ must be $y - (-2) = 3(x - 0)$, so the value of the tangent line at 2 (which must also be the value of $g(2)$) is 4. So $g(2) = 4$.

a. $y'(2) = f'(2) + g'(2) = 4 + 3 = 7$. The line contains the point $(2, f(2) + g(2)) = (2, 13)$. Thus, the equation of the tangent line is $y - 13 = 7(x - 2)$, or $y = 7x - 1$.

b. $y'(2) = f'(2) - 2g'(2) = 4 - 2 \cdot 3 = -2$. The line contains the point $(2, f(2) - 2g(2)) = (2, 1)$. Thus, the equation of the tangent line is $y - 1 = -2(x - 2)$, or $y = -2x + 5$.

c. $y'(2) = 4f'(2) = 4 \cdot 4 = 16$. The line contains the point $(2, 4f(2)) = (2, 36)$. Thus, the equation of the tangent line is $y - 36 = 16(x - 2)$, or $y = 16x + 4$.

3.3.77 For $f(x) = x^2 + bx + c$ we have $f'(x) = 2x + b$, so $f'(1) = 2 + b$. Because the slope of $4x + 2$ is 4, we require $2 + b = 4$, so $b = 2$. Also, because the value of $4x + 2$ at $x = 1$ is 6, we must have $f(1) = 1 + 2 + c = 6$, so $c = 3$. Thus the curve $f(x) = x^2 + 2x + 3$ has $y = 4x + 2$ as its tangent line at $x = 1$.

3.3.79 $G'(2) = 3f'(2) - g'(2) = 3(-3) - 1 = -10$.

3.3.81 $G'(5) = 3f'(5) - g'(5) = 3 \cdot 1 - (-1) = 4$.

3.3.83

a. Let $f(x) = x + e^x$ and $a = 0$. Then $\lim_{x \to 0} \frac{f(x) - f(0)}{x} = \lim_{x \to 0} \frac{x + e^x - 1}{x} = f'(0)$.

b. $f'(x) = 1 + e^x$, so $f'(0) = 1 + e^0 = 2$, so $\lim_{x \to 0} \frac{x + e^x - 1}{x} = 2$.

3.3.85

a. Let $f(x) = \sqrt{x}$ and $a = 9$. Then $\lim_{h \to 0} \frac{f(a+h) - f(a)}{h} = \lim_{h \to 0} \frac{\sqrt{9+h} - \sqrt{9}}{h} = f'(9)$.

b. Because $f'(x) = \frac{1}{2\sqrt{x}}$, we have $f'(9) = \frac{1}{6}$, so this is the value of the original limit.

3.3.87

a. Let $f(x) = e^x$, and $a = 3$. Then $\lim_{h \to 0} \frac{f(a+h) - f(a)}{h} = \lim_{h \to 0} \frac{e^{3+h} - e^3}{h} = f'(3)$.

b. Because $f'(x) = e^x$, we have $f'(3) = e^3$, so this is the value of the original limit.

3.3.89 $\lim\limits_{x \to 0} \dfrac{e^{3x} - 1}{x} = 3$.

3.3.91 $\lim\limits_{x \to 0^+} x^x = 1$.

3.3.93

$$\dfrac{d}{dx} x^n = \lim_{h \to 0} \dfrac{(x+h)^n - x^n}{h} = \lim_{h \to 0} \left(\dfrac{x^n + nx^{n-1}h + \frac{n(n-1)}{2} x^{n-2} h^2 + \cdots + nxh^{n-1} + h^n - x^n}{h} \right)$$

$$= \lim_{h \to 0} \left(nx^{n-1} + \dfrac{n(n-1)}{2} x^{n-2} h + \cdots + nxh^{n-2} + h^{n-1} \right) = nx^{n-1} + 0 + 0 + \cdots + 0 = nx^{n-1}$$

3.3.95

a. $\dfrac{d}{dx}(\sqrt{x}) = \dfrac{d}{dx} x^{1/2} = \dfrac{1}{2} \cdot x^{-1/2} = \dfrac{1}{2\sqrt{x}}$.

b.

$$\dfrac{d}{dx} x^{3/2} = \lim_{h \to 0} \dfrac{(x+h)^{3/2} - x^{3/2}}{h} = \lim_{h \to 0} \dfrac{((x+h)^{3/2} - x^{3/2})((x+h)^{3/2} + x^{3/2})}{h((x+h)^{3/2} + x^{3/2})}$$

$$= \lim_{h \to 0} \dfrac{(x+h)^3 - x^3}{h((x+h)^{3/2} + x^{3/2})} = \lim_{h \to 0} \dfrac{x^3 + 3x^2 h + 3xh^2 + h^3 - x^3}{h((x+h)^{3/2} + x^{3/2})}$$

$$= \lim_{h \to 0} \dfrac{3x^2 + 3xh + h^2}{((x+h)^{3/2} + x^{3/2})} = \dfrac{3x^2 + 0 + 0}{x^{3/2} + x^{3/2}} = \dfrac{3x^2}{2x^{3/2}} = \dfrac{3}{2} x^{1/2}.$$

c.

$$\dfrac{d}{dx} x^{5/2} = \lim_{h \to 0} \dfrac{(x+h)^{5/2} - x^{5/2}}{h} = \lim_{h \to 0} \dfrac{((x+h)^{5/2} - x^{5/2})((x+h)^{5/2} + x^{5/2})}{h((x+h)^{5/2} + x^{5/2})}$$

$$= \lim_{h \to 0} \dfrac{(x+h)^5 - x^5}{h((x+h)^{5/2} + x^{5/2})} = \lim_{h \to 0} \dfrac{x^5 + 5x^4 h + 10x^3 h^2 + 10x^2 h^3 + 5xh^4 + h^5 - x^5}{h((x+h)^{5/2} + x^{5/2})}$$

$$= \lim_{h \to 0} \dfrac{5x^4 + 10x^3 h + 10x^2 h^2 + 5xh^3 + h^4}{((x+h)^{5/2} + x^{5/2})}$$

$$= \dfrac{5x^4 + 0 + 0 + 0 + 0}{x^{5/2} + x^{5/2}} = \dfrac{5x^4}{2x^{5/2}} = \dfrac{5}{2} x^{3/2}.$$

d. It appears that $\dfrac{d}{dx} x^{n/2} = \dfrac{n}{2} \cdot x^{(n/2)-1}$.

3.3.97

a. $\dfrac{d}{dx}(e^{2x}) = \lim\limits_{h \to 0} \dfrac{e^{2(x+h)} - e^{2x}}{h} = \lim\limits_{h \to 0} \dfrac{e^{2x}(e^{2h} - 1)}{h} = e^{2x} \lim\limits_{h \to 0} \dfrac{e^{2h} - 1}{h}$.

b. Let $z = 2h$. Then $\lim\limits_{h \to 0} \dfrac{e^{2h} - 1}{h} = \lim\limits_{z \to 0} \dfrac{e^z - 1}{z/2} = 2 \lim\limits_{z \to 0} \dfrac{e^z - 1}{z} = 2$.

c. $\dfrac{d}{dx}(e^{2x}) = e^{2x} \lim\limits_{h \to 0} \dfrac{e^{2h} - 1}{h} = 2e^{2x}$.

3.4 The Product and Quotient Rules

3.4.1 The derivative of the product fg with respect to x is given by $f'(x)g(x) + f(x)g'(x)$.

3.4.3 $\dfrac{d}{dx}((x+1)(3x+2)) = \left(\dfrac{d}{dx}(x+1)\right)(3x+2) + (x+1)\dfrac{d}{dx}(3x+2) = 1(3x+2) + (x+1)3 = 6x+5.$

3.4.5 $\dfrac{d}{dx}\left(\dfrac{x-1}{3x+2}\right) = \dfrac{(3x+2)\frac{d}{dx}(x-1) - (x-1)\frac{d}{dx}(3x+2)}{(3x+2)^2} = \dfrac{3x+2 - 3(x-1)}{(3x+2)^2} = \dfrac{5}{(3x+2)^2}.$

3.4.7

a. Using the product rule: $f'(x) = 1(x-1) + x(1) = 2x-1$.

b. By expanding first: $f(x) = x^2 - x$ so $f'(x) = 2x-1$.

3.4.9

a. By the product rule: $y' = (2t+7)(3t-4) + (t^2+7t) \cdot 3 = 9t^2 + 34t - 28$.

b. By expanding first: $y' = \dfrac{d}{dt}(3t^3 + 17t^2 - 28t) = 9t^2 + 34t - 28$.

3.4.11

a. By the quotient rule:
$$f'(w) = \dfrac{w(3w^2-1) - (w^3-w) \cdot 1}{w^2} = \dfrac{2w^3}{w^2} = 2w \text{ for } w \neq 0.$$

b. For $w \neq 0$ this simplifies as $w^2 - 1$. $f'(w) = \dfrac{d}{dw}(w^2-1) = 2w$.

3.4.13

a. By the quotient rule: $y' = \dfrac{(x-a)(2x) - (x^2-a^2)}{(x-a)^2} = \dfrac{2x^2 - 2ax - x^2 + a^2}{(x-a)^2} = \dfrac{x^2 - 2ax + a^2}{(x-a)^2} = \dfrac{(x-a)^2}{(x-a)^2} = 1$ for $x \neq a$.

b. For $x \neq a$, this simplifies as $y = \dfrac{(x+a)(x-a)}{x-a} = x+a$. $y' = \dfrac{d}{dx}(x+a) = 1$.

3.4.15 $\dfrac{d}{dx}(f(x)g(x))\bigg|_{x=1} = f'(1)g(1) + f(1)g'(1) = 4 \cdot 2 + 5 \cdot 3 = 23$.

$\dfrac{d}{dx}\left(\dfrac{f(x)}{g(x)}\right)\bigg|_{x=1} = \dfrac{g(1)f'(1) - f(1)g'(1)}{g(1)^2} = \dfrac{2 \cdot 4 - 5 \cdot 3}{2^2} = -\dfrac{7}{4}.$

3.4.17 By the quotient rule, $f'(x) = \dfrac{(x+6) \cdot 1 - x \cdot 1}{(x+6)^2} = \dfrac{6}{(x+6)^2}$, so $f'(3) = \dfrac{6}{81} = \dfrac{2}{27}$ and $f'(-2) = \dfrac{3}{8}$.

3.4.19 $f'(x) = 12x^3(2x^2-1) + 3x^4 \cdot 4x = 24x^5 - 12x^3 + 12x^5 = 36x^5 - 12x^3$.

3.4.21 $f'(x) = \dfrac{(x+1) \cdot 1 - x \cdot 1}{(x+1)^2} = \dfrac{1}{(x+1)^2}.$

3.4.23 $f'(t) = \dfrac{5}{3}t^{2/3}e^t + t^{5/3}e^t = e^t t^{2/3}\left(\dfrac{5}{3} + t\right).$

3.4.25 $f'(x) = \dfrac{(e^x+1)e^x - e^x(e^x)}{(e^x+1)^2} = \dfrac{e^x}{(e^x+1)^2}.$

3.4.27 $f'(x) = (1)e^{-x} + x(-e^{-x}) = e^{-x}(1-x)$.

3.4.29 $y' = \dfrac{d}{dt}\left(\dfrac{3t-1}{2t-2}\right) = \dfrac{(2t-2)\cdot 3 - (3t-1)\cdot 2}{(2t-2)^2} = -\dfrac{4}{(2t-2)^2} = -\dfrac{1}{(t-1)^2}$.

3.4.31 $h'(x) = (1)(x^3+x^2+x+1) + (x-1)(3x^2+2x+1) = x^3+x^2+x+1+3x^3+2x^2+x-3x^2-2x-1 = 4x^3$.

3.4.33 $g'(w) = e^w(w^3-1) + e^w\cdot 3w^2 = e^w(w^3+3w^2-1)$.

3.4.35 $f'(t) = e^t(t^2-2t+2) + e^t(2t-2) = e^t(t^2-2t+2+2t-2) = t^2 e^t$.

3.4.37 $g'(x) = \dfrac{(x^2-1)\cdot e^x - e^x\cdot 2x}{(x^2-1)^2} = \dfrac{e^x(x^2-2x-1)}{(x^2-1)^2}$.

3.4.39 $f'(x) = (-9)\cdot 3\cdot x^{-9-1} = -27x^{-10}$.

3.4.41 $g'(t) = \dfrac{d}{dt}(3t^2 + 6t^{-7}) = 6t - 42t^{-8}$.

3.4.43 $g'(t) = \dfrac{d}{dt}(1 + 3t^{-1} + t^{-2}) = -3t^{-2} - 2t^{-3}$.

3.4.45 $g'(x) = \dfrac{(x-2)((x+1)e^x + e^x) - (x+1)e^x}{(x-2)^2} = \dfrac{e^x}{(x-2)^2}\cdot \dfrac{(x-2)(x+2) - (x+1)}{1} = \dfrac{e^x}{(x-2)^2}\cdot (x^2 - x - 5)$.

3.4.47 $h'(x) = \dfrac{(x+1)e^x - xe^x}{(x+1)^2} = \dfrac{e^x}{(x+1)^2}$.

3.4.49

$$g'(w) = \dfrac{(\sqrt{w}-w)(\tfrac{1}{2\sqrt{w}}+1) - (\sqrt{w}+w)(\tfrac{1}{2\sqrt{w}}-1)}{(\sqrt{w}-w)^2} = \dfrac{\tfrac{1}{2}+\sqrt{w}-\tfrac{1}{2}\sqrt{w}-w - (\tfrac{1}{2}-\sqrt{w}+\tfrac{1}{2}\sqrt{w}-w)}{(\sqrt{w}-w)^2}$$

$$= \dfrac{2\sqrt{w}-\sqrt{w}}{(\sqrt{w}-w)^2} = \dfrac{\sqrt{w}}{(\sqrt{w}-w)^2}.$$

3.4.51 $h'(w) = \dfrac{(w^{5/3}+1)\tfrac{5}{3}w^{2/3} - w^{5/3}(\tfrac{5}{3}w^{2/3})}{(w^{5/3}+1)^2} = \dfrac{\tfrac{5}{3}w^{2/3}}{(w^{5/3}+1)^2} = \dfrac{5w^{2/3}}{3(w^{5/3}+1)^2}$.

3.4.53 $f'(x) = \dfrac{d}{dx}\left(4x^2 - \dfrac{2x}{5x+1}\right) = 8x - \dfrac{(5x+1)2 - (2x)(5)}{(5x+1)^2} = 8x - \dfrac{2}{(5x+1)^2}$.

3.4.55

$$h'(r) = \dfrac{(r+1)(-1-\tfrac{1}{2\sqrt{r}}) - (2-r-\sqrt{r})\cdot 1}{(r+1)^2} = \dfrac{-r - \tfrac{\sqrt{r}}{2} - 1 - \tfrac{1}{2\sqrt{r}} - 2 + r + \sqrt{r}}{(r+1)^2}$$

$$= \dfrac{\tfrac{\sqrt{r}}{2} - \tfrac{1}{2\sqrt{r}} - 3}{(r+1)^2}\cdot \dfrac{2\sqrt{r}}{2\sqrt{r}} = \dfrac{r - 1 - 6\sqrt{r}}{2\sqrt{r}(r+1)^2}.$$

3.4.57 $h'(x) = (35x^6 + 5)(6x^3 + 3x^2 + 3) + (5x^7 + 5x)(18x^2 + 6x) = 15((7x^6+1)(2x^3+x^2+1) + (x^7+x)(6x^2+2x)) = 300x^9 + 135x^8 + 105x^6 + 120x^3 + 45x^2 + 15$.

3.4.59 $f(x) = \sqrt{(e^x + 4x^2)^2} = |e^x + 4x^2| = e^x + 4x^2$. Therefore, $f'(x) = e^x + 8x$.

3.4.61

a. $y' = \dfrac{(x-1)-(x+5)}{(x-1)^2} = -\dfrac{6}{(x-1)^2}$.

At $a=3$ we have $y' = -\dfrac{6}{4} = -\dfrac{3}{2}$ and $y=4$, so the equation of the tangent line is $y-4 = -\dfrac{3}{2} \cdot (x-3)$, or $y = -\dfrac{3}{2}x + \dfrac{17}{2}$.

b.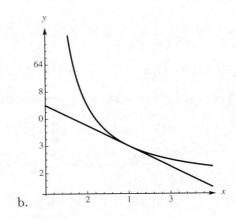

3.4.63

a. $y' = 2 + (1)e^x + xe^x$.
At $a=0$ we have $y' = 2+1+0 = 3$ and $y=1$. So the equation of the tangent line is $y-1 = 3(x-0)$, or $y = 3x+1$.

b.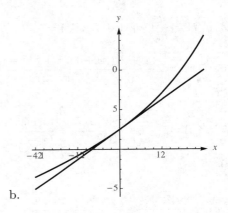

3.4.65

a. $p'(t) = \dfrac{(t+2)200 - 200t}{(t+2)^2} = \dfrac{400}{(t+2)^2}$.

b. $p'(5) = \dfrac{400}{49} \approx 8.16$.

c. The value of p' is as large as possible when its denominator is as small as possible, which is when $t=0$. The value of $p'(0)$ is 100.

d. $\lim\limits_{t \to \infty} p'(t) = \lim\limits_{t \to \infty} \dfrac{400}{(t+2)^2} = 0$. This means that the population eventually has a growth rate of 0, which means that the population approaches a steady state.

e.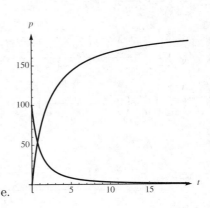

3.4.67

a. The instantaneous rate of change is $\dfrac{d}{dx}F(x) = -\dfrac{2kQq}{x^3}$ N/m $= -\dfrac{1.8 \times 10^{10} Qq}{x^3}$ N/m.

b. $\left[\dfrac{d}{dx}F(x)\right]\bigg|_{x=0.001} = -\dfrac{2(9 \times 10^9)}{(0.001)^3} = -\dfrac{18 \times 10^9}{10^{-9}} = -18 \times 10^{18} = -1.8 \times 10^{19}$ Newtons per meter.

c. Because the distance x appears in the denominator of $F'(x)$, the absolute value of the instantaneous rate of change decreases with the separation.

3.4.69

a. False. In fact, because e^5 is a constant, its derivative is zero.

b. False. It is certainly a reasonable way to proceed, but one could also write the given quantity as $x + 3 + 2x^{-1}$, and then proceed using the sum rule and the power rule and the extended power rule.

c. False. $\dfrac{d}{dx}\left(\dfrac{1}{x^5}\right) = \dfrac{d}{dx}\left(x^{-5}\right) = -5x^{-6} = -\dfrac{5}{x^6}$.

d. False. The derivative is $3x^2 e^x + x^3 e^x = x^2 e^x (3 + x)$.

3.4.71 $f'(x) = 2x(2 + x^{-3}) + x^2(-3x^{-4}) = 4x + 2x^{-2} - 3x^{-2} = 4x - x^{-2} = 4x - \dfrac{1}{x^2}$.

$f''(x) = 4 + 2x^{-3} = 2\left(2 + \dfrac{1}{x^3}\right)$.

$f'''(x) = -6x^{-4} = -\dfrac{6}{x^4}$.

3.4.73

$f'(x) = \dfrac{d}{dx}\left(\dfrac{x^2 - 7x}{x + 1}\right) = \dfrac{(x+1)(2x-7) - (x^2 - 7x) \cdot 1}{(x+1)^2} = \dfrac{x^2 + 2x - 7}{x^2 + 2x + 1} = \dfrac{x^2 + 2x - 7}{(x+1)^2}$.

$f''(x) = \dfrac{d}{dx}\left(\dfrac{x^2 + 2x - 7}{x^2 + 2x + 1}\right) = \dfrac{(x^2 + 2x + 1)(2x + 2) - (x^2 + 2x - 7)(2x + 2)}{(x+1)^4}$

$= \dfrac{(2x^2 + 4x + 2) - (2x^2 + 4x - 14)}{(x+1)^3} = \dfrac{16}{(x+1)^3}$.

3.4.75

$y' = -\dfrac{54x}{(x^2 + 9)^2}$. At $x = 2$, $y' = -\dfrac{108}{169}$ and $y = \dfrac{27}{13}$. Thus the tangent line is given by

a.
$$y - \dfrac{27}{13} = -\dfrac{108}{169}(x - 2),$$

or $y = -\dfrac{108}{169}x + \dfrac{567}{169}$.

b.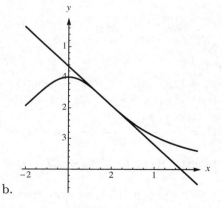

3.4.77 $\dfrac{d}{dx}\left[\dfrac{f(x)}{g(x)}\right]\bigg|_{x=2} = \dfrac{g(2)f'(2) - f(2)g'(2)}{(g(2))^2} = \dfrac{2 \cdot 5 - 4 \cdot 4}{4} = -\dfrac{3}{2}$.

3.4.79 $\dfrac{d}{dx}\left[\dfrac{f(x)}{x+2}\right]\bigg|_{x=4} = \dfrac{(4+2)f'(4) - f(4)}{36} = \dfrac{6 - 2}{36} = \dfrac{1}{9}$.

3.4.81 $\dfrac{d}{dx}\left[\dfrac{f(x)g(x)}{x}\right]\bigg|_{x=4} = \dfrac{4(f'(4)g(4) + f(4)g'(4)) - f(4)g(4)}{16} = \dfrac{4(1 \cdot 3 + 2 \cdot 1) - (2 \cdot 3)}{16} = \dfrac{14}{16} = \dfrac{7}{8}$.

3.4.83

a.

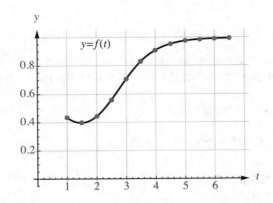

b. For $t \approx 3$.

c. $f'(t) = \dfrac{115.48(41.38) - 12.48(81.55)}{115.48^2} \approx 0.28$ mm/g per week. At a young age, the bird's wings are growing quickly relative to its age.

d. It appears that $f'(t) \approx 0$ for $t \geq 6.5$. $f'(6.5) = \dfrac{121.45(0.38) - 0.01(120.61)}{121.45^2} \approx 0.003$ mm/g per week. As the bird matures the bird's growth rate slows down, so the ratio of wing chord length to mass doesn't change much, which implies that the rate of change of wing chord length to mass, f', is almost 0.

3.4.85 $q'(x) = \dfrac{f'(x)g(x) - f(x)g'(x)}{g(x)^2}$, so $q'(3) = \dfrac{f'(3)g(3) - f(3)g'(3)}{g(3)^2} = \dfrac{5(2) - 2(-10)}{4} = \dfrac{30}{4} = \dfrac{15}{2}$.

3.4.87 $\dfrac{d}{dx}(f(x)g(x))\Big|_{x=4} = f'(4)g(4) + f(4)g'(4) = \dfrac{1}{2}(1) + 3(-1) = -\dfrac{5}{2}$.

3.4.89 $\dfrac{d}{dx}(xg(x))\Big|_{x=2} = g(2) + 2g'(2) = 3 + 2(-1) = 1$.

3.4.91

a. $y' = (2x - 3)h(x) + (x^2 - 3x)h'(x)$ so $y'(4) = 5(2) + 4(-3) = -2$. Also, $y(4) = 4(2) = 8$. So the equation is $y - 8 = -2(x - 4)$ or $y = -2x + 16$.

b. $y' = \dfrac{(x+2)h'(x) - h(x)}{(x+2)^2}$, so $y'(4) = \dfrac{6(-3) - 2}{36} = -\dfrac{20}{36} = -\dfrac{5}{9}$. Also, $y(4) = \dfrac{1}{3}$. So the equation of the tangent line is $y - \dfrac{1}{3} = -\dfrac{5}{9}(x - 4)$ or $y = -\dfrac{5}{9}x + \dfrac{23}{9}$.

3.4.93 Following the hint, we note that $f(0) = 100$ and we have $f'(x) = 40x^7 + 30x^4 + 20x^3 + 6x + 20$, so $f'(0) = 20$. Also note that $g(0) = 2$ and $g'(x) = 100x^9 + 72x^8 + 30x^4 + 12x + 4$, so $g'(0) = 4$. Then

$$q'(0) = \dfrac{g(0)f'(0) - f(0)g'(0)}{g(0)^2} = \dfrac{2(20) - 100(4)}{4} = -90.$$

3.4.95

a. The tangent line at $x = a$ is $y - a^2 = 2a(x - a)$ and at $x = b$ is $y - b^2 = 2b(x - b)$. These intersect when $a^2 + 2ax - 2a^2 = b^2 + 2bx - 2b^2$, or $(2a - 2b)x = a^2 - b^2$, which is met when $x = \dfrac{a+b}{2}$. So $c = \dfrac{a+b}{2}$.

b. The tangent line at $x = a$ is $y - \sqrt{a} = \dfrac{1}{2\sqrt{a}}(x-a)$ and at $x = b$ is $y - \sqrt{b} = \dfrac{1}{2\sqrt{b}}(x-b)$.

These intersect when $\sqrt{a} + \dfrac{1}{2\sqrt{a}}(x-a) = \sqrt{b} + \dfrac{1}{2\sqrt{b}}(x-b)$, or $\left(\dfrac{1}{2\sqrt{a}} - \dfrac{1}{2\sqrt{b}}\right)x = \dfrac{\sqrt{b} - \sqrt{a}}{2}$, which is met when $x = \sqrt{ab}$. So $c = \sqrt{ab}$.

c. The tangent line at $x = a$ is $y - \dfrac{1}{a} = -\dfrac{1}{a^2}(x-a)$ and at $x = b$ is $y - \dfrac{1}{b} = -\dfrac{1}{b^2}(x-b)$.

These intersect when $\dfrac{1}{a} - \dfrac{1}{a^2}(x-a) = \dfrac{1}{b} - \dfrac{1}{b^2}(x-b)$, or $\left(\dfrac{2}{a} - \dfrac{x}{a^2}\right) = \left(\dfrac{2}{b} - \dfrac{x}{b^2}\right)$, which is met when $x\left(\dfrac{1}{b^2} - \dfrac{1}{a^2}\right) = \dfrac{2}{b} - \dfrac{2}{a}$, or $x \cdot \left(\dfrac{a^2 - b^2}{a^2 b^2}\right) = \dfrac{2(a-b)}{ab}$. Thus we arrive at $x = \dfrac{2ab}{a+b}$. So $c = \dfrac{2ab}{a+b}$.

d. The tangent line at $x = a$ is $y - f(a) = f'(a)(x-a)$ and at $x = b$ is $y - f(b) = f'(b)(x-b)$.

These intersect when $f(a) + f'(a)(x-a) = f(b) + f'(b)(x-b)$, or $(f'(a) - f'(b))x = f(b) - f(a) - f'(b)b + f'(a)a$. Solving for x yields $x = \dfrac{f(b) - f(a) - f'(b)b + f'(a)a}{f'(a) - f'(b)}$ provided $f'(a) \neq f'(b)$ (which occurs when the tangent lines are parallel and don't intersect.)

3.4.97

$$\dfrac{d^2}{dx^2}(f(x)g(x)) = \dfrac{d}{dx}(f'(x)g(x) + f(x)g'(x)) = f''(x)g(x) + f'(x)g'(x) + f'(x)g'(x) + f(x)g''(x)$$
$$= f''(x)g(x) + 2f'(x)g'(x) + f(x)g''(x).$$

3.4.99

a.

$$\dfrac{d}{dx}[(f(x)g(x))h(x)] = \dfrac{d}{dx}[f(x)g(x)] \cdot h(x) + f(x)g(x) \cdot \dfrac{d}{dx}h(x)$$
$$= [f'(x)g(x) + f(x)g'(x)]h(x) + f(x)g(x)h'(x)$$
$$= f'(x)g(x)h(x) + f(x)g'(x)h(x) + f(x)g(x)h'(x).$$

b. $\dfrac{d}{dx}[e^x(x-1)(x+3)] = e^x(x-1)(x+3) + e^x(x+3) + e^x(x-1) = e^x(x^2 + 2x - 3 + x + 3 + x - 1) = e^x(x^2 + 4x - 1)$.

3.5 Derivatives of Trigonometric Functions

3.5.1 A direct substitution would yield the quotient of zero with itself, which isn't defined

3.5.3 Because $\tan x = \dfrac{\sin x}{\cos x}$, and $\cot x = \dfrac{\cos x}{\sin x}$, we can use the quotient rule to compute these derivatives, because we know the derivatives of $\sin x$ and of $\cos x$.

3.5.5 $f'(x) = \cos x$ and $f'(\pi) = \cos \pi = -1$.

3.5.7 $y' = \cos x$, so $y'(0) = \cos 0 = 1$. So the equation of the tangent line is $y - 0 = 1(x-0)$ or $y = x$.

3.5.9 $\dfrac{d}{dx}(\sin x + \cos x) = \cos x - \sin x$, so $\dfrac{d^2}{dx^2}(\sin x + \cos x) = \dfrac{d}{dx}(\cos x - \sin x) = -\sin x - \cos x = -(\sin x + \cos x)$.

3.5.11 $\lim\limits_{x \to 0} \dfrac{\sin 3x}{x} = \lim\limits_{x \to 0} \dfrac{3 \sin 3x}{3x} = 3 \lim\limits_{x \to 0} \dfrac{\sin 3x}{3x} = 3 \cdot 1 = 3$.

3.5.13 $\lim\limits_{x \to 0} \dfrac{\sin 7x}{\sin 3x} = \lim\limits_{x \to 0} \dfrac{\dfrac{7\sin 7x}{7x}}{\dfrac{3\sin 3x}{3x}} = \dfrac{7}{3} \cdot \lim\limits_{x \to 0} \dfrac{\dfrac{\sin 7x}{7x}}{\dfrac{\sin 3x}{3x}} = \dfrac{7}{3} \cdot \dfrac{1}{1} = \dfrac{7}{3}.$

3.5.15 $\lim\limits_{x \to 0} \dfrac{\tan 5x}{x} = \lim\limits_{x \to 0} \dfrac{5\sin 5x}{5x \cos 5x} = 5 \lim\limits_{x \to 0} \dfrac{\sin 5x}{5x} \cdot \lim\limits_{x \to 0} \dfrac{1}{\cos 5x} = 5 \cdot 1 \cdot 1 = 5.$

3.5.17 $\lim\limits_{x \to 0} \dfrac{\tan 7x}{\sin x} = \lim\limits_{x \to 0} \dfrac{\sin 7x}{\cos 7x \cdot \sin x} = \lim\limits_{x \to 0} \left(\dfrac{1}{\cos 7x} \cdot \dfrac{x}{\sin x} \cdot \dfrac{7 \sin 7x}{7x} \right) =$
$7 \cdot \lim\limits_{x \to 0} \dfrac{1}{\cos 7x} \cdot \lim\limits_{x \to 0} \dfrac{x}{\sin x} \cdot \lim\limits_{x \to 0} \dfrac{\sin 7x}{7x} = 7 \cdot 1 \cdot 1 \cdot 1 = 7.$

3.5.19 $\lim\limits_{x \to 2} \dfrac{\sin(x-2)}{x^2 - 4} = \lim\limits_{x \to 2} \left(\dfrac{1}{x+2} \cdot \dfrac{\sin(x-2)}{x-2} \right) = \lim\limits_{x \to 2} \dfrac{1}{x+2} \cdot \lim\limits_{x \to 2} \dfrac{\sin(x-2)}{x-2} = \dfrac{1}{4} \cdot 1 = \dfrac{1}{4}.$

3.5.21 $\lim\limits_{x \to 0} \dfrac{\sin ax}{\sin bx} = \lim\limits_{x \to 0} \dfrac{a \sin ax}{ax} \cdot \dfrac{bx}{b \sin bx} = \dfrac{a}{b} \lim\limits_{x \to 0} \dfrac{\sin ax}{ax} \cdot \lim\limits_{x \to 0} \dfrac{bx}{\sin bx} = \dfrac{a}{b} \cdot 1 \cdot 1 = \dfrac{a}{b}.$

3.5.23 $\dfrac{dy}{dx} = \cos x - \sin x.$

3.5.25 $\dfrac{dy}{dx} = -e^{-x} \sin x + e^{-x} \cos x = e^{-x}(\cos x - \sin x).$

3.5.27 $\dfrac{dy}{dx} = \sin x + x \cos x.$

3.5.29
$$\dfrac{dy}{dx} = \dfrac{(\sin x + 1)(-\sin x) - (\cos x)(\cos x)}{(1 + \sin x)^2} = \dfrac{-1(\sin^2 x + \cos^2 x) - \sin x}{(1 + \sin x)^2}$$
$$= \dfrac{-1(1 + \sin x)}{(1 + \sin x)^2} = -\dfrac{1}{1 + \sin x}.$$

3.5.31 $\dfrac{dy}{dx} = \cos x \cos x + \sin x \cdot (-\sin x) = \cos^2 x - \sin^2 x = \cos(2x).$

3.5.33 $\dfrac{dy}{dx} = -\sin x \cos x + \cos x(-\sin x) = -2\sin x \cos x = -\sin(2x).$

3.5.35 $\dfrac{dy}{dw} = 2w \sin w + w^2 \cos w + 2\cos w - 2w \sin w - 2 \cos w = w^2 \cos w.$

3.5.37 $\dfrac{dy}{dx} = \cos x \sin x + x(-\sin x)\sin x + x \cos x \cos x = \sin x \cos x - x \sin^2 x + x \cos^2 x = \dfrac{1}{2}\sin 2x + x \cos 2x.$

3.5.39 $\dfrac{dy}{dx} = \dfrac{(1 + \cos x)\cos x - \sin x(-\sin x)}{(1 + \cos x)^2} = \dfrac{1 + \cos x}{(1 + \cos x)^2} = \dfrac{1}{1 + \cos x}.$

3.5.41 $\dfrac{dy}{dx} = \dfrac{(1 + \cos x)\sin x - (1 - \cos x)(-\sin x)}{(1 + \cos x)^2} = \dfrac{2 \sin x}{(1 + \cos x)^2}.$

3.5.43 $\dfrac{dy}{dx} = \sec x \tan x - \csc x \cot x.$

3.5.45 $\dfrac{dy}{dx} = e^x \csc x + e^x(-\csc x \cot x) = e^x \csc x(1 - \cot x).$

3.5.47
$$\frac{dy}{dx} = \frac{(1+\csc x)(-\csc^2 x) - \cot x(-\csc x \cot x)}{(1+\csc x)^2} = \frac{-\csc^2 x - \csc^3 x + \csc x(\csc^2 x - 1)}{(1+\csc x)^2}$$
$$= \frac{-\csc x(1+\csc x)}{(1+\csc x)^2} = -\frac{\csc x}{1+\csc x}$$

3.5.49
$$\frac{dy}{dz} = \frac{0 - (\sec z \tan z \csc z - \sec z \csc z \cot z)}{\sec^2 z \csc^2 z} = \frac{\sec z \csc z(\cot z - \tan z)}{\sec^2 z \csc^2 z}$$
$$= \frac{\cot z - \tan z}{\sec z \csc z} = \cos^2 z - \sin^2 z = \cos(2z).$$

3.5.51 $\dfrac{dy}{dx} = 1 - (-\sin x \sin x + \cos x \cos x) = 1 + \sin^2 x - \cos^2 x = 1 + \sin^2 x - (1 - \sin^2 x) = 2\sin^2 x.$

3.5.53 $\dfrac{d}{dx}(\sec x) = \dfrac{d}{dx}\left(\dfrac{1}{\cos x}\right) = \dfrac{0-(-\sin x)}{\cos^2 x} = \dfrac{1}{\cos x} \cdot \dfrac{\sin x}{\cos x} = \sec x \tan x.$

3.5.55

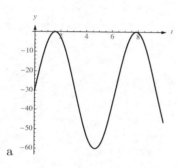

a.

b. $v(t) = y'(t) = 30\cos t$ cm per second.

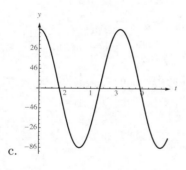

c.

d. $v(t) = 30\cos t = 0$ when $t = \dfrac{2n+1}{2}\cdot\pi$ where n is a non-negative integer. At those times, the position is given by $\begin{cases} 0 & \text{if } n \text{ is even} \\ -60 & \text{if } n \text{ is odd.} \end{cases}$

e. The maximum velocity is 30 cm per second because $|\cos t| \leq 1$ for all t. We have $\cos t = 1$ for $t = 2n\pi$ for a positive integer n. At those times, $y(2n\pi) = -30$.

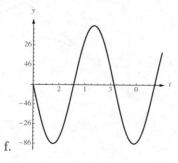

f.

$a(t) = v'(t) = -30\sin t.$

3.5.57 $y' = \sin x + x\cos x$, so $y'' = \cos x + \cos x + -x\sin x = 2\cos x - x\sin x$.

3.5.59 $y' = e^x \sin x + e^x \cos x$, so $y'' = e^x \sin x + e^x \cos x + e^x \cos x + e^x(-\sin x) = 2e^x \cos x$.

3.5.61 $y' = -\csc^2 x$ and $y'' = -((-\csc x \cot x)\csc x + \csc x(-\csc x \cot x)) = 2\cot x \csc^2 x$.

3.5.63

$$y' = \sec x \tan x \csc x - \sec x \csc x \cot x = \sec x \csc x(\tan x - \cot x) = \sec^2 x - \csc^2 x.$$

$$y'' = \sec x(\sec x \tan x) + (\sec x \tan x)\sec x - ((-\csc x \cot x)\csc x + \csc x(-\csc x \cot x))$$
$$= 2\sec^2 x \tan x + 2\csc^2 x \cot x.$$

3.5.65

a. False. $\dfrac{d}{dx}\sin^2 x = \sin x \cos x + \cos x \sin x = 2\sin x \cos x \neq \cos^2 x$.

b. False. $\dfrac{d^2}{dx^2}\sin x = \dfrac{d}{dx}\cos x = -\sin x \neq \sin x$.

c. True. $\dfrac{d^4}{dx^4}\cos x = \dfrac{d^3}{dx^3}(-\sin x) = \dfrac{d^2}{dx^2}(-\cos x) = \dfrac{d}{dx}\sin x = \cos x$.

d. True. In fact, $\pi/2$ isn't even in the domain of $\sec x$.

3.5.67 $\displaystyle\lim_{x \to \pi/4} \dfrac{\tan x - 1}{x - \pi/4} = \lim_{x \to \pi/4} \dfrac{\tan x - \tan \pi/4}{x - \pi/4} = \dfrac{d}{dx}\tan x \bigg|_{x=\pi/4} = \sec^2 \pi/4 = 2$.

3.5.69 $\displaystyle\lim_{h \to 0} \dfrac{\cos(\pi/6 + h) - (\sqrt{3}/2)}{h} = \lim_{h \to 0} \dfrac{\cos(\pi/6 + h) - (\cos \pi/6)}{h} = \dfrac{d}{dx}\cos x \bigg|_{x=\pi/6} = -\sin(\pi/6) = -1/2$.

3.5.71 $\displaystyle\lim_{h \to 0} \dfrac{\tan(5\pi/6 + h) + (1/\sqrt{3})}{h} = \lim_{h \to 0} \dfrac{\tan(5\pi/6 + h) - (\tan 5\pi/6)}{h} = \dfrac{d}{dx}\tan x \bigg|_{x=\pi/6} =$
$\sec^2(5\pi/6) = 4/3$.

3.5.73

a. $y' = 2\cos x$, so $y'(\pi/6) = \sqrt{3}$. $y(\pi/6) = 2$. The tangent line is thus given by $y - 2 = \sqrt{3}(x - \pi/6)$, or $y = \sqrt{3}x + 2 - \dfrac{\pi\sqrt{3}}{6}$.

b.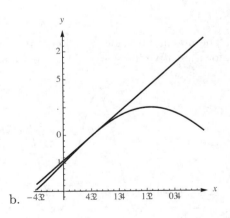

3.5.75

a. $y' = \dfrac{(1-\cos x)(-\sin x) - \cos x \sin x}{(1-\cos x)^2} = -\dfrac{\sin x}{(1-\cos x)^2}$, so $y'(\pi/3) = -2\sqrt{3}$. $y(\pi/3) = 1$. The tangent line is thus given by $y - 1 = -2\sqrt{3}(x - \pi/3)$, or $y = -2\sqrt{3}x + \dfrac{2\sqrt{3}\pi}{3} + 1$.

b.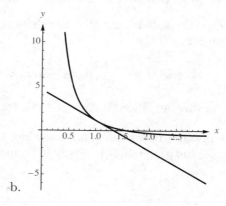

3.5.77 For a horizontal tangent line we need $f'(x) = 1 + 2\sin x = 0$, or $\sin x = -\dfrac{1}{2}$. This occurs for $x = \dfrac{7\pi}{6} + 2n\pi$ where n is any integer, or for $x = \dfrac{11\pi}{6} + 2n\pi$ where n is any integer.

3.5.79

a. $y'(t) = A\cos t$, $y''(t) = -A\sin t$, so $y''(t) + y(t) = -A\sin t + A\sin t = 0$ for all A and all t.

b. $y'(t) = -B\sin t$, $y''(t) = -B\cos t$, so $y''(t) + y(t) = -B\cos t + B\cos t = 0$ for all B and all t.

c. $y' = A\cos t - B\sin t$, $y'' = -A\sin t - B\cos t$, so $y''(t) + y(t) = -A\sin t - B\cos t + A\sin t + B\cos t = 0$ for all A, B, t.

3.5.81

$$\lim_{x\to 0} \dfrac{\cos x - 1}{x} = \lim_{x\to 0} \dfrac{(\cos x - 1)(\cos x + 1)}{x(\cos x + 1)} = \lim_{x\to 0} \dfrac{\cos^2 x - 1}{x(\cos x + 1)} = \lim_{x\to 0} -\dfrac{\sin^2 x}{x(\cos x + 1)}$$

$$= \lim_{x\to 0} \dfrac{\sin x}{x} \cdot \lim_{x\to 0} -\dfrac{\sin x}{(\cos x + 1)} = 1 \cdot \dfrac{0}{2} = 0.$$

3.5.83

$$\dfrac{d}{dx}\cos x = \lim_{h\to 0} \dfrac{\cos(x+h) - \cos x}{h} = \lim_{h\to 0} \dfrac{\cos x \cos h - \sin x \sin h - \cos x}{h}$$

$$= \cos x \left(\lim_{h\to 0} \dfrac{\cos h - 1}{h}\right) - \sin x \left(\lim_{h\to 0} \dfrac{\sin h}{h}\right) = \cos x \cdot 0 - \sin x \cdot 1 = -\sin x.$$

3.5.85 g is continuous at 0 if and only if $\lim_{x\to 0} g(x) = g(0)$. Because $\lim_{x\to 0} g(x) = \lim_{x\to 0} \dfrac{1-\cos x}{2x} = \dfrac{1}{2} \cdot 0 = 0$, we require $a = 0$ in order for g to be continuous.

3.5.87

a. $\dfrac{d}{dx}\sin^2 x = \sin x \cos x + \cos x \sin x = 2\sin x \cos x.$

b. $\dfrac{d}{dx}\sin^3 x = \dfrac{d}{dx}(\sin^2 x)(\sin x) = (2\sin x \cos x)\sin x + \sin^2 x \cdot \cos x = 3\sin^2 x \cos x.$

c. $\dfrac{d}{dx}\sin^4 x = \dfrac{d}{dx}(\sin^3 x)(\sin x) = (3\sin^2 x \cos x)(\sin x) + (\sin^3 x)(\cos x) = 4\sin^3 x \cos x.$

d. We guess that $\dfrac{d}{dx}\sin^n x = n\sin^{n-1} x \cos x$.

We have already seen that the claim is valid for $n = 2$. Suppose our guess is valid for a given positive integer n. Then

$$\dfrac{d}{dx}\sin^{n+1} x = \dfrac{d}{dx}(\sin^n x)(\sin x) = (n\sin^{n-1} x \cos x)(\sin x) + \sin^n x \cos x = (n+1)\sin^n x \cos x.$$

Thus by induction, the result holds for all n.

3.5.89 Because D is a difference quotient, and because $h = 0.01$ is small, D is a good approximation to f'. Therefore, the graph of D is nearly indistinguishable from the graph of $f'(x) = \cos x$.

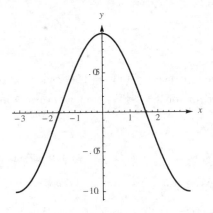

3.6 Derivatives as Rates of Change

3.6.1 The average rate of change is $\dfrac{f(x + \Delta x) - f(x)}{\Delta x}$, whereas the instantaneous rate of change is the limit of this quotient as $\Delta x \to 0$.

3.6.3 If $\dfrac{dy}{dx}$ is small, then small changes in x will result in relatively small changes in the value of y.

3.6.5 At 15 weeks, the puppy grows at a rate of 1.75 lb/week.

3.6.7 Acceleration is the instantaneous rate of change of the velocity; that is, if $s(t)$ is the position of an object at time t, then $s''(t) = \dfrac{d}{dt}(v(t)) = a(t)$ is the acceleration of the object at time t.

3.6.9 $V'(T) = 0.6$. The speed of sound increases by 0.6 m/s for each increase of 1° in temperature.

3.6.11

a. $v_{\text{avg}} = \dfrac{f(0.75) - f(0)}{0.75} = \dfrac{30 - 0}{0.75} = 40$ mph.

b. $v_{\text{avg}} = \dfrac{f(0.75) - f(0.25)}{0.75 - 0.25} = \dfrac{30 - 10}{0.5} = 40$ mph.

This is a pretty good estimate, since the graph is nearly linear over that time interval.

c. $v_{\text{avg}} = \dfrac{f(2.25) - f(1.75)}{2.25 - 1.75} = \dfrac{-14 - 16}{0.5} = -60$ mph.

At 11 a.m. the velocity is $v(2) \approx -60$ mph. The car is moving south with a speed of approximately 60 mph.

d. From 9 a.m. until about 10:08 a.m., the car moves north, away from the station. Then it moves south, passing the station at approximately 11:02 a.m., and continues south until about 11:40 a.m. Then the car drives north until 12:00 noon stopping south of the station.

3.6.13 Each of the first 200 stoves cost on average $70 to produce, while the 201st stove costs $65 to produce.

3.6.15

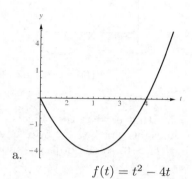

a. $f(t) = t^2 - 4t$ b. $f'(t) = 2t - 4$

b. $f'(t) = 0$ when $t = 2$ – that is when the object is stationary. For $0 \leq t < 2$ we have $f'(t) < 0$ so the object is moving to the left. For $2 < t \leq 5$ we have $f'(t) > 0$ so the object is moving to the right.

c. $f'(1) = -2$ ft/sec and $f''(t) = 2\,\text{ft/sec}^2$, so in particular, $f''(1) = 2\,\text{ft/sec}^2$.

d. $f'(t) = 0$ when $t = 2$ and $f''(2) = 2\,\text{ft/sec}^2$.

e. On the interval $(2, 5]$ the velocity and acceleration are both positive, so the object's speed is increasing.

3.6.17

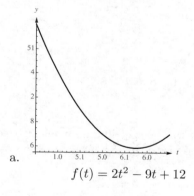

a. $f(t) = 2t^2 - 9t + 12$ b. $f'(t) = 4t - 9$

b. $f'(t) = 0$ when $t = 9/4$ – that is when the object is stationary. For $0 \leq t < 9/4$ we have $f'(t) < 0$ so the object is moving to the left. For $9/4 < t \leq 3$ we have $f'(t) > 0$ so the object is moving to the right.

c. $f'(1) = -5$ ft/sec and $f''(t) = 4\,\text{ft/sec}^2$, so in particular, $f''(1) = 4\,\text{ft/sec}^2$.

d. $f'(t) = 0$ when $t = 9/4$ and $f''(9/4) = 4\,\text{ft/sec}^2$.

e. On the interval $(9/4, 3]$ both the velocity and acceleration are positive, so the object's speed is increasing.

3.6.19

a.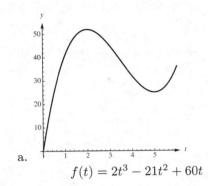
$f(t) = 2t^3 - 21t^2 + 60t$

b.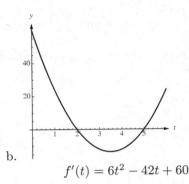
$f'(t) = 6t^2 - 42t + 60$

b. $f'(t) = 0$ when $6(t-2)(t-5) = 0$, which is at $t = 2$ and $t = 5$ – that is when the object is stationary. For $0 \le t < 2$ we have $f'(t) > 0$ so the object is moving to the right. For $2 < t < 5$ we have $f'(t) < 0$ so the object is moving to the left. For $5 < t \le 8$ we have $f'(t) > 0$, so the object is moving to the right again.

c. $f'(1) = 24$ ft/sec and $f''(t) = 12t - 42$, so $f''(1) = -30$ ft/sec².

d. $f'(t) = 0$ when $t = 2$ and $t = 5$. We have $f''(2) = -18$ ft/sec² and $f''(5) = 18$ ft/sec².

e. $f''(t) = 12t - 42$ is positive for $t > \frac{42}{12} = \frac{7}{2}$ and negative for $t < \frac{7}{2}$. So f' and f'' are both positive on $(5, 6]$ and are both negative on $(2, 3.5)$, so that is where the object is speeding up.

3.6.21 To find out when it hits the water, we set $s(t) = -16t^2 + 64$ equal to 0. This gives $-16t^2 = -64$, or $t^2 = 4$, so $t = 2$ is the time the stone hits the water. The velocity at time t is $v(t) = s'(t) = -32t$, so $v(2) = -64$, so the speed is $|-64| = 64$ ft/sec.

3.6.23

a. $v(t) = s'(t) = -32t + 32$.

b. $v(t) = 0$ when $-32t + 32 = 0$, so at $t = 1$ s.

c. $s(1) = -16 + 32 + 48 = 64$ ft.

d. $s(t) = 0$ when $-16t^2 + 32t + 48 = 0$, or $-16(t+1)(t-3) = 0$, so at $t = 3$ s.

e. $v(3) = -96 + 32 = -64$ ft/s.

f. Note that the acceleration is always negative, so the stone will be speeding up when its velocity is also negative, which occurs on the interval $(1, 3)$. This corresponds to the downward portion of the stone's trip.

3.6.25

a. $v(t) = s'(t) = -32t + 64$ ft/sec.

b. $v(t) = 0$ when $-32t + 64 = 0$, which is at $t = 2$.

c. $s(2) = -16 \cdot 4 + 64 \cdot 2 + 32 = 96$ feet.

d. $s(t) = 0$ when $-16t^2 + 64t + 32 = 0$. Using the quadratic formula we see that this occurs when $t = 2 + \sqrt{6} \approx 4.45$ seconds.

e. The velocity when the stone hits the ground is $v(2 + \sqrt{6}) = -32(2 + \sqrt{6}) + 64 = -32\sqrt{6} \approx -78.38$ feet per second.

3.6. Derivatives as Rates of Change

f. The acceleration due to gravity is always negative, so the object is speeding up when its velocity is negative; that is, on its downward journey during the interval $(2, 2 + \sqrt{6})$.

3.6.27 Because the maximum height is $\dfrac{v_0^2}{64}$ (see the previous problem), we need $\dfrac{v_0^2}{64} = 128$, so $v_0 = \sqrt{128 \cdot 64} \approx 90.5$ ft/s.

3.6.29

a. The average cost function is given by $\overline{C}(x) = \dfrac{C(x)}{x} = \dfrac{1000}{x} + .1$. The marginal cost function is given by $M(x) = C'(x) = .1$.

b. At $a = 2000$ we have $\overline{C}(2000) = \dfrac{1000}{2000} + .1 = .6$, and $M(2000) = .1$.

c. The average cost per item when producing 2000 items is $0.60. The cost of producing the next item is $0.10.

3.6.31

a. The average cost function is given by $\overline{C}(x) = \dfrac{C(x)}{x} = \dfrac{100}{x} + 40 - 0.01x$. The marginal cost function is given by $M(x) = C'(x) = 40 - 0.02x$.

b. At $a = 1000$ we have $\overline{C}(1000) = \dfrac{100}{1000} + 40 - (.01)(1000) = 30.1$, and $M(1000) = 20$.

c. The average cost per item when producing 1000 items is $30.10. The cost of producing the next item is $20.00.

3.6.33

a. $D(10) = 40 - 20 = 20$ DVDs per day.

b. Demand is zero when $D(p) = 40 - 2p = 0$, which occurs for $p = 20$ dollars.

c. The elasticity is $E(p) = \dfrac{dD}{dp}\dfrac{p}{D} = -2\left(\dfrac{p}{40 - 2p}\right) = \dfrac{p}{p - 20}$.

d. This quantity satisfies $-1 < E(p) < 0$ when $-1 < \dfrac{p}{p-20} < 0$ which occurs when $p < 20 - p$, or $p < 10$. So for prices in the interval $(0, 10)$ the demand is inelastic, while for prices in the interval $(10, 20)$ the demand is elastic.

e. If the price goes up from 10 to 10.25, that is a $\dfrac{.25}{10} = .025 = 2.5\%$ increase in price.

f. If the price goes up from 10 to 10.25, the demand goes from $D(10) = 40 - 20 = 20$ to $D(10.25) = 40 - 20.5 = 19.5$, which is a $\dfrac{.5}{20} = 2.5\%$ decrease.

3.6.35

a. False. For example, when a ball is thrown up in the air near the surface of the earth, its acceleration is constant (due to gravity) but its velocity changes during its trip.

b. True. If the rate of change of velocity is zero, then velocity must be constant.

c. False. If the velocity is constant over an interval, then the average velocity is equal to the instantaneous velocity over the interval.

d. True. For example, a ball dropped from a tower has negative acceleration and increasing speed as it falls toward the earth.

3.6.37 In each case, the stone reaches its maximum height when its velocity is zero.

On Mars, this occurs when $v(t) = s'(t) = 96 - 12t = 0$, or when $t = 8$ seconds. So the maximum height on Mars is $s(8) = 384$ feet.

On Earth, this occurs when $v(t) = s'(t) = 96 - 32t = 0$, or when $t = 3$ seconds. So the maximum height on Earth is $s(3) = 144$ feet.

The stone will travel $384 - 144 = 240$ feet higher on Mars.

3.6.39 The first stone reaches its maximum height when $f'(t) = -32t + 32 = 0$, so after 1 second, and its maximum height is therefore $f(1) = -16 + 32 + 48 = 64$ feet.

The second stone reaches its maximum height when $g'(t) = -32t + v_0 = 0$, so when $t = \dfrac{v_0}{32}$. Its height at that time is $g(v_0/32) = -16(v_0/32)^2 + (v_0^2/32) = \dfrac{v_0^2}{64}$. This is equal to 64 when $v_0 = 64$ feet per second.

3.6.41

a. The velocity is zero at $t = 1, 2,$ and 3.

b. The object is moving in the positive direction when the slope of s is positive, so from $t = 0$ to $t = 1$, and from $t = 2$ to $t = 3$. It is moving in the negative direction from $t = 1$ to $t = 2$, and for $t > 3$.

c.

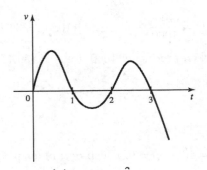

d. The speed is increasing on $(0, 1/2)$ as the velocity is positive and the acceleration is positive there. On $(1/2, 1)$ the speed is decreasing as the velocity is positive but the acceleration is negative. On $(1, 3/2)$ the speed is increasing as the velocity is negative and the acceleration is negative, but on $(3/2, 2)$ the speed is decreasing as the velocity is negative but the acceleration is positive. On approximately $(2, 2.6)$ the speed is increasing as the velocity is positive and the acceleration is positive, but on about $(2.6, 3)$ the velocity is positive but the acceleration is negative, so the speed is decreasing. On $(3, \infty)$ both the velocity and the acceleration are negative, so the object is speeding up.

3.6.43

a. $P(x) = xp(x) - C(x) = 100x + 0.02x^2 - 50x - 100 = 0.02x^2 + 50x - 100$.

b. The average profit is $\overline{P}(x) = \dfrac{P(x)}{x} = 0.02x + 50 - \dfrac{100}{x}$. The marginal profit is $P'(x) = .04x + 50$.

c. $\overline{P}(500) = 59.8$. $P'(500) = 70$.

d. The average profit for the first 500 items sold is \$59.80, while the profit on the 501st item is \$70.00.

3.6.45

a. $P(x) = xp(x) - C(x) = 100x + 0.04x^2 - 800$.

3.6. Derivatives as Rates of Change

b. The average profit is $\overline{P}(x) = \dfrac{P(x)}{x} = .04x + 100 - \dfrac{800}{x}$. The marginal profit is $P'(x) = .08x + 100$.

c. $\overline{P}(1000) = 139.2$. $P'(1000) = 180$.

d. The average profit for the first 1000 items sold is $139.20, while the profit on the 1001st item is $180.00.

3.6.47 Since 1925, the population appears to have grown most slowly in about 1935. Using a straight edge and drawing the (approximate) tangent line, we obtain the following figure. The tangent line appears to pass through the points $(5, 100)$ and $(50, 140)$. Therefore the slope of the tangent line is approximately $\dfrac{(140 - 100)}{(50 - 5)} = 0.89$. So the population was growing at about 890,000 people per year in 1935 (answers will vary — between about 700,000 and 1,000,000 people per year are acceptable).

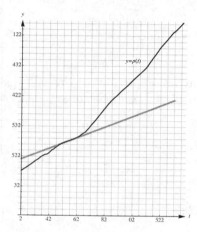

3.6.49

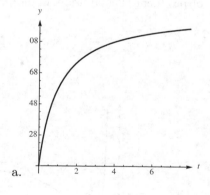

a.

b. $v(t) = s'(t) = \dfrac{(t+1)100 - 100t \cdot 1}{(t+1)^2} = \dfrac{100}{(t+1)^2}$.

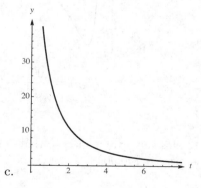

c.

The velocity of the marble is decreasing.

d. $s(t) = 80$ when $\dfrac{100t}{t+1} = 80$, or $100t = 80t + 80$, which occurs when $t = 4$ seconds.

e. $v(t) = 50$ when $\dfrac{100}{(t+1)^2} = 50$, or $(t+1)^2 = 2$. This occurs for $t = \sqrt{2} - 1 \approx 0.414$ seconds.

3.6.51

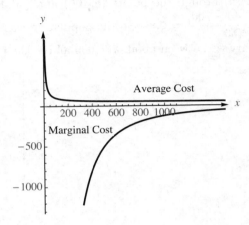

a. The average cost function is $\overline{C}(x) = \dfrac{C(x)}{25000} = 50 + \dfrac{5000}{x} + 0.00006x$. The marginal cost function is $C'(x) = -\dfrac{125000000}{x^2} + 1.5$.
The average cost decreases to about 50 per unit as the batch size increases, while the marginal cost is negative but increases.

b. $\overline{C}(5000) = 51.3$, $C'(5000) = -3.5$.

c. If the batch size is 5000, then the average cost of producing 25000 items is $51.30 per item. If the batch size is increased from 5000 to 5001, then the cost of producing 25000 items would decrease by about $3.50.

3.6.53

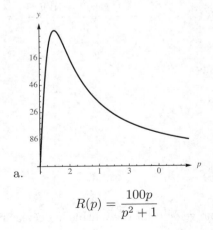

a. $R(p) = \dfrac{100p}{p^2 + 1}$

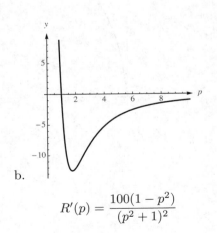

b. $R'(p) = \dfrac{100(1 - p^2)}{(p^2 + 1)^2}$

c. $R'(p)$ is zero at $p = 1$, and the maximum of $R(p)$ occurs at this same value of p, so that is the price to charge in order to maximize revenue. The revenue at this price is $50.00.

3.6. Derivatives as Rates of Change

3.6.55

a. The mass oscillates about the equilibrium point.

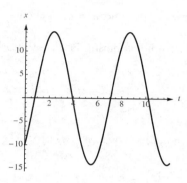

b. $\dfrac{dx}{dt} = 10\cos t + 10\sin t$ is the velocity of the mass at time t.

c. $\dfrac{dx}{dt} = 0$ when $\sin t = -\cos t$, which occurs when $t = \dfrac{4n+3}{4} \cdot \pi$ where n is any positive integer.

d. The model is unrealistic as it ignores the effects of friction and gravity. In reality, the amplitude would decrease as the mass oscillates.

3.6.57

a. Juan starts out faster, but slows toward the end, while Jean starts slower but increases her speed toward the end.

b. Because both start and finish at the same time, they finish with the same average angular velocity.

c. It is a tie.

d. Jean's velocity is given by $\theta'(t) = \dfrac{\pi t}{4}$. At $t = 2$, $\theta'(2) = \dfrac{\pi}{2}$ radians per minute. Her velocity is greatest at $t = 4$.

e. Juan's velocity is given by $\phi'(t) = \pi - \dfrac{\pi t}{4}$. At $t = 2$, $\phi'(2) = \dfrac{\pi}{2}$ radians per minute as well. His velocity is greatest at $t = 0$.

3.6.59

a. $v(t) = y'(t) = -15e^{-t}\cos t - 15e^{-t}\sin t$, so $v(1) \approx -7.625$ meters per second, and $v(3) \approx .63$ meters per second.

b. She is moving down for approximately 2.4 seconds, and then up until about 5.5 seconds, and then down again until about 8.6 seconds, and then up again.

c. The maximum velocity going up appears to be about about 0.65 meters per second.

3.6.61

a. $T'(t) = 160 - 80x$, so $T'(1) = 80$, so the heat flux at 1 is -80. At $x = 3$ we have $T'(3) = -80$, so the heat flux at 3 is 80.

b. The heat flux $-T'(x)$ is negative for $0 \leq x < 2$ and positive for $2 < x \leq 4$.

c. At any point other than the midpoint of the rod, heat flows toward the closest end of the rod, and "out the end."

3.7 The Chain Rule

3.7.1 If $y = f(x)$ and $u = g(x)$ then $\dfrac{dy}{dx} = \dfrac{dy}{du} \cdot \dfrac{du}{dx}$. Alternatively, we have $\dfrac{d}{dx}(f(g(x))) = f'(g(x))g'(x)$.

3.7.3 The inner function is $u = x^3 + x + 1$ and the outer function is $y = u^4$. So
$$\frac{dy}{dx} = \frac{dy}{du} \cdot \frac{du}{dx} = 4u^3(3x^2 + 1) = 4(x^3 + x + 1)^3(3x^2 + 1).$$

3.7.5

a. $u = g(x) = \cos x$, $y = f(u) = u^3$. So $\dfrac{dy}{dx} = \dfrac{dy}{du} \cdot \dfrac{du}{dx} = 3\cos^2 x \cdot (-\sin x) = -3\cos^2 x \sin x$.

b. $u = g(x) = x^3$, $y = f(u) = \cos u$. So $\dfrac{dy}{dx} = \dfrac{dy}{du} \cdot \dfrac{du}{dx} = -\sin x^3 \cdot 3x^2 = -3x^2 \sin x^3$.

3.7.7 The derivative of $f(g(x))$ equals f' evaluated at $g(x)$ multiplied by g' evaluated at x.

3.7.9 Let $g(x) = 4x + 1$ and $f(x) = \sqrt{x}$ so that $f(g(x)) = \sqrt{4x+1}$. Then $\dfrac{d}{dx} f(g(x)) = f'(g(x))g'(x) = \dfrac{1}{2\sqrt{g(x)}} \cdot 4 = \dfrac{2}{\sqrt{4x+1}}$.

3.7.11 $h'(3) = f'(g(3))g'(3) = f'(4)g'(3) = 10 \cdot 5 = 50$. Note that the value of $f(4)$ isn't needed to calculate this.

3.7.13 $y' = e^{kx} \cdot k = ke^{kx}$.

3.7.15 With $u = 3x + 7$ and $y = u^{10}$ we have $\dfrac{dy}{dx} = \dfrac{dy}{dx} \cdot \dfrac{du}{dx} = 10u^9 \cdot 3 = 30(3x+7)^9$.

3.7.17 With $u = \sin x$ and $y = u^5$ we have $\dfrac{dy}{dx} = \dfrac{dy}{du} \cdot \dfrac{du}{dx} = 5u^4 \cdot \cos x = 5\sin^4 x \cos x$.

3.7.19 With $u = x^2 + 1$ and $y = \sqrt{u}$ we have $\dfrac{dy}{dx} = \dfrac{dy}{du} \cdot \dfrac{du}{dx} = \dfrac{1}{2\sqrt{u}} \cdot (2x) = \dfrac{x}{\sqrt{x^2+1}}$.

3.7.21 With $u = 4x^2 + 1$ and $y = e^u$, we have $\dfrac{dy}{dx} = \dfrac{dy}{du} \dfrac{du}{dx} = e^u \cdot 8x = 8xe^{4x^2+1}$.

3.7.23 With $u = 5x^2$ and $y = \tan u$ we have $\dfrac{dy}{dx} = \dfrac{dy}{du} \cdot \dfrac{du}{dx} = \sec^2 u \cdot (10x) = 10x \sec^2 5x^2$.

3.7.25

a. $h'(3) = f'(g(3))g'(3) = f'(1) \cdot 20 = 5 \cdot 20 = 100$.

b. $h'(2) = f'(g(2))g'(2) = f'(5) \cdot 10 = -10 \cdot 10 = -100$.

c. $p'(4) = g'(f(4))f'(4) = g'(1) \cdot (-8) = 2 \cdot (-8) = -16$.

d. $p'(2) = g'(f(2))f'(2) = g'(3) \cdot 2 = 20 \cdot 2 = 40$.

e. $h'(5) = f'(g(5))g'(5) = f'(2) \cdot 20 = 2 \cdot 20 = 40$.

3.7.27 With $g(x) = 3x^2 + 7x$ and $f(u) = u^{10}$ we have $\dfrac{d}{dx}[f(g(x))] = f'(g(x))g'(x) = 10(3x^2 + 7x)^9(6x + 7)$.

3.7.29 With $g(x) = 10x + 1$ and $f(u) = \sqrt{u}$, we have
$$\frac{d}{dx}[f(g(x))] = f'(g(x))g'(x) = \frac{1}{2\sqrt{10x+1}} \cdot 10 = \frac{5}{\sqrt{10x+1}}.$$

3.7. The Chain Rule

3.7.31 With $g(x) = 7x^3 + 1$ and $f(u) = 5u^{-3}$ we have
$$\frac{d}{dx}[f(g(x))] = f'(g(x))g'(x) = -15(7x^3 + 1)^{-4}(21x^2) = -315(7x^3 + 1)^{-4} \cdot x^2.$$

3.7.33 With $g(x) = 3x + 1$ and $f(u) = \sec u$, we have
$$\frac{d}{dx}[f(g(x))] = f'(g(x))g'(x) = \sec(3x+1)\tan(3x+1) \cdot 3 = 3\sec(3x+1)\tan(3x+1).$$

3.7.35 With $g(x) = e^x$ and $f(u) = \tan u$ we have $\frac{d}{dx}[f(g(x))] = f'(g(x))g'(x) = \sec^2 u \cdot e^x = e^x \sec^2 e^x$.

3.7.37 With $g(x) = 4x^3 + 3x + 1$ and $f(u) = \sin u$ we have
$$\frac{d}{dx}[f(g(x))] = f'(g(x))g'(x) = \cos u \cdot (12x^2 + 3) = (12x^2 + 3) \cdot \cos(4x^3 + 3x + 1).$$

3.7.39 With $g(x) = 5x + 1$ and $f(u) = x^{2/3}$ we have
$$\frac{d}{dx}(f(g(x)) = f'(g(x)g'(x) = \frac{2}{3}(5x + 1)^{-\frac{1}{3}} \cdot 5 = \frac{10}{3(5x+1)^{\frac{1}{3}}}.$$

3.7.41 With $u = \dfrac{2x}{4x-3}$ and $f(u) = u^{1/4}$ we have
$$\frac{dy}{dx} = \frac{1}{4}\left(\frac{2x}{4x-3}\right)^{-\frac{3}{4}} \cdot \left(\frac{2(4x-3) - 2x \cdot 4}{(4x-3)^2}\right) = -\frac{3}{2}\left(\frac{4x-3}{2x}\right)^{\frac{3}{4}} \cdot \frac{1}{(4x-3)^2} = -\frac{3}{2^{7/4}x^{3/4}(4x-3)^{5/4}}.$$

3.7.43 With $g(x) = \sec x + \tan x$ and $f(u) = u^5$ we have $\frac{d}{dx}[f(g(x))] = f'(g(x))g'(x) = 5u^4 \cdot (\sec x \tan x + \sec^2 x) = 5(\sec x + \tan x)^4(\sec x \tan x + \sec^2 x) = 5\sec x(\sec x + \tan x)^5$.

3.7.45 Take $g(x) = 2x^6 - 3x^3 + 3$, and $n = 25$. Then $y' = n(g(x))^{n-1}g'(x) = 25(2x^6 - 3x^3 + 3)^{24}(12x^5 - 9x^2)$.

3.7.47 Take $g(x) = 1 + 2\tan u$, and $n = 4.5$. Then
$$y' = n(g(x))^{n-1}g'(x) = 4.5(1 + 2\tan u)^{3.5}(2\sec^2 u) = 9(1 + 2\tan u)^{3.5}\sec^2 u.$$

3.7.49
$$\frac{d}{dx}\sqrt{1 + \cot^2 x} = \frac{1}{2\sqrt{1 + \cot^2 x}} \cdot \frac{d}{dx}(1 + \cot^2 x) = \frac{1}{2\sqrt{1 + \cot^2 x}} \cdot 2\cot x \cdot \frac{d}{dx}\cot x$$
$$= \frac{1}{2\sqrt{1 + \cot^2 x}} \cdot 2\cot x \cdot (-\csc^2 x) = -\frac{\cot x \csc^2 x}{\sqrt{1 + \cot^2 x}}.$$

3.7.51 Note that $\frac{d}{dx}e^{-x} = -e^{-x}$. Then we have $y' = \dfrac{2e^x - 3e^{-x}}{3}$.

3.7.53
$$\frac{d}{dx}\sin(\sin(e^x)) = \cos(\sin(e^x))\frac{d}{dx}\sin(e^x)$$
$$= \cos(\sin(e^x)) \cdot \cos(e^x) \cdot e^x$$

3.7.55
$$\frac{d}{dx}\sin^5(\cos 3x) = 5\sin^4(\cos 3x) \cdot \frac{d}{dx}(\sin(\cos 3x))$$
$$= 5\sin^4(\cos 3x) \cdot \cos(\cos 3x) \cdot \frac{d}{dx}\cos 3x$$
$$= 5\sin^4(\cos 3x) \cdot \cos(\cos 3x) \cdot (-\sin 3x) \cdot 3$$
$$= -15\sin^4(\cos 3x)\cos(\cos 3x)\sin 3x.$$

3.7.57
$$\frac{d}{dt}\left(\frac{e^{2t}}{1+e^{2t}}\right) = \frac{(1+e^{2t})2e^{2t} - e^{2t}\cdot 2e^{2t}}{(1+e^{2t})^2}$$
$$= \frac{2e^{2t}(1+e^{2t}-e^{2t})}{(1+e^{2t})^2}$$
$$= \frac{2e^{2t}}{(1+e^{2t})^2}.$$

3.7.59 $\dfrac{d}{dx}\sqrt{x+\sqrt{x}} = \dfrac{1}{2\sqrt{x+\sqrt{x}}} \cdot \dfrac{d}{dx}(x+\sqrt{x}) = \dfrac{1}{2\sqrt{x+\sqrt{x}}} \cdot \left(1+\dfrac{1}{2\sqrt{x}}\right).$

3.7.61 $\dfrac{d}{dx}f(g(x^2)) = f'(g(x^2)) \cdot \dfrac{d}{dx}(g(x^2)) = f'(g(x^2)) \cdot g'(x^2) \cdot 2x.$

3.7.63 $y' = 5\left(\dfrac{x}{x+1}\right)^4 \cdot \dfrac{(x+1)(1) - x(1)}{(x+1)^2} = \dfrac{5x^4}{(x+1)^6}.$

3.7.65 $y' = e^{x^2+1}(2x)\sin x^3 + e^{x^2+1}(\cos x^3)3x^2 = xe^{x^2+1}(2\sin x^3 + 3x\cos x^3).$

3.7.67 $\dfrac{dy}{d\theta} = 2\theta\sec 5\theta + \theta^2(5\sec 5\theta\tan 5\theta) = \theta\sec 5\theta(2 + 5\theta\tan 5\theta).$

3.7.69 $y' = 4((x+2)(x^2+1))^3 \cdot ((1)(x^2+1) + (x+2)(2x)) = 4((x+2)(x^2+1))^3(3x^2+4x+1) = 4(x+2)^2(x^2+1)^3(3x+1)(x+1).$

3.7.71 $y' = \dfrac{1}{5}(x^4+\cos 2x)^{-4/5}(4x^3 - 2\sin 2x) = \dfrac{4x^3 - 2\sin 2x}{5(x^4+\cos 2x)^{4/5}}.$

3.7.73
$$y' = 2(p+3)^1\sin p^2 + (p+3)^2(\cos p^2)(2p) = (p+3)(2\sin p^2 + 2p^2\cos p^2 + 6p\cos p^2)$$
$$= 2(p+3)(\sin p^2 + p^2\cos p^2 + 3p\cos p^2).$$

3.7.75 $\dfrac{d}{dx}\sqrt{f(x)} = \dfrac{1}{2\sqrt{f(x)}} \cdot f'(x).$

3.7.77

a. True. The product rule alone will suffice.

b. True. This function is the composition of e^x with $\sqrt{x+1}$.

c. True. The derivative of the composition $f(g(x))$ is the product of $f'(g(x))$ with $g'(x)$, so it is the product of two derivatives.

d. False. In fact, $\dfrac{d}{dx}P(Q(x)) = P'(Q(x))Q'(x).$

3.7.79 Note that $a(70) = 13330$.

$$\dfrac{d}{dt}p(a(t))|_{t=70} = p'(a(70))a'(70) \approx \dfrac{738-765}{14330-13330} \cdot \dfrac{13440-13330}{80-70} = -0.297\,\text{hPa per minute}.$$

3.7.81 $m'(t) = 64e^{0.004t} \cdot (0.004) = 0.256e^{0.004t}$. $m'(65) = 0.256e^{0.26} \approx 0.33$. 65 days after the diet switch, the mass of the tortoise is increasing at about $1/3$ of a gram per day.

3.7. The Chain Rule

3.7.83

a. After 10 years we will have $A(10) = 200e^{0.0398} \approx \$297.77$.

b. The growth rate is $A'(t) = 200 \cdot (0.0398e^{0.0398t}) = 7.96e^{0.0398t}$. After 10 years, the growth rate is $A'(10) = 7.96e^{0.0398} \approx 11.85\,\text{dollars/year}$.

c. The tangent line is given by $y - 297.77 = 11.85(t - 10)$, or $y = 11.85t + 179.27$.

3.7.85

a. The slope is $f'(x) = e^{2x} + 2xe^{2x}$. This is zero when $e^{2x}(1 + 2x) = 0$, which occurs when $x = -\frac{1}{2}$.

b. The graph of f has a horizontal tangent line at $x = -1/2$.

3.7.87

$$\frac{d^2}{dx^2}\sin x^2 = \frac{d}{dx}(2x\cos x^2) = 2(\cos x^2 - 2x^2\sin x^2)$$
$$= 2\cos x^2 - 4x^2\sin x^2.$$

Note that in the middle of this calculation we used a result from the middle of the previous problem – namely the derivative of $x\cos x^2$.

3.7.89 $\dfrac{d^2}{dx^2}e^{-2x^2} = \dfrac{d}{dx}\left(-4xe^{-2x^2}\right) = -4e^{-2x^2} + 16x^2e^{-2x^2} = 4e^{-2x^2}(4x^2 - 1)$.

3.7.91 $y' = \dfrac{(x^3 - 6x - 1)(2)(x^2 - 1)(2x) - (x^2 - 1)^2(3x^2 - 6)}{(x^3 - 6x - 1)^2}$,

so $y'(0) = \dfrac{(-1)(2)(-1)(0) - (-1)^2(-6)}{(-1)^2} = 6$. The equation of the tangent line is thus $y + 1 = 6(x - 0)$, or $y = 6x - 1$.

3.7.93

a. $g'(4) = 3$, $g(4) = 3 \cdot 4 - 5 = 7$. $f'(7) = -2$, $f(7) = -2 \cdot 7 + 23 = 9$. Thus, $h(4) = f(g(4)) = f(7) = 9$, and $h'(4) = f'(g(4))g'(4) = f'(7) \cdot 3 = -2 \cdot 3 = -6$.

b. The tangent line to h at $(4, 9)$ is given by $y - 9 = -6(x - 4)$, or $y = -6x + 33$.

3.7.95 $y'(x) = 2e^{2x}$, so $y'\left(\frac{\ln 3}{2}\right) = 2e^{\ln 3} = 6$. Also, $y\left(\frac{\ln 3}{2}\right) = e^{\ln 3} = 3$. The tangent line is therefore given by $y - 3 = 6\left(x - \frac{\ln 3}{2}\right)$, or $y = 6x + 3 - 3\ln 3$.

3.7.97 First, note that $g'(x) = \cos(\pi f(x)) \cdot \pi f'(x)$.

a. $g'(0) = \cos(\pi \cdot f(0)) \cdot \pi f'(0) = \cos(-3\pi) \cdot 3\pi = -3\pi$.

b. $g'(1) = \cos(\pi \cdot f(1)) \cdot \pi f'(1) = \cos(3\pi) \cdot 5\pi = -5\pi$.

3.7.99

a. $\dfrac{d^2y}{dt^2} = \dfrac{d}{dt}\left(-y_0\sqrt{\dfrac{k}{m}}\sin\left(t\sqrt{\dfrac{k}{m}}\right)\right) = -y_0 \cdot \dfrac{k}{m} \cdot \cos\left(t\sqrt{\dfrac{k}{m}}\right)$.

b. $-\dfrac{k}{m}y = -\dfrac{k}{m}\left(y_0\cos\left(t\sqrt{\dfrac{k}{m}}\right)\right) = \dfrac{d^2y}{dt^2}$.

Copyright © 2019 Pearson Education, Inc.

3.7.101

a.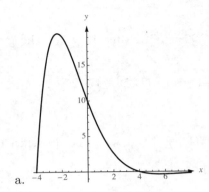

b.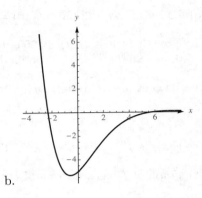

$$\frac{dy}{dt} = -5e^{-t/2}\cos\left(\frac{\pi t}{8}\right) - \frac{5\pi}{4}e^{-t/2}\sin\left(\frac{\pi t}{8}\right).$$

c. The velocity is zero at about -2.3 and at about 5.7, and the displacement has a maximum and a minimum at these points.

3.7.103

a. Assuming a non leap year, March 1st corresponds to $t = 59$. We have $D(59) = 12 - 3\cos\left(\frac{2\pi(69)}{365}\right) \approx 10.88$ hours.

b. $\dfrac{d}{dt}D(t) = 3 \cdot \dfrac{2\pi}{365}\sin\left(\dfrac{2\pi(t+10)}{365}\right)$ hours per day.

c. $D'(59) \approx 0.048$ hours per day ≈ 2 minutes and 52 seconds per day. This means that on March 1st, the days are getting longer by just shy of 3 minutes per day.

d.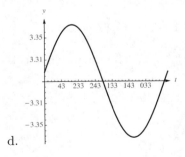

e. The largest increase in the length of the days appears to be at about $t = 81$, and the largest decrease at about $t = 265$. These correspond to March 22nd and to September 22nd. The least rapid changes occur at about $t = 172$ and $t = 355$. These correspond to June 21st and December 21st.

3.7.105

a. $E'(t) = 400 + 200\cos\left(\dfrac{\pi t}{12}\right)$ MW.

b. Because the maximum value of $\cos\theta$ is 1, the maximum value of $E'(t)$ will be 600 MW, where $\cos\left(\frac{\pi t}{12}\right) = 1$, which is where $t = 0$, which corresponds to noon.

3.7. The Chain Rule

c. Because the minimum value of $\cos\theta$ is -1, the minimum value of $E'(t)$ will be 200 MW, where $\cos\left(\frac{\pi t}{12}\right) = -1$, which is where $\frac{\pi t}{12} = \pi$, or $t = 12$, which corresponds to midnight.

d.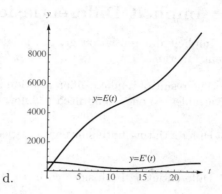

3.7.107

$$\frac{d}{dx}\left(f(x)(g(x))^{-1}\right) = f'(x)(g(x))^{-1} + f(x)(-(g(x))^{-2}g'(x)) = \frac{f'(x)}{g(x)} - \frac{f(x)g'(x)}{(g(x))^2}$$
$$= \frac{g(x)f'(x) - f(x)g'(x)}{(g(x))^2}.$$

3.7.109

a. $h(x) = (x^2 - 3)^5$, $a = 2$.

b. $h'(x) = 5(x^2 - 3)^4(2x) = 10x(x^2 - 3)^4$, so the value of this limit is $h'(2) = 20$.

3.7.111

a. $h(x) = \sin x^2$, $a = \frac{\pi}{2}$.

b. $h'(x) = (\cos x^2)(2x)$, so the value of this limit is $h'\left(\frac{\pi}{2}\right) = \pi \cdot \cos\left(\frac{\pi^2}{4}\right) \approx -2.45$.

3.7.113 $\lim\limits_{x\to 5}\dfrac{f(x^2) - f(25)}{x - 5} = \dfrac{d}{dx}\left[f(x^2)\right]_{x=5} = 2\cdot 5\cdot f'(25) = 10f'(25)$.

3.7.115

a. $\lim\limits_{v\to u} H(v) = \lim\limits_{v\to u}\left(\dfrac{f(v) - f(u)}{v - u} - f'(u)\right) = \lim\limits_{v\to u}\left(\dfrac{f(v) - f(u)}{v - u}\right) - f'(u) = f'(u) - f'(u) = 0$.

b. Suppose $u = v$. Then clearly both sides of the given expression are 0, so they are equal. Suppose $u \ne v$. Then $H(v) = \dfrac{f(v) - f(u)}{v - u} - f'(u)$, so $H(v) + f'(u) = \dfrac{f(v) - f(u)}{v - u}$, so the result holds by multiplying both sides of this equation by $v - u$.

c.
$$h'(a) = \lim_{x\to a}\frac{f(g(x)) - f(g(a))}{x - a} = \lim_{x\to a}\frac{H(g(x)) + f'(g(a))}{x - a}\cdot(g(x) - g(a))$$
$$= \lim_{x\to a}\left[(H(g(x)) + f'(g(a)))\cdot\frac{g(x) - g(a)}{x - a}\right].$$

d. $h'(a) = \lim\limits_{x\to a}\left[(H(g(x)) + f'(g(a)))\cdot\dfrac{g(x) - g(a)}{x - a}\right] = (0 + f'(g(a)))\cdot g'(a) = f'(g(a))g'(a)$.

3.8 Implicit Differentiation

3.8.1 Implicit differentiation gives a single unified derivative, whereas solving for y explicitly yields two different functions.

3.8.3 The result of implicit differentiation is often an expression involving both the dependent and independent variables, so one would need to know both in order to calculate the value of the derivative.

3.8.5 Differentiating both sides with respect to x gives $1 = 2y\frac{dy}{dx}$, so $\frac{dy}{dx} = \frac{1}{2y}$.

3.8.7 Differentiating both sides with respect to x gives $\cos y \cdot \frac{dy}{dx} + 0 = 1$, so $\frac{dy}{dx} = \frac{1}{\cos y}$.

3.8.9

    a. When $x = 0$ we have $-y + y^3 = 0$, so $y(y^2 - 1) = 0$, so y can be either 0 or ± 1. So the y-intercepts are $(0,0)$, $(0,1)$, and $(0,-1)$.

    b. Differentiating both sides with respect to x gives $2 - \frac{dy}{dx} + 3y^2\frac{dy}{dx} = 0$. Thus $2 = \frac{dy}{dx} - 3y^2\frac{dy}{dx}$, so $2 = \frac{dy}{dx}(1 - 3y^2)$ and $\frac{dy}{dx} = \frac{2}{1 - 3y^2}$.

    c. At $(0,0)$ we have $\frac{dy}{dx} = 2$, at $(0,1)$ and at $(0,-1)$ we have $\frac{dy}{dx} = \frac{2}{1-3} = -1$.

3.8.11 Differentiating both sides with respect to x gives $1 = 3y^2\frac{dy}{dx}$. Thus $\frac{dy}{dx} = \frac{1}{3y^2} = \frac{1}{3}y^{-2}$. Differentiating both sides again gives $\frac{d^2y}{dx^2} = -\frac{2}{3}y^{-3}\frac{dy}{dx} = -\frac{2}{3}y^{-3} \cdot \frac{1}{3}y^{-2} = -\frac{2}{9y^5}$.

3.8.13

    a. $4x^3 + 4y^3\frac{dy}{dx} = 0$. Thus $4y^3\frac{dy}{dx} = -4x^3$, so $\frac{dy}{dx} = -\frac{x^3}{y^3}$.

    b. When $x = 1$ and $y = -1$, we have $\frac{dy}{dx} = \frac{-1}{-1} = 1$.

3.8.15

    a. $2y\frac{dy}{dx} = 4$, so $\frac{dy}{dx} = \frac{2}{y}$.

    b. $\left.\frac{dy}{dx}\right|_{(1,2)} = \frac{2}{2} = 1$.

3.8.17

    a. $\frac{dy}{dx}\cos y = 20x^3$, so $\frac{dy}{dx} = \frac{20x^3}{\cos y}$.

    b. $\left.\frac{dy}{dx}\right|_{(1,\pi)} = \frac{20}{\cos \pi} = -20$.

3.8.19

    a. $-\frac{dy}{dx}\sin y = 1$, so $\frac{dy}{dx} = -\frac{1}{\sin y} = -\csc y$.

    b. $\left.\frac{dy}{dx}\right|_{(0,\pi/2)} = -\csc(\pi/2) = -1$.

3.8.21

    a. $1 \cdot y + x\frac{dy}{dx} = 0$, so $\frac{dy}{dx} = -\frac{y}{x}$.

    b. $\left.\frac{dy}{dx}\right|_{(1,7)} = -\frac{7}{1} = -7$.

3.8. Implicit Differentiation

3.8.23

a. $\frac{1}{3}x^{-\frac{2}{3}} + \frac{4}{3}y^{\frac{1}{3}}\frac{dy}{dx} = 0$, so $\frac{4}{3}y^{1/3}\frac{dy}{dx} = -\frac{1}{3x^{2/3}}$, so $\frac{dy}{dx} = -\frac{1}{4x^{2/3}y^{1/3}}$.

b. $\left.\frac{dy}{dx}\right|_{(1,1)} = -\frac{1}{4}$.

3.8.25

a. $y^{1/3} + \frac{1}{3}xy^{-2/3}\frac{dy}{dx} + \frac{dy}{dx} = 0$, so $\frac{1}{3}xy^{-2/3}\frac{dy}{dx} + \frac{dy}{dx} = -y^{1/3}$ and therefore $\frac{dy}{dx}\left(\frac{1}{3}xy^{-2/3} + 1\right) = -y^{1/3}$, so $\frac{dy}{dx} = -\frac{y^{1/3}}{\frac{1}{3}xy^{-2/3} + 1} = -\frac{3y}{x + 3y^{2/3}}$.

b. $\left.\frac{dy}{dx}\right|_{(1,8)} = -\frac{24}{13}$.

3.8.27 $\cos x + \cos y \frac{dy}{dx} = \frac{dy}{dx}$, so $\cos x = \frac{dy}{dx} - \cos y \frac{dy}{dx}$, and therefore $\cos x = \frac{dy}{dx}(1 - \cos y)$, and thus $\frac{dy}{dx} = \frac{\cos x}{1 - \cos y}$.

3.8.29 $1 + \frac{dy}{dx} = -\sin y \cdot \frac{dy}{dx}$, so $\frac{dy}{dx} + (\sin y)\frac{dy}{dx} = -1$, and $\frac{dy}{dx} = -\frac{1}{1 + \sin y}$.

3.8.31 $\left(y + x\frac{dy}{dx}\right)\cos(xy) = 1 + \frac{dy}{dx}$, so $y\cos(xy) + x\frac{dy}{dx}\cos(xy) = 1 + \frac{dy}{dx}$. If we rearrange terms in order to have the terms with a factor of $\frac{dy}{dx}$ all on the same side, we obtain $y\cos(xy) - 1 = \frac{dy}{dx} - x\frac{dy}{dx}\cos(xy)$. Factoring out the $\frac{dy}{dx}$ factor gives $y\cos(xy) - 1 = \frac{dy}{dx}(1 - x\cos(xy))$, so $\frac{dy}{dx} = \frac{y\cos(xy) - 1}{1 - x\cos(xy)}$.

3.8.33 $-2y\frac{dy}{dx}\sin y^2 + 1 = \frac{dy}{dx}e^y$, which we can write as $1 = \frac{dy}{dx}e^y + 2y\frac{dy}{dx}\sin y^2$, or $1 = \frac{dy}{dx}(e^y + 2y\sin y^2)$. Thus, $\frac{dy}{dx} = \frac{1}{e^y + 2y\sin y^2}$.

3.8.35

$$3x^2 = \frac{(x-y)(1 + \frac{dy}{dx}) - (x+y)(1 - \frac{dy}{dx})}{(x-y)^2}$$

$$3x^2(x-y)^2 = x + x\frac{dy}{dx} - y - y\frac{dy}{dx} - x + x\frac{dy}{dx} - y + y\frac{dy}{dx}$$

$$3x^2(x-y)^2 + 2y = 2x\frac{dy}{dx}$$

$$\frac{dy}{dx} = \frac{3x^2(x-y)^2 + 2y}{2x}$$

3.8.37

$$18x^2 + 21\frac{dy}{dx}y^2 = 13\left(y + x\frac{dy}{dx}\right)$$

$$21\frac{dy}{dx}y^2 - 13x\frac{dy}{dx} = 13y - 18x^2$$

$$\frac{dy}{dx} = \frac{13y - 18x^2}{21y^2 - 13x}.$$

3.8.39

$$\frac{4x^3 + 2y\frac{dy}{dx}}{2\sqrt{x^4+y^2}} = 5 + 6y^2\frac{dy}{dx}$$

$$y\frac{dy}{dx} - 6\frac{dy}{dx}y^2\sqrt{x^4+y^2} = 5\sqrt{x^4+y^2} - 2x^3$$

$$\frac{dy}{dx} = \frac{5\sqrt{x^4+y^2} - 2x^3}{y - 6y^2\sqrt{x^4+y^2}}.$$

3.8.41

a. $1280 = 40L^{1/3}K^{2/3}$, so $0 = \frac{40}{3}L^{-2/3}K^{2/3} + \frac{80}{3}L^{1/3}K^{-1/3}\cdot\frac{dK}{dL}$. Multiplying both sides by $\frac{3}{40}L^{2/3}K^{1/3}$ yields

$0 = K + 2L\frac{dK}{dL}$, so $\frac{dK}{dL} = -\frac{1}{2}\frac{K}{L}$.

b. With $L = 8$ and $K = 64$, $\frac{dK}{dL} = -\frac{64}{16} = -4$.

3.8.43

a. $V = \frac{\pi h^2(3r-h)}{3} = \frac{5\pi}{3}$. So

$$\frac{1}{3}[2\pi h(3r-h) + \pi h^2(3r'-1)] = 0,$$

$$6rh - 2h^2 + 3h^2r' - h^2 = 0,$$

so $r' = 1 - \frac{2r}{h}$.

b. At $r = 2$ and $h = 1$, we have $r' = 1 - 4 = -3$.

3.8.45

a. $\sin 0 + 5\cdot 0 = 0^2$, so the point $(0,0)$ does lie on the curve.

b. $y'\cos y + 5 = 2yy'$, so $5 = y'(2y - \cos y)$, so $y' = \frac{5}{2y - \cos y}$. At the given point we have $y' = -5$. The equation of the tangent line is therefore $y - 0 = -5(x - 0)$, or $y = -5x$.

3.8.47

a. $2^2 + 2\cdot 1 + 1^2 = 7$, so the point $(2,1)$ does lie on the curve.

b. $2x + y + xy' + 2yy' = 0$, which can be written $(x+2y)y' = -2x - y$. Solving for y' yields $y' = \frac{-2x - y}{x + 2y}$. Thus, at the point $(2,1)$ we have $y' = -\frac{5}{4}$. The equation of the tangent line is therefore $y - 1 = -\frac{5}{4}(x-2)$, or $y = -\frac{5}{4}x + \frac{7}{2}$.

3.8.49

a. $\cos(\pi/2 - \pi/4) + \sin(\pi/4) = (\sqrt{2}/2) + (\sqrt{2}/2) = \sqrt{2}$, so the point $(\pi/2, \pi/4)$ does lie on the curve.

b. $(1 - y')(-\sin(x-y)) + y'\cos y = 0$, which can be written as $y'(\cos y + \sin(x-y)) = \sin(x-y)$, so $y' = \frac{\sin(x-y)}{\cos y + \sin(x-y)}$. At the given point we have $y' = 1/2$. The equation of the tangent line is therefore $y - (\pi/4) = (1/2)(x - \pi/2)$, or $y = \frac{1}{2}x$.

3.8.51 $1 + 2yy' = 0$, so $y' = -\dfrac{1}{2y}$. Differentiating again, we obtain

$$y'' = -\dfrac{1}{2} \cdot \dfrac{-y'}{y^2} = \dfrac{y'}{2y^2} = \left(-\dfrac{1}{2y}\right) \cdot \dfrac{1}{2y^2} = -\dfrac{1}{4y^3}.$$

3.8.53 $1 + \dfrac{dy}{dx} = (\cos y)\dfrac{dy}{dx}$, so $1 = \dfrac{dy}{dx}(\cos y - 1)$, and thus $\dfrac{dy}{dx} = \dfrac{1}{\cos y - 1}$. Thus

$$\dfrac{d^2y}{dx^2} = -\dfrac{1}{(\cos y - 1)^2} \cdot \left(-\sin y \dfrac{dy}{dx}\right) = \dfrac{\sin y}{(\cos y - 1)^2} \cdot \dfrac{1}{(\cos y - 1)} = \dfrac{\sin y}{(\cos y - 1)^3}.$$

3.8.55 $2y'e^{2y} + 1 = y'$, so $y' = \dfrac{1}{1 - 2e^{2y}}$. Differentiating again, we obtain

$$y'' = -\left(1 - 2e^{2y}\right)^{-2}\left(-4e^{2y}y'\right) = \dfrac{4e^{2y}}{(1 - 2e^{2y})^3}.$$

3.8.57

a. False. For example, the equation $y\cos(xy) = x$, cannot be solved explicitly for y in terms of x.

b. True. We have $2x + 2yy' = 0$, and the result follows by solving for y'.

c. False. The equation $x = 1$ doesn't represent any sort of function – it is either just a number, or perhaps a vertical line, but it doesn't represent a differentiable function.

d. False. $y + xy' = 0$, so $y' = -\dfrac{y}{x}$, $x \neq 0$.

3.8.59

a. $y + xy' + \dfrac{3}{2}x^{1/2}y^{-1/2} - \dfrac{1}{2}x^{3/2}y^{-3/2}y' = 0$. Multiplying through by $2y^{3/2}$ gives

$$2y^{5/2} + 2xy^{3/2}y' + 3x^{1/2}y - x^{3/2}y' = 0.$$

Then

$$2xy^{3/2}y' - x^{3/2}y' = -2y^{5/2} - 3x^{1/2}y$$
$$y'(2xy^{3/2} - x^{3/2}) = -(2y^{5/2} + 3x^{1/2}y)$$
$$y' = -\dfrac{2y^{5/2} + 3x^{1/2}y}{2xy^{3/2} - x^{3/2}}$$
$$= \dfrac{y(2y^{3/2} + 3\sqrt{x})}{x(\sqrt{x} - 2y^{3/2})}.$$

b. At $(1,1)$, $y' = -5$.

3.8.61

a. There are two points on the curve associated with $x = 1$. When $x = 1$, we have $1 + y^2 - y = 1$, so $y(y-1) = 0$. The two points are thus $(1, 0)$ and $(1, 1)$. Differentiating yields $1 + 2yy' - y' = 0$, so $y' = \dfrac{1}{1 - 2y}$.
At $(1, 0)$, we have $y' = 1$, so the tangent line is given by $y = x - 1$.
At $(1, 1)$, we have $y' = -1$, so the tangent line is given by $y - 1 = -1(x - 1)$, or $y = -x + 2$.

b.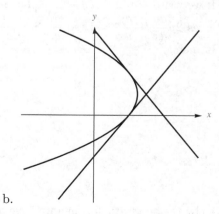

3.8.63

a. $y(2x) + (x^2 + 4)y' = 0$, so $y' = -\dfrac{2xy}{x^2 + 4}$.

b. At $y = 1$ we have $x^2 + 4 = 8$, so $x = \pm 2$. At the point $(2, 1)$ we have $y' = -\dfrac{4}{8} = -\dfrac{1}{2}$. At the point $(-2, 1)$ we have $y' = \dfrac{4}{8} = \dfrac{1}{2}$. Thus, the equations of the tangent lines are given by $y - 1 = -\dfrac{1}{2}(x - 2)$ and $y - 1 = \dfrac{1}{2}(x + 2)$, or $y = -\dfrac{1}{2}x + 2$ and $y = \frac{1}{2}x + 2$.

c. $y = \dfrac{8}{x^2 + 4}$, so $y' = \dfrac{0 - 8 \cdot 2x}{(x^2 + 4)^2} = -\dfrac{16x}{(x^2 + 4)^2}$.

d. $y' = -\dfrac{16x}{(x^2 + 4)^2} = -\dfrac{2x}{x^2 + 4} \cdot \dfrac{8}{x^2 + 4} = -\dfrac{2x}{x^2 + 4} \cdot y = -\dfrac{2xy}{x^2 + 4}$.

3.8.65

a. From exercise 61, we have that $y' = \dfrac{1}{1 - 2y}$. A vertical tangent would occur at a point whose y value would make $1 - 2y$ equal to zero. So we are looking for where $2y = 1$ or $y = \dfrac{1}{2}$.

If $y = \dfrac{1}{2}$, then $x + \dfrac{1}{4} - \dfrac{1}{2} = 1$, so $x = \dfrac{5}{4}$, and there is a vertical tangent at $\left(\dfrac{5}{4}, \dfrac{1}{2}\right)$.

b. Because y' is never zero, there are no horizontal tangent lines.

3.8.67 Differentiating with respect to x gives $18x + 2y\dfrac{dy}{dx} - 36 + 6\dfrac{dy}{dx} = 0$, so

$$2y\dfrac{dy}{dx} + 6\dfrac{dy}{dx} = -18x + 36$$
$$(2y + 6)\dfrac{dy}{dx} = -18x + 36$$
$$\dfrac{dy}{dx} = \dfrac{-9x + 18}{y + 3}.$$

This is zero when $x = 2$. Using the original equation, we have $36 + y^2 - 72 + 6y + 36 = 0$, or $y^2 + 6y = 0$, or $y = 0$ and $y = -6$. Thus there are horizontal tangent lines at $(2, 0)$ and $(2, -6)$. So $y = 0$ and $y = -6$ are horizontal tangent lines.

This curve has vertical tangent lines when $y = -3$. Using the original equation, we have $9x^2 + 9 - 36x - 18 + 36 = 0$ or $9x^2 - 36x + 27 = 0$, or $x^2 - 4x + 3 = 0$. This factors as $(x - 3)(x - 1) = 0$, so the corresponding values of x are 3 and 1. Thus the vertical tangent lines occur at $(3, -3)$ and $(1, -3)$.

3.8.69

a. If we write $y^3 - 1 = xy - x$, we have $(y - 1)(y^2 + y + 1) = x(y - 1)$, so $y^2 + y + 1 = x$. Differentiating gives $2y\dfrac{dy}{dx} + \dfrac{dy}{dx} = 1$, so $\dfrac{dy}{dx} = \dfrac{1}{1 + 2y}$. Note also that $y = 1$ satisfies the equation; $y' = 0$ on this branch.

b.
$$y^3 - 1 = x(y - 1)$$
$$(y - 1)(y^2 + y + 1) = x(y - 1)$$
$$y^2 + y + 1 = x$$
$$y^2 + y + (1 - x) = 0,$$

so by the quadratic formula we have $y = \dfrac{-1 \pm \sqrt{4x - 3}}{2}$. Note that this means that $\pm\sqrt{4x - 3} = 2y + 1$.

c.

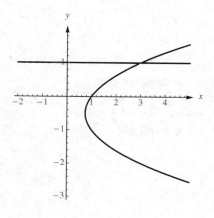

3.8.71

a. $4x^3 = 4x - 4yy'$, so $y' = \dfrac{x - x^3}{y}$.

b. $y = \pm\sqrt{x^2 - \dfrac{x^4}{2}}$.

c.

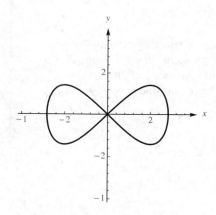

3.8.73

The slope of the normal line is the negative reciprocal of the slope of the tangent line. From 45: $y' = -5$, so the slope of the normal line is $\dfrac{1}{5}$. At the point (0,0), we have the normal line $y = \dfrac{1}{5}x$.

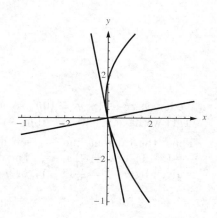

3.8.75

The slope of the normal line is the negative reciprocal of the slope of the tangent line. From 47: $y' = -\frac{5}{4}$, so the slope of the normal line is $\frac{4}{5}$. At the point $(2,1)$ we have the line $y - 1 = \frac{4}{5}(x-2)$, or $y = \frac{4}{5}x - \frac{3}{5}$.

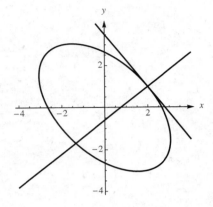

3.8.77

The slope of the normal line is the negative reciprocal of the slope of the tangent line. From 49: $y' = \frac{1}{2}$, so the slope of the normal line is -2. At the point $(\frac{\pi}{2}, \frac{\pi}{4})$ we have the line $y - \frac{\pi}{4} = -2(x - \frac{\pi}{2})$, or $y = -2x + \frac{5\pi}{4}$.

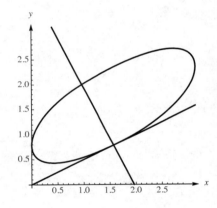

3.8.79

a. We have $9x^2 + 21y^2 y' = 10y'$, so at the point $(1,1)$ we have $9 + 21y' = 10y'$, so $y' = -\frac{9}{11}$.

Thus, the tangent line is given by $y - 1 = -\frac{9}{11}(x-1)$, or $y = \frac{-9}{11}x + \frac{20}{11}$. The normal line is given by $y - 1 = \frac{11}{9}(x-1)$, or $y = \frac{11}{9}x - \frac{2}{9}$.

b.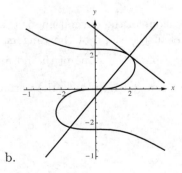

3.8. Implicit Differentiation

3.8.81

a. We have $2(x^2+y^2-2x)(2x+2yy'-2) = 4x + 4yy'$, so at the point $(2,2)$ we have $2(4+4-4)(4+4y'-2) = 8+8y'$, so $16+32y' = 8+8y'$, so $y' = -\frac{1}{3}$.

Thus, the tangent line is given by $y - 2 = -\frac{1}{3}(x-2)$, or $y = \frac{-1}{3}x + \frac{8}{3}$. The normal line is given by $y - 2 = 3(x-2)$, or $y = 3x - 4$.

b.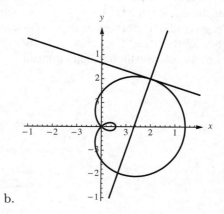

3.8.83 Note for $y = mx$, $\frac{dy}{dx} = m = \frac{y}{x}$, and for $x^2 + y^2 = a^2$, $\frac{dy}{dx} = -\frac{x}{y}$. So for any point (x, y), we have $\frac{y}{x}$ and $-\frac{x}{y}$ are negative reciprocals.

3.8.85 For $xy = a$ we have $xy' + y = 0$, so $y' = -\frac{y}{x}$. For $x^2 - y^2 = b$, we have $2x - 2yy' = 0$, so $y' = \frac{x}{y}$. Let (c, d) be a point on both curves. Then the slope of the normal line to the first curve is $\frac{c}{d}$, but that is the slope of the tangent line to the second curve. Thus the two curves are orthogonal at any points of intersection.

3.8.87

$$(2x + 2yy')(x^2 + y^2 + x) + (x^2 + y^2)(2x + 2yy' + 1) = 8y^2 + 16xyy'$$

$$2yy'(x^2 + y^2 + x) + (x^2 + y^2)2yy' - 16xyy' = 8y^2 - 2x(x^2 + y^2 + x) - (x^2 + y^2)(2x + 1)$$

$$y' = \frac{8y^2 - 2x(x^2 + y^2 + x) - (x^2 + y^2)(2x + 1)}{2y(x^2 + y^2 + x) + 2y(x^2 + y^2) - 16xy}$$

$$= \frac{8y^2 - 2x^3 - 2xy^2 - 2x^2 - 2x^3 - x^2 - 2xy^2 - y^2}{2y(x^2 + y^2 + x + x^2 + y^2 - 8x)}$$

$$= \frac{7y^2 - 3x^2 - 4xy^2 - 4x^3}{2y(2x^2 + 2y^2 - 7x)}.$$

3.8.89 $\frac{y'}{2\sqrt{y}} + y + xy' = 0$, so $y' + 2x\sqrt{y}y' = -2y\sqrt{y}$, so $y' = \frac{-2y\sqrt{y}}{2x\sqrt{y}+1} = -\frac{2y^{3/2}}{2x\sqrt{y}+1}$.
Differentiating again we obtain

$$y'' = -\frac{(2x\sqrt{y}+1)(3\sqrt{y}y') - 2y^{3/2}\left(2\sqrt{y} + \frac{xy'}{\sqrt{y}}\right)}{(2x\sqrt{y}+1)^2} = \frac{-(2x\sqrt{y}+1)(3\sqrt{y}y') + 4y^2 + 2xyy'}{(2x\sqrt{y}+1)^2}$$

$$= \left(\frac{-(2x\sqrt{y}+1)(3\sqrt{y})\left(\frac{-2y\sqrt{y}}{2x\sqrt{y}+1}\right) + 4y^2 + 2xy\left(\frac{-2y\sqrt{y}}{1+2x\sqrt{y}}\right)}{(2x\sqrt{y}+1)^2}\right) \cdot \frac{2x\sqrt{y}+1}{2x\sqrt{y}+1}$$

$$= \frac{(2x\sqrt{y}+1)(6y^2) + 4y^2(1 + 2x\sqrt{y}) - 4xy^{5/2}}{(2x\sqrt{y}+1)^3} = \frac{10y^2 + 16xy^2\sqrt{y}}{(2x\sqrt{y}+1)^3}.$$

3.8.91 Differentiating with respect to x yields $2x(3y^2 - 2y^3) + x^2(6y - 6y^2)\frac{dy}{dx} = 0$. Solving for $\frac{dy}{dx}$ yields $\frac{dy}{dx} = \frac{-2x(3y^2 - 2y^3)}{6x^2(y - y^2)} = \frac{3y^2 - 2y^3}{3x(y^2 - y)} = \frac{3y - 2y^2}{3x(y - 1)}$. The numerator is zero when $y = 0$ or when $y = \frac{3}{2}$. Using the original equation, there are no points where $y = 0$, and there are also no points where $y = \frac{3}{2}$ because we obtain the equation $x^2(\frac{27}{4} - \frac{27}{4}) = 4$, which has no solutions. So there are no horizontal tangent lines.

There could be vertical tangent lines where $x = 0$ or $y = 1$; in fact, when $y = 1$ the original equation becomes $x^2(3-2) = 4$, so $x = \pm 2$. There are vertical tangents at $(2, 1)$ and $(-2, 1)$. Letting $x = 0$ doesn't yield any points in the original equation.

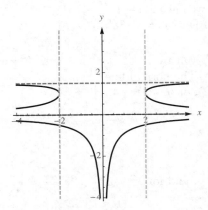

3.8.93 Differentiating with respect to x yields $(1 - y^2) + x(-2yy') + 3y^2 y' = 0$. Solving for y' yields $y' = \dfrac{y^2 - 1}{3y^2 - 2xy} = \dfrac{y^2 - 1}{y(3y - 2x)}$. There could be a horizontal tangent line where $y = \pm 1$, but the original equation with $y = \pm 1$ yields $\pm 1 = 0$, so there are no horizontal tangent lines.

There could be a vertical tangent line for $y = 0$ or for $y = \frac{2}{3}x$. Using the original equation, letting $y = 0$ yields $x = 0$, and letting $y = \frac{2}{3}x$ yields $x\left(1 - \dfrac{4x^2}{9}\right) + \dfrac{8x^3}{27} = 0$. For $x \neq 0$, this yields $27 - 12x^2 + 8x^2 = 0$, or $x^2 = \frac{27}{4}$, so $x = \pm\dfrac{\sqrt{27}}{2} = \pm\dfrac{3\sqrt{3}}{2}$. The corresponding y values are $y = \pm\dfrac{2}{3}x = \pm\sqrt{3}$. So there are vertical tangent lines at $(0, 0)$ and at $(3\sqrt{3}/2, \sqrt{3})$ and $(-3\sqrt{3}/2, -\sqrt{3})$.

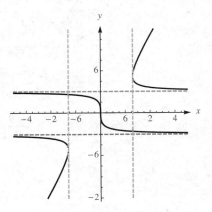

3.9 Derivatives of Logarithmic and Exponential Functions

3.9.1 $y = \ln x$ if and only if $x = e^y$. Differentiating implicitly yields $1 = e^y \cdot y'$, so $y' = \frac{1}{e^y} = \frac{1}{e^{\ln x}} = \frac{1}{x}$ for $x > 0$.

3.9.3 $\dfrac{d}{dx} \ln(kx) = \dfrac{1}{kx} \cdot k = \dfrac{1}{x}$. This is valid for $x > 0$ if $k > 0$ and $x < 0$ if $k < 0$. Also, we can write $\ln(kx) = \ln(k) + \ln(x)$, so its derivative is $0 + \dfrac{1}{x} = \dfrac{1}{x}$.

3.9.5 $\dfrac{d}{dx}\log_b x = \dfrac{1}{x\ln b}$ for $b>0$, $b\neq 1$ and $x>0$. If $b=e$, we have $\dfrac{d}{dx}\log_e x = \dfrac{1}{x\ln e} = \dfrac{1}{x}$.

3.9.7 $e^{x\ln(x^2+1)} = e^{\ln(x^2+1)^x} = (x^2+1)^x$.

3.9.9 $\dfrac{d}{dx}\left(\left(\ln\sqrt{x^2+1}\right)\right) = \dfrac{d}{dx}\left(\ln(x^2+1)^{1/2}\right) = \dfrac{d}{dx}\left(\dfrac{1}{2}\ln(x^2+1)\right) = \dfrac{1}{2}\dfrac{2x}{(x^2+1)} = \dfrac{x}{x^2+1}$.

3.9.11 $f(x) = e^{\ln(g(x)^{h(x)})} = e^{h(x)\cdot\ln(g(x))}$.

3.9.13 $\dfrac{d}{dx}(\ln(xe^x)) = \dfrac{d}{dx}(\ln x + \ln e^x) = \dfrac{d}{dx}(\ln x + x) = \dfrac{1}{x} + 1 = \dfrac{1+x}{x}$.

3.9.15 $\dfrac{d}{dx}(\ln(7x)) = \dfrac{d}{dx}(\ln 7 + \ln x) = 0 + \dfrac{1}{x} = \dfrac{1}{x}$.

3.9.17 $\dfrac{d}{dx}(\ln(x^2)) = \dfrac{d}{dx}(2\ln x) = \dfrac{2}{x}$.

3.9.19 $\dfrac{d}{dx}(\ln|\sin x|) = \dfrac{1}{\sin x}\cdot(\cos x) = \cot x$.

3.9.21 $\dfrac{d}{dx}\ln(x^4+1) = \dfrac{1}{x^4+1}\cdot 4x^3 = \dfrac{4x^3}{x^4+1}$.

3.9.23 $\dfrac{d}{dx}\left(\ln\left(\dfrac{x+1}{x-1}\right)\right) = \dfrac{(x-1)}{(x+1)}\left(\dfrac{(x-1)-(x+1)}{(x-1)^2}\right) = -\dfrac{2}{(x+1)(x-1)} = \dfrac{2}{1-x^2}$.

3.9.25 $\dfrac{d}{dx}((x^2+1)\ln x) = 2x\ln x + \dfrac{x^2+1}{x}$.

3.9.27 $\dfrac{d}{dx}(x^2(1-\ln x^2)) = 2x(1-\ln x^2) + x^2\left(\dfrac{-1}{x^2}\right)2x = 2x - 2x\ln x^2 - 2x = -2x\ln x^2$.

3.9.29 $\dfrac{d}{dx}\ln(\ln x) = \dfrac{1}{\ln x}\cdot\dfrac{1}{x}$.

3.9.31 $\dfrac{d}{dx}\left(\dfrac{\ln x}{\ln x + 1}\right) = \dfrac{(\ln x+1)(1/x) - (\ln x)(1/x)}{(\ln x+1)^2} = \dfrac{1}{x(\ln x+1)^2}$.

3.9.33 $y' = ex^{e-1}$.

3.9.35 $y' = \pi(2^x+1)^{\pi-1}2^x\ln 2$.

3.9.37 $y' = 8^x\ln 8$.

3.9.39 $y' = 5\cdot\dfrac{d}{dx}4^x = 5\cdot\ln 4\cdot 4^x$.

3.9.41 $y' = 2^{3+\sin x}(\ln 2)\cos x$

3.9.43 $y' = 3x^2 3^x + x^3 3^x\ln 3 = 3^x x^2(3+x\ln 3)$.

3.9.45 $\dfrac{dA}{dt} = 250(1.045)^{4t}\cdot\ln(1.045)\cdot 4 = 1000\cdot\ln 1.045\cdot(1.045^{4t})$.

3.9.47 $f'(x) = \dfrac{(2^x+1)2^x\ln 2 - 2^x(2^x\ln 2)}{(2^x+1)^2} = \dfrac{2^x\ln 2}{(2^x+1)^2}$.

3.9.49 Let $y = x^{\cos x}$. Then $\ln y = \cos x \ln x$. Differentiating both sides gives

$$\frac{1}{y} \cdot y' = (-\sin x) \ln x + \cos x \cdot \frac{1}{x}.$$

Therefore,

$$y' = x^{\cos x} \left(\frac{\cos x}{x} - \sin x \ln x \right).$$

At $\pi/2$ we have $y'(\pi/2) = (\pi/2)^0 (0/(\pi/2) - \ln(\pi/2)) = -\ln(\pi/2)$.

3.9.51 Let $y = x^{\sqrt{x}}$. Then $\ln y = \sqrt{x} \cdot \ln x$. Differentiating both sides gives

$$\frac{1}{y} \cdot y' = \frac{1}{2\sqrt{x}} \ln x + \frac{\sqrt{x}}{x}.$$

Therefore,

$$y' = x^{\sqrt{x}} \left(\frac{\ln x + 2}{2\sqrt{x}} \right).$$

At 4 we have $y'(4) = 4^2 \left(\frac{\ln 4 + 2}{4} \right) = 4 \ln 4 + 8$.

3.9.53 Because $f(x) = (\sin x)^{\ln x}$, we have $\ln f(x) = \ln x \ln \sin x$. Differentiating both sides gives

$$\frac{1}{f(x)} f'(x) = \frac{1}{x} \cdot \ln \sin x + \ln x \cdot \frac{1}{\sin x} \cos x.$$

Therefore,

$$f'(x) = (\sin x)^{\ln x} \left(\frac{\ln \sin x + x \ln x \cot x}{x} \right).$$

We have $f'(\pi/2) = 0$ because $\cot \pi/2 = 0$ and $\ln \sin(\pi/2) = \ln 1 = 0$.

3.9.55 Because $f(x) = (4 \sin x + 2)^{\cos x}$, we have $\ln f(x) = \cos x \ln(4 \sin x + 2)$. Differentiating both sides gives

$$\frac{1}{f(x)} f'(x) = -\sin x \cdot \ln(4 \sin x + 2) + \cos x \cdot \frac{4 \cos x}{(4 \sin x + 2)}.$$

Therefore,

$$f'(x) = (4 \sin x + 2)^{\cos x} \left(-\sin x \ln(4 \sin x + 2) + \frac{2 \cos^2 x}{(2 \sin x + 1)} \right).$$

We have $f'(\pi) = 2^{-1} \left(0 + \frac{2}{1} \right) = \frac{1}{2} \cdot \frac{2}{1} = 1$.

3.9.57

a. $T = 10 \cdot 2^{-0.274 \cdot 16}$ minutes ≈ 28.7 seconds.

b. $\dfrac{\Delta T}{\Delta a} = \dfrac{10 \cdot 2^{-0.274 \cdot 8} - 10 \cdot 2^{-0.274 \cdot 2}}{8 - 2} \approx -0.78$ minutes per 1000 feet, which is about -46.512 seconds per 1000 feet.

c. $\dfrac{dT}{da} = -2.74 \cdot 2^{-0.274 \cdot a} \cdot \ln 2$. At $a = 8$ we have $\dfrac{dT}{da} = -2.74 \cdot 2^{-0.274 \cdot 8} \cdot \ln 2 \approx -0.42$ minutes per 1000 feet. Every 1000 feet the airplane climbs, leaves about .42 minutes less time of consciousness, which corresponds to about 24.94 seconds.

3.9.59 Let $y = x^{\sin x}$. Then $\ln y = \sin x \ln x$, so $\dfrac{1}{y} \cdot y' = \cos x \ln x + \dfrac{\sin x}{x}$. At the point $(1, 1)$ we have $y' = \sin 1$, so the tangent line is given by $y - 1 = (\sin 1)(x - 1)$, or $y = (\sin 1)x + 1 - \sin 1$.

3.9.61 Let $y = (x^2)^x = x^{2x}$. Then $\ln y = x \ln x^2$ and $\frac{1}{y} \cdot y' = \ln x^2 + 2$, so $y' = x^{2x}(\ln x^2 + 2)$. This quantity is zero when $\ln x^2 = -2$, or $x^2 = e^{-2}$. Thus there are horizontal tangents at $|x| = e^{-1}$, so for $x = \pm\frac{1}{e}$. The two tangent lines are given by $y = \frac{1}{e^{2/e}}$ (at $\left(\frac{1}{e}, \frac{1}{e^{2/e}}\right)$) and $y = e^{2/e}$ (at $\left(-\frac{1}{e}, e^{2/e}\right)$.)

3.9.63 $y' = 4 \cdot \frac{2x}{(x^2 - 1) \cdot \ln 3} = \frac{8x}{(x^2 - 1) \cdot \ln 3}$.

3.9.65 $y' = -\sin x (\ln(\cos^2 x)) + \cos x \cdot \left(\frac{2\cos x(-\sin x)}{\cos^2 x}\right) = (-\sin x)(\ln(\cos^2 x) + 2)$.

3.9.67 $y' = \frac{d}{dx}(\log_4 x)^{-1} = -(\log_4 x)^{-2} \cdot \frac{1}{x \ln 4} = -\frac{1}{x(\ln 4)(\log_4 x)^2} = -\frac{\ln 4}{x \ln^2 x}$.

3.9.69 $f'(x) = \frac{d}{dx}(4\ln(3x + 1)) = \frac{4}{3x + 1} \cdot 3 = \frac{12}{3x + 1}$.

3.9.71 $f'(x) = \frac{d}{dx}\left(\frac{1}{2}\ln 10x\right) = \frac{d}{dx}\frac{1}{2}[\ln 10 + \ln x] = \frac{1}{2x}$.

3.9.73 $f'(x) = \frac{d}{dx}(\ln(2x - 1) + 3\ln(x + 2) - 2\ln(1 - 4x)) = \frac{2}{2x - 1} + \frac{3}{x + 2} + \frac{8}{1 - 4x}$.

3.9.75 Let $y = x^{10x}$. Then $\ln y = 10x \ln x$, so $\frac{1}{y} \cdot y' = 10 \ln x + 10$, and $y' = x^{10x}(10)(\ln x + 1)$.

3.9.77 Let $y = \frac{(x + 1)^{10}}{(2x - 4)^8}$, so $\ln y = \ln\left(\frac{(x + 1)^{10}}{(2x - 4)^8}\right) = 10\ln(x + 1) - 8\ln(2x - 4)$. Then

$$\frac{1}{y} \cdot y' = \frac{10}{x + 1} - \frac{8}{2x - 4} \cdot 2,$$

$$y' = \frac{(x + 1)^{10}}{(2x - 4)^8} \cdot \left(\frac{10}{x + 1} - \frac{8}{x - 2}\right).$$

3.9.79 Let $y = x^{\ln x}$. Then $\ln y = (\ln x)^2$. Thus $\frac{1}{y} \cdot y' = 2\ln x \cdot \frac{1}{x}$, so $y' = x^{\ln x}\left(\frac{2\ln x}{x}\right)$.

3.9.81 Let $y = \frac{(x + 1)^{3/2}(x - 4)^{5/2}}{(5x + 3)^{2/3}}$. Then $\ln y = \ln\left(\frac{(x + 1)^{3/2}(x - 4)^{5/2}}{(5x + 3)^{2/3}}\right) = \frac{3}{2}\ln(x + 1) + \frac{5}{2}\ln(x - 4) - \frac{2}{3}\ln(5x + 3)$. Then

$$\frac{1}{y} \cdot y' = \frac{3}{2(x + 1)} + \frac{5}{2(x - 4)} - \frac{10}{3(5x + 3)},$$

$$y' = \frac{(x + 1)^{3/2}(x - 4)^{5/2}}{(5x + 3)^{2/3}} \cdot \left(\frac{3}{2(x + 1)} + \frac{5}{2(x - 4)} - \frac{10}{3(5x + 3)}\right).$$

3.9.83 Let $y = (\sin x)^{\tan x}$, and assume $0 < x < \pi$, $x \neq \frac{\pi}{2}$. Then $\ln y = (\tan x)\ln(\sin x)$. Then

$$\frac{1}{y} \cdot y' = (\sec^2 x)\ln(\sin x) + \frac{\tan x \cos x}{\sin x},$$

$$y' = (\sin x)^{\tan x}\left((\sec^2 x)\ln(\sin x) + 1\right).$$

3.9.85 Let $y = \left(1 + \frac{1}{x}\right)^x$. Then $\ln y = x\ln\left(1 + \frac{1}{x}\right)$, so $\frac{1}{y} \cdot y' = \ln\left(1 + \frac{1}{x}\right) + x\left(\frac{-1/x^2}{1 + 1/x}\right) = \ln\left(1 + \frac{1}{x}\right) - \frac{1}{x + 1}$. Therefore, $y' = \left(1 + \frac{1}{x}\right)^x \left(\ln\left(1 + \frac{1}{x}\right) - \frac{1}{x + 1}\right)$.

3.9.87

a. False. $\log_2 9$ is a constant, so its derivative is 0.

b. False. If $x < -1$, then the right-hand side is defined while the left-hand side isn't.

c. False. The correct way to write that function would be $e^{(x+1)\ln 2}$.

d. False. $\dfrac{d}{dx}(\sqrt{2})^x = (\sqrt{2})^x \ln(\sqrt{2})$.

e. True. This follows from the generalized power rule.

f. True. Note that $\ln((4x+1)^{\ln x}) = (\ln x)\ln(4x+1) = \ln(4x+1)(\ln x) = \ln(x^{\ln(4x+1)})$. Then we have $\ln((4x+1)^{\ln x}) = \ln(x^{\ln(4x+1)})$, and exponentiating both sides gives the desired result.

3.9.89 $\dfrac{d^2}{dx^2}(\log x) = \dfrac{d}{dx}\left(\dfrac{1}{x \ln 10}\right) = -\dfrac{1}{x^2 \ln 10}$.

3.9.91 $\dfrac{d^3}{dx^3}(x^2 \ln x) = \dfrac{d^2}{dx^2}(2x \ln x + x) = \dfrac{d}{dx}(2\ln x + 2 + 1) = \dfrac{2}{x}$.

3.9.93

i. $y' = \dfrac{d}{dx}\left(e^{x \ln 3}\right) = (e^{x \ln 3}) \cdot \ln 3 = 3^x \ln 3$.

ii. Let $y = 3^x$. Then $\ln y = x \ln 3$. So $\dfrac{1}{y} \cdot y' = \ln 3$, and $y' = 3^x \ln 3$.

3.9.95

$y' = \dfrac{d}{dx} e^{\sin x \ln 2} = (\cos x)(\ln 2) 2^{\sin x}$. At $x = \pi/2$ we have $y' = 0$, so the tangent line is given by $y = 2$.

3.9.97

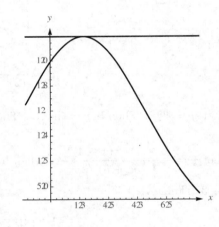

a. We used a graphing rectangle of $[0, 25] \times [0, 8000]$.

3.9. Derivatives of Logarithmic and Exponential Functions

b. To find when $P(t)$ hits 5000, we solve $5000 = \frac{400000}{50+7950e^{-0.5t}}$, or $50 + 7950e^{-0.5t} = 80$. This leads to $7950e^{-0.5t} = 30$, or $-0.5t = \ln\left(\frac{30}{7950}\right)$. So we have $t = 2\ln(265) \approx 11.16$ years.

The carrying capacity is $\lim_{t \to \infty} P(t) = \frac{400,000}{50} = 8000$. Ninety percent of 8000 is 7200, so we seek the time when $P(t) = 7200$. We have $7200 = \frac{400000}{50+7950e^{-0.5t}}$, or $50 + 7950e^{-0.5t} = \frac{500}{9}$. This leads to $7950e^{-0.5t} = \frac{50}{9}$, or $-0.5t = \ln\left(\frac{50}{71550}\right)$. So we have $t = 2\ln\left(\frac{71550}{50}\right) \approx 14.53$ years.

c. $\frac{dP}{dt} = -\frac{400000}{(50+7950e^{-0.5t})^2} \cdot (7950)(-0.5)e^{-0.5t}$.

At $t = 0$ we have $P'(0) = \frac{400,000 \cdot 7950 \cdot .5}{8000^2} = \frac{1,590,000,000}{8000^2} \approx 25$ fish per year.

At $t = 5$ we have $P'(5) = \frac{1,590,000,000 e^{-5/2}}{(50+7950e^{-5/2})^2} \approx 264$ fish per year.

d. The maximum is at about $t = 10$ years.

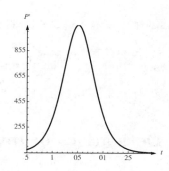

3.9.99

a. $\ln(P(t)) = \ln(3 \cdot 10^{10}) - \ln(2 + 3e^{-0.025t})$. $\frac{d}{dt}\ln(P(t)) = \frac{P'(t)}{P(t)} = r(t) = \frac{0.075 \cdot e^{-0.025t}}{2 + 3e^{-0.025t}}$. $r(0) = \frac{0.075}{5} = 0.015$, so the population is growing at 1.5% per year in 1999.

b. $r(11) = \frac{0.075e^{-0.275}}{2 + 3e^{-0.275}} \approx 0.0133$.

$r(21) = \frac{0.075e^{-0.525}}{2 + 3e^{-0.525}} \approx 0.0118$.

The relative growth rate decreases over time.

c. $\lim_{t \to \infty} r(t) = \lim_{t \to \infty} \frac{0.075}{3 + 2e^{0.025t}} = 0$, because the denominator increases without bound. The relative growth rate becomes smaller and smaller as the population nears the carrying capacity.

3.9.101

a.

t	$A(t)$
5	$17,442.50
15	$72,704.68
25	$173,248.49
35	$356,177.57

Average growth on $[5, 15]$ is $\frac{A(15) - A(5)}{10} \approx \$5526$ per year.

Average growth on $[15, 25]$ is $\frac{A(25) - A(15)}{10} \approx \$10,054$ per year.

Average growth on $[25, 35]$ is $\frac{A(35) - A(25)}{10} \approx \$18,293$ per year.

b. $A(40) \approx \$497{,}872.68$.

c. $A'(t) = 50{,}000 \cdot 12 \cdot (1.005)^{12t} \cdot \ln(1.005) \approx 2993 \cdot (1.005)^{12t}$. The rate of growth of the investment increases over time, so the earlier you start saving, the higher the rate of increase will be when you retire.

3.9.103 We search for a solution to $x = p^x$. If the two curves will have only one point of intersection, then they should be tangent at the point of intersection. So we need $1 = p^x \ln p$, or $\dfrac{1}{\ln p} = p^x = x$. So $\ln p = \dfrac{1}{x}$ and $p = e^{1/x}$. Then we have $x = p^x = (e^{1/x})^x = e$. So the point of intersection is (e, e) and the value of p is $e^{1/e} \approx 1.44467$.

3.9.105 Let $f(x) = \ln x$ and $a = e$. Then $f'(e) = \lim\limits_{x \to e} \dfrac{\ln x - 1}{x - e} = \dfrac{1}{e}$.

3.9.107 Let $f(x) = x^x$ and $a = 3$. Then $f'(3) = \lim\limits_{h \to 0} \dfrac{(3+h)^{3+h} - 27}{h} = 3^3 \cdot (\ln 3 + 1) = 27(1 + \ln 3)$.

3.9.109 Let $y = u(x)^{v(x)}$. Then $\ln y = v(x) \ln u(x)$, so $\dfrac{1}{y} \cdot y' = v'(x) \ln u(x) + v(x) \cdot \dfrac{u'(x)}{u(x)}$. Thus we have $y' = u(x)^{v(x)} \cdot \left(v'(x) \ln u(x) + v(x) \cdot \dfrac{u'(x)}{u(x)} \right) = u(x)^{v(x)} \cdot \left(\dfrac{v(x)}{u(x)} \dfrac{du}{dx} + \ln u(x) \dfrac{dv}{dx} \right)$.

3.10 Derivatives of Inverse Trigonometric Functions

3.10.1
$\dfrac{d}{dx} \sin^{-1} x = \dfrac{1}{\sqrt{1 - x^2}}$, $-1 < x < 1$.
$\dfrac{d}{dx} \tan^{-1} x = \dfrac{1}{1 + x^2}$, $-\infty < x < \infty$.
$\dfrac{d}{dx} \sec^{-1} x = \dfrac{1}{|x|\sqrt{x^2 - 1}}$, $|x| > 1$.

3.10.3 $y' = \dfrac{1}{1 + x^2}$. At $x = -2$ we have $y'(-2) = \dfrac{1}{1 + 4} = \dfrac{1}{5}$.

3.10.5 $(f^{-1})'(8) = \dfrac{1}{f'(2)} = \dfrac{1}{4}$.

3.10.7

a. Note that $f(0) = 4$, so $f^{-1}(4) = 0$. $(f^{-1})'(4) = \dfrac{1}{f'(0)} = \dfrac{1}{2}$.

b. Note that $f(1) = 6$, so $f^{-1}(6) = 1$. $(f^{-1})'(6) = \dfrac{1}{f'(1)} = \dfrac{1}{3/2} = \dfrac{2}{3}$.

c. Note that there is no given x so that $f(x) = 1$, so the desired derivative cannot be determined.

d. From the table directly, $f'(1) = \dfrac{3}{2}$.

3.10.9 Because $(f^{-1})'(8) = \dfrac{1}{f'(3)} = \dfrac{1}{7}$, the equation of the tangent line at $x = 8$ is $y - 3 = \frac{1}{7}(x - 8)$ or $y = \dfrac{1}{7}x + \dfrac{13}{7}$.

3.10.11 Let $f(x) = \sin x$. Then $f'(\pi/6) = \cos(\pi/6) = \dfrac{\sqrt{3}}{2}$, which implies that the slope of the curve $y = \sin^{-1} x$ at $(1/2, \pi/6)$ is $\dfrac{2}{\sqrt{3}}$.

3.10. Derivatives of Inverse Trigonometric Functions

3.10.13 $\dfrac{d}{dx}\sin^{-1}(2x) = \dfrac{2}{\sqrt{1-4x^2}}.$

3.10.15 $\dfrac{d}{dw}\cos(\sin^{-1}(2w)) = (-\sin(\sin^{-1}(2w)))\cdot \dfrac{d}{dw}\left(\sin^{-1}(2w)\right) = -2w\cdot \dfrac{2}{\sqrt{1-4w^2}} = -\dfrac{4w}{\sqrt{1-4w^2}}.$

3.10.17 $\dfrac{d}{dx}\sin^{-1}(e^{-2x}) = \dfrac{1}{\sqrt{1-e^{-4x}}}\cdot \dfrac{d}{dx}e^{-2x} = -\dfrac{2e^{-2x}}{\sqrt{1-e^{-4x}}}.$

3.10.19 $f'(x) = \dfrac{1}{1+100x^2}\cdot 10 = \dfrac{10}{100x^2+1}.$

3.10.21 $\dfrac{d}{dy}\tan^{-1}(2y^2-4) = \dfrac{1}{1+(2y^2-4)^2}\cdot \dfrac{d}{dy}(2y^2-4) = \dfrac{4y}{1+(2y^2-4)^2}.$

3.10.23 $\dfrac{d}{dz}\cot^{-1}\sqrt{z} = -\dfrac{1}{1+\sqrt{z}^2}\cdot \dfrac{d}{dz}\sqrt{z} = -\dfrac{1}{1+z}\cdot \dfrac{1}{2\sqrt{z}} = -\dfrac{1}{2\sqrt{z}(1+z)}.$

3.10.25
$$f'(x) = 2x + 6x^2\cot^{-1}x - \dfrac{2x^3}{1+x^2} - \dfrac{2x}{1+x^2} = 2x + 6x^2\cot^{-1}x - \dfrac{2x^3+2x}{1+x^2}$$
$$= 2x + 6x^2\cot^{-1}x - \dfrac{2x(x^2+1)}{1+x^2} = 6x^2\cot^{-1}x.$$

3.10.27 $f'(2) = 2w - \dfrac{2w}{1+w^4} = \dfrac{2w+2w^5}{1+w^4} - \dfrac{2w}{1+w^4} = \dfrac{2w^5}{1+w^4}.$

3.10.29 $\dfrac{d}{dx}\cos^{-1}\dfrac{1}{x} = -\dfrac{1}{\sqrt{1-\dfrac{1}{x^2}}}\cdot \left(-\dfrac{1}{x^2}\right) = \dfrac{1}{x^2\sqrt{\dfrac{x^2-1}{x^2}}} = \dfrac{|x|}{x^2\sqrt{x^2-1}} = \dfrac{1}{|x|\sqrt{x^2-1}},$ for $|x| > 1.$

3.10.31 $\dfrac{d}{du}\csc^{-1}(2u+1) = -\dfrac{1}{|2u+1|\sqrt{(2u+1)^2-1}}\cdot 2 = -\dfrac{2}{|2u+1|\sqrt{(2u+1)^2-1}} = -\dfrac{1}{|2u+1|\sqrt{u^2+u}}.$

3.10.33 $\dfrac{d}{dy}\cot^{-1}\left(\dfrac{1}{1+y^2}\right) = \left(-\dfrac{1}{1+\left(\dfrac{1}{1+y^2}\right)^2}\right)\cdot \left(-\dfrac{2y}{(1+y^2)^2}\right) = \dfrac{2y}{(1+y^2)^2+1}.$

3.10.35 $\dfrac{d}{dx}\sec^{-1}(\ln x) = \dfrac{1}{|\ln x|\sqrt{(\ln x)^2-1}}\cdot \dfrac{1}{x} = \dfrac{1}{x|\ln x|\sqrt{(\ln x)^2-1}}.$

3.10.37 $\dfrac{d}{dx}\csc^{-1}(\tan e^x) = -\dfrac{1}{|\tan e^x|\sqrt{(\tan e^x)^2-1}}\cdot (\sec^2 e^x)\cdot e^x.$

3.10.39 $\dfrac{d}{ds}\cot^{-1}(e^s) = -\dfrac{1}{1+e^{2s}}\cdot e^s = -\dfrac{e^s}{1+e^{2s}}.$

3.10.41 $f'(x) = \dfrac{1}{1+4x^2}\cdot 2,$ so $f'(1/2) = \dfrac{1}{1+1}\cdot 2 = 1.$ Thus the equation of the tangent line is $y - \pi/4 = 1(x-1/2),$ or $y = x + \dfrac{\pi}{4} - \dfrac{1}{2}.$

3.10.43 $f'(x) = -\dfrac{1}{\sqrt{1-x^4}}\cdot 2x = -\dfrac{2x}{\sqrt{1-x^4}},$ so $f'(1/\sqrt{2}) = -\dfrac{\sqrt{2}}{\sqrt{1-(1/4)}} = -\dfrac{2\sqrt{2}}{\sqrt{3}}.$ Thus the equation of the tangent line is $y - \pi/3 = -\dfrac{2\sqrt{2}}{\sqrt{3}}\left(x - 1/\sqrt{2}\right),$ or $y = -\dfrac{2\sqrt{2}}{\sqrt{3}}x + \dfrac{\pi}{3} + \dfrac{2}{\sqrt{3}}.$

3.10.45

a. $\dfrac{x}{150} = \cot\theta$, so $\theta = \cot^{-1}\left(\dfrac{x}{150}\right)$. Then $\dfrac{d\theta}{dx} = -\dfrac{1}{1+\left(\dfrac{x}{150}\right)^2} \cdot \dfrac{1}{150} = -\dfrac{150}{(150)^2 + x^2}$. When $x = 500$, we have $\dfrac{d\theta}{dx} = -\dfrac{150}{150^2 + 500^2} \approx -0.00055$ radians per meter.

b. The most rapid change is at $x = 0$ where $\dfrac{d\theta}{dx} = -\dfrac{1}{150} \approx -0.0067$ radians per meter.

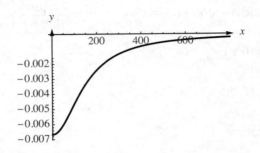

3.10.47 $f(4) = 16$ so $(f^{-1})'(16) = \dfrac{1}{f'(4)} = \dfrac{1}{3}$.

3.10.49 $f(0) = 1$ so $(f^{-1})'(1) = \dfrac{1}{f'(0)} = \dfrac{1}{5/e} = \dfrac{e}{5}$.

3.10.51 $f\left(\dfrac{\pi}{4}\right) = 1$ so $(f^{-1})'(1) = \dfrac{1}{f'\left(\dfrac{\pi}{4}\right)} = \dfrac{1}{\sec^2\left(\dfrac{\pi}{4}\right)} = \dfrac{1}{2}$.

3.10.53 $f(4) = 2$ so $(f^{-1})'(2) = \dfrac{1}{f'(4)} = \dfrac{1}{(1/2\sqrt{4})} = 4$.

3.10.55 $f(4) = 36$ and $(f^{-1})'(36) = \dfrac{1}{f'(4)} = \dfrac{1}{2(4+2)} = \dfrac{1}{12}$.

3.10.57 Note that $f(1) = 3$. So $(f^{-1})'(3) = \dfrac{1}{f'(1)} = \dfrac{1}{4}$.

3.10.59 $(f^{-1})'(4) = \dfrac{1}{f'(7)} = \dfrac{4}{5}$, so $f'(7) = \dfrac{5}{4}$.

3.10.61

a. True, because $\dfrac{d}{dx}\sin^{-1} x = -\dfrac{d}{dx}\cos^{-1} x$.

b. False. $\dfrac{d}{dx}\tan^{-1} x = \dfrac{1}{1+x^2}$ for all x, and this doesn't equal $\sec^2 x$ anywhere except at the origin (one curve is always less than or equal to one, and the other is always greater than or equal to one).

c. True. $\dfrac{d}{dx}\sin^{-1} x = \dfrac{1}{\sqrt{1-x^2}}$, and this is minimal when its denominator is as big as possible, which occurs when $x = 0$. So the smallest possible slope of a tangent line for this function on $(-1, 1)$ is $\dfrac{1}{\sqrt{1-0^2}} = 1$.

d. **True.** $\dfrac{d}{dx}\sin x = \cos x$ and $\cos x = 1$ for $x = 0$ and $-1 \leq \cos x \leq 1$ for all x. Thus 1 is the largest possible slope for a tangent line to the sine function.

e. **True.** This follows because the function $\dfrac{1}{x}$ is its own inverse. (Note that $f(f(x)) = \dfrac{1}{f(x)} = \dfrac{1}{1/x} = x$.) Thus, the derivative of the inverse of f is the derivative of f, which is $-\dfrac{1}{x^2}$.

3.10.63

a.

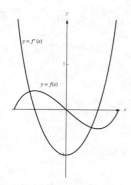

b. $f'(x) = 2x\sin^{-1}(x) + \dfrac{x^2 - 1}{\sqrt{1 - x^2}}$.

c. Note that f' is zero and f has a horizontal tangent line at about $x = -0.61$ and at about $x = 0.61$.

3.10.65

a. ,

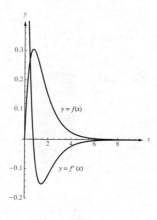

b. $f'(x) = -e^{-x}\tan^{-1}x + e^{-x}\dfrac{1}{1 + x^2}$.

c. Note that f' is zero and f has a horizontal tangent line at about $x = 0.75$.

3.10.67 Let $f(y) = 3y - 4$. Then $f'(y) = 3$ for all y in the domain of f. Let $y = f^{-1}(x)$. $(f^{-1})'(x) = \dfrac{1}{f'(y)} = \dfrac{1}{3}$.

3.10.69 Let $x = f(y) = y^2 - 4$ for $y > 0$. Note that this means that $y = \sqrt{x + 4}$. Then $f'(y) = 2y$. So $(f^{-1})'(x) = \dfrac{1}{f'(y)} = \dfrac{1}{2y} = \dfrac{1}{2\sqrt{x + 4}}$.

3.10.71 $y = e^{3x+1}$, so $\ln y = 3x + 1$, and thus $x = \dfrac{\ln y - 1}{3}$. Then $f^{-1}(x) = \dfrac{\ln x - 1}{3}$, so $(f^{-1})'(x) = \dfrac{1}{3x}$.

3.10.73 $y = 10^{12x-6}$ so $\log_{10} y = 12x - 6$, so $x = \dfrac{\log_{10} y + 6}{12}$. Then $f^{-1}(x) = \dfrac{\log_{10} x + 6}{12}$ and $(f^{-1})'(x) = \dfrac{1}{12x \ln 10}$.

3.10.75 For $y \geq -2$, let $x = \sqrt{y+2}$. Note that it then follows that $x \geq 0$. Then $1 = \frac{y'}{2\sqrt{y+2}}$, and $x^2 = y+2$, so $y = x^2 - 2$. Thus we have $(f^{-1})'(x) = y' = 2\sqrt{x^2 - 2 + 2} = 2|x| = 2x$, because $x \geq 0$.

3.10.77 For $y > 0$, let $x = y^{-1/2}$. Then $1 = -\frac{1}{2}y^{-3/2}y'$, so $y' = -2y^{3/2} = -2(x^{-2})^{3/2} = -2x^{-3}$ where $x > 0$.

3.10.79

a. Because $\frac{l}{10} = \csc(\theta)$, $\theta = \csc^{-1}\left(\frac{l}{10}\right)$, and $\frac{d\theta}{dl} = -\frac{1}{(l/10)\sqrt{(l/10)^2 - 1}} \cdot \frac{1}{10} = -\frac{10}{l\sqrt{l^2 - 100}}$.

b. $\left.\frac{d\theta}{dl}\right|_{l=50} = -\frac{10}{50\sqrt{2500 - 100}} \approx -0.0041$ radians per foot.

$\left.\frac{d\theta}{dl}\right|_{l=20} = -\frac{10}{20\sqrt{400 - 100}} \approx -0.029$ radians per foot.

$\left.\frac{d\theta}{dl}\right|_{l=11} = -\frac{10}{11\sqrt{121 - 100}} \approx -0.198$ radians per foot.

c. $\lim_{l \to 10^+} -\frac{10}{l\sqrt{l^2 - 100}} = -\infty$. The angle changes very quickly as we approach the dock.

d. $\frac{d\theta}{dl}$ is negative because this measures the change in θ as l increases – but when the boat is approaching the dock, l is decreasing.

3.10.81

a. $\sin\theta = \frac{c}{D}$, so $\theta = \sin^{-1}\left(\frac{c}{D}\right)$. Thus $\frac{d\theta}{dc} = \frac{1/D}{\sqrt{1 - \left(\frac{c}{D}\right)^2}} = \frac{1}{\sqrt{D^2 - c^2}}$.

b. $\left.\frac{d\theta}{dc}\right|_{c=0} = \frac{1}{\sqrt{D^2}} = \frac{1}{D}$.

3.10.83 $(f^{-1})'(y_0) = \frac{1}{f'(x_0)}$ where $y_0 = f(x_0)$.

$\frac{d}{dx}\sin^{-1}x = \frac{1}{\cos(\sin^{-1}x)} = \frac{1}{\sqrt{1 - \sin^2(\sin^{-1}x)}} = \frac{1}{\sqrt{1 - x^2}}$.

3.10.85 Using the identity $\cot^{-1}x + \tan^{-1}x = \frac{\pi}{2}$, we have the $\frac{d}{dx}\cot^{-1}x + \frac{d}{dx}\tan^{-1}x = 0$, so $\frac{d}{dx}\cot^{-1}x = -\frac{d}{dx}\tan^{-1}x$. Likewise, because $\csc^{-1}x + \sec^{-1}x = \frac{\pi}{2}$, we have $\frac{d}{dx}\csc^{-1}x + \frac{d}{dx}\sec^{-1}x = 0$, so $\frac{d}{dx}\csc^{-1}x = -\frac{d}{dx}\sec^{-1}x$.

3.10.87 $\cos(\sin^{-1}(x)) = \sqrt{1 - \sin^2(\sin^{-1}(x))} = \sqrt{1 - x^2}$ for $-1 \leq x \leq 1$.

3.10.89 $\tan(2\tan^{-1}(x)) = \frac{2\tan(\tan^{-1}(x))}{1 - \tan^2(\tan^{-1}(x))} = \frac{2x}{1 - x^2}$ for $-1 < x < 1$.

3.11 Related Rates

3.11.1 The area of a circle of radius r is $A(r) = \pi r^2$. If the radius $r = r(t)$ changes with time, then the area of the circle is a function of r and r is a function of t, so ultimately A is a function of t. If the radius changes at rate $\dfrac{dr}{dt}$, then the area changes at rate $2\pi r \dfrac{dr}{dt}$.

3.11.3 Because area is width times height, if one increases, the other must decrease in order for the area to remain constant.

3.11.5

a. $V = 200h$, so $\dfrac{dV}{dt} = 200\dfrac{dh}{dt}$.

b. $\dfrac{dV}{dt} = 200 \cdot \dfrac{1}{4} = 50\,\text{ft}^3/\text{min}$.

c. $\dfrac{dh}{dt} = \dfrac{dV/dt}{200} = \dfrac{10}{200} = \dfrac{1}{20}\,\text{ft/min}$.

3.11.7

a. $V = \dfrac{4}{3}\pi r^3$ so $\dfrac{dV}{dt} = 4\pi r^2 \dfrac{dr}{dt}$.

b. $\dfrac{dV}{dt} = 4\pi(4)^2(2) = 128\,\text{in}^3/\text{min}$.

c. $\dfrac{dr}{dt} = \dfrac{dV/dt}{4\pi r^2} = \dfrac{10}{4\pi 5^2} = \dfrac{1}{10\pi}\,\text{in/min}$.

3.11.9 $\dfrac{dz}{dt} = \dfrac{dx}{dt} + 3y^2 \dfrac{dy}{dt}$, so at the aforementioned time, $\dfrac{dz}{dt} = -1 + 3(2)^2(5) = 59$.

3.11.11

$A(x) = x^2$, $\dfrac{dx}{dt} = 2$ meters per second.

a. $\dfrac{dA}{dt} = 2x\dfrac{dx}{dt}$, so at $x = 10$ we have $\dfrac{dA}{dt} = 2 \cdot 10\text{m} \cdot 2\text{m/s} = 40\text{m}^2/\text{s}$.

b. At $x = 20\text{m}$ we have $\dfrac{dA}{dt} = 2 \cdot 20\text{m} \cdot 2\text{m/s} = 80\text{m}^2/\text{s}$.

3.11.13

a. Let x be the length of a leg of an isosceles right triangle. Then $\dfrac{dx}{dt} = 2$ meters per second. The area is given by $A(x) = \frac{1}{2}x^2$. Thus, $\dfrac{dA}{dt} = \dfrac{dA}{dx}\dfrac{dx}{dt} = x \cdot 2 = 2x$ square meters per second. When $x = 2$, we have $\dfrac{dA}{dx} = 4$, so the area is increasing at 4 square meters per second.

b. When the hypotenuse is 1 meter long, the legs are $1/\sqrt{2}$ meters long. So $A'(1/\sqrt{2}) = 2 \cdot \dfrac{1}{\sqrt{2}} = \sqrt{2}$, so the area is increasing at $\sqrt{2}$ square meters per second.

c. If h is the length of the hypotenuse, then $x^2 + x^2 = h^2$, so $h = \sqrt{2}x$. So $\dfrac{dh}{dt} = \dfrac{dh}{dx}\dfrac{dx}{dt} = \sqrt{2}\dfrac{dx}{dt} = \sqrt{2} \cdot 2 = 2\sqrt{2}$ meters per second.

3.11.15

a. Let r be the radius of the circle and A the area, and note that we are given $\frac{dA}{dt} = 1$ square cm per second. Because $A = \pi r^2$, we have $\frac{dA}{dt} = \frac{dA}{dr}\frac{dr}{dt}$, so

$$1 = 2\pi r \frac{dr}{dt},$$

and thus

$$\frac{dr}{dt} = \frac{1}{2\pi r}.$$

When $r = 2$, we have $\frac{dr}{dt} = \frac{1}{4\pi}$ cm per second.

b. When $c = 2\pi r = 2$, we have $r = 1/\pi$. At this time, $\frac{dr}{dt} = \frac{1}{2\pi r} = \frac{1}{2\pi(1/\pi)} = \frac{1}{2}$ cm per second.

3.11.17 $A(x) = \pi x^2$, so $\frac{dA}{dt} = 2\pi x \frac{dx}{dt}$. At $x = 10$ ft and $\frac{dx}{dt} = -2$ ft/min we have $\frac{dA}{dt} = 2\pi \cdot 10 \cdot (-2) = -40\pi$ ft^2/min.

3.11.19 $V(r) = \frac{4}{3}\pi r^3$, so $\frac{dV}{dt} = 4\pi r^2 \frac{dr}{dt} = 15$ in^3/min. At $r = 10$ inches we have $4\pi(10\,\text{in})^2 \frac{dr}{dt} = 15$ in^3/min. Thus, $\frac{dr}{dt} = \frac{3}{80\pi}$ in/min ≈ 0.012 in/min.

3.11.21 $V(r) = \frac{4}{3}\pi r^3$, and $S(r) = 4\pi r^2$. $\frac{dV}{dt} = 4\pi r^2 \frac{dr}{dt} = k \cdot 4\pi r^2$, so $\frac{dr}{dt} = k$, the constant of proportionality.

3.11.23 Let x be the distance the westbound airliner has traveled between noon and t hours after 1:00, and let y be the distance the northbound airliner has traveled t hours after 1:00, and let D be the distance between the planes. We have $D^2 = x^2 + y^2$, so $2D\frac{dD}{dt} = 2x\frac{dx}{dt} + 2y\frac{dy}{dt}$. We are given that $\frac{dx}{dt} = 500$ mph and $\frac{dy}{dt} = 550$ mph. At 2:30, we have that $x = 500 + 500 \cdot 1.5 = 1250$, and $y = 550 \cdot 1.5 = 825$ miles. $D = \sqrt{2243125} \approx 1497.7$ miles. Thus $\frac{dD}{dt} \approx \frac{1250 \cdot 500 + 825 \cdot 550}{1497.7} \approx 720.27$ miles per hour.

3.11.25

Let D be the length of the rope from the boat to the capstan, and let x be the horizontal distance from the boat to the dock. By the Pythagorean Theorem, $x^2 + 25 = D^2$, so $2x\frac{dx}{dt} = 2D\frac{dD}{dt}$, so $\frac{dx}{dt} = \frac{D}{x}\frac{dD}{dt}$. We are given that $\frac{dD}{dt} = -3$ feet per second, so when $x = 10$, we have $\frac{dx}{dt} = \frac{\sqrt{125}}{10} \cdot (-3) = -\frac{3\sqrt{5}}{2}$ feet per second. The boat is approaching the dock at $\frac{3\sqrt{5}}{2}$ feet per second.

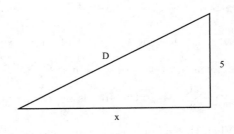

3.11.27 Let x be the distance the motorcycle has traveled since the instant it went under the balloon, and let y be the height of the balloon above the ground t seconds after the motorcycle went under it. We have $x^2 + y^2 = D^2$ where D is the distance between the motorcycle and the balloon. Thus, $2x\frac{dx}{dt} + 2y\frac{dy}{dt} = 2D\frac{dD}{dt}$, and we are given that $\frac{dy}{dt} = 10$ feet per second, and $\frac{dx}{dt} = 40\,\text{mph} = \frac{176}{3}\,\text{ft/s}$. After 10 seconds have passed, we have that $y = 150 + 100 = 250\,\text{ft}$, $x = \frac{1760}{3}\,\text{ft}$ and $D = \sqrt{250^2 + \left(\frac{1760}{3}\right)^2} \approx 638\,\text{ft}$. Thus, $\frac{dD}{dt} \approx \frac{1}{638}\left(\frac{1760}{3} \cdot \frac{176}{3} + 2500\right) \approx 57.86$ feet per second.

3.11.29

Let x be the distance between the fish and the fisherman's feet, and let D be the distance between the fish and the tip of the pole. Then $D^2 = x^2 + 144$, so $2D(dD/dt) = 2x(dx/dt)$. Note that $dD/dt = -1/3\,\text{ft/sec}$, so when $x = 20\,\text{ft}$, we have $dx/dt = \sqrt{400 + 144}/20 \cdot (-1/3) \approx -0.3887\,\text{ft/sec} \approx -4.66\,\text{in/sec}$. The fish is moving toward the fisherman at about 4.66 in/sec.

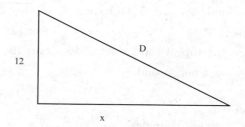

3.11.31 Let $h(t)$ be the height of the water in the tank at time t. Then the volume of the water in the tank at time t is given by $V = \pi r^2 h = \pi h$. We are seeking $\frac{dV}{dt}$ when $\frac{dh}{dt} = -1/2$ foot per minute. Because $\frac{dV}{dt} = \frac{dV}{dh}\frac{dh}{dt} = -\frac{1}{2}\pi$, the volume of the water in the tank is decreasing at $\pi/2$ cubic feet per minute, so the water is draining out at $\pi/2$ cubic feet per minute.

3.11.33 Let x be the distance from the bottom of the cylinder to the position of the piston. Let $V(x)$ be the volume of the cylinder when the piston is at position x. $V(x) = 25\pi x$, so $\frac{dV}{dt} = 25\pi \frac{dx}{dt}$. Because $\frac{dx}{dt} = -3\,\text{cm/s}$ we have $\frac{dV}{dt} = 25\pi(-3)\,\text{cm}^3/\text{s} = -75\pi\,\text{cm}^3/\text{s}$.

3.11.35 $V = \frac{1}{3}\pi r^2 h$ where $r = 3h$, so $V = 3\pi h^3$. We have that $\frac{dV}{dt} = 9\pi h^2 \frac{dh}{dt}$, and we given that $\frac{dh}{dt} = 2$ at the moment when $h = 12$, so at that time, $\frac{dV}{dt} = 9\pi \cdot 144\,\text{cm}^2 \cdot 2\,\text{cm/sec} = 2592\pi\,\text{cm}^3/\text{s}$. This is the rate at which the volume of the sandpile is increasing, so it must also be the rate at which the sand is leaving the bin, because there is no other sand involved.

3.11.37 Let h be the depth of the water in the tank at time t, and let r be the radius of the cone-shaped water at time t. By similar triangles, we have that $\frac{h}{r} = \frac{12}{6}$, so $h = 2r$. The volume of the water in the tank is given by $V = \frac{1}{3}\pi r^2 h = \frac{1}{3}\pi \frac{h^2}{4} \cdot h = \frac{\pi h^3}{12}$. Thus, $\frac{dV}{dt} = \frac{\pi h^2}{4}\frac{dh}{dt}$. When $\frac{dh}{dt} = -1$, we have $\frac{dV}{dt} = -\frac{\pi h^2}{4}$. when $h = 6$, we have $\frac{dV}{dt} = -9\pi$, so the water is draining from the tank at 9π cubic feet per minute.

3.11.39 The volume of a segment of water of height h within a hemisphere of radius 10 is given by $V = \frac{1}{3}\pi h^2(30 - h) = 10\pi h^2 - \frac{1}{3}\pi h^3$. We have that $\frac{dV}{dt} = 20\pi h \frac{dh}{dt} - \pi h^2 \frac{dh}{dt}$. We are given that $\frac{dV}{dt} = 3\,\text{m}^3/\text{min}$, so when $h = 5$ we have $3 = (100\pi - 25\pi)\frac{dh}{dt}$, so $\frac{dh}{dt} = \frac{3}{75\pi} = \frac{1}{25\pi}$ meters per minute.

3.11.41

Let h be the vertical distance from the ground to the top of the ladder, and let x be the horizontal distance from the wall to the bottom of the ladder. By the Pythagorean Theorem, we have that $x^2 + h^2 = 169$. Thus, $2x\dfrac{dx}{dt} + 2h\dfrac{dh}{dt} = 0$, so $\dfrac{dh}{dt} = -\dfrac{x}{h}\dfrac{dx}{dt}$, and we are given that $\dfrac{dx}{dt} = 0.5$ feet per second. At $x = 5$ we have $h = \sqrt{169 - 25} = 12$ feet. Thus, $\dfrac{dh}{dt} = -\dfrac{5}{12} \cdot \dfrac{1}{2} = -\dfrac{5}{24}$ feet per second. So the top of the ladder slides down the wall at $\dfrac{5}{24}$ feet per second.

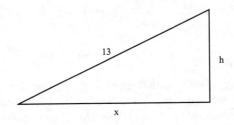

3.11.43

By similar triangles, $\dfrac{x+y}{20} = \dfrac{y}{5}$, so $x + y = 4y$, so $x = 3y$, and $\dfrac{dx}{dt} = 3\dfrac{dy}{dt}$. Because we are given that $\dfrac{dx}{dt} = -8$, we have $\dfrac{dy}{dt} = -\dfrac{8}{3}$ feet per second. The tip of her shadow is therefore moving at $-8 - \dfrac{8}{3} = -\dfrac{32}{3}$ feet per second.

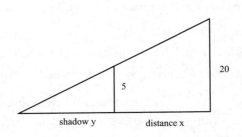

3.11.45

Let h be the vertical distance between the point on the elevator shaft positioned directly opposite the observer and the point on the elevator shaft that the observer is observing. So $h > 0$ corresponds to $\theta > 0$ and $h < 0$ corresponds to $\theta < 0$. We have $\dfrac{h}{20} = \tan\theta$, so $\dfrac{1}{20}\dfrac{dh}{dt} = \sec^2\theta\dfrac{d\theta}{dt}$. We are given that $\dfrac{dh}{dt} = 5\,\text{m/s}$. At $h = -10$, we have $\tan\theta = -.5$, so $\sec^2\theta = 1 + \tan^2\theta = 1 + (.5)^2 = 1.25$. So $\dfrac{d\theta}{dt} = \dfrac{1}{20 \cdot 1.25} \cdot 5 = \dfrac{1}{5}$ radian per second.

When $h = 20$, we have that $\tan\theta = 1$, so $\sec^2\theta = 1 + 1^2 = 2$, and thus $\dfrac{d\theta}{dt} = \dfrac{1}{20 \cdot 2} \cdot 5 = \dfrac{1}{8}$ radian per second.

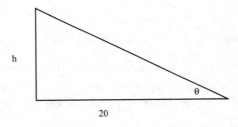

3.11.47 Let α be the angle between the line of sight to the bottom of the screen and the line of sight to the point 3 feet below where the floor and the wall meet. Note that $\cot\alpha = \dfrac{x}{3}$ and $\cot(\alpha + \theta) = \dfrac{x}{10}$, so $\alpha = \cot^{-1}(\dfrac{x}{3})$ and $\alpha + \theta = \cot^{-1}\left(\dfrac{x}{10}\right)$. Thus, $\theta = \cot^{-1}\left(\dfrac{x}{10}\right) - \cot^{-1}\left(\dfrac{x}{3}\right)$. So

$$\dfrac{d\theta}{dt} = -\dfrac{10x'}{100 + x^2} + \dfrac{3x'}{9 + x^2},$$

and at $x = 30$ feet, and with $\dfrac{dx}{dt} = 3$ feet per second, we have $\dfrac{d\theta}{dt} = -\dfrac{30}{1000} + \dfrac{9}{909} \approx -0.0201$ radians per second.

3.11. Related Rates

3.11.49

Let x be the distance the surface ship has traveled and D the depth of the submarine. We have $\frac{dx}{dt} = 10$ km/hr. Note that $\frac{D}{x} = \tan 20°$, so $D = x \cdot \tan 20° \approx 0.364x$. We have $\frac{dD}{dt} = 0.364 \frac{dx}{dt} = 3.64$ km/hr. The depth of the submarine is increasing at a rate of 3.64 km/hr.

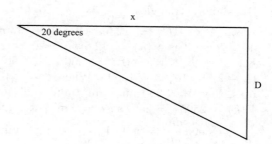

3.11.51

a. Let A be the point where the dragster started, let B be the point where camera 1 is located and let $C = y(t)$ be the position of the car at time t. Let θ be angle ABC. Note that $\tan \theta = \frac{y}{50}$, so $\sec^2 \theta \cdot \frac{d\theta}{dt} = \frac{1}{50} \frac{dy}{dt}$. At time $t = 2$, we have that $\tan^2 \theta = 4$, so $\sec^2 \theta = \tan^2 \theta + 1 = 5$. So $\frac{dy}{dt} = 5 \cdot 50 \cdot .75 = 187.5$ feet per second.

b. Let D be the point where camera 2 is located, and let ϕ be angle ADC. The $\phi = \tan^{-1}\left(\frac{y}{100}\right)$, so $\frac{d\phi}{dt} = \frac{1}{100(1 + \left(\frac{y}{100}\right)^2)} \cdot \frac{dy}{dt}$. After 2 seconds, we know that $y = 100$ and $\frac{dy}{dt} = 187.5$. Thus $\frac{d\phi}{dt} = \frac{100}{20,000} \cdot 187.5 = .9375$ radians per second.

3.11.53

a. $P = \frac{dE}{dt} = \frac{1}{2}v^2 \frac{dm}{dt}$.

b. $P = \frac{1}{2}v^2 \frac{dm}{dt} = \frac{1}{2}v^2(\rho A v) = \frac{1}{2}\rho A v^3$.

c. $P = \frac{1}{2}\rho A v^3 = \frac{1}{2}(1.23)(\pi)(3^2)(10^3) \approx 388.7$ W

d. $17388.7(0.25) \approx 4347.2$ W.

3.11.55 Let θ be the angle between the hands of the clock, and D the distance between the tips of the hands. By the Law of Cosines, $D^2 = 2.5^2 + 3^2 - 15\cos\theta$. So $2D\frac{dD}{dt} = 15\sin\theta\frac{d\theta}{dt}$. At 9:00 AM, we have $D^2 = 6.25 + 9$, so $D = \sqrt{15.25}$. Also, $\theta = \pi/2$ so $\sin\theta = 1$. Thus, $\frac{dD}{dt} = \frac{15}{2\sqrt{15.25}}\frac{d\theta}{dt}$. Now $\frac{d\theta}{dt} = \frac{d\theta_1}{dt} - \frac{d\theta_2}{dt}$ where $\frac{d\theta_1}{dt}$ is the angular change of the minute hand and $\frac{d\theta_2}{dt}$ is the angular change of the hour hand. We have $\frac{d\theta_1}{dt} = \frac{\pi}{30}$ radians per minute and $\frac{d\theta_2}{dt} = \frac{\pi}{360}$ radians per minute, so $\frac{d\theta}{dt} = \frac{11\pi}{360}$ radians per minute. Thus $\frac{dD}{dt} = \frac{15}{2\sqrt{15.25}} \cdot \frac{11\pi}{360} \approx 0.18436$ meters per minute, or about 11.06 meters per hour.

3.11.57
By similar triangles, $\dfrac{2}{50} = \dfrac{h}{b}$, so $b = 25h$. Also, $A = \tfrac{1}{2}bh = 12.5h^2$, so the volume for $0 \le h \le 2$ is $V(h) = 12.5 \cdot h^2 \cdot 20 = 250h^2$. For $2 < h \le 3$, $V(h) = 250 \cdot 2^2 + 50 \cdot 20 \cdot (h-2) = 1000h - 1000$. When $t = 250$ minutes, then $V = 250\,\text{min} \cdot 1\,\text{m}^3/\text{min} = 250\,\text{m}^3$. So $V(h) = 250h^2 = 250$, so $h = 1\,\text{m}$. At that time $\dfrac{dV}{dt} = 500h\dfrac{dh}{dt} = 500 \cdot 1 \cdot \dfrac{dh}{dt} = 1\,\text{m}^3/\text{min}$. So $\dfrac{dh}{dt} = \dfrac{1}{500}\,\text{m/min} = 0.002\,\text{m/min} = 2\,\text{mm/min}$.

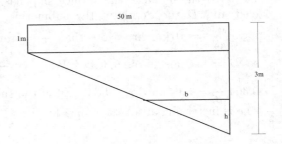

Fill time: The volume of the entire swimming pool is 2000 cubic meters, so at 1 cubic meter per minute, it will take 2000 minutes.

3.11.59 By the Law of Sines, $\dfrac{\sin\theta}{s} = \dfrac{\sin\left(\tfrac{3\pi}{4} - \theta\right)}{2}$, so

$$2\sin\theta = s\sin\left(\tfrac{3\pi}{4} - \theta\right) = s\left(\sin\left(\tfrac{3\pi}{4}\right)\cos\theta - \cos\left(\tfrac{3\pi}{4}\right)\sin\theta\right).$$

We have

$$2\sin\theta = \tfrac{\sqrt{2}}{2}s(\sin\theta + \cos\theta)$$

$$2\tan\theta = \tfrac{\sqrt{2}}{2}s(\tan\theta + 1)$$

$$\tan\theta = \dfrac{(\sqrt{2}/2)\cdot s}{2 - (\sqrt{2}/2)s} = \dfrac{\sqrt{2}s}{4 - \sqrt{2}s}$$

$$\theta = \tan^{-1}\left(\dfrac{\sqrt{2}s}{4 - \sqrt{2}s}\right).$$

Thus, $\dfrac{d\theta}{dt} = \dfrac{\sqrt{2}\cdot\tfrac{ds}{dt}}{4 - 2\sqrt{2}s + s^2}$. When $\dfrac{ds}{dt} = 15$ and $s = 7.5$ we arrive at $\dfrac{d\theta}{dt} = 0.54$ radians per hour.

3.11.61 Let x be the distance the eastbound boat has traveled at time t and let s be the distance the northeastbound boat has traveled. Note the diagram shown. By the Law of Sines, $\dfrac{\sin\left(\tfrac{\pi}{2} - \theta\right)}{s} = \dfrac{\sin\left(\tfrac{\pi}{4} + \theta\right)}{x}$. Thus,

$$x\left(\sin\left(\tfrac{\pi}{2}\right)\cos\theta - \cos\left(\tfrac{\pi}{2}\right)\sin\theta\right) = s\left(\sin\left(\tfrac{\pi}{4}\right)\cos\theta + \cos\left(\tfrac{\pi}{4}\right)\sin\theta\right)$$

So

$$x\cos\theta = \tfrac{\sqrt{2}}{2}\cdot s\cdot\cos\theta + \tfrac{\sqrt{2}}{2}\cdot s\cdot\sin\theta,$$

$$x = \tfrac{\sqrt{2}}{2}\cdot s + \tfrac{\sqrt{2}}{2}\cdot s\cdot\tan\theta,$$

and thus $\tan\theta = \dfrac{x - \tfrac{\sqrt{2}}{2}s}{\tfrac{\sqrt{2}}{2}s} = \dfrac{\sqrt{2}x - s}{s}$, and therefore $\theta = \tan^{-1}\left(\dfrac{\sqrt{2}x - s}{s}\right).$

3.11. Related Rates

We have

$$\frac{d\theta}{dt} = \frac{1}{1+\left(\frac{\sqrt{2}x-s}{s}\right)^2} \cdot \frac{\left(\sqrt{2}\left(\frac{dx}{dt}\right)-\left(\frac{ds}{dt}\right)\right)\cdot s - (\sqrt{2}x-s)\cdot \frac{ds}{dt}}{s^2} = \frac{\sqrt{2}\left(s\frac{dx}{dt}-x\frac{ds}{dt}\right)}{s^2+(\sqrt{2}x-s)^2}.$$

At time t, we have $s(t) = 15t$ and $x(t) = 12t$. Note that

$$s\frac{dx}{dt} - x\frac{ds}{dt} = 15t \cdot 12 - 12t \cdot 15 = 0.$$

Thus $\theta' = 0$ for every value of t, so that the angle is constant.

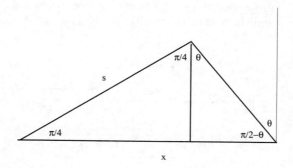

3.11.63

a. The volume of the water in the tank (as a function of h – the depth of the water in the tank) is given by 5 times the area of the segment of water in a cross-sectional circle. For a tank of radius 1, the formula for such a segment is $\cos^{-1}(1-h) - (1-h)\sqrt{2h-h^2}$. Thus the volume of the water in the tank is given by $V = 5(\cos^{-1}(1-h) - (1-h)\sqrt{2h-h^2})$. We have

$$\frac{dV}{dt} = 5 \cdot \left(-\frac{1}{\sqrt{1-(1-h)^2}} \cdot \left(-\frac{dh}{dt}\right) + \frac{dh}{dt}\sqrt{2h-h^2} - \frac{(1-h)^2}{\sqrt{2h-h^2}}\frac{dh}{dt}\right)$$

$$= 5\left(\sqrt{2h-h^2} + \frac{1-(1-h)^2}{\sqrt{2h-h^2}}\right)\frac{dh}{dt}$$

$$= 5\left(\frac{2h-h^2+1-1+2h-h^2}{\sqrt{2h-h^2}}\right)\frac{dh}{dt}$$

$$= 5\left(\frac{2(2h-h^2)}{\sqrt{2h-h^2}}\right)\frac{dh}{dt}$$

$$= 10\sqrt{2h-h^2} \cdot \frac{dh}{dt}$$

When $h = 0.5$, we have $-\frac{3}{2} = \frac{dV}{dt} = 5\sqrt{3}\frac{dh}{dt}$, so $\frac{dh}{dt} = -\frac{\sqrt{3}}{10}$ meters per hr.

b. The surface area of the water is given by $S = 5 \cdot 2\sqrt{2h-h^2}$. So $\frac{dS}{dt} = 10 \cdot \frac{2-2h}{2\sqrt{2h-h^2}} \cdot \frac{dh}{dt}$, so at $h = .5$, we have $\frac{5}{\sqrt{3/4}} \cdot -\frac{\sqrt{3}}{10} = -1$ square meter per hr.

Chapter Three Review

1

a. False. This function is not differentiable at $x = -\frac{1}{2}$. It is possible for a function to be continuous at a point and not differentiable at that point.

b. False. For example, $f(x) = x^2 + 3$ and $g(x) = x^2 + 100$ have the same derivative, but aren't the same function.

c. False. For example, $\frac{d}{dx}|e^{-x}| = \frac{d}{dx}e^{-x} = -e^{-x} \neq |-e^{-x}|$.

d. False. For example, the function $f(x) = |x|$ has no derivative at 0, but there is no vertical tangent there.

e. True. For example, a ball dropping from a high tower has acceleration due to gravity which is negative, but it is speeding up as it falls because the velocity (which is negative also) is in the same direction as the acceleration.

3

$$g'(x) = \lim_{h \to 0} \frac{g(x+h) - g(x)}{h} = \lim_{h \to 0} \frac{\frac{1}{(x+h)^2+5} - \frac{1}{x^2+5}}{h} =$$
$$= \lim_{h \to 0} \frac{x^2 + 5 - (x^2 + 2xh + h^2 + 5)}{h((x+h)^2+5)(x^2+5)}$$
$$= \lim_{h \to 0} \frac{-h(2x+h)}{h((x+h)^2+5)(x^2+5)}$$
$$= \lim_{h \to 0} \frac{-(2x+h)}{((x+h)^2+5)(x^2+5)}$$
$$= -\frac{2x}{(x^2+5)^2}.$$

5 $\frac{d}{dx}(\tan^3 - 3\tan x + 3x) = 3\tan^2 x \sec^2 x - 3\sec^2 x + 3 = 3\sec^2 x(\tan^2 x - 1) + 3 = 3(\tan^2 x + 1)(\tan^2 x - 1) + 3 = 3\tan^4 x - 3 + 3 = 3\tan^4 x.$

7
$$\frac{d}{dx}(x^4 - \ln(x^4 + 1)) = 4x^3 - \frac{4x^3}{x^4+1} = \frac{4x^3(x^4+1)}{x^4+1} - \frac{4x^3}{x^4+1} = \frac{4x^7 + 4x^3 - 4x^3}{x^4+1} = \frac{4x^7}{x^4+1}.$$

9 $y' = 2x^2 + 2\pi x + 7.$

11 $y' = 2^x (\ln 2).$

13 $y' = e^{2\theta} \cdot 2 = 2e^{2\theta}.$

15 $y' = \frac{3}{2}\sqrt{1+x^4}(4x^3) = 6x^3\sqrt{1+x^4}.$

17 $y' = 10t \sin t + 5t^2 \cos t.$

19 $y' = -e^{-x}(x^2 + 2x + 2) + e^{-x}(2x + 2) = -e^{-x}(x^2 + 2x + 2 - 2x - 2) = -x^2 e^{-x}.$

21 $y' = \frac{((\sec 2w + 1)2 \sec 2w \tan 2w - 2 \sec 2w \sec 2w \tan 2w)}{(\sec 2w + 1)^2} = \frac{2 \sec 2w \tan 2w}{(\sec 2w + 1)^2}.$

23 $y' = \frac{3}{\sec 3x} \sec 3x \tan 3x = 3 \tan 3x.$

Chapter Three Review

155

25 $y' = 100(5t^2 + 10)^{99}(10t) = 1000t(5t^2 + 10)^{99}$.

27 $y' = \dfrac{1}{\sin x^3} \cdot \cos x^3 \cdot 3x^2 = \dfrac{3x^2 \cos x^3}{\sin x^3} = 3x^2 \cot x^3$.

29 $y' = \dfrac{1}{(\sqrt{t^2-1})^2 + 1} \cdot \dfrac{1}{2}(t^2-1)^{-1/2} \cdot 2t = \dfrac{2t}{2t^2\sqrt{t^2-1}} = \dfrac{1}{t\sqrt{t^2-1}}$.

31 $y' = (4\sec^2(\theta^2 + 3\theta + 2)) \cdot (2\theta + 3) = (8\theta + 12)\sec^2(\theta^2 + 3\theta + 2)$.

33 $y' = \dfrac{w^5 \cdot \frac{1}{w} - \ln w \cdot 5w^4}{w^{10}} = \dfrac{w^4(1 - 5\ln w)}{w^{10}} = \dfrac{1 - 5\ln w}{w^6}$.

35 $y' = \dfrac{(8u+1)(8u+1) - (4u^2+u)(8)}{(8u+1)^2} = \dfrac{64u^2 + 16u + 1 - 32u^2 - 8u}{(8u+1)^2} = \dfrac{32u^2 + 8u + 1}{(8u+1)^2}$.

37 $y' = \sec^2(\sin\theta) \cdot \cos\theta$.

39 $y' = \cos\sqrt{\cos^2 x + 1} \cdot \dfrac{1}{2\sqrt{\cos^2 x + 1}} \cdot -2\cos x \sin x = -\dfrac{\cos\sqrt{\cos^2 x + 1}\cos x \sin x}{\sqrt{\cos^2 x + 1}}$.

41 $y' = \dfrac{d}{dt}\ln\sqrt{e^t + 1} = \dfrac{d}{dt}\left(\dfrac{1}{2}\ln(e^t + 1)\right) = \dfrac{1}{2} \cdot \dfrac{1}{e^t + 1} \cdot e^t = \dfrac{e^t}{2(e^t + 1)}$.

43 $y' = 2x + 2\tan^{-1}(\cot x) + 2x \cdot \dfrac{1}{1 + \cot^2 x} \cdot (-\csc^2 x) = 2x + 2\tan^{-1}(\cot x) + \dfrac{2x}{\csc^2 x} \cdot (-\csc^2 x) = 2\tan^{-1}(\cot x)$.

45 $y' = \ln^2 x + x \cdot 2\ln x \cdot \left(\dfrac{1}{x}\right) = \ln x \cdot (\ln x + 2)$.

47 $y' = 2^{x^2-x} \cdot \ln 2 \cdot (2x - 1)$.

49 Let $y = (x^2 + 1)^{\ln x}$. Then $\ln y = \ln x \ln(x^2 + 1)$. Then

$$\dfrac{1}{y}\dfrac{dy}{dx} = \dfrac{1}{x}\ln(x^2 + 1) + (\ln x)\dfrac{1}{x^2 + 1} \cdot 2x$$
$$= \dfrac{\ln(x^2 + 1)}{x} + \dfrac{2x\ln x}{x^2 + 1},$$

so

$$\dfrac{dy}{dx} = (x^2 + 1)^{\ln x}\left(\dfrac{\ln(x^2+1)}{x} + \dfrac{2x\ln x}{x^2+1}\right).$$

51 $y' = \dfrac{1}{\sqrt{1 - \left(\frac{1}{x}\right)^2}} \cdot -\dfrac{1}{x^2} = -\dfrac{1}{|x|\sqrt{x^2 - 1}}$.

53 $y' = 6\cot^{-1} 3x - \dfrac{18x}{9x^2 + 1} + \dfrac{18x}{9x^2 + 1} = 6\cot^{-1} 3x$.

55 Differentiate implicitly. $1 = -\sin(x - y)(1 - y')$, so $y' - 1 = \dfrac{1}{\sin(x - y)}$, so $y' = \dfrac{1}{\sin(x-y)} + 1 = \csc(x - y) + 1$.

57 Because

$$y' = \dfrac{(1 + \sin x)y'e^y - e^y \cos x}{(1 + \sin x)^2},$$

collecting terms gives

$$y'\left(1 - \frac{e^y}{1+\sin x}\right) = -\frac{e^y \cos x}{(1+\sin x)^2},$$

so

$$y'(1-y) = -\frac{\cos x}{1+\sin x} \cdot y.$$

Thus $y' = -\dfrac{y \cos x}{(1-y)(1+\sin x)}$. This can also be written as $y' = \dfrac{y \cos x}{e^y - 1 - \sin x}$.

59 $y'\sqrt{x^2+y^2} + y \cdot \dfrac{x+yy'}{\sqrt{x^2+y^2}} = 0$, and thus $y'\left(\sqrt{x^2+y^2} + \dfrac{y^2}{\sqrt{x^2+y^2}}\right) = -\dfrac{xy}{\sqrt{x^2+y^2}}$. This can be written as $y'\left(\dfrac{x^2+2y^2}{\sqrt{x^2+y^2}}\right) = -\dfrac{xy}{\sqrt{x^2+y^2}}$, so $y' = -\dfrac{xy}{x^2+2y^2}$.

61 Use logarithmic differentiation. $\ln y = 10\ln(3x+5) + \frac{1}{2}\ln(x^2+5) - 50\ln(x^3+3)$. Then

$$\frac{1}{y}\frac{dy}{dx} = \frac{30}{3x+5} + \frac{x}{x^2+5} - \frac{150x^2}{x^3-3}.$$

So

$$\frac{dy}{dx} = \frac{(3x+5)^{10}\sqrt{x^2+5}}{(x^3+1)^{50}}\left(\frac{30}{3x+5} + \frac{x}{x^2+5} - \frac{150x^2}{x^3-3}\right).$$

63 $f'(x) = \sec^{-1} x + \dfrac{1}{\sqrt{x^2-1}}$. So $f'(2/\sqrt{3}) = \dfrac{\pi}{6} + \sqrt{3}$.

65 $\dfrac{d}{dx}x^{1/x} = \dfrac{d}{dx}e^{\frac{\ln x}{x}} = e^{\frac{\ln x}{x}} \cdot \left(\dfrac{1-\ln x}{x^2}\right) = x^{1/x}\left(\dfrac{1-\ln x}{x^2}\right)$. So $\dfrac{d}{dx}x^{1/x}\bigg|_{x=1} = 1 \cdot \dfrac{1-0}{1^2} = 1$.

67 $y' = 2^x x \ln 2 + 2^x = 2^x(x\ln 2 + 1)$, so $y'' = 2^x \ln 2(x\ln 2 + 1) + 2^x \ln 2 = 2^x \ln 2(x\ln 2 + 2)$.

69 $y' = \dfrac{x^2 \cdot \frac{1}{x} - 2x \ln x}{x^4} = \dfrac{1 - 2\ln x}{x^3}$. Then

$$y'' = \frac{x^3(\frac{-2}{x}) - (1-2\ln x)3x^2}{x^6} = \frac{-2x^2 - 3x^2 + 6x^2 \ln x}{x^6} = \frac{6\ln x - 5}{x^4}.$$

71 Differentiating implicitly: $y + xy' + 2yy' = 0$, so $y'(x+2y) = -y$, so $y' = -\dfrac{y}{x+2y}$. Then $y'' =$

$$-\frac{(x+2y)y' - y(1+2y')}{(x+2y)^2} = \frac{y - xy'}{(x+2y)^2} = \frac{y + x\left(\frac{y}{x+2y}\right)}{(x+2y)^2} = \frac{y(x+2y) + xy}{(x+2y)^3} = \frac{2xy + 2y^2}{(x+2y)^3} = \frac{2}{(x+2y)^3}.$$

Note that for the last equality we are using the fact (from the original equation) that $xy + y^2 = 1$.

73 $y' = 9x^2 + \cos x$. At $x = 0$, $y' = 1$. So the tangent line is given by $y - 0 = 1(x - 0)$, or $y = x$.

75 $y' + \dfrac{y+xy'}{2\sqrt{xy}} = 0$. At the point $(1,4)$, we have $y' + \dfrac{4+y'}{4} = 0$, so $y' = -\dfrac{4}{5}$. The tangent line is given by $y - 4 = -\dfrac{4}{5}(x-1)$, or $y = -\dfrac{4}{5}x + \dfrac{24}{5}$.

77 $\dfrac{d}{dx}[x^2 f(x)] = 2xf(x) + x^2 f'(x)$.

79 $\dfrac{d}{dx}\left(\dfrac{xf(x)}{g(x)}\right) = \dfrac{(f(x) + xf'(x))g(x) - xf(x)g'(x)}{g(x)^2}$.

Chapter Three Review

81

a. This has (D) as its derivative. Note that it consists of two pieces each of which are linear with the same slope. So its derivative is constant – but at $x = 2$ the derivative doesn't exist. We can easily know that this is true because the function isn't continuous at $x = 2$, so it can't be differentiable there.

b. This has (C) as its derivative. The slope of the tangent line is positive for $x < 2$ and negative for $x > 2$ and doesn't exist at $x = 2$. Also, near $x = 2$ the slope is near zero.

c. This has (B) as its derivative. Note that the slope of the tangent line is always positive, and gets infinitely steep at $x = 2$.

d. This has (A) as its derivative. Note that the slope of the tangent line is positive for $x < 2$, negative for $x > 2$, and is infinitely steep at $x = 2$ where the cusp occurs.

83

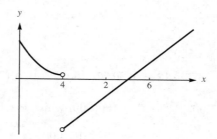

85

a. $\dfrac{d}{dx}[f(x) + 2g(x)]_{x=3} = f'(3) + 2g'(3) = 9 + 2 \cdot 9 = 27.$

b. $\dfrac{d}{dx}\left[\dfrac{f(x)}{g(x)}\right]_{x=1} = \dfrac{g(1)f'(1) - f(1)g'(1)}{g(1)^2} = \dfrac{9(7) - 3(5)}{9^2} = \dfrac{48}{81} = \dfrac{16}{27}.$

c. $\dfrac{d}{dx}(f(x)g(x))\bigg|_{x=3} = f'(3)g(3) + f(3)g'(3) = 9(7) + 1(9) = 72.$

d. $\dfrac{d}{dx}(f(x))^3\bigg|_{x=5} = 3f(5)^2 f'(5) = 3(9)^2 \cdot 5 = 1215.$

e. $(g^{-1})'(7) = \dfrac{1}{g'(3)} = \dfrac{1}{9}.$

87 Note that for $x = 2$, we have $y = \sqrt{8 + 2 - 1} = 3$. $\left(f^{-1}(x)\right)'\bigg|_{x=f(2)} = \dfrac{1}{f'(2)} = \dfrac{1}{\frac{3(2^2)+1}{2\sqrt{2^3+2-1}}} = \dfrac{6}{13}.$

89 If $f(x) = x^{-1/3}$, then $f^{-1}(x) = x^{-3}$. So $\left(f^{-1}\right)'(x) = -3x^{-4}$ for $x \neq 0$.

91

a. Because $f^{-1}(7) = 3$, we have $\left(f^{-1}\right)'(7) = \dfrac{1}{f'(3)} = \dfrac{1}{4}.$

b. Because $f^{-1}(3) = 1$, we have $\left(f^{-1}\right)'(3) = \dfrac{1}{f'(1)} = 1.$

c. $\left(f^{-1}\right)'(f(2)) = \dfrac{1}{f'(2)} = \dfrac{1}{3}.$

93 $\dfrac{d}{dx}(f(g(x)))\bigg|_{x=5} = f'(g(5))g'(5) = f'(3)g'(5) = 4(6) = 24$. Also, $f(g(5)) = f(3) = 2$. So the tangent line is given by $y - 2 = 24(x - 5)$, or $y = 24x - 118$.

95

a. $\dfrac{ds}{dt} = v(t) = 96 - 12t$, so $v(1) = 84$ ft/s.

b. $v(t) = 12$ when $96 - 12t = 12$, or $12t = 84$, so $t = 7$ s.

c. $v(t) = 0$ when $96 - 12t = 0$, or $t = 8$ s. The height at $t = 8$ is $s(8) = 96(8) - 6(64) = 384$ ft.

d. $s(t) = 0$ when $96t - 6t^2 = 0$, or $t(16 - t) = 0$, so $t = 0$ and $t = 16$. The velocity at time $t = 16$ is $v(16) = -96$, so it strikes the ground with a speed of 96 ft/s.

97

a. $A(20) = 40{,}000(1.0075^{240} - 1) \approx \$200{,}366$.

$A'(t) = 40{,}000(1.0075^{12t})\ln(1.0075) \cdot 12 \approx 3586.57(1.0075^{12t})$. $A'(20) \approx \$21{,}552$ dollars/year.

b. $A(t) = 100{,}000$ when $40{,}000(1.0075^{12t} - 1) = 100{,}000$, which occurs when
$$1.0075^{12t} - 1 = \dfrac{100{,}000}{40{,}000} = 2.5,$$
so when
$$1.0075^{12t} = 3.5.$$
This occurs when $12t\ln(1.0075) = \ln 3.5$, or $t = \dfrac{\ln 3.5}{12\ln(1.0075)} \approx 13.97 \approx 14$ years. At that time we have $A'(13.97) \approx \$12{,}551$ per year.

99

a. Average growth is $\dfrac{p(60) - p(50)}{10} = 2.7$ million people per year.

b. The curve is pretty straight between $t = 50$ and $t = 60$, so the secant line between these two points is approximately as steep as the tangent line at a point in between.

c. A reasonable estimate to the instantaneous grow rate at 1985 would be the slope of the secant line between $t = 80$ and $t = 90$. This is $\dfrac{p(90) - p(80)}{10} = 2.217$ million people per year.

101

a. $v(15) \approx \dfrac{400-200}{5} = 40$ meters per second.

b. Because the graph is a straight line for $t \geq 30$, $v(70) = \dfrac{D(90)-D(60)}{30} = \dfrac{1600-1400}{30} = \dfrac{20}{3}$ meters per second. The points at 60 and 90 were chosen because it is easier to detect the function values at those points using the given grid.

c. The average velocity is $\dfrac{D(90) - D(20)}{70} \approx \dfrac{1600 - 550}{70} = 15$ meters per second.

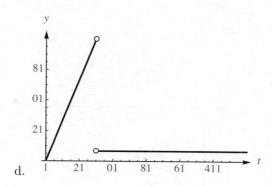

d.

e. The parachute was deployed.

103 We are looking for values of x so that $y'(x) = 0$. We have $y' = \sqrt{6-x} - \frac{x}{2\sqrt{6-x}}$, and this quantity is zero when $2(6-x) - x = 0$, or $12 - 3x = 0$, so when $x = 4$. So at the point $(4, 4\sqrt{2})$ there is a horizontal tangent line. There is a vertical tangent line at $x = 6$, because $\lim_{x \to 6^-} y'(x) = -\infty$.

105 Let $a = 5$ and $f(x) = \tan(\pi\sqrt{3x-11})$. Note that $f'(5) = \frac{3\pi \sec^2(2\pi)}{2} = \frac{3\pi}{4}$.

So $\lim_{x \to 5} \frac{f(x) - f(5)}{x - 5} = \lim_{x \to 5} \frac{\tan(\pi\sqrt{3x-11}) - 0}{x - 5} = f'(5) = \frac{3\pi}{4}$.

107

a. The average cost is $\frac{C(3000)}{3000} = \frac{1025000}{3000} \approx \$341.67$. The marginal cost is $C'(3000) = -0.04(3000) + 400 = \$280$.

b. The average cost of producing 3000 lawnmowers is \$341.67 per mower. The cost of producing the 3001st lawnmower is approximately \$280.

109

a. The average growth rate is $\frac{p(50) - p(0)}{50} = \frac{407500 - 80000}{50} = 6550$ people per year.

b. The growth rate in 1990 is $p'(40) = -5.1(40^2) + 144 \cdot 40 + 7200 = 4800$ people per year.

111

Let x be the distance the eastbound boat has traveled, and y the distance the southbound boat has traveled. By the Pythagorean Theorem, $D^2 = x^2 + y^2$, so $2D\frac{dD}{dt} = 2x\frac{dx}{dt} + 2y\frac{dy}{dt}$, so $\frac{dD}{dt} = \frac{x \cdot x' + y \cdot y'}{D}$. We are given that $x' = 40$, $y' = 30$, and at $t = .5$ hours, we have $x = 20$, $y = 15$, and $D = 25$. Thus, $\frac{dD}{dt} = \frac{20 \cdot 40 + 30 \cdot 15}{25} = 50$ mph.

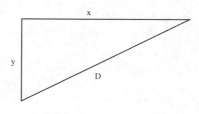

113 Let h be the elevation of the balloon, and s the length of the rope. We have $h = s\sin(65°)$, so $h' = s'\sin(65°) = -5 \cdot \sin(65°) \approx -4.53$ feet per second.

115

Let x be the distance the jet has flown since it went over the spectator. Let θ be the angle of elevation between the ground and the line from the spectator to the jet. Note that θ is also the angle pictured, and that $\cot\theta = \dfrac{x}{500}$. Thus, $\theta = \cot^{-1}\left(\dfrac{x}{500}\right)$. We are given that $x' = 450\,\text{mph} = 660\,\text{ft/sec}$.

$\theta' = -\dfrac{x'}{500 \cdot \left(1 + \left(\frac{x}{500}\right)^2\right)} = -\dfrac{500x'}{250,000 + x^2}$. After 2 seconds, $x = 1320$ feet, so at this time $\theta' = -\dfrac{500 \cdot 660}{250,000 + (1320)^2} \approx -0.166$ radians per second.

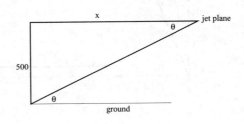

117 Let x equal the length of her shadow and let y equal her distance from the pole. Using similar triangles, we have $\dfrac{x}{5} = \dfrac{x+y}{15}$, which implies that $3x = x+y$ or $x = \dfrac{y}{2}$. It follows that the rate at which her shadow is increasing is $\dfrac{dx}{dt} = \dfrac{1}{2}\dfrac{dy}{dt} = \dfrac{1}{2}(3) = 1.5$ ft/s.

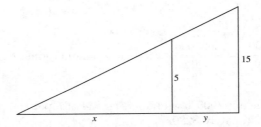

119

a. $(f^{-1})'\left(\dfrac{1}{\sqrt{2}}\right) = \dfrac{1}{f'\left(\frac{\pi}{4}\right)} = \dfrac{1}{\cos\left(\frac{\pi}{4}\right)} = \sqrt{2}.$

b. $\dfrac{d}{dx}\sin^{-1}(x)\bigg|_{x=1/\sqrt{2}} = \dfrac{1}{\sqrt{1-(1/2)}} = \dfrac{1}{\sqrt{1/2}} = \sqrt{2}.$

Chapter 4

Applications of the Derivative

4.1 Maxima and Minima

4.1.1 A number $M = f(c)$ where $c \in [a,b]$ with the property that $f(x) \leq M$ for all $x \in [a,b]$ is an absolute maximum for f on $[a,b]$, and a number $m = f(d)$ where $d \in [a,b]$ with the property that $f(x) \geq m$ for all $x \in [a,b]$ is an absolute minimum for f on $[a,b]$.

4.1.3 The function must be a continuous function defined on a closed interval.

4.1.5 The function shown has no absolute minimum on $[0,3]$ because $\lim_{x \to 0^+} f(x) = -\infty$. It has an absolute maximum near $x = 1$ and a local minimum near $x = 2.5$.

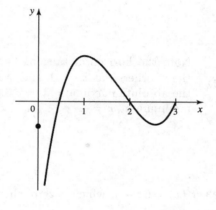

4.1.7 Note the existence of a horizontal tangent line at $x = 0$ where the maximum occurs.

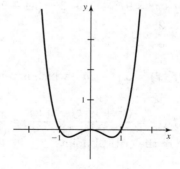

4.1.9 First find all the critical points by seeking all points x in the domain of f so that $f'(x) = 0$ or $f'(x)$ doesn't exist. Now compare the y-values of all of these points, together with the y-values of the endpoints. The largest y-value from among these is the maximum, and the smallest is the minimum.

4.1.11 $y = h(x)$ has an absolute maximum at $x = b$ and an absolute minimum at $x = c_2$.

4.1.13 $y = g(x)$ has no absolute maximum, but has an absolute minimum at $x = a$.

4.1.15 $y = f(x)$ has an absolute maximum at $x = b$ and an absolute minimum at $x = a$. It has local maxima at $x = p$ and $x = r$, and local minima at $x = q$ and $x = s$.

4.1.17 $y = g(x)$ has an absolute minimum at $x = b$ and an absolute maximum at $x = p$. It has local maxima at $x = p$ and $x = r$. It has a local minimum at $x = q$.

4.1.19 Note the horizontal tangent lines at 1 and 2, and the minimum at 0 and the maximum at 4.

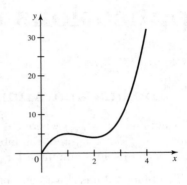

4.1.21 Note the horizontal tangent line at $x = 2$, and the "corners" at $x = 1$ and $x = 3$. Also note the absolute maximum at $x = 3$ and the absolute minimum at $x = 4$.

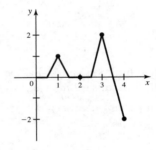

4.1.23 $f'(x) = 6x - 4$, which is zero when $x = \dfrac{2}{3}$. So that is the only critical point.

4.1.25 $f'(x) = x^2 - 9$, which is zero for $x = \pm 3$. So the critical points are $x = \pm 3$.

4.1.27 $f'(x) = 9x^2 + 3x - 2 = (3x+2)(3x-1)$, which is zero for $x = -\dfrac{2}{3}$ and $x = \dfrac{1}{3}$. So the critical points are $x = -\dfrac{2}{3}$ and $x = \dfrac{1}{3}$.

4.1.29 $f'(x) = 3x^2 - 4a^2$, which is zero for $3x^2 = 4a^2$, or $x^2 = \dfrac{4a^2}{3}$. So the critical points are $x = \pm\dfrac{2a}{\sqrt{3}}$.

4.1.31 $f'(t) = \dfrac{(t^2+1)(1) - t(2t)}{(t^2+1)^2} = \dfrac{1-t^2}{(t^2+1)^2}$. This quantity is zero exactly when $1 - t^2 = 0$, so $t = 1$ and $t = -1$ are the critical points.

4.1.33 $f'(x) = \dfrac{e^x - e^{-x}}{2}$, which is zero when $e^x = e^{-x}$ or $x = -x$, so only for $x = 0$.

4.1.35 $f'(x) = -\dfrac{1}{x^2} + \dfrac{1}{x} = \dfrac{x-1}{x^2}$. There is a critical point at $x = 1$.

4.1. Maxima and Minima

4.1.37 $f'(x) = 2x\sqrt{x+5} + x^2 \cdot \dfrac{1}{2\sqrt{x+5}} = \dfrac{4x(x+5)}{2\sqrt{x+5}} + \dfrac{x^2}{2\sqrt{x+5}} = \dfrac{5x^2+20x}{2\sqrt{x+5}}$. This is zero when $5x^2+20x = 5x(x+4)$ is zero, which occurs for $x = 0$ and $x = -4$. The critical points are $x = 0$ and $x = -4$.

4.1.39 $f'(x) = \sqrt{x-a} + \dfrac{x}{2\sqrt{x-a}} = \dfrac{2x-2a+x}{2\sqrt{x-a}} = \dfrac{3x-2a}{2\sqrt{x-a}}$. This expression is zero when $x = \dfrac{2a}{3}$; however, that number is not in the domain of f if $a > 0$. However, if $a < 0$, then $\dfrac{2a}{3}$ is in the domain, and thus gives a critical point.

4.1.41 $f'(t) = t^4 - a^4$, which is zero when $t^4 = a^4$, or $|t| = a$. So there are critical points at $t = a$ and at $t = -a$.

4.1.43 $f'(x) = 2x$, which is zero for $x = 0$. We have that $f(-2) = -6$, $f(0) = -10$, and $f(3) = -1$, so the maximum value of f on this interval is -1 and the minimum is -10.

4.1.45 $f'(x) = 3x^2 - 6x = 3x(x-2)$. This is zero for $x = 0$ and $x = 2$. Checking endpoints and critical points we have $f(-1) = -4$, $f(0) = 0$, $f(2) = -4$, and $f(3) = 0$. So there is an absolute maximum of 0 at $x = 0$ and $x = 3$, and an absolute minimum of -4 at $x = -1$ and $x = 2$.

4.1.47 $f'(x) = 15x^4 - 75x^2 + 60 = 15(x^4 - 5x^2 + 4) = 15(x^2-4)(x^2-1) = 15(x-2)(x+2)(x-1)(x+1)$. So the critical points are $x = 1, -1, 2, -2$. Checking the endpoints and critical points we have $f(-2) = -16$, $f(-1) = -38$, $f(0) = 0$, $f(1) = 38$, $f(2) = 16$, and $f(3) = 234$. There is an absolute minimum of -38 at $x = -1$, and an absolute maximum of 234 at $x = 3$.

4.1.49 $f'(x) = -2\cos x \sin x$, which is zero for $x = 0$, $x = \pi/2$, and $x = \pi$. Because there are endpoints at $x = 0$ and $x = \pi$, only $(\pi/2, 0)$ is a critical point. We have that $f(0) = 1$, $f(\pi/2) = 0$, and $f(\pi) = 1$, so the maximum value of f on this interval is 1 and the minimum is 0.

4.1.51 $f'(x) = 3\cos 3x$, which is zero when

$$3x = \ldots, -\pi/2, \pi/2, 3\pi/2, \ldots,$$

so when $x = \ldots, -\pi/6, \pi/6, \pi/2, \ldots$. The only such values on the given interval are $x = -\pi/6$ and $x = \pi/6$. We have

$$f(-\pi/4) = -\sqrt{2}/2 \approx -0.707,$$

$f(-\pi/6) = -1$, $f(\pi/6) = 1$, and $f(\pi/3) = 0$, so the absolute maximum of f is 1 at $x = \pi/6$ and the absolute minimum is -1 at $x = -\pi/6$.

4.1.53 Let $y = (2x)^x$, so that $\ln y = x\ln(2x)$. Then $\dfrac{1}{y}y' = \ln(2x) + \dfrac{x}{2x} \cdot 2 = 1 + \ln(2x)$. Thus $y' = (2x)^x(1 + \ln(2x))$. This quantity is zero when $1 + \ln(2x) = 0$, which occurs when $\ln(2x) = -1$, or $x = \dfrac{1}{2e} \approx 0.184$. We have $f(0.1) \approx 0.851$, $f\left(\dfrac{1}{2e}\right) = e^{-(1/2e)} \approx 0.832$, and $f(1) = 2$. So the absolute minimum is $e^{-(1/2e)}$ and the absolute maximum is 2.

4.1.55 $f'(x) = 2x - \dfrac{1}{\sqrt{1-x^2}} = \dfrac{2x\sqrt{1-x^2}-1}{\sqrt{1-x^2}}$. This is zero on $(-1,1)$ when the numerator is zero, which is when $2x\sqrt{1-x^2} = 1$, so when $(4x^2)(1-x^2) = 1$, or $4x^4 - 4x^2 + 1 = 0$. This factors as $(2x^2-1)(2x^2-1) = 0$, so we have solutions for $x = \pm\sqrt{1/2}$. We have $f(-1) = 1 + \pi$, $f(-1/\sqrt{2}) = \dfrac{1}{2} + \dfrac{3\pi}{4}$, $f(1/\sqrt{2}) = \dfrac{1}{2} + \dfrac{\pi}{4}$, and $f(1) = 1$. So the maximum for f is $1 + \pi$ and the minimum is 1.

4.1.57 $f'(x) = 6x^2 - 30x + 24 = 6(x^2 - 5x + 4) = 6(x-4)(x-1)$. This is zero at $x = 4$ and $x = 1$. $f(1) = 11$ and $f(4) = -16$. At the endpoints we have $f(0) = 0$ and $f(5) = -5$. The absolute maximum is 11 and the absolute minimum is -16.

4.1.59 $f'(x) = 4x^2 + 10x - 6 = 2(2x^2 + 5x - 3) = 2(x+3)(2x-1)$. This is zero when $x = -3$ and when $x = 1/2$. $f(-3) = 27$ and $f(1/2) = -19/12$. At the endpoints we have $f(-4) = 56/3 \approx 18.7$, and $f(1) = 1/3$. The absolute maximum is 27 and the absolute minimum is $-19/12$.

4.1.61 $f'(x) = \dfrac{(x^2+9)^5 - x \cdot 5(x^2+9)^4(2x)}{(x^2+9)^{10}} = \dfrac{(x^2+9)^4(x^2+9-10x^2)}{(x^2+9)^{10}} = \dfrac{9 - 9x^2}{(x^2+9)^6} = \dfrac{9(1-x)(1+x)}{(x^2+9)^2}$.

This expression is zero for $x = \pm 1$. Checking the endpoints and critical points gives $f(-2) = -\dfrac{2}{13^5}$, $f(-1) = -\dfrac{1}{10^5}$, $f(1) = \dfrac{1}{10^5}$ and $f(2) = \dfrac{2}{13^5}$. The absolute maximum is $\dfrac{1}{100000}$ at $x = 1$ and the absolute minimum is $-\dfrac{1}{100000}$ at $x = -1$.

4.1.63 $f'(x) = \sec x \tan x$ which is zero when $\tan x = 0$ (since $\sec x$ is never zero.) So we are looking for where $\dfrac{\sin x}{\cos x} = 0$, which is when $\sin x = 0$, which is at $x = 0$. $f(-\pi/4) = \sqrt{2} = f(\pi/4)$ and $f(0) = 1$. So the absolute maximum for f is $\sqrt{2}$ and the absolute minimum is 1.

4.1.65 $f'(x) = 3x^2 e^{-x} + x^3 \cdot (-e^{-x}) = e^{-x} \cdot (3x^2 - x^3) = e^{-x} \cdot x^2 \cdot (3 - x)$. This expression is zero when $x = 0$ and when $x = 3$, so $(0,0)$ and $(3, (27/e^3))$ are the critical points. $f(-1) = -e$, $f(0) = 0$, $f(3) = \dfrac{27}{e^3} \approx 1.344$, and $f(5) \approx 0.8422$. So the absolute maximum of f on the given interval is about 1.344, and the absolute minimum is $-e \approx -2.718$.

4.1.67 $f'(x) = x^{2/3}(-2x) + (4 - x^2) \cdot \dfrac{2}{3\sqrt[3]{x}} = \dfrac{-6x^2 + 8 - 2x^2}{3\sqrt[3]{x}} = \dfrac{8 - 8x^2}{3\sqrt[3]{x}}$. This quantity is zero when $x = \pm 1$ and it doesn't exist at $x = 0$. So there are critical points at $(0,0)$, $(-1, 3)$, and $(1, 3)$. Checking the endpoints, we have $f(\pm 2) = 0$. So the absolute minimum is 0 at $x = \pm 2$ and $x = 0$, and the absolute maximum is 3 at $x = \pm 1$.

4.1.69

a. The wind either passes through the blade without slowing down or the turbine blades slow down the wind so that $0 \leq v_2 \leq v_1$, and because $v_1 > 0$, $0 \leq \dfrac{v_2}{v_1} \leq 1$.

b. $R(1) = \dfrac{1}{2}(1+1)(1-1) = 0$. Given that $r = 1$, it is the case that $v_1 = v_2$ which means that the wind does not slow down after passing through the blades and therefore no power is captured by the turbine blades. It stands to reason that the blades should slow down the wind, at least a little bit, and therefore it seems unlikely or impossible for $r = 1$.

c. $R(0) = \dfrac{1}{2}$. A value of $r = 0$ means that $v_2 = 0$, so that the wind turbine stops the wind stream, which is unlikely or impossible to occur.

d. Finding the derivative of R and setting it equal to zero, we have

$$R'(r) = \dfrac{1}{2}((1 - r^2) + (1+r)(-2r)) = -\dfrac{1}{2}(3r^2 + 2r - 1) = -\dfrac{1}{2}(3r - 1)(r + 1) = 0.$$

Given that $R(0) = \dfrac{1}{2}$, $R\left(\dfrac{1}{3}\right) = \dfrac{1}{2}\left(\dfrac{4}{3}\right)\left(\dfrac{8}{9}\right) = \dfrac{16}{27} \approx 0.593$, R is maximized on $[0,1]$ when $r = \dfrac{1}{3}$. At best, a wind turbine can extract about 59.3% of the available power from the wind stream.

4.1.71 $s'(t) = 32 - 4t^3$ which is 0 when $t^3 = 8$, so $t = 2$. Because $s(0) = 0$, $s(2) = 48$, and $s(3) = 15$, the object is furthest to the right at time $t = 2$.

4.1.73 The stone will reach its maximum height when its velocity is zero, which occurs at the only critical point for this inverted parabola. We have that $v(t) = s'(t) = -32t + 64$, which is zero when $t = 2$. The height at this time is $s(2) = 256$, the maximum height.

4.1.75

a. Note that $P(n) = 50n - .5n^2 - 100$, so $P'(n) = 50 - n$, which is zero when $n = 50$. It is clear that this is a maximum, since the graph of P is an inverted parabola.

b. Given a domain of $[0, 45]$, since the only critical point is not in the domain, the maximum must occur at an endpoint. Because $P(0) = -100$, and $P(45) = \$1137.50$, he should take 45 people on the tour.

4.1.77

a. False. The derivative $f'(x) = \dfrac{1}{2\sqrt{x}}$ is never zero, and the function has no critical points.

b. False. For example, the function $f(x) = \begin{cases} \sin x & \text{if } -5 \leq x \leq 0, \\ -8 & \text{if } 0 < x \leq 5 \end{cases}$ is not continuous on $[-5, 5]$, but has an absolute maximum of 1.

c. False. For example, the function $f(x) = (x-2)^3$ satisfies $f'(2) = 0$, but it has neither a maximum nor a minimum at $x = 2$.

d. True. This follows from the theorems in this section.

4.1.79

a. $f'(x) = 2^x \cdot \ln 2 \cdot \sin x + 2^x \cos x = 2^x((\ln 2) \cdot \sin x + \cos x)$. Because 2^x is never zero, this expression is zero only when $(\ln 2) \cdot \sin x + \cos x = 0$, or $(\ln 2) \cdot \tan x = -1$, or $\tan x = \left(-\dfrac{1}{\ln 2}\right)$. So one solution is $x = \tan^{-1}\left(-\dfrac{1}{\ln 2}\right) \approx -0.9647$. And since the tangent function is periodic with period π, we also have solutions at approximately $-0.9647 + \pi \approx 2.1769$, and $-0.9647 + 2\pi \approx 5.3185$. These are the only solutions on the given interval.

b. $f(-2) \approx -0.2273$, $f(-0.9647) \approx -0.4211$, $f(2.1769) \approx 3.7164$, $f(5.3185) \approx -32.7968$, and $f(6) \approx -17.8826$. Thus the absolute maximum is about 3.7164 and the absolute minimum is about -32.7968.

c.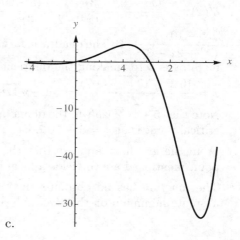

4.1.81

a. $f'(\theta) = 2\cos\theta - \sin\theta$, which is zero when $\tan\theta = 2$. So one critical point occurs at $\theta = \tan^{-1}(2) \approx 1.107$. And since the tangent function is periodic with period π, there are will also be solutions at this number plus or minus integer multiples of π. On the given interval, these are located at approximately $1.107 - 2\pi \approx -5.176$, at $1.107 - \pi \approx -2.034$, and at $1.107 + \pi \approx 4.249$.

b. From the graph, it appears that there is a local minimum at about $\theta = -2.034$ and at $\theta = 4.249$, and there is a local maximum at about $\theta = -5.176$, and at about $\theta = 1.107$.

c. From the graph, it appears that the local minimum at about $\theta = -2.034$ is also an absolute minimum, as is the one at $\theta = 4.249$. The local maximum at about $\theta = -5.176$, and at about $\theta = 1.107$ are also absolute maxima. The value of the absolute maximum appears to be about 2.24 and the value of the absolute minimum appears to be about -2.24.

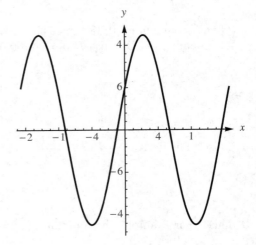

4.1.83

a. $h'(x) = \dfrac{(x^2 + 2x - 3)(-1) - (5-x)(2x+2)}{(x^2 + 2x - 3)^2} = \dfrac{x^2 - 10x - 7}{(x^2 + 2x - 3)^2}$. By the quadratic formula, the numerator is zero (making the quotient zero) when $x = \dfrac{10 \pm \sqrt{100 - 4(-7)}}{2} = 5 \pm \dfrac{1}{2}\sqrt{128} = 5 \pm 4\sqrt{2}$. Note that $5 + 4\sqrt{2}$ isn't in the domain, so the only critical point is at $x = 5 - 4\sqrt{2}$.

b. From the graph, it appears that the one critical point mentioned above yields a local maximum.

c. The function has no absolute maximum and no absolute minimum on the given interval.

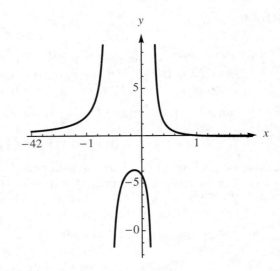

4.1.85

Note that

$$g(x) = \begin{cases} 3 - x + 2x + 2 = x + 5 & \text{if } -2 \leq x \leq -1, \\ 3 - x - 2x - 2 = 1 - 3x & \text{if } -1 \leq x \leq 3. \end{cases}$$

There is an absolute maximum of 4 and an absolute minimum of -8. The absolute maximum occurs at $x = -1$, and the absolute minimum occurs at a $x = 3$.

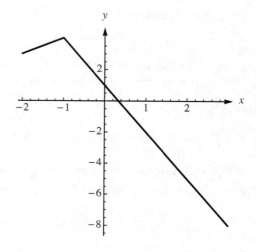

4.1.87

a. Because distance is rate times time, the time will be distance over rate. The swim distance is given by $\sqrt{2500 + x^2}$ meters, so the time for swimming is $\dfrac{\sqrt{2500 + x^2}}{2}$. For running, the distance is $50 - x$, so the time is $\dfrac{50 - x}{4}$. Thus we have $T(x) = \dfrac{\sqrt{2500 + x^2}}{2} + \dfrac{50 - x}{4}$.

b. $T'(x) = \dfrac{1}{2} \cdot \dfrac{1}{2} \left(x^2 + 2500\right)^{-1/2} \cdot 2x - \dfrac{1}{4} = \dfrac{x}{2\sqrt{x^2 + 2500}} - \dfrac{1}{4}$. This expression is zero when $\dfrac{x^2}{x^2 + 2500} = \dfrac{1}{4}$, so when $4x^2 = x^2 + 2500$, which occurs when $x^2 = \dfrac{2500}{3}$. So $x = \sqrt{\dfrac{2500}{3}} \approx 28.868$.

c. $T(0) = 37.5$, $T(28.868) \approx 34.151$, and $T(50) = 25\sqrt{2} \approx 35.355$. The absolute minimum occurs at the only critical point. The minimal crossing time is approximately 34.151 seconds.

d.

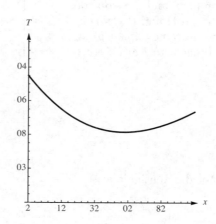

4.1.89

a. Note that since there is a local extreme value at 2 for f and since f is differentiable everywhere, we must have $f'(2) = 0$.

$g(2) = 2f(2) + 1 = 1.$

$h(2) = 2f(2) + 2 + 1 = 3$.

$g'(2) = 2 \cdot f'(2) + f(2) = 0$.

$h'(2) = 2f'(2) + f(2) + 1 = 1$.

b. h doesn't, since its derivative isn't zero at $x = 2$. However g might: for example, if $f(x) = (x-2)^2$ then $g(x) = x(x-2)^2 + 1$ has a local minimum at $x = 2$.

4.1.91

a. If $f(c)$ is a local maximum, then when x is near c but not equal to c, $f(c) \geq f(x)$, so $f(x) - f(c) \leq 0$.

b. When x is near to c but a little bigger than c, $x - c > 0$. So in this case, $\dfrac{f(x) - f(c)}{x - c} \leq 0$, since the numerator is negative (or 0) and the denominator is positive.

Thus, $\lim\limits_{x \to c^+} \dfrac{f(x) - f(c)}{x - c} = f'(c) \leq 0$

c. When x is near to c but a little smaller than c, $x - c < 0$. So in this case, $\dfrac{f(x) - f(c)}{x - c} \geq 0$, since the numerator is negative (or 0) and the denominator is negative, making the quotient positive (or 0).

Thus, $\lim\limits_{x \to c^-} \dfrac{f(x) - f(c)}{x - c} = f'(c) \geq 0$.

d. From the above, we have that $f'(c) \leq 0$ and $f'(c) \geq 0$, so $f'(c) = 0$.

4.2 Mean Value Theorem

4.2.1 If f is a continuous function on the closed interval $[a, b]$ and is differentiable on (a, b) and the slope of the secant line that joins $(a, f(a))$ and $(b, f(b))$ is zero, then there is at least one value c in (a, b) at which the slope of the line tangent to f at $(c, f(c))$ is also zero.

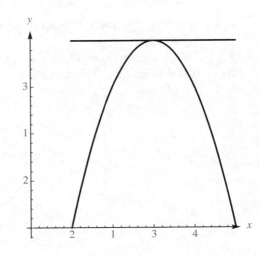

4.2.3 The function $f(x) = |x|$ is not differentiable at 0.

4.2.5

a. It appears that at $c = 1$ the tangent line is parallel to the secant line shown.

b. $f'(x) = \dfrac{x}{2}$. The slope of the secant line is $\dfrac{f(4) - f(-2)}{4 - (-2)} = \dfrac{5 - 2}{6} = \dfrac{1}{2}$. We have $\dfrac{x}{2} = \dfrac{1}{2}$ at $x = 1$, as conjectured.

4.2.7

a. It appears that at a point a little bigger than 1 (but less than 1.5) the tangent line is parallel to the secant line shown, and then again at a point between -1 and -1.5.

b. $f'(x) = \dfrac{5x^4}{16}$. The slope of the secant line is $\dfrac{f(2) - f(-2)}{2 - (-2)} = \dfrac{2 - (-2)}{2 - (-2)} = 1$. We have $\dfrac{5x^4}{16} = 1$ when $x^4 = \dfrac{16}{5}$, or $c = \pm \dfrac{2}{\sqrt[4]{5}} \approx \pm 1.34$.

4.2.9 We seek a function over an interval for which it isn't true that there is a tangent line parallel to the secant line between the endpoints.

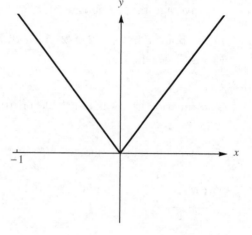

4.2.11 The function f is differentiable on $[0, 1]$ and $f(0) = f(1) = 0$, so Rolle's theorem applies. We wish to find a point x in $(0, 1)$ such that $f'(x) = 0$; we have $f'(x) = (x-1)^2 + 2x(x-1) = (x-1)(3x-1)$, so $x = 1/3$ satisfies the conclusion of Rolle's theorem.

4.2.13 The function f is differentiable on $[\pi/8, 3\pi/8]$ and $f(\pi/8) = f(3\pi/8) = 0$, so Rolle's theorem applies. We wish to find a point x in $(\pi/8, 3\pi/8)$ such that $f'(x) = 0$; we have $f'(x) = -4\sin 4x$, so $x = \pi/4$ satisfies the conclusion of Rolle's theorem.

4.2.15 The function f is not differentiable at $x = 0$, so Rolle's theorem does not apply.

4.2.17 g is continuous on $[-1, 3]$ and differentiable on $(-1, 3)$, and $g(-1) = 0 = g(3)$, so Rolle's theorem does apply. $g'(x) = 3x^2 - 2x - 5 = (x+1)(3x-5)$. This is zero for $x = -1$ (which is not on $(-1, 3)$) and for $x = 5/3$ (which is on $(-1, 3)$.) So $x = 5/3$ satisfies the conclusion of Rolle's theorem.

4.2.19 The average rate of change of the temperature from 3.2 km to 6.1 km is $\dfrac{-10.3 - 8.0}{6.1 - 3.2} \approx -6.3°/\text{km}$. Based on this, we cannot conclude that the lapse rate exceeds the critical value of $7°/\text{km}$.

4.2.21

a. The function f is continuous on $[-1, 2]$ and differentiable on $(-1, 2)$. so the Mean Value Theorem applies.

b. The average rate of change of f on $[-1, 2]$ is $\dfrac{f(2) - f(-1)}{2 - (-1)} = \dfrac{3 - 6}{3} = -1$. We wish to find a point c in $(-1, 2)$ such that $f'(c) = -1$, or equivalently $-2c = -1$ which gives $c = 1/2$.

4.2.23

a. The MVT does not apply because f is not differentiable at $x = 0$. To see this note that $\lim\limits_{h \to 0^+} \dfrac{f(0+h) - f(0)}{h} = 1$ and $\lim\limits_{h \to 0^-} \dfrac{f(0+h) - f(0)}{h} = -2$.

4.2.25

a. The function f is differentiable on $(0,1)$ and continuous on $[0,1]$ so the Mean Value Theorem applies.

b. The average rate of change of f on $[0,1]$ is $\dfrac{f(1)-f(0)}{1-0} = \dfrac{e-1}{1} = e-1$. We wish to find a point c in $(0,1)$ such that $f'(c)=e-1$, or equivalently $e^c = e-1$ which gives $c = \ln(e-1) \approx 0.541$.

4.2.27

a. f is continuous on $[-\pi/2, \pi/2]$ and differentiable on $(-\pi/2, \pi/2)$, so the Mean Value Theorem applies.

b. The average rate of change of f on $[-\pi/2,\pi/2]$ is $\dfrac{f(\pi/2)-f(-\pi/2)}{\pi/2-(-\pi/2)} = \dfrac{1-(-1)}{\pi} = \dfrac{2}{\pi}$. We wish to find a point c in $(-\pi/2, \pi/2)$ such that $f'(c) = \dfrac{2}{\pi}$, or equivalently $\cos c = \dfrac{2}{\pi}$ which gives $c \approx \pm 0.881$ (using a root finder).

4.2.29

a. The function f is differentiable on $(0,1/2)$ and continuous on $[0,1/2]$ so the Mean Value Theorem applies.

b. The average rate of change of f on $[0,1/2]$ is $\dfrac{f(1/2)-f(0)}{\frac{1}{2}-0} = \dfrac{\frac{\pi}{6}-0}{\frac{1}{2}} = \dfrac{\pi}{3}$.

We wish to find a point c in $(0,1/2)$ such that $f'(c) = \pi/3$, or equivalently $\dfrac{1}{\sqrt{1-c^2}} = \dfrac{\pi}{3}$, so $c = \sqrt{1-\dfrac{9}{\pi^2}}$.

4.2.31

a. The Mean Value Theorem does not apply because the function f is not differentiable at $x=0$.

b. Even though the Mean Value Theorem doesn't apply, it still happens to be the case that there are numbers c between -8 and 8 where the tangent line has slope $\dfrac{f(8)-f(-8)}{8-(-8)} = \dfrac{1}{2}$. This occurs where $\dfrac{2}{3}c^{-2/3} = 1/2$, which gives $c = \pm\dfrac{8}{9}\cdot\sqrt{3}$.

4.2.33

a. False. The function f is not differentiable at $x=0$.

b. True. If $f(x)-g(x) = c$ is constant, then $f'(x)-g'(x) = 0$.

c. False. If $f'(x)=0$ then we can conclude that $f(x)=c$ for some constant.

4.2.35

a.
$$\dfrac{d}{dx}\tan^{-1}(2/x^2) = \dfrac{1}{1+4/x^4}\cdot -\dfrac{4}{x^3} = \dfrac{1}{1+4/x^4}\cdot -\dfrac{4}{x^3}\cdot\dfrac{x}{x} = -\dfrac{4x}{x^4+4}.$$

$$\begin{aligned}\dfrac{d}{dx}\left(\tan^{-1}(x+1)-\tan^{-1}(x-1)\right) &= \dfrac{1}{1+(x+1)^2}-\dfrac{1}{1+(x-1)^2}\\ &= \dfrac{x^2-2x+2-(x^2+2x+2)}{1+(x+1)^2+(x-1)^2+(x^2-1)^2}\\ &= -\dfrac{4x}{1+x^2+2x+1+x^2-2x+1+x^4-2x^2+1}\\ &= -\dfrac{4x}{x^4+4}.\end{aligned}$$

Because these two functions have the same derivative, they differ by a constant. So for any x, $\tan^{-1}(2/x^2) - (\tan^{-1}(x+1) - \tan^{-1}(x-1))$ is a constant.

b. Because the two function in part (a) differ by a constant, we can compute the constant by evaluating for a specific number x. Choosing $x = 1$, we have $\tan^{-1} 2 - (\tan^{-1} 2 - \tan^{-1} 0) = 0$, so the constant is 0, and we have
$$\tan^{-1}(2/x^2) = \left(\tan^{-1}(x+1) - \tan^{-1}(x-1)\right).$$

4.2.37 The functions $h(x)$ and $p(x)$ have the same derivative as $f(x)$ because they differ from $f(x)$ by a constant.

4.2.39 The secant line between the endpoints has slope $\dfrac{f(4) - f(-4)}{4 - (-4)} = \dfrac{4-1}{8} = \dfrac{3}{8}$. The slope of the tangent line to the graph appears to have this value at approximately -2.5 and at about 2.6. These are eyeballed estimates, so your personal estimate may differ.

4.2.41 Because $f(1) \approx 2$ and $f(3) \approx 2$, the average rate of change of f on $[1,3]$ is $\dfrac{f(3) - f(1)}{3 - 1} = 0$. However, the tangent to f between $x = 1$ and $x = 2$ is the graph of f itself, which has slope 2, and the tangent to f between $x = 2$ and $x = 3$ is also the graph of f, which has slope 1. So there is no point in $(1, 3)$ where the tangent line has slope 0. This does not contradict the Mean Value Theorem because f is not continuous everywhere on $[1, 3]$, nor differentiable everywhere on $(1, 3)$ both of these hypotheses fail at x = 2.

4.2.43 The average speed of the car over the 28 minute period $(= 28/60$ hr$)$ is $\dfrac{30 - 0}{28/60} \approx 64.3$mi/hr, so the officer can conclude by the Mean Value Theorem that at some point the car exceeded the speed limit.

4.2.45 The runner's average speed is $6.2/(32/60) \approx 11.6$ mi /hr. By the Mean Value Theorem, the runner's speed was 11.6 mi/hr at least once. By the intermediate value theorem, all speeds between 0 and 11.6 mi/hr were reached. Because the initial and final speed was 0 mi/hr, the speed of 11 mi/hr was reached at least twice.

4.2.47 Observe that
$$\frac{f(b) - f(a)}{b - a} = \frac{A(b^2 - a^2) + B(b - a)}{b - a} = A(a + b) + B$$
and $f'(c) = 2Ac + B$, so the point c that satisfies the conclusion of the Mean Value Theorem is $c = (a+b)/2$.

4.2.49 Note that $f'(x) = 2 \tan x \sec^2 x$ and $g'(x) = 2 \sec x \sec x \tan x = 2 \tan x \sec^2 x$, so $f'(x) = g'(x)$. This implies that $f - g$ is a constant, which also follows from the trigonometric identity $\sec^2 x = \tan^2 x + 1$.

4.2.51 Let $f(x) = \sec^{-1} x$ and $g(x) = \cos^{-1}\left(\dfrac{1}{x}\right)$. Then
$$f'(x) = \frac{1}{|x|\sqrt{x^2 - 1}}$$
and
$$g'(x) = -\frac{1}{\sqrt{1 - (1/x)^2}} \cdot \left(-\frac{1}{x^2}\right) = \frac{1}{|x|^2 \sqrt{1 - (1/x)^2}} = \frac{1}{|x|\sqrt{x^2}\sqrt{1 - (1/x)^2}} = \frac{1}{|x|\sqrt{x^2 - 1}}.$$
Because $f'(x) = g'(x)$ for $x > 0$, we must have that $f(x) - g(x) = C$ for $x > 0$. Let $x = 1$. Then $C = f(1) - g(1) = 0 - 0 = 0$, so $f(x) - g(x) = 0$, so $f(x) = g(x)$ for $x > 0$. Similarly, $f(x) - g(x) = C$ for $x < 0$, and letting $x = -1$ gives $f(x) - g(x) = 0$ so $f(x) = g(x)$ for $x < 0$. Thus $\sec^{-1} x = \cos^{-1}\left(\dfrac{1}{x}\right)$ for all $x \neq 0$.

4.2.53 By the Mean Value Theorem, there is a number c in $(2, 4)$ so that $\dfrac{f(4) - f(2)}{4 - 2} = f'(c)$, or $f(4) - 7 = 2f'(c)$. Because $f'(c) < 2$, we must have $f(4) - 7 < 4$, which implies that $f(4) < 11$.

4.2.55 Let $a > 0$ and let $f(x) = \sqrt{1+x}$ on $[0, a]$. By the Mean Value Theorem, there is a number c in $(0, a)$ so that $\dfrac{f(a) - f(0)}{a - 0} = f'(c)$, which implies that

$$\frac{\sqrt{1+a} - 1}{a} = \frac{1}{2\sqrt{1+c}}.$$

Because $\sqrt{1+c} > 1$ for $c > 0$, $\dfrac{1}{2\sqrt{1+c}} < \dfrac{1}{2}$. So

$$\frac{\sqrt{1+a} - 1}{a} = \frac{1}{2\sqrt{1+c}} < \frac{1}{2}$$

so

$$\frac{\sqrt{1+a} - 1}{a} < \frac{1}{2}.$$

Rewriting this inequality gives

$$1 + \frac{a}{2} > \sqrt{1+a}$$

for $a > 0$.

4.2.57

a. If $g(x) = x$ then $g'(x) = 1$ and hence $\dfrac{f(b) - f(a)}{g(b) - g(a)} = \dfrac{f(b) - f(a)}{b - a} = \dfrac{f'(c)}{g'(c)} = f'(c)$.

b. We have $\dfrac{f(b) - f(a)}{g(b) - g(a)} = \dfrac{0 - (-1)}{6 - 2} = \dfrac{1}{4}$; $\dfrac{f'(c)}{g'(c)} = \dfrac{2c}{4} = \dfrac{c}{2}$; so $c = 1/2$.

4.3 What Derivatives Tell Us

4.3.1 If f' is positive on an interval, f is increasing on that interval. If f' is negative on an interval, f is decreasing on that interval.

4.3.3

a. $x - 3 = 0$ for $x = 3$, so $x = 3$ is a critical point of f.

b. $f'(x) > 0$ for $x > 3$, so f is increasing on $(3, \infty)$. $f' < 0$ on $(-\infty, 3)$, so f is decreasing on $(-\infty, 3)$.

4.3.5

One such example is $f(x) = x^3$ at $x = 0$.

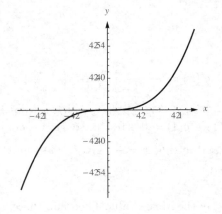

4.3.7

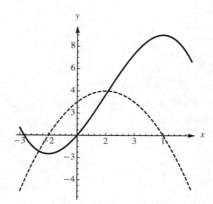

f' is dotted and possible f is solid.

4.3.9

Such a function would be decreasing until $x = 2$, then increasing until $x = 5$, and then decreasing again after that.

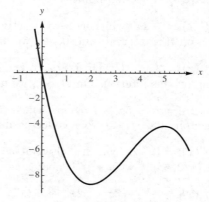

4.3.11

Such a function has extrema (minima) at 0 and 4, where the y value is zero. The function should never go below the x axis.

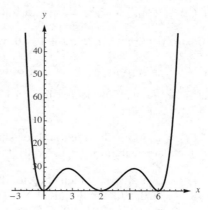

4.3.13

a. $g'' > 0$ on $(-\infty, 2)$ so g is concave up there; $g'' < 0$ on $(2, \infty)$, so g is concave down there.

b. There is an inflection point at $x = 2$.

4.3.15

Yes, for example, consider $f(x) = 100 - x^2$ on the interval $(-8, 0)$. It is above the x axis, increasing, and concave down on that interval.

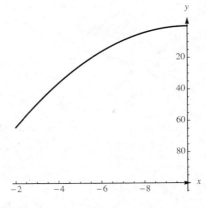

4.3.17 $f(x) = x^4$ has this property at 0. Note that $f''(x) = 12x^2$, which is 0 at $x = 0$, but the function doesn't have an inflection point there.

4.3.19 $f'(x) = -2x$, which is zero exactly when $x = 0$. On $(-\infty, 0)$ we note that $f' > 0$, so that f is increasing on this interval. On $(0, \infty)$, we note that $f' < 0$, so f is decreasing on this interval.

4.3.21 $f'(x) = 2(x - 1)$, which is zero exactly when $x = 1$. On $(-\infty, 1)$ we note that $f' < 0$, so that f is decreasing on this interval. On $(1, \infty)$, we note that $f' > 0$, so f is increasing on this interval.

4.3.23 $f'(x) = x^2 - 5x + 4 = (x-4)(x-1)$, which is 0 for $x = 1$ and $x = 4$. On $(-\infty, 1)$ $f' > 0$ so f is increasing; on $(1, 4)$ $f' < 0$ so f is decreasing; and on $(4, \infty)$ $f' > 0$ so f is increasing.

4.3.25 $f'(x) = 1 - 2x$, which is 0 when $x = 1/2$. On $(-\infty, 1/2)$ $f' > 0$ so f is increasing on this interval, while on $(1/2, \infty)$ $f' < 0$, so f is deceasing on this interval.

4.3.27 $f'(x) = -x^3 + 3x^2 - 2x = -x(x^2 - 3x + 2) = -x(x-1)(x-2)$. This is zero when $x = 0$, $x = 1$, and $x = 2$. Note that $f'(-1) > 0$, and $f'(1.5) > 0$, while $f'(.5) < 0$, and $f'(3) < 0$. So f is increasing on $(-\infty, 0)$ and on $(1, 2)$, while it is decreasing on $(0, 1)$ and on $(2, \infty)$.

4.3.29 $f'(x) = 2x \ln x^2 + x^2 \cdot \dfrac{1}{x^2} \cdot 2x = 2x(\ln x^2 + 1)$. This is undefined for $x = 0$, and when $\ln x^2 + 1 = 0$. For $x > 0$, this occurs when $2 \ln x + 1 = 0$, which occurs when $\ln x = -\dfrac{1}{2}$, or $x = \dfrac{1}{\sqrt{e}}$. By symmetry, we also have that $f'(x)$ is zero for $x = -\dfrac{1}{\sqrt{e}}$. Note that $\dfrac{1}{\sqrt{e}} \approx 0.6$, and that $f'(-1) < 0$, $f'(-1/2) > 0$, $f'(1/2) < 0$, and $f'(1) > 0$. Thus, f is decreasing on $(-\infty, -1/\sqrt{e})$ and on $(0, 1/\sqrt{e})$, and is increasing on $(-1/\sqrt{e}, 0)$ and on $(1/\sqrt{e}, \infty)$.

4.3.31 $f'(x) = 2 \sin x - 1$ which is 0 for $\sin x = 1/2$, or (on the given interval) $x = \pi/6, 5\pi/6$. On $(0, \pi/6)$ $f' < 0$ so f is decreasing; on $(\pi/6, 5\pi/6)$ $f' > 0$ so f is increasing; on $(5\pi/6, 2\pi)$ $f' < 0$ so f is decreasing.

4.3.33 $f'(x) = -9 \sin 3x$, which is 0 for $3x = -3\pi, -2\pi, -\pi, 0, \pi, 2\pi,$ and 3π, which corresponds to $x = -\pi, -2\pi/3, -\pi/3, 0, \pi/3, 2\pi/3$ and π. Note that $f'(-5\pi/6) = 9 > 0$, $f'(-\pi/2) = -9 < 0$, $f'(-\pi/6) = 9 > 0$, $f'(\pi/6) = -9 < 0$, $f'(\pi/2) = 9 > 0$, and $f'(5\pi/6) = -9 < 0$. Thus f is increasing on $(-\pi, -2\pi/3)$, on $(-\pi/3, 0)$, and on $(\pi/3, 2\pi/3)$, while f is decreasing on $(-2\pi/3, -\pi/3)$, on $(0, \pi/3)$, and on $(2\pi/3, \pi)$.

4.3.35 $f'(x) = (2/3)x^{-1/3}(x^2 - 4) + x^{2/3} \cdot 2x = \dfrac{2(x^2 - 4)}{3x^{1/3}} + \dfrac{2x^{5/3}}{1} \cdot \dfrac{3x^{1/3}}{3x^{1/3}} = \dfrac{8x^2 - 8}{3x^{1/3}} = \dfrac{8(x+1)(x-1)}{3x^{1/3}}$. This is zero for $x = \pm 1$ and is undefined for $x = 0$. Note that $f'(-8) = -84 < 0$, $f'(-1/8) = \dfrac{21}{4} > 0$, $f'(1/8) = -\dfrac{21}{4} < 0$, and $f''(8) = 84 > 0$. Thus f is decreasing on $(-\infty, -1)$ and on $(0, 1)$, while f is increasing on $(-1, 0)$ and on $(1, \infty)$.

4.3. What Derivatives Tell Us

4.3.37 $f'(x) = \dfrac{1}{2\sqrt{9-x^2}} \cdot (-2x) + \dfrac{1}{3\sqrt{1-\frac{x^2}{9}}} = \dfrac{-x}{\sqrt{9-x^2}} + \dfrac{1}{\sqrt{9-x^2}} = \dfrac{1-x}{\sqrt{9-x^2}}$. This is 0 for $x = 1$ and is undefined at the endpoints of the domain which are 3 and -3. On $(-3, 1)$ $f' > 0$ so f is increasing; on $(1, 3)$ $f' < 0$ so f is decreasing.

4.3.39 $f'(x) = -60x^4 + 300x^3 - 240x^2 = -60x^2(x^2 - 5x + 4) = -60x^2(x-4)(x-1)$. This is 0 for $x = 0$, $x = 1$, and $x = 4$. Note that $f'(-1) = -600 < 0$, $f'(1/2) = -26.25 < 0$, $f'(2) = 480 > 0$, and $f'(5) = -6000 < 0$. Thus f is increasing on $(1, 4)$ and is decreasing on $(-\infty, 1)$ and on $(4, \infty)$.

4.3.41 $f'(x) = -8x^3 + 2x = -2x(4x^2 - 1) = -2x(2x+1)(2x-1)$. This is zero for $x = 0$ and $x = \pm 1/2$. Note that $f'(-1) > 0$, $f'(-1/4) < 0$, $f'(1/4) > 0$, and $f'(1) < 0$, so f is increasing on $(-\infty, -1/2)$ and on $(0, 1/2)$, while it is decreasing on $(-1/2, 0)$ and on $(1/2, \infty)$.

4.3.43 We have $f'(x) = e^{-x^2/2} + xe^{-x^2/2} \cdot (-x) = (1 - x^2)e^{-x^2/2}$. This is zero only when $x = \pm 1$. Note that $f'(-2) = -3e^{-2} < 0$, $f'(0) = 1 > 0$, and $f'(2) = -3e^{-2} < 0$. Thus f is decreasing on $(-\infty, -1)$ and on $(1, \infty)$, and is increasing on $(-1, 1)$.

4.3.45

a. $f'(x) = 2x$, so $x = 0$ is the only critical point.

b. Note that $f' < 0$ for $x < 0$ and $f' > 0$ for $x > 0$, so f has a local minimum of $f(0) = 3$ at $x = 0$.

c. Note that $f(-3) = 12$, $f(0) = 3$ and $f(2) = 7$, so the absolute maximum is 12 and the absolute minimum is 3.

4.3.47

a. $f'(x) = x \cdot \dfrac{1}{2}(4 - x^2)^{-1/2} \cdot (-2x) + \sqrt{4-x^2} \cdot 1 = \dfrac{4 - 2x^2}{\sqrt{4-x^2}}$, which exists everywhere on $(-2, 2)$ and is zero only for $x = \pm\sqrt{2}$, so those are the only critical points.

b. Note that $f'(-1.5) < 0$, $f'(0) > 0$ and $f'(1.5) < 0$, so f has a local minimum of $f(-\sqrt{2}) = -2$ and a local maximum of $f(\sqrt{2}) = 2$.

c. Note that $f(-2) = 0 = f(2)$. So the absolute maximum is 2 at $x = \sqrt{2}$ and the absolute minimum is -2 at $x = -\sqrt{2}$.

4.3.49

a. $f'(x) = -3x^2 + 9$, which is zero when $9 = 3x^2$, or $x^2 = 3$. So the critical points are at $x = \pm\sqrt{3}$.

b. Note that $f'(-2) < 0$, $f'(0) > 0$, and $f'(2) < 0$, so there is a local minimum of $f(-\sqrt{3}) = -6\sqrt{3}$ and a local maximum of $f(\sqrt{3}) = 6\sqrt{3}$.

c. There is an absolute maximum of 28 at $x = -4$ and an absolute minimum of $-6\sqrt{3}$ at $x = -\sqrt{3}$.

4.3.51

a. $f'(x) = x^{2/3} + (x-5) \cdot \dfrac{2}{3}x^{-1/3} = \dfrac{5x - 10}{3x^{1/3}}$, which is undefined at $x = 0$ and is 0 at $x = 2$. So these are the two critical points.

b. Note that $f'(-1) > 0$ and $f'(1) < 0$, and $f'(3) > 0$ so f has a local maximum at $x = 0$ of $f(0) = 0$ and a local minimum at $x = 2$ of $-3\sqrt[3]{4} \approx -4.762$.

c. Note that $f(-5) = -10\sqrt[3]{25} \approx -29.24$, $f(0) = 0$, and $f(5) = 0$, so the absolute maximum of f on $[-5, 5]$ is 0 and the absolute minimum is $-10\sqrt[3]{25}$.

4.3.53

a. $f'(x) = \dfrac{\sqrt{x}}{x} + \dfrac{\ln x}{2\sqrt{x}} = \dfrac{2 + \ln x}{2\sqrt{x}}$. This is defined everywhere on $(0, \infty)$ and is 0 only at $x = e^{-2}$.

b. Note that $f' < 0$ on $\left(0, \dfrac{1}{e^2}\right)$ and $f' > 0$ on $\left(\dfrac{1}{e^2}, \infty\right)$, so there is a local minimum at $x = \dfrac{1}{e^2}$.

c. Because there is only one critical point, the local minimum at $x = \dfrac{1}{e^2}$ yields an absolute minimum of $f(1/e^2) = -\dfrac{2}{e} \approx -0.736$. There is no absolute maximum because f increases without bound as $x \to \infty$.

4.3.55 $f'(x) = -xe^{-x} + e^{-x} = e^{-x}(1 - x)$, which is 0 only for $x = 1$. f is continuous on $(-\infty, \infty)$ and contains only one critical point. Note that $f' > 0$ for $x < 1$ and $f' < 0$ for $x > 1$. So there is a local maximum of $f(1) = 1/e$ at $x = 1$. The local maximum of $1/e$ at $x = 1$ is an absolute maximum. There is no absolute minimum, because the function is unbounded in the negative direction as $x \to -\infty$.

4.3.57 A is continuous on $(0, \infty)$. $A'(r) = -\dfrac{24}{r^2} + 4\pi r = \dfrac{4\pi r^3 - 24}{r^2}$, which is 0 for $r = \sqrt[3]{6/\pi}$, so there is only one critical point on the stated interval. Note that $A' < 0$ on $(0, \sqrt[3]{6/\pi})$ and $A' > 0$ on $(\sqrt[3]{6/\pi}, \infty)$, so there is a local minimum of $A(\sqrt[3]{6/\pi}) = 36\sqrt[3]{\pi/6}$. The local minimum mentioned above is an absolute minimum. There is no absolute maximum, because A is unbounded as $r \to \infty$.

4.3.59

The function sketched should be increasing and concave up everywhere.

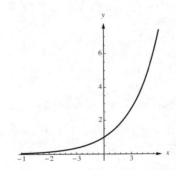

4.3.61

The function sketched should be decreasing everywhere, concave down for $x < 0$, and concave up for $x > 0$.

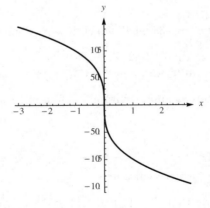

4.3.63 $f'(x) = 4x^3 - 6x^2$, so $f''(x) = 12x^2 - 12x = 12x(x - 1)$. Note that f'' is zero when $x = 0$ and $x = 1$, so these are potential inflection points. Also note that $f''(-1) > 0$, $f''(0.5) < 0$, and $f''(2) > 0$, so f is concave up on $(-\infty, 0)$ and on $(1, \infty)$, and is concave down on $(0, 1)$. There are inflection points at $(0, 1)$ and $(1, 0)$.

4.3.65 $f'(x) = 20x^3 - 60x^2$, and $f''(x) = 60x^2 - 120x = 60x(x-2)$. This is 0 for $x = 0$ and for $x = 2$. Note that $f''(-1) > 0$, $f''(1) < 0$, and $f''(3) > 0$. So f is concave up on $(-\infty, 0)$, concave down on $(0, 2)$, and concave up on $(2, \infty)$. There are inflection points at $x = 0$ and $x = 2$.

4.3.67 $f'(x) = e^x(x-3) + e^x = e^x(x-3+1) = e^x(x-2)$. $f''(x) = e^x(x-2) + e^x = e^x(x-2+1) = e^x(x-1)$. Note that f'' is zero only at $x = 1$. Also note that $f''(0) < 0$ and $f''(2) > 0$, so f is concave down on $(-\infty, 1)$ and is concave up on $(1, \infty)$. The point $(1, -2e)$ is an inflection point.

4.3.69 $g'(t) = \dfrac{6t}{3t^2 + 1}$, and $g''(t) = \dfrac{(3t^2+1) \cdot 6 - 6t(6t)}{(3t^2+1)^2} = \dfrac{6 - 18t^2}{(3t^2+1)^2}$. Note that g'' is 0 for $t = \pm\sqrt{1/3}$. Also, $g''(-1) < 0$, $g''(0) > 0$, and $g''(1) < 0$, so g is concave down on $(-\infty, -\sqrt{1/3})$ and on $(\sqrt{1/3}, \infty)$, and is concave up on $(-\sqrt{1/3}, \sqrt{1/3})$. There are inflection points at $t = \pm\sqrt{1/3}$.

4.3.71 $f'(x) = -xe^{-x^2/2}$, and $f''(x) = (-x)(-xe^{-x^2/2}) + e^{-x^2/2} \cdot -1 = e^{-x^2/2}(x^2 - 1)$. $f''(x)$ is 0 for $x = \pm 1$. Also $f'' > 0$ on $(-\infty, -1)$ and on $(1, \infty)$, so f is concave up there, while on $(-1, 1)$ f is concave down because $f'' < 0$ on that interval. The inflection points are at $(\pm 1, e^{-1/2})$.

4.3.73 $f'(x) = \sqrt{x}/x + (\ln x)\left(\dfrac{1}{2\sqrt{x}}\right) = \dfrac{2 + \ln x}{2\sqrt{x}}$. $f''(x) = \dfrac{2\sqrt{x}/x - (2 + \ln x)/\sqrt{x}}{(2\sqrt{x})^2} = -\dfrac{\ln x}{4\sqrt{x^3}}$. Note that f'' is 0 only at $x = 1$. On $(0, 1)$ we note that $f'' > 0$ so f is concave up, and on $(1, \infty)$ we note that $f'' < 0$ so f is concave down. There is an inflection point at $(1, 0)$.

4.3.75 $g'(t) = 15t^4 - 120t^3 + 240t^2$, and $g''(t) = 60t^3 - 360t^2 + 480t = 60t(t-2)(t-4)$. Note that g'' is 0 for $t = 0, 2$, and 4. Note also that $g'' < 0$ on $(-\infty, 0)$ and on $(2, 4)$, so g is concave down on those intervals, while $g'' > 0$ on $(0, 2)$ and on $(4, \infty)$, so g is concave up there. There are inflection points at $t = 0, 2$, and 4.

4.3.77 $f'(x) = 3x^2 - 6x = 3x(x-2)$. This is zero when $x = 0$ and when $x = 2$, and these are the critical points. $f''(x) = 6x - 6$. Note that $f''(0) < 0$ and $f''(2) > 0$. Thus by the Second Derivative Test, there is a local maximum at $x = 0$ and a local minimum at $x = 2$.

4.3.79 $f'(x) = -2x$, so $x = 0$ is a critical point. $f''(x) = -2$, so $f''(0) = -2$ and the critical point yields a local maximum.

4.3.81 $f'(x) = e^x(x-7) + e^x = e^x(x-6)$. This is zero when $x = 6$, and this is a critical point. $f''(x) = e^x(x-6) + e^x = e^x(x-5)$. Note that $f''(6) > 0$, so there is a local minimum at $x = 6$.

4.3.83 $f'(x) = 6x^2 - 6x = 6x(x-1)$, so $x = 0$ and $x = 1$ are critical points. $f''(x) = 12x - 6$, so $f''(1) = 6 > 0$, so the critical point at $x = 1$ yields a local minimum. Also, $f''(0) = -6 < 0$, so the critical point at 0 yields a local maximum.

4.3.85 $f'(x) = x^2 \cdot (-e^{-x}) + e^{-x} \cdot 2x = e^{-x}(2x - x^2)$, which is zero for $x = 0$ and $x = 2$, so these are the critical points. $f''(x) = e^{-x}(2 - 2x) + (2x - x^2)(-e^{-x}) = e^{-x}(2 - 4x + x^2)$. Note that $f''(0) = 2 > 0$, so there is a local minimum at $x = 0$. Also, $f''(2) = -2e^{-2} < 0$, so there is a local maximum at $x = 2$.

4.3.87 $f'(x) = 4x \ln x + 2x^2 \cdot \frac{1}{x} - 22x = 4x \ln x - 20x = 4x(\ln x - 5)$. This is zero for $x = e^5$, so that is the critical point. $f''(x) = 4 \ln x + 4x \cdot \dfrac{1}{x} - 20 = 4 \ln x - 16$. Note that $f''(e^5) > 0$, so there is a local minimum at e^5.

4.3.89 $p'(t) = 6t^2 + 6t - 36 = 6(t+3)(t-2)$, which is 0 at $t = -3$ and $t = 2$. Note that $p''(t) = 12t + 6$, so $p''(-3) = -30 < 0$ and $p''(2) = 30 > 0$, so there is a local maximum at $t = -3$ and a local minimum at $t = 2$.

4.3.91 $f'(x) = 4(x+a)^3$, which is 0 for $x = -a$. Note that $f''(x) = 12(x+a)^2$, which is 0 at $x = -a$, so the test is inconclusive. The first derivative test shows that there is a local minimum at $x = -a$.

4.3.93 $f'(x) = 24x^3 \ln x^2 + 6x^4 \left(\dfrac{2}{x}\right) - 28x^3 = 24x^3 \ln x^2 + 12x^3 - 28x^3 = 24x^3 \ln x^2 - 16x^3 = 8x^3(3\ln x^2 - 2)$.

Note that 0 is not in the domain of f, so the only values of x that makes f' 0 are those for which $\ln x^2 = \dfrac{2}{3}$, or $x^2 = e^{2/3}$, so $x = \pm e^{1/3}$.

$$f''(x) = 24x^2(3\ln x^2 - 2) + 8x^3\left(\dfrac{6}{x}\right) = 24x^2(3\ln x^2 - 2 + 2) = 72x^2 \ln x^2.$$

A check of f'' evaluated at either $\pm e^{1/3}$ shows a positive value, so there are local minimums at both $x = e^{1/3}$ and $x = -e^{1/3}$.

4.3.95

a. True. $f'(x) > 0$ implies that f is increasing, and $f''(x) < 0$ implies that f' is decreasing. So f is increasing, but at a decreasing rate.

b. False. In fact, if $f'(c)$ exists and isn't zero, then there isn't any kind of local extrema at $x = c$.

c. True. In fact, if two functions differ by a constant, then all of their derivatives are the same.

d. False. For example, consider $f(x) = x$ and $g(x) = x - 10$. Both are increasing, but $f(x)g(x) = x^2 - 10x$ is decreasing on $(-\infty, 5)$.

e. False. A continuous function with two local maxima must have a local minimum in between.

4.3.97

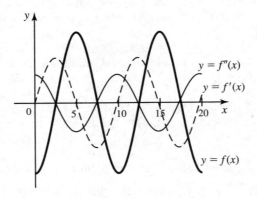

4.3.99 The graphs match as follows: (a) – (f) – (g); (b) – (e) –(i); (c) – (d) –(h). Note that (a) is always increasing, so its derivative must be always positive, and (f) switches from decreasing to increasing at 0, so its derivative must be negative for $x < 0$ and positive for $x > 0$.

Note that (b) has three extrema where there are horizontal tangent lines, so its derivative must cross the x-axis three times, and (e) has two extrema, so its derivative must cross the x-axis two times.

4.3.101

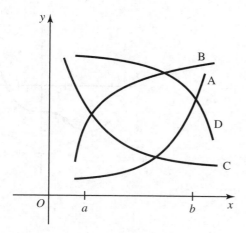 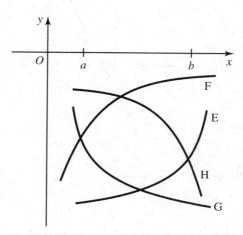

4.3.103

The graph sketched must have a flat tangent line at $x = -3/2$, $x = 0$, and $x = 1$, and must contain the points $(-2, 0)$, $(0, 0)$, and $(1, 0)$. The example to the right is only one possible such graph.

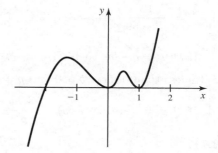

4.3.105

The graph sketched must be concave up on $(-\infty, -2)$ and on $(1, 3)$, and concave down on $(-2, 1)$ and on $(3, \infty)$. The example to the right is only one possible such graph.

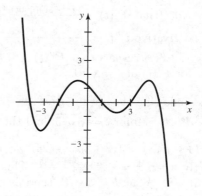

4.3.107

a. f is increasing on $(-2, 2)$. It is decreasing on $(-3, -2)$.

b. There are critical points of f at $x = -2$ and at $x = 0$. There is a local minimum at $x = -2$ and no extremum at $x = 0$.

c. There are inflection points of f at $x = -1$ and at $x = 0$.

d. f is concave up on $(-3, -1)$ and on $(0, 2)$, while it is concave down on $(-1, 0)$.

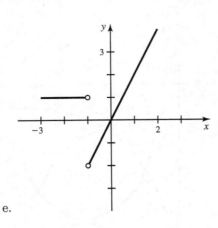
e.

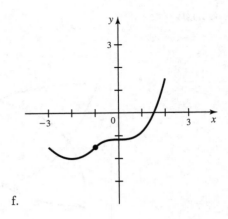
f.

4.3.109

a. $E = \dfrac{dD}{dp} \cdot \dfrac{p}{D} = -10 \dfrac{p}{500 - 10p} = \dfrac{p}{p - 50}$.

b. $E = \dfrac{12}{12 - 50} \cdot .045 = -1.42\%$.

c. If $D(p) = a - bp$, then $E(p) = -b \cdot \dfrac{p}{a - bp} = \dfrac{bp}{bp - a}$. So $E'(p) = \dfrac{(bp - a)b - bpb}{(bp - a)^2} = -\dfrac{ab}{(bp - a)^2}$, which is less than 0 for $a, b > 0$ and $p \ne a/b$.

d. If $D(p) = \dfrac{a}{p^b}$, then $E(p) = -\dfrac{ab}{p^{b+1}} \cdot \dfrac{p}{a/p^b} = -b$.

4.3.111

a. $\lim\limits_{t\to\infty} \dfrac{300t^2}{t^2 + 30} \cdot \dfrac{1/t^2}{1/t^2} = \lim\limits_{t\to\infty} \dfrac{300}{1 + (30/t^2)} = 300$.

b. Note that $P'(t) = \dfrac{(t^2+30)(600t) - 300t^2(2t)}{(t^2+30)^2} = \dfrac{18000t}{(t^2+30)^2}$. We want to maximize this, so we compute its derivative $P''(t) = \dfrac{(t^2+30)^2 \cdot 18000 - 18000t \cdot 2(t^2+30) \cdot 2t}{(t^2+30)^4} = \dfrac{54000(10-t^2)}{(t^2+30)^3}$. This is 0 for $t = \sqrt{10}$, and an analysis of $P''(t)$ reveals that $P''(t) > 0$ for $t < \sqrt{10}$ and $P''(t) < 0$ for $t > \sqrt{10}$ so there is a local maximum for $P'(t)$ at $t = \sqrt{10}$.

c. Following the outline from the previous problem, we see that $P'(t) = \dfrac{2bKt}{(t^2+b)^2}$, and $P''(t) = \dfrac{2bK(b-3t^2)}{(t^2+b)^3}$. $P''(t)$ is 0 for $t = \sqrt{b/3}$, and the first derivative test reveals that this is a local maximum.

4.3.113 $f'(x) = 4x^3 + 3ax^2 + 2bx + c$, and $f''(x) = 12x^2 + 6ax + 2b = 2(6x^2 + 3ax + b)$. Note that $f''(x) = 0$ exactly when $x = \dfrac{-3a \pm \sqrt{9a^2 - 24b}}{12}$. This represents no real solutions when $9a^2 - 24b < 0$, which occurs when $b > 3a^2/8$. When $b = 3a^2/8$, there is one root, but in this case the sign of f'' doesn't change at the double root $x = -a/4$, so there are no inflection points for f. In the case $b < 3a^2/8$, there are two roots of f'', both of which yield inflection points of f, as can be seen by the change in sign of f'' at its two roots.

4.3.115

a. $f'(x) = 3x^2 + 2ax + b$, which is 0 when $x = \dfrac{-2a \pm \sqrt{4a^2 - 12b}}{6} = \dfrac{-a \pm \sqrt{a^2 - 3b}}{3}$. These solutions represent distinct real numbers when $a^2 > 3b$. Let the two distinct roots be $r_1 < r_2$. Note that f' is negative on the interval (r_1, r_2) and positive on $(-\infty, r_1)$ and on (r_2, ∞) so there is a maximum at r_1 and a minimum at r_2.

b. If $a^2 < 3b$, then there are no real critical points, so there are no extreme values.

4.4 Graphing Functions

4.4.1 Because the intervals of increase and decrease and the intervals of concavity must be subsets of the domain, it is helpful to know what the domain is at the outset.

4.4.3 No. Polynomials are continuous everywhere, so they have no vertical asymptotes. Also, polynomials in x always tend to $\pm\infty$ as $x \to \pm\infty$.

4.4.5 The maximum and minimum must occur at either an endpoint or a critical point. So to find the absolute maximum and minimum, it suffices to find all the critical points, and then compare the values of the function at those points and at the endpoints. The largest such value is the maximum and the smallest is the minimum.

4.4.7

The function sketched should be decreasing and concave down for $x < 3$ and decreasing and concave up for $x > 3$.

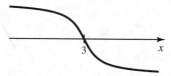

4.4.9

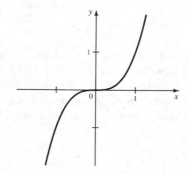

4.4.11

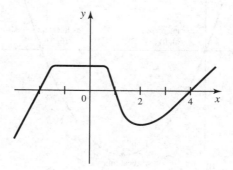

4.4.13

a. $f'(x) = 1 \cdot (x+3)^2 + (x-3)(2)(x+3) = (x+3)(x+3+2x-6) = (x+3)(3x-3) = 3(x-1)(x+3)$.
Also, $f''(x) = 3(x+3) + 3(x-1) = 3x + 9 + 3x - 3 = 6x + 6 = 6(x+1)$.

b. $f'(x) = 0$ for $x = 1$ and $x = -3$ (so those are the critical points), while $f''(x) = 0$ for $x = -1$, so that is the location of a potential inflection point.

c. $f' > 0$ on $(-\infty, -3)$ and $(1, \infty)$, so f is increasing there. $f' < 0$ on $(-3, 1)$ so f is decreasing there.

d. $f'' > 0$ on $(-1, \infty)$ so f is concave up there; $f'' < 0$ on $(-\infty, -1)$, so f is concave down there.

e. f has a local max of $f(-3) = 0$ at $x = -3$ and a local minimum of $f(1) = -32$ at $x = 1$. The point $(-1, -16)$ is an inflection point.

f. The y-intercept is $f(0) = -27$. The x-intercepts are $x = 3, -3$.

g.

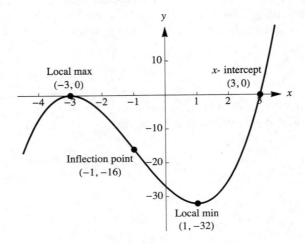

4.4.15 Because f is a polynomial, its domain is $(-\infty, \infty)$. There is neither even nor odd symmetry because $f(-x) \neq f(x)$ and $f(-x) \neq -f(x)$. $f'(x) = 2x - 6 = 2(x - 3)$ which is zero for $x = 3$. We have $f' > 0$ on $(3, \infty)$ so f is increasing there, while $f' < 0$ on $(-\infty, 3)$, so f is decreasing there, and there is a local minimum at $x = 3$. $f''(x) = 2$ which is always positive, so f is concave up on $(-\infty, \infty)$ and there are no inflection points. The local minimum value is $f(3) = -9$, the y-intercept is $(0, 0)$, which is also an x-intercept, and $(6, 0)$ is also an x-intercept.

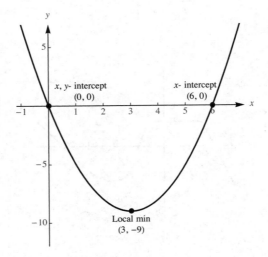

4.4.17 The domain of f is $(-\infty, \infty)$, and there is no symmetry. The y intercept is $f(0) = 0$, and the x-intercepts are 0 and 3 because $f(x) = x(x-3)^2$. $f'(x) = 3x^2 - 12x + 9 = 3(x^2 - 4x + 3) = 3(x-3)(x-1)$. This is zero when $x = 1$ and $x = 3$. Note that $f'(0) > 0$, $f'(2) < 0$ and $f'(4) > 0$, so f is increasing on $(-\infty, 1)$ and on $(3, \infty)$. It is decreasing on $(1, 3)$. Note that $f''(x) = 6x - 12$ which is zero at $x = 2$. Because $f''(1) < 0$ and $f''(3) > 0$, we conclude that f is concave down on $(-\infty, 2)$ and concave up on $(2, \infty)$. There is an inflection point at $(2, 2)$, a local maximum at $(1, 4)$ and a local minimum at $(3, 0)$.

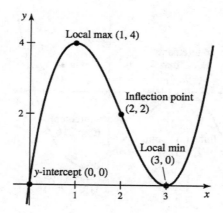

4.4.19 The domain of f is $(-\infty, \infty)$, and there is even symmetry, because $f(-x) = f(x)$. $f'(x) = 4x^3 - 12x = 4x(x^2 - 3)$. This is 0 when $x = \pm\sqrt{3}$ and when $x = 0$. $f''(x) = 12x^2 - 12 = 12(x^2 - 1)$, which is 0 when $x = \pm 1$. Note that $f'(-2) < 0$, $f'(-1) > 0$, $f'(1) < 0$, and $f'(2) > 0$. So f is decreasing on $(-\infty, -\sqrt{3})$ and on $(0, \sqrt{3})$. It is increasing on $(-\sqrt{3}, 0)$ and on $(\sqrt{3}, \infty)$. There is a local maximum of 0 at $x = 0$ and local minima of -9 at $x = \pm\sqrt{3}$. Note also that $f''(x) > 0$ for $x < -1$ and for $x > 1$ and $f''(x) < 0$ for $-1 < x < 1$, so there are inflection points at $x = \pm 1$. Also, f is concave down on $(-1, 1)$ and concave up on $(-\infty, -1)$ and on $(1, \infty)$. There is a y-intercept at $f(0) = 0$ and x-intercepts where $f(x) = x^4 - 6x^2 = x^2(x^2 - 6) = 0$, which is at $x = \pm\sqrt{6}$ and $x = 0$.

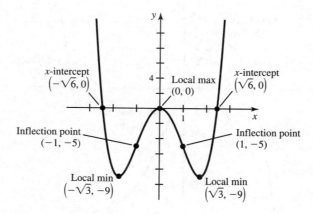

4.4.21 The domain of f is $(-\infty, \infty)$, and there is no symmetry. The y-intercept is $f(0) = -216$. The x-intercepts are 6 and -6.

$f'(x) = (x+6)^2 + (x-6)2(x+6) = (x+6)(x+6+2x-12) = (x+6)(3x-6) = 3(x+6)(x-2)$. The critical numbers are -6 and 2. Note that $f'(-7) > 0$, $f'(-2) < 0$, and $f'(3) > 0$, so f is increasing on $(-\infty, -6)$ and on $(2, \infty)$. It is decreasing on $(-6, 2)$. There is a local maximum of 0 at -6 and a local minimum of -256 at $x = 2$.

$f''(x) = 3(x-2) + 3(x+6) = 3(x-2+x+6) = 3(2x+4) = 6(x+2)$, which is zero for $x = -2$. Note that $f''(x) < 0$ for $x < -2$ and $f''(x) > 0$ for $x > -2$, so f is concave down on $(-\infty, -2)$ and concave up on $(-2, \infty)$. The point $(-2, -128)$ is an inflection point.

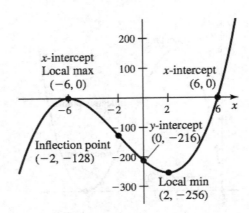

4.4.23 The domain of f is $(-\infty, \infty)$ and there is no symmetry. There are no asymptotes because f is a polynomial.

$f'(x) = 3x^2 - 12x - 135 = 3(x-9)(x+5)$, which is 0 for $x = 9$ and $x = -5$. $f'(x) > 0$ on $(-\infty, -5)$ and on $(9, \infty)$, so f is increasing on those intervals. $f'(x) < 0$ on $(-5, 9)$, so f is decreasing on that interval. There is a local maximum at $x = -5$ and a local minimum at $x = 9$.

$f''(x) = 6x - 12$, which is 0 for $x = 2$. $f''(x) > 0$ on $(2, \infty)$, so f is concave up on that interval. $f''(x) < 0$ on $(-\infty, 2)$, so f is concave down on that interval. There is a point of inflection at $x = 2$. The y-intercept is 0 and the x-intercepts are -9 and 15.

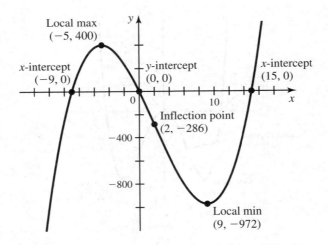

4.4.25 The domain of f is $(-\infty, \infty)$ and there is no symmetry. There are no asymptotes because f is a polynomial.

$f'(x) = 3x^2 - 6x - 144 = 3(x+6)(x-8)$, which is 0 for $x = -6$ and $x = 8$. $f'(x) > 0$ on $(-\infty, -6)$ and on $(8, \infty)$, so f is increasing on those intervals. $f'(x) < 0$ on $(-6, 8)$, so f is decreasing on that interval. There is a local maximum at $x = -6$ and a local minimum at $x = 8$.

$f''(x) = 6x - 6$, which is 0 for $x = 1$. $f''(x) > 0$ on $(1, \infty)$, so f is concave up on that interval. $f''(x) < 0$ on $(-\infty, 1)$, so f is concave down on that interval. There is a point of inflection at $x = 1$. The y-intercept is -140 and the x-intercepts are at -10, -1, and 14.

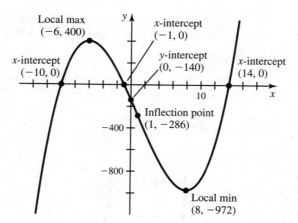

4.4.27 The domain of f is $[0, \infty)$ and there is no symmetry. $f'(x) = 1 - \dfrac{1}{\sqrt{x}} = \dfrac{\sqrt{x} - 1}{\sqrt{x}}$ which is zero for $x = 1$. $f' > 0$ on $(1, \infty)$, so f is increasing there, while $f' < 0$ on $(0, 1)$, so f is decreasing there. There is a local minimum of -1 at $x = 1$.

If we write $f'(x)$ as $f'(x) = 1 - x^{-1/2}$, then we see that $f''(x) = \dfrac{1}{2}x^{-3/2} = \dfrac{1}{2x^{3/2}}$. This is never zero, and is always positive. So f is concave up and has no inflection points. The x- and y-intercept is $(0, 0)$ and there is another x-intercept at $(4, 0)$.

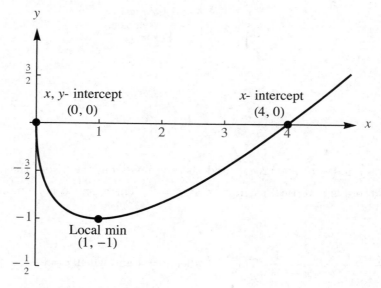

4.4.29 The domain of f is $(-\infty, -1) \cup (-1, 1) \cup (1, \infty)$, and there is odd symmetry, because $f(-x) = \dfrac{3(-x)}{(-x)^2 - 1} = -\dfrac{3x}{x^2 - 1} = -f(x)$. The only intercept is $(0, 0)$ which is both the y- and x-intercept.

Note that $\lim\limits_{x \to -1^+} f(x) = \infty$ and $\lim\limits_{x \to -1^-} f(x) = -\infty$, so there is a vertical asymptote at $x = -1$. Also, $\lim\limits_{x \to 1^+} f(x) = \infty$ and $\lim\limits_{x \to 1^-} f(x) = -\infty$, so there is a vertical asymptote at $x = 1$.

Note that $\lim\limits_{x \to \pm\infty} \dfrac{3x}{x^2 - 1} \cdot \dfrac{1/x^2}{1/x^2} = \lim\limits_{x \to \pm\infty} \dfrac{3/x}{1 - (1/x^2)} = 0$, so $y = 0$ is a horizontal asymptote.

$f'(x) = \dfrac{(x^2 - 1) \cdot 3 - 3x \cdot 2x}{(x^2 - 1)^2} = \dfrac{-3x^2 - 3}{(x^2 - 1)^2} = -\dfrac{3(x^2 + 1)}{(x^2 - 1)^2}$. This is never 0, and is in fact negative wherever it is defined. Thus, f is decreasing on $(-\infty, -1)$, on $(-1, 1)$, and on $(1, \infty)$. There are no extrema.

$f''(x) = \dfrac{(x^2 - 1)^2 (-6x) + 3(x^2 + 1) \cdot 2 \cdot (x^2 - 1) \cdot 2x}{(x^2 - 1)^4} = \dfrac{-6x^3 + 6x + 12x^3 + 12x}{(x^2 - 1)^3} = \dfrac{6x(x^2 + 3)}{(x^2 - 1)^3}$. This is 0 for $x = 0$. The point $(0, 0)$ is an point of inflection, because it is an interior point on the domain, and the second derivative changes from positive to negative there. The other concavity changes take place at the asymptotes. Note that f is concave down on $(-\infty, -1)$ and on $(0, 1)$, and concave up on $(-1, 0)$ and on $(1, \infty)$.

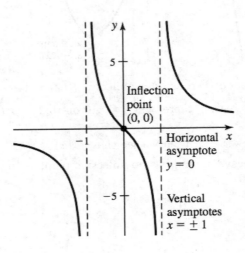

4.4.31 The domain of f is $(-\infty, 2) \cup (2, \infty)$, and there is no symmetry. Note that $\lim_{x \to 2^+} f(x) = \infty$ and $\lim_{x \to 2^-} f(x) = -\infty$, so there is a vertical asymptote at $x = 2$. There isn't a horizontal asymptote, because $\lim_{x \to \pm\infty} f(x) = \pm\infty$.

$$f'(x) = \frac{(x-2) \cdot 2x - x^2}{(x-2)^2} = \frac{x(x-4)}{(x-2)^2}. \text{ This is } 0 \text{ when } x = 4 \text{ and when } x = 0.$$

$$f''(x) = \frac{(x-2)^2(2x-4) - (x^2-4x) \cdot 2 \cdot (x-2)}{(x-2)^4} = \frac{8}{(x-2)^3}. \text{ This is never } 0.$$

Note that $f'(-1) > 0$, $f'(1) < 0$, $f'(3) < 0$ and $f'(5) > 0$. So f is decreasing on $(0, 2)$ and on $(2, 4)$. It is increasing on $(-\infty, 0)$ and on $(4, \infty)$. There is a local maximum of 0 at $x = 0$ and a local minimum of 8 at $x = 4$.

Note that $f''(x) > 0$ for $x > 2$ and $f''(x) < 0$ for $x < 2$, So f is concave up on $(2, \infty)$ and concave down on $(-\infty, 4)$. There are no inflection points, because the only change in concavity occurs at a vertical asymptote. The only intercept is $(0, 0)$.

By long division, we can write f as $x + 2 + \dfrac{4}{x-2}$. Therefore the line $y = x + 2$ is a slant asymptote.

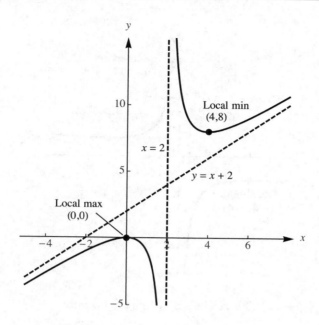

4.4.33 The domain of f is $(-\infty, -1/2) \cup (-1/2, \infty)$, and there is no symmetry.

Because $\displaystyle\lim_{x \to \pm\infty} \frac{x^2 + 12}{2x + 1} \cdot \frac{1/x}{1/x} = \lim_{x \to \pm\infty} \frac{x + (12/x)}{2 + (1/x)} = \pm\infty$, there is no horizontal asymptote. However, there is a slant asymptote of $y = \dfrac{x}{2} - \dfrac{1}{4}$, because we can write f as $f(x) = \dfrac{x}{2} - \dfrac{1}{4} + \dfrac{49/4}{2x + 1}$ by long division.

Also, because $\displaystyle\lim_{x \to (-1/2)^-} f(x) = -\infty$ and $\displaystyle\lim_{x \to (-1/2)^+} f(x) = \infty$, there is a vertical asymptote at $x = -1/2$.

$f'(x) = \dfrac{(2x+1) \cdot 2x - (x^2 + 12) \cdot 2}{(2x+1)^2} = \dfrac{2x^2 + 2x - 24}{(2x+1)^2} = \dfrac{2(x+4)(x-3)}{(2x+1)^2}$. This is 0 for $x = -4$ and $x = 3$.

$f''(x) = \dfrac{(2x+1)^2(4x+2) - (2x^2 + 2x - 24) \cdot 2(2x+1) \cdot 2}{(2x+1)^4} = \dfrac{98}{(2x+1)^3}$, which is never 0.

Note that $f'(x) > 0$ on $(-\infty, -4)$ and on $(3, \infty)$. So f is increasing on $(-\infty, -4)$ and on $(3, \infty)$. Also, $f'(x) < 0$ on $(-4, -1/2)$ and on $(-1/2, 3)$. So f is decreasing on those intervals. There is a local maximum of -4 at $x = -4$ and a local minimum of 3 at $x = 3$.

Note also that $f''(x) < 0$ for $x < -1/2$, and $f''(x) > 0$ for $x > -1/2$, so f is concave down on $(-\infty, -1/2)$ and is concave up on $(-1/2, \infty)$. There are no inflection points because the only change in concavity occurs at the vertical asymptote. There are no x-intercepts because $x^2 + 12 > 0$ for all x, and the y-intercept is $f(0) = 12$.

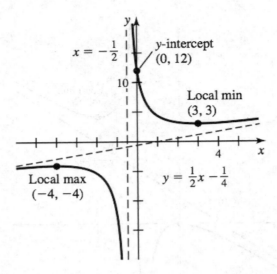

4.4.35 The domain of f is $(-\infty, \infty)$. There is even symmetry because

$$f(-x) = \tan^{-1}\left(\frac{(-x)^2}{\sqrt{3}}\right) = \tan^{-1}\left(\frac{x^2}{\sqrt{3}}\right) = f(x).$$

The only intercept is $(0,0)$. $f'(x) = \dfrac{1}{x^4/3+1} \cdot \dfrac{2x}{\sqrt{3}} = \dfrac{2\sqrt{3}x}{x^4+3}$. This is zero for $x = 0$, and $f' > 0$ on $(0, \infty)$ so f is increasing there, while $f' < 0$ on $(-\infty, 0)$ so f is deceasing there. There is a local minimum of 0 at $x = 0$.

$f''(x) = \dfrac{(x^4+3)(2\sqrt{3}) - 2\sqrt{3}x(4x^3)}{(x^4+3)^2} = \dfrac{2\sqrt{3}(3-3x^4)}{(x^4+3)^2} = \dfrac{6\sqrt{3}(1-x)(1+x)(1+x^2)}{(x^4+3)^2}$. This is zero for $x = 1$ and $x = -1$. $f'' < 0$ on $(-\infty, -1)$ and on $(1, \infty)$ so f is concave down there, while $f'' > 0$ on $(-1, 1)$, so f is concave up there. There are inflection points at $(1, \pi/6)$ and $(-1, \pi/6)$.

$\lim\limits_{x \to \pm\infty} f(x) = \dfrac{\pi}{2}$, so $y = \dfrac{\pi}{2}$ is a horizontal asymptote.

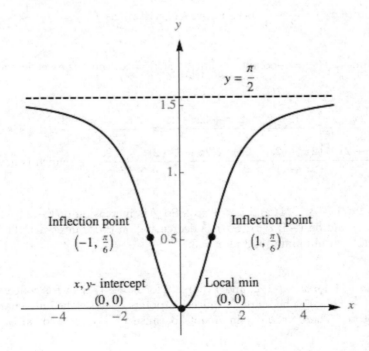

4.4.37 The domain of f is given to be $[-2\pi, 2\pi]$, and there is no symmetry, and no vertical asymptotes. There are no horizontal asymptotes to consider on this restricted domain.

$f'(x) = 1 - 2\sin x$. This is 0 when $\sin x = 1/2$, which occurs on the given interval for $x = -11\pi/6$, $-7\pi/6, \pi/6$, and $5\pi/6$. $f''(x) = -2\cos x$, which is 0 for $x = -3\pi/2, -\pi/2, \pi/2$, and $3\pi/2$.

Note that $f'(x) > 0$ on $(-2\pi, -11\pi/6)$, and on $(-7\pi/6, \pi/6)$, and on $(5\pi/6, 2\pi)$. So f is increasing on those intervals, while $f'(x) < 0$ on $(-11\pi/6, -7\pi/6))$ and on $(\pi/6, 5\pi/6)$, so f is decreasing there. f has local maxima at $x = -11\pi/6$ and at $x = \pi/6$ and local minima at $x = -7\pi/6$ and at $x = 5\pi/6$. Note also that $f''(x) < 0$ on $(-2\pi, -3\pi/2)$ and on $(-\pi/2, \pi/2)$ and on $(3\pi/2, 2\pi)$, so f is concave down on those intervals, while $f''(x) > 0$ on $(-3\pi/2, -\pi/2)$ and on $(\pi/2, 3\pi/2)$, so f is concave up there and there are inflection points at $x = \pm 3\pi/2$ and $x = \pm \pi/2$. The y-intercept is $f(0) = 2$ and the x-intercept is at approximately -1.030.

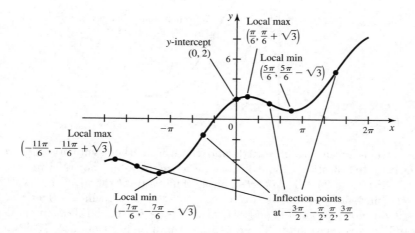

4.4.39 The domain of f is $(-\infty, \infty)$. There are no asymptotes. There are x-intercepts at $(0,0)$ and $(\pm 3\sqrt{3}, 0)$. f does have odd symmetry, because $f(-x) = -x - 3((-x)^{1/3}) = -(x - 3x^{1/3}) = -f(x)$.

$f'(x) = 1 - \dfrac{1}{\sqrt[3]{x^2}} = \dfrac{\sqrt[3]{x^2} - 1}{\sqrt[3]{x^2}}$. This is undefined at $x = 0$, and is equal to zero at ± 1. Note that $f'(-2) > 0$, $f'(-1/2) < 0$, $f'(1/2) < 0$, $f'(2) > 0$. Thus, f is increasing on $(-\infty, -1)$ and on $(1, \infty)$. Because f is continuous at 0 (even though f' doesn't exist there), we can combine the intervals $(-1, 0)$ and $(0, 1)$ and state that f is decreasing on $(-1, 1)$. There is a local maximum at $(-1, 2)$ and a local minimum at $(1, -2)$.

$f''(x) = \dfrac{2}{3\sqrt[3]{x^5}}$, which is never zero, but is undefined at $x = 0$. Note that $f''(x) < 0$ for $x < 0$ and $f''(x) > 0$ for $x > 0$, so f is concave down on $(-\infty, 0)$ and is concave up on $(0, \infty)$. There is an inflection point at $(0, 0)$.

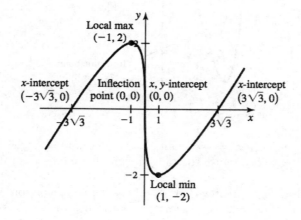

4.4.41 The domain of f is given to be $[0, 2\pi]$, so questions about symmetry and horizontal asymptotes aren't relevant. There are no vertical asymptotes.

$f'(x) = \cos x - 1$. This is never 0 on $(0, 2\pi)$. $f''(x) = -\sin x$, which is 0 on the given interval only for $x = \pi$. Note that $f'(x) < 0$ on $(0, 2\pi)$, so f is decreasing on the given interval and there are no relative extrema. Note also that $f''(x) < 0$ on $(0, \pi)$ and $f''(x) > 0$ on $(\pi, 2\pi)$, so f is concave down on $(0, \pi)$ and is concave up on $(\pi, 2\pi)$, and there is an inflection point at $x = \pi$. The only intercept is the origin $(0, 0)$.

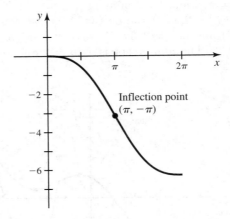

4.4.43 The domain of g is given to be $[-\pi, \pi]$, and there is neither symmetry nor vertical asymptotes. Because the domain is finite, questions about horizontal asymptotes are not relevant.

$g'(t) = e^{-t}\cos t + \sin t \cdot (-e^{-t}) = e^{-t}(\cos t - \sin t)$. This is 0 on the given interval for $t = -3\pi/4$ and $t = \pi/4$. $g''(t) = e^{-t}(-\sin t - \cos t) + (\cos t - \sin t)(-e^{-t}) = -2e^{-t}\cos t$, which is 0 for $t = -\pi/2$ and $t = \pi/2$.

Note that $g'(t) < 0$ on $(-\pi, -3\pi/4)$ and on $(\pi/4, \pi)$, so g is decreasing on those intervals. On $(-3\pi/4, \pi/4)$ we have $g'(t) > 0$ and so g is increasing. There is a local minimum of about -7.460 at $t = -3\pi/4$ and a local maximum of about 0.322 at $t = \pi/4$.

Note also that $g''(t) > 0$ on $(-\pi, -\pi/2)$ and on $(\pi/2, \pi)$, while $g''(t) < 0$ on $(-\pi/2, \pi/2i)$, so g is concave down on $(-\pi/2, \pi/2)$ and is concave up on $(\pi/2, \pi)$ and on $(-\pi, -\pi/2)$. There are inflection points at $t = \pm\pi/2$. The origin is both the y-intercept and an x-intercept. The endpoints are x-intercepts as well.

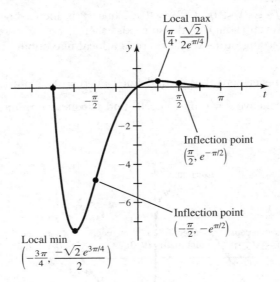

4.4.45 Note that $f(-x) = -x + \tan(-x) = -x - \tan x = -(x + \tan x) = -f(x)$, so f has odd symmetry. f has vertical asymptotes at $x = \pm\pi/2$, because the tangent function increases or decreases without bound as x approaches these values.

$f'(x) = 1 + \sec^2 x$ which is always greater than 0. Thus f is increasing on each interval on which it is defined, and it has no extrema. $f''(x) = 2\sec x \cdot \sec x \tan x = 2\sec^2 x \tan x$. This is 0 at $x = 0$.

Note that $f''(x)$ is positive on $(0, \pi/2)$, so f is concave up there. Also, $f''(x)$ is negative on $(-\pi/2, 0)$, so f is concave down there. There is a point of inflection at $(0,0)$.

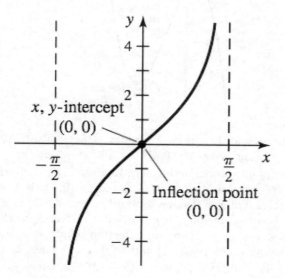

4.4.47 $f'(x)$ is 0 at $x = 1$ and $x = 3$.

$f'(x) > 0$ on $(0, 1)$ and on $(3, 4)$, so f is increasing on those intervals. $f'(x) < 0$ on $(1, 3)$, so f is decreasing on that interval. There is a local maximum at $x = 1$ and a local minimum at $x = 3$.

$f''(x)$ changes sign at $x = 2$ from negative to positive, so $x = 2$ is an inflection point where the concavity of f changes from down to up. An example of such a function is sketched.

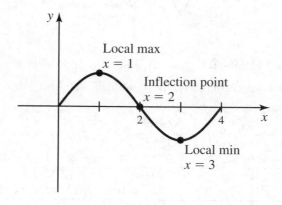

4.4.49 The domain of f is $(-\infty, \infty)$, and there is no symmetry. $f'(x) = x^2 - 4x - 5 = (x-5)(x+1)$. This is 0 when $x = -1, 5$. $f''(x) = 2x - 4$, which is 0 when $x = 2$. Note that $f'(-2) > 0$, $f'(0) < 0$, and $f'(6) > 0$. So f is increasing on $(-\infty, -1)$ and on $(5, \infty)$. It is decreasing on $(-1, 5)$. There is a local maximum of $14/3$ at $x = -1$ and a local minimum of $-94/3$ at $x = 5$. Note also that $f''(x) < 0$ for $x < 2$ and $f''(x) > 0$ for $x > 2$, so there is an inflection point at $(2, -40/3)$, and f is concave down on $(-\infty, 2)$ and concave up on $(2, \infty)$. The y intercept is $f(0) = 2$.

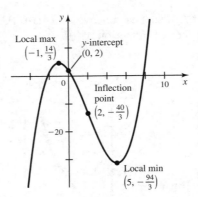

4.4.51 The domain of f is $(-\infty, \infty)$, and there is no symmetry. $f'(x) = 12x^3 + 12x^2 - 24x = 12x(x+2)(x-1)$. This is 0 when $x = -2$, when $x = 1$, and when $x = 0$. $f''(x) = 36x^2 + 24x - 24 = 12(3x^2 + 2x - 2)$, which is 0 when $x = \frac{-1 \pm \sqrt{7}}{3}$. These values are at approximately -1.215 and 0.549. Note that $f'(-3) < 0$, $f'(-1) > 0$, $f'(.5) < 0$, and $f'(2) > 0$. So f is decreasing on $(-\infty, -2)$ and on $(0, 1)$. It is increasing on $(-2, 0)$ and on $(1, \infty)$. There is a local maximum of 0 at $x = 0$ and a local minimum of -32 at $x = -2$ and a local minimum of -5 at $x = 1$. Let $r_1 < r_2$ be the two roots of $f''(x)$ mentioned above. Note that $f''(x) > 0$ for $x < r_1$ and for $x > r_2$ and $f''(x) < 0$ for $r_1 < x < r_2$, so there are inflection points at $x = r_1$ and at $x = r_2$. Also, f is concave down on (r_1, r_2) and concave up on $(-\infty, r_1)$ and on (r_2, ∞). There is a y-intercept at $f(0) = 0$ and x-intercepts where $f(x) = 3x^4 + 4x^3 - 12x^2 = x^2(3x^2 + 4x - 12) = 0$, which is at $x = \frac{-2 \pm 2\sqrt{10}}{3}$ and $x = 0$.

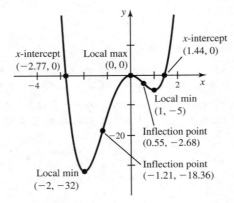

4.4.53 The domain of f is $(-\infty, -1) \cup (-1, 1) \cup (1, \infty)$, and there is no symmetry. Note that $\lim_{x \to -1^+} f(x) = \infty$ and $\lim_{x \to -1^-} f(x) = -\infty$, so there is a vertical asymptote at $x = -1$. Also, $\lim_{x \to 1^+} f(x) = -\infty$ and $\lim_{x \to 1^-} f(x) = \infty$, so there is a vertical asymptote at $x = 1$

Note that $\lim_{x \to \pm\infty} \frac{3x-5}{x^2-1} \cdot \frac{1/x^2}{1/x^2} = \lim_{x \to \pm\infty} \frac{3/x}{1 - (1/x^2)} = 0$, so $y = 0$ is a horizontal asymptote.

$f'(x) = \frac{(x^2-1) \cdot 3 - (3x-5) \cdot 2x}{(x^2-1)^2} = \frac{-3x^2 + 10x - 3}{(x^2-1)^2} = \frac{(-3x+1)(x-3)}{(x^2-1)^2}$. This is 0 when $x = 3$ and when $x = 1/3$. $f''(x) = \frac{(x^2-1)^2(-6x+10) - (-3x^2+10x-3)(2)(x^2-1) \cdot 2x}{(x^2-1)^4} = \frac{2(3x^3 - 15x^2 + 9x - 5)}{(x^2-1)^3}$.
This is 0 for $x \approx 4.405$. Let r be this root of $f''(x)$.

Note that $f'(-2) < 0$, $f'(-1/2) < 0$, $f'(1/2) > 0$, $f'(2) > 0$ and $f'(4) < 0$. So f is decreasing on $(-\infty, -1)$, on $(-1, 1/3)$ and on $(3, \infty)$. It is increasing on $(1/3, 1)$ and on $(1, 3)$. There is a local maximum of $1/2$ at $x = 3$ and a local minimum of $9/2$ at $x = 1/3$.

Note that $f''(x) < 0$ for $x < -1$ and $f''(x) < 0$ for $1 < x < r$, while $f''(x) > 0$ for $-1 < x < 1$, and for $x > r$. Thus f is concave up on $(-1, 1)$ and on (r, ∞) and concave down on $(-\infty, -1)$ and on $(1, r)$. There is an inflection point at r. There is a y-intercept at $f(0) = 5$ and an x-intercept at $(5/3, 0)$.

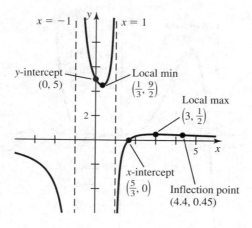

4.4.55

a. False. Maxima and minima can also occur at points where $f'(x)$ doesn't exist. Also, it is possible to have a zero of f' which doesn't correspond to an extreme point.

b. False. Inflection points can also occur at points where $f''(x)$ doesn't exist, and a zero of f'' might not correspond to an inflection point.

c. False. For example, $f(x) = \dfrac{(x^2 - 9)(x^2 - 16)}{(x+3)(x-4)}$ doesn't have a vertical asymptote at $x = -3$ or $x = 4$.

d. True. The limit of a rational function as $x \to \infty$ is a finite number when the degree of the denominator is greater than or equal to that of the numerator. If they both have the same degree, the limit is the ratio of the leading coefficients, and this is also true of the limit as $x \to -\infty$. In the case where the denominator has greater degree than the numerator, the limit is 0 as $x \to -\infty$ and as $x \to \infty$.

4.4.57 $f'(x)$ is 0 on the interior of the given interval at $x = \pm 3\pi/2$, $x = \pm \pi$, $x = \pm \pi/2$, and at $x = 0$. $f'(x) > 0$ on $(-2\pi, -3\pi/2)$, $(-\pi, -\pi/2)$, $(0, \pi/2)$, and on $(\pi, 3\pi/2)$, so f is increasing on those intervals. $f'(x) < 0$ on $(-3\pi/2, -\pi)$, $(-\pi/2, 0)$, $(\pi/2, \pi)$, and on $(3\pi/2, 2\pi)$, so f is decreasing on those intervals. There are local maxima at $x = \pm 3\pi/2$ and $x = \pm \pi/2$, and local minima at $x = 0$ and at $x = \pm \pi$. An example of such a function is sketched.

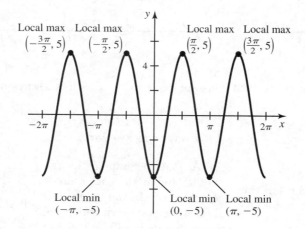

4.4.59 $f'(x)$ is 0 at $x = 0$, $x = -2$, and $x = 1$. $f'(x) > 0$ on $(-\infty, -2)$ and on $(1, \infty)$, so f is increasing on those intervals. $f'(x) < 0$ on $(-2, 0)$ and on $(0, 1)$, so f is decreasing on those intervals. There is a local maximum at $x = -2$ and a local minimum at $x = 1$. There isn't an extremum at $x = 0$.

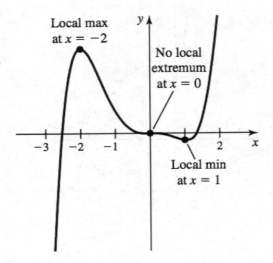

4.4.61

a.

(a) 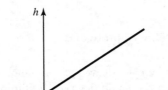   (b) 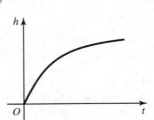   (c)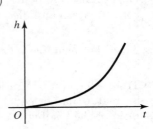

(d) 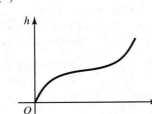   (e) 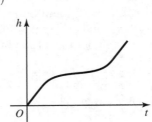   (f)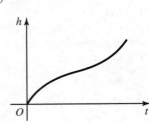

b. The water is being poured in at a constant rate, so the depth is always increasing, so $y = h(t)$ is an increasing function.

c. (a) No concavity   (b) Always concave down.   (c) Always concave up.

(d) Concave down for for the first half and concave up for the second half.

(e) At the beginning, in the middle, and at the end, there is no concavity. In the lower middle it is concave down and in the upper middle it will be concave up.

(f) This is concave down for the first half, and concave up for the second half.

d. (a) $h'(t)$ is constant, so there is no local max/min.   (b) $h'(t)$ is maximal at $t = 0$.   (c) $h'(t)$ is maximal at $t = 10$.

(d) $h'(t)$ is maximal at $t = 0$ and $t = 10$. (e) $h'(t)$ is maximal on the first and last straight parts of $h(t)$. (f) $h'(t)$ is maximal at $t = 0$ and $t = 10$.

4.4.63 f can be written as $f(x) = e^{(\ln x)/x}$. $f'(x) = e^{(\ln x)/x}\left(\dfrac{1-\ln x}{x^2}\right)$. This is 0 for $x = e$, and is positive on $(0, e)$ and negative on (e, ∞). There is a local maximum at $x = e$ of $e^{1/e}$.

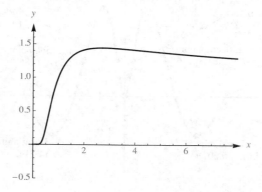

4.4.65

If $f''(x) > 0$ on $(-\infty, 0)$ and on $(0, \infty)$, then $f'(x)$ is increasing on both of those intervals. But if there is a local max at 0, the function f must be switching from increasing to decreasing there. This means that f' must be switching from positive to negative. But if f' is switching from positive to negative, but increasing, there must be a cusp at $x = 0$, so $f'(0)$ does not exist.

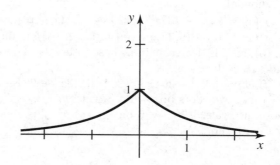

4.4.67 The domain of f is given to be $[-2\pi, 2\pi]$. There are no vertical asymptotes. Note that $f(-x) = \dfrac{(-x)(\sin(-x))}{((-x)^2 + 1)} = \dfrac{x \sin x}{x^2 + 1} = f(x)$, f has even symmetry. Questions about horizontal asymptotes aren't relevant because the given domain is an interval with finite length.

$f'(x) = \dfrac{(x^2 + 1)(x \cos x + \sin x) - x \sin x \cdot (2x)}{(x^2 + 1)^2}$, which can be simplified to

$$\dfrac{x(x^2 + 1)\cos x + (1 - x^2)\sin x}{(x^2 + 1)^2},$$

and with the aid of a computer algebra system, the roots of this expression can be found to be approximately ± 4.514 and ± 1.356, as well as $x = 0$. We will call the non-zero roots $\pm r_1$ and $\pm r_2$ where $0 < r_1 < r_2$. Note that $f'(x) < 0$ on $(-2\pi, -r_2)$ and on $(-r_1, 0)$ and (r_1, r_2), so f is decreasing there, while $f'(x) > 0$ on $(-r_2, -r_1)$, on $(0, r_1)$, and on $(r_2, 2\pi)$, so f is increasing on these intervals. There are local maxima at $x = \pm r_1$ and local minima at $x = 0$ and at $x = \pm r_2$.

$f''(x)$ has numerator $(x^2 + 1)^2((x^3 + x)(-\sin x) + \cos x(3x^2 + 1) + (1 - x^2)(\cos x) + \sin x(-2x)) - ((x^3 + x)\cos x + (1 - x^2)\sin x)(4x)(x^2 + 1)$ and denominator $(x^2 + 1)^4$. This simplifies to

$$f''(x) = \dfrac{(-x^5 - 7x)\sin x + (-2x^4 + 2)\cos x}{(x^2 + 1)^3},$$

which is 0 at approximately ± 5.961 and ± 2.561 and ± 0.494. We will call these 6 roots $\pm r_3$, $\pm r_4$ and $\pm r_5$ where $0 < r_3 < r_4 < r_5$. Note that $f''(x) < 0$ on $(-2\pi, -r_5)$ and on $(-r_4, -r_3)$, and on (r_3, r_4), and on

$(r_5, 2\pi)$, so f is concave down on these intervals, while $f''(x) > 0$ on $(-r_5, -r_4)$, and on $(-r_3, r_3)$, and on (r_4, r_5), so f is concave up on these intervals. There are points of inflection at each of $\pm r_3$, $\pm r_4$, and $\pm r_5$. There is an x-intercept at $(0, 0)$, which is also the y-intercept, as well as x-intercepts at $\pm 2\pi$.

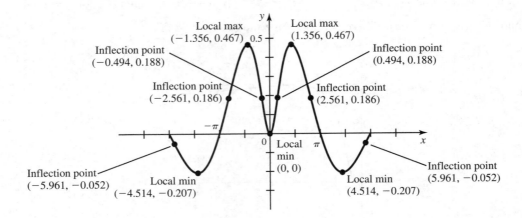

4.4.69 The domain of f is $(-\infty, \infty)$ and there is no symmetry. There are no asymptotes because f is a polynomial.

$f'(x) = 12x^3 - 132x^2 + 120x = 12x(x - 10)(x - 1)$, which is 0 for $x = 0$, $x = 1$, and $x = 10$.

$f'(x) > 0$ on $(0, 1)$ and on $(10, \infty)$, so f is increasing on those intervals. $f'(x) < 0$ on $(-\infty, 0)$ and on $(1, 10)$, so f is decreasing on those intervals. There is a local maximum at $x = 1$, and local minima at $x = 0$ and at $x = 10$.

$f''(x) = 36x^2 - 264x + 120 = 12(3x^2 - 22x + 10)$. This is 0 at approximately $x = .487$ and $x = 6.846$. Let these two roots be r_1 and r_2 with $r_1 < r_2$. $f''(x) > 0$ on $(-\infty, r_1)$ and on (r_2, ∞), so f is concave up on those intervals. $f''(x) < 0$ on (r_1, r_2), so f is concave down on that interval. There are points of inflection at $x = r_1$ and $x = r_2$.

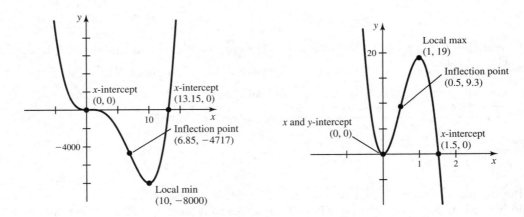

4.4.71 The domain of f is $(-\infty, \infty)$. There are no vertical asymptotes. Note that $f(-x) = \dfrac{\tan^{-1}(-x)}{(-x)^2 + 1} = -\dfrac{\tan^{-1}(x)}{x^2 + 1} = -f(x)$, so f has odd symmetry. Because $\lim\limits_{x \to \pm\infty} \tan^{-1}(x) = \pm\pi/2$, $\lim\limits_{x \to \pm\infty} f(x) = 0$, so $y = 0$ is a horizontal asymptote.

$f'(x) = \dfrac{(x^2 + 1)(1/(x^2 + 1)) - \tan^{-1} x \cdot 2x}{(x^2 + 1)^2} = \dfrac{1 - 2x \tan^{-1} x}{(x^2 + 1)^2}$. Using a computer algebra system shows that the numerator has two roots at approximately ± 0.765. Let the roots be $\pm r_1$ where $r_1 > 0$. Note that $f'(x) < 0$ on $(-\infty, -r_1)$ and on (r_1, ∞), so f is decreasing there, while $f'(x) > 0$ on $(-r_1, r_1)$, so f is increasing on that interval. There is a local minimum at $-r_1$ and a local maximum at r_1.

$$f''(x) = \frac{(x^2+1)^2\left[(-2x)(1/(x^2+1)) - \tan^{-1}x \cdot 2\right] - (1 - 2x\tan^{-1}x) \cdot 2(x^2+1)(2x)}{(x^2+1)^4}$$

$$= \frac{(6x^2 - 2)\tan^{-1}x - 6x}{(x^2+1)^3}.$$

Again, using a computer algebra system reveals roots at approximately ± 1.330 in addition to the root at 0. Let the non-zero roots of the numerator be $\pm r_2$ where $r_2 > 0$. We see that $f''(x) < 0$ on $(-\infty, -r_2)$, and on $(0, r_2)$, so f is concave down on those intervals, while $f''(x) > 0$ on $(-r_2, 0)$ and on (r_2, ∞), so f is concave up on those intervals, and there are points of inflection at $-r_2$, 0, and r_2.

There is an x-intercept at $(0,0)$, which is also the y-intercept.

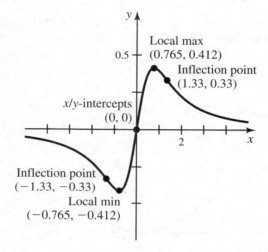

4.4.73 The equation is valid only for $|x| \leq 1$ and $|y| \leq 1$. Using implicit differentiation, we have $(2/3)x^{-1/3} + (2/3)y^{-1/3}y' = 0$, so $y' = \dfrac{-y^{1/3}}{x^{1/3}}$. This is 0 for $y = 0$ (in which case $x = \pm 1$) and doesn't exist for $x = 0$ (in which case $x = \pm 1$.) In the first quadrant the curve is decreasing, in the 2nd it is increasing, in the 3rd it is decreasing, and in the 4th it is increasing. Differentiating y' yields $y'' = \dfrac{x^{1/3}(-1/3)y^{-2/3}y' + y^{1/3}(1/3)(x^{-2/3})}{x^{2/3}} = \dfrac{y^{2/3} + x^{2/3}}{3x^{4/3}y^{1/3}}$, which is positive when y is positive and negative when y is negative, so the curve is concave up in the first and 2nd quadrants, and concave down in the 3rd and 4th.

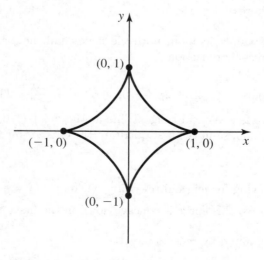

4.4.75 First note that the expression is symmetric when x and y are switched, so the curve should be symmetric about the line $y = x$. Also, if $y = x$, then $2x^3 = 3x^2$, so either $x = 0$ or $x = 3/2$, so this is where the curve intersects the line $y = x$.

Differentiating implicitly yields $3x^2 + 3y^2 y' = 3xy' + 3y$, so $y' = \dfrac{y - x^2}{y^2 - x}$. This is 0 when $y = x^2$, but this occurs on the curve when $x^3 + x^6 = 3x^3$, which yields $x = 0$ (and $y = 0$), or $x^3 = 2$, so $x = \sqrt[3]{2} \approx 1.260$. Note also that the derivative doesn't exist when $x = y^2$, which again yields $(0,0)$ and $y^6 + y^3 = 3y^3$, or $y = \sqrt[3]{2}$. So there should be a flat tangent line at approximately $(1.260, 1.587)$ and a vertical tangent line at about $(1.587, 1.260)$.

Differentiating again and solving for y'' yields $y''(x) = \dfrac{2xy\left(x^3 - 3xy + y^3 + 1\right)}{(x - y^2)^3} = \dfrac{2xy}{(x - y^2)^3}$. In the first quadrant, when $x > y^2$, the curve is concave up, when $x < y^2$, the curve is concave down. In both the 2nd and 4th quadrants, the curve is concave up.

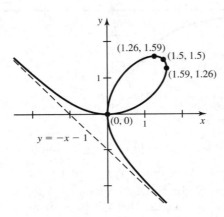

4.4.77 Note that the curve requires $-1 \leq x < 1$.

Note also that the curve is symmetric about both the x-axis and the y-axis, so we can just consider the first quadrant, and obtain the rest by reflection.

Differentiating implicitly yields $4x^3 - 2x + 2yy' = 0$, so $y' = \dfrac{x - 2x^3}{y}$. This is 0 in the first quadrant for $x = \sqrt{2}/2$. Note also that there is a vertical tangent line at the point $(1, 0)$. The derivative is positive on $(0, \sqrt{2}/2)$ and negative on $(\sqrt{2}/2, 1)$, so in the first quadrant the curve is increasing on that first interval and decreasing on the second.

Differentiating again and solving for y'' (and rewriting) yields $y''(x) = \dfrac{x^4(2x^2 - 3)}{y}$, which is negative in the first quadrant for $0 < x < 1$, so this curve is concave down in the first quadrant.

The rest of the curve can be found by reflection.

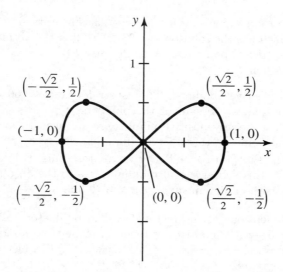

4.4.79 As n increases, the curves retain their symmetry, but move "outward." That is, the curves enclose a greater area. It appears that the figures approach the 2×2 square centered at the origin with sides parallel to the coordinate axes.

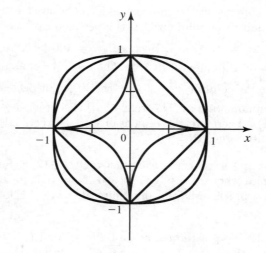

4.5 Optimization Problems

4.5.1 ...objective... constraints ....

4.5.3 The constraint is $x + y = 10$, so we can express $y = 10 - x$ or $x = 10 - y$. Therefore the objective function can be expressed $Q = x^2(10 - x)$ or $Q = (10 - y)^2 y$.

4.5.5

a. From the constraint, we have $y = 100 - 10x$. Substituting this into the objective function, we have $P(x) = x(100 - 10x) = 100x - 10x^2$.

b. $P'(x) = 100 - 20x$, which is zero for $x = 5$. Because $P''(x) = -20$, P is concave down everywhere, so the critical point $x = 5$ yields an absolute maximum. The value of $P(5)$ is $100(5) - 10(25) = 250$.

4.5.7 Let x and y be the two non-negative numbers. The constraint is $x + y = 23$, which gives $y = 23 - x$. The objective function to be maximized is the product of the numbers, $P = xy$. Using $y = 23 - x$, we have $P = xy = x(23 - x) = 23x - x^2$. Now x must be at least 0, and cannot exceed 23 (otherwise $y < 0$).

Therefore we need to maximize $P(x) = 23x - x^2$ for $0 \leq x \leq 23$. The critical points of the objective function satisfy $P'(x) = 23 - 2x = 0$, which has the solution $x = 23/2$. To find the absolute maximum of P, we check the endpoints of $[0, 23]$ and the critical point $x = 23/2$. Because $P(0) = P(23) = 0$ and $P(23/2) = (23/2)^2$, the absolute maximum occurs when $x = y = 23/2$.

4.5.9 Let x and y be the two positive numbers. The constraint is $xy = 50$, which gives $y = 50/x$. The objective function to be minimized is the sum of the numbers, $S = x + y$. Using $y = 50/x$, we have $S = x + y = x + \frac{50}{x}$. Now x can be any positive number, so we need to maximize $S(x) = x + 50/x$ on the interval $(0, \infty)$. The critical points of the objective function satisfy $S'(x) = 1 - \frac{50}{x^2} = 0$, which has the solution $x = \sqrt{50} = 5\sqrt{2}$. By the First (or Second) Derivative Test, this critical point corresponds to a local minimum, and by Theorem 4.5, this solitary local minimum is also the absolute minimum on the interval $(0, \infty)$. Therefore the numbers with minimum sum are $x = 5\sqrt{2}$ and $y = \frac{50}{5\sqrt{2}} = \frac{10}{\sqrt{2}} = 5\sqrt{2}$, so $x = y = 5\sqrt{2}$.

4.5.11 Let x and y be the dimensions of the rectangle. The perimeter is $2x + 2y$, so the constraint is $2x + 2y = 10$, which gives $y = 5 - x$. The objective function to be maximized is the area of the rectangle, $A = xy$. Thus we have $A = xy = x(5 - x) = 5x - x^2$. We have $x, y \geq 0$, which also implies $x \leq 5$ (otherwise $y < 0$). Therefore we need to maximize $A(x) = 5x - x^2$ for $0 \leq x \leq 5$. The critical points of the objective function satisfy $A'(x) = 5 - 2x = 0$, which has the solution $x = 5/2$. To find the absolute maximum of A, we check the endpoints of $[0, 5]$ and the critical point $x = 5/2$. Because $A(0) = A(5) = 0$ and $A(5/2) = 25/4$, the absolute maximum occurs when $x = y = 5/2$, so width = length = $5/2$ m.

4.5.13 Let x and y be be the dimensions of the rectangle. The area is $xy = 100$, so the constraint is $y = 100/x$. The objective function to be minimized is the perimeter of the rectangle, $P = 2x + 2y$. Using $y = 100/x$, we have $P = 2x + 2y = 2x + \dfrac{200}{x}$. Because $xy = 100 > 0$ we must have $x > 0$, so we need to minimize $P(x) = 2x + 200/x$ on the interval $(0, \infty)$. The critical points of the objective function satisfy $P'(x) = 2 - \dfrac{200}{x^2} = 0$, which has the solution $x = 10$. By the First (or Second) Derivative Test, this critical point corresponds to a local minimum, and by Theorem 4.5, this solitary local minimum is also the absolute minimum on the interval $(0, \infty)$. Therefore the dimensions of the rectangle with minimum perimeter are $x = 10$ and $y = \frac{100}{10} = 10$, so width = length = 10.

4.5.15 We seek to minimize $S = 2x + y$ subject to the constraint $y = 12/x$. Substituting gives $S = 2x + 12/x$, so $S'(x) = 2 - 12/x^2$. This is zero when $x^2 = 6$, or $x = \sqrt{6}$. Note that for $0 < x < \sqrt{6}$ we have $S'(x) < 0$, and for $x > \sqrt{6}$ we have $S'(x) > 0$, so we have a minimum at $x = \sqrt{6}$. Note that when $x = \sqrt{6}$, we have $y = 12/x = 12/\sqrt{6} = 2\sqrt{6}$.

4.5.17 Let the coordinates of the base of the rectangle be $(x, 0)$ and $(-x, 0)$ where $0 \leq x \leq 5$. Then the width of the rectangle is $2x$ and the height is $\sqrt{25 - x^2}$, so the area A is given by $A(x) = 2x\sqrt{25 - x^2}$. The critical points of this function satisfy $A'(x) = 2\sqrt{25 - x^2} + \dfrac{2x \cdot (-x)}{\sqrt{25 - x^2}} = \dfrac{2(25 - 2x^2)}{\sqrt{25 - x^2}} = 0$, which has unique solution $x = 5/\sqrt{2}$ in $(0, 5)$. We have $A(0) = A(5) = 0$, so the rectangle of maximum area has width $2x = 10/\sqrt{2}$ cm, height $y = \sqrt{25 - (25/2)} = 5/\sqrt{2}$ cm.

4.5.19 Let x be the length of the sides of the base of the box and y be the height of the box. The volume is $x \cdot x \cdot y = 8$, so the constraint is $x^2 y = 8$, which gives $y = \dfrac{8}{x^2}$. The objective function to be minimized is the surface area S of the box, which consists of $2x^2$ (for the top and base) + $4xy$ (for the 4 sides); therefore $S = 2x^2 + 4xy$. Using $y = \dfrac{8}{x^2}$, we have $S = 2x^2 + 4xy = 2x^2 + 4x \cdot \dfrac{8}{x^2} = 2x^2 + \dfrac{32}{x}$. The base side length can be any $x > 0$, so we need to maximize $S(x) = 2x^2 + \dfrac{32}{x}$ on the interval $(0, \infty)$. The critical points of the objective function satisfy $S'(x) = 4x - \dfrac{32}{x^2} = 0$; clearing denominators gives $4x^3 = 32$ so $x^3 = 8$, so $x = 2$. By the First (or Second) Derivative Test, this critical point corresponds to a local minimum, and by Theorem 4.5, this solitary local minimum is also the absolute minimum on the interval $(0, \infty)$. When $x = 2$ we have $y = \dfrac{8}{4} = 2$, so the box with the smallest surface area is $2 \times 2 \times 2$.

4.5. Optimization Problems

4.5.21 Let x be the length of the sides of the base of the box and y be the height of the box. The volume of the box is $x \cdot x \cdot y = x^2y$, so the constraint is $x^2y = 16$, which gives $y = 16/x^2$. Let c be the cost per square foot of the material used to make the sides. Then the cost to make the base is $2cx^2$, the cost to make the 4 sides is $4cxy$, and the cost to make the top is $\frac{1}{2}cx^2$. The objective function to be minimized is the total cost, which is $C = 2cx^2 + 4cxy + \frac{1}{2}cx^2 = \frac{5}{2}cx^2 + 4cx \cdot \frac{16}{x^2} = c\left(\frac{5x^2}{2} + \frac{64}{x}\right)$. The base side length can be any $x > 0$, so we need to maximize $C(x) = c(5x^2/2 + 64/x)$ on the interval $(0, \infty)$. The critical points of the objective function satisfy $5x - \frac{64}{x^2} = 0$, which gives $x^3 = 64/5$ or $x = 4/\sqrt[3]{5}$. By the First (or Second) Derivative Test, this critical point corresponds to a local minimum, and by Theorem 4.5, this solitary local minimum is also the absolute minimum on the interval $(0, \infty)$. Therefore the box with minimum cost has base $4/\sqrt[3]{5}$ ft by $4/\sqrt[3]{5}$ ft and height $y = 16/(4/\sqrt[3]{5})^2 = 5^{2/3}$ ft.

4.5.23 A point on the parabola $y = 1 - x^2$ has the form $(x, 1 - x^2)$, which has distance L to the point $(1, 1)$ given by $L^2 = (x-1)^2 + \left(1 - (1-x^2)\right)^2 = x^4 + x^2 - 2x + 1$. Because L is positive, it suffices to minimize L^2. The critical points of L^2 satisfy $\frac{dL^2}{dx} = 4x^3 + 2x - 2 = 2(2x^3 + x - 1) = 0$ This cubic equation has a unique root $x \approx 0.590$, so the point closest to $(1, 1)$ on this parabola is approximately $(0.590, 0.652)$.

4.5.25 The distance between $(x, 3x)$ and $(50, 0)$ is $d(x) = \sqrt{(3x-0)^2 + (x-50)^2}$. Instead of working with the distance, we can instead work with the square of the distance, because these two functions have minima which occur at the same place. So consider

$$(d(x))^2 = D(x) = (3x)^2 + (x-50)^2 = 9x^2 + x^2 - 100x + 2500 = 10(x^2 - 10x + 250).$$

$\frac{dD}{dx} = 10(2x - 10)$, which is zero for $x = 5$. Because $\frac{d^2D}{dx^2} = 20 > 0$, we see that the critical point at $x = 5$ is a minimum. So the minimum of D (and d) occurs at $x = 5$. The value of d at the point $(5, 15)$, is $d(5) = \sqrt{15^2 + (-45)^2} = 15\sqrt{10} \approx 47.4$.

4.5.27

a. Let x be the distance from the point on the shoreline nearest to the boat to the point where the woman lands on shore; then the remaining distance she must travel on shore is $6 - x$. By the Pythagorean Theorem, the distance the woman must row is $\sqrt{x^2 + 16}$. So the time for the rowing leg is $\frac{\text{distance}}{\text{rate}} = \frac{\sqrt{x^2+16}}{2}$ and the time for the walking leg is $\frac{\text{distance}}{\text{rate}} = \frac{6-x}{3}$. The total travel time for the trip is the objective function $T(x) = \frac{\sqrt{x^2+16}}{2} + \frac{6-x}{3}$. We wish to minimize this function for $0 \le x \le 6$. The critical points of the objective function satisfy $T'(x) = \frac{x}{2\sqrt{x^2+16}} - \frac{1}{3} = 0$, which when simplified gives $5x^2 = 64$, so $x = 8/\sqrt{5}$ is the only critical point in $(0, 6)$. From the First Derivative Test we see that T has a local minimum at this point, so $x = 8/\sqrt{5}$ must give the minimum value of T on $[0, 6]$.

b. Let $v > 0$ be the woman's rowing speed. Then the total travel time is now given by $T(x) = \frac{\sqrt{x^2+16}}{v} + \frac{6-x}{3}$. The derivative of the objective function is $T'(x) = \frac{x}{v\sqrt{x^2+16}} - \frac{1}{3}$. If we try to solve the equation $T'(x) = 0$ as in part (a) above, we see that there is at most one solution $x > 0$. Therefore there can be at most one critical point of T in the interval $(0, 6)$. Observe also that $T'(0) = -1/3 < 0$ so the absolute minimum of T on $[0, 6]$ cannot occur at $x = 0$. So one of two things must happen: there is a unique critical point for T in $(0, 6)$ which is the absolute minimum for T on $[0, 6]$, and then $T'(6) > 0$; or, T is decreasing on $[0, 6]$, and then $T'(6) \le 0$ (the quickest way to the restaurant is to row directly in this case). The condition $T'(6) \le 0$ is equivalent to $\frac{6}{\sqrt{6^2+16}} \le \frac{v}{3}$ which gives $v \ge 9/\sqrt{13}$ mi/hr.

4.5.29 Let x be the distance from the point on shore nearest the island to the point where the underwater cable meets the shore, and let y be the be the length of the underwater cable. In terms of the angle θ in the figure, $\tan\theta = 3.5/x$ so $x = 3.5\cot\theta$, and $\sin\theta = 3.5/y$ so $y = 3.5\csc\theta$. The objective function to be minimized is the cost given by $C(\theta) = 2400 \cdot 3.5\csc\theta + 1200 \cdot (8 - 3.5\cot\theta) = 8400\csc\theta - 4200\cot\theta + 9600$. The angle θ must be between $\tan^{-1}(3.5/8)$ (≈ 0.412) and $\pi/2$. The critical points of $C(\theta)$ satisfy $C'(\theta) = 8400(-\csc\theta\cot\theta) - 4200(-\csc^2\theta) = 4200\csc^2\theta(1 - 2\cos\theta)$, which has unique solution $\theta = \pi/3$ in the interval under consideration. By the First Derivative Test, this critical point corresponds to a local minimum, and by Theorem 4.5, this solitary local minimum is also the absolute minimum on the interval $[\tan^{-1}(3.5/8), \pi/2]$. Therefore the optimal point on shore has distance $x = 3.5\cot(\pi/3) = 7\sqrt{3}/6$ mi from the point on shore nearest the island, in the direction of the power station.

4.5.31 Let L be the ladder length and x be the distance between the foot of the ladder and the fence. The Pythagorean Theorem gives the relationship $L^2 = (x+5)^2 + b^2$, where b is the height of the top of the ladder. We see that $b/(x+5) = 8/x$ by similar triangles, which gives $b = 8(x+5)/x$. Substituting in the expression for L^2 above gives $L^2 = (x+5)^2 + 64\dfrac{(x+5)^2}{x^2} = (x+5)^2\left(1 + \dfrac{64}{x^2}\right)$. It suffices to minimize L^2 instead of L. However in this case x and b must satisfy $x, b \leq 20$. Solving $20 = 8(x+5)/x$ for x gives $x = 10/3$, so the condition $b \leq 20$ corresponds to $x \geq 10/3$, and we see that we must minimize L^2 for $10/3 \leq x \leq 20$. We have $\dfrac{d}{dx}L^2 = (x+5)^2\left(-\dfrac{128}{x^3}\right) + 2(x+5)\left(1 + \dfrac{64}{x^2}\right) = \dfrac{2(x+5)(x^3 - 320)}{x^3}$. Because $x > 0$, the only critical point is $x = \sqrt[3]{320} \approx 6.840$. By the First Derivative Test, this critical point corresponds to a local minimum, and by Theorem 4.5, this solitary local minimum is also the absolute minimum on the interval $[10/3, 20]$. Substituting $x \approx 6.840$ in the expression for L^2 we find the length of the shortest ladder $L \approx 18.220$ ft.

4.5.33 If we remove a sector of angle θ from a circle of radius 20, the remaining circumference is $2\pi\cdot 20 - \theta\cdot 20 = 20(2\pi - \theta)$, so the base of the cone formed has radius $r = \dfrac{20(2\pi - \theta)}{2\pi} = \dfrac{10(2\pi - \theta)}{\pi}$. As θ varies from 0 to 2π, the radius ranges from 0 to 20, but all possible cones formed have side length 20. The height h of the cone is given by the Pythagorean Theorem: $h^2 + r^2 = 20^2$, so $h = \sqrt{400 - r^2}$. The volume of the cone given by $V = \dfrac{\pi}{3}r^2 h = \dfrac{\pi}{3}r^2\sqrt{400 - r^2}$.

Thus

$$V'(r) = \frac{2\pi}{3}r\sqrt{400 - r^2} + \frac{\pi}{3}r^2(400 - r^2)^{-1/2}\cdot\frac{1}{2}\cdot(-2r)$$
$$= \frac{\pi}{3}\cdot\frac{2r(400 - r^2) - r^3}{\sqrt{400 - r^2}}$$
$$= \frac{\pi}{3}\cdot\frac{800r - 3r^3}{\sqrt{400 - r^2}}.$$

The only positive critical point occurs where $r = \sqrt{\frac{800}{3}} = 20\sqrt{\frac{2}{3}}$. An application of the First Derivative Test shows that this is a maximum. So

$$h = \sqrt{400 - \left(20\sqrt{\frac{2}{3}}\right)^2} = \sqrt{400 - 400\cdot\frac{2}{3}} = 20\sqrt{\frac{1}{3}}.$$

4.5.35

a. Let r and h be the radius and height of the can. The volume of the can is $V = \pi r^2 h$, which gives the constraint $\pi r^2 h = 354$ or $h = 354/(\pi r^2)$. The objective function to be minimized is the surface area, which consists of $2\pi r^2$ (for the top and bottom of the can) and $2\pi rh$ (for the side of the can). Therefore the objective function to be minimized is $A = 2\pi r^2 + 2\pi rh = 2\pi\left(r^2 + r\left(\dfrac{354}{\pi r^2}\right)\right) = 2\pi\left(r^2 + \dfrac{354}{\pi r}\right)$.

We need to minimize $A(r)$ for $r > 0$. The critical points of $A(r)$ satisfy $A'(r) = 2\pi\left(2r - \dfrac{354}{\pi r^2}\right) = 0$,

4.5. Optimization Problems

which gives $r = \sqrt[3]{(177/\pi)} \approx 3.834$ cm. The corresponding value of h is $h = \dfrac{354}{\pi r^2} = \dfrac{354r}{\pi r^3} = 2r \cdot \dfrac{177}{\pi r^3} = 2r$, so $h = 2\sqrt[3]{(177/\pi)} \approx 7.667$ cm. By the First (or Second) Derivative Test, this critical point corresponds to a local minimum, and by Theorem 4.5, this solitary local minimum is also the absolute minimum on the interval $(0, \infty)$.

b. We modify the objective function in part (a) above to account for the fact that the top and bottom of the can have double thickness: $A = 4\pi r^2 + 2\pi rh = 2\pi \left(2r^2 + r\left(\dfrac{354}{\pi r^2}\right)\right) = 4\pi \left(r^2 + \dfrac{177}{\pi r}\right)$. We need to minimize $A(r)$ for $r > 0$. The critical points of $A(r)$ satisfy $A'(r) = 4\pi \left(2r - \dfrac{177}{\pi r^2}\right) = 0$, which gives $r = \sqrt[3]{(177/2\pi)} \approx 3.043$ cm. The corresponding value of h is $h = \dfrac{354}{\pi r^2} = \dfrac{354r}{\pi r^3} = 4r \cdot \dfrac{177}{2\pi r^3} = 4r$, so $h = 4\sqrt[3]{(177/2\pi)} \approx 12.171$ cm. These dimensions are closer to those of a real soda can.

4.5.37 Let x and y be the dimensions of the flower garden; the area of the flower garden is 30, so we have the constraint $xy = 30$ which gives $y = 30/x$. The dimensions of the garden and borders are $x + 4$ and $y + 2$, so the objective function to be minimized for $x > 0$ is $A = (x+4)(y+2) = (x+4)\left(\dfrac{30}{x} + 2\right) = 2x + \dfrac{120}{x} + 38$. The critical points of $A(x)$ satisfy $A'(x) = 2 - \dfrac{120}{x^2} = 0$, which has unique solution $x = \sqrt{60} = 2\sqrt{15}$. By the First (or Second) Derivative test, this critical point gives a local minimum, which by Theorem 4.5 must be the absolute minimum of A over $(0, \infty)$. The corresponding value of y is $30/2\sqrt{15} = \sqrt{15}$, so the dimensions are $\sqrt{15}$ by $2\sqrt{15}$ m.

4.5.39 Because the length of the box (in inches) is $18 - 2x$, the width is $9 - x$, and the height is x, the volume of the box is $V(x) = x(9-x)(18-2x) = 2x^3 - 36x^2 + 162x$, where $0 \leq x \leq 9$. Taking the derivative of V, we have $V'(x) = 6x^2 - 72x + 162 = 6(x^2 - 12x + 27) = 6(x-3)(x-9)$. Setting the derivative equal to 0 and solving for x, we find that $x = 3$ and $x = 9$. Because $V(0) = V(9) = 0$ and $V(3) = 216$, the box of maximum volume is $12 \times 6 \times 3$ and has a volume of 216 in^3.

4.5.41 Let x and y equal the width and height, respectively, of the rectangular part of the window. Then the perimeter is $x + 2y + \dfrac{\pi x}{2} = 20$ and the area of the window is $A = xy + \dfrac{1}{2}\pi\left(\dfrac{x}{2}\right)^2 = xy + \dfrac{\pi x^2}{8}$. Solving the perimeter equation for y, we have $y = \dfrac{1}{2}\left(20 - x - \dfrac{\pi x}{2}\right) = 10 - \dfrac{x}{2} - \dfrac{\pi x}{4}$. By substitution, $A(x) = x\left(10 - \dfrac{x}{2} - \dfrac{\pi x}{4}\right) + \dfrac{1}{8}\pi x^2 = \left(\dfrac{\pi}{8} - \dfrac{1}{2} - \dfrac{\pi}{4}\right)x^2 + 10x = \left(-\dfrac{1}{2} - \dfrac{\pi}{8}\right)x^2 + 10x$. Taking the derivative of A, we have $A'(x) = \left(-1 - \dfrac{2\pi}{8}\right)x + 10$, which is 0 for $x = \dfrac{10}{1 + 2\pi/8} = \dfrac{40}{4 + \pi} \approx 5.6$ ft. At this x value, $y \approx 2.8$ ft. Because $A''\left(\dfrac{40}{4+\pi}\right) < 0$, the Second Derivative Test implies that the area is maximized when the dimensions of the rectangular pane are approximately 5.6 ft wide by 2.8 ft high.

4.5.43 The radius r and height h of the barrel satisfy the constraint $r^2 + h^2 = d^2$, which we can rewrite as $r^2 = d^2 - h^2$. The volume of the barrel is given by $V = \pi r^2 h = \pi(d^2 - h^2)h = \pi(d^2h - h^3)$. The height h must satisfy $0 \leq h \leq d$, so we need to maximize $V(h)$ on the interval $[0, d]$. The critical points of V satisfy $V'(h) = \pi(d^2 - 3h^2) = 0$. The only critical point in $(0, d)$ is $h = d/\sqrt{3}$, which gives the maximum volume because at the endpoints $V(0) = V(d) = 0$. The corresponding r value satisfies $r^2 = d^2 - d^2/3 = 2d^2/3$, so $r = \sqrt{2}d/\sqrt{3}$ and we see that the ratio r/h that maximizes the volume is $\sqrt{2}$.

4.5.45 Let r and h be the radius and height of the cylinder. The distance d from the centroid of the cylinder (the midpoint of the cylinder's axis of rotation) to any point on the top or bottom edge satisfies $d^2 = r^2 + \left(\dfrac{h}{2}\right)^2$ so the constraint is $r^2 + (h/2)^2 = R^2$. The volume of the cylinder is given by $V = \pi r^2 h = \pi\left(R^2 - \left(\dfrac{h}{2}\right)^2\right)h = \pi\left(R^2 h - \dfrac{h^3}{4}\right)$. Because $r, h \geq 0$ we must have $0 \leq h \leq 2R$. We wish to maximize

$V(h)$ on this interval. The critical points of $V(h)$ satisfy $V'(h) = \pi \left(R^2 - \dfrac{3h^2}{4} \right) = 0$ which gives $h = 2R/\sqrt{3}$, and from the constraint we obtain $r = \sqrt{2}R/\sqrt{3}$. The volume $V(h) = 0$ at the endpoints $h = 0$ and $h = 2R$, so the maximum volume must occur at this critical point.

4.5.47 Let R and H be the radius and height of the larger cone and let r and h be the radius and height of the smaller inscribed cone. The region that lies above the smaller cone inside the larger cone is a cone with radius r and height $H - h$; by similar triangles we have $\dfrac{H-h}{r} = \dfrac{H}{R}$ so $h = \dfrac{H}{R}(R - r)$. The volume of the smaller cone is $V = \dfrac{\pi}{3} r^2 h = \dfrac{\pi H}{3R}\left(Rr^2 - r^3\right)$, which we must maximize over $0 \leq r \leq R$. The critical points of $V(r)$ satisfy $V'(r) = \dfrac{\pi H}{3R}(2Rr - 3r^2) = 0$ which has unique solution $r = 2R/3$ in $(0, R)$. Because $V(r) = 0$ at the endpoints $r = 0$ and $r = R$, the smaller cone with maximum volume has radius $r = 2R/3$ and height $h = H/3$, so the optimal ratio of the heights is 3:1.

4.5.49

a. We find $g(0) = 0$, $g(40) = 30$ and $g(60) = 25$ miles per gallon. The value at $v = 0$ is reasonable because when a car first starts moving it needs a lot of power from its engine, so the gas mileage is very low. The decline from 30 to 25 mi/gal as v increases from 40 mi/hr to 60 mi/hr reflects the fact that gas mileage tends to decrease at speeds over 55 mi/hr.

b. The quadratic function $g(v) = (85v - v^2)/60$ takes its maximum value at $v = 85/2 = 42.5$ mi/hr.

c. At speed v the amount of gas needed to drive one mile is $1/g(v)$ and the time it takes is $1/v$. Hence the cost of gas for one mile is $p/g(v)$ and the cost for the driver is w/v, and so the cost for L miles is $C(v) = Lp/g(v) + Lw/v$.

d. We have $C(v) = 400\left(\dfrac{4}{g(v)} + \dfrac{20}{v}\right) = 1600\left(\dfrac{1}{g(v)} + \dfrac{5}{v}\right)$. The critical points of $C(v)$ satisfy $\dfrac{g'(v)}{g(v)^2} + \dfrac{5}{v^2} = 0$, which simplifies to $v^2 g'(v) + 5g(v)^2 = 0$. Substituting the formula for $g(v)$ above and using $g'(v) = (85 - 2v)/60$, we can factor out v^2 and reduce to the quadratic equation $v^2 - 194v + 8245 = 0$, which has roots $v \approx 62.883, 131.117$. The First (or Second) Derivative Test shows that $C(v)$ has a local minimum at $v \approx 62.9$, which is the unique critical point for $0 \leq v \leq 131$. Therefore the cost is minimized at this value of v.

e. Because L is a constant factor in the cost function $C(v)$, changing L will not change the critical points of $C(v)$.

f. The critical points of $C(v)$ now satisfy the equation $\dfrac{4.2 g'(v)}{g(v)^2} + \dfrac{20}{v^2} = 0$, which simplifies to $4.2 v^2 g'(v) + 20 g(v)^2 = 0$. As above, substituting the formula for $g(v)$ above and using $g'(v) = (85 - 2v)/60$, we can factor out v^2 and reduce to the quadratic equation $v^2 - 195.2v + 8296 = 0$, which has roots $v \approx 62.532, 132.668$. As in part (d), the minimum cost occurs for $v \approx 62.532$, slightly less than the speed in part (d).

g. The critical points of $C(v)$ now satisfy the equation $\dfrac{4 g'(v)}{g(v)^2} + \dfrac{15}{v^2} = 0$, which simplifies to $4 v^2 g'(v) + 15 g(v)^2 = 0$. As above, substituting the formula for $g(v)$ above and using $g'(v) = (85 - 2v)/60$, we can factor out v^2 and reduce to the quadratic equation $v^2 - 202v + 8585 = 0$, which has roots $v \approx 60.800, 141.200$. As in part (d), the minimum cost occurs for $v \approx 60.8$, less than the speed in part (d).

4.5.51 The viewing angle θ is given by $\theta = \cot^{-1}\left(\dfrac{x}{10}\right) - \cot^{-1}\left(\dfrac{x}{3}\right)$, and we wish to maximize this function for $x > 0$. The critical points satisfy $\theta'(x) = -\dfrac{1}{1 + \left(\frac{x}{10}\right)^2} \cdot \dfrac{1}{10} - (-)\dfrac{1}{1 + \left(\frac{x}{3}\right)^2} \cdot \dfrac{1}{3} = \dfrac{3}{x^2 + 3^2} - \dfrac{10}{x^2 + 10^2} = 0$

which simplifies to $3(x^2 + 100) = 10(x^2 + 9)$ or $x^2 = 30$. Therefore $x = \sqrt{30} \approx 5.477$ ft is the only critical point in $(0, \infty)$. By the First (or Second) Derivative Test, this critical point corresponds to a local maximum, and by Theorem 4.5, this solitary local maximum must be the absolute maximum on the interval $(0, \infty)$.

4.5.53 Let x be the distance between the point and the weaker light source; then $12 - x$ is the distance to the stronger light source. The intensity is proportional to $I(x) = \dfrac{1}{x^2} + \dfrac{2}{(12-x)^2}$, so we can take this as our objective function to be minimized for $0 < x < 12$. The critical points of $I(x)$ satisfy $I'(x) = -\dfrac{2}{x^3} + \dfrac{4}{(12-x)^3} = 0$ which gives $\left(\dfrac{12-x}{x}\right)^3 = 2$, or $\dfrac{12-x}{x} = \sqrt[3]{2}$, or $x = \dfrac{12}{\sqrt[3]{2}+1} \approx 5.310$. By the First (or Second) Derivative Test, this critical point corresponds to a local minimum, and by Theorem 4.5, this solitary local minimum is also the absolute minimum on the interval $(0, 12)$. Therefore the intensity is weakest at the point $12/(\sqrt[3]{2}+1) \approx 5.310$ m from the weaker source.

4.5.55

a. Let $x, d-x$ be the distances from the point where the rope meets the ground to the poles of height m, n respectively. Then the rope has length $L(x) = \sqrt{x^2 + m^2} + \sqrt{(d-x)^2 + n^2}$. We wish to minimize this function for $0 \leq x \leq d$. The critical points of $L(x)$ satisfy $L'(x) = \dfrac{x}{\sqrt{x^2+m^2}} - \dfrac{d-x}{\sqrt{(d-x)^2+n^2}} = 0$, which is equivalent to $\dfrac{x}{\sqrt{x^2+m^2}} = \dfrac{d-x}{\sqrt{(d-x)^2+n^2}}$, or in terms of the angles θ_1 and θ_2 in the figure, $\sec\theta_1 = \sec\theta_2$ and therefore $\theta_1 = \theta_2$. Observe that $L'(0) < 0$ and $L'(d) > 0$, so the minimum value of $L(x)$ must occur at some $x \in (0, d)$. There must be exactly one critical point, because as x ranges from 0 to d, θ_1 decreases and θ_2 increases, and so $\theta_1 = \theta_2$ can occur for at most one value of x.

b. Because the speed of light is constant, travel time is minimized when distance is minimized, which we saw in part (a) occurs when $\theta_1 = \theta_2$.

4.5.57 Let h be the height of the cylindrical tower and r the radius of the dome. The cylinder has volume $\pi r^2 h$, and the hemispherical dome has volume $2\pi r^3/3$ (half the volume of a sphere of radius r). The total volume is 750, so we have the constraint $\pi r^2 h + \dfrac{2\pi r^3}{3} = 750$ which gives $h = \dfrac{750}{\pi r^2} - \dfrac{2r}{3}$. We must have $h \geq 0$, which is equivalent to $r \leq \sqrt[3]{1125/\pi}$. The objective function to be maximized is the cost of the metal to make the silo, which is proportional to the surface area of the cylinder $(= 2\pi rh)$ plus 1.5 times the surface area of the hemisphere $(= 2\pi r^2)$. So we can take as objective function

$$C = 2\pi rh + 1.5 \cdot 2\pi r^2 = 2\pi r\left(\dfrac{750}{\pi r^2} - \dfrac{2r}{3}\right) + 3\pi r^2 = \dfrac{1500}{r} + \dfrac{5}{3}\pi r^2.$$

The critical points of $C(r)$ satisfy $C'(r) = -\dfrac{1500}{r^2} + \dfrac{10}{3}\pi r = 0$, which gives $\pi r^3 = 450$ and hence $r = \sqrt[3]{450/\pi}$. The corresponding value of h is

$$h = \dfrac{750}{\pi r^2} - \dfrac{2r}{3} = \dfrac{750 r}{\pi r^3} - \dfrac{2r}{3} = \left(\dfrac{750}{450} - \dfrac{2}{3}\right) r = r.$$

By the First (or Second) Derivative Test, this critical point corresponds to a local minimum, and by Theorem 4.5, this solitary local minimum is also the absolute minimum on the interval $[0, \sqrt[3]{1125/\pi}]$. Therefore the dimensions that minimize the cost are $r = h = \sqrt[3]{450/\pi}$ m.

4.5.59

a. $f'(x) = 2(x - a_1) + 2(x - a_2) = 4x - 2(a_1 + a_2)$. This is zero for $x = \dfrac{a_1 + a_2}{2}$. Because $f'(x) < 0$ for $x < \dfrac{a_1 + a_2}{2}$ and $f'(x) > 0$ for $x > \dfrac{a_1 + a_2}{2}$, we have a minimum at $x = \dfrac{a_1 + a_2}{2}$.

b. $f'(x) = 2(x-a_1) + 2(x-a_2) + 2(x-a_3) = 6x - 2(a_1+a_2+a_3)$. This is zero for $x = \dfrac{a_1+a_2+a_3}{3}$.
Because $f'(x) < 0$ for $x < \dfrac{a_1+a_2+a_3}{3}$ and $f'(x) > 0$ for $x > \dfrac{a_1+a_2+a_3}{3}$, we have a minimum at $x = \dfrac{a_1+a_2+a_3}{3}$.

c. $f'(x) = 2\sum_{k=1}^{n}(x-a_k) = 2nx - 2\sum_{k=1}^{n} a_k$. This is zero when $x = \dfrac{\sum_{k=1}^{n} a_k}{n}$. An application of the First Derivative Test shows that this value of x yields a minimum.

4.5.61 The cross-section is a trapezoid with height $3\sin\theta$; the larger of the parallel sides has length $3 + 2\cdot 3\cos\theta = 3 + 6\cos\theta$ and the smaller parallel side has length 3. The area of this trapezoid is given by

$$A(\theta) = \frac{1}{2}(3 + (3+6\cos\theta))\cdot 3\sin\theta = 9(1+\cos\theta)\sin\theta = 9\left(\sin\theta + \frac{\sin 2\theta}{2}\right),$$

using the identity $\sin 2\theta = 2\sin\theta\cos\theta$. We wish to maximize this function for $0 \le \theta \le \pi/2$. The critical points of $A(\theta)$ satisfy $\cos\theta + \cos 2\theta = \cos\theta + 2\cos^2\theta - 1 = 0$, using the identity $\cos 2\theta = 2\cos^2\theta - 1$. Therefore $x = \cos\theta$ satisfies the quadratic equation $2x^2 + x - 1 = 0$, which has roots $x = 1/2$ and -1. So the only critical point in $(0,\pi/2)$ is $\theta = \cos^{-1}(1/2) = \pi/3$, which by the First (or Second) Derivative Test and Theorem 4.5 gives the maximum area.

4.5.63 Let the radius of the Ferris wheel have length r, and let α be the angle the specific seat on the Ferris wheel makes with the center of the wheel (see the figure in the text). This point has coordinates $(r\cos\alpha, r + r\sin\alpha)$ so the distance from the seat to the base of the wheel is

$$d = \sqrt{r^2\cos^2\alpha + r^2(1+\sin\alpha)^2} = \sqrt{2}r\sqrt{1+\sin\alpha}.$$

Therefore the observer's angle satisfies $\tan\theta = \dfrac{r\sqrt{2}}{20}\sqrt{1+\sin\alpha}$. Think of θ and α as functions of time t and differentiate: $\sec^2\theta\,\dfrac{d\theta}{dt} = \dfrac{r\sqrt{2}}{20}\cdot\dfrac{\cos\alpha}{2\sqrt{1+\sin\alpha}}\dfrac{d\alpha}{dt} = \dfrac{\pi r\sqrt{2}}{40}\cdot\dfrac{\cos\alpha}{\sqrt{1+\sin\alpha}}$. Therefore

$$\frac{d\theta}{dt} = \frac{\pi r\sqrt{2}}{40}\cdot\frac{\cos^2\theta\cos\alpha}{\sqrt{1+\sin\alpha}}.$$

Observe that $\left|\dfrac{d\theta}{dt}\right| = \dfrac{\pi r\sqrt{2}}{40}\dfrac{\cos^2\theta|\cos\alpha|}{\sqrt{1+\sin\alpha}}\dfrac{\sqrt{1-\sin\alpha}}{\sqrt{1-\sin\alpha}}$, which can be written as $\dfrac{\pi r\sqrt{2}}{40}\cos^2\theta\sqrt{1-\sin\alpha}$. When the seat on the Ferris wheel is at its lowest point we have $\theta = 0$ and $\alpha = -\pi/2$, which gives $\cos^2\theta = 1$ and $\sqrt{1-\sin\alpha} = \sqrt{2}$. At any other point on the wheel we have $\cos^2\theta \le 1$ and $\sqrt{1-\sin\alpha} < \sqrt{2}$, so θ is changing most rapidly when the seat is at its lowest point.

4.5.65 The critical points of the function $a(\theta)$ satisfy

$$a'(\theta) = \omega^2 r\left(-\sin\theta - \frac{2r\sin 2\theta}{L}\right) = -\omega^2 r\sin\theta\left(1 + \frac{4r\cos\theta}{L}\right) = 0,$$

using the identity $\sin 2\theta = 2\sin\theta\cos\theta$. There are two cases to consider separately: (a) $0 < L < 4r$ and (b) $L \ge 4r$. In case (a) the critical points in $[0, 2\pi]$ are $\theta = 0, \pi, 2\pi$ and also $\theta = \cos^{-1}(-L/(4r))$ and $2\pi - \cos^{-1}(-L/(4r))$. Comparing the values of $a(\theta)$ at these points shows that the maximum acceleration occurs at $\theta = 0$ and 2π and the minimum occurs at $\theta = \cos^{-1}(-L/(4r))$ and $2\pi - \cos^{-1}(-L/(4r))$. (There is a local maximum at $\theta = \pi$.) In case (b) the only critical points are $\theta = 0, \pi$ and 2π, and comparing the values of $a(\theta)$ at these points shows that the maximum acceleration occurs at $\theta = 0$ and 2π as in case (a), whereas the minimum occurs at $\theta = \pi$ in this case.

4.5.67

a. Using the Pythagorean Theorem, we find that the height of this triangle is 2. Let x be the distance from the point P to the base of the triangle; then the distance from P to the top vertex is $2 - x$ and the distance to each of the base vertices is $\sqrt{x^2 + 4}$, again by the Pythagorean Theorem. Therefore the sum of the distances to the three vertices is given by $S(x) = 2\sqrt{x^2 + 4} + 2 - x$. We wish to minimize this function for $0 \leq x \leq 2$. The critical points of $S(x)$ satisfy $S'(x) = \dfrac{2x}{\sqrt{x^2 + 4}} - 1 = 0$, which has unique solution $x = 2/\sqrt{3}$ in $(0, 2)$. By the First Derivative Test, this critical point corresponds to a local minimum, and by Theorem 4.5, this solitary local minimum is also the absolute minimum on the interval $[0, 2]$. Therefore the optimal location for P is $2/\sqrt{3}$ units above the base.

b. In this case the objective function to be minimized is $S(x) = 2\sqrt{x^2 + 4} + h - x$ where $0 \leq x \leq h$. Exactly as above, we find that the only critical point $x > 0$ is $x = 2/\sqrt{3}$. This will give the absolute minimum on $[0, h]$ as long as $h \geq 2/\sqrt{3}$. When $h < 2/\sqrt{3}$, $S(x)$ is decreasing on $[0, h]$ and the minimum occurs at the endpoint $x = h$.

4.5.69 Let x be the distance between the point on the track nearest your initial position to the point where you catch the train. If you just catch the back of the train, then the train will have travelled $x + 1/3$ miles, which will require time $T = \dfrac{\text{distance}}{\text{rate}} = \dfrac{x + \frac{1}{3}}{20}$. The distance you must run is $\sqrt{x^2 + 1/(16)}$, so your running speed must be $v = \dfrac{\text{distance}}{\text{time}} = \dfrac{20\sqrt{x^2 + \frac{1}{16}}}{x + \frac{1}{3}}$. We wish to minimize this function for $x \geq 0$. The derivative of $v(x)$ can be written $v'(x) = \left(\dfrac{x}{x^2 + \frac{1}{16}} - \dfrac{1}{x + \frac{1}{3}} \right) v(x)$, so the critical points of $v(x)$ satisfy $\dfrac{x}{x^2 + \frac{1}{16}} = \dfrac{1}{x + \frac{1}{3}}$ so $x \left(x + \dfrac{1}{3} \right) = x^2 + \dfrac{1}{16}$ which gives $x = 3/16$ mi. By the First (or Second) Derivative Test, this critical point corresponds to a local minimum, and by Theorem 4.5, this solitary local minimum is also the absolute minimum on the interval $[0, \infty)$. The minimum running speed is $v\left(\dfrac{3}{16}\right) = \dfrac{20\sqrt{\left(\frac{3}{16}\right)^2 + \frac{1}{16}}}{\frac{3}{16} + \frac{1}{3}} = \dfrac{60\sqrt{9 + 16}}{9 + 16} = 12$ mph.

4.5.71

a. Let r and h be the radius and height of the inscribed cylinder. The region that lies above the cylinder inside the cone is a cone with radius r and height $H - h$; by similar triangles we have $\dfrac{H - h}{r} = \dfrac{H}{R}$ so $h = \dfrac{H}{R}(R - r)$. The volume of the cylinder is $V = \pi r^2 h = \dfrac{\pi H}{R}\left(Rr^2 - r^3\right)$, which we must maximize over $0 \leq r \leq R$. The critical points of $V(r)$ satisfy $V'(r) = \dfrac{\pi H}{R}\left(2Rr - 3r^2\right) = 0$, which has unique solution $r = 2R/3$ in $(0, R)$. Because $V(r) = 0$ at the endpoints $r = 0$ and $r = R$, the cylinder with maximum volume has radius $r = 2R/3$, height $h = H/3$ and volume $V = \pi r^2 h = \dfrac{4\pi}{27}R^2 H = \dfrac{4}{9} \cdot \dfrac{\pi}{3} R^2 H$; i.e. 4/9 the volume of the cone.

b. The lateral surface area of the cylinder is $A = 2\pi r h = 2\pi r \cdot \dfrac{H}{R}(R - r) = \dfrac{2\pi H}{R} r(R - r)$. This function takes its maximum over $0 \leq r \leq R$ at $r = R/2$, so the cylinder with maximum lateral surface area has dimensions $r = R/2$ and $h = H/2$.

4.5.73 Following the hint, place two points P and Q above the midpoint of the base of the square, at distances x and y to the sides (see figure), where $0 \leq x, y \leq 1/2$. Then join the bottom vertices of the square to P, the upper vertices to Q and join P to Q. This road system has total length $L = 2\sqrt{x^2 + \dfrac{1}{4}} +$

$2\sqrt{y^2 + \frac{1}{4}} + (1 - x - y) = 1 + \left(\sqrt{4x^2 + 1} - x\right) + \left(\sqrt{4y^2 + 1} - y\right)$. We can minimize the contributions from x and y separately; the critical points of the function $f(x) = \sqrt{4x^2 + 1} - x$ satisfy $f'(x) = \dfrac{4x}{\sqrt{4x^2 + 1}} - 1 = 0$ which gives $\sqrt{4x^2 + 1} = 4x$, so $12x^2 = 1$ and $x = 1/(2\sqrt{3})$. By the First (or Second) Derivative Test, this critical point corresponds to a local minimum, and by Theorem 4.5, this solitary local minimum is also the absolute minimum on the interval $[0, 1/2]$. The minimum value of $f(x)$ on this interval is $f(1/(2\sqrt{3})) = \sqrt{3}/2$, so the shortest road system has length $L = 1 + 2 \cdot \dfrac{\sqrt{3}}{2} = 1 + \sqrt{3} \approx 2.732$.

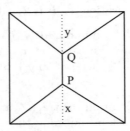

4.5.75 A point on the curve $y = \sqrt{x}$ has the form $(x, \sqrt{x})$, which has distance L to the point $(p, 0)$ given by $L^2 = (x - p)^2 + \sqrt{x}^2 = x^2 + (1 - 2p)x + p^2$. Because L is positive, it suffices to minimize L^2 for $x \geq 0$. This quadratic function takes its minimum at $x = -(1 - 2p)/2 = p - 1/2$, so in case (i) the minimum occurs at the point $(p - 1/2, \sqrt{p - 1/2})$ and in case (ii) there are no critical points for $x > 0$, the function L^2 is increasing on $[0, \infty)$ so the minimum occurs at $(0, 0)$.

4.5.77 Let the angle of the cuts with the horizontal be ϕ_1 and ϕ_2, where $\phi_1 + \phi_2 = \theta$. The volume of the notch is proportional to $\tan\phi_1 + \tan\phi_2 = \tan\phi_1 + \tan(\theta - \phi_1)$, so it suffices to minimize the objective function $V(\phi_1) = \tan\phi_1 + \tan(\theta - \phi_1)$ for $0 \leq \phi_1 \leq \theta$. The critical points of $V(\phi_1)$ satisfy $\sec^2\phi_1 - \sec^2(\theta - \phi_1) = 0$, which is equivalent to the condition $\cos\phi_1 = \cos(\theta - \phi_1)$. This is satisfied if $\phi_1 = \theta - \phi_1$ which gives $\phi_1 = \theta/2$. There are no other solutions in $(0, \theta)$, because $\cos\phi_1$ is decreasing and $\cos(\theta - \phi_1)$ is increasing on $(0, \theta)$ and therefore can intersect at most once. So the only critical point occurs when $\phi_1 = \phi_2 = \theta/2$, and the First Derivative Test shows that this critical point is a local minimum; by Theorem 4.5, this must be the absolute minimum on $[0, \theta]$.

4.5.79 Let x and y be the lengths of the sides of the pen, with y the side parallel to the barn. The diagonal has length $\sqrt{x^2 + y^2}$, by the Pythagorean Theorem. Therefore the constraint is $2x + y + \sqrt{x^2 + y^2} = 200$, which we rewrite as $2x + y = 200 - \sqrt{x^2 + y^2}$. Square both sides to obtain $4x^2 + 4xy + y^2 = 40{,}000 - 400\sqrt{x^2 + y^2} + x^2 + y^2$, which simplifies to $3x^2 + 4xy = 40{,}000 - 400\sqrt{x^2 + y^2}$. Now substitute $\sqrt{x^2 + y^2} = 200 - 2x - y$ in this equation and simplify to obtain $(3x - 200)(x - 200) = 4(100 - x)y$ so $y = \dfrac{(3x - 200)(x - 200)}{4(100 - x)}$. The objective function to be maximized is the area of the pen, $A = xy$. Using the expression above for y in terms of x, we have

$$A = xy = \dfrac{x(3x - 200)(x - 200)}{4(100 - x)} = -\dfrac{1}{4} \cdot \dfrac{x(3x - 200)(x - 200)}{(x - 100)}.$$

The length x must be at least 0, and because the diagonal is at least as long as x, we must have $3x \leq 200$; so x cannot exceed $200/3$. Therefore we need to maximize the function $A(x)$ defined above for $0 \leq x \leq 200/3$. We have

$$A'(x) = -\dfrac{1}{4} \cdot \left(\dfrac{(3x - 200)(x - 200)}{(x - 100)} + \dfrac{x \cdot 3(x - 200)}{(x - 100)} + \dfrac{x(3x - 200)}{(x - 100)} - \dfrac{x(3x - 200)(x - 200)}{(x - 100)^2}\right)$$

$$= \left(\dfrac{1}{x} + \dfrac{3}{3x - 200} + \dfrac{1}{x - 200} - \dfrac{1}{x - 100}\right) A(x).$$

Because $A(x) > 0$ for $0 < x < 200/3$, the critical points of the objective function satisfy $\dfrac{1}{x} + \dfrac{3}{3x - 200} + \dfrac{1}{x - 200} = \dfrac{1}{x - 100}$ which when simplified gives the equation $6x^3 - 1700x^2 + 160{,}000x - 4{,}000{,}000 = 0$. Using

4.6 Linear Approximation and Differentials

4.6.1

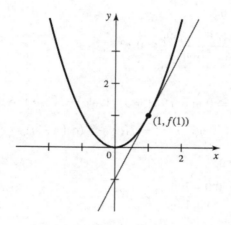

4.6.3 If f is differentiable at the point, then near that point, f is approximately linear, so the function nearly coincides with the tangent line at that point.

4.6.5 $L(x) = f(1) + f'(1)(x-1) = 2 + 3(x-1) = 3x - 1$, so $f(1.1) \approx L(1.1) = 3(1.1) - 1 = 2.3$.

4.6.7 $L(x) = f(4) + f'(4)(x-4) = 3 + 2(x-4)$. So $f(3.85) \approx L(3.85) = 3 + 2(3.85 - 4) = 3 - .3 = 2.7$.

4.6.9 The relationship is given by $dy = f'(x)dx$, which is the linear approximation of the change Δy in $y = f(x)$ corresponding to a change dx in x.

4.6.11 $\Delta y \approx f'(a)\Delta x$, so $f'(a) \approx \dfrac{\Delta y}{\Delta x}$. Therefore $f'(5) \approx \dfrac{f(5.01) - f(5)}{5.01 - 5} = \dfrac{0.25}{0.01} = 25$.

4.6.13 The approximate average speed is $L(-1) = 60 - (-1) = 61$ miles per hour. The exact speed is $\dfrac{3600}{59}$ miles per hour which is about 61.02 miles per hour.

4.6.15 Let $T(x) = \dfrac{60D}{60+x}$. Then $T'(x) = -\dfrac{60D}{(60+x)^2}$, so $T'(0) = -\dfrac{D}{60}$. The linear approximation is given by $L(x) = T(0) - \dfrac{D}{60}(x - 0)$, or $L(x) = D\left(1 - \dfrac{x}{60}\right)$.

4.6.17 With $D = 80$, we have $T(-3) \approx L(-3) = 80(1 - \dfrac{-3}{60}) = 80 + 4 = 84$ minutes. The exact time required is $T(-3) = \dfrac{60 \cdot 80}{60 - 3} \approx 84.211$ minutes.

4.6.19 $f'(x) = 8x + 1$, so $f'(1) = 9$.
$$L(x) = f(1) + f'(1)(x-1) = 5 + 9(x-1) = 9x - 4.$$

4.6.21 $g'(t) = \dfrac{2}{2\sqrt{2t+9}} = \dfrac{1}{\sqrt{2t+9}}$, so $g'(-4) = 1$.
$$L(t) = g(-4) + g'(-4)(t+4) = 1 + 1(t+4) = t + 5.$$

4.6.23 $f'(x) = 3e^{3x-6}$, so $f'(2) = 3$.
$$L(x) = f(2) + f'(2)(x-2) = 1 + 3(x-2) = 3x - 5.$$

4.6.25

a. Note that $f(a) = f(2) = 8$ and $f'(a) = -2a = -4$, so the linear approximation has equation
$$y = L(x) = f(a) + f'(a)(x-a) = 8 + (-4)(x-2) = -4x + 16.$$

b. We have $f(2.1) \approx L(2.1) = 7.6$.

c. The percentage error is $100 \cdot \dfrac{|7.6 - 7.59|}{7.59} \approx 0.13\%$.

4.6.27

a. Note that $f(a) = f(0) = \ln 1 = 0$ and $f'(a) = 1/(1+a) = 1$, so the linear approximation has equation
$$y = L(x) = f(a) + f'(a)(x-a) = x.$$

b. We have $f(0.9) \approx L(0.9) = 0.9$.

c. The percentage error is $100 \cdot \dfrac{|0.9 - \ln 1.9|}{|\ln 1.9|} \approx 40\%$.

4.6.29

a. Note that $f(a) = f(0) = \cos 0 = 1$ and $f'(a) = -\sin a = 0$, so the linear approximation has equation
$$y = L(x) = f(a) + f'(a)(x-a) = 1.$$

b. We have $f(-0.01) \approx L(-0.01) = 1$.

c. The percentage error is $100 \cdot \dfrac{|1 - \cos(-0.01)|}{\cos(-0.01)} \approx 0.005\%$.

4.6.31

a. Note that $f(a) = 8^{-1/3} = 1/2$ and $f'(a) = (-1/3)(8)^{-4/3} = -1/48$, so the linear approximation has equation
$$y = L(x) = f(a) + f'(a)(x-a) = \frac{1}{2} + \frac{-1}{48}x.$$

b. We have $f(-0.1) \approx L(-0.1) \approx 0.50208333$

c. The percentage error is $100 \cdot \dfrac{|7.9^{-1/3} - .050208333|}{(7.9)^{-1/3}} \approx 0.003\%$.

4.6.33

a. Note that $f(a) = f(0) = 1$ and $f'(a) = -1/(1+a)^2 = -1$, so the linear approximation has equation
$$y = L(x) = f(a) + f'(a)(x-a) = 1 - x.$$

b. The linear approximation to $1/1.1$ is $\dfrac{1}{1.1} \approx L(0.1) = 0.9$.

c. The percentage error is $100 \cdot \dfrac{|0.9 - \frac{1}{1.1}|}{\frac{1}{1.1}} = 1\%$.

4.6. Linear Approximation and Differentials

4.6.35

a. Note that $f(a) = f(0) = 1$ and $f'(a) = -e^{-a} = -1$, so the linear approximation has equation
$$y = L(x) = f(a) + f'(a)(x-a) = 1 - x.$$

b. The linear approximation to $e^{-0.03}$ is $e^{-0.03} \approx L(0.03) = 0.97$.

c. The percentage error is $100 \cdot \dfrac{|0.97 - e^{-0.03}|}{e^{-0.03}} \approx 0.046\%$.

4.6.37 Let $f(x) = 1/x$, $a = 200$. Then $f(a) = 0.005$ and $f'(a) = -1/a^2 = -0.000025$, so the linear approximation to f near $a = 200$ is
$$L(x) = f(a) + f'(a)(x - a) = 0.005 - 0.000025(x - 200).$$

Therefore
$$\frac{1}{203} = f(203) \approx L(203) = 0.004925.$$

4.6.39 Let $f(x) = \sqrt{x}$, $a = 144$. Then $f(a) = 12$ and $f'(a) = 1/(2\sqrt{a}) = 1/24$, so the linear approximation to f near $a = 144$ is
$$L(x) = f(a) + f'(a)(x - a) = 12 + \frac{1}{24}(x - 144).$$

Therefore
$$\sqrt{146} = f(146) \approx L(146) = \frac{145}{12}.$$

4.6.41 Let $f(x) = \ln x$, $a = 1$. Then $f(a) = 0$ and $f'(a) = 1/a = 1$, so the linear approximation to f near $a = 1$ is
$$L(x) = f(a) + f'(a)(x - a) = x - 1.$$

Therefore
$$\ln(1.05) = f(1.05) \approx L(1.05) = 0.05.$$

4.6.43 Let $f(x) = e^x$, $a = 0$. Then $f(a) = 1$ and $f'(a) = e^a = 1$, so the linear approximation to f near $a = 0$ is
$$L(x) = f(a) + f'(a)(x - a) = 1 + x.$$

Therefore
$$e^{0.06} = f(0.06) \approx L(0.06) \approx 1.060.$$

4.6.45 Let $f(x) = 1/\sqrt[3]{x}$, $a = 512$. Then $f(a) = 1/8$ and $f'(a) = -1/(3a^{4/3}) = -1/12{,}288$, so the linear approximation to f near $a = 512$ is
$$L(x) = f(a) + f'(a)(x - a) = \frac{1}{8} - \frac{1}{12{,}288}(x - 512).$$

Therefore
$$\frac{1}{\sqrt[3]{510}} = f(510) \approx L(510) = \frac{769}{6144} \approx 0.1252.$$

4.6.47

a. With $f(x) = \dfrac{2}{x}$ and $a = 1$, we have $f(a) = 2$ and $f'(a) = -\dfrac{2}{a^2} = -2$. Thus the linear approximation to $f(x)$ at $x = 1$ is $L(x) = f(1) + f'(1)(x - 1) = 2 + -2(x - 1) = -2x + 4$.

b. A plot of f with L:

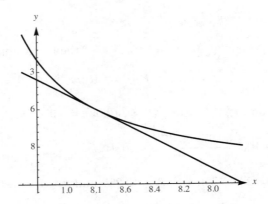

c. The linear approximation in part (b) appears to be an underestimate everywhere, because it lies below the graph of f.

d. Because $f'(x) = -\dfrac{2}{x^2}$, we have $f''(x) = \dfrac{4}{x^3}$, so that $f''(1) > 0$, and f is concave up at $x = 1$. This is consistent with L being an underestimate near $x = 1$.

4.6.49

a. With $f(x) = e^{-x}$ and $a = \ln 2$, we have $f(a) = \frac{1}{2}$ and $f'(a) = -e^{-a} = -\frac{1}{2}$. Thus the linear approximation to f at $x = \ln 2$ is

$$L(x) = f(\ln 2) + f'(\ln 2)(x - \ln 2) = \frac{1}{2} + -\frac{1}{2}(x - \ln 2) = -\frac{1}{2}x + \frac{1}{2}(1 + \ln 2).$$

b. A plot of f with L:

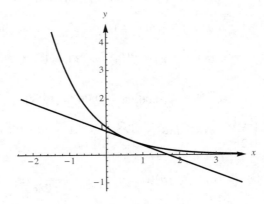

c. The linear approximation in part (b) appears to be an underestimate everywhere, because it lies below the graph of f.

d. Because $f'(x) = -e^{-x}$, we have $f''(x) = e^{-x}$, so that $f''(x) > 0$ for all values of x and f is therefore concave up everywhere. This is consistent with L being an underestimate.

4.6.51 $E(x) = |L(x) - s(x)| = \left| 60 - x - \dfrac{3600}{60 + x} \right|$. A graph is shown below, for x from -7.26 to 8.26.

4.6. Linear Approximation and Differentials

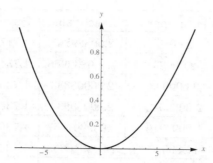

$E(x) \leq 1$ when $-7.26 \leq x \leq 8.26$, which corresponds to driving times for 1 mi from about 53 s to 68 s. Therefore, $L(x)$ gives approximations to $s(x)$ that are within 1 mi/hr of the true value when you drive 1 mile in t seconds, where $53 < t < 68$.

4.6.53

a. True. Note that $f(0) = 0$ and $f'(0) = 0$, so the linear approximation at 0 is in fact $L(x) = 0$.

b. False. The function $f(x) = |x|$ is not differentiable at $x = 0$, so there is no good linear approximation at 0.

c. True. For linear functions, the linear approximation at any point and the function are equal.

d. True. Note that $f'(x) = \frac{1}{x}$, so that $f''(x) = \frac{-1}{x^2}$, and $f''(e) < 0$, so f is concave down near $x = e$. Thus L is an overestimate of f.

4.6.55 Note that $V'(r) = 4\pi r^2$, so $\Delta V \approx V'(a)\Delta r = 4\pi a^2 \Delta r$. Substituting $a = 5$ and $\Delta r = 0.1$ gives $\Delta V \approx 4\pi \cdot 25 \cdot 0.1 = 10\pi \approx 31.416 \text{ ft}^3$.

4.6.57 Note that V is a linear function of h with $V'(h) = \pi r^2 = 400\pi$, so $\Delta V = V'(a)\Delta r = 400\pi \Delta r$. Substituting $\Delta r = -0.1$ gives $\Delta V = -40\pi \approx -125.664 \text{ cm}^3$.

4.6.59 Note that $S'(r) = \pi\sqrt{r^2 + h^2} + \pi r \cdot \frac{r}{\sqrt{r^2 + h^2}} = \pi \frac{2r^2 + h^2}{\sqrt{r^2 + h^2}}$, so $\Delta S \approx S'(a)\Delta r = \pi \frac{2a^2 + h^2}{\sqrt{a^2 + h^2}}\Delta r$. Substituting $h = 6$, $a = 10$ and $\Delta r = -0.1$ gives $\Delta S \approx \pi \frac{236}{\sqrt{136}}(-0.1) = \frac{-59\pi}{5\sqrt{34}} \approx -6.358 \text{ m}^2$.

4.6.61 We have $f'(x) = 2$, so $dy = 2\,dx$.

4.6.63 We have $f'(x) = -3/x^4$, so $dy = -\frac{3}{x^4}dx$.

4.6.65 We have $f'(x) = a\sin x$, so $dy = a\sin x\,dx$.

4.6.67 We have $f'(x) = 9x^2 - 4$, so $dy = (9x^2 - 4)\,dx$.

4.6.69 We have $f'(x) = \sec^2 x$, so $dy = \sec^2 x\,dx$.

4.6.71 Note that $f(a) = f(8) = 2$ and $f'(a) = (1/3)a^{-2/3} = 1/12$, so the linear approximation has equation

$$y = L(x) = f(a) + f'(a)(x - a) = 2 + \frac{1}{12}(x - 8) = \frac{x}{12} + \frac{4}{3}.$$

x	Linear approx	Exact value	Percent error
8.1	$2.008\overline{3}$	2.00829885	1.717×10^{-3}
8.01	$2.00083\overline{3}$	2.000832986	1.734×10^{-5}
8.001	$2.000083\overline{3}$	2.00008333	1.736×10^{-7}
8.0001	$2.0000083\overline{3}$	2.00000833	1.736×10^{-9}
7.9999	$1.9999991\overline{6}$	1.999991667	1.736×10^{-9}
7.999	$1.999991\overline{6}$	1.999916663	1.736×10^{-7}
7.99	$1.99991\overline{6}$	1.999166319	1.738×10^{-5}
7.9	$1.991\overline{6}$	1.991631701	1.756×10^{-3}

The percentage errors become extremely small as x approaches 8 In fact, each time we decrease Δx by a factor of 10, the percentage error decreases by a factor of 100.

4.6.73

a. The linear approximation near $x = 1$ is more accurate for f because the rate at which f' is changing at 1 is smaller than the rate at which g' is changing at 1. The graph of f bends away from the linear function more slowly than the graph of g.

b. The larger the value of $|f''(a)|$, the greater the deviation of the curve $y = f(x)$ from the tangent line at points near $x = a$.

4.7 L'Hôpital's Rule

4.7.1 If $\lim_{x \to a} f(x) = 0$ and $\lim_{x \to a} g(x) = 0$, then we say $\lim_{x \to a} \frac{f(x)}{g(x)}$ is of indeterminate form $\frac{0}{0}$.

4.7.3 Take the limit of the quotient of the derivatives of the numerator and denominator.

4.7.5

a. Let $f(x) = x^2$ and $g(x) = \frac{1}{x^2}$. Then $\lim_{x \to 0} f(x) = 0$ and $\lim_{x \to 0} g(x) = \infty$, but $\lim_{x \to 0} f(x)g(x) = \lim_{x \to 0} x^2 \cdot \frac{1}{x^2} = \lim_{x \to 0} 1 = 1$.

b. Let $f(x) = 2x^2$ and $g(x) = \frac{1}{x^2}$. Then $\lim_{x \to 0} f(x) = 0$ and $\lim_{x \to 0} g(x) = \infty$, but $\lim_{x \to 0} f(x)g(x) = \lim_{x \to 0} 2x^2 \cdot \frac{1}{x^2} = \lim_{x \to 0} 2 = 2$.

4.7.7 If $\lim_{x \to a} f(x)g(x)$ has the form $0 \cdot \infty$, then $\lim_{x \to a} \frac{f(x)}{1/g(x)}$ has the indeterminate form $\frac{0}{0}$ or $\frac{\infty}{\infty}$.

4.7.9 $\lim_{x \to 0} (\tan^{-1} x) \left(\frac{1}{5x} \right) = \lim_{x \to 0} \frac{\tan^{-1} x}{5x} = \lim_{x \to 0} \frac{1/(1+x^2)}{5} = \frac{1}{5}$.

4.7.11 If $\lim_{x \to a} f(x) = 1$ and $\lim_{x \to a} g(x) = \infty$, then $f(x)^{g(x)} \to 1^\infty$ as $x \to a$, which is meaningless; so direct substitution does not work.

4.7.13 This means $\lim_{x \to \infty} \frac{g(x)}{f(x)} = 0$.

4.7.15 By Theorem 4.14, we have $\ln x, x^3, 2^x, x^x$ in order of increasing growth rates.

4.7. L'Hôpital's Rule

4.7.17 L'Hôpital's rule gives $\displaystyle\lim_{x\to 2}\frac{x^2-2x}{8-6x+x^2}=\lim_{x\to 2}\frac{2x-2}{-6+2x}=\frac{2}{-2}=-1$.

4.7.19 $\displaystyle\lim_{x\to 1}\frac{x^2+2x}{x+3}=\frac{1^2+2}{1+3}=\frac{3}{4}$. Note that this is not an indeterminate form.

4.7.21 L'Hôpital's rule gives $\displaystyle\lim_{x\to 1}\frac{\ln x}{4x-x^2-3}=\lim_{x\to 1}\frac{1/x}{4-2x}=\frac{1}{2}$.

4.7.23 Apply L'Hôpital's rule three times:
$$\lim_{x\to\infty}\frac{3x^4-x^2}{6x^4+12}=\lim_{x\to\infty}\frac{12x^3-2x}{24x^3}=\lim_{x\to\infty}\frac{36x^2-2}{72x^2}=\lim_{x\to\infty}\frac{72x}{144x}=\frac{1}{2}.$$

4.7.25 L'Hôpital's rule gives $\displaystyle\lim_{x\to e}\frac{\ln x-1}{x-e}=\lim_{x\to e}\frac{1/x}{1}=\frac{1}{e}$.

4.7.27 Apply L'Hôpital's rule:
$$\lim_{x\to 0^+}\frac{1-\ln x}{1+\ln x}=\lim_{x\to 0^+}\frac{-1/x}{1/x}=\lim_{x\to 0^+}(-1)=-1.$$

4.7.29 L'Hôpital's rule gives $\displaystyle\lim_{x\to 0}\frac{3\sin 4x}{5x}=\lim_{x\to 0}\frac{12\cos 4x}{5}=\frac{12}{5}$.

4.7.31 L'Hôpital's rule gives $\displaystyle\lim_{u\to\pi/4}\frac{\tan u-\cot u}{u-\pi/4}=\lim_{u\to\pi/4}\frac{\sec^2 u+\csc^2 u}{1}=2+2=4$.

4.7.33 Apply L'Hôpital's rule twice: $\displaystyle\lim_{x\to 0}\frac{1-\cos 3x}{8x^2}=\lim_{x\to 0}\frac{3\sin 3x}{16x}=\lim_{x\to 0}\frac{9\cos 3x}{16}=\frac{9}{16}$.

4.7.35 Apply L'Hôpital's rule twice:
$$\lim_{x\to\pi}\frac{\cos x+1}{(x-\pi)^2}=\lim_{x\to\pi}\frac{-\sin x}{2(x-\pi)}=\lim_{x\to\pi}\frac{-\cos x}{2}=\frac{1}{2}.$$

4.7.37 Apply L'Hôpital's rule three times:

$$\lim_{x\to\pi/2^-}\frac{\tan x}{3/(2x-\pi)}=\lim_{x\to\pi/2^-}\frac{\sec^2 x}{-6/(2x-\pi)^2}=-(1/6)\lim_{x\to\pi/2^-}\frac{(2x-\pi)^2}{\cos^2 x}=(-1/6)\lim_{x\to\pi/2^-}\frac{4(2x-\pi)}{2\cos x(-\sin x)}=$$

$$(-1/6)\lim_{x\to\pi/2^-}\frac{8x-4\pi}{-\sin(2x)}=(-1/6)\lim_{x\to\pi/2^-}\frac{8}{-2\cos(2x)}=-\frac{2}{3}.$$

4.7.39 Apply L'Hôpital's rule twice:
$$\lim_{x\to 0}\frac{e^x-\sin x-1}{x^4+8x^3+12x^2}=\lim_{x\to 0}\frac{e^x-\cos x}{4x^3+24x^2+24x}=\lim_{x\to 0}\frac{e^x+\sin x}{12x^2+48x+24}=\frac{1}{24}.$$

4.7.41 L'Hôpital's rule gives:
$$\lim_{x\to\infty}\frac{e^{1/x}-1}{1/x}=\lim_{x\to\infty}\frac{e^{1/x}(-1/x^2)}{(-1/x^2)}=\lim_{x\to\infty}e^{1/x}=1.$$

4.7.43 Apply L'Hôpital's rule twice:
$$\lim_{x\to-1}\frac{x^3-x^2-5x-3}{x^4+2x^3-x^2-4x-2}=\lim_{x\to-1}\frac{3x^2-2x-5}{4x^3+6x^2-2x-4}=\lim_{x\to-1}\frac{6x-2}{12x^2+12x-2}=4.$$

4.7.45 Applying L'Hôpital's rule twice gives:
$$\lim_{x\to\infty}\frac{\ln(3x+5)}{\ln(7x+3)+1} = \lim_{x\to\infty}\frac{3/(3x+5)}{7/(7x+3)} = \frac{3}{7}\lim_{x\to\infty}\frac{7x+3}{3x+5} = \frac{3}{7}\lim_{x\to\infty}\frac{7}{3} = 1.$$

4.7.47 L'Hôpital's rule gives $\displaystyle\lim_{v\to 3}\frac{v-1-\sqrt{v^2-5}}{v-3} = \lim_{v\to 3}\frac{1-\dfrac{v}{\sqrt{v^2-5}}}{1} = -\frac{1}{2}$.

4.7.49 Apply L'Hôpital's rule twice:
$$\lim_{x\to 2}\frac{x^2-4x+4}{\sin^2\pi x} = \lim_{x\to 2}\frac{2x-4}{2\pi(\sin\pi x)(\cos\pi x)} = \lim_{x\to 2}\frac{2x-4}{\pi\sin 2\pi x} = \lim_{x\to 2}\frac{2}{2\pi^2\cos 2\pi x} = \frac{2}{2\pi^2} = \frac{1}{\pi^2}.$$

4.7.51 Applying L'Hôpital's rule twice gives:
$$\lim_{x\to\infty}\frac{x^2-\ln(2/x)}{3x^2+2x} = \lim_{x\to\infty}\frac{2x+(1/x)}{6x+2} = \lim_{x\to\infty}\frac{2-1/x^2}{6} = \frac{2}{6} = \frac{1}{3}.$$

4.7.53 Observe that $\displaystyle\lim_{x\to 0}x\csc x = \lim_{x\to 0}\frac{x}{\sin x} = \lim_{x\to 0}\frac{1}{\cos x} = 1$, by L'Hôpital's rule.

4.7.55 By L'Hôpital's rule: $\displaystyle\lim_{x\to 0}\frac{\sin 7x}{\sin 6x} = \lim_{x\to 0}\frac{7\cos 7x}{6\cos 6x} = \frac{7}{6}$.

4.7.57 Observe that $\displaystyle\lim_{x\to(\pi/2)^-}\left(\frac{\pi}{2}-x\right)\sec x = \lim_{x\to(\pi/2)^-}\frac{\pi/2-x}{\cos x} = \lim_{x\to(\pi/2)^-}\frac{-1}{-\sin x} = 1$ by L'Hôpital's rule.

4.7.59
$$\lim_{x\to 1^+}\left(\frac{1}{x-1}-\frac{1}{\ln x}\right) = \lim_{x\to 1^+}\frac{\ln x-x+1}{(x-1)\ln x}$$
$$= \lim_{x\to 1^+}\frac{1/x-1}{\ln x+(x-1)(1/x)}$$
$$= \lim_{x\to 1^+}\frac{1/x-1}{\ln x+1-1/x} = \lim_{x\to 1^+}\frac{-1/x^2}{1/x+1/x^2} = -\frac{1}{2},$$
where L'Hôpital's rule was used twice.

4.7.61 Observe that $\displaystyle\lim_{x\to 0}\left(\cot x-\frac{1}{x}\right) = \lim_{x\to 0}\left(\frac{\cos x}{\sin x}-\frac{1}{x}\right) = \lim_{x\to 0}\frac{x\cos x-\sin x}{x\sin x}$. Apply L'Hôpital's rule twice:
$$\lim_{x\to 0}\frac{x\cos x-\sin x}{x\sin x} = \lim_{x\to 0}\frac{\cos x-x\sin x-\cos x}{\sin x+x\cos x}$$
$$= -\lim_{x\to 0}\frac{x\sin x}{\sin x+x\cos x} = -\lim_{x\to 0}\frac{\sin x+x\cos x}{\cos x+\cos x-x\sin x} = -\frac{0}{2} = 0.$$

4.7.63 Make the substitution $x=\dfrac{1}{t}$. Then
$$\lim_{x\to\infty}(x^2-\sqrt{x^4+16x^2}) = \lim_{t\to 0^+}\left(\frac{1}{t^2}-\sqrt{\frac{1}{t^4}+\frac{16}{t^2}}\right)$$
$$= \lim_{t\to 0^+}\frac{1-\sqrt{1+16t^2}}{t^2} = \lim_{t\to 0^+}\frac{\frac{-32t}{2\sqrt{1+16t^2}}}{2t}$$
$$= \lim_{t\to 0^+}\frac{-8}{\sqrt{1+16t^2}} = -8.$$

4.7. L'Hôpital's Rule

4.7.65 Observe that
$$(\sqrt{x-2} - \sqrt{x-4}) \cdot \frac{\sqrt{x-2}+\sqrt{x-4}}{\sqrt{x-2}+\sqrt{x-4}} = \frac{x-2-(x-4)}{\sqrt{x-2}+\sqrt{x-4}} = \frac{2}{\sqrt{x-2}+\sqrt{x-4}},$$
so
$$\lim_{x\to\infty} \sqrt{x-2} - \sqrt{x-4} = \lim_{x\to\infty} \frac{2}{\sqrt{x-2}+\sqrt{x-4}} = 0.$$

4.7.67 Use the identity $1+2+\cdots+n = \frac{n(n+1)}{2}$; then
$$\lim_{n\to\infty} \frac{1+2+\cdots+n}{n^2} = \lim_{n\to\infty} \frac{n(n+1)}{2n^2} = \lim_{n\to\infty} \frac{n+1}{2n} = \frac{1}{2}.$$

4.7.69
$$\lim_{x\to\infty} \frac{\log_2 x}{\log_3 x} = \lim_{x\to\infty} \frac{\frac{1}{x\ln 2}}{\frac{1}{x\ln 3}} = \frac{\ln 3}{\ln 2}.$$

4.7.71 By L'Hôpital's rule, $\displaystyle\lim_{x\to 6} \frac{(5x+2)^{1/5} - 2}{x^{-1} - 6^{-1}} = \lim_{x\to 6} \frac{(5x+2)^{-4/5}}{-x^{-2}} = -\frac{9}{4}.$

4.7.73 Make the substitution $t = 1/x$; then
$$\lim_{x\to\infty} x^3\left(\frac{1}{x} - \sin\frac{1}{x}\right) = \lim_{t\to 0^+} \frac{t - \sin t}{t^3} = \lim_{t\to 0^+} \frac{1 - \cos t}{3t^2} = \lim_{t\to 0^+} \frac{\sin t}{6t} = \lim_{t\to 0^+} \frac{\cos t}{6} = \frac{1}{6},$$
using L'Hôpital's rule.

4.7.75 Note that $\ln x^{2x} = 2x\ln x$, so we evaluate
$$L = \lim_{x\to 0^+} 2x\ln x = 2\lim_{x\to 0^+} \frac{\ln x}{1/x} = 2\lim_{x\to 0^+} \frac{1/x}{-1/x^2} = 2\lim_{x\to 0^+} (-x) = 0$$
by L'Hôpital's rule. Therefore $\displaystyle\lim_{x\to 0^+} x^{2x} = e^L = 1.$

4.7.77 Note that $\ln(\tan\theta)^{\cos\theta} = \cos\theta \ln\tan\theta$, so we evaluate
$$L = \lim_{\theta\to\pi/2^-} \cos\theta \ln\tan\theta = \lim_{\theta\to\pi/2^-} \frac{\ln\tan\theta}{\sec\theta}.$$
L'Hôpital's rule gives
$$\lim_{\theta\to\pi/2^-} \frac{\ln\tan\theta}{\sec\theta} = \lim_{\theta\to\pi/2^-} \frac{\sec^2\theta/\tan\theta}{\sec\theta\tan\theta} = \lim_{\theta\to\pi/2^-} \frac{\sec\theta}{\tan^2\theta} = \lim_{\theta\to\pi/2^-} \frac{\cos\theta}{\sin^2\theta} = 0,$$
so $\displaystyle\lim_{\theta\to\pi/2^-} (\tan\theta)^{\cos\theta} = e^L = 1.$

4.7.79 Note that $\ln(1+x)^{\cot x} = \cot x\ln(1+x)$, so we evaluate
$$L = \lim_{x\to 0^+} \cot x\ln(1+x) = \lim_{x\to 0^+} \frac{\ln(1+x)}{\tan x} = \lim_{x\to 0^+} \frac{1/(1+x)}{\sec^2 x} = \lim_{x\to 0^+} \frac{\cos^2 x}{1+x} = 1$$
by L'Hôpital's rule. Therefore $\displaystyle\lim_{x\to 0^+} (1+x)^{\cot x} = e^L = e.$

4.7.81 Note that $\ln(e^{ax} + x)^{1/x} = \frac{1}{x}\ln(e^{ax} + x)$, so we evaluate
$$L = \lim_{x\to 0} \frac{\ln(e^{ax} + x)}{x} = \lim_{x\to 0} \frac{(ae^{ax} + 1)/(e^{ax} + x)}{1} = \frac{(a+1)/1}{1} = a+1.$$
Therefore, $\displaystyle\lim_{x\to 0}(e^{ax} + x)^{1/x} = e^{a+1}.$

4.7.83 Note that $\ln(x+\cos x)^{1/x} = (\ln(x+\cos x))/x$, so we evaluate

$$L = \lim_{x\to 0} \frac{\ln(x+\cos x)}{x} = \lim_{x\to 0} \frac{(x+\cos x)^{-1}(1-\sin x)}{1} = 1$$

by L'Hôpital's rule. Therefore $\lim_{x\to 0}(x+\cos x)^{1/x} = e^L = e$.

4.7.85

a. After each year the balance increases by the factor $1+r$; therefore the balance after t years is $B(t) = P(1+r)^t$.

b. Let $L = \lim_{m\to\infty} mt\ln\left(1+\frac{r}{m}\right) = t\lim_{m\to\infty}\frac{\ln\left(1+\frac{r}{m}\right)}{1/m}$. By L'Hôpital's rule we have

$$L = t\lim_{m\to\infty}\frac{\frac{1}{1+\frac{r}{m}}\left(\frac{-r}{m^2}\right)}{-1/m^2} = t\lim_{m\to\infty}\frac{r}{1+\frac{r}{m}} = rt,$$

so $\lim_{m\to\infty} B(t) = Pe^L = Pe^{rt}$.

4.7.87 By factoring, we have

$$\lim_{x\to 0}\frac{(e^x-1)(e^x+5)}{(e^x+1)(e^x-1)} = \lim_{x\to 0}\frac{e^x+5}{e^x+1} = \frac{6}{2} = 3.$$

Or, using L'Hôpital's rule, we have

$$\lim_{x\to 0}\frac{2e^{2x}+4e^x}{2e^{2x}} = \frac{2+4}{2} = 3.$$

4.7.89

$$\lim_{x\to 1}\frac{x\ln x - x + 1}{x\ln^2 x} = \lim_{x\to 1}\frac{\ln x + x\cdot\frac{1}{x}-1}{\ln^2 x + 2x\ln x\cdot\frac{1}{x}} = \lim_{x\to 1}\frac{\ln x}{\ln^2 x + 2\ln x} = \lim_{x\to 1}\frac{1}{\ln x + 2} = \frac{1}{2},$$

where the first equality follows from L'Hôpital's rule and the penultimate follows by dividing the numerator and denominator by $\ln x$.

4.7.91 Let $z = \ln x^{\frac{1}{1+\ln x}}$. Then $z = \frac{\ln x}{1+\ln x}$, and $\lim_{x\to 0^+} z = \lim_{x\to 0^+}\frac{\ln x}{1+\ln x} = \lim_{x\to 0^+}\frac{1/x}{1/x} = \lim_{x\to 0^+} 1 = 1$. Then $\lim_{x\to 0^+} e^z = \lim_{x\to 0^+} x^{\frac{1}{1+\ln x}} = e^1 = e$.

4.7.93 Apply L'Hôpital's rule: $\lim_{x\to 0}\frac{a^x-b^x}{x} = \lim_{x\to 0}\frac{(\ln a)a^x-(\ln b)b^x}{1} = \ln a - \ln b$.

4.7.95 By Theorem 4.14, $e^{0.01x}$ grows faster than x^{10} as $x\to\infty$.

4.7.97 Note that $\ln x^{20} = 20\ln x$, so $\ln x^{20}$ and $\ln x$ have comparable growth rates as $x\to\infty$.

4.7.99 By Theorem 4.14, x^x grows faster than 100^x as $x\to\infty$.

4.7.101 By Theorem 4.14, 1.00001^x grows faster than x^{20} as $x\to\infty$.

4.7.103 Note that $\lim_{x\to\infty}\frac{e^{x^2}}{e^{10x}} = \lim_{x\to\infty} e^{x^2-10x} = \infty$, so e^{x^2} grows faster than e^{10x} as $x\to\infty$.

4.7. L'Hôpital's Rule

4.7.105

a. False. $\lim_{x\to 2} x^2 - 1 = 3$, so L'Hôpital's rule does not apply. In fact, $\lim_{x\to 2} \dfrac{x-2}{x^2-1} = \dfrac{0}{3} = 0$.

b. False. L'Hôpital's rule does not say $\lim_{x\to a} f(x)g(x) = \lim_{x\to a} f'(x) \lim_{x\to a} g'(x)$. In fact, $\lim_{x\to 0} x \sin x = 0 \cdot 0 = 0$.

c. False. This limit has the form $0^\infty = 0$.

d. False. This limit has the indeterminate form 1^∞ which is not always 1.

e. True. $\ln x^{100} = 100 \ln x$.

f. True. Note that $\lim_{x\to\infty} \dfrac{e^x}{2^x} = \lim_{x\to\infty} \left(\dfrac{e}{2}\right)^x = \infty$ because $e/2 > 1$.

4.7.107

The domain of f is $(0, \infty)$, so questions about symmetry aren't relevant. There are no asymptotes. To see the behavior as $x \to 0^+$ we compute

$$\lim_{x\to 0^+} \frac{\ln x}{x^{-1}} = \lim_{x\to 0^+} \frac{1/x}{-x^{-2}} = \lim_{x\to 0^+} -x = 0.$$

$f'(x) = x \cdot 1/x + \ln x = 1 + \ln x$. This is 0 for $x = 1/e$. Note that $f'(x) < 0$ for $0 < x < 1/e$ and $f'(x) > 0$ for $x > 1/e$, so f is decreasing on $(0, 1/e)$ and increasing on $(1/e, \infty)$ and there is a local minimum (which is also an absolute minimum) at $x = 1/e$.

$f''(x) = 1/x$, which is always positive on the domain, so f is concave up on its domain and there are no inflection points.

There is an x-intercept at $x = 1$.

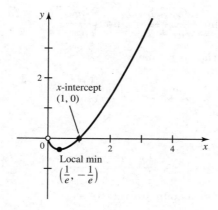

4.7.109

The domain of p is $(-\infty, \infty)$. There are no vertical asymptotes. Note that $p(-x) = -xe^{-(-x)^2/2} = -(xe^{-x^2/2}) = -p(x)$, so p has odd symmetry. Note that

$$\lim_{x\to\pm\infty} p(x) = \lim_{x\to\pm\infty} \frac{x}{e^{x^2/2}} = \lim_{x\to\pm\infty} \frac{1}{xe^{x^2/2}} = 0,$$

so $y = 0$ is a horizontal asymptote.

$$p'(x) = x \cdot (-xe^{-x^2/2}) + e^{-x^2/2} \cdot 1 = e^{-x^2/2}(1 - x^2) = e^{-x^2/2}(1-x)(1+x).$$

This is 0 for $x = \pm 1$. Note that $p'(x) < 0$ on $(-\infty, -1)$ and on $(1, \infty)$, so p is decreasing on those intervals, and $p'(x) > 0$ on $(-1, 1)$, so p is increasing on that interval. There is a local maximum at $x = 1$ and a local minimum at $x = -1$.

$$p''(x) = e^{-x^2/2}(-2x) + (1-x^2) \cdot (-x)e^{-x^2/2} = e^{-x^2/2}x(x^2 - 3),$$

which is 0 at $x = 0$ and at $x = \pm\sqrt{3}$. Note that $p''(x) > 0$ on $(-\sqrt{3}, 0)$ and on $(\sqrt{3}, \infty)$, so p is concave up on those intervals, while $p''(x) < 0$ on $(-\infty, -\sqrt{3})$ and on $(0, \sqrt{3})$, so p is concave down on those intervals. There are inflection points at each of $x = \pm\sqrt{3}$ and at $x = 0$.

There is an x-intercept at $(0, 0)$, which is also the y-intercept.

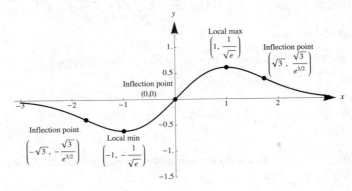

4.7.111 L'Hôpital's rule gives
$$\lim_{x \to \infty} \frac{\sqrt{ax+b}}{\sqrt{cx+d}} = \lim_{x \to \infty} \frac{a}{\sqrt{ax+b}} \cdot \frac{\sqrt{cx+d}}{c} = \frac{a}{c} \lim_{x \to \infty} \frac{\sqrt{cx+d}}{\sqrt{ax+b}},$$
which is the same form as the original limit, so L'Hôpital's rule fails in this case. We can evaluate this limit as follows: first observe that
$$\lim_{x \to \infty} \frac{ax+b}{cx+d} = \frac{a}{c}$$
by L'Hôpital's rule; therefore
$$\lim_{x \to \infty} \frac{\sqrt{ax+b}}{\sqrt{cx+d}} = \lim_{x \to \infty} \sqrt{\frac{ax+b}{cx+d}} = \sqrt{\frac{a}{c}}.$$

4.7.113 Let $t = b^x$, then $x = \ln t / \ln b$ and we have
$$\lim_{x \to \infty} \frac{x^p}{b^x} = \lim_{t \to \infty} \frac{\ln^p t}{t \ln^p b} = 0,$$
by Theorem 4.14.

4.7.115 Note that $\log_a x = \ln x / \ln a$, so
$$\frac{\log_a x}{\log_b x} = \frac{\ln b}{\ln a},$$
and therefore $\log_a x$ and $\log_b x$ grow at a comparable rate as $x \to \infty$.

4.7.117 The triangle ABP has base $1 - \cos\theta$ and height $\sin\theta$, so its area is
$$f(\theta) = \frac{1}{2} \sin\theta (1 - \cos\theta).$$

The sector OBP has area $\theta/2$, and the triangle OBP has base 1 and height $\sin\theta$; therefore
$$g(\theta) = \frac{1}{2}(\theta - \sin\theta).$$

We have
$$\lim_{\theta \to 0} \frac{g(\theta)}{f(\theta)} = \lim_{\theta \to 0} \frac{\theta - \sin\theta}{\sin\theta(1 - \cos\theta)} = \lim_{\theta \to 0} \frac{\theta - \sin\theta}{\sin\theta - (1/2)\sin 2\theta}.$$

Three applications of L'Hôpital's rule gives
$$\lim_{\theta \to 0} \frac{g(\theta)}{f(\theta)} = \lim_{\theta \to 0} \frac{1 - \cos\theta}{\cos\theta - \cos 2\theta} = \lim_{\theta \to 0} \frac{\sin\theta}{2\sin 2\theta - \sin\theta} = \lim_{\theta \to 0} \frac{\cos\theta}{4\cos 2\theta - \cos\theta} = \frac{1}{3}.$$

4.7.119 Note that
$$\ln\left(1+\frac{a}{x}\right)^x = \frac{\ln(1+a/x)}{1/x}$$
so we evaluate
$$L = \lim_{x\to\infty} \frac{\ln(1+a/x)}{1/x} = \lim_{x\to\infty} \frac{1}{1+a/x}\cdot -\frac{a}{x^2}\cdot\frac{1}{-1/x^2} = \lim_{x\to\infty}\frac{a}{1+a/x} = a$$
by L'Hôpital's rule. Therefore
$$\lim_{x\to\infty}\left(1+\frac{a}{x}\right)^x = e^L = e^a.$$

4.7.121

a. Observe that $\lim_{x\to\infty}\frac{b^x}{e^x} = \lim_{x\to\infty}\left(\frac{b}{e}\right)^x$. This limit is ∞ exactly when $b > e$.

b. Observe that $\lim_{x\to\infty}\frac{e^{ax}}{e^x} = \lim_{x\to\infty} e^{(a-1)x}$. This limit is ∞ exactly when $a > 1$.

4.8 Newton's Method

4.8.1 Newton's method generates a sequence of x-intercepts of lines tangent to the graph of f to approximate the roots of f.

4.8.3 Starting at $x_0 = 3$ on the x-axis, travel with your eye downward until you hit the point on the curve at $(3,-3)$. Then follow the tangent line at that point until you hit the x-axis, and you will be at the point $x_1 = 2$. Repeating, we start at the point $(2,0)$ and go down to the point on the curve which is $(2,-1)$. Then following the tangent line until it hits the x-axis again, we find ourselves at $x_2 = 1$. Repeating, we go down to the point on the curve at about $(1,-1/3)$, and then follow the tangent line until hitting the x-axis at $x_3 = 0$.

4.8.5 It is hard to find these exactly by hand, but you should find (approximately) $x_1 = 0.75$.

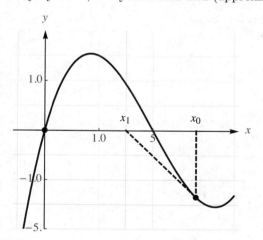

4.8.7 Generally, if two successive Newton approximations agree in their first p digits, then those approximations have p digits of accuracy. The method is terminated when the desired accuracy is reached.

4.8.9 Because $f'(x_n) = 2x_n$, we have
$$x_{n+1} = x_n - \frac{x_n^2 - 6}{2x_n} = \frac{2x_n^2 - x_n^2 + 6}{2x_n} = \frac{x_n^2 + 6}{2x_n}.$$
$x_1 = 2.5$ and $x_2 = 2.45$.

4.8.11 Because $f'(x_n) = -e^{-x_n} - 1$, we have

$$x_{n+1} = x_n - \frac{e^{-x_n} - x_n}{-e^{-x_n} - 1} = \frac{(-e^{-x_n} - 1)x_n - (e^{-x_n} - x_n)}{-e^{-x_n} - 1} = \frac{-e^{-x_n}x_n - e^{-x_n}}{-e^{-x_n} - 1} = \frac{x_n + 1}{e^{x_n} + 1}.$$

$x_1 = 0.564382$ and $x_2 = 0.567142$.

4.8.13 Because $f'(x_n) = 2x_n$, we have

$$x_{n+1} = x_n - \frac{x_n^2 - 10}{2x_n} = \frac{2x_n^2 - (x_n^2 - 10)}{2x_n} = \frac{x_n^2 + 10}{2x_n}.$$

n	x_n
0	3.0
1	3.16667
2	3.16228
3	3.16228

4.8.15 Because $f'(x_n) = \cos(x_n) + 1$, we have

$$x_{n+1} = x_n - \frac{\sin x_n + x_n - 1}{\cos x_n + 1}.$$

n	x_n
0	0.5
1	0.51096
2	0.51097
3	0.51097

4.8.17 Because $f'(x_n) = \sec^2(x_n) - 2$, we have

$$x_{n+1} = x_n - \frac{\tan x_n - 2x_n}{\sec^2(x_n) - 2}.$$

n	x_n
0	1.2
1	1.16934
2	1.16561
3	1.16556
4	1.16556

4.8.19 Because $f'(x_n) = \frac{-1}{\sqrt{1-x_n^2}} - 1$, we have

$$x_{n+1} = x_n - \frac{\cos^{-1}(x_n) - x_n}{\frac{-1}{\sqrt{1-x_n^2}} - 1}.$$

n	x_n
0	0.75
1	0.73915
2	0.73909
3	0.73909

4.8.21 A preliminary sketch of f indicates that there are two roots, near $x = -0.4$. and $x = 1.3$

The Newton's method recursion for f is given by

$$x_{n+1} = x_n - \frac{\cos(2x_n) - x_n^2 + 2x_n}{-2\sin(2x_n) - 2x_n + 2}.$$

We have:

4.8. Newton's Method

n	x_n
0	-0.4
1	-0.337825
2	-0.335412
3	-0.335408
4	-0.335408

n	x_n
0	1.3
1	1.33256
2	1.33306
3	1.33306

The roots are approximately -0.335408 and 1.33306.

4.8.23 A preliminary sketch of f indicates that there is one root, near $x = 0.2$
The Newton's method recursion for f is given by

$$x_{n+1} = x_n - \frac{e^{-x_n} - (x_n + 4)/5}{-e^{-x_n} - 1/5}.$$

We have:

n	x_n
0	0.2
1	0.179122
2	0.179295
3	0.179295

The root is approximately 0.179295.

4.8.25 A preliminary sketch of f indicates that there are two roots, near $x = .5$ and near $x = 3$.
The Newton's method recursion for f is given by

$$x_{n+1} = x_n - \frac{\ln x_n - x_n^2 + 3x_n - 1}{(1/x_n) - 2x_n + 3}.$$

We have:

n	x_n
0	0.5
1	0.610787
2	0.620655
3	0.620723
4	0.620723

n	x_n
0	3
1	3.03698
2	3.03645
3	3.03645

The roots are approximately 0.620723 and 3.03645.

4.8.27 A preliminary sketch of the two curves seems to indicate that they intersect once near $x = 2$.
Let $f(x) = \sin x - x/2$. Then $f'(x_n) = \cos x_n - 1/2$. The Newton's method formula becomes

$$x_{n+1} = x_n - \frac{\sin x_n - x_n/2}{\cos x_n - 1/2}.$$

If we use an initial estimate of $x_0 = 2$, we obtain $x_1 = 1.901$, $x_2 = 1.89551$, $x_3 = 1.89549$ and $x_4 = 1.89549$, so the point of intersection appears to be at approximately $x = 1.89549$.

4.8.29 A preliminary sketch of the two curves seems to indicate that they intersect three times, once between -2.5 and -2, once between 0 and $1/2$, and once between 1.5 and 2.

Let $f(x) = 4 - x^2 - 1/x$. Then $f'(x_n) = -2x_n + (1/x_n^2)$. The Newton's method formula becomes

$$x_{n+1} = x_n - \frac{4 - x_n^2 - (1/x_n)}{-2x_n + (1/x_n^2)}.$$

If we use an initial estimate of $x_0 = -2.25$, we obtain $x_1 = -2.11843$, $x_2 = -2.11491$, $x_3 = -2.11491$, so there appears to be a point of intersection near $x = -2.115$.

If we use an initial estimate of $x_0 = .25$, we obtain $x_1 = .254032$, $x_2 = .254102$, so there appears to be a point of intersection near $x = .254$.

If we use an initial estimate of $x_0 = 1.75$, we obtain $x_1 = 1.86535$, $x_2 = 1.86081$, so there appears to be another point of intersection near $x = 1.86$.

4.8.31 A preliminary sketch of the two curves seems to indicate that they intersect twice, once just to the right of 0, and once between 2 and 2.5.

Let $f(x) = 4\sqrt{x} - (x^2 + 1)$. Then $f'(x_n) = 2/\sqrt{x_n} - 2x_n$. The Newton's method formula becomes

$$x_{n+1} = x_n - \frac{4\sqrt{x_n} - (x_n^2 + 1)}{2/\sqrt{x_n} - 2x_n}.$$

If we use an initial estimate of $x_0 = .1$, we obtain $x_1 = .0583788$, $x_2 = .0629053$, $x_3 = .0629971$, so there appears to be a point of intersection near $x = .06299$.

If we use an initial estimate of $x_0 = 2.25$, we obtain $x_1 = 2.23026$, $x_2 = 2.23012$, $x_3 = 2.23012$, so there appears to be a point of intersection near $x = 2.23012$.

4.8.33 The tumor decreases in size, then it starts growing again. Because the initial size of the tumor is 1, we need to find when $V(t) = \dfrac{1}{2}$. Using Newton's Method on the function $g(t) = V(t) - 1/2$, with an initial estimate of $t_0 = 6$, we find $t_1 = 6.40065$, $t_2 = 6.41774$, $t_3 = 6.41777$, $t_4 = 6.41777$. So the tumor reaches half its original size after about 6.4 days.

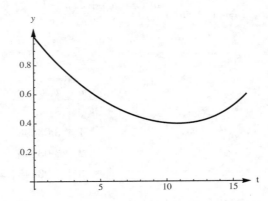

4.8.35

a. We need $A = 1,000,000$, so $\dfrac{10,000((1+r)^{30} - 1)}{r} = 1,000,000$. Multiplying both sides by r yields $10,000((1+r)^{30} - 1) = 1,000,000 r$, or

$$1,000,000 r - 10,000(1+r)^{30} + 10,000 = 0.$$

b. Using Newton's Method with an initial estimate of 0.05 gives

4.8. Newton's Method

n	r_n
0	0.05
1	0.121455
2	0.0969022
3	0.0811171
4	0.0743737
5	0.073191
6	0.0731527

The interest rate required is about 7.3%.

4.8.37

a. We are seeking the time t when first $y(t) = 2.5e^{-t}\cos 2t$ is zero. This occurs first for $t = \pi/4$.

b. We are seeking the minimum value for y. We have $y'(t) = -2.5e^{-t}\cos 2t + 2.5e^{-t}(-2\sin 2t) = -2.5e^{-t}(\cos 2t + 2\sin 2t)$. This is zero when $\cos 2t = -2\sin 2t$, or $\tan 2t = \frac{-1}{2}$. Let $f(t) = \tan 2t + \frac{1}{2}$. If we apply Newton's method to $f(t)$ with a starting point of $t_0 = 1$, we obtain a root of 1.33897 after five iterations. An application of the First Derivative Test shows that there is a local minimum for y at this number. The displacement at this time is -0.586107. This local minimum is in fact an absolute minimum.

c. The second time that $y(t) = 2.5e^{-t}\cos 2t$ is zero is when $2t = \frac{3\pi}{2}$, or $t = 3\pi/4$.

d. Following our work in part b, we look for a root of $f(t) = \tan 2t + \frac{1}{2}$ that is bigger than 1.33897. From the graph, we are looking near $t = 3$. Applying Newton's method to $f(t)$ with an initial value of $x_0 = 3$ gives a root 2.90977 after three iterations. Applying the First Derivative Test, we see that there is a local maximum of 0.12181 at $x = 2.90977$.

4.8.39

Let $p(x) = x^4 - 7$. Because $p'(x_n) = 4x_n^3$, we have

$$x_{n+1} = x_n - \frac{x_n^4 - 7}{4x_n^3} = \frac{3}{4}x_n + \frac{7}{4} \cdot \frac{1}{x_n^3}.$$

We use an initial guess of $x_0 = 1.7$.

n	x_n
0	1.7
1	1.6312
2	1.6266
3	1.62658
4	1.62658

It appears that $\sqrt[4]{7} \approx 1.62658$.

4.8.41

Let $p(x) = x^3 + 9$. Because $p'(x_n) = 3x_n^2$, we have

$$x_{n+1} = x_n - \frac{x_n^3 + 9}{3x_n^2} = \frac{2}{3}x_n - \frac{3}{x_n^2}.$$

We use an initial guess of $x_0 = -2$.

n	x_n
0	-2.0
1	-2.08333
2	-2.08008
3	-2.08008

It appears that $\sqrt[3]{-9} \approx -2.08008$.

4.8.43

$$f'(x) = \frac{-x\sin x - \cos x}{x^2},$$

which is zero when $x \sin x + \cos x = 0$. Note that $f'(1) < 0$ and $f'(\pi) > 0$, so there must be a local minimum on the interval $(1, \pi)$. Let $g(x) = x \sin x + \cos x$. Then $g'(x_n) = \sin x_n + x_n \cos x_n - \sin x_n = x_n \cos x_n$, and the Newton's method formula becomes

$$x_{n+1} = x_n - \frac{x_n \sin x_n + \cos x_n}{x_n \cos x_n}.$$

If we use an initial estimate of $x_0 = 2.5$, we obtain $x_1 = 2.84702$, $x_2 = 2.79918$, $x_3 = 2.79839$, $x_4 = 2.79839$, so the smallest local minimum of f on $(0, \infty)$ occurs at approximately 2.79839.

4.8.45 $f'(x) = 9x^4 - 30x^3 + 7x^2 + 60x$ and $f''(x) = 36x^3 - 90x^2 + 14x + 60 = 2(18x^3 - 45x^2 + 7x + 30)$. We are seeking roots of $f''(x)$. If we apply Newton's method to f'' we obtain the recursion

$$x_{n+1} = x_n - \frac{36x_n^3 - 90x_n^2 + 14x_n + 60}{108x_n^2 - 180x_n + 14}.$$

Starting with an initial estimate of $x_0 = 1$, we obtain $x_1 = 1.34483$, $x_2 = 1.45527$, $x_3 = 1.49284$, $x_4 = 1.49974$, and $x_5 \approx 1.5$. We check directly that 1.5 is a root of f'', so $2x - 3$ is a factor of f'', and using long division, we see that $f''(x) = 2(2x - 3)(9x^2 - 9x - 10) = 2(2x - 3)(3x + 2)(3x - 5)$. So the potential inflection points of f are located at $x = -2/3$, $x = 3/2$, and $x = 5/3$. A check of the sign of f'' on the various intervals confirms that these are all the locations of inflection points.

4.8.47

a. True.

b. False. The quadratic formula gives exact values.

c. False. It sometime fails depending on factors such as the shape of the curve and the closeness of the initial estimate.

4.8.49 A preliminary sketch suggests there is 1 intersection point, near $x = 0.75$. Using Newton's Method on the function $g(x) = f(x) - x = \cos x - x$ gives.

n	x_n
0	0.75
1	0.73911
2	0.73908
3	0.73908

The fixed point is approximately 0.73908.

4.8.51 Let $g(x) = 2x \cos x - x$. Fixed points of f are roots of g. Clearly $x = 0$ is a root of g. The Newton's method recursion for g is given by

$$x_{n+1} = x_n - \frac{2x_n \cos x_n - x_n}{2 \cos x_n - 2x \sin x_n - 1}.$$

A preliminary sketch of g indicates that there is only one nonzero root on $[0, 2]$, near $x = 1$. We have:

n	x_n
0	1
1	1.0503
2	1.04721
3	1.0472
4	1.0472

The fixed points are 0 and approximately 1.0472.

4.8.53

a.

n	x_n
0	2
1	5.33333
2	11.0553
3	22.2931

b.

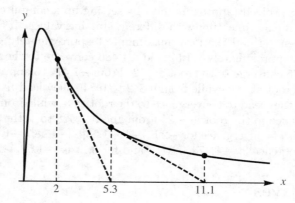

c. The tangent lines intersect the x-axis farther and farther away from the root r.

4.8.55

a. The Newton's method formula would be:

$$x_{n+1} = x_n - \frac{1/x_n - a}{-1/x_n^2} = x_n + x_n - ax_n^2 = 2x_n - ax_n^2 = (2 - ax_n)x_n.$$

b. The approximation to $1/7$ is 0.14285714.

n	x_n
0	0.1
1	0.13
2	0.1417
3	0.14284777
4	0.14285714
5	0.14285714

4.8.57 Let $f(\lambda) = \tan(\pi\lambda) - \lambda$. We are looking for the first three positive roots of f. A preliminary sketch indicates that they are located near 1.4, 2.4, and 3.4. The Newton's method recursion is given by

$$x_{n+1} = x_n - \frac{\tan(\pi x_n) - x_n}{\pi \sec^2(\pi x_n) - 1}.$$

We obtain the following results:

n	x_n
0	1.4
1	1.34741
2	1.30555
3	1.29121
4	1.29012
5	1.29011
6	1.29011

n	x_n
0	2.4
1	2.37876
2	2.37331
3	2.37305
4	2.37305

n	x_n
0	3.4
1	3.4101
2	3.40919
3	3.40918
4	3.40918

The first three positive eigenvalues are approximately 1.29011, 2.37305, and 3.40918.

4.8.59 This problem can be solved (approximately) by setting up a computer or calculator program to run Newton's method, and then experimenting with different starting values. If this is done, it can be seen that any initial estimate between -4 and the local maximum at approximately -1.53 converges to the root at -2. Initial values between approximately -1.52 and -1.486 converge to the root at 3, while starting values between -1.485 and -1.475 converge to the root at -2. From -1.474 to approximately 0.841, starting values lead to convergence to the root at -1, while from 0.842 to 0.846 they lead to convergence to the root at -2. From about 0.847 to 0.862 they lead to convergence to the root at 3, while from 0.863 to the local minimum at 1.528 they lead to convergence to the root at -2. From about 1.528 to 4, the convergence is to the root at 3. Thus the approximate basis of convergence for -2 is $[-4, -1.53] \cup [-1.485, -1.475] \cup [0.842, 0.846]$. For -1 the approximate basis of convergence is $[-1.474, 0.841]$, and for 3, it is $[-1.52, -1.486] \cup [0.847, 0.862] \cup [1.53, 4]$.

4.9 Antiderivatives

4.9.1 Derivative, antiderivative.

4.9.3 $x + C$, where C is any constant.

4.9.5 $\dfrac{x^{p+1}}{p+1} + C$, where C is any real number and $p \neq -1$.

4.9.7 $\ln|x| + C$, where C is any constant.

4.9.9 Observe that $F(-1) = 4 + C = 4$, so $C = 0$.

4.9.11 The antiderivatives of $5x^4$ are $x^5 + C$. Check: $\dfrac{d}{dx}(x^5 + C) = 5x^4$.

4.9.13 The antiderivatives of $2\sin x + 1$ are $-2\cos x + x + C$. Check:
$$\frac{d}{dx}(-2\cos x + x + C) = -2(-\sin x) + 1 = 2\sin x + 1.$$

4.9.15 The antiderivatives of $3\sec^2 x$ are $3\tan x + C$. Check: $\dfrac{d}{dx}(3\tan x + C) = 3\sec^2 x$.

4.9.17 The antiderivatives of $-2/y^3 = -2y^{-3}$ are $y^{-2} + C$. Check: $\dfrac{d}{dy}(y^{-2} + C) = -2y^{-3}$.

4.9.19 The antiderivatives of e^x are $e^x + C$. Check: $\dfrac{d}{dx}(e^x + C) = e^x$.

4.9.21 The antiderivatives of $\dfrac{1}{s^2+1}$ are $\tan^{-1} s + C$. Check: $\dfrac{d}{ds}(\tan^{-1}(s) + C) = \dfrac{1}{s^2+1}$.

4.9. Antiderivatives

4.9.23 $\int (3x^5 - 5x^9)\, dx = 3 \cdot \dfrac{x^6}{6} - 5 \cdot \dfrac{x^{10}}{10} + C = \dfrac{1}{2}x^6 - \dfrac{1}{2}x^{10} + C$. Check: $\dfrac{d}{dx}(\dfrac{1}{2}x^6 - \dfrac{1}{2}x^{10} + C) = 3x^5 - 5x^9$.

4.9.25 $\int \left(4\sqrt{x} - \dfrac{4}{\sqrt{x}}\right) dx = \int (4x^{1/2} - 4x^{-1/2})\, dx = 4 \cdot \dfrac{x^{3/2}}{3/2} - 4 \cdot \dfrac{x^{1/2}}{1/2} + C = \dfrac{8}{3}x^{3/2} - 8x^{1/2} + C$. Check: $\dfrac{d}{dx}(\dfrac{8}{3}x^{3/2} - 8x^{1/2} + C) = 4\sqrt{x} - \dfrac{4}{\sqrt{x}}$.

4.9.27 $\int (5s + 3)^2\, ds = \int (25s^2 + 30s + 9)\, ds = 25\dfrac{s^3}{3} + 30\dfrac{s^2}{2} + 9s + C = \dfrac{25s^3}{3} + 15s^2 + 9s + C$. Check: $\dfrac{d}{ds}\left(\dfrac{25s^3}{3} + 15s^2 + 9s + C\right) = 25s^2 + 30s + 9 = (5s+3)^2$.

4.9.29 $\int (3x^{1/3} + 4x^{-1/3} + 6)\, dx = 3 \cdot \dfrac{3}{4}x^{4/3} + 4 \cdot \dfrac{3}{2}x^{2/3} + 6x + C = \dfrac{9}{4}x^{4/3} + 6x^{2/3} + 6x + C$. Check: $\dfrac{d}{dx}\left(\dfrac{9}{4}x^{4/3} + 6x^{2/3} + 6x + C\right) = 3x^{1/3} + 4x^{-1/3} + 6$.

4.9.31 $\int (3x+1)(4-x)\, dx = \int (12x - 3x^2 + 4 - x)\, dx = \int (-3x^2 + 11x + 4)\, dx = -x^3 + \dfrac{11}{2}x^2 + 4x + C$. Check: $\dfrac{d}{dx}\left(\dfrac{9}{2}x^{4/3} + C\right) = 6\sqrt[3]{x}$.

4.9.33 $\int (3x^{-4} + 2 - 3x^{-2})\, dx = -x^{-3} + 2x + 3x^{-1} + C$. Check: $\dfrac{d}{dx}(-x^{-3} + 2x + 3x^{-1} + C) = 3x^{-4} + 2 - 3x^{-2}$.

4.9.35 $\int \dfrac{4x^4 - 6x^2}{x}\, dx = \int \left(\dfrac{4x^4}{x} - \dfrac{6x^2}{x}\right) dx = \int (4x^3 - 6x)\, dx = x^4 - 3x^2 + C$. Check: $\dfrac{d}{dx}(x^4 - 3x^2 + C) = 4x^3 - 6x$.

4.9.37 $\int \dfrac{x^2 - 36}{x-6}\, dx = \int \dfrac{(x-6)(x+6)}{x-6}\, dx = \int (x+6)\, dx = \dfrac{x^2}{2} + 6x + C$. Check: $\dfrac{d}{dx}\left(\dfrac{x^2}{2} + 6x + C\right) = x + 6 = \dfrac{x^2 - 36}{x - 6}$.

4.9.39 $\int (\csc^2 \theta + 2\theta^2 - 3\theta)\, d\theta = -\cot\theta + \dfrac{2}{3}\theta^3 - \dfrac{3}{2}\theta^2 + C$. Check: $\dfrac{d}{d\theta}\left(-\cot\theta + \dfrac{2}{3}\theta^3 - \dfrac{3}{2}\theta^2 + C\right) = \csc^2\theta + 2\theta^2 - 3\theta$.

4.9.41 $\int \dfrac{2 + 3\cos y}{\sin^2 y}\, dy = \int \left(\dfrac{2}{\sin^2 y} + \dfrac{3\cos y}{\sin^2 y}\right) dy = \int (2\csc^2 y + 3\cot y \csc y)\, dy = -2\cot y - 3\csc y + C$. Check: $\dfrac{d}{dy}(-2\cot y - 3\csc y + C) = -2(-\csc^2 y) - 3(-\csc y \cot y) = \dfrac{2}{\sin^2 y} + \dfrac{3\cos y}{\sin^2 y} = \dfrac{2 + 3\cos y}{\sin^2 y}$.

4.9.43 $\int (\sec^2 x - 1)\, dx = \tan x - x + C$. Check: $\dfrac{d}{dx}(\tan x - x + C) = \sec^2 x - 1$.

4.9.45 $\int (\sec^2 \theta + \sec\theta \tan\theta)\, d\theta = \tan\theta + \sec\theta + C$. Check: $\dfrac{d}{d\theta}(\tan\theta + \sec\theta + C) = \sec^2\theta + \sec\theta \tan\theta$.

4.9.47 $\int (3t^2 + 2\csc^2 t)\, dt = t^3 - 2\cot t + C$. Check: $\dfrac{d}{dt}(t^3 - 2\cot t + C) = 3t^2 + 2\csc^2 t$.

4.9.49 $\int \sec\theta (\tan\theta + \sec\theta + \cos\theta)\, d\theta = \int (\sec\theta \tan\theta + \sec^2\theta + 1)\, d\theta = \sec\theta + \tan\theta + \theta + C$. Check: $\dfrac{d}{d\theta}(\sec\theta + \tan\theta + \theta + C) = \sec\theta \tan\theta + \sec^2\theta + 1 = \sec\theta(\tan\theta + \sec\theta + \cos\theta)$.

4.9.51 $\int \dfrac{1}{2y}\,dy = \dfrac{1}{2}\int y^{-1}\,dy = \dfrac{1}{2}\ln|y| + C.$ Check: $\dfrac{d}{dy}\left(\dfrac{1}{2}\ln|y| + C\right) = \dfrac{1}{2y}.$

4.9.53 $\int \dfrac{6}{\sqrt{4-4x^2}}\,dx = \dfrac{6}{2}\int \dfrac{1}{\sqrt{1-x^2}}\,dx = 3\sin^{-1}x + C.$ Check: $\dfrac{d}{dx}(3\sin^{-1}x + C) = \dfrac{3}{\sqrt{1-x^2}} = \dfrac{6}{2\sqrt{1-x^2}} = \dfrac{6}{\sqrt{4-4x^2}}.$

4.9.55 $\int \dfrac{4}{x\sqrt{x^2-1}}\,dx = 4\sec^{-1}|x| + C.$ Check: $\dfrac{d}{dx}(4\sec^{-1}|x| + C) = \dfrac{4}{x\sqrt{x^2-1}}.$

4.9.57 $\int \dfrac{1}{x\sqrt{36x^2-36}}\,dx = \dfrac{1}{6}\int \dfrac{1}{x\sqrt{x^2-1}}\,dx = \dfrac{1}{6}\sec^{-1}|x| + C.$ Check: $\dfrac{d}{dx}\left(\dfrac{1}{6}\sec^{-1}|x| + C\right) = \dfrac{1}{6}\cdot\dfrac{1}{x\sqrt{x^2-1}} = \dfrac{1}{x\sqrt{36x^2-36}}.$

4.9.59 $\int \dfrac{t+1}{t}\,dt = \int \left(\dfrac{t}{t} + \dfrac{1}{t}\right)dt = \int \left(1 + \dfrac{1}{t}\right)dt = t + \ln|t| + C.$ Check: $\dfrac{d}{dt}(t + \ln|t| + C) = 1 + \dfrac{1}{t} = \dfrac{t+1}{t}.$

4.9.61 $\int e^{x+2}\,dx = \int e^2 e^x\,dx = e^2\int e^x\,dx = e^2 e^x + C = e^{x+2} + C.$ Check: $\dfrac{d}{dx}(e^{x+2} + C) = e^{x+2}.$

4.9.63 $\int \dfrac{e^{2w} - 5e^w + 4}{e^w - 1}\,dw = \int \dfrac{(e^w - 1)(e^w - 4)}{e^w - 1}\,dw = \int (e^w - 4)\,dw = e^w - 4w + C.$ Check: $\dfrac{d}{dw}(e^w - 4w + C) = e^w - 4 = \dfrac{(e^w - 4)(e^w - 1)}{e^w - 1} = \dfrac{e^{2w} - 5e^w + 4}{e^w - 1}.$

4.9.65 $\int \dfrac{1+\sqrt{x}}{x}\,dx = \int (x^{-1} + x^{-1/2})\,dx = \ln|x| + 2x^{1/2} + C = \ln|x| + 2\sqrt{x} + C.$ Check: $\dfrac{d}{dx}(\ln|x| + 2\sqrt{x} + C) = \dfrac{1}{x} + \dfrac{1}{\sqrt{x}} = \dfrac{1+\sqrt{x}}{x}.$

4.9.67 $\int \sqrt{x}(2x^6 - 4\sqrt[3]{x})\,dx = \int (2x^{13/2} - 4x^{5/6})\,dx = 2\cdot\dfrac{2}{15}x^{15/2} - 4\cdot\dfrac{6}{11}x^{11/6} + C = \dfrac{4}{15}x^{15/2} - \dfrac{24}{11}x^{11/6} + C.$ Check: $\dfrac{d}{dx}\left(\dfrac{4}{15}x^{15/2} - \dfrac{24}{11}x^{11/6} + C\right) = 2x^{13/2} - 4x^{5/6} = \sqrt{x}(2x^6 - 4\sqrt[3]{x}).$

4.9.69 We have $F(x) = \int (x^5 - 2x^2 + 1)\,dx = \dfrac{x^6}{6} - \dfrac{2x^3}{3} + x + C$; substituting $F(0) = 1$ gives $C = 1$, and thus $F(x) = \dfrac{x^6}{6} - \dfrac{2x^3}{3} + x + 1.$

4.9.71 $F(x) = \int (8x^3 + \sin x)\,dx = 2x^4 - \cos x + C$; substituting $F(0) = 2$ gives $-1 + C = 2$, so $C = 3$, and thus $F(x) = 2x^4 - \cos x + 3.$

4.9.73 We have $F(v) = \int \sec v \tan v\,dv = \sec v + C$; substituting $F(0) = 2$ gives $\sec 0 + C = 1 + C = 2$, so $C = 1$, and thus $F(v) = \sec v + 1, -\pi/2 < t < \pi/2.$

4.9.75 We have $F(y) = \int \dfrac{3y^3 + 5}{y}\,dy = \int \left(\dfrac{3y^3}{y} + \dfrac{5}{y}\right)dy = \int \left(3y^2 + \dfrac{5}{y}\right)dy = y^3 + 5\ln|y| + C$; substituting $F(1) = 3$ gives $1 + 0 + C = 3$, so $C = 2$, and thus $F(y) = y^3 + 5\ln y + 2, y > 0.$

4.9.77 We have $f(x) = \int (2x - 3)\,dx = x^2 - 3x + C$; substituting $f(0) = 4$ gives $C = 4$, so $f(x) = x^2 - 3x + 4.$

4.9. Antiderivatives

4.9.79 We have $g(x) = \int 7x\left(x^6 - \frac{1}{7}\right)dx = \int (7x^7 - x)\,dx = \frac{7}{8}x^8 - \frac{x^2}{2} + C$; substituting $g(1) = 2$ gives $\frac{7}{8} - \frac{1}{2} + C = 2$, so $C = \frac{13}{8}$, and thus $g(x) = \frac{7}{8}x^8 - \frac{x^2}{2} + \frac{13}{8}$.

4.9.81 $f(u) = \int 4(\cos u - \sin u)\,du = 4\sin u + 4\cos u + C$; substituting $f(\pi/2) = 0$ gives $4 + 0 + C = 0$, so $C = -4$, and thus $h(u) = 4\sin u + 4\cos u - 4$.

4.9.83 $y(t) = \int \left(\frac{3}{t} + 6\right)dt = 3\ln|t| + 6t + C$; substituting $y(1) = 8$ gives $0 + 6 + C = 8$, so $C = 2$, and thus $y(t) = 3\ln t + 6t + 2,\ t > 0$.

4.9.85 $y(\theta) = \int \frac{\sqrt{2}\cos^3\theta + 1}{\cos^2\theta}\,d\theta = \int \left(\frac{\sqrt{2}\cos^3\theta}{\cos^2\theta} + \frac{1}{\cos^2\theta}\right)d\theta = \int \left(\sqrt{2}\cos\theta + \sec^2\theta\right)d\theta = \sqrt{2}\sin\theta + \tan\theta + C$; substituting $y(\pi/4) = 3$ gives $1 + 1 + C = 3$, so $C = 1$, and thus $y(\theta) = \sqrt{2}\sin\theta + \tan\theta + 1$, $-\pi/2 < \theta < \pi/2$.

4.9.87 We have $f(x) = \int (2x - 5)\,dx = x^2 - 5x + C$; substituting $f(0) = 4$ gives $C = 4$, so $f(x) = x^2 - 5x + 4$.

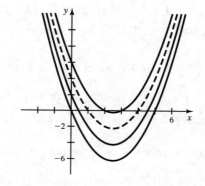

4.9.89 We have $f(x) = \int (3x + \sin x)\,dx = \frac{3}{2}x^2 - \cos x + C$; substituting $f(0) = 3$ gives $C = 4$, so $f(x) = \frac{3}{2}x^2 - \cos x + 4$.

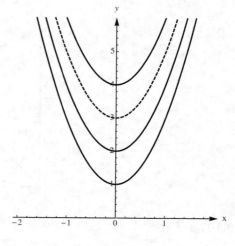

4.9.91 We have $s(t) = \int (2t + 4)\,dt = t^2 + 4t + C$; substituting $s(0) = 0$ gives $C = 0$, so $s(t) = t^2 + 4t$.

4.9.93 We have $s(t) = \int 2\sqrt{t}\,dt = 2 \cdot \frac{2}{3}t^{3/2} + C = \frac{4}{3}t^{3/2} + C$; substituting $s(0) = 1$ gives $C = 1$, so $s(t) = \frac{4}{3}t^{3/2} + 1$.

4.9.95 We have $s(t) = \int (6t^2 + 4t - 10)\,dt = 2t^3 + 2t^2 - 10t + C$; substituting $s(0) = 0$ gives $C = 0$, so $s(t) = 2t^3 + 2t^2 - 10t$.

4.9.97 $v(t) = \int a(t)\,dt = \int -32\,dt = -32t + C_1$. Because $v(0) = 20$, we have $0 + C_1 = 20$, so $v(t) = -32t + 20$.

$s(t) = \int v(t)\,dt = \int (-32t + 20)\,dt = -16t^2 + 20t + C_2$. Because $s(0) = 0$, we have $C_2 = 0$, and thus $s(t) = -16t^2 + 20t$.

4.9.99 $v(t) = \int a(t)\,dt = \int 0.2t\,dt = 0.1t^2 + C_1$. Because $v(0) = 0$, we have $C_1 = 0$.

$s(t) = \int v(t)\,dt = \int 0.1t^2\,dt = \frac{1}{30}t^3 + C_2$. Because $s(0) = 1$, we have $C_2 = 1$, and thus $s(t) = \frac{1}{30}t^3 + 1$.

4.9.101 $v(t) = \int a(t)\,dt = \int (2 + 3\sin t)\,dt = 2t - 3\cos t + C_1$. Because $v(0) = 1$, we have $0 - 3 + C_1 = 1$, so $C_1 = 4$.

$s(t) = \int v(t)\,dt = \int (2t - 3\cos t + 4)\,dt = t^2 - 3\sin t + 4t + C_2$. Because $s(0) = 10$, we have $C_2 = 10$, so $s(t) = t^2 + 4t - 3\sin t + 10$.

4.9.103 $a(t) = 16$, so $v(t) = \int 16\,dt = 16t + C_1$. Because $v(0) = 0$, we have $C_1 = 0$, so $v(t) = 16t$.

$s(t) = \int v(t)\,dt = \int 16t\,dt = 8t^2 + C_2$, where we can assume $s(0) = 0$ so $C_2 = 0$. Then the distance traveled in the first 5 seconds is $s(5) = 200$ feet.

4.9.105 Runner A has position function $s(y) = \int \sin t\,dt = -\cos t + C$; the initial condition $s(0) = 0$ gives $C = 1$, so $s(t) = 1 - \cos t$. Runner B has position function $S(t) = \int \cos t\,dt = \sin t + C$; the initial condition $S(0) = 0$ gives $C = 0$, so $S(t) = \sin t$. The smallest $t > 0$ where $s(t) = S(t)$ is $t = \pi/2$.

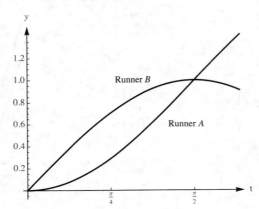

4.9.107

a. We have $v(t) = -9.8t + v_0$ and $v_0 = 30$, so $v(t) = -9.8t + 30$.

b. The height of the softball above ground is given by $s(t) = \int (-9.8t + 30)\,dt = -4.9t^2 + 30t + s_0 = -4.9t^2 + 30t$.

c. The ball reaches its maximum height when $v(t) = -9.8t + 30 = 0$, which gives $t = 30/9.8 \approx 3.06$ s; the maximum height is $s(30/9.8) \approx 45.92$ m.

d. The ball strikes the ground when $s(t) = 0$ (and $t > 0$), which gives $t(30 - 4.9t) = 0$, so $t = 30/4.9 \approx 6.12$ s.

4.9.109

a. We have $v(t) = -9.8t + v_0$ and $v_0 = 10$, so $v(t) = -9.8t + 10$.

b. The height of the payload above ground is given by $s(t) = \int (-9.8t + 10)\,dt = -4.9t^2 + 10t + s_0 = -4.9t^2 + 10t + 400$.

c. The payload reaches its maximum height when $v(t) = -9.8t + 10 = 0$, which gives $t = 10/9.8 \approx 1.02$ s; the maximum height is $s(10/9.8) \approx 405.10$ m.

d. The payload strikes the ground when $s(t) = 0$ (and $t > 0$), which gives $-4.9t^2 + 10t + 400 = 0$, so $t \approx 10.11$ s.

4.9.111

a. True, because $F'(x) = G'(x)$.

b. False; f is the derivative of F.

c. True; $\int f(x)\,dx$ is the most general antiderivative of $f(x)$, which is $F(x) + C$.

d. False; a function cannot have more than one derivative.

e. False; one can only conclude that $F(x)$ and $G(x)$ differ by a constant.

4.9.113 We have $F'(x) = \int \cos x\,dx = \sin x + C$; $F'(0) = 3$ so $C = 3$. Then $F(x) = \int (\sin x + 3)\,dx = -\cos x + 3x + C$; $F(\pi) = 4$ gives $1 + 3\pi + C = 4$ so $C = 3 - 3\pi$ and $F(x) = -\cos x + 3x + 3 - 3\pi$.

4.9.115 We have $F''(x) = \int (672x^5 + 24x)\,dx = 672 \cdot \dfrac{x^6}{6} + 24 \cdot \dfrac{x^2}{2} + C$; $F''(0) = 0$ so $C = 0$. Next $F'(x) = \int (112x^6 + 12x^2)\,dx = 112 \cdot \dfrac{x^7}{7} + 12 \cdot \dfrac{x^3}{3} + C$; $F'(0) = 2$ so $C = 2$. Finally $F(x) = \int (16x^7 + 4x^3 + 2)\,dx = 16 \cdot \dfrac{x^8}{8} + 4 \cdot \dfrac{x^4}{4} + 2x + C$; $F(0) = 1$ so $C = 1$ and $F(x) = 2x^8 + x^4 + 2x + 1$.

4.9.117

a. We have
$$Q(t) = \int 0.1(100 - t^2)\,dt = 0.1\left(100t - \dfrac{t^3}{3}\right) + C;$$
$Q(0) = 0$, so $C = 0$ and $Q(t) = 10t - t^3/30$ gal.

b.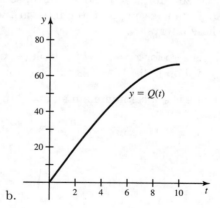

c. $Q(10) = 200/3 \approx 67$ gal.

4.9.119 Check that $\dfrac{d}{dx}(2\sin\sqrt{x}) = 2\cos\sqrt{x}\left(\dfrac{1}{2\sqrt{x}}\right) = \dfrac{\cos\sqrt{x}}{\sqrt{x}}$.

4.9.121 Check that $\dfrac{d}{dx}\left(\dfrac{1}{3}\sin x^3\right) = \dfrac{1}{3}\cos x^3 (3x^2) = x^2 \cos x^3$.

Chapter Four Review

1

a. False. The point $(c, f(c))$ is a critical point for f, but is not necessarily a local maximum or minimum. Example: $f(x) = x^3$ at $c = 0$.

b. False. The fact that $f''(c) = 0$ does not necessarily imply that f changes concavity at c. Example: $f(x) = x^4$ at $c = 0$.

c. True. Both are antiderivatives of $2x$.

d. True. The function has a maximum on the closed interval determined by the two local minima, and the only way the maximum can occur at the endpoints is if the function is constant, in which case every point is a local max and min.

e. True. The slope of the linearization is given by $f'(0) = \cos(0) = 1$, and the line has y-intercept $(0, f(0)) = (0, 0)$.

f. False. For example, $\lim_{x \to \infty} x^2 = \infty$ and $\lim_{x \to \infty} x = \infty$, but $\lim_{x \to \infty} (x^2 - x) = \infty$.

3

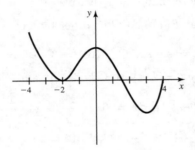

5

a. f' is zero for $x = 0$ and $x = \pm 10$, so those are the critical points. $f' > 0$ on $(-10, 0)$ and on $(10, \infty)$, so f is increasing on those intervals, while $f' < 0$ on $(-\infty, -10)$ and on $(0, 10)$, so f is decreasing on those intervals.

b. f'' is zero for $x = \pm 6$. $f'' > 0$ on $(-\infty, -6)$ and on $(6, \infty)$, so f is concave up on those intervals, while $f' < 0$ on $(-6, 6)$, so f is concave down on that interval.

c. There is a local minimum at $x = -10$ and $x = 10$, and a local maximum at $x = 0$.

d.

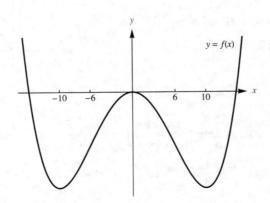

7 $f'(x) = 12x^3 - 12x = 12x(x^2 - 1) = 12x(x-1)(x+1)$, which is zero for $x = 0, -1, 1$, so those are the critical points. $f(0) = 9$, $f(-1) = f(1) = 6$, and checking endpoints, we have $f(\pm 2) = 33$. So the absolute maximum is 33 at the endpoints and the absolute minimum is 6 at $x = \pm 1$.

9 $f'(x) = 6x^2 - 6x - 36 = 6(x-3)(x+2)$, which is zero for $x =$ and $x = -2$, so the critical points are $x = 3, -2$. Because $\lim_{x \to \infty} f(x) = \infty$ and $\lim_{x \to -\infty} f(x) = -\infty$, this function has no absolute max or min on $(-\infty, \infty)$.

11 $f'(x) = \dfrac{2x-2}{x^2 - 2x + 2}$, which is zero for $x = 1$. The only critical point is $x = 1$. $f(0) = f(2) = \ln 2$, while $f(1) = 0$. The absolute maximum is $\ln 2$ at the endpoints, while the absolute minimum is 0 at $x = 1$.

13

$$g'(x) = -\frac{1}{2}\cos x + \frac{1}{2}\cos x \cos x - \frac{1}{2}\sin x \sin x = \frac{1}{2}\left(-\cos x + \cos^2 x - (1 - \cos^2 x)\right)$$
$$= \frac{1}{2}\left(2\cos^2 x - \cos x - 1\right) = \frac{1}{2}\left(2\cos x + 1\right)\left(\cos x - 1\right).$$

This is zero on the interior of the given interval where $\cos x = -\dfrac{1}{2}$, which is $x = \dfrac{2\pi}{3}$ and $x = \dfrac{4\pi}{3}$, so those are the critical points. We have $g(2\pi/3) = -\dfrac{3\sqrt{3}}{8}$ and $g(4\pi/3) = \dfrac{3\sqrt{3}}{8}$, while $g(0) = g(2\pi) = 0$. So the absolute minimum is $-\dfrac{3\sqrt{3}}{8}$ and the absolute maximum is $\dfrac{3\sqrt{3}}{8}$.

15 The critical points satisfy $f'(x) = 2\ln x + 2x \cdot \dfrac{1}{x} = 2\ln x + 2 = 0$, which has solution $x = 1/e$. The Second Derivative Test shows that this critical point is a local minimum, so by Theorem 4.5 the absolute minimum value on the interval $(0, \infty)$ is $f\left(\dfrac{1}{e}\right) = -\dfrac{2}{e} + 10$. Because $\lim_{x \to \infty} x \ln x = \infty$, this function does not have an absolute maximum on $(0, \infty)$.

17

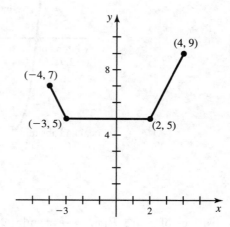

All points in the interval $[-3, 2]$ are critical points. The absolute max occurs at $(4, 9)$; there are no local maxima. All points $(x, 5)$ for x in the interval $[-3, 2]$ are absolute and local minima.

19

a. $f'(x) = x^8 + 15x^4 - 16 = (x^4 + 16)(x^4 - 1) = (x^4 + 16)(x^2 + 1)(x^2 - 1) = (x^4 + 16)(x^2 + 1)(x+1)(x-1)$, which is zero for $x = \pm 1$. $f' > 0$ on $(-\infty, -1)$ and on $(1, \infty)$ so f is increasing there, while $f' < 0$ on $(-1, 1)$ so f is decreasing there.

b. $f''(x) = 8x^7 + 60x^3 = 4x^3(x^4 + 15)$, which is zero only for $x = 0$. $f'' < 0$ on $(-\infty, 0)$ so f is concave down there, while $f'' > 0$ on $(0, \infty)$ so f is concave up there.

21 $f'(x) = 10x^4 - 40x^3 + 60x^2 + 1$, so $f''(x) = 40x^3 - 120x^2 + 120x = 40x(x^2 - 3x + 3)$. The quadratic $x^2 - 3x + 3$ has no real roots, so $x = 0$ is the only possible inflection point. The sign of $f''(x)$ changes at $x = 0$ so an inflection point occurs at $(0, 1)$.

23 $f'(x) = (x+a)^3 + (x-a)(3)(x+a)^2 = (x+a)^2(x+a+3x-3a) = (x+a)^2(4x-2a) = 2(x+a)^2(2x-a)$. This is zero for $x = \dfrac{a}{2}$ and $x = -a$, so those are the critical points. $f''(x) = 4(x+a)(2x-a) + 2(x+a)^2(2) = 4(x+a)(2x-a+x+a) = 4(x+a)(3x) = 12x(x+a)$. This is zero for $x = 0$ and $x = -a$, and an analysis of the sign of f'' shows that these are the locations of inflection points.

25 The derivatives of f are $f'(x) = 2x^3 - 6x + 4$ and $f''(x) = 6x^2 - 6$. Observe that $f'(x) = 2(x-1)^2(x+2)$, so we have critical points $x = 1, -2$. Solving $f''(x) = 6(x^2 - 1) = 0$ gives possible inflection points at $x = \pm 1$. Testing the sign of $f'(x)$ shows that f is decreasing on the interval $(-\infty, -2)$ and is increasing on $(-2, \infty)$. The First Derivative Test shows that a local minimum occurs at $x = -2$, and that the critical point $x = 1$ is neither a local max or min.

Testing the sign of $f''(x)$ shows that f is concave down on the interval $(-1, 1)$ and is concave up on the intervals $(-\infty, -1)$ and $(1, \infty)$. Therefore inflection points occur at $x = \pm 1$. Using a numerical solver, we see that the graph has x-intercepts at $x \approx -2.917, -0.215$. We also observe that $\lim\limits_{x \to \pm\infty} f(x) = \infty$, so f has no absolute maximum and an absolute minimum at $x = -2$.

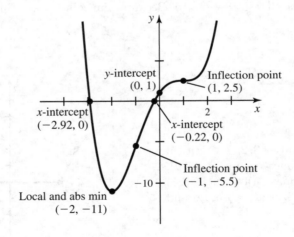

27 The derivatives of f are $f'(x) = -4\pi \sin[\pi(x-1)]$ and $f''(x) = -4\pi^2 \cos[\pi(x-1)]$. Solving $f'(x) = 0$ gives critical point $x = 1$, and solving $f''(x) = 0$ gives possible inflection points at $x = 1/2, 3/2$. Testing the sign of $f'(x)$ shows that f is decreasing on the interval $(1, 2)$ and increasing on $(0, 1)$. The First Derivative Test shows that a local maximum occurs at $x = 1$. By Theorem 4.5, this solitary local maximum must be the absolute maximum for f on the interval $[0, 2]$.

Testing the sign of $f''(x)$ shows that f is concave down on the interval $(1/2, 3/2)$ and concave up on the intervals $(0, 1/2)$ and $(3/2, 2)$. Therefore inflection points occur at $x = 1/2, 3/2$. These points are also the x-intercepts of the graph. We also observe that $f(0) = f(2) = -4$, so f takes its absolute minimum at these points.

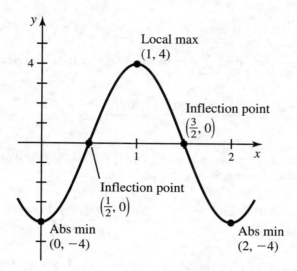

29 The domain of f is $(-\infty, 1) \cup (1, \infty)$. There is a vertical asymptote at $x = 1$. By long division, we can write $f(x) = x + 1 + \dfrac{4}{x-1}$, so $y = x + 1$ is a slant asymptote. $f(-x) = \dfrac{x^2 + 3}{-x - 1} = -\dfrac{x^2 + 3}{x + 1}$ which is not the same as either $f(x)$ or $-f(x)$, so f has neither even nor odd symmetry.

$$f'(x) = \frac{(x-1)(2x) - (x^2 + 3)}{(x-1)^2} = \frac{x^2 - 2x - 3}{(x-1)^2} = \frac{(x+1)(x-3)}{(x-1)^2},$$

which is zero for $x = -1$ and $x = 3$, so these are the critical points. $f' > 0$ on $(-\infty, -1)$ and $(3, \infty)$, so f is increasing on those intervals, while $f' < 0$ on $(-1, 1)$ and on $(1, 3)$ so f is decreasing on those intervals. There is a local minimum of 6 at $x = 3$ and a local maximum of -2 at $x = -1$.

$$f''(x) = \frac{(x-1)^2(2x-2) - (x^2 - 2x - 3)(2(x-1))}{(x-1)^4} = \frac{(2)(x-1)(x^2 - 2x + 1 - (x^2 - 2x - 3))}{(x-1)^4} = \frac{8}{(x-1)^3}.$$

$f'' > 0$ on $(1, \infty)$ so f is concave up there, and $f'' < 0$ so f is concave down there. There are no inflection points as the only concavity change occurs at a vertical asymptote. The only intercept is the y-intercept at $f(0) = -3$.

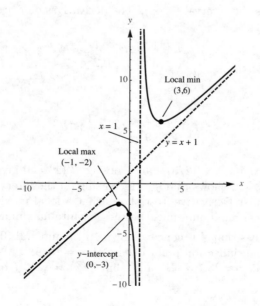

31 The domain of f is $(-\infty, \infty)$ and because $f(-x) = \dfrac{10(-x)^2}{(-x)^2+3} = f(x)$, f is symmetric with respect to the y-axis. Because $\lim\limits_{x\to\pm\infty} \dfrac{10x^2}{x^3+3} = \lim\limits_{x\to\pm\infty} \dfrac{20x}{2x} = 10$, the line $y=10$ is a horizontal asymptote.

$$f'(x) = \frac{(x^2+3)(20x) - 10x^2(2x)}{(x^2+3)^2} = \frac{60x}{(x^2+3)^2},$$

which is zero for $x = 0$. $f' > 0$ on $(0, \infty)$, so f is increasing there, while $f' < 0$ on $(-\infty, 0)$, so f is decreasing there. There is a local (and absolute) minimum of 0 at $x = 0$.

$$f''(x) = \frac{(x^2+3)^2(60) - 60x(2)(x^2+3)(2x)}{(x^2+3)^4} = \frac{60(x^2+3)(x^2+3-4x^2)}{(x^2+3)^4} = \frac{180(1-x)(1+x)}{(x^2+3)^3},$$

which is zero for $x = -1$ and $x = 1$. $f'' < 0$ on $(-\infty, -1)$ and on $(1, \infty)$, so f is concave down there, while $f'' > 0$ on $(-1, 1)$, so f is concave up there. There are inflection points at $(\pm 1, 5/2)$. The only intercept is $(0, 0)$, which is both an x- and a y-intercept.

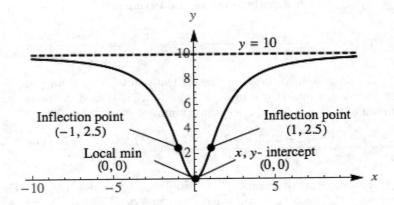

33 Note that the domain of f is the interval $(0, \infty)$. The derivatives of f are

$$f'(x) = \frac{1}{3}x^{-2/3} - \frac{1}{2}x^{-1/2},$$

and

$$f''(x) = -\frac{2}{9}x^{-5/3} + \frac{1}{4}x^{-3/2}.$$

Solving $f'(x) = 0$ gives critical point $x = (2/3)^6$, and solving $f''(x) = 0$ gives a possible inflection point at $x = (8/9)^6$. Testing the sign of $f'(x)$ shows that f is increasing on the interval $(0, (2/3)^6)$ and decreasing on the interval $((2/3)^6, \infty)$. The First Derivative Test shows that a local maximum occurs at $x = (2/3)^6$ and By Theorem 4.5 this solitary local maximum must be the absolute maximum of f on the interval $(0, \infty)$.

Testing the sign of $f''(x)$ shows that f is concave down on the interval $(0, (8/9)^6)$ and concave up on the interval $((8/9)^6, \infty)$. Therefore an inflection point occurs at $x = (8/9)^6$. Using a numerical solver, we find that the graph has x-intercept at $x \approx 23.767$. Because $\lim\limits_{x\to\infty} f(x) = -\infty$, f has no absolute minimum on the interval $(0, \infty)$.

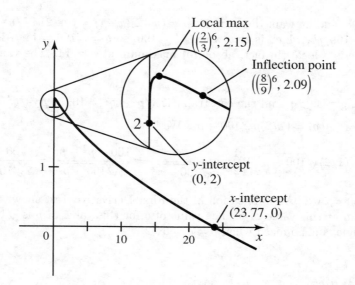

35 The volume of the box (where x is measured in inches) is given by $V(x) = x(6-x)(12-x)$, for $0 \le x \le 6$. Using the product rule for three functions, we have

$$V'(x) = (6-x)(12-x) + x(-1)(12-x) + x(6-x)(-1) = 72 - 6x - 12x + x^2 - 12x + x^2 - 6x + x^2 = 3(x^2 - 12x + 24).$$

By the quadratic formula, this is zero on the domain for $x = 6 - 2\sqrt{3} \approx 2.5$. Using the Second Derivative Test, we have $V''(a) = 3(2a - 12)$ which is negative for $a \approx 2.5$, so the solitary critical point on the domain gives a maximum. Therefore, the largest box is approximately $2.5 \times 3.5 \times 9.5$.

37 Let x equal your distance from the louder speaker; then your distance from the other speaker is $100 - x$. So we want to minimize the function $I(x) = \dfrac{3}{x^2} + \dfrac{1}{(100-x)^2}$, for $0 < x < 100$. Then

$$\begin{aligned}I'(x) &= -\frac{6}{x^3} + \frac{2}{(100-x)^3} = \frac{-6(100-x)^3 + 2x^3}{x^3(100-x)^3} \\ &= \left(\frac{-6000000 + 180000x - 1800x^2 + 6x^3 + 2x^3}{x^3(100-x)^3}\right) \\ &= \frac{8(x^3 - 225x^2 + 22{,}500x - 750{,}000)}{(x^3)(100-x)^3}.\end{aligned}$$

Using a computer algebra system or a root finder on a calculator, we find that $I' = 0$ for $x \approx 59$. Because $I'(1) < 0$ and $I'(99) > 0$, we see that the critical number yields a minimum for $0 < x < 100$. So you should stand 59 m away from the louder speaker.

39 The objective function to be maximized is the volume of the cone, given by $V = \pi r^2 h / 3$. By the Pythagorean Theorem, r and h satisfy the constraint $h^2 + r^2 = 16$, which gives $r^2 = 16 - h^2$. Therefore

$$V(h) = \frac{\pi}{3} h(16 - h^2) = \frac{\pi}{3}(16h - h^3).$$

We must maximize this function for $0 \le h \le 4$. The critical points of $V(h)$ satisfy $V'(h) = \dfrac{\pi}{3}(16 - 3h^2) = 0$, which has unique solution $h = 4/\sqrt{3} = 4\sqrt{3}/3$ in $(0, 4)$. Because $V(0) = V(4) = 0$, $h = 4\sqrt{3}/3$ gives the maximum value of $V(h)$ on $[0, 4]$. The corresponding value of r satisfies $r^2 = 16 - \dfrac{16}{3} = \dfrac{32}{3}$, so $r = \dfrac{4\sqrt{2}}{\sqrt{3}} = \dfrac{4\sqrt{6}}{3}$.

41 We have that $xy = 98$, and we want to maximize $p = (y-2)(x-1) = (y-2)(98/y-1) = 98-y-196/y+2$. Note that $p'(y) = -1 + 196/y^2$, which is zero when $y^2 = 196$, so $y = \sqrt{196} = 14$. Also note that $p'(13) > 0$ and $p'(15) < 0$, so there is a local (in fact, absolute) maximum at $y = 14$. The value of x when $y = 14$ is $x = 98/14 = 7$.

43 The area of the triangle is $\frac{1}{2}pq$, and the constraint is $\sqrt{p^2 + q^2} = 10$, or $p^2 + q^2 = 100$. So we can write the area of the triangle as $A(p) = (1/2)p\sqrt{100 - p^2}$. We have

$$A'(p) = (1/2)\sqrt{100 - p^2} + \frac{1}{2}p \cdot \frac{-p}{\sqrt{100 - p^2}} = \frac{100 - p^2 - p^2}{2\sqrt{100 - p^2}} = \frac{100 - 2p^2}{2\sqrt{100 - p^2}}.$$

This is zero for $p = \sqrt{50} = 5\sqrt{2}$. An application of the First Derivative Test shows that there is a local (in fact, absolute) maximum at this value of p. The value of q for this value of p is $\sqrt{100 - 50} = 5\sqrt{2}$ as well. So the area of the triangle is maximized when $p = q = 5\sqrt{2}$.

45

a. $f'(x) = \frac{2}{3}x^{-1/3}$, so $f'(27) = \frac{2}{3\sqrt[3]{27}} = \frac{2}{9}$. Thus

$$L(x) - 9 = \frac{2}{9}(x - 27),$$

or $L(x) = 9 + \frac{2}{9}(x - 27) = \frac{2}{9}x + 3$.

b. $f(29) \approx L(29) = 9 + \frac{4}{9} \approx 9.44$. This is an overestimate, because $f''(27) < 0$. Note that the calculator value of $f(29)$ is about 9.43913.

47 Let $f(x) = 1/x^2$ and let $a = 4$. Then $f'(x) = \frac{-2}{x^3}$ so $f'(4) = \frac{-2}{64} = \frac{-1}{32}$. The linearization is

$$L(x) = \frac{1}{16} + (-1/32)(x - 4).$$

Then $f(4.2) = 1/(4.2)^2 \approx L(4.2) = \frac{1}{16} - \frac{1}{32} \cdot \frac{2}{10} = 0.05625$.

49 $\Delta h \approx h'(a)\Delta t = -32 \cdot 5(0.7) = -112$ feet.

51 The value of c appears to be between 2 and 3. We are seeking c so that

$$f'(c) = \frac{f(b) - f(a)}{b - a} = \frac{10 - 0}{10 - 0} = 1,$$

so we want $\frac{\sqrt{10}}{2\sqrt{c}} = 1$. Thus $10 = 4c$ and $c = 2.5$, as projected.

53

a. The average rate of change of $P(t)$ on the interval $[0, 8]$ is

$$\frac{P(8) - P(0)}{8 - 0} = \frac{800/9 - 0}{8} = \frac{100}{9} \text{cells/week}.$$

b. We solve

$$P'(t) = \frac{100}{(t+1)^2} = \frac{100}{9}$$

which gives $(t + 1)^2 = 9$, so $t = 2$ weeks.

Chapter Four Review

55 It is possible to note that 1 is a root by inspection. Then by long division by $x - 1$, we have $f(x) = (x - 1)(3x^2 - x - 1)$. We apply Newton's method to the function $g(x) = 3x^2 - x - 1$.

The Newton's method recursion is given by $x_{n+1} = x_n - \dfrac{3x_n^2 - x_n - 1}{6x_n - 1}$. Applying this recursion to the initial estimates of -0.5 and 0.8 yields:

n	x_n
0	-0.5
1	-0.4375
2	-0.434267
3	-0.434259
4	-0.434259

n	x_n
0	0.8
1	0.768421
2	0.767592
3	0.767592
4	0.767592

The roots are thus 1, and approximately -0.434259 and 0.767592.

57 First note that $f'(x) = 10x^4 - 18x^2 - 4$ and $f''(x) = 40x^3 - 36x = 4x(10x^2 - 9)$. This is clearly 0 when $x = 0$, and when $10x^2 - 9 = 0$. Applying Newton's method to the function $g(x) = 10x^2 - 9$ with initial estimates of -1 and 1 yields:

n	x_n
0	-1
1	-0.95
2	-0.948684
3	-0.948683
4	-0.948683
5	-0.948683

n	x_n
0	1
1	0.95
2	0.948684
3	0.948683
4	0.948683
5	0.948683

Checking the signs of $f''(x)$ on the appropriate intervals leads to the conclusion that these are all the locations of inflection points of f. So the inflection points of f are located at 0 and approximately ± 0.948683.

59 By L'Hôpital's rule:

$$\lim_{x \to \infty} \frac{2x^5 - x + 1}{5x^6 + x} = \lim_{x \to \infty} \frac{10x^4 - 1}{30x^5 + 1} = \lim_{x \to \infty} \frac{40x^3}{150x^4} = \lim_{x \to \infty} \frac{4}{15x} = 0.$$

61 L'Hôpital's rule gives $\lim\limits_{t \to 0} \dfrac{1 - \cos 6t}{2t} = \lim\limits_{t \to 0} \dfrac{6 \sin 6t}{2} = 0.$

63 Observe that $\lim\limits_{\theta \to 0} \dfrac{3 \sin^2 2\theta}{\theta^2} = 3 \left(\lim\limits_{\theta \to 0} \dfrac{\sin 2\theta}{\theta} \right)^2$; L'Hôpital's rule gives $\lim\limits_{\theta \to 0} \dfrac{\sin 2\theta}{\theta} = \lim\limits_{\theta \to 0} \dfrac{2 \cos 2\theta}{1} = 2$, so $\lim\limits_{\theta \to 0} \dfrac{3 \sin^2 2\theta}{\theta^2} = 3 \cdot 2^2 = 12.$

65 Observe that $2\theta \cot 3\theta = 2 \cos 3\theta \cdot \dfrac{\theta}{\sin 3\theta}$; $\lim\limits_{\theta \to 0} \dfrac{\theta}{\sin 3\theta} = \lim\limits_{\theta \to 0} \dfrac{1}{3 \cos 3\theta} = \dfrac{1}{3}$ by L'Hôpital's rule, so

$$\lim_{\theta \to 0} 2\theta \cot 3\theta = 2 \cdot 1 \cdot \frac{1}{3} = \frac{2}{3}.$$

67 Make the change of variables $x = 1/y$; then $\ln^{10} y = (\ln y)^{10} = (-\ln x)^{10} = \ln^{10} x$, and so $\lim\limits_{y \to 0^+} \dfrac{\ln^{10} y}{\sqrt{y}} = \lim\limits_{x \to \infty} \sqrt{x} \ln^{10} x = \infty.$

69 Apply L'Hôpital's rule twice:
$$\lim_{x\to 1}\frac{x^4-x^3-3x^2+5x-2}{x^3+x^2-5x+3}=\lim_{x\to 1}\frac{4x^3-3x^2-6x+5}{3x^2+2x-5}=\lim_{x\to 1}\frac{12x^2-6x-6}{6x+2}=0.$$

71 $\lim_{x\to 0}\csc x\sin^{-1}x=\lim_{x\to 0}\frac{\sin^{-1}x}{\sin x}=\lim_{x\to 0}\frac{1/\sqrt{1-x^2}}{\cos x}=1$, by L'Hôpital's rule.

73 Observe that $\lim_{x\to\infty}\frac{x+1}{x-1}=1$, by L'Hôpital's rule. Hence $\lim_{x\to\infty}\ln\left(\frac{x+1}{x-1}\right)=\ln 1=0$.

75 Note that $\ln(\sin x)^{\tan x}=\tan x\ln\sin x$, so we evaluate $L=\lim_{x\to\pi/2^-}\tan x\ln\sin x=\lim_{x\to\pi/2^-}\frac{\ln\sin x}{\cot x}=$
$\lim_{x\to\pi/2^-}\frac{\cot x}{-\csc^2 x}=\lim_{x\to\pi/2^-}(-\cos x\sin x)=0$ by L'Hôpital's rule. Therefore $\lim_{x\to\pi/2^-}(\sin x)^{\tan x}=e^L=1$.

77 Note that for $0<x<1$, we have $\ln x<0$, so $|\ln x|=-\ln x$. Then $|\ln x|^x=(-\ln x)^x$. Consider the natural logarithm of this quantity, $x\ln(-\ln x)=\frac{\ln(-\ln x)}{\frac{1}{x}}$. Using L'Hôpital's rule we have

$$\lim_{x\to 0^+}\frac{\ln(-\ln x)}{\frac{1}{x}}=\lim_{x\to 0^+}\frac{\frac{1}{x\ln x}}{\frac{-1}{x^2}}=\lim_{x\to 0^+}\frac{-x}{\ln x}=0.$$

Thus $\lim_{x\to 0^+}|\ln x|^x=e^0=1$.

79 Note that $\ln\left(1-\frac{3}{x}\right)^x=x\ln\left(1-\frac{3}{x}\right)=\frac{\ln(1-3/x)}{1/x}$, so we evaluate

$$L=\lim_{x\to\infty}\frac{\ln(1-3/x)}{1/x}=\lim_{x\to\infty}\frac{(1-3/x)^{-1}(3/x^2)}{-1/x^2}=-3$$

by L'Hôpital's rule. Therefore $\lim_{x\to\infty}\left(1-\frac{3}{x}\right)^x=e^L=e^{-3}$.

81 Let $y=(x-1)^{\sin\pi x}$. Then $\ln y=\sin\pi x\ln(x-1)=\frac{\ln(x-1)}{\csc\pi x}$. Then

$$\lim_{x\to 1}\ln y=\lim_{x\to 1}\frac{\ln(x-1)}{\csc\pi x}=\lim_{x\to 1}\frac{\frac{1}{x-1}}{-\pi\csc\pi x\cot\pi x}=\lim_{x\to 1}\frac{\sin^2\pi x}{-\pi(x-1)\cos\pi x}$$
$$=\lim_{x\to 1}\frac{1}{-\pi\cos\pi x}\cdot\lim_{x\to 1}\frac{\sin^2\pi x}{x-1}=\frac{1}{\pi}\lim_{x\to 1}\frac{2\pi\sin\pi x\cos\pi x}{1}=\frac{1}{\pi}\cdot 0=0.$$

Then $\lim_{x\to 1}y=\lim_{x\to 1}e^{\ln y}=e^0=1$.

83 Observe that $\lim_{x\to\infty}\frac{x^{1/2}}{x^{1/3}}=\lim_{x\to\infty}x^{1/6}=\infty$, so $x^{1/2}$ grows faster than $x^{1/3}$ as $x\to\infty$.

85 By Theorem 4.14, $\sqrt{x}$ grows faster than $\ln^{10}x$ as $x\to\infty$.

87 Observe that $\lim_{x\to\infty}\frac{e^x}{3^x}=\lim_{x\to\infty}\left(\frac{e}{3}\right)^x=0$ because $e/3<1$. Therefore 3^x grows faster than e^x as $x\to\infty$.

89 Observe that $4^{x/2}=(4^{1/2})^x=2^x$, so these functions are identical and hence have comparable growth rates as $x\to\infty$.

91 $\int(2x+1)^2\,dx=\int(4x^2+4x+1)\,dx=\frac{4}{3}x^3+2x^2+x+C$.

93 $\displaystyle\int \left(\frac{1}{x^2} - \frac{2}{x^{5/2}}\right) dx = \int (x^{-2} - 2x^{-5/2}) dx = -x^{-1} - 2 \cdot \left(-\frac{2}{3}\right) x^{-3/2} = -\frac{1}{x} + \frac{4}{3} x^{-3/2} + C.$

95 $\displaystyle\int (1 + 3\cos\theta)\, d\theta = \theta + 3\sin\theta + C.$

97 $\displaystyle\int \frac{dx}{1 - \sin^2 x} = \int \frac{dx}{\cos^2 x} = \int \sec^2 x\, dx = \tan x + C.$

99 $\displaystyle\int \frac{12}{x}\, dx = 12 \ln |x| + C.$

101 $\displaystyle\int \frac{x^2}{x^4 + x^2}\, dx = \int \frac{1}{x^2 + 1}\, dx = \tan^{-1} x + C.$

103 $\displaystyle\int \left(\sqrt[4]{x^3} + \sqrt{x^5}\right) dx = \int (x^{3/4} + x^{5/2}) dx = \frac{4}{7} x^{7/4} + \frac{2}{7} x^{7/2} + C.$

105 We have $f(t) = \displaystyle\int (\sin t + 2t)\, dt = -\cos t + t^2 + C;\ f(0) = -1 + C = 5$, so $C = 6$ and $f(t) = -\cos t + t^2 + 6.$

107 Note that by long division, we can write $\dfrac{x^4 - 2}{1 + x^2} = x^2 - 1 - \dfrac{1}{1 + x^2}$. Therefore
$f(x) = \displaystyle\int \left(x^2 - 1 - \frac{1}{1 + x^2}\right) dx = \frac{x^3}{3} - x - \tan^{-1} x + C.$ Because $f(1) = \dfrac{2}{3}$, we have $\dfrac{1}{3} - \dfrac{\pi}{4} + C = -\dfrac{2}{3}$, so $C = \dfrac{\pi}{4}$. Thus $f(x) = \dfrac{x^3}{3} - x - \tan^{-1} x + \dfrac{\pi}{4}.$

109 The velocity of the rocket is given by

$$v(t) = -9.8t + v_0 = -9.8t + 120,$$

and the position function of the rocket is

$$s(t) = \int (-9.8t + 120)\, dt = -4.9t^2 + 120t + s_0 = -4.9t^2 + 120t + 125.$$

The rocket reaches its maximum height when $v(t) = 0$, which occurs at $t = 120/9.8 \approx 12.24$ s; the maximum height is $s(120/9.8) \approx 859.69$ m. The rocket hits the ground when $s(t) = 0$; solving this quadratic equation gives $t \approx 25.49$ s.

111

  a. We have $v(t) = -32t + v_0$ and $v_0 = 64$, so $v(t) = -32t + 64.$

  b. The height of the ball above the river is given by $s(t) = \displaystyle\int (-32t + 64)\, dt = -16t^2 + 64t + s_0.$, and because $s(0) = 128$, we have $s_0 = 128$. So $s(t) = -16t^2 + 64t + 128.$

  c. The ball reaches its maximum height when $v(t) = -32t + 64 = 0$, which occurs for $t = 2$. The height at this time is $s(2) = 192.$

  d. The ball strikes the river when $s(t) = 0$ (and $t > 0$), which gives $-16t^2 + 64t + 128 = -16(t^2 - 4t - 8) = 0.$ Using the quadratic formula, we see that this occurs for $t = 2(1 + \sqrt{3})$. The velocity with which the ball strikes the river is therefore $v((2(1 + \sqrt{3})) = -64\sqrt{3} \approx -110.9$ ft per s.

113 For the second limit, let $y = x^2$:

$$\lim_{x \to 0} \frac{x^2}{1 - e^{-x^2}} = \lim_{y \to 0} \frac{y}{1 - e^{-y}} = \lim_{y \to 0} \frac{1}{e^{-y}} = 1$$

by L'Hôpital's rule. For the first, observe that for $x > 0$

$$\frac{x}{\sqrt{1 - e^{-x^2}}} = \left(\frac{x^2}{1 - e^{-x^2}}\right)^{1/2},$$

so $\lim\limits_{x \to 0^+} \dfrac{x}{\sqrt{1 - e^{-x^2}}} = 1$ as well.

115 Note that $\lim\limits_{x \to 0^+} x^x = 1$, as shown in Section 4.6. Therefore $\lim\limits_{x \to 0^+} f(x) = \lim\limits_{x \to 0^+} (x^x)^x = 1^0 = 1$, and $\lim\limits_{x \to 0} g(x) = \lim\limits_{x \to 0} x^{(x^x)} = 0^1 = 0$.

117

a. First, observe that

$$\lim_{x \to \infty} \left(1 + \frac{1}{x}\right)^{x+a} = \lim_{x \to \infty} \left(1 + \frac{1}{x}\right)^x \left(1 + \frac{1}{x}\right)^a = e \cdot 1 = e.$$

Therefore $\lim\limits_{x \to \infty} \ln g(x) = 1$. It suffices to determine whether $\ln g(x) - 1$ is positive or negative as $x \to \infty$. To do this, consider

$$\lim_{x \to \infty} x(\ln g(x) - 1) = \lim_{t \to 0} \frac{(1 + at)\ln(1 + t) - t}{t^2},$$

where we make the change of variables $t = 1/x$. This limit can be evaluated by using L'Hôpital's rule twice:

$$\lim_{t \to 0} \frac{(1 + at)\ln(1 + t) - t}{t^2} = a - \frac{1}{2}.$$

Therefore when $a > 1/2$ we have $g(x) > e$ as $x \to \infty$,

b. Using the analysis above, for $0 < a < 1/2$, $g(x) < e$ as $x \to \infty$. In the case $a = 1/2$ we consider the limit

$$\lim_{x \to \infty} x^2(\ln g(x) - 1) = \lim_{t \to 0} \frac{(1 + at)\ln(1 + t) - t}{t^3},$$

which can be evaluated by using L'Hôpital's three times:

$$\lim_{t \to 0} \frac{(1 + t/2)\ln(1 + t) - t}{t^3} = \frac{1}{12}.$$

Therefore $g(x) > e$ as $x \to \infty$ in this case as well.

Chapter 5

Integration

5.1 Approximating Areas under Curves

5.1.1

In the first 2 seconds, the object moves $15 \cdot 2 = 30$ meters. In the next three seconds, the object moves $25 \cdot 3 = 75$ meters, so the total displacement is $75 + 30 = 105$ meters.

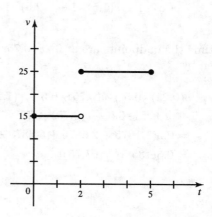

5.1.3

a. The displacement is approximately $40(4) + 70(4) = 440$ ft.

b. The displacement is approximately $30(2) + 50(2) + 80(2) + 40(2) = 400$ ft.

5.1.5

a. The displacement is approximately $60(2) + 30(2) + 80(2) = 340$ ft.

b. The displacement is approximately $70(1) + 60(1) + 50(1) + 30(1) + 40(1) + 80(1) = 330$ ft.

5.1.7 Subdivide the interval from 0 to $\pi/2$ into subintervals. On each subinterval, pick a sample point (like the left endpoint, or the right endpoint, or the midpoint, for example) and call the first sample point x_1 and the second sample point x_2 and so on. For each sample point x_i, calculate the area of the rectangle which lies over the subinterval and has height $f(x_i) = \cos x_i$. Do this for each subinterval, and add the areas of the corresponding rectangles together. This will give an approximation to the area under the curve, with generally a better approximation occurring as n increases – where n is the number of subintervals used.

5.1.9 We have $\Delta x = \dfrac{7-1}{6} = 1$. The left Riemann sum is given by $1(10 + 9 + 7 + 5 + 2 + 1) = 34$ and the right Riemann sum is given by $1(9 + 7 + 5 + 2 + 1 + 0) = 24$.

245

5.1.11 Because the interval $[1,3]$ has length 2, if we subdivide it into 4 subintervals, each will have length $\Delta x = \dfrac{2}{4} = \dfrac{1}{2}$. The grid points will be $x_0 = 1$, $x_1 = 1 + \Delta x = 1.5$, $x_2 = 1 + 2\Delta x = 2$, $x_3 = 1 + 3\Delta x = 2.5$, and $x_4 = 1 + 4\Delta x = 3$.

If we use the left-hand side of each subinterval, we will use 1, 1.5, 2, and 2.5.

If we use the right-hand side of each subinterval, we will use 1.5, 2, 2.5, and 3.

If we use the midpoint of each subinerval, we will use 1.25, 1.75, 2.25, and 2.75.

5.1.13 It is an underestimate. If we use the right-hand side of each subinterval to determine the height of the rectangles, the height of each rectangle will be the minimum of f over the subinterval, so the sum of the areas of the rectangles will be less than the corresponding area under the curve.

5.1.15

a. On the first subinterval, the midpoint is 0.5, and $v(0.5) = 1.75$. On the 2nd subinterval, the midpoint is 1.5 and $v(1.5) = 7.75$. Continuing in this manner, we obtain the estimate to the displacement of

$$v(.5) \cdot 1 + v(1.5) \cdot 1 + v(2.5) \cdot 1 + v(3.5) \cdot 1 = 1.75 + 7.75 + 19.75 + 37.75 = 67 \text{ ft}.$$

b. This time the midpoints are at 0.25, 0.75, 1.25 .... Each subinterval has length $\frac{1}{2}$. Thus, the estimate is given by

$$v(0.25) \cdot 0.5 + v(0.75) \cdot 0.5 + v(1.25) \cdot 0.5 + v(1.75) \cdot 0.5 + v(2.25) \cdot 0.5 + v(2.75) \cdot 0.5$$
$$+ v(3.25) \cdot 0.5 + v(3.75) \cdot 0.5$$
$$= 0.5(1.1875 + 2.6875 + 5.6875 + 10.1875 + 16.1875 + 23.6875 + 32.6875 + 43.1875)$$
$$= 0.5(135.5) = 67.75 \text{ ft}..$$

5.1.17

The left-hand grid points are 0 and 4. The length of each subinterval is $8/2 = 4$. So the left Riemann sum is given by $v(0) \cdot 4 + v(4) \cdot 4 = 4 \cdot (1 + 9) = 40$ m.

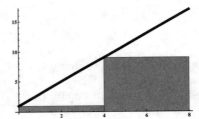

5.1.19

The left-hand grid points are 0, 2, 4, and 6. The length of each subinterval is 2. So the left Riemann sum is given by $v(0) \cdot 2 + v(2) \cdot 2 + v(4) \cdot 2 + v(6) \cdot 2 = 2 \cdot \left(\dfrac{1}{1} + \dfrac{1}{5} + \dfrac{1}{9} + \dfrac{1}{13}\right) \approx 2.776$ m.

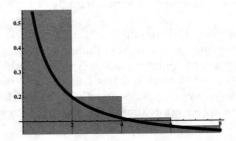

5.1. Approximating Areas under Curves

5.1.21

The left-hand grid points are 0, 3, 6, 9, and 12. The length of each subinterval is 3. So the left Riemann sum is given by $\sum_{k=1}^{5} v(3(k-1)) \cdot 3 = 12 + 24 + 12\sqrt{7} + 12\sqrt{10} + 12\sqrt{13} \approx 148.963$ mi.

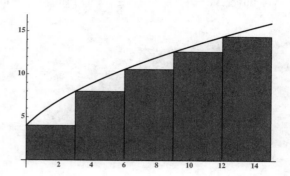

5.1.23 The left Riemann sum is given by $f(1) + f(2) + f(3) + f(4) + f(5) = 2 + 3 + 4 + 5 + 6 = 20$. The right Riemann sum is given by $f(2) + f(3) + f(4) + f(5) + f(6) = 3 + 4 + 5 + 6 + 7 = 25$.

5.1.25

a.

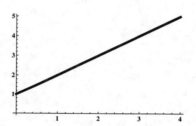

b. We have $\Delta x = \dfrac{4-0}{4} = 1$. The grid points are $x_0 = 0$, $x_1 = 1$, $x_2 = 2$, $x_3 = 3$, and $x_4 = 4$.

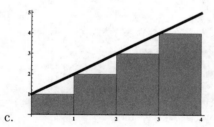

c.

d. The left Riemann sum is $1 \cdot 1 + 2 \cdot 1 + 3 \cdot 1 + 4 \cdot 1 = 10$, which is an underestimate of the area under the curve. The right Riemann sum is $2 \cdot 1 + 3 \cdot 1 + 4 \cdot 1 + 5 \cdot 1 = 14$ which is an overestimate of the area under the curve.

5.1.27

a.

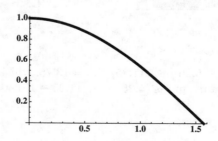

b. We have $\Delta x = \dfrac{\pi/2 - 0}{4} = \dfrac{\pi}{8}$. The grid points are $x_0 = 0$, $x_1 = \dfrac{\pi}{8}$, $x_2 = \dfrac{\pi}{4}$, $x_3 = \dfrac{3\pi}{8}$, and $x_4 = \dfrac{\pi}{2}$.

c.

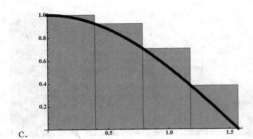
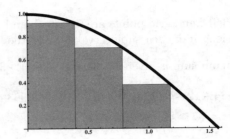

d. The left Riemann sum is $1 \cdot \dfrac{\pi}{8} + \cos(\pi/8) \cdot \dfrac{\pi}{8} + \cos(\pi/4) \cdot \dfrac{\pi}{8} + \cos(3\pi/8) \cdot \dfrac{\pi}{8} \approx 1.185$, which is an overestimate of the area under the curve. The right Riemann sum is $\cos(\pi/8) \cdot \dfrac{\pi}{8} + \cos(\pi/4) \cdot \dfrac{\pi}{8} + \cos(3\pi/8) \cdot \dfrac{\pi}{8} + 0 \cdot \dfrac{\pi}{8} \approx 0.791$ which is an underestimate of the area under the curve.

5.1.29

a.

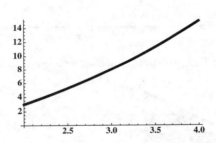

b. We have $\Delta x = \dfrac{4-2}{4} = \dfrac{1}{2}$. The grid points are $x_0 = 2$, $x_1 = 2.5$, $x_2 = 3$, $x_3 = 3.5$, and $x_4 = 4$.

c.

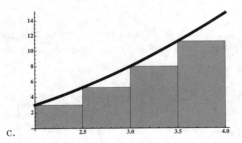
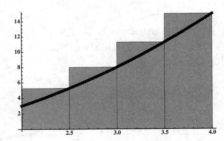

d. The left Riemann sum is $(3 + 5.25 + 8 + 11.25) \cdot 0.5 = 13.75$, which is an underestimate of the area under the curve. The right Riemann sum is $(5.25 + 8 + 11.25 + 15) \cdot 0.5 = 19.75$ which is an overestimate of the area under the curve.

5.1.31

a.

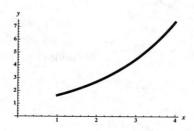

b. We have $\Delta x = \dfrac{4-1}{6} = \dfrac{1}{2}$. The grid points are $x_0 = 1$, $x_1 = 1.5$, $x_2 = 2$, $x_3 = 2.5$, $x_4 = 3$, $x_5 = 3.5$, and $x_6 = 4$.

c.

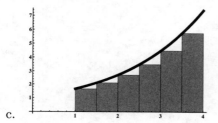 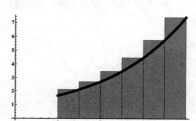

d. The left Riemann sum is $\frac{1}{2}\left(e^{1/2} + e^{3/4} + e + e^{5/4} + e^{3/2} + e^{7/4}\right) \approx 10.105$, which is an underestimate of the area under the curve. The right Riemann sum is $\dfrac{1}{2}\left(e^{3/4} + e + e^{5/4} + e^{3/2} + e^{7/4} + e^2\right) \approx 12.975$ which is an overestimate of the area under the curve.

5.1.33 We have $\Delta x = 2$, so the midpoints are 1, 3, 5, 7, and 9. So the midpoint Riemann sum is $2(f(1) + f(3) + f(5) + f(7) + f(9)) = 670$.

5.1.35

a. The left Riemann sum is $315(6) + 350(6) + 365(6) + 370(6) + 350(6) = 10{,}500$ m. The right Riemann sum is $350(6) + 365(6) + 370(6) + 350(6) + 290(6) = 10{,}350$ m. Answers will vary.

b. The left Riemann sum is more accurate.

c. More accurate approximations are obtained by increasing the number of subintervals.

5.1.37

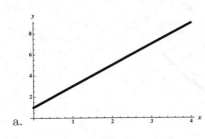
a.

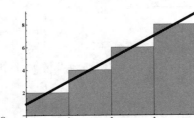
c.

b. We have $\Delta x = \dfrac{4-0}{4} = 1$. The gridpoints are $x_0 = 0$, $x_1 = 1$, $x_2 = 2$, $x_3 = 3$, and $x_4 = 4$, so the midpoints are .5, 1.5, 2.5, and 3.5.

d. The midpoint Riemann sum is $1(2+4+6+8) = 20$.

5.1.39

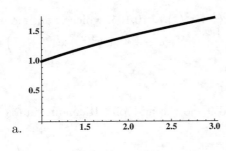
a.

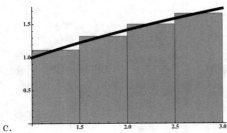
c.

b. We have $\Delta x = \dfrac{3-1}{4} = \dfrac{1}{2}$. So the midpoints are 1.25, 1.75, 2.25, and 2.75.

d. The midpoint Riemann sum is $.5(\sqrt{1.25} + \sqrt{1.75} + \sqrt{2.25} + \sqrt{2.75}) \approx 2.800$.

5.1.41

a.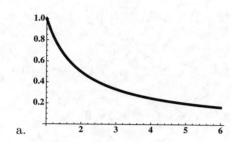

b. We have $\Delta x = \dfrac{6-1}{5} = 1$. So the midpoints are 1.5, 2.5, 3.5, 4.5, and 5.5.

c.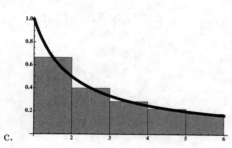

d. The midpoint Riemann sum is $2/3 + 2/5 + 2/7 + 2/9 + 2/11 \approx 1.756$.

5.1.43 Note that $\Delta x = \dfrac{2-0}{4} = .5$. So the left Riemann sum is given by $0.5(5 + 3 + 2 + 1) = 5.5$ and the right Riemann sum is given by $0.5(3 + 2 + 1 + 1) = 3.5$.

5.1.45

a.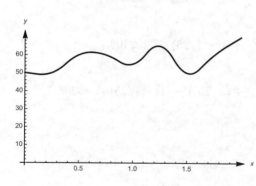

b. With $n = 2$, we have $\Delta x = \dfrac{2-0}{2} = 1$, so the midpoints are .5 and 1.5. So the midpoint Riemann sum is $60 + 50 = 110$. For $n = 4$, we have $\Delta x = \dfrac{2-0}{4} = \dfrac{1}{2}$, so the midpoints are 0.25, 0.75, 1.25, and 1.75. The midpoint Riemann sum in this case is $0.5(50 + 60 + 65 + 60) = \dfrac{235}{2} = 117.5$.

5.1.47

a. $\displaystyle\sum_{k=1}^{5} k$.

b. $\displaystyle\sum_{k=1}^{6} (k+3)$.

c. $\displaystyle\sum_{k=1}^{4} k^2$.

d. $\displaystyle\sum_{k=1}^{4} \dfrac{1}{k}$.

5.1.49

a. $\sum_{k=1}^{10} k = 1 + 2 + 3 + \ldots + 10 = 55.$

b. $\sum_{k=1}^{6} (2k+1) = 3 + 5 + 7 + 9 + 11 + 13 = 48.$

c. $\sum_{k=1}^{4} k^2 = 1 + 4 + 9 + 16 = 30.$

d. $\sum_{n=1}^{5} (1 + n^2) = 2 + 5 + 10 + 17 + 26 = 60.$

e. $\sum_{m=1}^{3} \frac{2m+2}{3} = \frac{4}{3} + \frac{6}{3} + \frac{8}{3} = 6.$

f. $\sum_{j=1}^{3} (3j - 4) = -1 + 2 + 5 = 6.$

g. $\sum_{p=1}^{5} (2p + p^2) = 3 + 8 + 15 + 24 + 35 = 85.$

h. $\sum_{n=0}^{4} \sin \frac{n\pi}{2} = 0 + 1 + 0 + (-1) + 0 = 0.$

5.1.51 Note that $\Delta x = \frac{1}{10}$, and $x_k = a + k\Delta x = \frac{k}{10}$.

a. The left Riemann sum is given by $\sum_{k=0}^{39} 3\sqrt{k/10} \cdot (1/10) \approx 15.6809.$

The right Riemann sum is given by $\sum_{k=1}^{40} 3\sqrt{k/10} \cdot (1/10) \approx 16.2809.$

The midpoint Riemann sum is given by $\sum_{k=0}^{39} 3\sqrt{(1/20) + (k/10)} \cdot (1/10) \approx 16.0055.$

b. It appears that the actual area is about 16.

5.1.53 Note that $\Delta x = \frac{1}{25}$, and $x_k = 2 + k\Delta x = 2 + (k/25)$. So $f(x_k) = (2 + (k/25))^2 - 1$.

a. The left Riemann sum is given by $\sum_{k=0}^{74} [(2 + (k/25))^2 - 1] \cdot (1/25) \approx 35.5808.$

The right Riemann sum is given by $\sum_{k=1}^{75} [(2 + (k/25))^2 - 1] \cdot (1/25) \approx 36.2408.$

The midpoint Riemann sum is given by $\sum_{k=0}^{74} [((61/30) + (k/25))^2 - 1] \cdot (1/25) \approx 35.9996.$

b. It appears that the actual area is about 36.

5.1.55 $\Delta x = \frac{1 - (-1)}{n} = \frac{2}{n}$. $x_k = -1 + k\Delta x = -1 + \frac{2k}{n}$. The right Riemann sum is given by

$$\sum_{k=1}^{n} f(x_k)\Delta x = \frac{2}{n} \sum_{k=1}^{n} \left(12 - 3\left(-1 + \frac{2k}{n}\right)^2\right).$$

n	Right Riemann Sum
10	21.96
30	21.9956
60	21.9989
80	21.9994

5.1. Approximating Areas under Curves

The sum approaches 22.

5.1.57 $\Delta x = \dfrac{\pi - (-\pi)}{n} = \dfrac{2\pi}{n}$. $x_k = -\pi + k\Delta x = -\pi + \dfrac{2\pi k}{n}$. The right Riemann sum is given by

$$\sum_{k=1}^{n} f(x_k)\Delta x = \dfrac{2\pi}{n} \sum_{k=1}^{n} \left(\dfrac{1 - \cos\left(-\pi + \frac{2\pi k}{n}\right)}{2} \right).$$

n	Right Riemann Sum
10	3.14159
30	3.14159
60	3.14159
80	3.14159

The sum approaches π.

5.1.59

a. True. Because the curve is a straight line, the region under the curve and over each subinterval is a trapezoid. The formula for the area of such a trapezoid over $[x_i, x_{i+1}]$ is $\dfrac{f(x_i) + f(x_{i+1})}{2} \cdot \Delta x = \dfrac{2x_i + 5 + 2x_{i+1} + 5}{2} \cdot \Delta x = (x_i + x_{i+1} + 5)\Delta x$ and the area given by using the midpoint formula is $f\left(\dfrac{x_i + x_{i+1}}{2}\right)\Delta x = (x_i + x_{i+1} + 5)\Delta x$. So the area under the curve is exactly given by the midpoint Riemann sum. Note that this holds for any straight line.

b. False. The left Riemann sum will underestimate the area under an increasing function.

c. True. The value of f at the midpoint will always be between the value of f at the endpoints, if f is monotonic increasing or monotonic decreasing.

5.1.61 $\displaystyle\sum_{k=1}^{50} \left(\dfrac{2k}{25} + 1\right) \cdot \dfrac{2}{25} = 12.16.$

5.1.63 $\displaystyle\sum_{k=0}^{31} \left(3 + \dfrac{1}{8} + \dfrac{k}{4}\right)^3 \cdot \dfrac{1}{4} \approx 3639.125.$

5.1.65 This is the right Riemann sum for f on the interval $[1, 5]$ for $n = 4$.

5.1.67 This is the midpoint Riemann sum for f on the interval $[2, 6]$ for $n = 4$.

5.1.69 For all of the calculations below, we have $\Delta x = \dfrac{1}{2}$, and grid points $x_0 = 0$, $x_1 = 0.5$, $x_2 = 1$, $x_3 = 2.5$, and $x_4 = 2$.

a.

The left Riemann sum is given by $\frac{1}{2}(f(0) + f(0.5) + f(1) + f(1.5))$ which is equal to $\frac{1}{2}(2 + 2.25 + 3 + 4.25) = 5.75$.

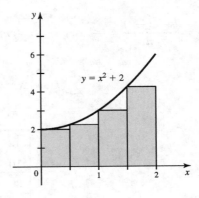

b.

The midpoint Riemann sum is given by $\frac{1}{2}(f(.25) + f(.75) + f(1.25) + f(1.75))$ which is equal to $\frac{1}{2}(2.0625 + 2.5625 + 3.5625 + 5.0625) = 6.625$.

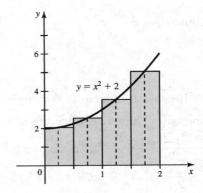

c.

The right Riemann sum is given by $\frac{1}{2}(f(0.5) + f(1) + f(1.5) + f(2))$ which is equal to $\frac{1}{2}(2.25 + 3 + 4.25 + 6) = 7.75$.

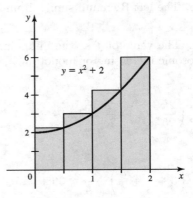

5.1.71

a. During the first second, the velocity steadily increases from 0 to 20, then it remains constant until $t = 3$. From $t = 3$ until $t = 5$ it steadily decreases, and then remains constant until $t = 6$.

b. Between $t = 0$ and $t = 2$ the area under the curve is $\frac{1}{2} \cdot 1 \cdot 20 + 1 \cdot 20 = 30$.

c. Between $t = 2$ and $t = 5$ the displacement is the sum of the area of a rectangle with area 20 and a trapezoid with area 30, so the displacement is 50 meters.

5.1. Approximating Areas under Curves

d. Between $t = 0$ and $t = 5$ the displacement is 80. Between $t = 5$ and any time $t \geq 5$ the displacement is $10(t-5)$ so the displacement between $t = 0$ and $t \geq 5$ is $80 + 10(t-5)$.

5.1.73

a. Between 0 and 5, the area under the curve is given by the area of a square of area 4 and the area of a trapezoid of area 10.5, so the total area is 14.5.

b. Between 5 and 10, the area under the curve is given by the area of a trapezoid of area 5.5 and the area of a rectangle of area $4 \cdot 6 = 24$, so the total area is 29.5.

c. The mass of the entire rod would be the total area under the curve from 0 to 10, which would be $14.5 + 29.5 = 44$ grams.

d. At $x = \dfrac{19}{3}$ there is a mass of 22 on each side. Note that from 0 to 6 the mass is 20 grams, so the center of mass is a little greater than 6.

5.1.75 If $0 \leq t \leq 2$, the displacement is $30t$. If $2 \leq t \leq 2.5$, the displacement is $60 + 50(t-2)$. If $2.5 \leq t \leq 3$, the displacement is $85 + 44(t - 2.5)$.

Thus, $d(t) = \begin{cases} 30t & \text{if } 0 \leq t \leq 2, \\ 50t - 40 & \text{if } 2 \leq t \leq 2.5 \\ 44t - 25 & \text{if } 2.5 \leq t \leq 3. \end{cases}$

5.1.77 Using the left Riemann sum

$$\sum_{k=0}^{n-1} \left| 1 - \left(-1 + \frac{3k}{n}\right)^3 \right| \cdot \frac{3}{n},$$

we have

n	16	32	64
A_n	4.33054	4.52814	4.63592

It appears that the areas are approaching 4.75.

5.1.79 The midpoint Riemann sum gives

$$\sum_{k=0}^{n-1} f\left(a + \frac{b-a}{2n} + \frac{k(b-a)}{n}\right) \cdot \frac{b-a}{n} = \sum_{k=0}^{n-1} \left(m\left(a + \frac{b-a}{2n} + \frac{k(b-a)}{n}\right) + c\right) \cdot \frac{b-a}{n} =$$

$$m \cdot a \cdot n \cdot \frac{b-a}{n} + \frac{m(b-a)^2 n}{2n^2} + \frac{(n-1)n}{2} \cdot \frac{m(b-a)^2}{n^2} + \frac{cn(b-a)}{n} =$$

$$m \cdot a \cdot (b-a) + \frac{m(b-a)^2}{2n} + \frac{m(b-a)^2}{2} - \frac{m(b-a)^2}{2n} + c(b-a) =$$

$$m \cdot a \cdot (b-a) + \frac{m(b-a)^2}{2} + c(b-a) = (b-a) \cdot \left(\frac{m(a+b)}{2} + c\right).$$

This proves that the midpoint Riemann sum is independent of n. Because the region in question is a trapezoid, we know that the exact area is given by the width of the subinterval times the average value at the endpoints, which is

$$(b-a)\left(\frac{f(a) + f(b)}{2}\right) = (b-a)\left(\frac{ma + c + mb + c}{2}\right) = (b-a)\left(\frac{m(a+b)}{2} + c\right).$$

5.1.81 For a function that is concave up and increasing, each rectangle of the right Riemann sum will have its top edge above the curve, since the value of the function at the right edge of the rectangle will be larger than at any other point in the rectangle. Thus this will be an overestimate. For a function that is concave up and decreasing, however, each rectangle will lie wholly below the curve, since the value of the function at the right edge will be smaller than at any other point in the rectangle. Thus this will be an underestimate. For a function that is concave down and increasing, each rectangle of the right Riemann sum will have its top edge above the curve, since the value of the function at the right edge of the rectangle will be larger than at any other point in the rectangle. Thus this will be an overestimate. Finally, for a function that is concave down and decreasing, however, each rectangle will lie wholly below the curve, since the value of the function at the right edge will be smaller than at any other point in the rectangle. Thus this will be an underestimate. Graphs of each of the four situations are below:

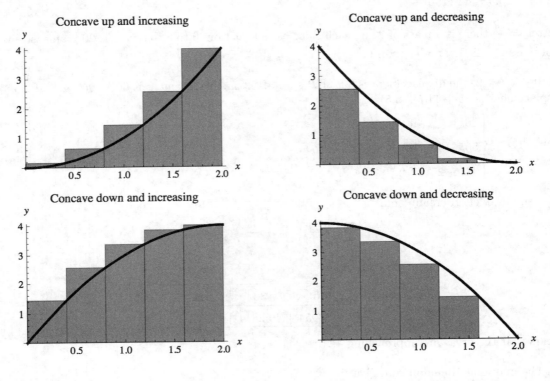

So the answer is

	Increasing on $[a,b]$	Decreasing on $[a,b]$
Concave up on $[a,b]$	Overestimate	Underestimate
Concave down on $[a,b]$	Overestimate	Underestimate

5.2 Definite Integrals

5.2.1 The net area is the difference between the area above the x-axis and below the curve, and below the x-axis and above the curve.

5.2.3 The left Riemann sum is: $40(1) + 30(1) + 20(1) + 0(1) - 10(1) - 20(1) = 60$. The right Riemann sum is $30(1) + 20(1) + 0(1) - 10(1) - 20(1) - 20(1) = 0$.

5.2.5 The left Riemann sum is: $2(2) - 3(2) - 5(2) = -12$. The right Riemann sum is $(-3)(2) - 5(2) - 1(2) = -18$. The midpoint Riemann sum is $-2(2) - 4(2) - 2(2) = -16$.

5.2.7 The integral represents the area of a 5 by 2 rectangle, so the value is 10.

5.2. Definite Integrals

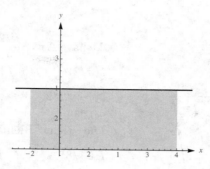

5.2.9 The integral represents the difference in areas of a triangle with base 2 and height 4 (so its area is 4) and one with base 1 and height 2 (so its area is 1), so the value of the integral is $4 - 1 = 3$.

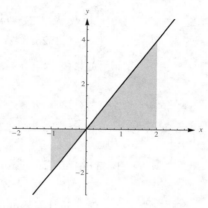

5.2.11 Because each of the functions $\sin x$ and $\cos x$ have the same amount of area above the x-axis as below between 0 and 2π, these both have value 0.

5.2.13 Because a region "from $x = a$ to $x = a$" has no width, its area is zero. This is akin to asking for the area of a one-dimensional object.

5.2.15 This integral represents the area under $y = x$ between $x = 0$ and $x = a$, which is a right triangle. The length of the base of the triangle is a and the height is a, so the area is $\frac{1}{2} \cdot a^2$, so $\int_0^a x\,dx = \frac{a^2}{2}$.

5.2.17

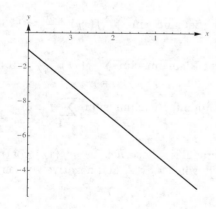

The left Riemann sum is $f(0) \cdot 1 + f(1) \cdot 1 + f(2) \cdot 1 + f(3) \cdot 1 = -1 - 3 - 5 - 7 = -16$.
The right Riemann sum is $f(1) \cdot 1 + f(2) \cdot 1 + f(3) \cdot 1 + f(4) \cdot 1 = -3 - 5 - 7 - 9 = -24$.
The midpoint Riemann sum is $f(.5) \cdot 1 + f(1.5) \cdot 1 + f(2.5) \cdot 1 + f(3.5) \cdot 1 = -2 - 4 - 6 - 8 = -20$.

5.2.19

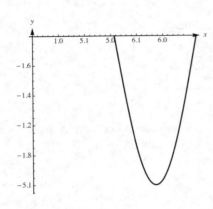

We have $\Delta x = \dfrac{\pi/2}{4} = \dfrac{\pi}{8}$. The left Riemann sum is
$f(\pi/2)\cdot\dfrac{\pi}{8}+f(5\pi/8)\cdot\dfrac{\pi}{8}+f(6\pi/8)\cdot\dfrac{\pi}{8}+f(7\pi/8)\cdot\dfrac{\pi}{8} =$
$\left(0 - \sqrt{2}/2 - 1 - \sqrt{2}/2\right)\cdot\dfrac{\pi}{8} = \dfrac{\pi}{8}\cdot(-1-\sqrt{2}) \approx$
-0.948.

The right Riemann sum is $f(5\pi/8)\cdot\dfrac{\pi}{8} +$
$f(6\pi/8)\cdot\dfrac{\pi}{8} + f(7\pi/8)\cdot\dfrac{\pi}{8} + f(\pi)\cdot\dfrac{\pi}{8} =$
$\left(-\sqrt{2}/2 - 1 - \sqrt{2}/2 - 0\right)\cdot\dfrac{\pi}{8} = \dfrac{\pi}{8}\cdot(-1-\sqrt{2}) \approx$
-0.948.

The midpoint Riemann sum is $f(9\pi/16)\cdot\dfrac{\pi}{8} +$
$f(11\pi/16)\cdot\dfrac{\pi}{8} + f(13\pi/16)\cdot\dfrac{\pi}{8} + f(15\pi/16)\cdot\dfrac{\pi}{8} \approx$
$\dfrac{\pi}{8}\cdot(-0.382683 - 0.92388 - 0.92388 - 0.382683) \approx$
-1.026.

5.2.21

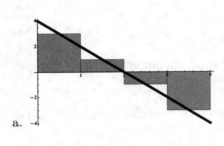

a.

b. The left Riemann sum $\sum_{k=0}^{3} f(x_k)\cdot 1 = 4$.

The right Riemann sum $\sum_{k=1}^{4} f(x_k)\cdot 1 = -4$.

The midpoint Riemann sum $\sum_{k=1}^{4} f(x_k^*)\cdot 1 = 0$.

c. The rectangles whose height is $f(x_k)$ contribute positively when $x_k < 2$ and negatively when $x_k > 2$.

5.2.23

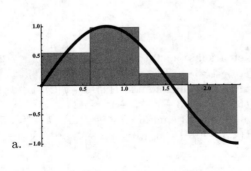

a.

b. The left Riemann sum $\sum_{k=0}^{3} f(x_k)\cdot \dfrac{3\pi}{16} \approx 0.7353$.

The right Riemann sum $\sum_{k=1}^{4} f(x_k)\cdot \dfrac{3\pi}{16} \approx 0.146$.

The midpoint Riemann sum $\sum_{k=1}^{4} f(x_k^*)\cdot \dfrac{3\pi}{16} \approx 0.530$.

c. The rectangles whose height is $f(x_k)$ contribute positively when $x_k < \dfrac{\pi}{2}$ and negatively when $x_k > \dfrac{\pi}{2}$.

5.2.25

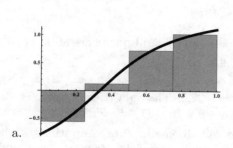

a.

b. The left Riemann sum $\sum_{k=0}^{3} f(x_k) \cdot \frac{1}{4} \approx 0.082$.

The right Riemann sum $\sum_{k=1}^{4} f(x_k) \cdot \frac{1}{4} \approx 0.555$.

The midpoint Riemann sum $\sum_{k=1}^{4} f(x_k^*) \cdot \frac{1}{4} \approx 0.326$.

c. The rectangles whose height is $f(x_k)$ contribute positively when $x_k > 1/3$ and negatively when $x_k < 1/3$.

5.2.27

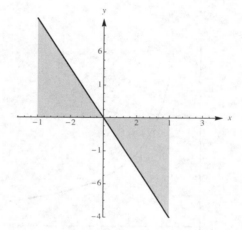

The region above the axis is a triangle with base 2 and height $f(-2) = 6$, and the region below the axis is a triangle with base 2 and height $-f(2) = 6$, so the net area is 0, and the area is 12.

5.2.29

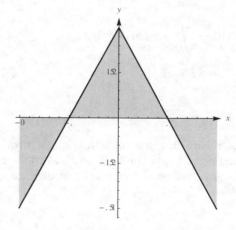

The region above the axis is a triangle with base 2 and height $f(0) = 1$, while the region below the axis consists of two triangles each with base 1 and height 1, so the net area is 0, and the area is 2.

5.2.31

a.
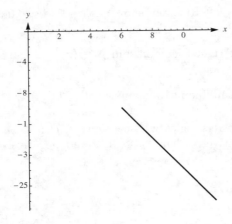

b. $\Delta x = \dfrac{1}{2}$, so the grid points are at 3, 3.5, 4, 4.5, 5, 5.5, and 6.

c. The left Riemann sum is $0.5(-5-6-7-8-9-10) = -22.5$. The right Riemann sum is $-.5(-6-7-8-9-10-11) = -25.5$.

d. The left Riemann sum overestimates the true value, while the right Riemann sum underestimates it.

5.2.33

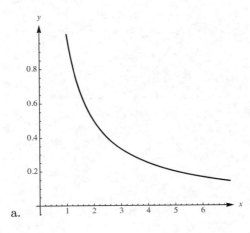

a.

b. $\Delta x = 1$, so the grid points are at 1, 2, 3, 4, 5, 6, and 7.

c. The left Riemann sum is approximately $1 + 0.5 + 0.333333 + 0.25 + -.2 + 0.166666 = 2.45$. The right Riemann sum is approximately $.5 + 0.333333 + 0.25 + 0.2 + 0.166666 + 0.142857 \approx 1.593$.

d. The left Riemann sum overestimates the true value, while the right Riemann sum underestimates it.

5.2.35 This is $\displaystyle\int_0^2 (x^2 + 1)\, dx$.

5.2.37 This is $\displaystyle\int_1^2 x \ln(x)\, dx$.

5.2.39

The region in question is a triangle with base 4 and height 8, so the area is $\frac{1}{2} \cdot 8 \cdot 4 = 16$, and this is therefore the value of the definite integral as well.

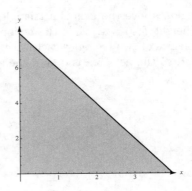

5.2. Definite Integrals

5.2.41

The region consists of two triangles, both below the axis. One has base 1 and height 1, the other has base 2 and height 2, so the net area is $-\frac{1}{2} \cdot 1 \cdot 1 - \frac{1}{2} \cdot 2 \cdot 2 = -2.5$.

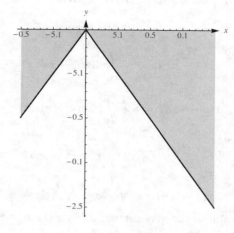

5.2.43

The region consists of a quarter circle of radius 4, situated above the axis. So the net area is $\frac{\pi \cdot 4^2}{4} = 4\pi$.

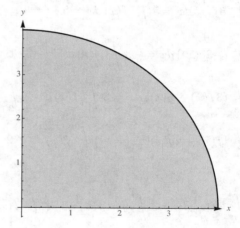

5.2.45

The region consists of a rectangle of area 10 above the axis, and a trapezoid of area 16 above the axis, so the net area is $10 + 16 = 26$.

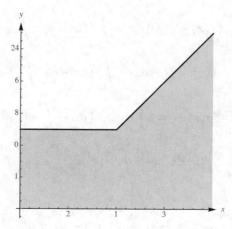

5.2.47 $\int_0^\pi x \sin x \, dx = A(R_1) + A(R_2) = 1 + \pi - 1 = \pi$.

5.2.49 $\int_0^{2\pi} x \sin x \, dx = A(R_1) + A(R_2) - A(R_3) - A(R_4) = 1 + \pi - 1 - \pi - 1 - 2\pi + 1 = -2\pi.$

5.2.51

a. $\int_4^0 3x(4-x) \, dx = -\int_0^4 3x(4-x) \, dx = -32.$

b. $\int_0^4 x(x-4) \, dx = -\frac{1}{3}\int_0^4 3x(4-x) \, dx = -\frac{1}{3} \cdot 32 = -\frac{32}{3}.$

c. $\int_4^0 6x(4-x) \, dx = -2 \cdot \int_0^4 3x(4-x) \, dx = -2 \cdot 32 = -64.$

d. $\int_0^8 3x(4-x) \, dx = \int_0^4 3x(4-x) \, dx + \int_4^8 3x(4-x) \, dx = 32 + \int_4^8 3x(4-x) \, dx.$ It is not possible to evaluate the given integral from the information given.

5.2.53

a. $\int_0^3 5f(x) \, dx = 5 \int_0^3 f(x) \, dx = 5 \cdot 2 = 10.$

b. $\int_3^6 (-3g(x)) \, dx = -3 \int_3^6 g(x) \, dx = -3 \cdot 1 = -3.$

c. $\int_3^6 (3f(x) - g(x)) \, dx = 3\int_3^6 f(x) \, dx - \int_3^6 g(x) \, dx = 3(-5) - 1 = -16.$

d. $\int_6^3 [f(x) + 2g(x)] \, dx = -\left[\int_3^6 f(x) \, dx + 2\int_3^6 g(x) \, dx\right] = -[-5 + 2 \cdot 1] = 3.$

5.2.55

a. $\int_1^4 3f(x) \, dx = 3\int_1^4 f(x) \, dx = 3 \cdot \left(\int_1^6 f(x) \, dx - \int_4^6 f(x) \, dx\right) = 3 \cdot (10 - 5) = 15.$

b. $\int_1^6 (f(x) - g(x)) \, dx = \int_1^6 f(x) \, dx - \int_1^6 g(x) \, dx = 10 - 5 = 5.$

c. $\int_1^4 (f(x) - g(x)) \, dx = \int_1^4 f(x) \, dx - \int_1^4 g(x) \, dx = \left(\int_1^6 f(x) \, dx - \int_4^6 f(x) \, dx\right) - 2 = (10 - 5) - 2 = 3.$

d. $\int_4^6 (g(x) - f(x)) \, dx = \int_4^6 g(x) \, dx - \int_4^6 f(x) \, dx = \left(\int_1^6 g(x) \, dx - \int_1^4 g(x) \, dx\right) - 5 = (5 - 2) - 5 = -2.$

e. $\int_4^6 8g(x) \, dx = 8\left(\int_1^6 g(x) \, dx - \int_1^4 g(x) \, dx\right) = 8(5 - 2) = 24.$

f. $\int_4^1 2f(x) \, dx = -2\int_1^4 f(x) \, dx = -2 \cdot \left(\int_1^6 f(x) \, dx - \int_4^6 f(x) \, dx\right) = -2(10 - 5) = -10.$

5.2.57

a. $\int_0^1 (4x - 2x^3) \, dx = -2\int_0^1 (x^3 - 2x) \, dx = -2 \cdot \left(-\frac{3}{4}\right) = \frac{3}{2}.$

b. $\int_1^0 (2x - x^3) \, dx = \int_0^1 (x^3 - 2x) \, dx = -\frac{3}{4}.$

5.2. Definite Integrals

5.2.59 $\int_0^a f(x)\,dx = 16$.

5.2.61 $\int_a^c f(x)\,dx = 11 - 5 = 6$.

5.2.63 $\int_0^c |f(x)|\,dx = 16 + 5 + 11 = 32$.

5.2.65 $\int_a^0 f(x)\,dx = -\int_0^a f(x)\,dx = -16$.

5.2.67 $\int_0^1 (2x + \sqrt{1-x^2} + 1)\,dx = \int_0^1 2x\,dx + \int_0^1 \sqrt{1-x^2}\,dx + \int_0^1 1\,dx$. The first integral in this sum represents the area of a triangle with base 1 and height 2 (which has value 1), the second represents the area of a quarter of a circle of radius 1 (which has value $\frac{\pi}{4}$, and the third represents the area of a 1×1 square (which has value 1). So the integral's value is $1 + \frac{\pi}{4} + 1 = 2 + \frac{\pi}{4}$.

5.2.69

a. True. See problem 78 in the previous section for a proof.

b. True. See problem 79 in the previous section for a proof.

c. True. Because both of those function are periodic with period $\frac{2\pi}{a}$, and both have the same amount of area above the axis as below for one period, the net area of each between 0 and $\frac{2\pi}{a}$ is zero.

d. False. For example $\int_0^{2\pi} \sin x\,dx = 0 = \int_{2\pi}^0 \sin x\,dx$, but $\sin x$ is not a constant function.

e. False. Because x is not a constant, it can not be factored outside of the integral. For example $\int_0^1 x \cdot 1\,dx \neq x \int_0^1 1\,dx$.

5.2.71

a. $\sum_{k=1}^{20} \left(\left(\frac{k-1}{20}\right)^2 + 1\right) \cdot \frac{1}{20} \approx 1.309$. $\sum_{k=1}^{20} \left(\left(\frac{k}{20}\right)^2 + 1\right) \cdot \frac{1}{20} \approx 1.359$.

$\sum_{k=1}^{50} \left(\left(\frac{k-1}{50}\right)^2 + 1\right) \cdot \frac{1}{50} \approx 1.323$. $\sum_{k=1}^{50} \left(\left(\frac{k}{50}\right)^2 + 1\right) \cdot \frac{1}{50} \approx 1.343$.

$\sum_{k=1}^{100} \left(\left(\frac{k-1}{100}\right)^2 + 1\right) \cdot \frac{1}{100} \approx 1.328$. $\sum_{k=1}^{100} \left(\left(\frac{k}{100}\right)^2 + 1\right) \cdot \frac{1}{100} \approx 1.338$.

b. It appears that the integral's value is about $\frac{4}{3}$.

5.2.73

a. $\sum_{k=1}^{20} \cos^{-1}((k-1)/20) \cdot \frac{1}{20} \approx 1.036$. $\sum_{k=1}^{20} \cos^{-1}(k/20) \cdot \frac{1}{20} \approx 0.958$.

$\sum_{k=1}^{50} \cos^{-1}((k-1)/50) \cdot \frac{1}{50} \approx 1.015$. $\sum_{k=1}^{50} \cos^{-1}(k/50) \cdot \frac{1}{50} \approx 0.983$.

$\sum_{k=1}^{100} \cos^{-1}((k-1)/100) \cdot \frac{1}{100} \approx 1.008$. $\sum_{k=1}^{100} \cos^{-1}(k/100) \cdot \frac{1}{100} \approx 0.992$.

b. It appears that the integral's value is 1.

5.2.75

a. $\sum_{k=1}^{n} 2\sqrt{1 + \dfrac{3}{2n} + \dfrac{3(k-1)}{n}} \cdot \dfrac{3}{n}.$

b.
n	20	50	100
Midpoint Sum	9.3338	9.33341	9.33335

It appears that the integral's value is about $\dfrac{28}{3}$.

5.2.77

a. $\sum_{k=1}^{n} \left(4\left(\dfrac{2}{n} + \dfrac{4(k-1)}{n}\right) - \left(\dfrac{2}{n} + \dfrac{4(k-1)}{n}\right)^2 \right) \cdot \dfrac{4}{n}.$

b.
n	20	50	100
Midpoint Sum	10.68	10.6688	10.6672

It appears that the integral's value is about $\dfrac{32}{3}$.

5.2.79

$$\int_0^2 (2x+1)\,dx = \lim_{n\to\infty} \sum_{k=1}^n f(x_k)\Delta x = \lim_{n\to\infty} \sum_{k=1}^n \left[2\left(\dfrac{2k}{n}\right)+1\right]\dfrac{2}{n}$$

$$= \lim_{n\to\infty} \left[\dfrac{8}{n^2}\sum_{k=1}^n k + \dfrac{2}{n}\sum_{k=1}^n 1\right]$$

$$= \lim_{n\to\infty} \left[\dfrac{8}{n^2}\cdot \dfrac{n(n+1)}{2} + \dfrac{2}{n}\cdot n\right]$$

$$= \lim_{n\to\infty} \left[\dfrac{4(n+1)}{n} + 2\right] = 4+2 = 6.$$

5.2.81

$$\int_3^7 (4x+6)\,dx = \lim_{n\to\infty} \sum_{k=1}^n f(x_k)\Delta x = \lim_{n\to\infty} \sum_{k=1}^n \left[4\left(3+\dfrac{4k}{n}\right)+6\right]\dfrac{4}{n}$$

$$= \lim_{n\to\infty} \left[\dfrac{64}{n^2}\sum_{k=1}^n k + \dfrac{72}{n}\sum_{k=1}^n 1\right]$$

$$= \lim_{n\to\infty} \left[\dfrac{64}{n^2}\cdot \dfrac{n(n+1)}{2} + \dfrac{72}{n}\cdot n\right]$$

$$= \lim_{n\to\infty} \left[\dfrac{32(n+1)}{n} + 72\right] = 104.$$

5.2.83

$$\int_1^4 (x^2 - 1)\,dx = \lim_{n\to\infty} \sum_{k=1}^n f(x_k)\Delta x = \lim_{n\to\infty} \sum_{k=1}^n \left[\left(1 + \frac{3k}{n}\right)^2 - 1\right]\frac{3}{n}$$

$$= \lim_{n\to\infty} \left[\frac{18}{n^2}\sum_{k=1}^n k + \frac{27}{n^3}\sum_{k=1}^n k^2\right]$$

$$= \lim_{n\to\infty} \left[\frac{18}{n^2}\cdot\frac{n(n+1)}{2} + \frac{27}{n^3}\cdot\frac{n(n+1)(2n+1)}{6}\right]$$

$$= \lim_{n\to\infty} \left[\frac{9(n+1)}{n} + \frac{18n^2 + 27n + 9}{2n^2}\right] = 9 + 9 = 18.$$

5.2.85

$$\int_0^1 (4x^3 + 3x^2)\,dx = \lim_{n\to\infty} \sum_{k=1}^n f(x_k)\Delta x = \lim_{n\to\infty} \frac{1}{n}\sum_{k=1}^n \left(4\left(\frac{k}{n}\right)^3 + 3\left(\frac{k}{n}\right)^2\right)$$

$$= \lim_{n\to\infty} \left(\frac{4}{n^4}\sum_{k=1}^n k^3 + \frac{3}{n^3}\sum_{k=1}^n k^2\right)$$

$$= \lim_{n\to\infty} \left(\frac{4}{n^4}\left(\frac{n^2(n+1)^2}{4}\right) + \frac{3}{n^3}\left(\frac{n(n+1)(2n+1)}{6}\right)\right)$$

$$= 1 + 1 = 2.$$

5.2.87

The region in question is a semicircle above the axis with radius 5, so the area is $\frac{1}{2}\pi\cdot 5^2 = \frac{25\pi}{2}$.

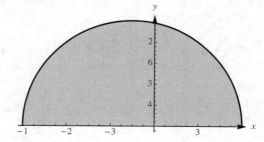

5.2.89 $\int_0^{10} f(x)\,dx = \int_0^5 2\,dx + \int_5^{10} 3\,dx = 10 + 15 = 25.$

5.2.91 $\int_1^5 x\lfloor x\rfloor\,dx = \int_1^2 x\lfloor x\rfloor\,dx + \int_2^3 x\lfloor x\rfloor\,dx + \int_3^4 x\lfloor x\rfloor\,dx + \int_4^5 x\lfloor x\rfloor\,dx = \int_1^2 x\,dx + \int_2^3 2x\,dx + \int_3^4 3x\,dx + \int_4^5 4x\,dx.$ Each of these integrals represents the area of a trapezoid with base 1. The value of the integral is $\frac{1+2}{2} + \frac{4+6}{2} + \frac{9+12}{2} + \frac{16+20}{2} = 35.$

5.2.93

$$\int_a^b cf(x)\,dx = \lim_{\Delta\to 0} \sum_{k=1}^n cf(x_k^*)\Delta x_k$$

$$= \lim_{\Delta\to 0} c\sum_{k=1}^n f(x_k^*)\Delta x_k$$

$$= c\lim_{\Delta\to 0} \sum_{k=1}^n f(x_k^*)\Delta x_k = c\int_a^b f(x)\,dx.$$

5.2.95 Let n be a positive integer. Let $\Delta x = \dfrac{1}{n}$. Note that each grid point $\dfrac{k}{n}$ for $0 \leq k \leq n$ where i is an integer is a rational number. So $f(x_k) = 1$ for each grid point. So the right Riemann sum is $\sum_{k=1}^{n} f(x_k) \dfrac{1}{n} = \dfrac{1}{n} \sum_{k=1}^{n} 1 = \dfrac{1}{n} \cdot n = 1$. The left Riemann sum calculation is similar, as is the midpoint Riemann sum calculation (because the grid midpoints are also rational numbers – they are the average of two rational numbers and hence are rational as well).

5.2.97

a. Note that for all values of $k = 1, 2, \ldots, n$, we have $x_{k-1} x_{k-1} \leq x_{k-1} x_k$, so $\sqrt{x_{k-1} x_{k-1}} \leq \sqrt{x_{k-1} x_k}$, and thus $x_{k-1} \leq \sqrt{x_{k-1} x_k}$. Similarly, $x_{k-1} x_k \leq x_k x_k$, so $\sqrt{x_{k-1} x_k} \leq \sqrt{x_k x_k} = x_k$, so $\sqrt{x_{k-1} x_k} \leq x_k$. Thus $x_{k-1} \leq \sqrt{x_{k-1} x_k} \leq x_k$ for all $k = 1, 2, \ldots, n$.

b. $\dfrac{1}{x_{k-1}} - \dfrac{1}{x_k} = \dfrac{x_k}{x_{k-1} x_k} - \dfrac{x_{k-1}}{x_{k-1} x_k} = \dfrac{x_k - x_{k-1}}{x_{k-1} x_k} = \dfrac{\Delta x_k}{x_{k-1} x_k}$, for all $k = 1, 2, \ldots, n$.

c. The Riemann sum is $\sum_{k=1}^{n} \dfrac{\Delta x_k}{\overline{x_k}^2}$. Using $\overline{x_k} = \sqrt{x_{k-1} x_k}$, we have

$$\sum_{k=1}^{n} \dfrac{\Delta x_k}{x_{k-1} x_k} = \sum_{k=1}^{n} \left(\dfrac{1}{x_{k-1}} - \dfrac{1}{x_k} \right),$$

where the last equality follows from part (b). Now note that the sum telescopes (that is, has many canceling terms).

$$\sum_{k=1}^{n} \left(\dfrac{1}{x_{k-1}} - \dfrac{1}{x_k} \right) = \left(\dfrac{1}{x_0} - \dfrac{1}{x_1} \right) + \left(\dfrac{1}{x_1} - \dfrac{1}{x_2} \right) + \cdots + \left(\dfrac{1}{x_{n-2}} - \dfrac{1}{x_{n-1}} \right) + \left(\dfrac{1}{x_{n-1}} - \dfrac{1}{x_n} \right)$$

$$= \dfrac{1}{x_0} + \left(-\dfrac{1}{x_1} + \dfrac{1}{x_1} \right) + \left(-\dfrac{1}{x_2} + \dfrac{1}{x_2} \right) + \cdots + \left(-\dfrac{1}{x_{n-1}} + \dfrac{1}{x_{n-1}} \right) - \dfrac{1}{x_n}$$

$$= \dfrac{1}{x_0} - \dfrac{1}{x_n}$$

$$= \dfrac{1}{a} - \dfrac{1}{b}.$$

d. The integral is the limit of the Riemann sum as $n \to \infty$. Thus we have

$$\int_a^b \dfrac{dx}{x^2} = \lim_{n \to \infty} \sum_{k=1}^{n} \dfrac{\Delta x_k}{\overline{x_k}^2} = \lim_{n \to \infty} \left(\dfrac{1}{a} - \dfrac{1}{b} \right) = \dfrac{1}{a} - \dfrac{1}{b}.$$

5.3 Fundamental Theorem of Calculus

5.3.1 A is also an antiderivative of f.

5.3.3 The fundamental theorem says that $\int_a^b f(x)\,dx = F(b) - F(a)$ where F is any antiderivative of f. So to evaluate $\int_a^b f(x)\,dx$, one could find an antiderivative $F(x)$, and then evaluate this at a and b and then subtract, obtaining $F(b) - F(a)$.

5.3. Fundamental Theorem of Calculus

5.3.5

$A(x) = \int_0^x (3-t)\,dt$ represents the area between 0 and x and below this curve. As x increases (but remains less than 3), the trapezoidal region's area increases, so the area function increases until x is 3.

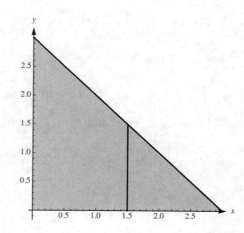

5.3.7 $\dfrac{d}{dx}\int_a^x f(t)\,dt = f(x)$, and $\int f'(x)\,dx = f(x) + C$.

5.3.9 $\dfrac{d}{dx}\int_a^x f(t)\,dt = f(x)$, and $\dfrac{d}{dx}\int_a^b f(t)\,dt = 0$. The latter is the derivative of a constant, the former follows from the Fundamental Theorem.

5.3.11 $\int_3^8 f'(t)\,dt = f(8) - f(3) = 20 - 4 = 16$.

5.3.13

a. $A(-2) = \int_{-2}^{-2} f(t)\,dt = 0$.

b. $F(8) = \int_4^8 f(t)\,dt = -9$.

c. $A(4) = \int_{-2}^4 f(t)\,dt = 8 + 17 = 25$.

d. $F(4) = \int_4^4 f(t)\,dt = 0$.

e. $A(8) = \int_{-2}^8 f(t)\,dt = 25 - 9 = 16$.

5.3.15

a. $A(x) = \int_0^x f(t)\,dt = \int_0^x 5\,dt = 5x$.

b. $A'(x) = 5 = f(x)$.

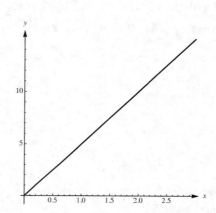

5.3.17

a. $A(2) = \int_0^2 t\, dt = 2$. $A(4) = \int_0^4 t\, dt = 8$. Because the region whose area is $A(x) = \int_0^x t\, dt$ is a triangle with base x and height x, its value is $\frac{1}{2}x^2$.

b. $F(4) = \int_2^4 t\, dt = 6$. $F(6) = \int_2^6 t\, dt = 16$. Because the region whose area is $A(x) = \int_2^x t\, dt$ is a trapezoid with base $x - 2$ and $h_1 = 2$ and $h_2 = x$, its value is $(x-2)\frac{2+x}{2} = \frac{x^2-4}{2} = \frac{x^2}{2} - 2$.

c. We have $A(x) - F(x) = \frac{x^2}{2} - (\frac{x^2}{2} - 2) = 2$, a constant.

5.3.19

a. The region is a triangle with base $x + 5$ and height $x + 5$, so its area is $A(x) = \frac{1}{2}(x+5)^2$.

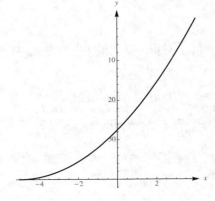

b. $A'(x) = x + 5 = f(x)$.

5.3.21

a. The region is a trapezoid with base $x - 2$ and heights $h_1 = f(2) = 7$ and $h_2 = f(x) = 3x + 1$, so its area is $A(x) = (x-2) \cdot \frac{7 + 3x + 1}{2} = (x-2) \cdot (\frac{3}{2}x + 4) = \frac{3}{2}x^2 + x - 8$.

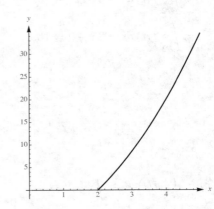

b. $A'(x) = 3x + 1 = f(x)$.

5.3.23 $\int_0^1 (x^2 - 2x + 3)\, dx = \left(\frac{x^3}{3} - x^2 + 3x\right)\Big|_0^1 = \frac{1}{3} - 1 + 3 - (0 - 0 + 0) = \frac{7}{3}$. It does appear that the area is between 2 and 3.

5.3.25

$\int_{-2}^{3}(x^2 - x - 6)\,dx = \left(\dfrac{x^3}{3} - \dfrac{x^2}{2} - 6x\right)\Big|_{-2}^{3} = -\dfrac{125}{6}.$

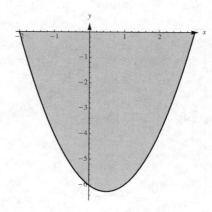

5.3.27

$\int_{0}^{5}(x^2 - 9)\,dx = \left(\dfrac{x^3}{3} - 9x\right)\Big|_{0}^{5} = \dfrac{125}{3} - 45 - (0 - 0) = -\dfrac{10}{3}.$

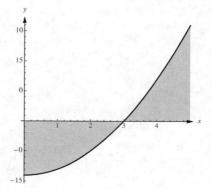

5.3.29 $\int_{0}^{2} 4x^3\,dx = x^4\Big|_{0}^{2} = 16 - 0 = 16.$

5.3.31 $\int_{1}^{8} 8x^{1/3}\,dx = 6x^{4/3}\Big|_{1}^{8} = 6(16 - 1) = 90.$

5.3.33 $\int_{0}^{1}(x + \sqrt{x})\,dx = \left(\dfrac{x^2}{2} + \dfrac{2x^{3/2}}{3}\right)\Big|_{0}^{1} = \dfrac{1}{2} + \dfrac{2}{3} - (0 + 0) = \dfrac{7}{6}.$

5.3.35 $\int_{1}^{9} \dfrac{2}{\sqrt{x}}\,dx = \int_{1}^{9} 2x^{-1/2}\,dx = 4x^{1/2}\Big|_{1}^{9} = 12 - 4 = 8.$

5.3.37 $\int_{-2}^{2}(x^2 - 4)\,dx = \left(\dfrac{x^3}{3} - 4x\right)\Big|_{-2}^{2} = \dfrac{8}{3} - 8 - \left(-\dfrac{8}{3} + 8\right) = \dfrac{16}{3} - 16 = -\dfrac{32}{3}.$

5.3.39 $\int_{1/2}^{1}(x^{-3} - 8)\,dx = \left(\dfrac{x^{-2}}{-2} - 8x\right)\Big|_{1/2}^{1} = -\dfrac{1}{2} - 8 - (-2 - 4) = -\dfrac{5}{2}.$

5.3.41 $\int_{1}^{4}(1 - x)(x - 4)\,dx = \int_{1}^{4}(-x^2 + 5x - 4)\,dx = \left(-\dfrac{x^3}{3} + \dfrac{5x^2}{2} - 4x\right)\Big|_{1}^{4} = \dfrac{9}{2}.$

5.3.43 $\int_{-2}^{-1} x^{-3}\,dx = \dfrac{x^{-2}}{-2}\Big|_{-2}^{-1} = -\dfrac{1}{2x^2}\Big|_{-2}^{-1} = -\dfrac{1}{2} - \left(-\dfrac{1}{8}\right) = -\dfrac{3}{8}.$

5.3.45 $\int_0^{\pi/4} \sec^2 \theta \, d\theta = \tan \theta \Big|_0^{\pi/4} = 1 - 0 = 1.$

5.3.47 $\int_1^2 \frac{3}{t} \, dt = 3 \ln|t| \Big|_1^2 = 3 \ln 2 - 3 \ln 1 = \ln 8.$

5.3.49 $\int_1^8 \sqrt[3]{y} \, dy = \frac{3}{4} y^{4/3} \Big|_1^8 = 12 - \frac{3}{4} = \frac{45}{4}.$

5.3.51
$$\int_1^4 \frac{x-2}{\sqrt{x}} \, dx = \int_1^4 \left(\frac{x}{\sqrt{x}} - \frac{2}{\sqrt{x}} \right) dx = \int_1^4 \left(x^{1/2} - 2x^{-1/2} \right) dx$$
$$= \left(\frac{2}{3} x^{3/2} - 4 x^{1/2} \right) \Big|_1^4 = \frac{16}{3} - 8 - \left(\frac{2}{3} - 4 \right) = \frac{14}{3} - \frac{12}{3} = \frac{2}{3}.$$

5.3.53 $\int_0^{\pi/3} \sec x \tan x \, dx = \sec x \Big|_0^{\pi/3} = 2 - 1 = 1.$

5.3.55 $\int_{\pi/4}^{3\pi/4} (\cot^2 x + 1) \, dx = \int_{\pi/4}^{3\pi/4} \csc^2 x \, dx = -\cot x \Big|_{\pi/4}^{3\pi/4} = -(-1 - 1) = 2.$

5.3.57 $\int_1^{\sqrt{3}} \frac{1}{1 + x^2} \, dx = \tan^{-1} \Big|_1^{\sqrt{3}} = \tan^{-1} \sqrt{3} - \tan^{-1} 1 = \pi/3 - \pi/4 = \pi/12.$

5.3.59 $\int_1^2 \frac{z^2 + 4}{z} \, dz = \int_1^2 \left(z + \frac{4}{z} \right) dz = \left(\frac{z^2}{2} + 4 \ln z \right) \Big|_1^2 = 2 + 4 \ln 2 - \left(\frac{1}{2} + 0 \right) = \ln 16 + \frac{3}{2}.$

5.3.61 $\int_0^\pi f(x) \, dx = \int_0^{\pi/2} (\sin x + 1) \, dx + \int_{\pi/2}^\pi (2 \cos x + 2) \, dx = (-\cos x + x) \Big|_0^{\pi/2} + (2 \sin x + 2x) \Big|_{\pi/2}^\pi =$
$(0 + \pi/2) - (-1 + 0) + (0 + 2\pi) - (2 + \pi) = 3\pi/2 - 1.$

5.3.63

The area (and net area) of this region is given by $\int_1^4 \sqrt{x} \, dx = \frac{2}{3} x^{3/2} \Big|_1^4 = \frac{16}{3} - \frac{2}{3} = \frac{14}{3}.$

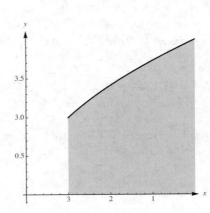

5.3.65

The net area of this region is given by
$$\int_{-2}^{2}(x^4-16)\,dx = \left(\frac{x^5}{5}-16x\right)\Big|_{-2}^{2} =$$
$$\frac{32}{5}-32-\left(-\frac{32}{5}+32\right) = \frac{64}{5}-64 = -\frac{256}{5}.$$
Thus the area is $\frac{256}{5}$.

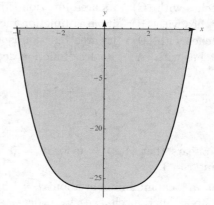

5.3.67 Because this region is below the axis, the area of it is given by $-\int_{2}^{4}(x^2-25)\,dx = -\left(\frac{x^3}{3}-25x\right)\Big|_{2}^{4} = -\left(\frac{64}{3}-100-\left(\frac{8}{3}-50\right)\right) = 50-\frac{56}{3} = \frac{94}{3}.$

5.3.69 Because this region is below the axis, the area of it is given by $-\int_{-2}^{-1}\frac{1}{x}\,dx = -\left(\ln|x|\Big|_{-2}^{-1}\right) = \ln 2 - \ln 1 = \ln 2.$

5.3.71 The area is given by

$$-\int_{-\pi/4}^{0}\sin x\,dx + \int_{0}^{3\pi/4}\sin x\,dx = \left(\cos x\Big|_{-\pi/4}^{0}\right) + \left(-\cos x\Big|_{0}^{3\pi/4}\right) = \left(1-\frac{\sqrt{2}}{2}\right) + \left(1+\frac{\sqrt{2}}{2}\right) = 2.$$

5.3.73 By a direct application of the Fundamental Theorem, this is $x^2 + x + 1$.

5.3.75 This is $-\dfrac{d}{dx}\int_{1}^{x}\sqrt{t^4+1}\,dt = -\sqrt{x^4+1}$.

5.3.77 By the Fundamental Theorem and the chain rule, this is $\dfrac{1}{x^6}\cdot 3x^2 = \dfrac{3}{x^4}$.

5.3.79 $\dfrac{d}{dx}\int_{0}^{\cos x}(t^4+6)\,dt = -(\cos^4 x + 6)\sin x.$

5.3.81 $\dfrac{d}{dz}\int_{\sin z}^{10}\dfrac{dt}{t^4+1} = -\dfrac{d}{dz}\int_{10}^{\sin z}\dfrac{dt}{t^4+1} = -\dfrac{1}{\sin^4+1}\cdot\cos z = -\dfrac{\cos z}{\sin^4 z+1}.$

5.3.83 $\dfrac{d}{dt}\left(\int_{1}^{t}\dfrac{3}{x}\,dx - \int_{t^2}^{1}\dfrac{3}{x}\,dx\right) = \dfrac{d}{dt}\int_{1}^{t}\dfrac{3}{x}\,dx + \dfrac{d}{dt}\int_{1}^{t^2}\dfrac{3}{x}\,dx = \dfrac{3}{t}+\dfrac{6t}{t^2} = \dfrac{9}{t}.$

5.3.85 This can be written as

$$\dfrac{d}{dx}\left(\int_{-x}^{0}\sqrt{1+t^2}\,dt + \int_{0}^{x}\sqrt{1+t^2}\,dt\right) = \dfrac{d}{dx}\left(-\int_{0}^{-x}\sqrt{1+t^2}\,dt + \int_{0}^{x}\sqrt{1+t^2}\,dt\right)$$
$$= -\sqrt{1+(-x)^2}(-1) + \sqrt{1+x^2} = 2\sqrt{1+x^2}.$$

5.3.87
 (a) matches with (C) – its area function is increasing linearly.
 (b) matches with (B) – its area function increases then decreases.
 (c) matches with (D) – its area function is always increasing on $[0, b]$, although not linearly.
 (d) matches with (A) – its area function decreases at first and then eventually increases.

5.3.89

a. It appears that $A(x) = 0$ for $x = 0$ and at about $x = 3$.

b. A has a local minimum at about $x = 1.5$ where the area function changes from decreasing to increasing, and a local max at around $x = 8.5$ where the area function changes from increasing to decreasing.

c.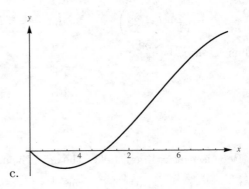

5.3.91

a. It appears that $A(x) = 0$ for $x = 0$ and $x = 10$.

b. A has a local maximum at $x = 5$ where the area function changes from increasing to decreasing.

c.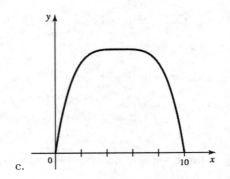

5.3.93 $A(2) = -\frac{1}{4}\pi \cdot 2^2 = -\pi$. $A(5) = -\pi + \frac{1}{2} \cdot 3 \cdot 3 = \frac{9}{2} - \pi$. $A(8) = \frac{9}{2} - \pi + \frac{1}{2} \cdot 3 \cdot 3 = 9 - \pi$. $A(12) = 9 - \pi - \frac{1}{2} \cdot 4 \cdot 2 = 5 - \pi$.

5.3.95

a. $A(x) = \int_0^x e^t \, dt = e^t \Big|_0^x = e^x - (1)$.

c. $A(\ln 2) = e^{\ln 2} - 1 = 2 - 1 = 1$. $A(\ln 4) = e^{\ln 4} - 1 = 4 - 1 = 3$. There is twice as much area under the curve between $\ln 2$ and $\ln 4$ as there is between 0 and $\ln 2$.

b.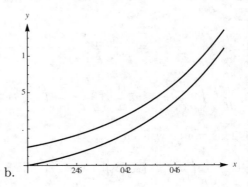

5.3.97

a. $A(x) = \int_0^x \cos t\, dt = \sin t \Big|_0^x = \sin x$.

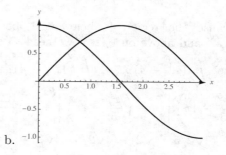

b.

c. $A(\pi/2) = 1$. $A(\pi) = 0$.

5.3.99 $f'(x) = x^2(x-3)(x-4)$, so the critical points of f are $x = 0$, $x = 3$, and $x = 4$. $f' > 0$ on $(-\infty, 0)$ and on $(0, 3)$ and on $(4, \infty)$, so f is increasing on those intervals, while $f' < 0$ on $(3, 4)$, so f is decreasing on that interval.

5.3.101

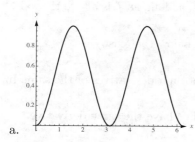

a.

b. $g'(x) = \sin^2 x$.

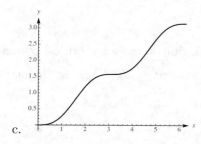

c.

Note that g' is always positive, so g is always increasing. There are inflection points where g' changes from increasing to decreasing, and vice versa.

5.3.103

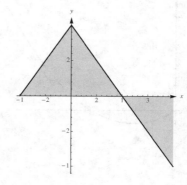

We can use geometry – there is a triangle with base 4 and height 2 and a triangle with base 2 and height 2, so the total area is $\frac{1}{2} \cdot 4 \cdot 2 + \frac{1}{2} \cdot 2 \cdot 2 = 6$.

5.3.105

Because the region is below the axis on $[1, \sqrt{2}]$ and above on $[\sqrt{2}, 4]$ we need to compute $\int_{\sqrt{2}}^{4} (x^4 - 4)\, dx - \int_{1}^{\sqrt{2}} (x^4 - 4)\, dx = \left(\dfrac{x^5}{5} - 4x\right)\Big|_{\sqrt{2}}^{4} - \left(\dfrac{x^5}{5} - 4x\right)\Big|_{1}^{\sqrt{2}} = \dfrac{1024}{5} - 16 - (4\sqrt{2}/5 - 4\sqrt{2}) - (4\sqrt{2}/5 - 4\sqrt{2}) + \dfrac{1}{5} - 4 = 185 + \dfrac{32\sqrt{2}}{5}$

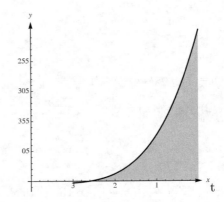

5.3.107

a. True. The net area under the curve increases as x increases, as long as f is above the axis.

b. True. The net area decreases as x increases, as long as f is below the axis.

c. False. These do not have the same derivative, so they are not antiderivatives of the same function.

d. True, because the two functions differ by a constant, and thus have the same derivative.

e. True, because the derivative of a constant is zero.

5.3.109 Because $\dfrac{d}{db}\int_{-1}^{b} x^2(3-x)\, dx = b^2(3-b)$ we see that this function of b has critical points at $b = 0$ and $b = 3$. Note also that the integrand is positive on $[0, 3]$, but is negative on $[3, \infty)$. So the maximum for this area function occurs at $b = 3$.

5.3.111

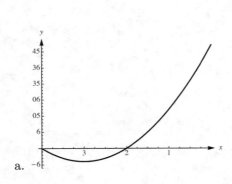

a.

b. We seek b so that $\int_{0}^{b}(x^2 - 4x) = 0$ for $b > 0$. We have $\left(\dfrac{x^3}{3} - 2x^2\right)\Big|_{0}^{b} = \dfrac{b^3}{3} - 2b^2 = 0$, which occurs for $\dfrac{b}{3} = 2$, or $b = 6$.

c. We seek b so that $\int_{0}^{b}(x^2 - ax) = 0$ for $b > 0$. We have $\left(\dfrac{x^3}{3} - \dfrac{ax^2}{2}\right)\Big|_{0}^{b} = \dfrac{b^3}{3} - \dfrac{ab^2}{2} = 0$, which occurs for $\dfrac{b}{3} = \dfrac{a}{2}$, or $b = \dfrac{3a}{2}$.

5.3.113 Differentiating both sides of the given equation yields $f(x) = -2\sin x + 3$.

Using a computer or calculator, we obtain:

x	500	1000	1500	2000
$S(x)$	1.5726	1.57023	1.57087	1.57098

5.3.115 This appears to be approaching $\frac{\pi}{2}$.
Note that between 0 and π, the area is approximately half the area of a rectangle with height 1 and base π, and then from π on there is approximately as much area above the axis as below.

5.3.117 By the Fundamental Theorem, $S'(x) = \sin x^2$, so $S''(x) = 2x \cos x^2$, so

$$S'(x)^2 + \left(\frac{S''(x)}{2x}\right)^2 = \sin^2 x^2 + \cos^2 x^2 = 1.$$

5.3.119

a. By definition of Reimann sums, $\int_a^b f'(x)\,dx$ is approximated by $\sum_{k=1}^{n} f'(x_{k-1})\Delta x$. But $f'(x_{k-1}) = \lim_{h \to 0} \frac{f(x_{k-1} + h) - f(x_{k-1})}{h}$. If $h = \Delta x$, then we have

$$f'(x_{k-1}) \approx \frac{f(x_{k-1} + \Delta x) - f(x_{k-1})}{\Delta x} = \frac{f(x_k) - f(x_{k-1})}{\Delta x},$$

so that

$$\int_a^b f'(x)\,dx \approx \sum_{k=1}^{n} \frac{f(x_k) - f(x_{k-1})}{\Delta x} \cdot \Delta x.$$

b. Canceling the Δx factors we obtain

$$\int_a^b f'(x)\,dx \approx \sum_{k=1}^{n} \frac{f(x_k) - f(x_{k-1})}{\Delta x} \cdot \Delta x$$
$$= \sum_{k=1}^{n} (f(x_k) - f(x_{k-1}))$$
$$= (f(x_1) - f(x_0)) + (f(x_2) - f(x_1)) + \cdots + (f(x_{n-1}) - f(x_{n-2})) + (f(x_n) - f(x_{n-1}))$$
$$= f(x_n) - f(x_0) = f(b) - f(a).$$

c. The analogy between the two situations is that both (a) the sum of difference quotients and (b) integral of a derivative are equal to the difference in function values at the endpoints.

5.4 Working with Integrals

5.4.1 If f is odd, it is symmetric about the origin, which guarantees that between $-a$ and a, there is as much area above the axis and under f as there is below the axis and above f, so the net area must be 0.

5.4.3

a. Because $\int_{-8}^{8} f(x)\,dx = 18 = 2\int_{0}^{8} f(x)\,dx$, we have $\int_{0}^{8} f(x)\,dx = \frac{18}{2} = 9$.

b. Because $xf(x)$ is an odd function when $f(x)$ is even, we have $\int_{-8}^{8} xf(x)\,dx = 0$.

5.4.5 The integrand can be written as $(5x^4 + 2x^2 + 1) + (3x^3 + x)$. The function $f(x) = 5x^4 + 2x^2 + 1$ is an even function and the function $g(x) = 3x^3 + x$ is an odd function. Thus,

$$\int_{-4}^{4} (f(x) + g(x))\,dx = \int_{-4}^{4} f(x)\,dx + \int_{-4}^{4} g(x)\,dx = 2\int_{0}^{4} f(x)\,dx + 0 = 2\int_{0}^{4} (5x^4 + 2x^2 + 1)\,dx.$$

5.4.7 $f(x) = x^{12}$ is an even function, because $f(-x) = (-x)^{12} = x^{12} = f(x)$. $g(x) = \sin x^2$ is also even, because $g(-x) = \sin((-x)^2) = \sin x^2 = g(x)$.

5.4.9 The average value of a continuous function on a closed interval $[a, b]$ will always be between the maximum and the minimum value of f on that interval. Because the function is continuous, the Intermediate Value Theorem assures us that the function will take on each value between the maximum and the minimum somewhere on the interval.

5.4.11 Because x^9 is an odd function, $\int_{-2}^{2} x^9\,dx = 0$.

5.4.13 $\int_{-2}^{2} (3x^8 - 2)\,dx = 2\int_{0}^{2} (3x^8 - 2)\,dx = 2\left(\dfrac{x^9}{3} - 2x\right)\Big|_{0}^{2} = \left(\dfrac{1024}{3}\right) - 8 = \dfrac{1000}{3}$.

5.4.15 $\int_{-2}^{2} (x^2 + x^3)\,dx = 2\int_{0}^{2} x^2\,dx = 2\left(\dfrac{x^3}{3}\right)\Big|_{0}^{2} = 2\left(\dfrac{8}{3} - 0\right) = \dfrac{16}{3}$.

5.4.17 Note that the first two terms of the integrand form an odd function, and the last two terms form an even function. $\int_{-2}^{2} (x^9 - 3x^5 + 2x^2 - 10)\,dx = 2\int_{0}^{2} (2x^2 - 10)\,dx = 2\left(\dfrac{2x^3}{3} - 10x\right)\Big|_{0}^{2} = \dfrac{32}{3} - 40 = -\dfrac{88}{3}$.

5.4.19 Because the integrand is an odd function and the interval is symmetric about 0, this integral's value is 0.

5.4.21 $\sec^2 x$ is even, so the value of this integral is $2\int_{0}^{\pi/4} \sec^2 x\,dx = 2(\tan x)\Big|_{0}^{\pi/4} = 2\cdot(1 - 0) = 2$.

5.4.23 The integrand is an odd function, so the value of this integral is zero.

5.4.25

The average value is $\dfrac{1}{1-(-1)}\int_{-1}^{1} x^3\,dx = \dfrac{1}{2}(x^4/4)\Big|_{-1}^{1} = 0$.

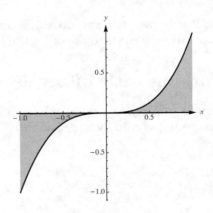

5.4. Working with Integrals

5.4.27

The average value is
$$\frac{1}{1-(-1)} \int_{-1}^{1} \frac{1}{x^2+1}\, dx = \frac{1}{2} \tan^{-1} x \Big|_{-1}^{1} = \frac{\pi/4 - (-\pi/4)}{2} = \frac{\pi}{4}.$$

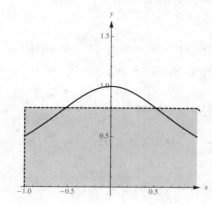

5.4.29

The average value is
$$\frac{1}{\pi} \int_{-\pi/2}^{\pi/2} \cos x\, dx = \frac{1}{\pi} (\sin x) \Big|_{-\pi/2}^{\pi/2}$$
$$= \frac{1}{\pi} \cdot (1 - -1) = \frac{2}{\pi} \approx 0.6366.$$

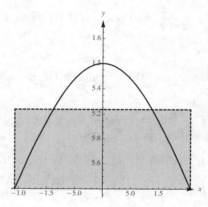

5.4.31

The average value is $\frac{1}{1} \int_0^1 x^n\, dx = \left(\frac{x^{n+1}}{n+1}\right)\Big|_0^1 = \frac{1}{n+1}$. The picture shown is for the case $n = 3$.

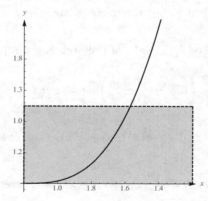

5.4.33 The average distance to the axis is given by $\frac{1}{20} \int_0^{20} 30x(20-x)\, dx$. This is equal to $\frac{1}{20} \int_0^{20} (600x - 30x^2)\, dx = \frac{1}{20} \left(300x^2 - 10x^3\right)\Big|_0^{20} = 2000.$

5.4.35 The average velocity is
$$\frac{1}{6-0} \int_0^6 (t^2 + 3t)\, dt = \frac{1}{6}\left(\frac{t^3}{3} + \frac{3t^2}{2}\right)\Big|_0^6 = \frac{1}{6}(72 + 54 - 0) = \frac{1}{6}(126) = 21\,\mathrm{m/s}.$$

5.4.37 The average height is $\dfrac{1}{\pi} \displaystyle\int_0^\pi 10 \sin x \, dx = \dfrac{1}{\pi}(-10 \cos x)\Big|_0^\pi = \dfrac{1}{\pi}(10 - -10) = \dfrac{20}{\pi}$.

5.4.39 The average value is $\dfrac{1}{4}\displaystyle\int_0^4 (8-2x)\,dx = \dfrac{1}{4}\left(8x - x^2\right)\Big|_0^4 = 4$. The function has a value of 4 when $8 - 2x = 4$, which occurs when $x = 2$.

5.4.41 The average value is $\dfrac{1}{a}\displaystyle\int_0^a \left(1 - \dfrac{x^2}{a^2}\right)dx = \dfrac{1}{a}\left(x - \dfrac{x^3}{3a^2}\right)\Big|_0^a = \dfrac{2}{3}$. The function attains this value when $\dfrac{2}{3} = 1 - \dfrac{x^2}{a^2}$, which is when $x^2 = \dfrac{a^2}{3}$, which on the given interval occurs for $x = \sqrt{3}a/3$.

5.4.43 The average value is $\dfrac{1}{2}\displaystyle\int_{-1}^1 (1 - |x|)\,dx = \dfrac{1}{2}\int_{-1}^0 (1+x)\,dx + \dfrac{1}{2}\int_0^1 (1-x)\,dx = \dfrac{1}{2}\left(x + \dfrac{x^2}{2}\right)\Big|_{-1}^0 + \dfrac{1}{2}\left(x - \dfrac{x^2}{2}\right)\Big|_0^1 = \dfrac{1}{4} + \dfrac{1}{4} = \dfrac{1}{2}$. The function attains this value twice, once on $[-1,0]$ when $1 + x = \dfrac{1}{2}$ which occurs when $x = -\dfrac{1}{2}$, and once on $[0,1]$ when $1 - x = \dfrac{1}{2}$ which occurs when $x = \dfrac{1}{2}$.

5.4.45

a. True. Because of the symmetry, the net area between 0 and 4 will be twice the net area between 0 and 2.

b. True. This follows because the symmetry implies that the net area from a to $a + 2$ is the opposite of the net area from $a - 2$ to a.

c. True. If $f(x) = cx + d$ on $[a,b]$ the value at the midpoint is $c \cdot \dfrac{a+b}{2} + d$, and the average value is
$$\dfrac{1}{b-a}\int_a^b (cx+d)\,dx = \dfrac{1}{b-a}\left(\dfrac{cx^2}{2} + dx\right)\Big|_a^b = \dfrac{1}{b-a}\left(\dfrac{cb^2}{2} + db - \left(\dfrac{ca^2}{2} + da\right)\right) = \dfrac{c}{2}\cdot(a+b) + d.$$

d. False, for example, when $a = 1$, we have that the maximum value of $x - x^2$ on $[0,1]$ occurs at $\dfrac{1}{2}$ and is equal to $\dfrac{1}{4}$, but the average value is $\displaystyle\int_0^1 (x - x^2)\,dx = \left(\dfrac{x^2}{2} - \dfrac{x^2}{3}\right)\Big|_0^1 = \dfrac{1}{2} - \dfrac{1}{3} = \dfrac{1}{6}$.

5.4.47 The average height of the arch is given by
$$\dfrac{1}{630}\int_{-315}^{315}\left(630 - \dfrac{630}{315^2}x^2\right)dx = \dfrac{630}{630}\left(x - \dfrac{x^3}{3\cdot 315^2}\right)\Big|_{-315}^{315} = (315 - 105 - (-315 + 105)) = 420 \text{ ft.}$$

5.4.49 $f(g(-x)) = f(g(x))$, so $f(g(x))$ is an even function, and $\displaystyle\int_{-a}^a f(g(x))\,dx = 2\int_0^a f(g(x))\,dx$.

5.4.51 $p(g(-x)) = p(g(x))$, so $p(g(x))$ is an even function, and $\displaystyle\int_{-a}^a p(g(x))\,dx = 2\int_0^a p(g(x))\,dx$.

5.4.53

a. The average value is $\displaystyle\int_0^1 (ax - ax^2)\,dx = \left(\dfrac{ax^2}{2} - \dfrac{ax^3}{3}\right)\Big|_0^1 = \dfrac{a}{2} - \dfrac{a}{3} = \dfrac{a}{6}$.

b. The function is equal to its average value when $\dfrac{a}{6} = ax - ax^2$ which occurs when $6x - 6x^2 = 1$, so when $6x^2 - 6x + 1 = 0$. On the given interval, this occurs for $x = \dfrac{6 \pm \sqrt{12}}{12} = \dfrac{3 \pm \sqrt{3}}{6}$.

5.4.55

a. The area of the triangle is $\frac{1}{2} \cdot 2a \cdot a^2 = a^3$. The area under the parabola is $\int_{-a}^{a} (a^2 - x^2)\, dx =$
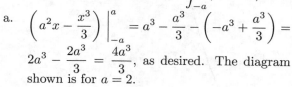
$= a^3 - \frac{a^3}{3} - \left(-a^3 + \frac{a^3}{3}\right) = 2a^3 - \frac{2a^3}{3} = \frac{4a^3}{3}$, as desired. The diagram shown is for $a = 2$.

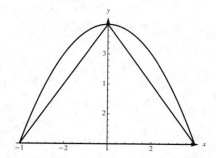

b. The area of the rectangle described is $2a \cdot a^2 = 2a^3$, and $\frac{2}{3}$ of this is $\frac{4a^3}{3}$, which is the area under the parabola derived above.

5.4.57 Suppose f is even, so that $f(-x) = f(x)$. Then $f^n(x) = f^n(-x)$, so that f^n is an even function, no matter what the parity of n is.

Suppose g is an odd function, so that $g(-x) = -g(x)$. Then $g^n(-x) = (-1)^n g^n(x)$, so g^n is even when n is even, and is odd when n is odd.

Summarizing, we have:

	f is even	f is odd
n is even	f^n is even	f^n is even
n is odd	f^n is even	f^n is odd

5.4.59

a. Note that h is continuous and differentiable on $[a, b]$, and that $h(a) = (a-b)\int_a^a f(t)\, dt + (a-a)\int_a^b g(t)\, dt = 0 + 0 = 0$. Also, $h(b) = (b-b)\int_a^b f(t)\, dt + (b-a)\int_b^b g(t)\, dt = 0 + 0 = 0$. So by Rolle's theorem, there exists c between a and b so that $h'(c) = 0$. By the Product and Sum Rules, and the Fundamental Theorem of Calculus, we have $h'(x) = \int_a^x f(t)\, dt + (x-b)f(x) + \int_x^b g(t)\, dt - (x-a)g(x)$.

So at the promised number c we have $h'(c) = \int_a^c f(t)\, dt + (c-b)f(c) + \int_c^b g(t)\, dt - (c-a)g(c) = 0$, so

$$\int_a^c f(t)\, dt + \int_c^b g(t)\, dt = (b-c)f(c) + (c-a)g(c).$$

b. Given f continuous on $[a, b]$, let g be the constant zero function, that is, $g(x) = 0$ for all x. Applying the result of part (a), we have that there exists a c between a and b with

$$\int_a^c f(t)\, dt + 0 = f(c)(b-c) + 0,$$

so

$$\int_a^c f(t)\, dt = f(c)(b-c),$$

as desired.

c. There exists a rectangle with base from c to b and height $f(c)$ so that the area of the rectangle is equal to the value of the integral of f from a to c.

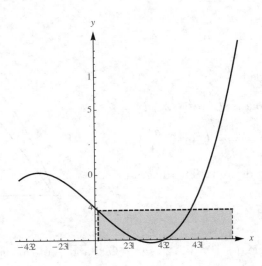

d. Given a function f continuous on $[a, b]$, let $g = f$. Then by part (a) there exists c between a and b so that
$$\int_a^c f(t)\,dt + \int_c^b f(t)\,dt = f(c)(b-c) + f(c)(c-a),$$
so
$$\int_a^b f(t)\,dt = f(c)(b - c + c - a) = f(c)(b-a),$$
so
$$\frac{1}{b-a}\int_a^b f(t)\,dt = f(c).$$

5.5 Substitution Rule

5.5.1 It is based on the Chain Rule for differentiation.

5.5.3 Typically u is substituted for the inner function, so $u = g(x)$.

5.5.5 The new integral is $\int_{g(a)}^{g(b)} f(u)\,du$.

5.5.7 Because $u = x^2 + 1$, $du = 2x\,dx$. Substituting yields $\int u^4\,du = \frac{u^5}{5} + C = \frac{(x^2+1)^5}{5} + C$.

5.5.9 Because $u = \sin x$, $du = \cos x\,dx$. Substituting yields $\int u^3\,du = \frac{u^4}{4} + C = \frac{\sin^4(x)}{4} + C$.

5.5.11 Let $u = x + 1$. Then $du = dx$, and $\int (x+1)^{12}\,dx = \int u^{12}\,du = \frac{u^{13}}{13} + C = \frac{(x+1)^{13}}{13} + C$. Check: $\frac{d}{dx}\left(\frac{(x+1)^{13}}{13} + C\right) = (x+1)^{12}$.

5.5.13 Let $u = 2x + 1$. Then $du = 2dx$ and $\int \sqrt{2x+1}\,dx = \frac{1}{2}\int \sqrt{u}\,du = \frac{1}{3}u^{3/2} + C = \frac{(2x+1)^{3/2}}{3} + C$. Check: $\frac{d}{dx}\left(\frac{(2x+1)^{3/2}}{3} + C\right) = \frac{3}{2}\cdot\frac{1}{3}\cdot(2x+1)^{1/2}\cdot 2 = \sqrt{2x+1}$.

5.5. Substitution Rule

5.5.15

a. $\int e^{10x}\,dx = \dfrac{1}{10}e^{10x} + C.$

b. $\int \sec 5x \tan 5x\,dx = \dfrac{1}{5}\sec 5x + C.$

c. $\int \sin 7x\,dx = -\dfrac{1}{7}\cos 7x + C.$

d. $\int \cos\dfrac{x}{7}\,dx = 7\sin\dfrac{x}{7} + C.$

e. $\int \dfrac{dx}{81 + 9x^2} = \dfrac{1}{9}\int\dfrac{dx}{9 + x^2} = \dfrac{1}{27}\tan^{-1}\dfrac{x}{3} + C.$

f. $\int \dfrac{dx}{\sqrt{36 - x^2}} = \sin^{-1}\dfrac{x}{6} + C.$

5.5.17 Let $u = x^2 - 1$. Then $du = 2x\,dx$. Substituting yields $\int u^{99}\,du = \dfrac{u^{100}}{100} + C = \dfrac{(x^2 - 1)^{100}}{100} + C.$

5.5.19 Let $u = 1 - 4x^3$. Then $du = -12x^2\,dx$, so $-\dfrac{1}{6}du = 2x^2\,dx$. Substituting yields $-\dfrac{1}{6}\int\dfrac{1}{\sqrt{u}}\,du = -\dfrac{1}{3}\cdot\sqrt{u} + C = -\dfrac{1}{3}\cdot\sqrt{1 - 4x^3} + C$

5.5.21 Let $u = x^2 + x$. Then $du = (2x + 1)\,dx$. Substituting yields $\int u^{10}\,du = \dfrac{u^{11}}{11} + C = \dfrac{(x^2 + x)^{11}}{11} + C.$

5.5.23 Let $u = x^4 + 16$. Then $du = 4x^3\,dx$, so $\dfrac{1}{4}du = x^3\,dx$. Substituting yields $\dfrac{1}{4}\int u^6\,du = \dfrac{1}{4}\cdot\dfrac{u^7}{7} + C = \dfrac{(x^4 + 16)^7}{28} + C.$

5.5.25 $\int \dfrac{dx}{\sqrt{36 - 4x^2}} = \dfrac{1}{2}\int\dfrac{dx}{\sqrt{9 - x^2}} = \dfrac{1}{2}\sin^{-1}\dfrac{x}{3} + C$ by equation 10 in Table 5.6.

5.5.27 Let $u = x^3$. Then $du = 3x^2\,dx$. Then $\int 6x^2 4^{x^3}\,dx = 2\int 4^u\,du = 2\cdot\dfrac{4^u}{\ln 4} + C = 2\cdot\dfrac{4^{x^3}}{\ln 4} + C = \dfrac{4^{x^3}}{\ln 2} + C.$

5.5.29 Let $u = x^6 - 3x^2$. Then $du = (6x^5 - 6x)\,dx$, so $\dfrac{1}{6}du = (x^5 - x)\,dx$. Substituting yields $\dfrac{1}{6}\int u^4\,du = \dfrac{1}{6}\cdot\dfrac{u^5}{5} + C = \dfrac{(x^6 - 3x^2)^5}{30} + C.$

5.5.31 Let $u = 5x$ so that $du = 5dx$. Substituting yields $\dfrac{3}{5}\int\dfrac{1}{\sqrt{1 - u^2}}\,du = \dfrac{3}{5}\sin^{-1}u + C = \dfrac{3}{5}\sin^{-1}x + C.$

5.5.33 The integral can be rewritten as $\int \dfrac{e^w}{36 + (e^w)^2}\,dw$. Let $u = e^w$, so that $du = e^w\,dw$. Substituting yields $\int \dfrac{du}{36 + u^2} = \dfrac{1}{6}\tan^{-1}\dfrac{u}{6} + C = \dfrac{1}{6}\tan^{-1}\dfrac{e^w}{6} + C.$

5.5.35 Let $u = x^2$ so that $du = 2x\,dx$. Substitution yields $\dfrac{1}{2}\int \csc u\cot u\,du = -\dfrac{1}{2}\csc u + C = -\dfrac{1}{2}\csc x^2 + C.$

5.5.37 Let $u = 10x + 7$ so that $du = 10\,dx$. Substituting yields $\dfrac{1}{10}\int \sec^2 u\,du = \dfrac{1}{10}\tan u + C = \dfrac{1}{10}\tan(10x + 7) + C.$

5.5.39 Let $u = 4t + 1$ so that $du = 4dt$. Substituting yields $\dfrac{1}{4}\displaystyle\int 10^u \, du = \dfrac{1}{4} \cdot \dfrac{10^u}{\ln 10} + C = \dfrac{10^u}{4\ln 10} + C$.

5.5.41 Let $u = \cot x$. Then $du = -\csc^2 x \, dx$. Substituting yields $-\displaystyle\int u^{-3} \, du = \dfrac{1}{2u^2} + C = \dfrac{1}{2\cot^2 x} + C$.

5.5.43 Note that $\sin x \sec^8 x = \dfrac{\sin x}{\cos^8 x}$. Let $u = \cos x$, so that $du = -\sin x \, dx$. Substituting yields $-\displaystyle\int u^{-8} \, du = \dfrac{1}{7u^7} + C = \dfrac{1}{7\cos^7 x} + C = \dfrac{\sec^7 x}{7} + C$.

5.5.45 $\displaystyle\int_0^{\pi/8} \cos 2x \, dx = \left(\dfrac{\sin 2x}{2}\right)\bigg|_0^{\pi/8} = \dfrac{\sqrt{2}/2 - 0}{2} = \dfrac{\sqrt{2}}{4}$.

5.5.47 Let $u = 4 - x^2$. Then $du = -2x \, dx$. Also, when $x = 0$ we have $u = 4$ and when $x = 1$ we have $u = 3$. Substituting yields $-\displaystyle\int_4^3 u \, du = \int_3^4 u \, du = \left(\dfrac{u^2}{2}\right)\bigg|_3^4 = 8 - 4.5 = 3.5$.

5.5.49 Let $u = 2^x + 4$ so that $du = 2^x \ln 2$. Also, when $x = 1$, $u = 6$ and when $x = 3$, $u = 12$. Substituting yields
$$\dfrac{1}{\ln 2}\int_6^{12} \dfrac{du}{u} = \dfrac{1}{\ln 2}\ln u \bigg|_6^{12} = \dfrac{1}{\ln 2}(\ln 12 - \ln 6) = \dfrac{1}{\ln 2}(\ln 2) = 1.$$

5.5.51 Let $u = \sin\theta$. Then $du = \cos\theta \, d\theta$. Also, when $\theta = 0$ we have $u = 0$ and when $\theta = \pi/2$ we have $u = 1$. Substituting yields $\displaystyle\int_0^1 u^2 \, du = \left(\dfrac{u^3}{3}\right)\bigg|_0^1 = \dfrac{1}{3}$.

5.5.53 Let $u = e^w$. Then $du = e^w \, dw$. Also, when $w = \ln \pi/4$, $u = \pi/4$, and when $w = \ln \pi/2$, $u = \pi/2$. Substituting yields
$$\int_{\pi/4}^{\pi/2} \cos u \, du = \sin u \bigg|_{\pi/4}^{\pi/2} = 1 - \dfrac{\sqrt{2}}{2} = \dfrac{2 - \sqrt{2}}{2}.$$

5.5.55 Let $u = x^3 + 1$. Then $du = 3x^2 \, dx$. Also, when $x = -1$ we have $u = 0$ and when $x = 2$ we have $u = 9$. Substituting yields $\dfrac{1}{3}\displaystyle\int_0^9 e^u \, du = \left(\dfrac{e^u}{3}\right)\bigg|_0^9 = \dfrac{e^9 - 1}{3}$.

5.5.57 Let $u = \sin x$. Then $du = \cos x \, dx$. Also, when $x = \pi/4$ we have $u = \sqrt{2}/2$ and when $x = \pi/2$ we have $u = 1$. Substituting yields $\displaystyle\int_{\sqrt{2}/2}^1 \dfrac{1}{u^2} \, du = \left(\dfrac{-1}{u}\right)\bigg|_{\sqrt{2}/2}^1 = \left(-1 - \left(-\dfrac{2}{\sqrt{2}}\right)\right) = \sqrt{2} - 1$.

5.5.59 Let $u = 5x$, so that $du = 5 \, dx$. Also, when $x = 2/(5\sqrt{3})$ we have $u = 2/\sqrt{3}$ and when $x = 2/5$ we have $u = 2$. Substituting yields $\displaystyle\int_{2/\sqrt{3}}^2 \dfrac{du}{u\sqrt{u^2 - 1}} = \sec^{-1} u \bigg|_{2/\sqrt{3}}^2 = \dfrac{\pi}{3} - \dfrac{\pi}{6} = \dfrac{\pi}{6}$.

5.5.61 Let $u = x^2 + 1$, so that $du = 2x \, dx$. Substituting yields $\dfrac{1}{2}\displaystyle\int_1^{17} \dfrac{1}{u} \, du = \dfrac{1}{2}\ln|u| \bigg|_1^{17} = \dfrac{\ln 17}{2}$.

5.5.63 Let $u = 3x$, so that $du = 3 \, dx$. Substituting yields $\dfrac{4}{3}\displaystyle\int_1^{3/\sqrt{3}} \dfrac{1}{u^2 + 1} \, du = \dfrac{4}{3}\tan^{-1} u \bigg|_1^{3/\sqrt{3}} = \dfrac{4}{3}\left(\dfrac{\pi}{3} - \dfrac{\pi}{4}\right) = \dfrac{4}{3} \cdot \dfrac{\pi}{12} = \dfrac{\pi}{9}$.

5.5.65 Let $u = 1 - x^2$. Then $du = -2x \, dx$. Also note that when $x = 0$ we have $u = 1$, and when $x = 1$ we have $u = 0$. Substituting yields $-\dfrac{1}{2}\displaystyle\int_1^0 \sqrt{u} \, du = \dfrac{1}{2}\int_0^1 \sqrt{u} \, du = \left(\dfrac{u^{3/2}}{3}\right)\bigg|_0^1 = \dfrac{1}{3}$.

5.5. Substitution Rule

5.5.67 Let $u = x^2 - 1$, so that $du = 2x\,dx$. Also note that when $x = 2$ we have $u = 3$, and when $x = 3$ we have $u = 8$. Substituting yields $\frac{1}{2}\int_3^8 u^{-1/3}\,du = \frac{1}{2}\left(\frac{3u^{2/3}}{2}\right)\Big|_3^8 = \frac{3}{4}\left(4 - \sqrt[3]{9}\right)$.

5.5.69 Let $u = 16 - x^4$. Then $du = -4x^3\,dx$. Also note that when $x = 0$ we have $u = 16$, and when $x = 2$ we have $u = 0$. Substituting yields $\frac{1}{4}\int_0^{16}\sqrt{u}\,du = \frac{1}{4}\left(\frac{2u^{3/2}}{3}\right)\Big|_0^{16} = \frac{32}{3}$.

5.5.71 Let $u = 2 + \cos x$ so that $du = -\sin x\,dx$. Note that when $x = -\pi$, $u = 1$ and when $x = 0$, $u = 3$. Substituting yields $\int_1^3 -\frac{1}{u}\,du = (-\ln|u|)\Big|_1^3 = -(\ln 3 - \ln 1) = -\ln 3$.

5.5.73 Let $u = 3x + 1$ so that $du = 3\,dx$. Note that $9x^2 + 6x + 1 = (3x+1)^2 = u^2$, and also that when $x = 1$, $u = 4$ and when $x = 2$, $u = 7$. Substituting yields $\frac{4}{3}\int_4^7 \frac{1}{u^2}\,du = \frac{4}{3}\left(-\frac{1}{u}\right)\Big|_4^7 = \frac{4}{3}\left(-\frac{1}{7} - \left(-\frac{1}{4}\right)\right) = \frac{4}{3}\left(\frac{3}{28}\right) = \frac{1}{7}$.

5.5.75 The average velocity is given by $\frac{1}{10 - 0}\int_0^{10}(8\sin \pi t + 2t)\,dt = \frac{1}{10}\left(-\frac{8}{\pi}\cos \pi t + t^2\right)\Big|_0^{10} = \frac{1}{10}\left(-\frac{8}{\pi}\cos 10\pi + 100 + \frac{8}{\pi}\cos 0 - 0\right) = 10$.

5.5.77

a. $\int_0^4 \frac{200}{(t+1)^2}\,dt = \left(\frac{-200}{t+1}\right)\Big|_0^4 = -40 + 200 = 160$.

b. $\int_0^6 \frac{200}{(t+1)^3}\,dt = \left(\frac{-200}{2(t+1)^2}\right)\Big|_0^6 = \frac{-100}{49} + 100 = \frac{4800}{49}$.

c. $\Delta P = \int_0^T \frac{200}{(t+1)^r}\,dt$. This decreases as r increases, because $\frac{200}{(t+1)^r} > \frac{200}{(t+1)^{r+1}}$.

d. Suppose $\int_0^{10}\frac{200}{(t+1)^r}\,dt = 350$. Then $\left(\frac{200(t+1)^{-r+1}}{1-r}\right)\Big|_0^{10} = 350$, so $11^{1-r} - 1 = \frac{350(1-r)}{200}$, and thus $\frac{11}{11^r} = \frac{7-7r}{4} + \frac{4}{4} = \frac{11-7r}{4}$, and $11^r = \frac{44}{11-7r}$. Using trial and error to find r, we arrive at $r \approx 1.278$.

e. $\int_0^T \frac{200}{(t+1)^3}\,dt = \left(-\frac{200}{2(t+1)^2}\right)\Big|_0^T = -\frac{100}{(T+1)^2} + 100$. As $T \to \infty$, this expression $\to 100$, so in the long run, the bacteria approaches a finite limit.

5.5.79 Let $u = x - 4$, so that $u + 4 = x$. Then $du = dx$. Substituting yields $\int \frac{u+4}{\sqrt{u}}\,du = \int\left(\frac{u}{\sqrt{u}} + \frac{4}{\sqrt{u}}\right)du = \int u^{1/2} + 4u^{-1/2}\,du = \frac{2}{3}u^{3/2} + 8u^{1/2} + C = \frac{2}{3}\cdot(x-4)^{3/2} + 8\sqrt{x-4} + C$.

5.5.81 Let $u = x + 4$, so that $u - 4 = x$. Then $du = dx$. Substituting yields

$$\int \frac{u-4}{\sqrt[3]{u}}\,du = \int\left(u^{2/3} - 4u^{-1/3}\right)du = \frac{3}{5}u^{5/3} + -6u^{2/3} + C$$
$$= \frac{3}{5}(x+4)^{5/3} - 6(x+4)^{2/3} + C.$$

5.5.83 Let $u = 2x + 1$. Then $du = 2dx$ and $x = \frac{u-1}{2}$. Substituting yields $\frac{1}{2}\int \frac{u-1}{2} \cdot \sqrt[3]{u}\, du = \frac{1}{4}\int (u^{4/3} - u^{1/3})\, du = \frac{1}{4}\left(\frac{3}{7}u^{7/3} - \frac{3}{4}u^{4/3}\right) + C = \frac{3(2x+1)^{7/3}}{28} - \frac{3(2x+1)^{4/3}}{16} + C = (2x+1)^{4/3}\left(\frac{3(2x+1)}{28} - \frac{3}{16}\right) = \frac{3}{112}(2x+1)^{4/3}(8x + 4 - 7) = \frac{3}{112}(2x+1)^{4/3}(8x - 3)$.

5.5.85 Let $u = x + 10$. Then $du = dx$ and $x = u - 10$. Substituting gives $\int (u-10)u^9\, du = \int (u^{10} - 10u^9)\, du = \frac{u^{11}}{11} - u^{10} + C = \frac{1}{11}(x+10)^{11} - (x+10)^{10} + C = (x+10)^{10}\left(\frac{x+10}{11} - 1\right) + C = \frac{(x+10)^{10}(x-1)}{11} + C$.

5.5.87 $\int_{-\pi}^{\pi} \cos^2 x\, dx = 2\int_0^{\pi} \frac{1 + \cos 2x}{2}\, dx = \left(x + \frac{\sin 2x}{2}\right)\Big|_0^{\pi} = \pi$.

5.5.89 $\int \sin^2\left(\theta + \frac{\pi}{6}\right) d\theta = \frac{1}{2}\int \left(1 - \cos\left(2\theta + \frac{\pi}{3}\right)\right) d\theta = \frac{\theta}{2} - \frac{\sin\left(2\theta + \frac{\pi}{3}\right)}{4} + C$.

5.5.91 $\int_{-\pi/4}^{\pi/4} \sin^2 2\theta\, d\theta = 2\int_0^{\pi/4} \sin^2 2\theta\, d\theta = 2\int_0^{\pi/4} \frac{1 - \cos 4\theta}{2}\, d\theta = \left(\theta - \frac{\sin 4\theta}{4}\right)\Big|_0^{\pi/4} = \frac{\pi}{4}$.

5.5.93 Let $u = \sin^2 y + 2$ so that $du = 2\sin y \cos y\, dy = \sin 2y\, dy$. Substituting yields $\int_2^{9/4} \frac{1}{u}\, du = (\ln|u|)\Big|_2^{9/4} = \ln(9/4) - \ln 2 = \ln(9/8)$.

5.5.95

a. True. This follows by substituting $u = f(x)$ to obtain the integral $\int u\, du = \frac{u^2}{2} + C = \frac{f(x)^2}{2} + C$.

b. True. Again, this follows from substituting $u = f(x)$ to obtain the integral $\int u^n\, du = \frac{u^{n+1}}{n+1} + C = \frac{(f(x))^{n+1}}{n+1} + C$ where $n \neq -1$.

c. False. If this were true, then $\sin 2x$ and $2\sin x$ would have to differ by a constant, which they do not. In fact, $\sin 2x = 2\sin x \cos x$.

d. False. The derivative of the right hand side is $(x^2 + 1)^9 \cdot 2x$ which is not the integrand on the left hand side.

e. False. If we let $u = f'(x)$, then $du = f''(x)\, dx$. Substituting yields $\int_{f'(a)}^{f'(b)} u\, du = \left(\frac{u^2}{2}\right)\Big|_{f'(a)}^{f'(b)} = \frac{(f'(b))^2}{2} - \frac{(f'(a))^2}{2}$.

5.5.97 $A(x) = \int_0^{\sqrt{\pi}} x \sin x^2\, dx$. Let $u = x^2$, so that $du = 2x\, dx$. Also, when $x = 0$ we have $u = 0$ and when $x = \sqrt{\pi}$ we have $u = \pi$. Substituting yields $\frac{1}{2}\int_0^{\pi} \sin u\, du = \frac{1}{2}(-\cos u)\Big|_0^{\pi} = 1$.

5.5.99 $A(a) = \int_0^a \left(\frac{1}{a} - \frac{x^2}{a^3}\right) dx = \left(\frac{x}{a} - \frac{x^3}{3a^3}\right)\Big|_0^a = 1 - \frac{1}{3} = \frac{2}{3}$. This is a constant function.

5.5. Substitution Rule

5.5.101

a. Let $u = \sin px$, so that $du = p\cos px\, dx$. Note that when $x = 0$, $u = 0$, and when $x = \pi/2p$, $u = 1$. Substituting yields $\dfrac{1}{p}\displaystyle\int_0^1 f(u)\, du = \dfrac{\pi}{p}$.

b. Let $u = \sin x$ so that $du = \cos x\, dx$. Note that when $x = -\pi/2$, $u = -1$ and when $x = \pi/2$, $u = 1$. Substituting yields $\displaystyle\int_{-1}^1 f(u)\, du = 0$, because f is an odd function. Alternatively, we could note that when f is odd, $\cos x \cdot f(\sin x)$ is also odd, because $\sin x$ is odd and $\cos x$ is even. Thus the given integral must be zero because it is the definite integral of an odd function over a symmetric interval about 0.

5.5.103 $\dfrac{1}{\pi/k - 0}\displaystyle\int_0^{\pi/k} \sin kx\, dx = \dfrac{k}{\pi}\cdot\left(\dfrac{-\cos kx}{k}\right)\Big|_0^{\pi/k} = \dfrac{1}{\pi}(1 - (-1)) = \dfrac{2}{\pi}$.

5.5.105 The area on the left is given by $\displaystyle\int_4^9 \dfrac{(\sqrt{x}-1)^2}{2\sqrt{x}}\, dx$. If we let $u = \sqrt{x} - 1$ so that $du = \dfrac{1}{2\sqrt{x}}\, dx$, we obtain the equivalent integral $\displaystyle\int_1^2 u^2\, du$ which represents the area on the right.

5.5.107 Let $u = f(x)$, so that $du = f'(x)\, dx$. Substituting yields

$$\int_4^5 (5u^3 + 7u^2 + u)\, du = \left(\dfrac{5u^4}{4} + \dfrac{7u^3}{3} + \dfrac{u^2}{2}\right)\Big|_4^5 = \dfrac{7297}{12}.$$

5.5.109 If we let $u = \sqrt{x+a}$, then $u^2 = x + a$ and $2u\, du = dx$. Substituting yields $\displaystyle\int_{\sqrt{a}}^{\sqrt{1+a}} (u^2 - a)\cdot u \cdot 2u\, du = \displaystyle\int_{\sqrt{a}}^{\sqrt{1+a}} (2u^4 - 2au^2)\, du = \left(\dfrac{2u^5}{5} - \dfrac{2au^3}{3}\right)\Big|_{\sqrt{a}}^{\sqrt{1+a}} = \dfrac{2(\sqrt{1+a})^5}{5} - \dfrac{2a(\sqrt{1+a})^3}{3} - \dfrac{2a^{5/2}}{5} + \dfrac{2a^{5/2}}{3}$.

If we let $u = x + a$, then $u - a = x$ and $du = dx$. Substituting yields $\displaystyle\int_a^{a+1} (u - a)\sqrt{u}\, du = \left(\dfrac{2u^{5/2}}{5} - \dfrac{2au^{3/2}}{3}\right)\Big|_a^{a+1} = \dfrac{2(a+1)^{5/2}}{5} - \dfrac{2a(a+1)^{3/2}}{3} - \dfrac{2a^{5/2}}{5} + \dfrac{2a^{5/2}}{3}$.

Note that the two results are the same.

5.5.111 If we let $u = \cos\theta$, then $du = -\sin\theta\, d\theta$. Substituting yields $\displaystyle\int -u^{-4}\, du = \dfrac{1}{3u^3} + C = \dfrac{1}{3\cos^3\theta} + C = \dfrac{\sec^3\theta}{3} + C$.

If we let $u = \sec\theta$, then $du = \sec\theta\tan\theta\, d\theta$. Substituting yields $\displaystyle\int u^2\, du = \dfrac{u^3}{3} + C = \dfrac{\sec^3\theta}{3} + C$. Note that the two results are the same.

5.5.113

a. Because $\sin 2x = 2\sin x\cos x$, we can write $(\sin x\cos x)^2 = \left(\dfrac{\sin 2x}{2}\right)^2 = \dfrac{\sin^2 2x}{4}$. Then we have $I = \dfrac{1}{4}\displaystyle\int \sin^2 2x\, dx = \dfrac{1}{4}\left(\dfrac{x}{2} - \dfrac{\sin 4x}{8}\right) + C = \dfrac{x}{8} - \dfrac{\sin 4x}{32} + C$. Note that we used the result of the previous problem during this derivation.

b. $I = \dfrac{1}{4}\displaystyle\int (1 - \cos 2x)(1 + \cos 2x)\, dx = \dfrac{1}{4}\displaystyle\int (1 - \cos^2 2x)\, dx = \dfrac{1}{4}\displaystyle\int \sin^2 2x\, dx = \dfrac{1}{4}\left(\dfrac{x}{2} - \dfrac{\sin 4x}{8}\right) + C = \dfrac{x}{8} - \dfrac{\sin 4x}{32} + C$.

c. The results are consistent. The work involved is similar in each method.

5.5.115

a. Let $u = cx$. Note that $du = c \cdot dx$. Substitution yields $\int_a^b f(cx)\,dx = \frac{1}{c}\int_{ac}^{bc} f(u)\,du$.

b.
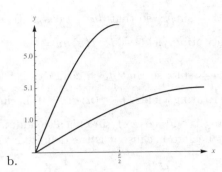

5.5.117 Let $u = \sqrt{x+1}$ so that $u^2 = x+1$. Then $2u\,du = dx$. Substituting yields $\int 2 \cdot \frac{u\,du}{\sqrt{1+u}}$. Now let $v = \sqrt{1+u}$ so that $v^2 = 1+u$ and $2v\,dv = du$. Now a substitution yields $4\int \frac{(v^2-1)v}{v}\,dv = 4\int (v^2-1)\,dv = \frac{4v^3}{3} - 4v + C = \frac{4}{3}(1+u)^{3/2} - 4\sqrt{1+u} + C = \frac{4}{3}\left(1+\sqrt{x+1}\right)^{3/2} - 4\sqrt{1+\sqrt{x+1}} + C = \frac{4}{3}\sqrt{1+\sqrt{x+1}}\left(1+\sqrt{x+1}-3\right) = \frac{4}{3}\sqrt{1+\sqrt{x+1}}(\sqrt{x+1}-2) + C$.

5.5.119 Let $u = \cos\theta$, so that $du = -\sin\theta\,d\theta$. This substitution yields $\int_0^1 \frac{u}{\sqrt{u^2+16}}\,du$. Now let $v = u^2+16$, so that $dv = 2u\,du$. Now a substitution yields $\frac{1}{2}\int_{16}^{17} v^{-1/2}\,dv = \sqrt{v}\Big|_{16}^{17} = \sqrt{17} - 4$.

Chapter Five Review

1

a. True. The antiderivative of a linear function is a quadratic function.

b. False. $A'(x) = f(x)$, not $F(x)$.

c. True. Note that f is an antiderivative of f', so this follows from the Fundamental Theorem.

d. True. Because $|f(x)| \geq 0$ for all x, this integral must be positive, unless f is constantly 0.

e. False. For example, the average value of $\sin x$ on $[0, 2\pi]$ is zero.

f. True. This is equal to $2\int_a^b f(x)\,dx - 3\int_a^b g(x)\,dx = 2\int_a^b f(x)\,dx + 3\int_b^a g(x)\,dx$.

g. True. The derivative of the right hand side is $f'(g(x))g'(x)$ by the Chain Rule.

3

a.

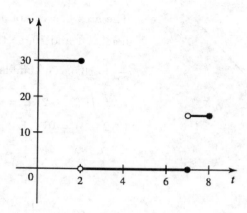

b. The area is $2 \cdot 30 + 0 + 15 \cdot 1 = 75$.

c. The area represents the distance that the diver ascends.

5 $\Delta x = \dfrac{4-1}{6} = \dfrac{1}{2}$. The left Riemann sum is given by $(f(1)+f(1.5)+f(2)+f(2.5)+f(3)+f(3.5))\left(\dfrac{1}{2}\right) = (\sqrt{5}+\sqrt{7}+\sqrt{9}+\sqrt{11}+\sqrt{13}+\sqrt{15})\left(\dfrac{1}{2}\right) \approx 9.34$. The right Riemann sum is given by $(f(1.5)+f(2)+f(2.5)+f(3)+f(3.5)+f(4))\left(\dfrac{1}{2}\right) = (\sqrt{7}+\sqrt{9}+\sqrt{11}+\sqrt{13}+\sqrt{15}+\sqrt{17})\left(\dfrac{1}{2}\right) \approx 10.28$. The midpoint Riemann sum is given by $(f(1.25)+f(1.75)+f(2.25)+f(2.75)+f(3.25)+f(3.75))\left(\dfrac{1}{2}\right) = (\sqrt{6}+\sqrt{8}+\sqrt{10}+\sqrt{12}+\sqrt{14}+\sqrt{16})\left(\dfrac{1}{2}\right) \approx 9.82$.

7

n	Midpoint Riemann sum
10	114.167
30	114.022
60	114.006

From the table, it appears that $\displaystyle\int_1^{25} \sqrt{2x-1}\,dx = 114$. This result can be verified using the Fundamental Theorem of Calculus and the substitution $u = 2x - 1$.

9

a. The right Riemann sum is $4 \cdot 1 + 7 \cdot 1 + 10 \cdot 1 = 21$.

b. The right Riemann sum is $\displaystyle\sum_{k=1}^{n}\left(3\left(1+\dfrac{3k}{n}\right)-2\right)\cdot\dfrac{3}{n}$.

c. The sum evaluates as $\displaystyle\sum_{k=1}^{n}\dfrac{3}{n} + \sum_{k=1}^{n}\dfrac{27k}{n^2} = 3 + \dfrac{27}{n^2}\cdot\dfrac{n(n+1)}{2}$. As $n \to \infty$, the limit of this expression is $3 + 13.5 = 16.5$.

d.

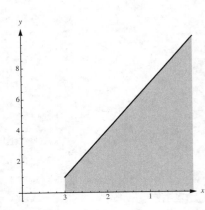

The area consists of a trapezoid with base 3 and heights 1 and 10, so the value is $(3)\left(\dfrac{1+10}{2}\right) = \dfrac{33}{2} = 16.5$. The Fundamental Theorem assures us that $\int_1^4 (3x-2)\,dx = \left(\dfrac{3x^2}{2} - 2x\right)\Big|_1^4 = (24-8) - (3/2 - 2) = 16 + 1/2 = 16.5$.

11 Let $\Delta x = \dfrac{2-0}{n} = \dfrac{2}{n}$. Let $x_k = 0 + k\Delta x = \dfrac{2k}{n}$. Then $f(x_k) = \dfrac{4k^2}{n^2} - 4$. Thus,

$$\lim_{n\to\infty}\sum_{k=1}^n f(x_k)\Delta x = \lim_{n\to\infty}\sum_{k=1}^n \left(\dfrac{4k^2}{n^2} - 4\right)\cdot\dfrac{2}{n} = \lim_{n\to\infty}\left(\dfrac{8}{n^3}\sum_{k=1}^n k^2 - \dfrac{8}{n}\sum_{k=1}^n 1\right) =$$

$$\lim_{n\to\infty}\left(\dfrac{8}{n^3}\cdot\dfrac{n(n+1)(2n+1)}{6} - 8\right) = \dfrac{8}{3} - 8 = \dfrac{-16}{3}.$$

13 Let $\Delta x = \dfrac{4-0}{n} = \dfrac{4}{n}$. Let $x_k = 0 + k\Delta x = 0 + \dfrac{4k}{n}$. Then $f(x_k) = \dfrac{64k^3}{n^3} - \dfrac{4k}{n}$. Thus,

$$\lim_{n\to\infty}\sum_{k=1}^n f(x_k)\Delta x = \lim_{n\to\infty}\sum_{k=1}^n \left(\dfrac{64k^3}{n^3} - \dfrac{4k}{n}\right)\cdot\dfrac{4}{n} = \lim_{n\to\infty}\left(\dfrac{256}{n^4}\sum_{k=1}^n k^3 - \dfrac{16}{n^2}\sum_{k=1}^n k\right) =$$

$$\lim_{n\to\infty}\left(\dfrac{256}{n^4}\cdot\dfrac{n^2(n+1)^2}{4} - \dfrac{16}{n^2}\cdot\dfrac{n(n+1)}{2}\right) = 64 - 8 = 56.$$

15

a. $\int_{-4}^4 f(x)\,dx = 2\int_0^4 f(x)\,dx = 2\cdot 10 = 20.$

b. $\int_{-4}^4 3g(x)\,dx = 3\cdot 0 = 0.$

c. $\int_{-4}^4 4f(x) - 3g(x)\,dx = 2\cdot 4\cdot\int_0^4 f(x)\,dx - 3\cdot 0 = 8\cdot 10 - 0 = 80.$

d. Let $u = 4x^2$, so that $du = 8x\,dx$. Substituting yields $\int_0^4 f(u)\,du = 10.$

e. Because f is an even function, $3xf(x)$ is an odd function. Thus, $\int_{-2}^2 3xf(x)\,dx = 0.$

17 $\int_1^4 3f(x)\,dx = 3\int_1^4 f(x)\,dx = 3\cdot 6 = 18.$

19 $\int_1^4 (3f(x) - 2g(x))\,dx = 3\int_1^4 f(x)\,dx - 2\int_1^4 g(x)\,dx = 3\cdot 6 - 2\cdot 4 = 18 - 8 = 10.$

21 There is not enough information to compute this integral.

23

a. This region can be divided up into a 4×2 rectangle and a right triangle with base and height equal to 1. Thus, the integral is equal to $8 + \frac{1}{2} = 8.5$.

b. $\int_6^4 f(x)\,dx = -\int_4^6 f(x)\,dx$. The region whose area is $\int_4^6 f(x)\,dx$ consists of a 1×3 rectangle, together with a right triangle with base 1 and height 3, so $\int_4^6 f(x)\,dx = 3 + \frac{3}{2} = 4.5$, and $\int_6^4 f(x)\,dx = -4.5$.

c. $\int_5^7 f(x)\,dx = \int_5^6 f(x)\,dx + \int_6^7 f(x)\,dx$. The region lying over $[5,6]$ is a right triangle with height 3 and base 1, so its area is $\frac{3}{2}$. The region lying under $[6,7]$ has the same area, but is below the x-axis, so $\int_6^7 f(x)\,dx = -\frac{3}{2}$. So $\int_5^7 f(x)\,dx = \frac{3}{2} + -\frac{3}{2} = 0$.

d. Note that $\int_4^5 f(x)\,dx = 3$, because the area represented is that of a 1×3 rectangle. Now by the work above, $\int_0^7 f(x)\,dx = \int_0^4 f(x)\,dx + \int_4^5 f(x)\,dx + \int_5^7 f(x)\,dx = 8.5 + 3 + 0 = 11.5$.

25 $\int_0^4 \sqrt{8x - x^2}\,dx = \int_0^4 \sqrt{16 - (x^2 - 8x + 16)}\,dx = \int_0^4 \sqrt{16 - (x-4)^2}\,dx$. This represents one quarter of the area inside the circle centered at $(4,0)$ with radius 4, so its value is $\frac{1}{4} \cdot 16\pi = 4\pi$.

27 It appears that B is the derivative of A, and C is the derivative of B. Thus we must have $A = \int_0^x f(t)\,dt$, $B = f(x)$, and $C = f'(x)$. Note that A is decreasing where B is negative and increasing where B is positive, and has a minimum where B is zero. Note also that B is increasing where C is positive, and is decreasing where C is negative, and has a maximum where C is zero.

29 $\dfrac{d}{dx}\int_7^x \sqrt{1 + t^4 + t^6}\,dt = \sqrt{1 + x^4 + x^6}$.

31 $\dfrac{d}{dx}\int_x^5 \sin w^6\,dw = -\dfrac{d}{dx}\int_5^x \sin w^6\,dw = -\sin x^6$.

33

$$\frac{d}{dx}\int_{-x}^x \frac{dt}{t^{10}+1} = \frac{d}{dx}\left(\int_{-x}^0 \frac{dt}{t^{10}+1} + \int_0^x \frac{dt}{t^{10}+1}\right)$$

$$= \frac{d}{dx}\left(-\int_0^{-x} \frac{dt}{t^{10}+1} + \int_0^x \frac{dt}{t^{10}+1}\right) = -\frac{1}{(-x)^{10}+1}\cdot(-1) + \frac{1}{x^{10}+1} = \frac{2}{x^{10}+1}.$$

Alternatively, if we realize that the integrand is an even function, we could write

$$\frac{d}{dx}2\int_0^x \frac{dt}{t^{10}+1} = \frac{2}{x^{10}+1}.$$

35 $f(x) = -\int_1^x (t-3)(t-6)^{11}\,dt$, so $f'(x) = -(x-3)(x-6)^{11}$. $f' = 0$ for $x = 3$ and $x = 6$. We have $f' > 0$ on $(3,6)$, so f is increasing there, while $f' < 0$ on $(-\infty, 3)$ and on $(6, \infty)$, so f is decreasing on those intervals.

37 Let $u = ax + b$. Then $du = a\,dx$, or $dx = \dfrac{du}{a}$. The substitution gives

$$\int f(ax+b)\,dx = \frac{1}{a}\int f(u)\,du = \frac{1}{a}F(u) + C = \frac{1}{a}F(ax+b) + C.$$

39 $\displaystyle\int_{-2}^{2}(3x^4 - 2x + 1)\,dx = \left(\dfrac{3x^5}{5} - x^2 + x\right)\Big|_{-2}^{2} = \dfrac{96}{5} - 4 + 2 - \left(-\dfrac{96}{5} - 4 - 2\right) = \dfrac{192}{5} + 4 = \dfrac{212}{5}.$

41 $\displaystyle\int(9x^8 - 7x^6)\,dx = x^9 - x^7 + C.$

43 $\displaystyle\int_0^1 (x + \sqrt{x})\,dx = \left(\dfrac{x^2}{2} + \dfrac{2x^{3/2}}{3}\right)\Big|_0^1 = \dfrac{1}{2} + \dfrac{2}{3} = \dfrac{7}{6}.$

45 $\displaystyle\int_{\pi/6}^{\pi/3}(\sec^2 t + \csc^2 t)\,dt = (\tan t - \cot t)\Big|_{\pi/6}^{\pi/3} = \left(\sqrt{3} - \dfrac{1}{\sqrt{3}}\right) - \left(\dfrac{1}{\sqrt{3}} - \sqrt{3}\right) = 2\sqrt{3} - \dfrac{2}{\sqrt{3}} = \dfrac{6-2}{\sqrt{3}} = \dfrac{4\sqrt{3}}{3}.$

47 $\displaystyle\int_{\sqrt{2}}^{2}\dfrac{1}{x\sqrt{x^2-1}}\,dx = \sec^{-1}x\Big|_{\sqrt{2}}^{2} = \sec^{-1}2 - \sec^{-1}\sqrt{2} = \dfrac{\pi}{3} - \dfrac{\pi}{4} = \dfrac{\pi}{12}.$

49 Let $u = \sin x$. Then $du = \cos x\,dx$. Then $\displaystyle\int\dfrac{\cos x}{\sin^{7/4}x}\,dx = \int u^{-7/4}\,du = -\dfrac{4}{3}u^{-3/4}\,du = -\dfrac{4}{3}(\sin x)^{-3/4} + C = -\dfrac{4}{3\sin^{3/4}x} + C.$

51 Let $u = x^3$, so that $du = 3x^2\,dx$. Substituting gives $\displaystyle\int x^2\cos x^3\,dx = \dfrac{1}{3}\int\cos u\,du = \dfrac{1}{3}\sin u + C = \dfrac{1}{3}\sin x^3 + C.$

53 Let $u = \sin 7w$. Then $du = 7\cos 7w$. Substituting yields $\dfrac{1}{7}\displaystyle\int\dfrac{du}{16+u^2} = \dfrac{1}{7}\cdot\dfrac{1}{4}\tan^{-1}\left(\dfrac{u}{4}\right) = \dfrac{1}{28}\tan^{-1}\left(\dfrac{\sin 7w}{4}\right) + C.$

55 Let $u = x^2 + 1$. Then $du = 2x\,dx$ and when $x = 0$, $u = 1$ and when $x = 1$, $u = 2$. Substituting gives

$$\dfrac{1}{2}\int_1^2 2^u\,du = \dfrac{2^u}{2\ln 2}\Big|_1^2 = \dfrac{2}{\ln 2} - \dfrac{1}{\ln 2} = \dfrac{1}{\ln 2}.$$

57 $\displaystyle\int_0^2(2x+1)^3\,dx = \left(\dfrac{(2x+1)^4}{8}\right)\Big|_0^2 = \dfrac{625}{8} - \dfrac{1}{8} = 78.$

59 Let $u = 3r^2 + 2$. Then $du = 6r\,dr$. When $r = 0$, $u = 2$ and when $r = 1$, $u = 5$. Substituting gives

$$\dfrac{5}{6}\int_2^5 e^u\,du = \dfrac{5}{6}e^u\Big|_2^5 = \dfrac{5}{6}(e^5 - e^2) = \dfrac{5}{6}e^2(e^3 - 1).$$

61 Let $u = e^x$, so that $du = e^x\,dx$. We can rewrite the given integral as $\displaystyle\int e^{e^x}e^x\,dx$, so that substituting gives $\displaystyle\int e^u\,du = e^u + C = e^{e^x} + C.$

63 Let $u = 2x$ so that $du = 2\,dx$. Substituting gives

$$\dfrac{1}{2}\int\dfrac{du}{\sqrt{1-u^2}} = \dfrac{1}{2}\sin^{-1}u + C = \dfrac{1}{2}\sin^{-1}2x + C.$$

65 Let $u = \dfrac{x}{6}$, then $du = \dfrac{1}{6} dx$. When $x = 0$, $u = 0$ and when $x = 2\pi$, $u = \pi/3$. Substituting gives $6 \displaystyle\int_0^{\pi/3} \cos^2 u\, du$. Using Example 6 of section 5.6, we have

$$6 \int_0^{\pi/3} \cos^2 u\, du = 6 \left(\dfrac{u}{2} + \dfrac{\sin 2u}{4} \right) \bigg|_0^{\pi/3} = \pi + \dfrac{3\sqrt{3}}{4}.$$

67 $\displaystyle\int_0^\pi \sin^2 5\theta\, d\theta = \int_0^\pi \dfrac{1 - \cos 10\theta}{2} d\theta = \left(\dfrac{\theta}{2} - \dfrac{\sin 10\theta}{20} \right) \bigg|_0^\pi = \dfrac{\pi}{2}.$

69 Let $u = x^3 + 3x^2 - 6x$, and note that $du = 3x^2 + 6x - 6\, dx = 3(x^2 + 2x - 2)\, dx$. Substituting yields $\dfrac{1}{3} \displaystyle\int_8^{36} \dfrac{1}{u} du = \left(\dfrac{1}{3} \ln u \right) \bigg|_8^{36} = \dfrac{1}{3}(\ln 36 - \ln 8) = \dfrac{1}{3} \ln \left(\dfrac{9}{2} \right).$

71 Note that the integrand is an odd function (as it is an odd function divided by an even function). Therefore $\displaystyle\int_{-5}^5 \dfrac{w^3}{\sqrt{w^{50} + w^{20} + 1}} = 0.$

73 Let $u = \dfrac{1}{x}$. Then $du = -\dfrac{1}{x^2} dx$. Substituting yields $-\displaystyle\int \sin u\, du = \cos u + C = \cos\left(\dfrac{1}{x} \right) + C.$

75 Let $u = \tan^{-1} x$. Then $du = \dfrac{1}{1 + x^2} dx$. Substituting yields $\displaystyle\int \dfrac{1}{u} du = \ln |u| + C = \ln |\tan^{-1} x| + C.$

77 Let $u = x + 3$. Then $du = dx$ and $u - 3 = x$. Substituting gives $\displaystyle\int (u - 3) u^{10}\, du = \int (u^{11} - 3u^{10})\, du = \dfrac{u^{12}}{12} - \dfrac{3u^{11}}{11} + C = u^{11} \left(\dfrac{u}{12} - \dfrac{3}{11} \right) = (x+3)^{11} \left(\dfrac{x}{12} - \dfrac{1}{44} \right) = (x+3)^{11} \left(\dfrac{11x - 3}{132} \right).$

79 Let $u = 25 - x^2$. Then $du = -2x\, dx$. Substituting yields $-\dfrac{1}{2} \displaystyle\int_{25}^{16} u^{-1/2} du = -\sqrt{u}\, \bigg|_{25}^{16} = 5 - 4 = 1.$

81 Let $u = 5x$ so that $du = 5 dx$. When $x = \sqrt{2}/5$, $u = \sqrt{2}$, and when $x = 2/5$, $u = 2$. Substituting then gives

$$\int_{\sqrt{2}}^2 \dfrac{du}{u\sqrt{u^2 - 1}} = \sec^{-1} |u| \, \bigg|_{\sqrt{2}}^2 = \left(\sec^{-1} 2 - \sec^{-1} \sqrt{2} \right) = \left(\dfrac{\pi}{3} - \dfrac{\pi}{4} \right) = \dfrac{\pi}{12}.$$

83 $\displaystyle\int_{-10}^{10} \dfrac{x}{\sqrt{200 - x^2}} dx = 0$ because the integrand is an odd function.

85 $\displaystyle\int_0^4 f(x)\, dx = \int_0^3 (2x + 1)\, dx + \int_3^4 (3x^2 + 2x - 8)\, dx = (x^2 + x) \bigg|_0^3 + (x^3 + x^2 - 8x) \bigg|_3^4 = 12 + (48 - 12) = 48.$

87 The area is given by $\displaystyle\int_{-4}^4 (16 - x^2)\, dx = \left(16x - \dfrac{x^3}{3} \right) \bigg|_{-4}^4 = 64 - \dfrac{64}{3} - \left(-64 + \dfrac{64}{3} \right) = \dfrac{256}{3}.$

89 The area is given by $2 \displaystyle\int_0^{2\pi} \sin(x/4)\, dx = 2 \left(-4 \cos(x/4) \right) \bigg|_0^{2\pi} = -8(0 - 1) = 8.$

91

i. $\displaystyle\int_{-1}^1 (x^4 - x^2)\, dx = \left(\dfrac{x^5}{5} - \dfrac{x^3}{3} \right) \bigg|_{-1}^1 = \left(\dfrac{1}{5} - \dfrac{1}{3} \right) - \left(-\dfrac{1}{5} - \left(-\dfrac{1}{3} \right) \right) = -\dfrac{4}{15}.$

ii. Because the region lies completely below the x-axis, the area bounded by the curve and the x-axis is
$-\int_{-1}^{1}(x^4 - x^2)\,dx = \dfrac{4}{15}$.

93 The average height of the arch is given by

$$\dfrac{1}{630}\int_{-315}^{315}\left(1260 - 315\left(e^{0.00418x} + e^{-.00418x}\right)\right)dx$$

$$= \dfrac{1}{630}\left(1260x - \dfrac{315}{0.00418}\left(e^{0.00418x} - e^{-.00418x}\right)\right)\Bigg|_{-315}^{315} \approx 431.5 \text{ ft.}$$

95 The displacement is $\int_{0}^{2} 5\sin\pi t\, dt = -\dfrac{5\cos\pi t}{\pi}\Big|_{0}^{2} = -\dfrac{5}{\pi}(\cos 2\pi - \cos 0) = 0$. The distance traveled is

$$\int_{0}^{1} 5\sin\pi t\, dt + \int_{1}^{2} -5\sin\pi t\, dt = -\dfrac{5\cos\pi t}{\pi}\Big|_{0}^{1} + \dfrac{5\cos\pi t}{\pi}\Big|_{1}^{2}$$

$$= \dfrac{5}{\pi} + \dfrac{5}{\pi} + \dfrac{5}{\pi} + \dfrac{5}{\pi} = \dfrac{20}{\pi}.$$

97 The average value is

$$\dfrac{1}{\ln 2}\int_{0}^{\ln 2} e^{2x}\, dx = \dfrac{1}{2\ln 2} e^{2x}\Big|_{0}^{\ln 2} = \dfrac{4-1}{2\ln 2} = \dfrac{3}{2\ln 2}.$$

99

a. The average value is 2.5. This is because for a straight line, the average value occurs at the midpoint of the interval, which is at the point $(3.5, 2.5)$, so $c = 3.5$.

b. The average value is 3 over the interval $[2,4]$ and 3 over the interval $[4,6]$, so is 3 over the interval $[2,6]$. The function takes on this value at $c = 3$ and $c = 5$.

101 Let $u = 2x$, so that $du = 2\,dx$. We have $\dfrac{1}{2}\int_{2}^{4} f'(u)\,du = \dfrac{1}{2}\cdot(f(4) - f(2)) = \dfrac{f(4)}{2} - 2$. Because we are given that this quantity is 10, we have $f(4) = 24$.

103

By the Fundamental Theorem, $f'(x) = \dfrac{1}{x}$, which is always positive for $x > 1$. Thus f is always increasing. Also, $f(1) = \int_{1}^{1} \dfrac{1}{t}\,dt = 0$.
Also, $f''(x) = -\dfrac{1}{x^2}$ which is always negative, so f is always concave down.

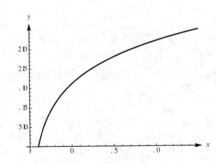

105

a. $F(-2) = \int_{-1}^{-2} f(t)\,dt = \int_{-1}^{-2} t\,dt = \dfrac{t^2}{2}\Big|_{-1}^{-2} = \dfrac{3}{2}$.

$F(2) = \int_{-1}^{2} f(t)\,dt = \int_{-1}^{0} f(t)\,dt + \int_{0}^{2} f(t)\,dt = \int_{-1}^{0} t\,dt + \int_{0}^{2} t^2\,dt = \dfrac{t^2}{2}\Big|_{-1}^{0} + \dfrac{t^3}{6}\Big|_{0}^{2} = \left(0 - \dfrac{1}{2}\right) + \left(\dfrac{8}{6} - 0\right) = \dfrac{5}{6}$.

b. By the Fundamental Theorem of Calculus, $F'(x) = f(x)$, so for $-2 \leq x < 0$ we have $F''(x) = x$.

c. By the Fundamental Theorem of Calculus, $F'(x) = f(x)$, so for $0 \leq x < 2$ we have $F''(x) = \frac{x^2}{2}$.

d. $F'(-1) = -1$ and $F'(1) = \frac{1}{2}$. These represent the rate of change of F at the given points. Because the graph in the exercise is the derivative of F, this is just the value of f at the given points.

e. Note that $F''(x) = \begin{cases} 1 & \text{if } -2 \leq x < 0, \\ x & \text{if } 0 \leq x \leq 2. \end{cases}$ Thus we have $F''(-1) = 1$ and $F''(1) = 1$.

f. The difference $F(x) - G(x) = \int_{-1}^{-2} f(t)\,dt = \frac{3}{2}$, as noted in part (a).

107 By L'hôpital's rule, we have
$$\lim_{x \to 2} \frac{\int_2^x e^{t^2}\,dt}{x-2} = \lim_{x \to 2} \frac{e^{x^2}}{1} = e^4.$$

109

Because x^n and $\sqrt[n]{x}$ are inverse functions of each other, they are symmetric in the square $[0,1] \times [0,1]$ about the line $y = x$. Together, the two regions completely fill up the 1×1 square, so these two areas add to one.

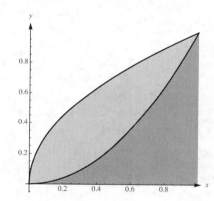

111 Factoring out $\frac{1}{b^2}$ gives $\frac{1}{b^2} \int \frac{dx}{(ax/b)^2 + 1}$. Now let $u = \frac{ax}{b}$, so that $du = \frac{a}{b}\,dx$. Substituting yields $\frac{1}{b^2} \cdot \frac{b}{a} \int \frac{du}{u^2+1} = \frac{1}{ab}\left(\tan^{-1}(u)\right) + C = \frac{1}{ab}\tan^{-1}\left(\frac{ax}{b}\right) + C$.

113 Note that $f'(x) = (x-1)^{15}(x-2)^9$, and that the zeros of f' are at $x = 1$ and $x = 2$.

a. f' is positive and thus f is increasing on $(-\infty, 1)$ and on $(2, \infty)$, while f' is negative and f is decreasing on $(1, 2)$.

b. $f''(x) = 15(x-1)^{14}(x-2)^9 + (x-1)^{15} \cdot 9(x-2)^8 = 3(x-1)^{14}(x-2)^8(8x - 13)$.

f is concave up on $\left(\frac{13}{8}, \infty\right)$ and concave down on $\left(-\infty, \frac{13}{8}\right)$.

c. f has a local maximum at $x = 1$ and a local minimum at $x = 2$.

d. f has an inflection point at $x = \frac{13}{8}$.

115 This follows by differentiating each side of the equation. $\frac{d}{dx}\left(u(x) + 2\int_0^x u(t)\,dt\right) = u'(x) + 2u(x)$, and $\frac{d}{dx}10 = 0$. The reverse is not true, because if $u(x) + 2\int_0^x u(t)\,dt = C$ for any constant C, then it would satisfy the second equation, even if $C \neq 10$.

117 $\int_0^c x(x-c)^2\,dx = \int_0^c (x^3 - 2cx^2 + c^2 x)\,dx = \left(\dfrac{x^4}{4} - 2c\dfrac{x^3}{3} + \dfrac{c^2 x^2}{2}\right)\bigg|_0^c = \dfrac{c^4}{4} - \dfrac{2c^4}{3} + \dfrac{c^4}{2} = \dfrac{c^4}{12}$. This is one when $c = \sqrt[4]{12}$.

Chapter 6

Applications of Integration

6.1 Velocity and Net Change

6.1.1 The position of an object is the coordinate of the object on the line at a given time, often denoted $s(t)$. The displacement over an interval $[a,b]$ is $s(b) - s(a)$, the difference of the object's ending position and beginning position. It can be written as $\int_a^b v(t)\,dt$ where $v(t)$ is the object's velocity at time t. The distance traveled by the object is $\int_a^b |v(t)|\,dt$, the sum of the distance traveled along the line to the right and the distance traveled along the line to the left over the given time interval.

6.1.3 The displacement is given by $\int_a^b v(t)\,dt$, because this quantity is equal to $s(b) - s(a)$.

6.1.5 The value of Q at time t will be given by $Q(t) = Q(0) + \int_0^t Q'(x)\,dx$.

6.1.7

a. The velocity is positive for $0 \leq t < 1$ and for $3 < t < 5$, so the object is moving in the positive direction on those intervals.

b. The displacement is $\int_0^3 v(t)\,dt = 12 - 16 = -4$.

c. The distance traveled is $\int_1^5 |v(t)|\,dt = 16 + 10 = 26$.

d. The displacement is $\int_0^5 v(t)\,dt = 12 - 16 + 10 = 6$.

e. After five hours the object's position is 6 miles from the original position in the positive direction.

6.1.9

a. The displacement is the net area, which is $\frac{1}{2} \cdot 2 \cdot 2 - \frac{1}{2} \cdot \frac{4}{3} \cdot 1 + \frac{1}{2} \cdot \frac{5}{3} \cdot 2 = 3$.

b. The distance traveled is $\frac{1}{2} \cdot 2 \cdot 2 + \frac{1}{2} \cdot \frac{4}{3} \cdot 1 + \frac{1}{2} \cdot \frac{5}{3} \cdot 2 = \frac{13}{3}$

c. $s(5) = s(0) + \int_0^5 v(t)\,dt = 0 + 3 = 3$.

295

d. $s(t) = \begin{cases} \int_0^t (-x+2)\, dx & \text{if } 0 \le t \le 3, \\ \dfrac{3}{2} + \int_3^t (3x-10)\, dx & \text{if } 3 < t \le 4, \\ 2 + \int_4^t (-2x+10)\, dx & \text{if } 4 < t \le 5 \end{cases} = \begin{cases} -\dfrac{t^2}{2} + 2t & \text{if } 0 \le t \le 3, \\ \dfrac{3t^2}{2} - 10t + 18 & \text{if } 3 < t \le 4, \\ -t^2 + 10t - 22 & \text{if } 4 < t \le 5. \end{cases}$

6.1.11

a. Note that $v(t) > 0$ on $[0, \pi/2]$, so the distance traveled is
$$\int_0^{\pi/2} 3\sin 2t\, dt = -\dfrac{3}{2}\cos 2t \Big|_0^{\pi/2} = -\dfrac{3}{2}(-1-1) = 3.$$

b. Note that the function is alternately positive then negative (with the same enclosed area) on each interval of length $\pi/2$. The displacement on $[0, \pi]$ is then $3 - 3 = 0$, the displacement on $[0, 3\pi/2]$ is $3 - 3 + 3 = 3$, and the displacement on $[0, 2\pi]$ is $3 - 3 + 3 - 3 = 0$.

c. The total distance traveled is $3 + 3 + 3 + 3 = 12$.

6.1.13

a. $v(t) = 3t(t-2)$ which is zero for $t = 0$ and $t = 2$. $v(t) > 0$ on $(2, 3)$ and is negative on $(0, 2)$.

b. The displacement is $\int_0^3 (3t^2 - 6t)\, dt = (t^3 - 3t^2)\Big|_0^3 = 3^3 - 3(3)^2 = 0$ m.

c. The total distance traveled is $\int_0^2 (6t - 3t^2)\, dt + \int_2^3 (3t^2 - 6t)\, dt = (3t^2 - t^3)\Big|_0^2 + (t^3 - 3t^2)\Big|_2^3 = (12 - 8) + (27 - 27) - (8 - 12) = 8$ m.

6.1.15

a. $v(t) = 3(t^2 - 6t + 8) = 3(t-4)(t-2)$, which is zero for $t = 4$ and $t = 2$. $v > 0$ on $(0, 2)$ and $(4, 5)$, while v is negative on $(2, 4)$.

b. The displacement is given by $\int_0^5 (3t^2 - 18t + 24)\, dt = (t^3 - 9t^2 + 24t)\Big|_0^5 = 5^3 - 9(5^2) + 24(5) = 20$ m.

c. On the time interval $(0, 2)$ the distance traveled is $\int_0^2 (3t^2 - 18t + 24)\, dt = (t^3 - 9t^2 + 24t)\Big|_0^2 = 8 - 36 + 48 = 20$ m. On the interval $(2, 4)$, the distance traveled is $-\int_2^4 (3t^2 - 18t + 24)\, dt = -(t^3 - 9t^2 + 24t)\Big|_2^4 = -(64 - 144 + 96) + (8 - 36 + 48) = 20 - 16 = 4$ m. On the time interval $(4, 5)$, the distance traveled is $\int_4^5 (3t^2 - 18t + 24)\, dt = (t^3 - 9t^2 + 24t)\Big|_4^5 = 20 - 16 = 4$ m. Therefore the total distance traveled is $20 + 4 + 4 = 28$ m.

6.1.17

a. $s(t) = \int \sin t\, dt = -\cos t + C$, and because $s(0) = 1$, we must have $C = 2$. Thus, $s(t) = 2 - \cos t$.

b. $s(t) = s(0) + \int_0^t \sin x\, dx = 1 + (-\cos x)\Big|_0^t = 2 - \cos t$.

6.1.19

a. $s(t) = \int (6-2t)\,dt = 6t - t^2 + C$, and because $s(0) = 0$, we must have $C = 0$. Thus, $s(t) = 6t - t^2$.

b. Also, $s(t) = s(0) + \int_0^t (6-2x)\,dx = (6x - x^2)\Big|_0^t = 6t - t^2$.

6.1.21

a. $s(t) = \int (9-t^2)\,dt = 9t - \dfrac{t^3}{3} + C$, and because $s(0) = -2$, we must have $C = -2$. Thus, $s(t) = 9t - \dfrac{t^3}{3} - 2$.

b. $s(t) = s(0) + \int_0^t (9-x^2)\,dx = -2 + \left(9x - \dfrac{x^3}{3}\right)\Big|_0^t = 9t - \dfrac{t^3}{3} - 2$.

6.1.23

a. $s(t) = s(0) + \int_0^t 2\pi \cos \pi x\, dx = 2\sin \pi x\big|_0^t = 2\sin \pi t$.

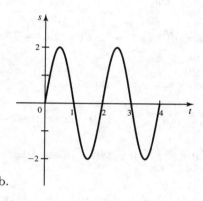

b.

c. The mass reaches its lowest point at $t = 1.5$, $t = 3.5$ and $t = 5.5$.

d. The mass reaches its highest point at $t = .5$, $t = 2.5$, and $t = 4.5$.

6.1.25

a. $s(t) = s(0) + \int_0^t 30(16 - x^2)\,dx = (480x - 10x^3)\Big|_0^t = 480t - 10t^3 = 10t(48 - t^2)$.

b. Because the velocity is positive, this is given by $s(2) - s(0) = 960 - 80 = 880$ miles.

c. The velocity is 400 when $480 - 30t^2 = 400$, or $t = \sqrt{8/3}$. At this point the plane has traveled $s(\sqrt{8/3}) = 480\sqrt{8/3} - 10\sqrt{8/3}^3 \approx 740.290$ miles.

6.1.27

a. The velocity has a maximum of 60 for $20 \leq t \leq 45$. The velocity is 0 at $t = 0$ and at $t = 60$.

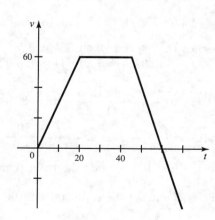

b. $\int_0^{20} 3t\,dt + \int_{20}^{30} 60\,dt = 1200$ m.

c. $1200 + \int_{30}^{45} 60\,dt + \int_{45}^{60} (240 - 4t)\,dt = 1200 + 900 + (240t - 2t^2)\Big|_{45}^{60} = 2550$ m.

d. At time $t = 60$ the automobile is at position 2550. In the following 15 seconds, it moves $\int_{60}^{75} (240 - 4t)\,dt = 450$ feet in the opposite direction, so it is at position $2550 - 450 = 2100$.

6.1.29 $v(t) = v(0) + \int_0^t a(x)\,dx = 70 + \int_0^t -32\,dx = 70 - 32x\Big|_0^t = -32t + 70.$

$s(t) = s(0) + \int_0^t (-32x + 70)\,dx = 10 + (-16x^2 + 70x)\Big|_0^t = -16t^2 + 70t + 10.$

6.1.31 $v(t) = v(0) + \int_0^t -9.8\,dx = 20 + -9.8x\Big|_0^t = -9.8t + 20.$

$s(t) = s(0) + \int_0^t (-9.8x + 20)\,dx = 0 + (-4.9x^2 + 20x)\Big|_0^t = -4.9t^2 + 20t.$

6.1.33 $v(t) = v(0) + \int_0^t -0.01x\,dx = 10 + (-0.005x^2)\Big|_0^t = 10 - 0.005t^2.$

$s(t) = s(0) + \int_0^t (10 - .005x^2)\,dx = \left(10x - \frac{1}{600}x^3\right)\Big|_0^t = 10t - \frac{1}{600}t^3.$

6.1.35 $v(t) = v(0) + \int_0^t \cos 2x\,dx = 5 + \left(\frac{1}{2}\sin 2x\right)\Big|_0^t = \frac{1}{2}\sin 2t + 5.$

$s(t) = s(0) + \int_0^t \left(\frac{1}{2}\sin 2x + 5\right) dx = 7 + \left(-\frac{1}{4}\cos 2x + 5x\right)\Big|_0^t = -\frac{1}{4}\cos 2t + 5t + \frac{29}{4}.$

6.1.37

a. The velocity is given by $v(0) + \int_0^t 88\,dx = 88x\Big|_0^t = 88t$ ft/s.

The position is given by $s(0) + \int_0^t 88x\,dx = 44x^2\Big|_0^t = 44t^2$ ft.

b. The car travels $s(4) = 44 \cdot 16 = 704$ feet.

c. Because a quarter mile is 1320 feet, we need $44t^2 = 1320$, so $t = \sqrt{30} \approx 5.477$ seconds.

d. We need $44t^2 = 300$, so $t \approx 2.611$ seconds.

e. It reaches that speed when $88t = 178$, or $t = 89/44$ seconds. At that time the racer has traveled $s(89/44) = 44(89/44)^2 = 89^2/44 \approx 180.023$ feet.

6.1.39 $v(t) = v(0) + \int_0^t -\frac{1280}{(1+8x)^3}\,dx = 80 + \left(\frac{80}{(1+8x)^2}\right)\Big|_0^t = \frac{80}{(1+8t)^2}.$

$s(t) = s(0) + \int_0^t \frac{80}{(1+8x)^2}\,dx = \left(-\frac{10}{1+8x}\right)\Big|_0^t = 10 - \frac{10}{1+8t}.$

Then in the first 0.2 seconds the train travels $s(0.2) - s(0) = 10 - \frac{10}{2.6} = 10 - \frac{50}{13} = \frac{80}{13} \approx 6.154$ miles.

Between time 0.2 and 0.4 the train travels $s(0.4) - s(0.2) = 10 - \frac{10}{4.2} - \left(10 - \frac{10}{2.6}\right) = \frac{50}{13} - \frac{50}{21} = \frac{400}{273} \approx 1.465$ miles.

6.1. Velocity and Net Change

6.1.41

a. $P(20) = 250 + \int_0^{20} (30 + 30\sqrt{t})\, dt = 250 + \left(30t + 20t^{3/2}\right)\Big|_0^{20} = 250 + 600 + 800\sqrt{5} = 850 + 800\sqrt{5} \approx 2639$ people.

b. $P(t) = 250 + \int_0^t (30 + 30\sqrt{x})\, dx = 250 + \left(30x + 20x^{3/2}\right)\Big|_0^t = 250 + 30t + 20t^{3/2}$ people.

6.1.43

a. $N(20) = 1500 + \int_0^{20} 100e^{-0.25t}\, dt = 1500 + \left(-400e^{-0.25t}\right)\Big|_0^{20} = 1500 + (-400e^{-5} + 400) = 1900 - \dfrac{400}{e^5} \approx 1897$ cells.

$N(40) = 1500 + \int_0^{40} 100e^{-0.25t}\, dt = 1500 + \left(-400e^{-0.25t}\right)\Big|_0^{40} = 1500 + (-400e^{-10} + 400) = 1900 - \dfrac{400}{e^{10}} \approx 1900$ cells.

b. $N(t) = 1500 + \int_0^t 100e^{-0.25x}\, dx = 1500 + \left(-400e^{-0.25x}\right)\Big|_0^t = 1500 + (-400e^{-0.25t} + 400) = -400e^{-0.25t} + 1900$ cells.

6.1.45

a. In the first 35 days the number of barrels produced is $\int_0^{30} 800\, dt + \int_{30}^{35} (2600 - 60t)\, dt = 24000 + 3250 = 27250$.

b. In the first 50 days the number of barrels produced is $27250 + \int_{35}^{40} (2600 - 60t)\, dt + \int_{40}^{50} 200\, dt = 27250 + 1750 + 2000 = 31000$.

c. A constant 200 barrels per day times 20 days yields 4000 barrels.

6.1.47

a. $Q(t) = Q(0) + \int_0^t 10^7 e^{-kx}\, dx = \left(\dfrac{10^7}{-k} e^{-kx}\right)\Big|_0^t = \dfrac{10^7(1 - e^{-kt})}{k}$.

b. $\lim_{t\to\infty} Q(t) = \lim_{t\to\infty} \dfrac{10^7(1 - e^{-kt})}{k} = \dfrac{10^7}{k}$. This represents the total number of barrels extracted if the nation extracts the oil indefinitely where it is assumed that the nation has at least $\dfrac{10^7}{k}$ barrels of oil in reserve.

c. We seek k so that $\dfrac{10^7}{k} = 2 \times 10^9$, which gives $k = \dfrac{1}{200} = 0.005$.

d. We want T so that $(2 \times 10^7) \int_0^T e^{-0.005t}\, dt = 2 \times 10^9$, so $\left(-200e^{-0.005t}\right)\Big|_0^T = 100$, so $1 - e^{-T/200} = 1/2$, so $T = 200 \ln 2 \approx 138.629$ years.

6.1.49

a. $\int_0^2 20(1 + \cos(\pi t/12))\, dt = \left(20t + \dfrac{240}{\pi} \sin(\pi t/12)\right)\Big|_0^2 = 40 + \dfrac{240}{\pi} \cdot \dfrac{1}{2} = 40 + \dfrac{120}{\pi} \approx 78.197\, \text{m}^3$.

b. $Q(t) = \int_0^t 20(1 + \cos(\pi x/12))\, dx$, which is equal to

$$\left(20x + \dfrac{240}{\pi} \sin(\pi x/12)\right)\Big|_0^t = 20t + \dfrac{240}{\pi} \cdot \sin(\pi t/12)\, \text{m}^3.$$

c. The reservoir is full when $20T + \dfrac{240}{\pi}\sin(\pi T/12) = 2500$, which occurs for $T \approx 122.6$ hours.

6.1.51

a. $V(t) = V(0) + \displaystyle\int_0^t -\dfrac{\pi}{2}\sin\dfrac{\pi x}{2}\,dx = 6 + \left(\cos\dfrac{\pi x}{2}\right)\Big|_0^t = 5 + \cos\dfrac{\pi t}{2}$.

b. Because $\sin\dfrac{\pi t}{2}$ is periodic with period 4, the breathing cycle repeats every 4 seconds, so there are $\dfrac{60}{4} = 15$ breaths per minute.

c. The lungs are full at $t = 0$, at which time $V(0) = 6$ L, so this is the capacity of the lungs. The tidal volume is the difference between this amount and the amount in the lungs after each exhalation, which is the minimum value of $V(t)$. This minimum occurs at the first positive zero of $V'(t)$, which is at $t=2$. Because $V(2) = 4$, the tidal volume is $6 - 4 = 2$L.

6.1.53

a. $E = \displaystyle\int_0^{24}(300 - 200\sin(\pi t/12))\,dt = \left(300t + \dfrac{2400}{\pi}\cos(\pi t/12)\right)\Big|_0^{24} = 7200$ MWh. This is equivalent to $7.2 \times 10^6 \cdot 3.6 \times 10^6 = 2.592 \times 10^{13}$ Joules.

b. For one day, $\dfrac{7.2 \times 10^6 \text{ KWh}}{450\,\text{Kwh/kg}} = 16,000$ kg coal needed.

For one year, $16000\,\text{kg} \times 365 = 5,840,000$ kg coal needed.

c. For one day, $\dfrac{7.2 \times 10^6 \text{ KWh}}{1.6 \times 10^4 \text{ Kwh/g}} = 450$ g U-235 need.

For one year, $450 \times 365 = 164,250$ g needed.

d. $\dfrac{7.2 \times 10^6 \text{ KWh/day}}{(200\,\text{KW/turbine}) \cdot (24\,\text{hours/day})} = 1500$ turbines.

6.1.55

a. The additional cost is $\displaystyle\int_{100}^{150}(2000 - .5x)\,dx = \left(2000x - \dfrac{x^2}{4}\right)\Big|_{100}^{150} = 96875$ dollars.

b. The additional cost is $\displaystyle\int_{500}^{550}(2000 - .5x)\,dx = \left(2000x - \dfrac{x^2}{4}\right)\Big|_{500}^{550} = 86875$ dollars.

6.1.57

a. The additional cost is $\displaystyle\int_{100}^{150}(300 + 10x - .01x^2)\,dx = \left(300x + 5x^2 - \dfrac{x^3}{300}\right)\Big|_{100}^{150} = 69583.33$ dollars.

b. The additional cost is $\displaystyle\int_{500}^{550}(300 + 10x - .01x^2)\,dx = \left(300x + 5x^2 - \dfrac{x^3}{300}\right)\Big|_{500}^{550} = 139583.33$ dollars.

6.1.59

a. False. This would only be the case if the motion was all in the same direction. If the object changes direction at all, then the distance traveled is greater than the displacement.

b. True. This is because $v(t) = |v(t)|$ in this case.

c. True. This is because $R(t) > 0$ for $0 < t < 10$, but $R(t) < 0$ for $t > 10$.

d. True. The cost of increasing production from A to B is given by $\int_A^B C'(t)\,dt$, which is geometrically the area under the curve $y = C'(x)$ from A to B. If C' is positive and decreasing, there is more area under the curve from A to $2A$ than from $2A$ to $3A$.

6.1.61 The distance traveled is $\int_0^4 \left(1 - \dfrac{t^2}{16}\right) dt = \left(t - \dfrac{t^3}{48}\right)\Big|_0^4 = \dfrac{8}{3}$. So the same distance could have been traveled over the given time period at a constant velocity of $\dfrac{8/3}{4} = \dfrac{2}{3}$.

6.1.63 The distance traveled is $\int_0^5 t\sqrt{25 - t^2}\,dt = \dfrac{1}{2}\int_0^{25} \sqrt{u}\,du = \left(\dfrac{1}{3} u^{3/2}\right)\Big|_0^{25} = \dfrac{125}{3}$. So the same distance could have been traveled over the given time period at a constant velocity of $\dfrac{125/3}{5} = \dfrac{25}{3}$.

6.1.65

a.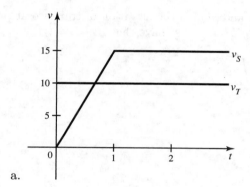

b. After 1 hour, Theo has ridden $1 \cdot 10 = 10$ miles, and Sasha has ridden $\frac{1}{2} \cdot 15 = 7.5$ miles, so Theo has ridden farther.

c. After 2 hours, Theo has ridden $2 \cdot 10 = 20$ miles, and Sasha has ridden $7.5 + 15 \cdot 1 = 22.5$ miles, so Sasha has ridden farther.

d. The times when they arrive at the various mile markers are in the following table:

	10	15	20
Theo	1	3/2	2
Sasha	7/6	3/2	11/6

Note that Theo hits the 10 mile marker first, then they are tied as they hit the 15 mile marker, and Sasha hits the 20 mile marker first. The area under v_S is the same as the area under v_T for $t = 1.5$, for $t < 1.5$ the area under v_T is greater, and for $t > 1.5$, the area under v_S is greater.

e. Theo will then hit the 20 mile mark in $18.8/10 = 1.88$ hours. Sasha hits the 20 mile mark at $t = 11/6 \approx 1.833$ hours, so Sasha will win.

f. A head start of 0.2 hours is equivalent for Theo of $10 \cdot 0.2 = 2$ miles. It will take him $18/10 = 1.8$ hours to ride the other 18 miles, while it still takes Sasha about 1.83 hours to cover 20 miles, so Theo will win.

6.1.67 Let $D(t)$ equal the depth of the snow t hours after it started snowing and let $t = T$ correspond to noon. This means that $D(0) = 0$ and because the snow piles up on the road at a constant rate, $D'(t) = k$ for some positive constant k. It follows that

$$D(t) = D(0) + \int_0^k = kx\Big|_0^t = kt.$$

Let $v(t)$ equal the velocity of the plow t hours after it started snowing. The plowing rate is inversely proportional to the depth of the snow which means that $v(t) = \dfrac{C}{kt}$ for some constant C and for $t \geq T$.

Because the plow goes twice as far from noon to 1 PM as from 1 PM to 2 PM,

$$\int_T^{T+1} \frac{C}{kt}\,dt = 2\int_{T+1}^{T+2} \frac{C}{kt}\,dt.$$

Evaluating the integrals, we have $\frac{C}{k}\ln\left(\frac{T+1}{T}\right) = \frac{2C}{k}\ln\left(\frac{T+2}{T+1}\right)$, or $\ln\left(\frac{T+1}{T}\right) = \ln\left(\frac{T+2}{T+1}\right)^2$. This implies that $(T+1)^3 = T(T+2)^2$, Expanding both sides and collecting terms gives the quadratic equation $T^2 + T - 1 = 0$, which has a positive root of $T = \dfrac{-1+\sqrt{5}}{2} \approx 0.618$ hours ≈ 37 minutes. So it started snowing at approximately 11:23 AM.

6.1.69 Using the Fundamental Theorem, we have

$$\int_a^b f'(x)\,dx = f(b) - f(a) = g(b) - g(a) = \int_a^b g'(x)\,dx.$$

6.1.71 If $f(x)$ is the elevation of trail one at position x and $g(x)$ is the elevation of trail two at position x, then because we are given that $f(a) = g(a)$ and $f(b) = g(b)$, we must have that

$$\int_a^b f'(x)\,dx = \int_a^b g'(x)\,dx,$$

so the two trails have the same net change in elevation.

6.2 Regions Between Curves

6.2.1 $\displaystyle\int_a^b (f(x) - g(x))\,dx + \int_b^c (g(x) - f(x))\,dx.$

6.2.3

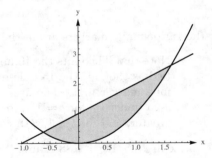

6.2.5

a. The area is given by $\displaystyle\int_0^1 ((2-x) - x)\,dx = \int_0^1 (2 - 2x)\,dx = (2x - x^2)\Big|_0^1 = 1.$

b. The area is that of a triangle with base 2 and height 1, so its area is $\dfrac{1}{2}\cdot 2 \cdot 1 = 1.$

6.2.7 $\displaystyle\int_0^1 y\,dy + \int_1^2 (2-y)\,dy.$

6.2.9 The curves intersect when $x = x^2 - 2$, or $(x+1)(x-2) = 0$, so at $x = -1$ and $x = 2$. The area is

$$\int_{-1}^2 (x - (x^2 - 2)) = \left(2x + \frac{x^2}{2} - \frac{x^3}{3}\right)\Big|_{-1}^2 = 4.5.$$

6.2.11 The curves intersect when $\dfrac{6}{x^2+1} = 3x^2$, or $3x^2(x^2+1) = 6$. This can be written as $3x^4+3x^2-6 = 0$, or $(x^4 + x^2 - 2) = 0$. Factoring gives $(x^2 + 2)(x^2 - 1) = (x^2 + 2)(x - 1)(x + 1) = 0$, so the curves intersect at $x = -1$ and $x = 1$. Using symmetry, the area is given by

$$2\int_0^1 \left(\frac{6}{x^2+1} - 3x^2\right) dx = 2\left(6\tan^{-1} x - x^3\right)\Big|_0^1 = 2\left(\frac{6\pi}{4} - 1\right) = 3\pi - 2.$$

6.2.13 By inspection, the curves intersect at $x = 1$. The area is given by

$$\int_0^1 (3 - x - 2^x)\, dx = \left(3x - x^2/2 - \frac{2^x}{\ln 2}\right)\Big|_0^1 = 3 - \frac{1}{2} - \frac{2}{\ln 2} - \left(-\frac{1}{\ln 2}\right) = \frac{5}{2} - \frac{1}{\ln 2}.$$

6.2.15 The curves intersect at $\pi/4$, so the area is given by $\displaystyle\int_0^{\pi/4} \sin x\, dx + \int_{\pi/4}^{\pi/2} \cos x\, dx = (-\cos x)\Big|_0^{\pi/4} + (\sin x)\Big|_{\pi/4}^{\pi/2} = (1 - \sqrt{2}/2) + (1 - \sqrt{2}/2) = 2 - \sqrt{2}.$

6.2.17 The curves intersect when $x^2 - 4x = -2x$, or $x^2 - 2x = x(x - 2) = 0$, so $x = 0$ and $x = 2$. The area is given by

$$\int_0^2 (-2x - (x^2 - 4x))\, dx = \int_0^2 (2x - x^2)\, dx = \left(x^2 - \frac{x^3}{3}\right)\Big|_0^2 = 4 - \frac{8}{3} = \frac{4}{3}.$$

6.2.19 The curves intersect when $2y = y^2 - 3$, or $y^2 - 2y - 3 = (y - 3)(y + 1) = 0$. This occurs for $y = $ and $y = -1$. The area is given by

$$\int_{-1}^3 (2y - (y^2 - 3))\, dy = \left(y^2 - \frac{y^3}{3} + 3y\right)\Big|_{-1}^3 = (9 - 9 + 9) - \left(1 + \frac{1}{3} - 3\right) = \frac{32}{3}.$$

6.2.21 Note that the two curves intersect where $y = 2 - y^2$, so for $y^2 + y - 2 = (y + 2)(y - 1) = 0$. So they intersect for $y = -2$ and $y = 1$. The area is given by

$$\int_{-2}^1 (4 - 2y - 2y^2)\, dy = \left(4y - y^2 - \frac{2}{3}y^3\right)\Big|_{-2}^1 = \left(4 - 1 - \frac{2}{3}\right) - \left(-8 - 4 + \frac{16}{3}\right) = 9.$$

6.2.23 The area is given by

$$\int_{-4}^5 \left(8 - y - \frac{(y-2)^2}{3}\right) dy = \left(8y - \frac{y^2}{2} - \frac{(y-2)^3}{9}\right)\Big|_{-4}^5 = 40 - \frac{25}{2} - 3 - (-32 - 8 + 24) = \frac{81}{2}.$$

6.2.25 The two curves meet at $x = 0$; at that point $y = 1$. So the integration is from $y = 0$ to $y = 1$. Solving for x gives the curves $x = 2y^2 - 2$ and $x = 1 - y^2$. Thus the area of the region is

$$\int_0^1 ((1 - y^2) - (2y^2 - 2))\, dy = \int_0^1 (-3y^2 + 3)\, dy = (-y^3 + 3y)\Big|_0^1 = 2.$$

6.2.27 The curves intersect when $x^2 - 2 = x$, or $x^2 - x - 2 = (x - 2)(x + 1) = 0$, or $x = -1$ and $x = 2$. The relevant point for the shaded region is $(-1, -1)$.

The area is

$$\int_{-1}^0 (y - (-\sqrt{y+2}))\, dy = \int_{-1}^0 (y + \sqrt{y+2})\, dy = \left(\frac{y^2}{2} + \frac{2(y+2)^{3/2}}{3}\right)\Big|_{-1}^0 = \frac{2\sqrt{8}}{3} - \left(\frac{1}{2} + \frac{2}{3}\right) = \frac{8\sqrt{2} - 7}{6}.$$

6.2.29 The curves $x = y^2 - 3y + 12$ and $x = -2y^2 - 6y + 30$ intersect where $y^2 - 3y + 12 = -2y^2 - 6y + 30$, or $3y^2 + 3y - 18 = 3(y+3)(y-2) = 0$. So the points of intersection are $y = -3$ and $y = 2$. Thus the area is

$$\int_{-3}^{2} ((-2y^2 - 6y + 30) - (y^2 - 3y + 12))\, dy = \int_{-3}^{2} (-3y^2 - 3y + 18)\, dy$$

$$= \left(-y^3 - \frac{3}{2}y^2 + 18y\right)\bigg|_{-3}^{2}$$

$$= -8 - 6 + 36 - 27 + \frac{27}{2} + 54 = \frac{125}{2}.$$

6.2.31

a. The area is given by $\int_0^1 (\sqrt{x} - x^3)\, dx$.

b. The area can also be written $\int_0^1 (\sqrt[3]{y} - y^2)\, dy$.

6.2.33 The area is $\int_0^1 \left(10\sqrt{t} - 7t\right) dt = \left(\frac{20}{3}t^{3/2} - \frac{7t^2}{2}\right)\bigg|_0^1 = \frac{20}{3} - \frac{7}{2} \approx 3.17$ km.

The area under the curve v_1 equals the distance traveled by the first runner and the area under the curve v_2 is the distance traveled by the second runner. So the area between the curves is the difference in the distance covered by the faster runner and the slower runner, which means that the faster runner ran about 3.17 km further than the slower runner.

6.2.35

a. The area of R_1 is given by $\int_0^1 6x(2 - x^2)^2\, dx$. Let $u = 2 - x^2$. Then $du = -2x\, dx$, and substituting yields

$$\int_2^1 -3u^2\, du = (-u^3)\bigg|_2^1 = -1 - (-8) = 7.$$

b. The area of R_1 minus the area of a triangle with base 1 and height 6 is equal to the area of R_2. Thus, the area of R_2 is $7 - 3 = 4$.

6.2.37 The nonvertical lines intersect when $4x + 4 = 6x + 6$, or $x = -1$. The vertical line $x = 4$ intersects both nonvertical lines when $x = 4$. The area is given by $\int_{-1}^{4} (6x + 6 - (4x + 4))\, dx =$

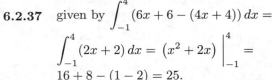

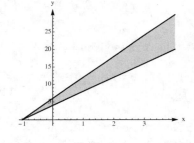

$\int_{-1}^{4} (2x + 2)\, dx = (x^2 + 2x)\bigg|_{-1}^{4} = 16 + 8 - (1 - 2) = 25.$

6.2.39 The curves e^x and e^{-2x} intersect when $e^x = e^{-2x}$, or $e^{3x} = 1$, which occurs only for $x = 0$. The vertical line $x = \ln 4$ clearly intersects the curves for $x = \ln 4$. The area is given by $\int_0^{\ln 4} (e^x -$

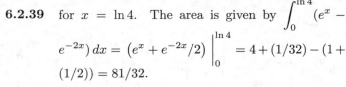

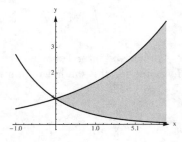

$e^{-2x})\, dx = \left(e^x + e^{-2x}/2\right)\bigg|_0^{\ln 4} = 4 + (1/32) - (1 + (1/2)) = 81/32.$

6.2.41 The curves intersect when $\frac{2}{1+x^2} = 1$, or $x^2 + 1 = 2$, or $x^2 - 1 = (x-1)(x+1) = 0$. So the intersections occur when $x = \pm 1$. Using symmetry, the area is $2\int_0^1 \left(\frac{2}{1+x^2} - 1\right) dx = 2\left(2\tan^{-1} x - x\right)\Big|_0^1 = \pi - 2$.

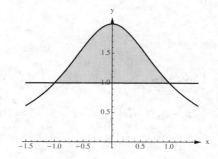

6.2.43 The curves intersect when $x = 1/x$, or $x^2 = 1$. This occurs in the first quadrant when $x = 1$. The area is given by $\int_0^1 x\,dx + \int_1^2 1/x\,dx = \left(x^2/2\right)\Big|_0^1 + (\ln x)\Big|_1^2 = 1/2 + \ln 2$.

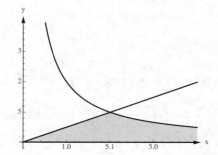

6.2.45 The curves intersect at $x = \pm 1$. Using symmetry, the area is
$$2\int_0^1 (2 - x - x^2)\,dx = 2\left(2x - \frac{1}{2}x^2 - \frac{1}{3}x^2\right)\Big|_0^1 = \frac{7}{3}.$$

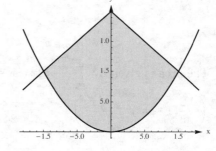

6.2.47 $\int_2^6 (x/2 - |x-3|)\,dx = \int_2^3 (x/2 - (3-x))\,dx + \int_3^6 (x/2 - (x-3))\,dx = \int_2^3 (3x/2 - 3)\,dx + \int_3^6 (3 - x/2)\,dx = (3x^2/4 - 3x)\Big|_2^3 + (3x - x^2/4)\Big|_3^6 = 27/4 - 9 - (3 - 6) + (18 - 9 - (9 - 9/4)) = 3/4 + 9/4 = 3$.

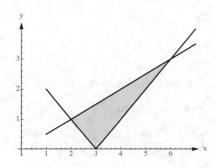

6.2.49

The area is given by $\int_0^8 (4 - x^{2/3})\,dx =$
$\left(4x - 3x^{5/3}/5\right)\Big|_0^8 = 32 - 96/5 = 64/5.$

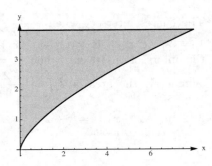

6.2.51

The area is given by $\int_0^{\ln 2} (2e^{-x} + 1 - e^x)\,dx =$
$\left(-2e^{-x} + x - e^x\right)\Big|_0^{\ln 2} = -1 + \ln 2 - 2 - (-2 - 1) = \ln 2.$

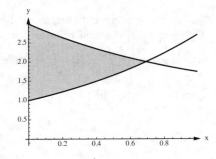

6.2.53

The area is given by $\int_0^{\sqrt{3}/2} (2x\sqrt{1-x^2} - x)\,dx =$
$\left(-\frac{2}{3}(1-x^2)^{3/2} - \frac{x^2}{2}\right)\Big|_0^{\sqrt{3}/2} = \frac{-2}{24} - \frac{3}{8} + \frac{2}{3} = \frac{5}{24}.$

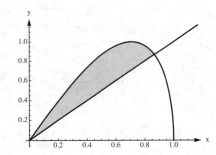

6.2.55

The curves intersect at the point $(9,3)$. The area is given by $\int_0^3 \left(\frac{y}{2} + \frac{15}{2} - y^2\right) dy =$
$\left(\frac{y^2}{4} + \frac{15y}{2} - \frac{y^3}{3}\right)\Big|_0^3 = \frac{9}{4} + \frac{45}{2} - 9 = \frac{63}{4}.$

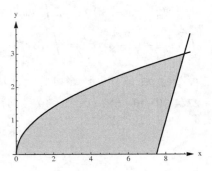

6.2.57

The curves intersect for $x^2 - 2x + 1 = 5x - 9$, or $x^2 - 7x + 10 = (x-5)(x-2) = 0$, so for $x = 2$ and $x = 5$. The area is given by $\int_2^5 (5x - 9 - (x-1)^2)\,dx = \left(5x^2/2 - 9x - (x-1)^3/3\right)\Big|_2^5 = 125/2 - 45 - 64/3 - (10 - 18 - (1/3)) = 4.5$.

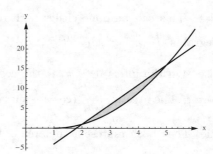

6.2.59

This is given by $\int_0^4 (3y - (y^2 - y))\,dy = \int_0^4 (4y - y^2)\,dy = \left(2y^2 - y^3/3\right)\Big|_0^4 = 32 - 64/3 = 32/3$.

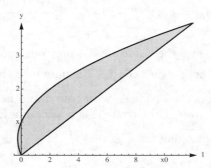

6.2.61

This area is given by $\int_{1/2}^2 (5/2 - 1/x - x)\,dx = \left(5x/2 - \ln x - x^2/2\right)\Big|_{1/2}^2 = 5 - \ln 2 - 2 - (5/4 - \ln(1/2) - 1/8) = 15/8 - 2\ln 2$.

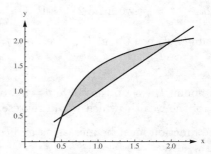

6.2.63 $y = 8x$ and $y = 9 - x^2$ intersect when $x^2 + 8x - 9 = (x+9)(x-1) = 0$, so at $x = 1$ in the first quadrant. $y = \frac{5}{2}x$ and $y = 9 - x^2$ intersect when $x^2 + \frac{5}{2}x - 9 = 0$, or $2x^2 + 5x - 18 = (2x+9)(x-2) = 0$, so at $x = 2$ in the first quadrant. Thus the area of the region is

$$A = \int_0^1 \left(8x - \frac{5}{2}x\right) dx + \int_1^2 \left(9 - x^2 - \frac{5}{2}x\right) dx$$
$$= \int_0^1 \frac{11}{2}x\,dx + \int_1^2 \left(9 - x^2 - \frac{5}{2}x\right) dx$$
$$= \left(\frac{11}{4}x^2\right)\Big|_0^1 + \left(9x - \frac{x^3}{3} - \frac{5}{4}x^2\right)\Big|_1^2$$
$$= \frac{11}{4} + 18 - \frac{8}{3} - 5 - 9 + \frac{1}{3} + \frac{5}{4} = \frac{17}{3}.$$

6.2.65

a. False. This can be done either with respect to x or with respect to y. For the latter, the relevant integral is $\int_0^1 (y - y^2)\,dy$.

b. False. On the interval $(0, \pi/4)$ the cosine function is greater, but on $(\pi/4, \pi/2)$ the sine function is greater. The area is $\int_0^{\pi/4} (\cos x - \sin x)\,dx + \int_{\pi/4}^{\pi/2} (\sin x - \cos x)\,dx$.

c. True. They both represent the area of the region in the first quadrant under $y = x$ and above $y = x^2$.

6.2.67 $A = 2 \int_0^1 y^2 \sqrt{1 - y^3}\,dy = \frac{2}{3} \int_0^1 \sqrt{u}\,du$ (where $u = 1 - y^3$). So $A = \frac{2}{3} \left(\frac{2}{3} u^{3/2} \right) \Big|_0^1 = \frac{4}{9}$.

6.2.69 $A_n = \int_0^1 (x^{1/n} - x)\,dx = \left(\frac{nx^{(n+1)/n}}{n+1} - x^2/2 \right) \Big|_0^1 = \frac{n}{n+1} - \frac{1}{2} = \frac{n-1}{2n+2}$.

6.2.71 Using the result of the previous problem, $\lim_{n \to \infty} A_n = \lim_{n \to \infty} \frac{n-1}{n+1} = 1$. As $n \to \infty$, the region in question approaches the 1×1 square over the interval $[0, 1]$, which has area 1.

6.2.73 This is a triangle with base 2 and height 1, so it has area 1. Note that for $0 \leq x \leq 1$ the line which forms the top of the triangle is $y = x$, and for $1 \leq x \leq 2$, the line is $y = 2 - x$. Integrating with respect to y, we are looking for k so that $\int_0^k (2 - y - y)\,dy = \frac{1}{2}$. Now,

$$\int_0^k (2 - 2y)\,dy = (2y - y^2) \Big|_0^k = 2k - k^2.$$

Setting $2k - k^2 = \frac{1}{2}$, we have $2k^2 - 4k + 1 = 0$, which has roots $k = \frac{2 \pm \sqrt{2}}{2}$. Only the negative sign choice gives a value between 0 and 1, so we want $k = \frac{2 - \sqrt{2}}{2} = 1 - \frac{\sqrt{2}}{2}$.

6.2.75

a. The point (n, n) on the curve $y = x$ would represent the notion that the lowest $p\%$ of the society owns $p\%$ of the wealth, which would represent a form of equality.

b. The function must be increasing and concave up because the poorest $p\%$ cannot own more than $p\%$ of the wealth.

c. $y = x^{1.1}$ is closest to $y = x$, and $y = x^4$ is furthest from $y = x$.

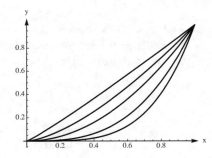

d. Note that $B = \int_0^1 L(x)\,dx$, and $A + B = 1/2$, so $A = \frac{1}{2} - \int_0^1 L(x)\,dx$. Then $G = \frac{A}{A+B} = \frac{A}{1/2} = 2A = 1 - 2\int_0^1 L(x)\,dx$.

e. For $L(x) = x^p$, we have $G = 1 - 2\int_0^1 x^p\,dx = 1 - 2\left(\dfrac{x^{p+1}}{p+1}\right)\Big|_0^1 = 1 - \dfrac{2}{p+1} = \dfrac{p-1}{p+1}$. So we have

p	1.1	1.5	2	3	4
G	1/21	1/5	1/3	1/2	3/5

f. For $p = 1$ we have $G = 0$. Because $\lim\limits_{p\to\infty} \dfrac{p-1}{p+1} = 1$, the largest possible value of G approaches 1.

g. For $L(x) = 5x^2/6 + x/6$, note that $L(0) = 0$, $L(1) = 1$, $L'(x) = 5x/3 + 1/6 > 0$ on $[0, 1]$, and $L''(x) = 5/3 > 0$ as well. The Gini index is

$$G = 1 - 2\int_0^1 (5x^2/6 + x/6)\,dx = 1 - 2\left(5x^3/18 + x^2/12\right)\Big|_0^1 = 1 - 2(5/18 + 1/12) = 1 - 5/9 - 1/6 = 5/18.$$

6.2.77

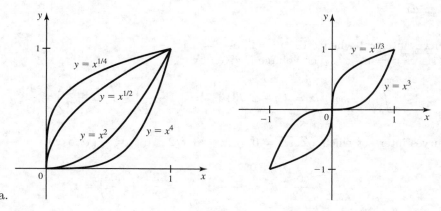

a.

b. $A_n(x)$ is the net area of the region between the graphs of f and g from 0 to x.

c. Note that $\int_0^1 (f(s) - g(s))\,ds = \left(\dfrac{s^{n+1}}{n+1} - \dfrac{s^{(1/n)+1}}{(1/n)+1}\right)\Big|_0^1 = \dfrac{1}{n+1} - \dfrac{n}{n+1} = \dfrac{1-n}{n+1} < 0$. So we seek the smallest c so that $\int_1^c (f(s) - g(s))\,ds = \dfrac{n-1}{n+1}$. This occurs when $\left(\dfrac{s^{n+1}}{n+1} - \dfrac{s^{(1/n)+1}}{(1/n)+1}\right)\Big|_1^c = \dfrac{n-1}{n+1}$, or $\dfrac{c^{n+1}}{n+1} = \dfrac{c^{(n+1)/n}}{(n+1)/n}$, or $c^{n+1} = nc^{(n+1)/n}$, or $c^{n-(1/n)} = n$, so $c = n^{n/(n^2-1)}$. As n increases, this root increases as well.

6.3 Volume by Slicing

6.3.1 $A(x)$ is the area of the cross section through the solid at the point x.

6.3.3

a. $A(x) = (\sqrt{3-x})^2 = 3 - x$.

b. $\int_0^2 (3-x)\,dx$.

6.3.5

a. The radius of a cross section is $\sqrt{\cos x}$.

b. $A(x) = \pi(\sqrt{\cos x})^2 = \pi \cos x$.

c. $V = \int_0^2 \pi \cos x\,dx$.

6.3.7

a. The outer radius is given by $\sqrt{x}+1$.

b. The inner radius is 1.

c. $A(x) = \pi\left(\left(\sqrt{x}+1\right)^2 - 1\right)$.

d. $V = \int_0^4 \pi\left(\left(\sqrt{x}+1\right)^2 - 1\right) dx$.

6.3.9

a. The radius is $\sqrt{x}$.

b. $A(x) = \pi(\sqrt{x})^2 = \pi x$.

c. $V = \int_0^4 \pi x\, dx$.

6.3.11 The curves intersect when $\sqrt{1-x^2} = 0$, or $x = \pm 1$. $A(x) = (\sqrt{1-x^2})^2 = 1 - x^2$. The volume is given by

$$\int_{-1}^1 (1-x^2)\, dx = (x - x^3/3)\Big|_{-1}^1 = \left(1 - \frac{1}{3}\right) - \left(-1 + \frac{1}{3}\right) = \frac{4}{3}.$$

6.3.13 The curves intersect when $\sqrt{\cos x} = 0$, or $x = \pm \pi/2$. $A(x) = \frac{1}{2}(\sqrt{\cos x})^2 = \frac{\cos x}{2}$. The volume is given by

$$\int_{-\pi/2}^{\pi/2} \frac{\cos x}{2}\, dx = \frac{\sin x}{2}\Big|_{-\pi/2}^{\pi/2} = \frac{1}{2} - \left(-\frac{1}{2}\right) = 1.$$

6.3.15 For each value of x, the height of the triangle is $2 - x$, which is the diameter of the semicircle, so the area of that semicircle is

$$A(x) = \frac{\pi}{2}\left(\frac{2-x}{2}\right)^2 = \frac{\pi}{8}(4 - 4x + x^2).$$

Thus the volume is

$$V = \int_0^2 A(x)\, dx = \int_0^2 \frac{\pi}{8}(4 - 4x + x^2)\, dx = \frac{\pi}{8}\left(4x - 2x^2 + x^3/3\right)\Big|_0^2 = \frac{\pi}{3}.$$

6.3.17 $V = \pi \int_0^3 4x^2\, dx = 4\pi\left(x^3/3\right)\Big|_0^3 = 36\pi$.

6.3.19 $V = \pi \int_0^{\ln 4} e^{-2x}\, dx = \frac{-\pi}{2}\left(e^{-2x}\right)\Big|_0^{\ln 4} = \frac{-\pi}{2}((1/16) - 1) = \frac{15\pi}{32}$.

6.3.21 $V = \pi \int_{-1}^1 \left(\frac{1}{\sqrt{1+y^2}}\right)^2 dy = \pi \tan^{-1} y\Big|_{-1}^1 = \pi(\pi/4 + \pi/4) = \pi^2/2$.

6.3.23 $V = \pi \int_0^4 ((2\sqrt{x})^2 - x^2)\, dx = \pi\left(2x^2 - x^3/3\right)\Big|_0^4 = \pi(32 - 64/3) = \frac{32\pi}{3}$.

6.3.25 $V = \pi \int_{\ln 2}^{\ln 3} ((e^{x/2})^2 - (e^{-x/2})^2)\, dx = \pi \int_{\ln 2}^{\ln 3} (e^x - e^{-x})\, dx = \pi\left(e^x + e^{-x}\right)\Big|_{\ln 2}^{\ln 3} = \pi((3 + 1/3) - (2 + 1/2)) = 5\pi/6$.

6.3.27 $V = \pi \int_0^1 (1 - y^{2/3}) \, dy = \pi \left(y - \dfrac{3}{5} y^{5/3} \right) \Big|_0^1 = \dfrac{2\pi}{5}$.

6.3.29 $V = \pi \int_0^{\pi/2} \cos^2 x \, dx = \dfrac{\pi}{2} \int_0^{\pi/2} (1 + \cos 2x) \, dx = \dfrac{\pi}{2} \left(x + \dfrac{\sin 2x}{2} \right) \Big|_0^{\pi/2} = \dfrac{\pi^2}{4}$.

6.3.31 $V = \pi \int_0^{\pi} \sin^2 x \, dx = \dfrac{\pi}{2} \int_0^{\pi} (1 - \cos 2x) \, dx = \dfrac{\pi}{2} \left(x - \dfrac{\sin 2x}{2} \right) \Big|_0^{\pi} = \dfrac{\pi^2}{2}$.

6.3.33 $V = \pi \int_0^{\pi/4} \sin^2 y \, dy = \dfrac{\pi}{2} \int_0^{\pi/4} (1 - \cos 2y) \, dy = \dfrac{\pi}{2} \left(y - \dfrac{\sin 2y}{2} \right) \Big|_0^{\pi/4} = \dfrac{\pi}{2} \left(\dfrac{\pi}{4} - \dfrac{1}{2} \right) = \dfrac{\pi(\pi - 2)}{8}$.

6.3.35 $V = \pi \int_0^{\pi/2} (\sin x - \sin^2 x) \, dx = \pi \int_0^{\pi/2} \left(\sin x - \dfrac{1}{2}(1 - \cos 2x) \right) dx = \pi \left(\dfrac{\sin 2x}{4} - \cos x - \dfrac{x}{2} \right) \Big|_0^{\pi/2}$
$= \pi \left(1 - \dfrac{\pi}{4} \right) = \dfrac{4\pi - \pi^2}{4}$.

6.3.37 $V = \pi \int_0^2 (16 - y^4) \, dy = \pi \left(16y - \dfrac{y^5}{5} \right) \Big|_0^2 = \dfrac{128\pi}{5}$.

6.3.39 $V = \pi \int_2^6 \left(\dfrac{1}{\sqrt{x}} \right)^2 dx = \pi (\ln x) \Big|_2^6 = \pi \ln 3$.

6.3.41 $V = \pi \int_0^2 e^{2x} \, dx = \dfrac{\pi}{2} (e^{2x}) \Big|_0^2 = \dfrac{\pi}{2}(e^4 - 1)$.

6.3.43 $V = \pi \int_0^{\ln 8} (e^{2y} - e^y) \, dy = \pi (e^{2y}/2 - e^y) \Big|_0^{\ln 8} = \pi \left(32 - 8 - \left(\dfrac{1}{2} - 1 \right) \right) = \dfrac{49\pi}{2}$.

6.3.45 About the x-axis: $V_x = \pi \int_0^5 4x^2 \, dx = \pi (4x^3/3) \Big|_0^5 = \dfrac{500\pi}{3}$.

About the y-axis: $V_y = \pi \int_0^{10} (25 - y^2/4) \, dy = \pi (25y - y^3/12) \Big|_0^{10} = \dfrac{500\pi}{3}$.

The volumes are the same.

6.3.47 About the x-axis: $V_x = \pi \int_0^1 (1 - x^3)^2 \, dx = \pi \int_0^1 (1 - 2x^3 + x^6) \, dx = \pi (x - x^4/2 + x^7/7) \Big|_0^1 = \dfrac{9\pi}{14}$.

About the y-axis: $V_y = \pi \int_0^1 (\sqrt[3]{1-y})^2 \, dy = -\pi \left(\dfrac{3}{5}(1-y)^{5/3} \right) \Big|_0^1 = -\pi(0 - 3/5) = \dfrac{3\pi}{5}$. The volume V_x is bigger.

6.3.49 $V = \pi \int_0^1 \left(1 - (1 - \sqrt{x}) \right)^2 dx = \pi \int_0^1 x \, dx = \pi \dfrac{x^2}{2} \Big|_0^1 = \dfrac{\pi}{2}$.

6.3.51 $V = \pi \int_0^{\pi/3} (2 - (2 - \sec y))^2 \, dy = \pi \int_0^{\pi/3} \sec^2 y \, dy = \pi \tan y \Big|_0^{\pi/3} = \pi \sqrt{3}$.

6.3.53 Using the disk method, a disk located at x has a radius of $1 - \sqrt{x}$ when revolved about the line $y = 1$, so the volume is

$V = \pi \int_0^1 (1 - \sqrt{x})^2 \, dx = \pi \int_0^1 (1 - 2\sqrt{x} + x) \, dx = \pi \left(x - \dfrac{4}{3} x^{3/2} + \dfrac{1}{2} x^2 \right) \Big|_0^1 = \pi \left(1 - \dfrac{4}{3} + \dfrac{1}{2} \right) = \dfrac{\pi}{6}$.

6.3.55 We use the washer method. For each x, the washer has outer radius $2 + 2\sin x$ and inner radius 2, so the volume is

$$V = \pi \int_0^\pi \left((2 + 2\sin x)^2 - 2^2\right) dx = \pi \int_0^\pi \left(8\sin x + 4\sin^2 x\right) dx$$

$$= \pi \int_0^\pi \left(8\sin x + 2(1 - \cos 2x)\right) dx = \pi(-8\cos x + 2x - \sin 2x)\Big|_0^\pi$$

$$= \pi(8 + 2\pi + 8) = 2\pi(\pi + 8).$$

6.3.57 We use the washer method. For each x, the outer radius is $1 + \sin x$ and the inner radius is $1 + (1 - \sin x) = 2 - \sin x$. Thus the volume is

$$V = \pi \int_{\pi/6}^{5\pi/6} \left((1 + \sin x)^2 - (2 - \sin x)^2\right) dx = \pi \int_{\pi/6}^{5\pi/6} (6\sin x - 3) dx$$

$$= \pi(-6\cos x - 3x)\Big|_{\pi/6}^{5\pi/6} = \pi\left(3\sqrt{3} - \frac{5\pi}{2} + 3\sqrt{3} + \frac{\pi}{2}\right) = \pi\left(6\sqrt{3} - 2\pi\right).$$

6.3.59 The lines $x = 2 - y$ and $x = 1 - \frac{y}{2}$ meet at the point $(0, 2)$. For each y, we obtain a washer whose outer radius is $3 - (1 - \frac{y}{2}) = 2 + \frac{y}{2}$ and whose inner radius is $3 - (2 - y) = y + 1$. Thus the volume is

$$V = \pi \int_0^2 \left(\left(2 + \frac{y}{2}\right)^2 - (y + 1)^2\right) dy = \pi \int_0^2 \left(-\frac{3}{4}y^2 + 3\right) dy = \pi\left(-\frac{1}{4}y^3 + 3y\right)\Big|_0^2 = 4\pi.$$

6.3.61

a. False. The cross sections are not disks or washers.

b. True. It is given by $V = \pi \int_0^R (R^2 - x^2) dx$.

c. True. This is because if we shift the sine function horizontally by $\pi/2$ units, we obtain the cosine function.

6.3.63 The volume V_S is $\pi \int_0^{\sqrt{a}} (y^2 - a)^2 dy = \pi \left(y^5/5 - 2ay^3/3 + a^2 y\right)\Big|_0^{\sqrt{a}} = \frac{8\pi}{15} a^{5/2}$.

The volume V_T is $\frac{1}{3}\pi a^2 \cdot a^{1/2} = \frac{1}{3}\pi a^{5/2}$. The ratio of $V_S/V_T = \frac{8/15}{1/3} = \frac{8}{5}$.

6.3.65

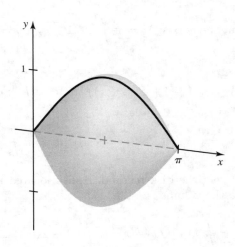

a. This comes from revolving the region in the first quadrant bounded by $\sin x$ and the line $y = 0$ between $x = 0$ and $x = \pi$ around the x axis.

b. This comes from revolving the region in the first quadrant bounded by $y = x + 1$ and the line $y = 0$ between $x = 0$ and $x = 2$ around the x axis.

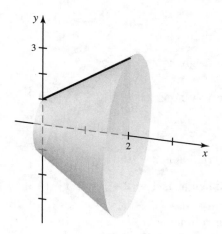

6.3.67 At a given point x in $[0, 4]$, let $A(x)$ equal the area of a cross section of the solid of revolution that is perpendicular to the x-axis. Use the formula $V \approx \sum_{k=1}^{n} A(x_k^*)\Delta x$ to estimate the volume.

Left Riemann sum: $V \approx \pi(1^2 + 4^2 + 7^2 + 10^2) = 166\pi$.
Right Riemann sum: $V \approx \pi(4^2 + 7^2 + 10^2 + 12^2) = 309\pi$.
Midpoint Riemann sum: $V \approx \pi(3^2 + 5^2 + 8^2 + 11^2) = 219\pi$.

6.3.69

a. Think of the cone as being obtained by revolving the region under $y = x$ in the first quadrant from $x = 0$ to $x = R$ around the x-axis. The volume is $\pi \int_0^R x^2\, dx = \pi R^3/3 = \frac{1}{3}V_C$.

b. Think of the hemisphere as being obtained by revolving the region under $y = \sqrt{R^2 - x^2}$ between $x = 0$ and $x = R$ around the x-axis. The volume is $\pi \int_0^R (R^2 - x^2)\, dx = \pi \left(R^2 x - x^3/3 \right) \Big|_0^R = \frac{2\pi R^3}{3} = \frac{2}{3}V_C$.

6.3.71 $V = \pi \int_{-2}^{2} ((3 + \sqrt{4 - y^2})^2 - (3 - \sqrt{4 - y^2})^2)\, dy = 24\pi \int_0^2 \sqrt{4 - y^2}\, dy = 24\pi(\pi) = 24\pi^2$. Note that the last integral evaluated represents $1/4$ the area of a circle of radius 2

6.3.73

a. By the general slicing method, $V(x) = \int_a^b A(x)\, dx$. Because the two figures have the same cross sections $A(x)$, they must therefore have the same volumes.

b. According to Cavelieri's principle, the volume is $V = \pi r^2 h$ (the angle $\pi/4$ is irrelevant).

6.4 Volume by Shells

6.4.1 $V = 2\pi \int_a^b x(f(x) - g(x))\, dx$.

6.4.3 ...revolved about the x axis. ...using the disk/washer method and integrating with respect to $\underline{x}$ or using the shell method and integrating with respect to $\underline{y}$.

6.4.5

a. The radius is x.

b. The height is $2 - x^2 - x$.

c. The volume is $2\pi \int_0^1 x(2 - x^2 - x)\,dx$.

6.4.7

a. The radius is $2 - y$.

b. The height is $4 - (2-y)^2 = 4 - (4 - 4y + y^2) = 4y - y^2$.

c. The volume is $2\pi \int_0^2 (2-y)(4y - y^2)\,dy$.

6.4.9 $V = 2\pi \int_0^1 x(x - x^2)\,dx = 2\pi \int_0^1 (x^2 - x^3)\,dx = 2\pi \left(\dfrac{x^3}{3} - \dfrac{x^4}{4}\right)\Big|_0^1 = 2\pi\left(\dfrac{1}{3} - \dfrac{1}{4}\right) = \dfrac{\pi}{6}$.

6.4.11 $V = 2\pi \int_0^1 x(3 - 3x)\,dx = 2\pi \left(3x^2/2 - x^3\right)\Big|_0^1 = \pi$.

The volume of a cone is $\dfrac{1}{3}\pi r^2 h = \dfrac{1}{3}\pi \cdot 1^2 \cdot 3 = \pi$.

6.4.13 $V = 2\pi \int_0^2 y(4 - y^2)\,dy = 2\pi\left(2y^2 - y^4/4\right)\Big|_0^2 = 8\pi$.

6.4.15 $V = 2\pi \int_2^4 y(4 - y)\,dy = 2\pi \left(2y^2 - \dfrac{y^3}{3}\right)\Big|_2^4 = \dfrac{32\pi}{3}$.

6.4.17 $V = 2\pi \int_0^{\sqrt{\pi/2}} x \cos x^2\,dx = \pi \left(\sin x^2\right)\Big|_0^{\sqrt{\pi/2}} = \pi$.

6.4.19 $V = 2\pi \int_0^1 x(1 - x^2)\,dx = 2\pi \int_0^1 (x - x^3)\,dx = 2\pi \left(\dfrac{x^2}{2} - \dfrac{x^4}{4}\right)\Big|_0^1 = 2\pi\left(\dfrac{1}{2} - \dfrac{1}{4}\right) = \dfrac{\pi}{2}$.

6.4.21 $V = 2\pi \int_0^3 y(y^2)\,dy = 2\pi \int_0^3 y^3\,dy = 2\pi \left(\dfrac{y^4}{4}\right)\Big|_0^3 = 81\pi/2$.

6.4.23 Note that the lines intersect at $(0,0)$, $(2,0)$ and $(1,1)$.
$V = 2\pi \int_0^1 y((2-y) - y)\,dy = 2\pi \int_0^1 (2y - 2y^2)\,dy = 2\pi\left(y^2 - \dfrac{2y^3}{3}\right)\Big|_0^1 = \dfrac{2\pi}{3}$.

6.4.25 $V = 2\pi \int_1^2 \dfrac{x}{(x^2 + 1)^2}\,dx$. Let $u = x^2 + 1$, so that $du = 2x\,dx$. Substituting gives

$$V = \pi \int_2^5 u^{-2}\,du = -\pi \cdot \dfrac{1}{u}\Big|_2^5 = -\pi\left(\dfrac{1}{5} - \dfrac{1}{2}\right) = \dfrac{3\pi}{10}.$$

6.4.27 $V = 2\pi \int_2^{16} y(2/y)^{2/3}\,dy = 2^{5/3}\pi \int_2^{16} y^{1/3}\,dy = 2^{5/3}\pi \left(3y^{4/3}/4\right)\Big|_2^{16} = 2^{5/3}\pi(3 \cdot 2^{10/3} - 3 \cdot 2^{-2/3}) = 3\pi(2^5 - 2) = 90\pi$.

6.4. Volume by Shells

6.4.29 $V = 2\pi \int_1^2 x(e^x/x)\,dx = 2\pi\, e^x \Big|_1^2 = 2\pi(e^2 - e).$

6.4.31 Note that $y = \sqrt{\cos^{-1} x}$ intersects the axes at $(0, \sqrt{\pi/2})$ and $(1,0)$. $V = 2\pi \int_0^{\sqrt{\pi/2}} y \cos y^2\, dy = \pi \sin y^2 \Big|_0^{\sqrt{\pi/2}} = \pi(1 - 0) = \pi.$

6.4.33 Note that the line $y = 1$ intersects the curve $y = x^3 - x^8 + 1$ at $x = 0$ and $x = 1$. $V = 2\pi \int_0^1 x(x^3 - x^8 + 1 - 1)\,dx = 2\pi \int_0^1 (x^4 - x^9)\,dx = 2\pi \left(\dfrac{x^5}{5} - \dfrac{x^{10}}{10}\right)\Big|_0^1 = 2\pi\left(\dfrac{1}{5} - \dfrac{1}{10}\right) = \dfrac{\pi}{5}.$

6.4.35 With washers we have
$$V = \pi \int_0^1 (x^{2/3} - x^2)\,dx = \pi\left(\dfrac{3x^{5/3}}{5} - \dfrac{x^3}{3}\right)\Big|_0^1 = \dfrac{4\pi}{15}.$$
With shells we have
$$V = 2\pi \int_0^1 y(y - y^3)\,dy = 2\pi\left(\dfrac{y^3}{3} - \dfrac{y^5}{5}\right)\Big|_0^1 = 2\pi\left(\dfrac{1}{3} - \dfrac{1}{5}\right) = \dfrac{4\pi}{15}.$$

6.4.37 Using washers we have
$$V = \pi \int_{-10}^{25} (\sqrt[3]{y+2} + 2)^2\,dy = \pi \int_{-8}^{27} (u^{1/3} + 2)^2\,du = \pi \int_{-8}^{27} (u^{2/3} + 4u^{1/3} + 4)\,du$$
$$= \pi\left(3u^{5/3}/5 + 3u^{4/3} + 4u\right)\Big|_{-8}^{27} = \pi(3^6/5 + 3^5 + 108 - (-96/5 + 48 - 32)) = \pi\left(\dfrac{2484}{5} + \dfrac{16}{5}\right) = 500\pi.$$
Using shells we have
$$V = 2\pi \int_0^5 x(25 - (x-2)^3 + 2)\,dx = 2\pi \int_0^5 (27x - x(x-2)^3)\,dx = 2\pi \int_0^5 -x^4 + 6x^3 - 12x^2 + 35x\,dx$$
$$= 2\pi\left(-\dfrac{x^5}{5} + \dfrac{3x^4}{2} - 4x^3 + \dfrac{35x^2}{2}\right)\Big|_0^5 = 500\pi$$

6.4.39 $V = 2\pi \int_0^1 (x+2)x^2\,dx = 2\pi \int_0^1 (x^3 + 2x^2)\,dx = \pi\left(\dfrac{x^4}{4} + \dfrac{2x^3}{3}\right)\Big|_0^1 = \dfrac{11\pi}{6}.$

6.4.41 $V = 2\pi \int_0^1 (2-x)x^2\,dx = 2\pi \int_0^1 (2x^2 - x^3)\,dx = 2\pi\left(\dfrac{2x^3}{3} - \dfrac{x^4}{4}\right)\Big|_0^1 = 2\pi\left(\dfrac{2}{3} - \dfrac{1}{4}\right) = \dfrac{5\pi}{6}.$

6.4.43
$$V = 2\pi \int_0^1 (y+2)(1 - \sqrt{y})\,dy = 2\pi \int_0^1 (y + 2 - y^{3/2} - 2y^{1/2})\,dy = 2\pi\left(\dfrac{y^2}{2} + 2y - \dfrac{2y^{5/2}}{5} - \dfrac{4y^{3/2}}{3}\right)\Big|_0^1$$
$$= 2\pi\left(\dfrac{1}{2} + 2 - \dfrac{2}{5} - \dfrac{4}{3}\right) = \dfrac{23\pi}{15}.$$

6.4.45 Using washers, we have
$$V = \pi \int_0^1 (3^2 - (x^2 + 2)^2)\,dx = \pi \int_0^1 (9 - x^4 - 4x^2 - 4)\,dx = \pi \int_0^1 (5 - x^4 - 4x^2)\,dx$$
$$= \pi\left(5x - \dfrac{x^5}{5} - \dfrac{4x^3}{3}\right)\Big|_0^1 = \pi\left(5 - \dfrac{1}{5} - \dfrac{4}{3}\right) = \dfrac{52\pi}{15}.$$

Using shells, we have

$$V = 2\pi \int_0^1 (y+2)\sqrt{y}\, dy = 2\pi \int_0^1 (y^{3/2} + 2y^{1/2})\, dy = 2\pi \left(\frac{2y^{5/2}}{5} + \frac{4y^{3/2}}{3}\right)\bigg|_0^1$$

$$= 2\pi \left(\frac{2}{5} + \frac{4}{3}\right) = \frac{52\pi}{15}.$$

6.4.47 Using washers, we have

$$V = \pi \int_0^1 ((6-x^2)^2 - 5^2)\, dx = \pi \int_0^1 (x^4 - 12x^2 + 11)\, dx = \pi \left(\frac{x^5}{5} - 4x^3 + 11x\right)\bigg|_0^1 = \frac{36\pi}{5}.$$

Using shells, we have

$$V = 2\pi \int_0^2 (6-y)\sqrt{y}\, dy = 2\pi \int_0^1 (6y^{1/2} - y^{3/2})\, dy = 2\pi \left(4y^{3/2} - \frac{2y^{5/2}}{5}\right)\bigg|_0^1 = 2\pi \left(4 - \frac{2}{5}\right) = \frac{36\pi}{5}.$$

6.4.49

a. The volume is $V = 2\pi \int_0^r y(2\sqrt{r^2 - y^2})\, dy$. Let $u = r^2 - y^2$, so that $du = -2y\, dy$. Substituting gives

$$2\pi \int_{r^2}^0 -\sqrt{u}\, du = 2\pi \int_0^{r^2} u^{1/2}\, du = 2\pi \frac{2u^{3/2}}{3}\bigg|_0^{r^2} = \frac{4}{3}\pi r^3.$$

b. The volume is

$$V = \pi \int_{-r}^r (\sqrt{r^2 - x^2})^2\, dx = \pi \int_{-r}^r (r^2 - x^2)\, dx = \pi \left(r^2 x - \frac{x^3}{3}\right)\bigg|_{-r}^r$$

$$= \pi \left(r^3 - \frac{r^3}{3} - \left(-r^3 - \frac{(-r)^3}{3}\right)\right) = 2\pi \left(r^3 - \frac{r^3}{3}\right) = \frac{4}{3}\pi r^3.$$

6.4.51

a. $V = 2\pi \int_1^5 2x\sqrt{4-(x-3)^2}\, dx.$

b. $V = 2\pi \int_0^2 ((3+\sqrt{4-y^2})^2 - (3-\sqrt{4-y^2})^2)\, dy = 24\pi \int_0^2 \sqrt{4-y^2}\, dy.$

c. Note that the last integral in part b represents $\frac{1}{4}$ of the area of the circle of radius 2 centered at the origin, so its value is $\frac{1}{4}\pi \cdot 4 = \pi$. So the volume is $24\pi \cdot \pi = 24\pi^2$.

6.4.53 First, note that the curve and the line intersect at $(0,0)$ and $(1,0)$.

We choose the washer method:

$$V = \pi \int_0^1 (x-x^4)^2\, dx = \pi \int_0^1 (x^2 - 2x^5 + x^8)\, dx$$

$$= \pi \left(x^3/3 - x^6/3 + x^9/9\right)\bigg|_0^1 = \pi/9.$$

6.4.55 Using washers:

$$V = 2\pi \int_0^1 ((2-x^2)^2 - (x^2)^2)\, dx = 2\pi \int_0^1 (4 - 4x^2)\, dx = 2\pi \left(4x - 4x^3/3\right)\bigg|_0^1 = \frac{16\pi}{3}.$$

6.4. Volume by Shells

6.4.57 Using shells:
$$V = 2\pi \int_2^6 x(2x+2-x)\,dx = 2\pi \int_2^6 (x^2+2x)\,dx = 2\pi\left(x^3/3+x^2\right)\bigg|_2^6 = \frac{608\pi}{3}.$$

6.4.59 Using shells:
$$V = 2\pi \int_0^1 y(e^{y^2} - e^{y^2/2})\,dy = 2\pi\left(e^{y^2}/2 - e^{y^2/2}\right)\bigg|_0^1 = \pi(e - 2\sqrt{e} - (1-2)) = \pi(\sqrt{e}-1)^2.$$

6.4.61 The curves intersect when $x^2 = 2-x$, or $x^2+x-2 = (x-1)(x+2) = 0$, so for $x=1$ and $x=-2$. Only $x=1$ is in the first quadrant; the corresponding y coordinate is $y=1$. Also, $2-x \geq x^2$ on $[0,1]$.

Using shells, the height of each shell is $2-x-x^2$, so the volume is
$$V = 2\pi \int_0^1 x(2-x-x^2)\,dx = 2\pi\left(x^2 - \frac{x^3}{3} - \frac{x^4}{4}\right)\bigg|_0^1 = \frac{5\pi}{6}.$$

6.4.63

a. True. Otherwise, we wouldn't have shells!

b. False. Either method can be used when revolving around either axis.

c. True.

6.4.65 Consider the region in the first quadrant bounded by the coordinate axes and the line $y = 8 - (8/3)x$. We can generate the desired cone by revolving this region around the y-axis. We then have
$$V = 2\pi \int_0^3 x(8 - (8/3)x)\,dx = \frac{16\pi}{3}\int_0^3 (3x-x^2)\,dx = \frac{16\pi}{3}\left(3x^2/2 - x^3/3\right)\bigg|_0^3 = 24\pi.$$

6.4.67 Consider the triangle in the first quadrant bounded by $y=0$, $x=3$, and $y = 9-(3/2)x$. The solid in question is formed by revolving this region around the y-axis. The volume is
$$V = 2\pi \int_3^6 x(9-(3/2)x)\,dx = 2\pi\left(9x^2/2 - x^3/2\right)\bigg|_3^6 = 54\pi.$$

6.4.69

a. $V_1 = \pi \int_0^1 (ax^2+1)^2\,dx = \pi \int_0^1 (a^2x^4 + 2ax^2 + 1)\,dx = \pi(a^2/5 + 2a/3 + 1).$

$V_2 = 2\pi \int_0^1 x(ax^2+1)\,dx = 2\pi(a/4 + 1/2) = \pi a/2 + \pi.$

b. These are equal when $\pi a/2 + \pi = \pi a^2/5 + 2a\pi/3 + \pi$, or when $\pi a^2/5 + a\pi/6 = 0$, which occurs when $a=0$ and when $a = -5/6$.

6.4.71 $V = 2\pi \int_0^2 xf(x^2)\,dx = \pi \int_0^4 f(u)\,du$, where $u = x^2$. Thus, $V = 10\pi$.

6.4.73

a. The longest diagonal of the cube is equal to the diameter of the sphere, which is $\sqrt{r^2+r^2+r^2} = \sqrt{3}r$. Thus, if R is the radius of the sphere, we must have $R = \frac{\sqrt{3}}{2}r$.

Now analyzing the cone which contains the sphere, we see that the height h of the cone is $3R$, so $h = \frac{3\sqrt{3}}{2}r$.

Now the volume of the cylinder is $V = \pi\left(\frac{3r}{2}\right)^2\left(\frac{3\sqrt{3}r}{2}\right) = \frac{27\sqrt{3}\pi r^3}{8}.$

b. Imagine the cone with its vertex up. Consider the plane which contains the vertex of the cone and two non-adjacent vertices of the cube's bottom face. The cross section of this plane with the cone and cube consists of an $r \times \sqrt{2}r$ rectangle (where r is the side length of the cube) and an isosceles triangle of base 2 and height 3, with the rectangle inscribed in the triangle, and the longer side of the rectangle lying on the base of the triangle. Using similar triangles, we have $\dfrac{r}{3} = \dfrac{1-(r/\sqrt{2})}{1}$, so $r = \dfrac{3\sqrt{2}}{3+\sqrt{2}}$, and the volume of the cube is $\dfrac{54\sqrt{2}}{(3+\sqrt{2})^3}$.

c. Imagine the sphere with the hole as being obtained by revolving the region pictured around the x-axis, where the relevant curves are $y = r$ and $y = \sqrt{R^2 - x^2}$ where r is the radius of the hole and R is the radius of the sphere. Note that if you draw the triangle with vertices $(0,0)$, $(5,0)$, and $(5,r)$, you have a right triangle with legs of length 5 and r, and hypotenuse of length R, so $25 + r^2 = R^2$.

The volume we are interested in is

$$V = 2\pi \int_0^5 \left((R^2 - x^2) - r^2\right) dx = 2\pi \int_0^5 (25 - x^2)\, dx = 2\pi \left(25x - x^3/3\right)\Big|_0^5 = \frac{500\pi}{3}.$$

Note that (surprisingly!), the result doesn't depend on the radius of the original sphere.

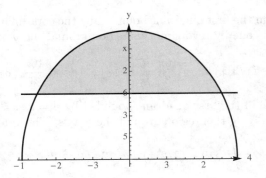

6.4.75

a. The cross sections of such an object are washers, with inner radius given b $g(x) - y_0$ and outer radius given by $f(x) - y_0$, so the volume is given by $\pi \int_a^b \left((f(x) - y_0)^2 - (g(x) - y_0)^2\right) dx$.

b. The cross section of this object are also washers, and this time the inner radius is $y_0 - f(x)$ and the outer radius is $y_0 - g(x)$, so the volume is given by $\pi \int_a^b \left((y_0 - g(x))^2 - (y_0 - f(x))^2\right) dx$.

6.5 Length of Curves

6.5.1 Given $f(x)$ and a and b, compute $f'(x)$ and then compute $\int_a^b \sqrt{1 + f'(x)^2}\, dx$.

6.5.3 Because $f'(x) = 3x^2$, the arc length is $\int_{-2}^5 \sqrt{1 + f'(x)^2}\, dx = \int_{-2}^5 \sqrt{1 + 9x^4}\, dx$.

6.5.5 Because $f'(x) = -2e^{-2x}$, the arc length is $\int_0^2 \sqrt{1 + f'(x)^2}\, dx = \int_0^2 \sqrt{1 + 4e^{-4x}}\, dx$.

6.5. Length of Curves

6.5.7 $L = \int_1^5 \sqrt{1+2^2}\,dx = 4\sqrt{5}$. Using the fact that this is a straight line segment between the points $(1,3)$ and $(5,11)$, we can calculate the arc length using the distance formula in the plane:
$$L = \sqrt{(5-1)^2 + (11-3)^2} = \sqrt{16+64} = \sqrt{80} = 4\sqrt{5}.$$

6.5.9 $L = \int_{-2}^6 \sqrt{1+(y')^2}\,dx = \int_{-2}^6 \sqrt{1+(-8)^2}\,dx = \int_{-2}^6 \sqrt{65}\,dx = 8\sqrt{65}.$

6.5.11 $y' = \sqrt{x}/2$, so $1+(y')^2 = 1 + x/4$. Thus,
$$L = \int_0^{60} \sqrt{1+(x/4)}\,dx = \int_0^{60} \frac{1}{2}\sqrt{x+4}\,dx = \left(\frac{1}{3}(x+4)^{3/2}\right)\bigg|_0^{60} = \frac{504}{3} = 168.$$

6.5.13 $y' = x(x^2+2)^{1/2}$, so $1+(y')^2 = 1 + x^2(x^2+2) = x^4 + 2x^2 + 1 = (x^2+1)^2$. Thus,
$$L = \int_0^1 (x^2+1)\,dx = (x^3/3 + x)\bigg|_0^1 = \frac{4}{3}.$$

6.5.15 $y' = x^3 - \frac{1}{4x^3}$, so $1+(y')^2 = 1 + x^6 - \frac{1}{2} + \frac{1}{16x^6} = \left(x^3 + \frac{1}{4x^3}\right)^2$. Thus,
$$L = \int_1^2 \left(x^3 + \frac{1}{4x^3}\right)dx = \left(x^4/4 + \frac{-1}{8}x^{-2}\right)\bigg|_1^2 = 4 - (1/32) - (1/4 - (1/8)) = \frac{123}{32}.$$

6.5.17 $\frac{dx}{dy} = y^3 - 1/(4y^3)$, so $1+\left(\frac{dx}{dy}\right)^2 = 1 + y^6 - 1/2 + 1/(16y^6) = \left(y^3 + \frac{1}{4y^3}\right)^2$. So
$$L = \int_1^2 \left(y^3 + \frac{1}{4}\cdot y^{-3}\right)dy = \left(y^4/4 - \frac{1}{8y^2}\right)\bigg|_1^2 = 4 - (1/32) - (1/4 - 1/8) = \frac{123}{32}.$$

6.5.19 $\frac{dx}{dy} = 2$, so $1 + \left(\frac{dx}{dy}\right)^2 = 1+4 = 5$, so $L = \int_{-3}^4 \sqrt{5}\,dy = 7\sqrt{5}.$

6.5.21

a. $y' = 2x$, so $1+(y')^2 = 1 + 4x^2$, so $L = \int_{-1}^1 \sqrt{1+4x^2}\,dx.$

b. $L = \int_{-1}^1 \sqrt{1+4x^2}\,dx \approx 2.958.$

6.5.23

a. $y' = 1/x$, so $1+(y')^2 = 1 + (1/x)^2 = \frac{x^2+1}{x^2}$, so $L = \int_1^4 \frac{\sqrt{x^2+1}}{x}\,dx.$

b. $L = \int_1^4 \frac{\sqrt{x^2+1}}{x}\,dx \approx 3.343.$

6.5.25

a. $x' = \frac{1}{2\sqrt{y-2}}$, so $1+(x')^2 = 1 + \frac{1}{4(y-2)} = \frac{4y-7}{4y-8}$, so $L = \int_3^4 \sqrt{\frac{4y-7}{4y-8}}\,dy.$

b. $L = \int_3^4 \sqrt{\frac{4y-7}{4y-8}}\,dy \approx 1.083.$

6.5.27

a. $y' = -2\sin 2x$, so $1 + (y')^2 = 1 + 4\sin^2 2x$, so $L = \int_0^\pi \sqrt{1 + 4\sin^2 2x}\, dx$.

b. $L = \int_0^\pi \sqrt{1 + 4\sin^2 2x}\, dx \approx 5.270$.

6.5.29

a. $y' = -1/x^2$, so $1 + (y')^2 = 1 + \dfrac{1}{x^4} = \dfrac{x^4 + 1}{x^4}$. Thus, $L = \int_1^{10} \dfrac{\sqrt{x^4 + 1}}{x^2}\, dx$.

b. $L = \int_1^{10} \dfrac{\sqrt{x^4 + 1}}{x^2}\, dx \approx 9.153$.

6.5.31 $f'(x) = 0.00074x$, so $L = \int_{-640}^{640} \sqrt{1 + (0.00074x)^2}\, dx \approx 1326.4$ meters.

6.5.33

a. False. For example, if $f(x) = x^2$, the first integrand is $\sqrt{1 + 4x^2}$ and the second is $1 + 2x$, which clearly yield different values for (for example) $a = 0$ and $b = 1$.

b. True. They are both equal to $\int_a^b \sqrt{1 + f'(x)^2}\, dx$.

c. False. Because $\sqrt{1 + f'(x)^2} > 0$, arc length can't be negative.

6.5.35

a. We are seeking functions $f(x)$ so that $f'(x) = \pm 4x^2$, so any function of the form $f(x) = \pm 4x^3/3 + C$ will work.

b. We are seeking functions $f(x)$ so that $f'(x) = \pm 6\cos(2x)$, so any function of the form $f(x) = \pm 3\sin(2x) + C$ will work.

6.5.37 The length of the parabola is $\int_{-1}^{1} \sqrt{1 + 4x^2}\, dx \approx 2.9597$.

The length of the given cosine function is $\int_{-1}^{1} \sqrt{1 + \dfrac{\pi^2}{4} \sin^2(\pi x/2)}\, dx \approx 2.924$, so the parabola is longer.

6.5.39

a. Let $u = 2x$. Then the given integral is equal to $\dfrac{1}{2} \int_a^b \sqrt{1 + f'(u)^2}\, du = \dfrac{L}{2}$.

b. Let $u = cx$. Then the given integral is equal to $\dfrac{1}{c} \int_a^b \sqrt{1 + f'(u)^2}\, du = \dfrac{L}{c}$.

6.5.41

a. $f'(x) = Aae^{ax} - \dfrac{1}{4Aa} e^{-ax}$, so $1 + f'(x)^2 = 1 + (Aae^{ax})^2 - \dfrac{1}{2} + \left(\dfrac{1}{4Aa} e^{-ax}\right)^2 = (Aae^{ax})^2 + \dfrac{1}{2} + \left(\dfrac{1}{4Aa} e^{-ax}\right)^2 = \left(Aae^{ax} + \dfrac{1}{4Aa} e^{-ax}\right)^2$. So the arc length for $c \leq x \leq d$ is

$$L = \int_c^d \left(Aae^{ax} + \dfrac{1}{4Aa} e^{-ax}\right) dx = \left(Ae^{ax} - \dfrac{1}{4Aa^2} e^{-ax}\right)\bigg|_c^d.$$

b. Applying the previous result with $c = 0$ and $d = \ln 2$, we have

$$L = \left(Ae^{ax} - \dfrac{1}{4Aa^2} e^{-ax}\right)\bigg|_0^{\ln 2} = A2^a - \dfrac{1}{4Aa^2} 2^{-a} - A + \dfrac{1}{4Aa^2} = A(2^a - 1) - \dfrac{1}{4a^2 A}(2^{-a} - 1).$$

6.6 Surface Area

6.6.1 $S = \pi r \sqrt{r^2 + h^2} = 3\pi\sqrt{9+16} = 15\pi$.

6.6.3 Evaluate $\int_a^b 2\pi f(x)\sqrt{1+f'(x)^2}\,dx$.

6.6.5

a. $S = \int_0^2 2\pi(2-x)\sqrt{1+(-1)^2}\,dx = 2\sqrt{2}\pi \int_0^2 (2-x)\,dx = 2\sqrt{2}\pi\left(2x - \frac{x^2}{2}\right)\Big|_0^2 = 4\sqrt{2}\pi$.

b. The surface is a cone with $\ell = 2\sqrt{2}$ and $r=2$; using the formula for the lateral surface area of a cone, we have $s = \pi r \ell = 4\sqrt{2}\pi$.

6.6.7 $S = 2\pi \int_0^6 (3x+4)\sqrt{1+9}\,dx = 2\sqrt{10}\pi\left(3x^2/2 + 4x\right)\Big|_0^6 = 2\sqrt{10}\pi(54+24) = 156\sqrt{10}\pi$.

6.6.9

$$S = 2\pi \int_9^{20} 8\sqrt{x}\sqrt{1+(16/x)}\,dx = 16\pi \int_9^{20} \sqrt{x+16}\,dx$$

$$= 16\pi\left(\frac{2(x+16)^{3/2}}{3}\right)\Big|_9^{20} = \frac{32\pi}{3}(216-125) = \frac{2912\pi}{3}.$$

6.6.11 Note that $x = y^3/3$ for $0 \le y \le 2$. The surface area is $S = 2\pi \int_0^2 (y^3/3)\sqrt{1+y^4}\,dy$. Let $u = 1+y^4$ so that $du = 4y^3\,dy$. Substituting yields $S = 2\pi \int_1^{17} \frac{1}{12} u^{1/2}\,du = \frac{\pi}{9}\left(u^{3/2}\right)\Big|_1^{17} = \frac{\pi}{9}(17\sqrt{17} - 1)$.

6.6.13

$$S = 2\pi \int_{-1/2}^{1/2} \sqrt{1-x^2}\sqrt{1+\left(\frac{-x}{\sqrt{1-x^2}}\right)^2}\,dx$$

$$= 4\pi \int_0^{1/2} \sqrt{1-x^2}\sqrt{1+\frac{x^2}{1-x^2}}\,dx = 4\pi \int_0^{1/2} \sqrt{1-x^2}\sqrt{\frac{1}{1-x^2}}\,dx = 4\pi \int_0^{1/2} dx = 2\pi.$$

6.6.15 Note that $x = \frac{y+1}{4}$ for $3 \le y \le 15$. The surface area is

$$S = 2\pi \int_3^{15} \frac{y+1}{4}\sqrt{1+\frac{1}{16}}\,dy = \frac{\pi\sqrt{17}}{8}\int_3^{15}(y+1)\,dy$$

$$= \frac{\pi\sqrt{17}}{8}\left(y^2/2 + y\right)\Big|_3^{15} = \frac{\pi\sqrt{17}}{8}(225/2 + 15 - (9/2 + 3)) = \frac{\pi\sqrt{17}}{8}(120) = 15\sqrt{17}\pi.$$

6.6.17

$$S = 2\pi \int_{-2}^{2} (1/4)(e^{2x}+e^{-2x})\sqrt{1 + \frac{(2e^{2x}-2e^{-2x})^2}{16}}\,dx = \frac{\pi}{2}\int_{-2}^{2}(e^{2x}+e^{-2x})\sqrt{\frac{16 + 4e^{4x} - 8 + 4e^{-4x}}{16}}\,dx$$

$$= \frac{\pi}{8}\int_{-2}^{2}(e^{2x}+e^{-2x})\sqrt{(2e^{2x}+2e^{-2x})^2}\,dx = \frac{\pi}{8}\int_{-2}^{2}(e^{2x}+e^{-2x})(2e^{2x}+2e^{-2x})\,dx$$

$$= \frac{\pi}{4}\int_{-2}^{2}(e^{4x}+2+e^{-4x})\,dx = \frac{\pi}{4}\left(\frac{e^{4x}}{4} + 2x - \frac{e^{-4x}}{4}\right)\Big|_{-2}^{2}$$

$$= \frac{\pi}{4}\left(\frac{e^8}{4} + 4 - \frac{e^{-8}}{4} - \left(\frac{e^{-8}}{4} - 4 - \frac{e^8}{4}\right)\right) = \frac{\pi}{4}\left(\frac{e^8}{2} + 8 - \frac{e^{-8}}{2}\right).$$

6.6.19

$$S = 2\pi \int_2^{10} \sqrt{12y - y^2}\sqrt{1 + \frac{(12-2y)^2}{4(12y-y^2)}}\, dy = 2\pi \int_2^{10} \sqrt{12y - y^2}\sqrt{\frac{48y - 4y^2 + 144 - 48y + 4y^2}{4(12y - y^2)}}\, dy$$

$$= 2\pi \int_2^{10} 6\, dy = 12\pi(10 - 2) = 96\pi$$

6.6.21 $S = 2\pi \int_1^7 \sqrt{8x - x^2}\sqrt{1 + \frac{(8 - 2x)^2}{4(8x - x^2)}}\, dx = 2\pi \int_1^7 \sqrt{8x - x^2}\sqrt{\frac{32x - 4x^2 + 64 - 32x + 4x^2}{4(8x - x^2)}}\, dx =$

$2\pi \int_1^7 \sqrt{8x - x^2}\sqrt{\frac{64}{4(8x - x^2)}}\, dx = 2\pi \int_1^7 4\, dx = 8\pi(7 - 1) = 48\pi$. Because the surface area is 48π square meters, the volume of paint required to cover the surface to a thickness of 0.0015 meters is $48\pi(.0015) \approx .226$ cubic meters. This is about 59.75 gallons.

6.6.23

a. False. One would need to find $x = f^{-1}(y)$ and compute the corresponding integral using f^{-1}.

b. False. For example, the curve given in number 14 above isn't one-to-one on the given interval, but the surface is still defined.

c. True. Because the curve is symmetric about the y-axis, the surface generated by revolving half the curve is half the surface generated by revolving the whole curve.

d. False. This curve is symmetric about the y-axis, so the surface generated by revolving the whole curve is the same as the surface generated by revolving the portion over $[0, 4]$.

6.6.25

a. $S = 2\pi \int_0^{\pi/2} \cos x \sqrt{1 + \sin^2 x}\, dx$.

b. $S \approx 7.21$.

6.6.27

a. $S = 2\pi \int_0^{\pi/4} \tan x \sqrt{1 + \sec^4 x}\, dx$.

b. $S \approx 3.84$.

6.6.29 Let $y = f(x) = (a^{2/3} - x^{2/3})^{3/2}$. Note that $f'(x) = -x^{-1/3}\sqrt{a^{2/3} - x^{2/3}}$, so

$$\sqrt{1 + f'(x)^2} = \sqrt{1 + \frac{a^{2/3} - x^{2/3}}{x^{2/3}}} = \sqrt{\frac{x^{2/3} + a^{2/3} - x^{2/3}}{x^{2/3}}} = \frac{a^{1/3}}{x^{1/3}}.$$

Thus (using symmetry) $S = 4\pi \int_0^a (a^{2/3} - x^{2/3})^{3/2}(a^{1/3} x^{-1/3})\, dx$. Let $u = (a^{2/3} - x^{2/3})$, so that $du = (-2x^{-1/3}/3)\, dx$. Substituting gives

$$S = -6\pi a^{1/3} \int_{a^{2/3}}^0 u^{3/2}\, du = 6\pi a^{1/3} \int_0^{a^{2/3}} u^{3/2}\, du = \frac{12\pi}{5} a^{1/3} \left(u^{5/2}\right)\Big|_0^{a^{2/3}} = \frac{12\pi a^2}{5}.$$

6.6. Surface Area

6.6.31

$$S = 2\pi \int_1^2 \left(x^{3/2} - \frac{\sqrt{x}}{3}\right)\sqrt{1 + (3\sqrt{x}/2 - x^{-1/2}/6)^2}\,dx = 2\pi \int_1^2 (x^{3/2} - \sqrt{x}/3)\sqrt{1 + \left(\frac{9x-1}{6\sqrt{x}}\right)^2}\,dx$$

$$= 2\pi \int_1^2 (x^{3/2} - \sqrt{x}/3)\sqrt{\frac{36x + 81x^2 - 18x + 1}{36x}}\,dx = 2\pi \int_1^2 (x^{3/2} - \sqrt{x}/3)\sqrt{\left(\frac{(1+9x)^2}{36x}\right)}\,dx$$

$$= 2\pi \int_1^2 \left(\frac{(1+9x)x}{6} - \frac{1+9x}{18}\right)dx = \frac{2\pi}{18}\int_1^2 (3x + 27x^2 - 1 - 9x)\,dx$$

$$= \frac{2\pi}{18}\int_1^2 (27x^2 - 6x - 1)\,dx = \frac{\pi}{9}\left(9x^3 - 3x^2 - x\right)\Big|_1^2 = \frac{\pi}{9}(72 - 12 - 2 - (9 - 3 - 1)) = \frac{53\pi}{9}$$

6.6.33

$$S = 2\pi \int_{1/2}^2 (x^3/3 + x^{-1}/4)\sqrt{1 + (x^2 - x^{-2}/4)^2}\,dx$$

$$= 2\pi \int_{1/2}^2 (x^3/3 + x^{-1}/4)\sqrt{1 + x^4 - 1/2 + x^{-4}/16}\,dx = 2\pi \int_{1/2}^2 (x^3/3 + x^{-1}/4)\sqrt{x^4 + 1/2 + x^{-4}/16}\,dx =$$

$$= 2\pi \int_{1/2}^2 (x^3/3 + x^{-1}/4)\sqrt{(x^2 + x^{-2}/4)^2}\,dx = 2\pi \int_{1/2}^2 (x^3/3 + x^{-1}/4)(x^2 + x^{-2}/4)\,dx$$

$$= 2\pi \int_{1/2}^2 (x^5/3 + x/12 + x/4 + x^{-3}/16)\,dx = 2\pi \int_{1/2}^2 (x^5/3 + x/3 + x^{-3}/16)\,dx$$

$$= 2\pi \left(x^6/18 + x^2/6 - x^{-2}/32\right)\Big|_{1/2}^2 = 2\pi(32/9 + 2/3 - 1/128 - (1/1152 + 1/24 - 1/8)) = \frac{275\pi}{32}$$

6.6.35

$$S = 2\pi \int_1^4 (4y^{3/2} - y^{1/2}/12)\sqrt{1 + (6y^{1/2} - y^{-1/2}/24)^2}\,dy$$

$$= 2\pi \int_1^4 (4y^{3/2} - y^{1/2}/12)\sqrt{1 + 36y - 1/2 + 1/(576y)}\,dy$$

$$= 2\pi \int_1^4 (4y^{3/2} - y^{1/2}/12)\sqrt{36y + 1/2 + 1/(576y)}\,dy$$

$$= 2\pi \int_1^4 (4y^{3/2} - y^{1/2}/12)\sqrt{(6y^{1/2} + y^{-1/2}/24)^2}\,dy$$

$$= 2\pi \int_1^4 (4y^{3/2} - y^{1/2}/12)(6y^{1/2} + y^{-1/2}/24)\,dy$$

$$= 2\pi \int_1^4 (24y^2 - y/3 - 1/288)\,dy$$

$$= 2\pi \left(8y^3 - y^2/6 - y/288\right)\Big|_1^4 = 2\pi \left(512 - \frac{8}{3} - \frac{1}{72} - \left(8 - \frac{1}{6} - \frac{1}{288}\right)\right) = \frac{48143\pi}{48}.$$

6.6.37 By symmetry, we can let $y = f(x) = \sqrt{r^2 - x^2}$, and imagine the surface obtained by revolving this curve around the x-axis for $a \leq x \leq a + h$. The surface area is

$$S = 2\pi \int_a^{a+h} \sqrt{r^2 - x^2}\sqrt{1 + \left(\frac{-x}{\sqrt{r^2 - x^2}}\right)^2}\,dx.$$

This can be written as

$$2\pi \int_a^{a+h} \sqrt{r^2 - x^2} \sqrt{\frac{r^2 - x^2 + x^2}{r^2 - x^2}} \, dx$$

$$= 2\pi \int_a^{a+h} \sqrt{r^2 - x^2} \frac{\sqrt{r^2}}{\sqrt{r^2 - x^2}} \, dx = 2\pi \int_a^{a+h} r \, dx = 2\pi r(a+h-a) = 2\pi rh.$$

6.6.39

a. Using R for the radius of the sphere, The ratio is $\frac{6R^2}{R^3} = \frac{6}{R}$.

b. The ratio is $\frac{4\pi R^2}{(4/3)\pi R^3} = \frac{3}{R}$.

c. We have an ellipsoid whose axes have lengths $2R$, R, and R, where $R = \ell\sqrt[3]{4}$. The volume of the ellipsoid is

$$V = \frac{4\pi}{3} \cdot R \cdot \frac{R}{2} \cdot \frac{R}{2} = \frac{\pi R^3}{3} = \frac{4\pi \ell^3}{3}.$$

The surface area is

$$S = 2\pi \cdot \frac{R}{2}\left(\frac{R}{2} + \frac{R^2}{\sqrt{R^2 - (R^2/4)}} \sin^{-1} \frac{\sqrt{R^2 - (R^2/4)}}{R}\right)$$

$$= \pi R \left(\frac{R}{2} + \frac{2R^2}{\sqrt{3}R} \sin^{-1}\left(\frac{\sqrt{3}R}{2R}\right)\right)$$

$$= \pi \ell \sqrt[3]{4} \left(\frac{\ell\sqrt[3]{4}}{2} + \frac{2\ell\sqrt[3]{4}}{\sqrt{3}} \sin^{-1}\left(\frac{\sqrt{3}}{2}\right)\right)$$

$$= \pi \ell \sqrt[3]{4} \left(\frac{\ell\sqrt[3]{4}}{2} + \frac{2\ell\sqrt[3]{4}}{\sqrt{3}} \cdot \frac{\pi}{3}\right)$$

$$= \pi \ell^2 4^{2/3} \left(\frac{1}{2} + \frac{2\pi}{3\sqrt{3}}\right).$$

The SAV ratio is therefore

$$\frac{3\pi \ell^2 4^{2/3}\left(\frac{1}{2} + \frac{2\pi}{3\sqrt{3}}\right)}{4\pi \ell^3} = \frac{3(3\sqrt{3} + 4\pi)}{4^{1/3}\ell 6\sqrt{3}} = \frac{9 + 4\sqrt{3}\pi}{6 \cdot 4^{1/3}\ell}.$$

d.

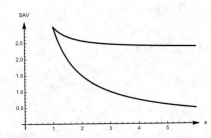

The lower curve is for the sphere and the upper for the ellipsoid.

e. More spherical is better for minimizing heat loss.

6.6.41

a. Note that $g(x) = cf(x)$, so that $g'(x) = cf'(x)$.

$$2\pi \int_a^b g(x)\sqrt{c^2 + g'(x)^2}\, dx = 2\pi \int_a^b cf(x)\sqrt{c^2 + c^2 f'(x)^2}\, dx = 2c^2\pi \int_a^b f(x)\sqrt{1 + f'(x)^2}\, dx = c^2 A.$$

b. Note that $h'(x) = cf'(cx)$. Consider

$$2\pi \int_{a/c}^{b/c} f(cx)\sqrt{c^2 + c^2 f'(cx)^2}\, dx = 2c\pi \int_{a/c}^{b/c} f(cx)\sqrt{1 + f'(cx)^2}\, dx.$$

Let $u = cx$ so that $du = c\, dx$. Then our integral is equal to

$$2\pi \int_a^b f(u)\sqrt{1 + f'(u)^2}\, du = A.$$

6.7 Physical Applications

6.7.1 $m = \rho_1 \cdot l_1 + \rho_2 \cdot l_2 = 1\,\text{g/cm} \cdot 50\,\text{cm} + 2\,\text{g/cm} \cdot 50\,\text{cm} = 150\,\text{g}.$

6.7.3 The work is the product of 5 Newtons and 5 meters, which is 25 J.

6.7.5 Different volumes of water are moved different distances.

6.7.7 $F = \rho g h = 1000 \cdot 9.8 \cdot 4 = 39,200\,\text{N/m}^2.$

6.7.9 $W = \int_5^{10} 25\pi \rho g (15 - y)\, dy.$

6.7.11 $W = \int_0^{10} 25\pi \rho g (10 - y)\, dy.$

6.7.13 $m = \int_0^\pi (1 + \sin x)\, dx = (x - \cos x)\Big|_0^\pi = \pi - (-1) - (0 - 1) = \pi + 2.$

6.7.15 $m = \int_0^2 (2 - x/2)\, dx = (2x - x^2/4)\Big|_0^2 = 3.$

6.7.17 $m = \int_0^1 x\sqrt{2 - x^2}\, dx = \left(\frac{-1}{3}(2 - x^2)^{3/2}\right)\Big|_0^1 = (-1/3) + (2\sqrt{2}/3) = \frac{2\sqrt{2} - 1}{3}.$

6.7.19 $m = \int_0^2 1\, dx + \int_2^4 (1 + x)\, dx = 2 + (x + x^2/2)\Big|_2^4 = 2 + (4 + 8) - (2 + 2) = 10.$

6.7.21 $W = \int_0^3 2x\, dx = x^2 \Big|_0^3 = 9\,\text{J}.$

6.7.23

a. Because $f(0.2) = 0.2k = 30$, we have $k = 150$.

b. $W = \int_0^{-0.4} 150x\, dx = 75x^2 \Big|_0^{-0.4} = 75 \cdot 0.16 = 12\,\text{J}.$

c. $W = \int_0^{0.3} 150x\, dx = 75x^2 \Big|_0^{0.3} = 75 \cdot 0.09 = 6.75\,\text{J}.$

d. $W = \int_{0.2}^{0.4} 150x\,dx = 75x^2 \Big|_{0.2}^{0.4} = 75(0.16 - 0.04) = 9$ J.

6.7.25

a. $f(x) = kx$, and $f(0.5) = 50$, so $k(0.5) = 50$, so $k = 100$.

Therefore $W = \int_0^{1.5} 100x\,dx = (50x^2)\Big|_0^{1.5} = 112.5$ J.

b. $\int_0^{-0.5} 100x\,dx = (50x^2)\Big|_0^{-0.5} = 12.5$ J.

6.7.27

a. $f(0.2) = 0.2k = 50$, so $k = 250$. $W = \int_0^{0.5} 250x\,dx = 125x^2 \Big|_0^{0.5} = 125 \cdot 0.25 = 31.25$ J.

b. $\int_0^{0.2} kx\,dx = kx^2/2 \Big|_0^{0.2} = 0.02k = 50$, so $k = 2500$. $W = \int_0^{0.5} 2500x\,dx = 1250x^2 \Big|_0^{0.5} = 1250 \cdot 0.25 = 312.5$ J.

6.7.29

a. We have $\int_0^{0.5} kx\,dx = \frac{k}{2}x^2 \Big|_0^{0.5} = 0.125k = 100$, so that $k = 800$. Then

$$W = \int_0^{1.25} 800x\,dx = 400x^2 \Big|_0^{1.25} = 625 \text{ J}.$$

b. $f(0.5) = 0.5k = 250$, so $k = 500$ and thus

$$W = \int_0^{1.25} 500x\,dx = 250x^2 \Big|_0^{1.25} = 390.625 \text{ J}.$$

6.7.31

a. $W_1 = \int_0^{30} \rho g(30 - y)\,dy = 5g\left(30y - y^2/2\right)\Big|_0^{30} = 2250g = 2250 \cdot 9.8 = 22050$ J.

b. $W = W_1 + W_2$, where W_2 is the work to just lift the block. $W_2 = 50g \cdot 30 = 1500g$, so $W = 2250g + 1500g = 3750g = 3750 \cdot 9.8 = 36750$ J.

6.7.33 The bottom half of the chain weighs 25 kg, so this problem is equivalent to determining the sum of the work done by the winch in winding up a 25-kg, 10-m chain and lifting a 25-kg mass a distance of 10 m. So

$$W = \int_{10}^{20} 2.5(9.8)(20-y)\,dy + 25(9.8)(10) = 2450 + 24.5\int_{10}^{20}(20-y)\,dy = 2450 + 24.5\left(20y - \frac{y^2}{2}\right)\Big|_{10}^{20}$$
$$= 2450 + 24.5\,(400 - 200 - (200 - 50)) = 2450 + 1225 = 3675 \text{ J}.$$

6.7.35 $W = \int_0^{2.5} \rho g A(y)(2.5 - y)\,dy = 1000 \cdot 9.8 \cdot 25 \cdot 15 \int_0^{2.5}(2.5 - y)\,dy = 3675000\left(2.5y - \frac{y^2}{2}\right)\Big|_0^{2.5} = 3675000 \cdot \frac{(2.5)^2}{2} = 11,484,375$ J.

6.7.37 $W = \int_0^4 \rho\pi g 2^2(10-y)\,dy = 4\pi\rho g\left(10y - \dfrac{y^2}{2}\right)\Big|_0^4 = 4\pi\rho g(40-8) = 128\pi\rho g \approx 3.941 \times 10^6$ J.

6.7.39

a. Let the vertex of the cone be at $(0,0)$, with the y-axis vertically oriented. Note that the area of a horizontal slice at height y is $\pi y^2/16$, and it must move $6-y$ meters to get to the top.

$$W = \int_0^6 \rho g \pi \dfrac{y^2}{16}(6-y)\,dy = \dfrac{\pi\rho g}{16}\int_0^6 (6y^2 - y^3)\,dy = \dfrac{\pi\rho g}{16}\left(2y^3 - \dfrac{y^4}{4}\right)\Big|_0^6 = \dfrac{\pi\rho g}{16}(108) = 66{,}150\pi \text{ J}.$$

b. Not true.

$$\int_0^3 \rho g \pi \dfrac{y^2}{16}(6-y)\,dy = \dfrac{\pi\rho g}{16}\int_0^3 (6y^2 - y^3)\,dy = \dfrac{\pi\rho g}{16}\left(2y^3 - \dfrac{y^4}{4}\right)\Big|_0^3 = \dfrac{\pi\rho g}{16} \cdot 33.75 \approx 20672\pi \text{ J},$$

less than half the amount from part (a). Note that while the water must be raised further than water in the top half, due to the shape of the tank, there is far less water in the bottom half than in the top.

6.7.41 A vertical cross section of the tank that passes through the center of the tank is a circle of radius 8. If we place the origin of the xy-plane at the center of this circle, then we have

$$W = \int_{-8}^8 1000(9.8)(64 - y^2)\pi(10+y)\,dy = 9800\pi \int_{-8}^8 (640 + 64y - 10y^2 - y^3)\,dy$$

$$= 9800\pi \left(640y + 32y^2 - \dfrac{10y^3}{3} - \dfrac{y^4}{4}\right)\Big|_{-8}^8$$

$$= 9800\pi \left(5120 + 2048 - \dfrac{5120}{3} - 1024 - \left(-5120 + 2048 + \dfrac{5120}{3} - 1024\right)\right)$$

$$= 9800\pi \left(10240 - \dfrac{10240}{3}\right) = 9800\pi \left(\dfrac{20480}{3}\right) \approx 210{,}176{,}737 \text{ J}.$$

6.7.43

a. Orient the axes so that the lower corners of the trough are at $(-0.25, 0)$ and at $(0.25, 0)$. Then the upper corners are at $(-0.5, 1)$ and at $(0.5, 1)$. Note that the line between $(0.25, 0)$ and $(0.5, 1)$ is given by $y = 4x - 1$. The area of a slice at height y is $2x \cdot 10 = 20 \cdot \frac{1}{4}(y+1) = 5(y+1)$. Thus,

$$W = \rho g \int_0^1 5(y+1)(1-y)\,dy = 5\rho g \int_0^1 (1-y^2)\,dy = 5\rho g\left(y - y^3/3\right)\Big|_0^1 = \dfrac{10\rho g}{3} \approx 32{,}667 \text{ J}.$$

b. Yes. If the length is doubled, the area of each slice is doubled, so the work integral is doubled as well.

6.7.45 Let the vertex of the cone be at $(0,0)$, with the y-axis vertically oriented. Note that the area of a horizontal slice at height y is $\pi y^2/16$, and it must move $3-y$ meters to get to the point 1 meter above the top.

$$W = \int_0^2 \rho g \pi \dfrac{y^2}{16}(3-y)\,dy = (\pi\rho g/16)\int_0^2 (3y^2 - y^3)\,dy$$

$$= (\pi\rho g/16)\left(y^3 - y^4/4\right)\Big|_0^2 = (\pi\rho g/16)(4) = \dfrac{\pi\rho g}{4} \approx 7696.9 \text{ J}.$$

6.7.47 Orient the axes so that the lower corners of the trapezoid are at $(5,0)$ and $(-5,0)$, and the upper corners are at $(10,15)$ and $(-10,15)$. Note that the line between the corners for $x > 0$ is given by $y = 3(x-5)$, so at level y, we have a width of $2x = \dfrac{2y}{3} + 10$.

$$F = \rho g \int_0^{15} (15-y)\left(\frac{2y+30}{3}\right) dy = \frac{2\rho g}{3} \int_0^{15} (225 - y^2)\, dy$$
$$= \frac{2\rho g}{3}\left(225y - y^3/3\right)\Big|_0^{15} = \frac{2\rho g}{3}\cdot 2250 = 1500\rho g = 1.470 \times 10^7\text{ N}.$$

6.7.49 Orient the axes so that the bottom vertex is at $(0,0)$. The other vertices are at $(\pm 10, 30)$, and the line between $(0,0)$ and $(10,30)$ is given by $y = 3x$. Thus, a slice at height y has width $2x = 2y/3$.

$$F = \int_0^{30} \rho g(30-y)\frac{2y}{3}\, dy = \frac{2\rho g}{3}\left(15y^2 - y^3/3\right)\Big|_0^{30} = \frac{2\rho g}{3}(4500) = 3000\rho g = 2.940 \times 10^7\text{ N}.$$

6.7.51 The width of the plate at depth y is $2-y$, so the force on the plate is

$$F = \int_1^2 \rho g(2-y)y\, dy = \rho g\left(y^2 - \frac{y^3}{3}\right)\Big|_1^2 = \rho g \cdot \frac{2}{3} \approx 6533\text{ N}.$$

6.7.53 $F = \displaystyle\int_1^{1.5} \rho g(4-y)\cdot 0.5\, dy = \rho g\left(2y - y^2/4\right)\Big|_1^{1.5} = \dfrac{11\rho g}{16} = 6737.5\text{ N}.$

6.7.55 $F = \displaystyle\int_0^{50} (150+2y)\cdot 80\, dy = 80\left(150y + y^2\right)\Big|_0^{50} = 8 \times 10^5\text{ N}.$

6.7.57

a. True. $m = \displaystyle\int_a^b \rho(x)\, dx = \dfrac{1}{b-a}\int_a^b \rho(x)\, dx \cdot (b-a) = \overline{\rho}\cdot L.$

b. True. $\displaystyle\int_0^L kx\, dx = \dfrac{kL^2}{2} = \int_0^{-L} kx\, dx.$

c. True. This follows because work is force times distance.

d. False. Although they have the same geometry, they are placed at different depths of the water, so the force is different.

6.7.59

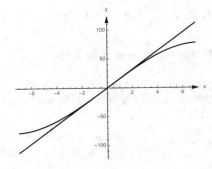

a. Compared to the linear spring $F(x) = 16x$, the restoring force is less for large displacements.

b. $W = \displaystyle\int_0^{1.5} (16x - 0.1x^3)\, dx = \left(8x^2 - .025x^4\right)\Big|_0^{1.5} = 17.87\text{ J}.$

c. $W = \displaystyle\int_0^{-2} (16x - 0.1x^3)\, dx = \left(8x^2 - .025x^4\right)\Big|_0^{-2} = 31.6\text{ J}.$

6.7. Physical Applications

6.7.61 Dividing the leak rate by the lifting rate, we find that the bucket is leaking at 20 kg/m. So the force exerted by the bucket after it has been raised y meters is $F(y) = 9.8(5000 - 20y)$. Therefore, the work required to lift the bucket 30 m is

$$W = \int_0^{30} 9.8(5000 - 20y)\,dy = 9.8\left(5000y - 10y^2\right)\Big|_0^{30} = 9.8\,(150{,}000 - 9000) = 1{,}381{,}800 \text{ J}.$$

6.7.63 Orient the axes so that $(0,0)$ is in the middle of the bottom of the cup. Note that the line between $(0.02, 0)$ and $(0.025, 0.15)$ is given by $y = 30x - 3/5$, so $x = \dfrac{y}{30} + \dfrac{1}{50}$. The area of a cross section at height y is given by $\pi x^2 = \pi \left(\dfrac{5y+3}{150}\right)^2$. Note that the distance the slice must travel is $0.2 - y$, because it must go 0.05 above the top of the glass. Thus,

$$W = \pi \rho g \int_0^{0.15} \left(\frac{5y+3}{150}\right)^2 (.2 - y)\,dy = \frac{\pi \rho g}{5 \cdot 150^2} \int_0^{0.15} (5y+3)^2 (1 - 5y)\,dy$$

$$= \frac{\pi \rho g}{5 \cdot 150^2} \int_0^{0.15} (-125y^3 - 125y^2 - 15y + 9)\,dy$$

$$= \frac{\pi \rho g}{5 \cdot 150^2} \left(-125y^4/4 - 125y^3/3 - 15y^2/2 + 9y\right)\Big|_0^{0.15}$$

$$= \frac{\pi \rho g}{5 \cdot 150^2} \cdot 1.0248 \approx 0.280 \text{ J}.$$

6.7.65

a. $F = \displaystyle\int_1^3 \rho g(4 - y) \cdot 2\,dy = \rho g\left(8y - y^2\right)\Big|_1^3 = 8\rho g = 78{,}400$ N. This is less than 90,000 N., so the window can withstand the force.

b. $F(h) = \displaystyle\int_1^3 \rho g(h - y) \cdot 2\,dy = \rho g\left(2hy - y^2\right)\Big|_1^3 = \rho g(6h - 9 - (2h - 1)) = 4\rho g(h - 2)$. This is less than or equal to 90,000 Newtons when $h \leq 4.296$ meters.

6.7.67 The plate on the left has more than half of its area below the horizontal line which is 3/2 below the surface, while the plate on the right has exactly half its area below that line, so the plate on the left should have more force than the plate on the right.

a. Note that the line in the first quadrant forom $(0, 0)$ to $(\sqrt{2}/2, \sqrt{2}/2)$ is given by $y = x$, while the line from $(\sqrt{2}/2, \sqrt{2}/2)$ to $(0, \sqrt{2})$ is given by $y = -x + \sqrt{2}$. So the width of a slice at height y in the lower part of the region is $2y$, and in the upper part of the region is $2(\sqrt{2} - y)$.

$$F = \rho g \int_0^{\sqrt{2}/2} (\sqrt{2} + 1 - y)(2y)\,dy + \rho g \int_{\sqrt{2}/2}^{\sqrt{2}} (\sqrt{2} + 1 - y)2(\sqrt{2} - y)\,dy$$

$$= \rho g \int_0^{\sqrt{2}/2} (-2y^2 + 2\sqrt{2}y + 2y)\,dy + \rho g \int_{\sqrt{2}/2}^{\sqrt{2}} (2y^2 - 4\sqrt{2}y - 2y + 2\sqrt{2} + 4)\,dy$$

$$= \rho g \left(-\frac{2y^3}{3} + \sqrt{2}y^2 + y^2\right)\Big|_0^{\sqrt{2}/2} + \rho g \left(2\left(\frac{y^3}{3} - \frac{1}{2}(1 + 2\sqrt{2})y^2 + \sqrt{2}y + 2y\right)\right)\Big|_{\sqrt{2}/2}^{\sqrt{2}}$$

$$= \rho g \left(\frac{1}{2} + \frac{\sqrt{2}}{3} + \frac{1}{2} + \frac{1}{3\sqrt{2}}\right) = \rho g \left(1 + \frac{\sqrt{2}}{2}\right) \text{ N}.$$

This is 16,730 N.

b. $F = \int_0^1 \rho g(2-y) \cdot 1 \, dy = \rho g \left(2y - y^2/2\right) \Big|_0^1 = \dfrac{3\rho g}{2}$ N. This is 14,700 N.

6.7.69

a.
$$W = \int_0^{2500000} \dfrac{GMm}{(x+R)^2} \, dx = GMm \left(\dfrac{-1}{x+R}\right) \Big|_0^{2500000}$$
$$= GMm \left(\dfrac{1}{R} - \dfrac{1}{R+2500000}\right) = \dfrac{GMm \, 2500000}{R(R+2500000)} \approx 8.87435 \times 10^9 \text{ J}.$$

b. $W(x) = \int_0^x \dfrac{GMm}{(t+R)^2} \, dt = GMm \left(\dfrac{-1}{t+R}\right)\Big|_0^x = GMm \left(\dfrac{1}{R} - \dfrac{1}{R+x}\right) = \dfrac{GMmx}{R(R+x)} = \dfrac{500 GMx}{R(R+x)}.$

c. $\lim\limits_{x \to \infty} \dfrac{GMmx}{R(R+x)} = \lim\limits_{x \to \infty} \dfrac{GMm}{R\left(\left(\frac{R}{x}\right)+1\right)} = \dfrac{GMm}{R}.$

d. Suppose $\dfrac{GMmx}{R(R+x)} = \dfrac{1}{2} mv^2$, then $v^2 = \dfrac{2GMx}{R(R+x)}$, and as $x \to \infty$ we have $v^2 = \dfrac{2GM}{R}$, so $v = \sqrt{\dfrac{2GM}{R}}$.

Chapter Six Review

1

a. True. A vertical slice would lead to shells, while a horizontal slice would lead to either disks or washers.

b. True. In order to find position, you would also need to know either its initial position, or at least its position at some time.

c. True. If dV/dt is constant, then V is a linear function of time.

3

a. $12t^2 - 30t + 12 = 6(2t^2 - 5t + 2) = 6(2t-1)(t-2)$, which is zero for $t = 2$ and $t = \dfrac{1}{2}$. $v(t) > 0$ for $0 \leq t < \dfrac{1}{2}$ and for $2 < t \leq 3$, so the motion is in the positive direction on those intervals, while $v < 0$ for $\dfrac{1}{2} < t < 2$, so the motion is in the negative direction on that interval.

b. The displacement is given by $\int_0^3 (12t^2 - 30t + 12) \, dt = \left(4t^3 - 15t^2 + 12t\right)\Big|_0^3 = 108 - 135 + 36 = 9 \text{ m}.$

c. The distance traveled is
$$\int_0^{1/2} (12t^2 - 30t + 12) \, dt + \int_2^3 (12t^2 - 30t + 12) \, dt - \int_{1/2}^2 (12t^2 - 30t + 12) \, dt$$
$$= \left(4t^3 - 15t^2 + 12t\right)\Big|_0^{1/2} + \left(4t^3 - 15t^2 + 12t\right)\Big|_2^3 - \left(4t^3 - 15t^2 + 12t\right)\Big|_{1/2}^2$$
$$= (1/2 - 15/4 + 6) + (108 - 135 + 36) - (32 - 60 + 24) - ((32 - 60 + 24) - (1/2 - 15/4 + 6))$$
$$= 22.5 \text{ m}.$$

d. $s(t) = s(0) + \int_0^t (12x^2 - 30x + 12)\,dx = 1 + (4x^3 - 15x^2 + 12x)\Big|_0^t = 4t^3 - 15t^2 + 12t + 1.$

With the antiderivative method, we would have

$$s(t) = \int v(t)\,dt = \int (12t^2 - 30t + 12)\,dt = 4t^3 - 15t^2 + 12t + C,$$

and because $s(0) = 1$, we would have $C = 1$. Therefore, $s(t) = 4t^3 - 15t^2 + 12t + 1$.

5 The position $s(t)$ and the displacement are the same, because the projective started on the ground at position 0.

$$s(t) = \int_0^t v(x)\,dx = \int_0^t (20 - 10x)\,dx = (20x - 5x^2)\Big|_0^t = 20t - 5t^2.$$

Note that the projectile is moving up for $0 < t < 2$ and down for $2 < t < 4$. Thus the distance traveled is equal to the position for $0 < t < 2$, but for $2 < t < 4$ the distance traveled is $20 + (20 - (20t - 5t^2)) = 40 - 20t + 5t^2$.

7

a. $v(t) = \int a(t)\,dt = \int 2\sin(\pi t/4)\,dt = -\dfrac{8}{\pi}\cos(\pi t/4) + C$, and because $v(0) = -\dfrac{8}{\pi}$, we have $C = 0$.
Thus, $v(t) = -\dfrac{8}{\pi}\cos(\pi t/4)$.

$s(t) = \int v(t)\,dt = \int -\dfrac{8}{\pi}\cos(\pi t/4)\,dt = -\dfrac{32}{\pi^2}\sin(\pi t/4) + D$, and because $s(0) = 0$ we have $D = 0$.
Thus, $s(t) = -\dfrac{32}{\pi^2}\sin(\pi t/4)$.

b. s is periodic with period 8, so we only consider $0 \le t \le 8$. There are critical numbers for s at $t = 2$ and $t = 6$. There is a maximum for s of $\dfrac{32}{\pi^2}$ at $t = 6$ and a minimum of $-\dfrac{32}{\pi^2}$ at $t = 2$.

c. The average velocity is $\dfrac{1}{8}\int_0^8 \dfrac{8}{\pi}\cos(\pi t/4)\,dt = -\dfrac{4}{\pi^2}(\sin(\pi t/4))\Big|_0^8 = 0.$

The average position is $\dfrac{1}{8}\int_0^8 -\dfrac{32}{\pi^2}\sin(\pi t/4)\,dt = \dfrac{16}{\pi^3}(\cos(\pi t/4))\Big|_0^8 = 0.$

9

a. For $0 \le t \le 8$ we have $R'(t) = 4t^{1/3}$, so $R(t) = 3t^{4/3} + C$, but $C = 0$, so $R(t) = 3t^{4/3}$.

b. Note that $R(8) = 48$, so for $t > 8$, $R(t) = 48 + \int_8^t 2\,dx = 48 + 2(t - 8)$.

We have $R(t) = \begin{cases} 3t^{4/3} & \text{if } 0 \le t \le 8; \\ 2t + 32 & \text{if } t > 8. \end{cases}$

c. The fuel runs out when $150 = 48 + 2(t - 8)$, which occurs for $t = 59$.

11

a.
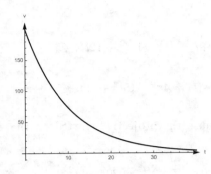

b. The velocity is 50 when $200e^{-t/10} = 50$, which occurs when $e^{t/10} = 4$, so when $t = 10\ln 4$.

c.
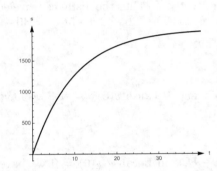

The position is given by $\int_0^t 200e^{-x/10}\,dx = -2000e^{-x/10}\Big|_0^t = 2000(1-e^{-t/10})$.

d. No. $\lim_{t\to\infty} s(t) = 2000 < 2500$.

13

a. Tom's position function is given by $\int_0^t 20e^{-2x}\,dx = \left(-10e^{-2x}\right)\Big|_0^t = 10(1-e^{-2t})$.
Sue's position is given by $\int_0^t 15e^{-x}\,dx = \left(-15e^{-x}\right)\Big|_0^t = 15(1-e^{-t})$.

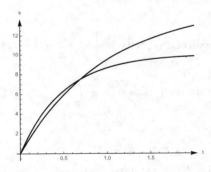

b. We are looking for t so that $10(1-e^{-2t}) = 15(1-e^{-t})$, which occurs when $10(e^{-t})^2 - 15e^{-t} + 5 = 0$, or $2u^2 - 3u + 1 = 0$ where $u = e^{-t}$. This quadratic factors as $(2u-1)(u-1)$, so if $e^{-t} = 1$ or $e^{-t} = 1/2$, so $t = 0$ or $t = \ln 2$.

c. Sue takes the lead and doesn't relinquish it at $t = \ln 2$.

15 The area is given by

$$\int_0^{\pi/4}(\sin 2x - 1 - (-\cos 2x))\,dx = \left(-\frac{\cos 2x}{2} - x + \frac{\sin 2x}{2}\right)\Big|_0^{\pi/4} = \left(0 - \frac{\pi}{4} + \frac{1}{2}\right) - \left(-\frac{1}{2} - 0 + 0\right) = 1 - \frac{\pi}{4}.$$

Chapter Six Review   333

17 The area is given by
$$\int_0^1 (e^y - 1)\, dy = (e^y - y)\Big|_0^1 = (e-1) - (1-0) = e - 2.$$

19 These two curves meet when $4x = x\sqrt{25 - x^2}$, so at $x = 0$ and when $4 = \sqrt{25 - x^2}$, which occurs for $x = 3$. On $[0, 3]$, we have $x\sqrt{25 - x^2} > 4x$ (for example check at $x = 1$), so the area is
$$A = \int_0^3 \left(x\sqrt{25 - x^2} - 4x \right) dx.$$

Note that $\int_0^3 -4x\, dx = -2x^2\Big|_0^3 = -18$. For the first term, use the substitution $u = 25 - x^2$, so that $du = -2x\, dx$ and the integration bounds become $u = 25$ to $u = 16$. Then we have
$$A = -\frac{1}{2}\int_{25}^{16} u^{1/2}\, du - 18 = -\frac{1}{2}\left(\frac{2}{3}u^{3/2}\right)\Big|_{25}^{16} - 18 = -\frac{64}{3} + \frac{125}{3} - 18 = \frac{7}{3}.$$

21 The curves intersect for $x^2 = 2x^2 - 4x$, or $x^2 - 4x = x(x-4) = 0$, so for $x = 0$ and $x = 4$. For $0 \le x \le 2$ the region is bounded above by x^2 and below by the x-axis, and for $2 \le x \le 4$ the region is bounded above by x^2 and below by $2x^2 - 4x$. The area is given by
$$\int_0^2 x^2\, dx + \int_2^4 (x^2 - 2x^2 + 4x)\, dx = \left(x^3/3 \right)\Big|_0^2 + \left(-x^3/3 + 2x^2 \right)\Big|_2^4 = (8/3) + -64/3 + 32 - (-8/3 + 8) = 8.$$

23 Note that we can write $1 - \left|\dfrac{x}{2} - 1\right|$ as $\begin{cases} \dfrac{x}{2} & \text{if } x \le 2; \\ 2 - \dfrac{x}{2} & \text{if } x \ge 2. \end{cases}$

For $0 \le x \le 2$, the region is bounded above by $\dfrac{x}{2}$ and below by $\dfrac{x}{6}$. For $2 \le x \le 3$, the region is bounded above by $2 - \dfrac{x}{2}$ and below by $\dfrac{x}{6}$. The area is then given by
$$\int_0^2 (x/2 - x/6)\, dx + \int_2^3 (2 - x/2 - x/6)\, dx = \left(x^2/6 \right)\Big|_0^2 + \left(2x - x^2/3 \right)\Big|_2^3 = \frac{2}{3} + (6 - 3) - (4 - 4/3) = 1.$$

25 The curve and the line intersect where $2x^2 - 6x + 5 = 1$, or $2x^2 - 6x + 4 = 2(x^2 - 3x + 2) = 2(x-1)(x-2) = 0$, which is for $x = 1$ and $x = 2$. The area is given by
$$\int_1^2 (1 - (2x^2 - 6x + 5))\, dx = \int_1^2 (-4 + 6x - 2x^2)\, dx = \left(-4x + 3x^2 - \frac{2x^3}{3} \right)\Big|_1^2$$
$$= -8 + 12 - \frac{16}{3} - \left(-4 + 3 - \frac{2}{3} \right) = \frac{1}{3}.$$

27 The area of R_1 is
$$\int_0^1 (3 - x - 2\sqrt{x})\, dx = \left(3x - \frac{x^2}{2} - \frac{4x^{3/2}}{3} \right)\Big|_0^1 = 3 - \frac{1}{2} - \frac{4}{3} = \frac{7}{6}.$$

The area of R_2 can be computed by subtracting the area of R_1 from the area of a triangle with base 3 and height 3, so it is
$$\frac{9}{2} - \frac{7}{6} = \frac{20}{6} = \frac{10}{3}.$$

The area of R_3 is the difference of $\int_1^3 2\sqrt{x}\, dx$ and the area of a triangle with base 2 and height 2, so it is
$$\int_1^3 2\sqrt{x}\, dx - 2 = \left(\frac{4x^{3/2}}{3} \right)\Big|_1^3 - 2 = \frac{12\sqrt{3}}{3} - \frac{4}{3} - 2 = 4\sqrt{3} - \frac{10}{3}.$$

Copyright © 2019 Pearson Education, Inc.

29
$$V = 2\pi \int_0^1 x(3 - x - 2\sqrt{x})\,dx = 2\pi \int_0^1 (3x - x^2 - 2x^{3/2})\,dx$$
$$= 2\pi \left(\frac{3x^2}{2} - \frac{x^3}{3} - \frac{4x^{5/2}}{5}\right)\Big|_0^1 = 2\pi \left(\frac{3}{2} - \frac{1}{3} - \frac{4}{5}\right) = \frac{11\pi}{15}.$$

31
$$V = 2\pi \int_0^2 y\left(3 - y - \frac{y^2}{4}\right) dy = 2\pi \int_0^2 \left(3y - y^2 - \frac{y^3}{4}\right) dy$$
$$= 2\pi \left(\frac{3y^2}{2} - \frac{y^3}{3} - \frac{y^4}{16}\right)\Big|_0^2 = 2\pi \left(6 - \frac{8}{3} - 1\right) = \frac{14\pi}{3}.$$

33 $V = 2\pi \int_1^3 (3 - x)\left(2\sqrt{x} - 3 + x\right) dx.$

35 $A = \int_0^1 1\,dx + \int_1^4 (2 - \sqrt{x})\,dx = 1 + \left(2x - \frac{2x^{3/2}}{3}\right)\Big|_1^4 = 1 + \left(8 - \frac{16}{3} - \left(2 - \frac{2}{3}\right)\right) = \frac{7}{3}.$

37 $V = \pi \int_0^1 (2 - y)^4\,dy = \pi \left(-\frac{(2-y)^5}{5}\right)\Big|_0^1 = \pi \left(-\frac{1}{5} + \frac{32}{5}\right) = \frac{31\pi}{5}.$

39 The area of region R_1 is $\int_0^{\pi/3} \sec^2 x\,dx = \tan x \Big|_0^{\pi/3} = \sqrt{3}$. The area of region R_2 can be calculated by subtracting the area of R_1 from the area of a rectangle with base $\frac{\pi}{3}$ and height 4. Thus the area of R_2 is $\frac{4\pi}{3} - \sqrt{3}.$

41 $A(a) = \int_0^{\sqrt[3]{a}} \left(\sqrt{x/a} - x^2/a\right) dx = \left(\frac{2}{3}\frac{x^{3/2}}{\sqrt{a}} - \frac{x^3}{3a}\right)\Big|_0^{\sqrt[3]{a}} = \frac{1}{3}.$

43 From $x = 0$ to $x = 1$ the area function is $A(x) = (\sqrt{x})^2 = x$, while from $x = 1$ to $x = 2$ it is $A(x) = (2 - x)^2.$
$$\int_0^1 x\,dx + \int_1^2 (2-x)^2\,dx = \frac{1}{2}x^2 \Big|_0^1 - \frac{1}{3}(2-x)^3 \Big|_1^2 = \frac{1}{2} + \frac{1}{3} = \frac{5}{6}.$$

45 Because the cross sections are parallel to the x-axis, solve for x to get $x = y^2$ and $x = 2 - y$. Then from $y = 0$ to $y = 1$ the area function is $A(y) = (y^2)^2 = y^4$ and from $y = 1$ to $y = 2$ it is $A(y) = (2 - y)^2$. The by the general slicing method, the volume is
$$\int_0^1 y^4\,dy + \int_1^2 (2-y)^2\,dy = \frac{y^5}{5}\Big|_0^1 - \frac{1}{3}(2-y)^3 \Big|_1^2 = \frac{1}{5} + \frac{1}{3} = \frac{8}{15}.$$

47

a. $V = 2\pi \int_0^1 x((1 + \sqrt{x}) - (1 - \sqrt{x}))\,dx = 4\pi \int_0^1 x^{3/2}\,dx = 4\pi \cdot \frac{2x^{5/2}}{5}\Big|_0^1 = \frac{8\pi}{5}.$

b. $V = \pi \int_0^2 (1^2 - ((1-y)^2)^2)\,dy = \pi \int_0^2 (1 - (y^4 - 4y^3 + 6y^2 - 4y + 1))\,dy = \pi \int_0^2 (-y^4 + 4y^3 - 6y^2 + 4y)\,dy =$
$\pi \left(-\frac{y^5}{5} + y^4 - 2y^3 + 2y^2\right)\Big|_0^2 = \frac{8\pi}{5}.$

49 It appears to be easiest to use shells, because then we won't have to split the integral up, and also the functions will integrate more easily. Solving for y we have $y = e^{x^2}$ and $y = e^{2-x^2}$. The height of each shell is then $e^{2-x^2} - e^{x^2}$ and the volume is

$$2\pi \int_0^1 x(e^{2-x^2} - e^{x^2})\, dx = 2\pi \left(-\frac{1}{2}e^{2-x^2} - \frac{1}{2}e^{x^2}\right)\bigg|_0^1 = \pi(-e - e + e^2 + 1) = \pi(e^2 - 2e + 1) = \pi(e-1)^2.$$

51 We use the shell method. $V = 2\pi \int_0^{\sqrt{3}/2} \dfrac{x}{\sqrt{1-x^2}}\, dx$. Let $u = 1 - x^2$ so that $du = -2x\, dx$. Substituting gives $V = 2\pi \int_1^{1/4} \dfrac{-1}{2\sqrt{u}}\, du = \pi \int_{1/4}^1 u^{-1/2}\, du = 2\pi \sqrt{u}\,\bigg|_{1/4}^1 = 2\pi \left(1 - \dfrac{1}{2}\right) = \pi.$

53 We use the disk method. The radius of each disk is $4 - (x-2)^2 = 4x - x^2$, so the volume is

$$V = \pi \int_0^4 (4x - x^2)^2\, dx = \pi \int_0^4 (x^4 - 8x^3 + 16x^2)\, dx$$
$$= \pi \left(\frac{x^5}{5} - 2x^4 + \frac{16}{3}x^3\right)\bigg|_0^4 = \pi \left(\frac{1024}{5} - 512 + \frac{1024}{3}\right) = \frac{512\pi}{15}.$$

55 The two non-horizontal lines intersect at $(2, 4)$. To revolve about $y = -2$, we use the shell method. First solve for x to obtain $x = \frac{y}{2}$ and $x = 6 - y$; then the height of each shell is $6 - y - \frac{y}{2} = 6 - \frac{3}{2}y$, and the radius at y is $y - (-2) = y + 2$. The volume is

$$2\pi \int_0^4 (y+2)\left(6 - \frac{3}{2}y\right) dy = 2\pi \int_0^4 \left(12 + 3y - \frac{3}{2}y^2\right) dy = 2\pi \left(12y + \frac{3}{2}y^2 - \frac{1}{2}y^3\right)\bigg|_0^4 = 80\pi.$$

To revolve about $x = -2$, we use the washer method. With the equations above, the outer radius of each washer is $6 - y - (-2) = 8 - y$ and the inner radius is $\frac{y}{2} - (-2) = 2 + \frac{y}{2}$. The volume is

$$V = \pi \int_0^4 \left((8-y)^2 - \left(2 + \frac{y}{2}\right)^2\right) dy = \pi \int_0^4 \left(\frac{3}{4}y^2 - 18y + 60\right) dy$$
$$= \pi \left(\frac{1}{4}y^3 - 9y^2 + 60y\right)\bigg|_0^4 = \pi(16 - 144 + 240) = 112\pi.$$

57 We use disks when revolving about the x-axis. The volume is

$$\pi \int_0^1 c^2 x^2 (1-x)^2\, dx = \pi c^2 \int_0^1 (x^2 - 2x^3 + x^4)\, dx = \pi c^2 \left(\frac{1}{3}x^3 - \frac{1}{2}x^4 + \frac{1}{5}x^5\right)\bigg|_0^1 = \pi c^2 \left(\frac{1}{3} - \frac{1}{2} + \frac{1}{5}\right) = \frac{\pi c^2}{30}.$$

We use shells when revolving about the y-axis. The volume is

$$2\pi \int_0^1 x \cdot cx(1-x)\, dx = 2\pi c \int_0^1 (x^2 - x^3)\, dx = 2\pi c \left(\frac{1}{3}x^3 - \frac{1}{4}x^4\right)\bigg|_0^1 = \frac{\pi c}{6}.$$

These are equal when $\dfrac{\pi c^2}{30} = \dfrac{\pi c}{6}$, which for $c \neq 0$ occurs when $c = 5$.

59 $L = \int_{\sqrt{2}}^{\sqrt{5}} \sqrt{1 + \dfrac{1}{x^2 - 1}}\, dx = \int_{\sqrt{2}}^{\sqrt{5}} \dfrac{x}{\sqrt{x^2 - 1}}\, dx.$ Let $u = x^2 - 1$ so that $du = 2x\, dx$. Substituting gives $\dfrac{1}{2}\int_1^4 u^{-1/2}\, du = (\sqrt{u})\bigg|_1^4 = 2 - 1 = 1.$

61 Note that $y' = 1/(2\sqrt{x}) - \sqrt{x}/2$, so $1 + y'^2 = (1/(2\sqrt{x}) + \sqrt{x}/2)^2.$ $L = \int_1^3 \sqrt{1 + y'^2}\, dy = \int_1^3 1/(2\sqrt{x}) + \sqrt{x}/2\, dx = \left(\sqrt{x} + x^{3/2}/3\right)\bigg|_1^3 = 2\sqrt{3} - 4/3.$

63 $y' = 2(x+1)$, so $\sqrt{1 + f'(x)^2} = \sqrt{1 + (2x+2)^2} = \sqrt{4x^2 + 8x + 5}$.
The arc length is $\int_2^4 \sqrt{4x^2 + 8x + 5}\, dx \approx 16.127$.

65 Note that $y' = 1/x$, so $1 + y'^2 = \dfrac{x^2+1}{x^2}$, and $\sqrt{1+y'^2} = \dfrac{\sqrt{x^2+1}}{x}$.
So
$$L = \int_1^b \frac{\sqrt{x^2+1}}{x}\, dx = \left(\sqrt{x^2+1} - \ln\left(\frac{1+\sqrt{x^2+1}}{x}\right)\right)\bigg|_1^b = \sqrt{b^2+1} - \sqrt{2} + \ln\left(\frac{(\sqrt{b^2+1}-1)(1+\sqrt{2})}{b}\right).$$
Using a computer algebra system, we see that this has value 2 for $b \approx 2.715$.

67

a. $S = 2\pi \int_0^3 \sqrt{3x - x^2}\sqrt{1 + \dfrac{(3-2x)^2}{4(3x-x^2)}}\, dx = 2\pi \int_0^3 \sqrt{3x - x^2}\sqrt{\dfrac{12x - 4x^2 + 9 - 12x + 4x^2}{4(3x-x^2)}}\, dx = \pi \int_0^3 3\, dx = 3\pi(3-0) = 9\pi$.

b. $V = \pi \int_0^3 (3x - x^2)\, dx = \pi\left(3x^2/2 - x^3/3\right)\bigg|_0^3 = \pi(27/2 - 9) = 9\pi/2$.

69

a. $S = 2\pi \int_1^2 \left(\dfrac{x^4}{2} + \dfrac{1}{16x^2}\right)\sqrt{1 + (2x^3 - (1/8x^3))^2}\, dx =$
$2\pi \int_1^2 \left(\dfrac{x^4}{2} + \dfrac{1}{16x^2}\right)\sqrt{1 + (4x^6 - 1/2 + (64/x^6))}\, dx = 2\pi \int_1^2 \left(\dfrac{x^4}{2} + \dfrac{1}{16x^2}\right)\sqrt{4x^6 + 1/2 + 64/x^6}\, dx =$
$2\pi \int_1^2 \left(\dfrac{x^4}{2} + \dfrac{1}{16x^2}\right)\sqrt{(2x^3 + (1/8x^3))^2}\, dx = 2\pi \int_1^2 \left(\dfrac{x^4}{2} + \dfrac{1}{16x^2}\right)\left(2x^3 + \dfrac{1}{8x^3}\right)\, dx = 2\pi \int_1^2 (x^7 + x/16 + x/8 + x^{-5}/128)\, dx = 2\pi \int_1^2 (x^7 + 3x/16 + x^{-5}/128)\, dx = 2\pi\left(x^8/8 + 3x^2/32 - x^{-4}/512\right)\bigg|_1^2 = 2\pi((32 + 3/8 - 1/8192) - (1/8 + 3/32 - 1/512)) = \dfrac{263439\pi}{4096}$.

b. Using the fact that $\sqrt{1 + f'(x)^2} = 2x^3 + 1/(8x^3)$ which was discovered during the previous calculation, we have $L = \int_1^2 (2x^3 + x^{-3}/8)\, dx = \left(x^4/2 - x^{-2}/16\right)\bigg|_1^2 = (8 - 1/64) - (1/2 - 1/16) = \dfrac{483}{64}$.

c. $V = 2\pi \int_1^2 (x^5/2 + x^{-1}/16)\, dx = 2\pi\left(x^6/12 + (1/16)\ln x\right)\bigg|_1^2 = 2\pi(16/3 + (\ln 2)/16 - (1/12 + 0)) = \dfrac{21\pi}{2} + \dfrac{\pi \ln 2}{8}$.

d.
$V = \pi \int_1^2 (x^4/2 + x^{-2}/16)^2\, dx = \pi \int_1^2 (x^8/4 + x^2/16 + x^{-4}/256)\, dx = \pi\left(x^9/36 + x^3/48 - x^{-3}/768\right)\bigg|_1^2$
$= \pi(512/36 + 8/48 - 1/6144 - (1/36 + 1/48 - 1/768)) = \dfrac{264341\pi}{18432}$.

71 $m = \int_0^3 150 e^{-x/3}\, dx = \left(-450 e^{-x/3}\right)\bigg|_0^3 = 450(1 - e^{-1})$ gm.

Chapter Six Review

73

a. Because $50 = \int_0^{.2} kx\, dx = \dfrac{kx^2}{2}\bigg|_0^{0.2} = \dfrac{k}{50}$, we must have $k = 2500$.

$$W = \int_{0.2}^{0.7} 2500x\, dx = 1250x^2 \bigg|_{0.2}^{0.7} = 562.5\text{ J}.$$

b. $f(0.2) = 0.2k = 50$, so $k = 250$, and thus

$$W = \int_{0.2}^{0.7} 250x\, dx = 125x^2 \bigg|_{0.2}^{0.7} = 56.25\text{ J}.$$

75

a. Note that the density of the chain is 2 kg/m.

$$W = \int_0^{10} 2(9.8)(10-y)\, dy = 19.6\left(10y - \dfrac{y^2}{2}\right)\bigg|_0^{10} = 19.6\,(100 - 50) = 980\text{ J}.$$

b. The lower 6 meters of the chain are lifted 4 meters. The work required for this is $6(2)(9.8)(4) = 470.4\text{ J}$. The work required to lift the upper four meters of the chain is

$$\int_6^{10} 2(9.8)(10-y)\, dy = 19.6\left(10y - \dfrac{y^2}{2}\right)\bigg|_6^{10} = 19.6(100 - 50 - (60-18)) = 156.8.$$

So the total work required is $470.4 + 156.8 = 672.2\text{ J}$.

77

a. $W = \int_0^6 1000(9.8)(8)(6-y)\, dy = 78,400\left(6y - \dfrac{y^2}{2}\right)\bigg|_0^6 = 78,400(36-18) = 1,411,200\text{ J}.$

b. $W = \int_0^2 1000(9.8)(8)(7-y)\, dy = 78,400\left(7y - \dfrac{y^2}{2}\right)\bigg|_0^2 = 78,400(14-2) = 940,800\text{ J}.$

79

a.
$$W = \int_0^6 1000(9.8)\dfrac{4}{9}\pi y^2(6-y)\, dy = \dfrac{39200}{9}\pi\int_0^6 (6y^2 - y^3)\, dy$$

$$= \dfrac{39200}{9}\pi\left(2y^3 - \dfrac{y^4}{4}\right)\bigg|_0^6 = \dfrac{39200}{9}\pi(432 - 324) \approx 1,477,805\text{ J}.$$

b. The work to pump out the top 3 feet is

$$\int_3^6 \dfrac{39200}{9}\pi(6y^2 - y^3)\, dy = \dfrac{39200}{9}\pi\left(2y^3 - \dfrac{y^4}{4}\right)\bigg|_3^6 = \dfrac{39200}{9}\pi(432 - 324 - (54 - 20.25)) \approx 1,015,592\text{ J}.$$

Therefore, the work to pump out the bottom 3 feet is about $1,477,805 - 1,015,592 = 461,814\text{ J}$.

81 $W = \int_0^9 1000(9.8)\pi y(10-y)\, dy = 9800\pi \int_0^9 (10y - y^2)\, dy = 9800\pi\left(5y^2 - \dfrac{y^3}{3}\right)\bigg|_0^9 = 9800\pi(405 - 243) \approx 4,987,592\text{ J}.$

83 $F = \int_0^1 1000(9.8)(3-y)\dfrac{y}{2}\, dy = 4900\int_0^1 (3y - y^2)\, dy = 4900\left(\dfrac{3y^2}{2} - \dfrac{y^3}{3}\right)\bigg|_0^1 = 4900\left(\dfrac{3}{2} - \dfrac{1}{3}\right) \approx 5716.7\text{ N}.$

85 Orient the semicircle so that the center is at the point $(0, 20)$. Then the force is given by

$$\int_0^{20} \rho g(20-y)(2)\sqrt{40y - y^2}\, dy.$$

Let $u = 40y - y^2$ so that $du = (40 - 2y)\, dy$. Then we have

$$\rho g \int_0^{400} u^{1/2}\, du = \rho g \left(\frac{2}{3} u^{3/2}\right)\bigg|_0^{400} \approx 5.2 \times 10^7.$$

Chapter 7

Logarithmic, Exponential, and Hyperbolic Functions

7.1 Logarithmic and Exponential Functions Revisited

7.1.1 The domain is $(0, \infty)$ and the range is $(-\infty, \infty)$.

7.1.3 $\int 4^x \, dx = \int e^{x \ln 4} \, dx = \dfrac{1}{\ln 4} e^{x \ln 4} + C = \dfrac{4^x}{\ln 4} + C.$

7.1.5 $3^x = e^{x \ln 3}$. $x^\pi = e^{\pi \ln x}$. $x^{\sin x} = e^{\sin x \cdot \ln x}$.

7.1.7 $\dfrac{d}{dx}(x \ln x^3) = x \cdot \dfrac{1}{x^3} \cdot 3x^2 + \ln x^3 = 3 + \ln x^3 = 3 + 3 \ln x = 3(1 + \ln x).$

7.1.9 $\dfrac{d}{dx} \sin(\ln x) = \cos(\ln x) \cdot \dfrac{1}{x} = \dfrac{\cos(\ln x)}{x}.$

7.1.11 $\dfrac{d}{dx}((\ln 2x)^{-5}) = -5(\ln 2x)^{-6}(1/x) = -\dfrac{5}{x \ln^6 2x}.$

7.1.13 $2x^{4x} = e^{4x \ln 2x}$, so

$$\dfrac{d}{dx} 2x^{4x} = \dfrac{d}{dx} e^{4x \ln 2x} = e^{4x \ln(2x)} \left(4x \cdot \dfrac{1}{x} + \ln 2x \cdot 4\right) = (2x)^{4x}(4 + 4 \ln 2x) = 4^{2x+1} x^{4x}(1 + \ln 2x).$$

7.1.15 $\dfrac{d}{dx} 2^{(x^2)} = 2^{(x^2)} \cdot \ln 2 \cdot (2x) = 2^{(x^2+1)} x \ln 2.$

7.1.17 $\dfrac{d}{dx} e^{2x \ln(x+1)} = e^{2x \ln(x+1)} \cdot \left(\dfrac{2x}{x+1} + 2 \ln(x+1)\right)$, which can be written as

$$(x+1)^{2x} \left(\dfrac{2x}{x+1} + 2 \ln(x+1)\right).$$

7.1.19 $\dfrac{d}{dy} e^{\sin y \ln y} = e^{\sin y \ln y} \left(\cos y \ln y + \dfrac{\sin y}{y}\right)$, which can be written as $y^{\sin y} \left(\cos y \ln y + \dfrac{\sin y}{y}\right).$

7.1.21 $\dfrac{d}{dx} e^{-10x^2} = e^{-10x^2} \cdot -20x = -20x e^{-10x^2}.$

7.1.23 $\dfrac{d}{dx} x^{2x} = \dfrac{d}{dx} e^{2x \ln x} = e^{2x \ln x}(2 + 2 \ln x) = 2x^{2x}(1 + \ln x).$

7.1.25 $\dfrac{d}{dx}(1/x)^x = \dfrac{d}{dx}e^{x\ln(1/x)} = \dfrac{d}{dx}e^{-x\ln x} = e^{-x\ln x}(-1-\ln x) = \left(\dfrac{1}{x}\right)^x(-1-\ln x).$

7.1.27
$$\dfrac{d}{dx}(1+(4/x))^x = \dfrac{d}{dx}e^{x\ln(1+(4/x))} = e^{x\ln(1+(4/x))}\left(\ln(1+(4/x)) + \dfrac{x}{1+(4/x)}\cdot -\dfrac{4}{x^2}\right)$$
$$= (1+(4/x))^x(\ln(1+(4/x)) - \dfrac{4}{x+4})$$

7.1.29 $\displaystyle\int_0^3 \dfrac{2x-1}{x+1}\,dx = \int_0^3 \left(2 - \dfrac{3}{x+1}\right)dx = (2x - 3\ln(x+1))\Big|_0^3 = 6(1-\ln 2).$

7.1.31 Let $u = \ln x$ so that $du = \dfrac{1}{x}dx$. Then
$$\int_e^{e^2} \dfrac{dx}{x\ln^3 x}\,dx = \int_1^2 u^{-3}\,du = \dfrac{-1}{2}(u^{-2})\Big|_1^2 = \dfrac{-1}{2}(1/4 - 1) = \dfrac{3}{8}.$$

7.1.33 Let $u = 4 + e^{2x}$ so that $du = 2e^{2x}\,dx$. Substituting gives
$$\dfrac{1}{2}\int \dfrac{du}{u} = \dfrac{1}{2}\ln|u| + C = \dfrac{1}{2}\ln(4+e^{2x}) + C.$$

7.1.35 Let $u = \ln(\ln x)$ so that $du = \dfrac{1}{x\ln x}\,dx$. Substituting yields
$$\int_{\ln 2}^{\ln 3} \dfrac{1}{u^2}\,du = \left(\dfrac{-1}{u}\right)\Big|_{\ln 2}^{\ln 3} = \dfrac{1}{\ln 2} - \dfrac{1}{\ln 3}.$$

7.1.37 Let $u = -x^2/2$ so that $du = -x\,dx$. Substituting yields
$$-4\int_0^{-2} e^u\,du = 4\int_{-2}^0 e^u\,du = 4\,(e^u)\Big|_{-2}^0 = 4(1-e^{-2}) = 4 - \dfrac{4}{e^2}.$$

7.1.39 Let $u = \sqrt{x}$ so that $du = \dfrac{1}{2\sqrt{x}}\,dx$. Then
$$\int \dfrac{e^{\sqrt{x}}}{\sqrt{x}}\,dx = 2\int e^u\,du = 2e^u + C = 2e^{\sqrt{x}} + C.$$

7.1.41 Let $u = e^x - e^{-x}$ so that $du = e^x + e^{-x}\,dx$. Then we have
$$\int \dfrac{e^x + e^{-x}}{e^x - e^{-x}}\,dx = \int \dfrac{1}{u}\,du = \ln|u| + C = \ln|e^x - e^{-x}| + C.$$

7.1.43 $\displaystyle\int_{-1}^1 10^x\,dx = \left(\dfrac{10^x}{\ln 10}\right)\Big|_{-1}^1 = \dfrac{10 - 10^{-1}}{\ln 10} = \dfrac{99}{10\ln 10}.$

7.1.45 $\displaystyle\int_1^2 (1+\ln x)x^x\,dx = (x^x)\Big|_1^2 = 4 - 1 = 3.$

7.1.47 Let $u = x^3 + 8$ so that $du = 3x^2\,dx$. Substituting gives
$$\dfrac{1}{3}\int 6^u\,du = \dfrac{6^u}{3\ln 6} + C = \dfrac{6^{x^3+8}}{3\ln 6} + C.$$

7.1. Logarithmic and Exponential Functions Revisited

7.1.49 Let $u = 3x^2 + 1$ so that $du = 6x\,dx$. Substituting yields
$$\frac{1}{6}\int e^u\,du = \frac{1}{6}e^u + C = \frac{1}{6}e^{3x^2+1} + C.$$

7.1.51 $\displaystyle\int 3^{-2x}\,dx = \int e^{-2x\ln 3}\,dx = e^{-2x\ln 3}\cdot\frac{1}{-2\ln 3} + C = -\frac{3^{-2x}}{2\ln 3} + C = \frac{-1}{9^x \ln 9} + C.$

7.1.53 Let $u = x^3$, so that $du = 3x^2\,dx$. Then we have
$$\int \frac{1}{3} 10^u\,du = \frac{10^u}{3\ln 10} + C = \frac{10^{x^3}}{3\ln 10} + C.$$

7.1.55 Let $u = \ln x$ so that $du = \frac{1}{x}dx$. Then we have
$$\int_0^{\ln(2)+1} 3^u\,du = \left(\frac{3^u}{\ln 3}\right)\Big|_0^{\ln(2)+1} = \frac{3\cdot 3^{\ln 2} - 1}{\ln 3}.$$

7.1.57 Let $u = \ln x$ so that $du = \frac{1}{x}dx$. Then we have
$$\int_0^2 u^5\,du = \left(\frac{u^6}{6}\right)\Big|_0^2 = \frac{32}{3}.$$

7.1.59 Let $u = e^{3x} + e^{-3x}$ so that $du = 3(e^{3x} - e^{-3x})\,dx$. Substituting gives
$$\frac{1}{3}\int_2^{65/8} \frac{1}{u}\,du = \frac{1}{3}(\ln u)\Big|_2^{65/8} = \frac{1}{3}(\ln(65/8) - \ln 2) = \frac{1}{3}\ln(65/16).$$

7.1.61 Let $u = 5 + \sqrt{x}$. Then $du = \dfrac{1}{2\sqrt{x}}$. Substituting yields
$$\int 2e^u\,du = 2e^u + C = 2e^{5+\sqrt{x}} + C.$$

7.1.63

h	$(1+2h)^{1/h}$	h	$(1+2h)^{1/h}$
10^{-1}	6.1917	-10^{-1}	9.3132
10^{-2}	7.2446	-10^{-2}	7.5404
10^{-3}	7.3743	-10^{-3}	7.4039
10^{-4}	7.3876	-10^{-4}	7.3905
10^{-5}	7.3889	-10^{-5}	7.3892
10^{-6}	7.3890	-10^{-6}	7.3891

Let $y = (1+2h)^{1/h}$. Then $\ln y = \dfrac{\ln(1+2h)}{h}$.
$\displaystyle\lim_{h\to 0}\ln y = \lim_{h\to 0}\frac{\ln(1+2h)}{h} = \lim_{h\to 0}\frac{2}{1+2h} = 2$, so the limit of y as $h \to 0$ is e^2.

7.1.65

x	$\dfrac{2^x-1}{x}$	x	$\dfrac{2^x-1}{x}$
10^{-1}	.71773	-10^{-1}	.66967
10^{-2}	.69556	-10^{-2}	.69075
10^{-3}	.69339	-10^{-3}	.69291
10^{-4}	.69317	-10^{-4}	.69312
10^{-5}	.69315	-10^{-5}	.69314
10^{-6}	.69315	-10^{-6}	.69315

$\displaystyle\lim_{x\to 0}\frac{2^x-1}{x} = \lim_{x\to 0}\frac{2^x\ln 2}{1} = \ln 2.$

7.1.67

a. True. This follows because $e^{\ln xy} = xy = e^{\ln x}e^{\ln y} = e^{\ln x + \ln y}$, and because the exponential function is one-to-one.

b. False. Zero is not in the domain of the natural logarithm function.

c. False. For example, $\ln(1+1) = \ln 2 \neq \ln(1) + \ln(1) = 0$.

d. False. $e^{2\ln x} = e^{\ln x^2} = x^2 \neq 2^x$.

e. False. $\int_0^e \frac{1}{x}\,dx = \lim_{b\to 0^+}\int_b^e \frac{1}{x}\,dx = \lim_{b\to 0^+}(\ln x)\Big|_b^e = 1 - \lim_{b\to 0^+}\ln b$, which does not exist.

7.1.69 The average value is given by $\dfrac{1}{p-1}\int_1^p \dfrac{1}{x}\,dx = \dfrac{1}{p-1}(\ln x)\Big|_1^p = \dfrac{\ln p}{p-1}$. Note that

$$\lim_{p\to\infty} \frac{\ln p}{p-1} = \lim_{p\to\infty} \frac{1/p}{1} = 0.$$

7.1.71

a. No. Let

$$h(a) = \int_{1-a}^{1+a} \frac{1-x}{x}\,dx = \int_{1-a}^{1} \frac{1-x}{x}\,dx + \int_{1}^{1+a} \frac{1-x}{x}\,dx = -\int_{1}^{1-a} \frac{1-x}{x}\,dx + \int_{1}^{1+a} \frac{1-x}{x}\,dx.$$

Then

$$h'(a) = \frac{1-(1-a)}{1-a} + \frac{1-(1+a)}{1+a} = \frac{a}{1-a} - \frac{a}{1+a} = \frac{a+a^2+(-a)+a^2}{1-a^2} = \frac{2a^2}{1-a^2} > 0.$$

Also, $h(a) = 0$. Because h is increasing on $(0,1)$, there are no other numbers on that interval where h is zero.

b. No. Let

$$g(a) = \int_{1/a}^{a} f(x)\,dx = \int_{1/a}^{1/2} f(x)\,dx + \int_{1/2}^{a} f(x)\,dx = -\int_{1/2}^{1/a} f(x)\,dx + \int_{1/2}^{a} f(x)\,dx.$$

Then

$$g'(a) = -f(1/a)\cdot -\frac{1}{a^2} + f(a) = \frac{1-(1/a)}{a} + \frac{1-a}{a} = \frac{2-(a+1/a)}{a} < 0,$$

because $a + 1/a > 2$ for $a > 1$. Because g has value 0 at $a=1$ and is decreasing on $(1,\infty)$ it is never 0 on that interval.

7.1.73

a. Let $\exp 0 = z$. Then $\ln\exp 0 = \ln z$, so $\ln z = 0$. Then because $\ln z = \ln 1$, we have $z = 1$, so $\exp 0 = 1$.

b. $\ln\left(\dfrac{\exp x}{\exp y}\right) = \ln(\exp x) - \ln(\exp y) = x - y$. Thus, $\exp(x-y) = \dfrac{\exp x}{\exp y}$.

c. $\ln((\exp x)^p) = p\ln(\exp x) = px$, so $\exp px = (\exp x)^p$.

7.1.75 Because $1/x$ is decreasing, we know that the left Riemann sum is an overestimate. Using $n=2$ subintervals, we have $\int_1^2 1/x\,dx = \ln 2 < \dfrac{1}{1}\cdot\dfrac{1}{2} + \dfrac{1}{3/2}\cdot\dfrac{1}{2} = \dfrac{5}{6} < 1$.

Because $1/x$ is decreasing, we know that the right Riemann sum is an underestimate. We will use $n=8$ subintervals. We have $\int_1^3 1/x\,dx = \ln 3 > 2\left(\dfrac{1}{9} + \dfrac{1}{11} + \cdots + \dfrac{1}{21}\right) > 1$.

7.1.77 Because $1/x$ is decreasing, the left Riemann sum is an overestimate to the integral. Over the interval $[1, n+1]$ with n subdivisions, we must have $\int_1^{n+1} \frac{1}{x}\,dx = \ln(n+1) < \frac{n+1-1}{n}\left(1 + \frac{1}{2} + \frac{1}{3} + \cdots + \frac{1}{n}\right) = \sum_{i=1}^{n} \frac{1}{i}$. Because $\ln(n+1) \to \infty$ as $n \to \infty$, it must follow that the harmonic partial sum $\left(1 + \frac{1}{2} + \frac{1}{3} + \cdots + \frac{1}{n}\right) \to \infty$ as well.

7.2 Exponential Models

7.2.1 Exponential growth occurs for a constant relative growth rate.

7.2.3 It is the time it takes the population to double in size.

7.2.5 For exponential growth modeled by $y = y_0 e^{kt}$, the doubling time T_2 is $T_2 = \frac{\ln 2}{k}$, where $k > 0$ is the growth constant.

7.2.7 The doubling time is $\frac{\ln 2}{k}$, so we solve $\frac{\ln 2}{k} = 20$, or $k = \frac{\ln 2}{20} \approx 0.03466$.

7.2.9 Compound interest and world population growth.

7.2.11 We have $y(t) = y_0 e^{kt}$ and $y(1) = y_0 e^k = 1.11 y_0$. Therefore $e^k = 1.11$, and $k = \ln 1.11 \approx 0.10436$.

7.2.13 For $f(t)$, $\frac{df}{dt} = 10.5$, so the absolute growth rate is constant. For $g(t)$, $\frac{dg}{dt} = 100 e^{t/10} \cdot \frac{1}{10} = 10 e^{t/10}$, so the growth rate is not constant but the relative growth rate $\frac{1}{g(t)} g'(t) = \frac{1}{10}$ is constant.

7.2.15

a. The growth is modeled by $p(t) = 90,000 e^{kt}$, with $t = 0$ corresponding to 2016, and time measured in years. Because $(1.024)(90,000) = 90,000 e^k$, we have $k = \ln(1.024) \approx 0.02373$.

b. $120,000 = 90,000 e^{t \ln 1.024}$, so $\frac{4}{3} = e^{t \ln 1.024}$, and $\ln \frac{4}{3} = t \ln 1.024$, so $t = \frac{\ln(4/3)}{\ln 1.024} \approx 12.1$ years. The population will reach 120,000 in 2028.

7.2.17

a. The growth is modeled by $y(t) = 50000 e^{kt}$. When $t = 10$, we have $p(10) = 50000 e^{10k} = 55000$, so $k = \frac{\ln 1.1}{10} \approx 0.00953$.

b. In 20 years, the population should be about $50000 e^{2 \ln 1.1} = 50000 (1.1)^2 = 60,500$.

7.2.19

a. The price of a cart of groceries is modeled by $y(t) = 100 e^{kt}$ where $t = 0$ corresponds to 2010, and t is measured in years and $k = \ln(1.016) \approx 0.01587$.

b. The price of the groceries in 2025 when $t = 15$ should be about $y(15) = 100 e^{15 \ln(1.016)} \approx \$126.88$.

7.2.21 $1200 = 1000 e^{5k}$, so $k = \frac{\ln 1.2}{5}$, and therefore $y(t) = 1000 e^{t(\ln 1.2)/5}$. In one year, the account grows by a factor of $e^{(\ln 1.2)/5} \approx 1.0371$, so the APY is about 3.71%.

7.2.23

a. The population is modeled by $P(t) = 334.4 e^{t \ln(1.0079)}$. The doubling time is $\frac{\ln 2}{\ln 1.0079} \approx 88.1$ years. The population in 2050 will be about $P(30) = 334.4 e^{30 \ln(1.0079)} \approx 423.44$ million.

b. If the growth rate is 0.7%, the doubling time is $\frac{\ln 2}{\ln 1.007} \approx 99.4$ years and the population in 2050 will be about $334.4e^{30\ln(1.007)} \approx 412.2$ million

7.2.25 The growth is modeled by $p(t) = 25.1e^{kt}$. When $t = 6$, we have $p(6) = 25.1e^{6k} = 26.47$, so $k = \frac{\ln(26.47/25.1)}{6} \approx 0.00886$. Thus $p(15) \approx 25.1e^{0.00886(15)} \approx 28.67$ million people.

7.2.27 The homicide rate is modeled by $H(t) = 800e^{kt}$, where $t = 0$ corresponds to 2018, and time is measured in years, and $k = \ln 0.97$. The rate will reach 600 when $\frac{600}{800} = e^{(\ln 0.97)t}$, or $t = \frac{\ln(6/8)}{\ln 0.97} \approx 9.44$ years. So it should achieve this rate in 2027.

7.2.29 The amount of Valium in the bloodstream is modeled by $a(t) = 20e^{kt}$. If the half-life is 36 hours, then $36 = \frac{\ln 2}{k}$, so $k = \frac{\ln 1/2}{36}$. So $a(12) = 20e^{-12\ln(2)/36} = 20e^{-\ln(2)/3} \approx 15.87$ mg.

The concentration of Valium will reach 2 mg when $0.1 = e^{-\ln(2)t/36}$, which is when $t = \frac{-36\ln(0.1)}{\ln 2} \approx 119.59$ hours.

7.2.31 The population is modeled by $p(t) = 1.853e^{kt}$. Because $p(6) = 1.853e^{6k} = 1.831$, we have $e^{6k} = \frac{1.831}{1.853}$, so $k = \frac{\ln(1.831/1.853)}{6}$. The population in 2025 is predicated to be $1.853e^{\frac{1}{6}\ln(1.831/1.853)15} \approx 1.798$ million. This may not be an appropriate long-term model because the downturn in the population of West Virginia may be temporary.

7.2.33 Let $y = 1000e^{kx}$ model the pressure at x feet above sea level. We know that $\frac{1}{3} = e^{30000k}$, so $k = \frac{\ln(1/3)}{30000}$. We want to know for what x does $\frac{1}{2} = e^{kx}$, so we are seeking $x = \frac{\ln 2}{k} = \frac{30000\ln(2)}{\ln 3} \approx 18,928$ feet above sea level.

The pressure is 1/100th of the sea-level pressure when $x = \frac{30000\ln(100)}{\ln 3} \approx 125,754$ feet.

7.2.35 The amount of U-238 in the rock is modeled by $a(t) = a_0 e^{kt}$ with $k = \frac{\ln(0.5)}{4.5}$, where time is measured in billions of years. We seek t so that $a_0 = a_0 e^{kt}$, so $t = \frac{\ln(0.85)}{k} = \frac{\ln(0.85)(4.5)}{\ln(0.5)} \approx 1.055$. So the cloth was painted about 1.055 billion years ago.

7.2.37 The amount of caffeine is modeled by $y(t) = e^{kt}$, and $0.8 = e^{2k}$, so $k = \frac{\ln 0.8}{2}$. The half-life is $\frac{\ln(0.5)}{k} = \frac{2\ln(0.5)}{\ln(0.8)} \approx 6.2$ hours.

7.2.39 Let $y(t)$ equal the cost of the light t years after 2018. Then $y(t) = 4e^{kt}$. Ten years from now, it is predicted that the light will cost 1/10 of its current cost, which means that $y(10) = 4/10 = 0.4$. Replace y with 0.4 and t with 10 in the equation $y(t) = 4e^{kt}$, we have $0.4 = 4e^{10k}$, or $e^{10k} = 0.1$. Solving for k we find that $k = (\ln 0.1)/10 \approx -0.2303$. This means that $y(t) = 4e^{-0.2302t}$ and the cost of the light in 2021 is predicted to be $y(3) = 4e^{-0.2302(3)} \approx \$2.01$.

7.2.41 The initial volume is $V_0 = \frac{4\pi}{3}\left(\frac{5}{10000}\right)^3 = \frac{\pi}{6 \times 10^9}$ cubic cm. $k = \frac{\ln 2}{35}$. Suppose $0.5 = V_0 e^{kt} = \frac{\pi}{6 \times 10^9} e^{(t\ln 2)/35}$. Then $\frac{t\ln 2}{35} = \ln\left(\frac{3 \times 10^9}{\pi}\right)$, and $t = \frac{35}{\ln 2}\ln\left(\frac{3 \times 10^9}{\pi}\right) \approx 1044$ days.

7.2.43

a. False. If that was the correct formula, then $y(1) = y_0 e^{.06} = (1.06184)y_0 \neq (1.06)y_0$.

7.2. Exponential Models

b. False. If it increases by ten precent per year, then after 3 years it increases by a factor of $(1.1)^3 = 1.331$ which corresponds to 33.1 percent.

c. True. The relative decay rate is constant, so the decay is exponential.

d. True. This follows because the doubling time is related to k by the equation $T_2 = \dfrac{\ln 2}{k}$.

e. True. This time would be the constant $\dfrac{\ln 10}{k}$.

7.2.45

a. $V_1(t) = 0.99(0.5)e^{kt}$, where $k = \dfrac{\ln(0.5)}{5.7} \approx -0.1216$. So $V_1(t) = 0.495e^{-0.1216t}$.

b. $V_2(t) = 0.01(0.5)e^{kt}$, where $k = \dfrac{\ln 2}{2.9} \approx 0.239$. So $V_2(t) = 0.005e^{0.239t}$.

c. $V(t) = V_1(t) + V_2(t) = 0.495e^{-0.1216t} + 0.005e^{0.239t}$.

d. The tumor initially shrinks significantly in size but eventually starts growing again.

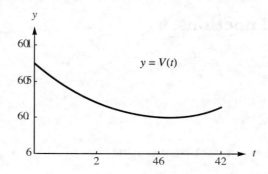

d. Solving the equation $V'(t) = 0$ for t, we find that $t \approx 10.9$ days. So it's best to treat the mouse again just before the end of the 10th day after the first treatment.

7.2.47

a. After 5 hours, Abe has run $\displaystyle\int_0^5 \dfrac{4}{t+1}\,dt = (4\ln(t+1))\big|_0^5 = 4\ln 6 \approx 7.17$ miles. After 5 hours, Bob has run $\displaystyle\int_0^5 4e^{-t/2}\,dt = \left(8e^{-t/2}\right)\Big|_5^0 = 8(1-e^{-5/2}) \approx 7.34$. So after 5 hours, Bob is ahead.

After 10 hours, Abe has run $\displaystyle\int_0^{10} \dfrac{4}{t+1}\,dt = (4\ln(t+1))\big|_0^{10} = 4\ln 11 \approx 9.59$ miles. After 10 hours, Bob has run $\displaystyle\int_0^{10} 4e^{-t/2}\,dt = \left(8e^{-t/2}\right)\Big|_{10}^0 = 8(1-e^{-5}) \approx 7.95$. So after 10 hours, Abe is ahead.

b. Bob's distance function is given by $8(1-e^{-t/2})$ and is bounded above by 8. Abe's distance function is given by $4\ln(t+1)$ and is unbounded.

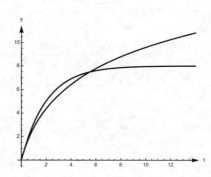

7.2.49 $(1 + 0.008)^{12} - 1 \approx 10.034\%$, which is more than $12 \cdot 0.008 \approx 9.6\%$.

7.2.51 As in the previous problem, $s(t) = \dfrac{v_0}{k}(1 - e^{-kt})$. Consider the equations $s(t_2) - s(t_1) = 1200 = \dfrac{v_0}{k}(e^{-kt_1} - e^{-kt_2})$ and $v(t_2) - v(t_1) = -100 = v_0(e^{-kt_2} - e^{-kt_1})$. Dividing these two equations yields $k = \dfrac{1}{12}$.

Now $v(t_1) = 1000 = v_0 e^{-t_1/12}$ and $v(t_2) = 900 = v_0 e^{-t_2/12}$, so $\dfrac{1000}{900} = e^{(-1/12)(t_1 - t_2)}$, so $t_1 - t_2 = -12\ln(10/9) \approx -1.2643$ seconds. The deceleration takes about 1.2643 seconds to occur.

7.2.53 If $y_0 e^{kt} = y_0(1+r)^t$, then $(e^k)^t = (1+r)^t$, so $e^k = 1+r$, and $k = \ln(1+r)$. If $y_0 e^{kt} = y_0 2^{t/T_2}$, then $(e^k)^t = (2^{1/T_2})^t$, so $e^k = 2^{1/T_2}$, and $k = \dfrac{\ln 2}{T_2}$, or $T_2 = \dfrac{\ln 2}{k}$. Also, we have $T_2 = \dfrac{\ln 2}{k} = \dfrac{\ln 2}{\ln(1+r)}$, so $T_2 \ln(1+r) = \ln 2$. Then $(1+r)^{T_2} = 2$, so $r = 2^{1/T_2} - 1$.

7.2.55 The time required to double occurs when $2y_0 = y_0 e^{kt}$, or when $t = \dfrac{\ln 2}{k}$. Therefore, the doubling time is $\dfrac{\ln 2}{k}$ which is constant as a function of t.

7.3 Hyperbolic Functions

7.3.1 $\cosh x = \dfrac{e^x + e^{-x}}{2}$; $\sinh x = \dfrac{e^x - e^{-x}}{2}$.

7.3.3 $\cosh^2 x - \sinh^2 x = 1$.

7.3.5 $\sinh^{-1} x = \ln(x + \sqrt{x^2 + 1})$.

7.3.7 Evaluate $\sinh^{-1}(1/5)$, which has the same value.

7.3.9 $\displaystyle\int \dfrac{dx}{16 - x^2} = \dfrac{1}{4}\coth^{-1}(x/4) + C$ for $|x| > 4$; in this case, the values in the interval of integration satisfy $|x| > 4$.

7.3.11 $\tanh x = \dfrac{\sinh x}{\cosh x} = \dfrac{\frac{e^x - e^{-x}}{2}}{\frac{e^x + e^{-x}}{2}} = \dfrac{e^x - e^{-x}}{e^x + e^{-x}} \cdot \dfrac{e^x}{e^x} = \dfrac{e^{2x} - 1}{e^{2x} + 1}$.

7.3.13

$$\cosh^2 x + \sinh^2 x = \left(\dfrac{e^x + e^{-x}}{2}\right)^2 + \left(\dfrac{e^x - e^{-x}}{2}\right)^2 = \dfrac{e^{2x} + 2 + e^{-2x}}{4} + \dfrac{e^{2x} - 2 + e^{-2x}}{4}$$

$$= \dfrac{2e^{2x} + 2e^{-2x}}{4} = \dfrac{e^{2x} + e^{-2x}}{2} = \cosh(2x).$$

7.3.15 $\cosh x + \sinh x = \dfrac{e^x + e^{-x}}{2} + \dfrac{e^x - e^{-x}}{2} = \dfrac{2e^x}{2} = e^x$.

7.3.17 $\dfrac{1 + \cosh 2x}{2} = \dfrac{1 + \cosh^2 x + \sinh^2 x}{2} = \dfrac{\cosh^2 x + (\sinh^2 x + 1)}{2} = \dfrac{\cosh^2 x + \cosh^2 x}{2} = \cosh^2 x$.
$\dfrac{\cosh 2x - 1}{2} = \dfrac{\cosh^2 x + \sinh^2 x - 1}{2} = \dfrac{\sinh^2 x + (\cosh^2 x - 1)}{2} = \dfrac{\sinh^2 x + \sinh^2 x}{2} = \sinh^2 x$.

7.3.19 $\dfrac{d}{dx}(\coth x) = \dfrac{d}{dx}\dfrac{\cosh x}{\sinh x} = \dfrac{\sinh^2 x - \cosh^2 x}{\sinh^2 x} = \dfrac{-1}{\sinh^2 x} = -\text{csch}^2 x$.

7.3.21 $\dfrac{d}{dx}(\text{csch } x) = \dfrac{d}{dx}\dfrac{1}{\sinh x} = \dfrac{\sinh x \cdot 0 - \cosh x}{\sinh^2 x} = -\text{csch } x \coth x$.

7.3. Hyperbolic Functions

7.3.23 $f'(x) = 2(\cosh x)(\sinh x)$.

7.3.25 $f'(x) = 2(\tanh x)(\text{sech}^2 x)$.

7.3.27 $f'(x) = \dfrac{1}{\text{sech } 2x}(-\text{sech } 2x \tanh 2x)(2) = -2\tanh 2x$.

7.3.29 $f'(x) = 2x\cosh^2 3x + x^2(2\cosh(3x)\sinh(3x))(3) = 2x\cosh(3x)(\cosh(3x) + 3x\sinh(3x))$.

7.3.31 $f'(x) = \dfrac{1}{\sqrt{16x^2 - 1}} \cdot 4 = \dfrac{4}{\sqrt{16x^2 - 1}}$.

7.3.33 $f'(v) = \dfrac{1}{\sqrt{v^4 + 1}} \cdot (2v) = \dfrac{2v}{\sqrt{v^4 + 1}}$.

7.3.35 $f'(x) = \sinh^{-1}(x) + \dfrac{x}{\sqrt{x^2+1}} - \dfrac{x}{\sqrt{x^2+1}} = \sinh^{-1}(x)$.

7.3.37 $\displaystyle\int \cosh 2x\, dx = \dfrac{1}{2}\sinh 2x + C$.

7.3.39 Let $u = 1 + \cosh x$ so that $du = \sinh x\, dx$. Substituting gives $\displaystyle\int \dfrac{1}{u}\, du = \ln|u| + C = \ln|1 + \cosh x| + C$.

7.3.41 Recall that $\tanh^2 x = 1 - \text{sech}^2 x$. So we have $\displaystyle\int (1 - \text{sech}^2 x)\, dx = x - \tanh x + C$.

7.3.43 Let $u = \cosh 3x$ so that $du = 3\sinh 3x\, dx$. Substituting gives $\dfrac{1}{3}\displaystyle\int_1^{\cosh 3} u^3\, du = \left. (u^4/12)\right|_1^{\cosh 3} = \cosh^4 3/12 - 1/12 \approx 856.034$.

7.3.45 $\displaystyle\int_0^{\ln 2} \tanh x\, dx = \int_0^{\ln 2} \dfrac{\sinh x}{\cosh x}\, dx$. Let $u = \cosh x$ so that $du = \sinh x\, dx$. Substituting gives $\displaystyle\int_1^{5/4} \dfrac{1}{u}\, du = \left. (\ln u)\right|_1^{5/4} = \ln(5/4)$.

7.3.47 We have $\dfrac{1}{8}\displaystyle\int \dfrac{dx}{1 - (x/\sqrt{8})^2}\, dx$. Let $u = x/\sqrt{8}$ so that $du = \dfrac{dx}{\sqrt{8}}$. Substituting gives $\dfrac{1}{\sqrt{8}}\displaystyle\int \dfrac{du}{1 - u^2} = \dfrac{1}{\sqrt{8}}\coth^{-1}(u) + C = \dfrac{1}{\sqrt{8}}\coth^{-1}(x/\sqrt{8}) + C$.

7.3.49 We have $\dfrac{1}{36}\displaystyle\int \dfrac{e^x}{1 - (e^x/6)^2}\, dx$.
Let $u = e^x/6$ so that $du = e^x/6\, dx$. Substituting gives

$$\dfrac{1}{6}\int \dfrac{du}{1 - u^2} = \dfrac{1}{6}\tanh^{-1}(u) + C = \dfrac{1}{6}\tanh^{-1}(e^x/6) + C.$$

7.3.51 We have $\dfrac{1}{2}\displaystyle\int \dfrac{dx}{x\sqrt{1 - (x^4/2)^2}}$. Let $u = x^4/2$ so that $du = 2x^3\, dx$. Substituting gives

$$\dfrac{1}{2}\int \dfrac{du}{4u\sqrt{1 - u^2}} = -\dfrac{1}{8}\text{sech}^{-1} u + C = -\dfrac{1}{8}\text{sech}^{-1}(x^4/2) + C.$$

7.3.53 $\displaystyle\int \dfrac{\cosh z}{\sinh^2 z}\, dz = \int \coth z\, \text{csch } z\, dx = -\text{csch } z + C$.

7.3.55 Let $u = \sinh^{-1} x$ so that $du = \frac{dx}{\sqrt{x^2+1}}$. Substituting gives

$$\int_{\sinh^{-1}(5/12)}^{\sinh^{-1}(3/4)} u\, du = \left. (u^2/2) \right|_{\sinh^{-1}(5/12)}^{\sinh^{-1}(3/4)} = \frac{1}{2}\left(\left(\sinh^{-1}(3/4)\right)^2 - \left(\sinh^{-1}(5/12)\right)^2 \right) \approx 0.158.$$

7.3.57

a. Note that $\sinh \ln x = \dfrac{e^{\ln x} - e^{-\ln x}}{2} = \dfrac{x - 1/x}{2} = x/2 - 1/(2x)$. So $\displaystyle\int \dfrac{\sinh \ln x}{x} = \int (1/2 - x^{-2}/2)\, dx = x/2 + x^{-1}/2 + C = \dfrac{x^2 + 1}{2x} + C.$

b. Let $u = \ln x$, so that $du = \dfrac{1}{x}\, dx$. Substituting gives $\displaystyle\int \sinh u\, du = \cosh u + C = \cosh \ln x + C = \dfrac{e^{\ln x} + e^{-\ln x}}{2} + C = \dfrac{x + 1/x}{2} + C = \dfrac{x^2 + 1}{2x} + C.$

7.3.59

a. The values of $\coth x$ on the interval $[5, 10]$ are very close to 1. So the area under the curve is very close to that of a 5×1 rectangle.

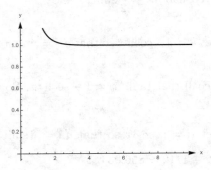

b. $\displaystyle\int_5^{10} \coth x\, dx = \int_5^{10} \dfrac{\cosh x}{\sinh x}\, dx = \left. (\ln|\sinh x|) \right|_5^{10} = \ln \sinh 10 - \ln \sinh 5 \approx 5.0000454.$ The absolute value of the error is about 0.0000454.

7.3.61

a. The curves intersect when $\dfrac{1}{\cosh x} = \dfrac{\sinh x}{\cosh x}$, or when $\sinh x = 1$, so at $x = \sinh^{-1}(1) = \ln(1 + \sqrt{2})$.

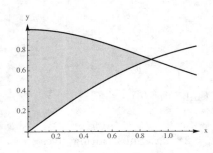

b. We need to compute $A = \displaystyle\int_0^{\ln(1+\sqrt{2})} \text{sech}\, x - \tanh x\, dx.$ Recall that $\displaystyle\int \text{sech}\, x\, dx = \tan^{-1}|\sinh x| + C$ and $\displaystyle\int \tanh x\, dx = \ln \cosh x + C.$

Thus, $A = \left. \left(\tan^{-1} \sinh x - \ln \cosh x \right) \right|_0^{\ln(1+\sqrt{2})} = \tan^{-1} \sinh(\ln(1 + \sqrt{2})) - \ln \cosh(\ln(1 + \sqrt{2})) - (0 - 0) \approx 0.44.$

7.3.63 Let $u = \ln x$ so that $du = \dfrac{dx}{x}$. Substituting gives

$$\int_0^2 \dfrac{du}{\sqrt{u^2 + 1}} = \left. \left(\sinh^{-1} u\right) \right|_0^2 = \sinh^{-1}(2) - 0 = \ln(2 + \sqrt{5}).$$

7.3. Hyperbolic Functions

7.3.65 $\int_{-2}^{2} \frac{dt}{t^2 - 3^2} = -\frac{1}{3} \left(\tanh^{-1}(x/3)\right) \Big|_{-2}^{2} = -\frac{1}{3} \left(\tanh^{-1}(2/3) - \tanh^{-1}(-2/3)\right) =$
$-\frac{1}{3} \left(\frac{1}{2} \ln((5/3)/(1/3)) - \frac{1}{2} \ln((1/3)/(5/3))\right) = -\frac{1}{3} \ln 5 \approx -0.54.$

7.3.67 Let $u = x^{1/3}$ so that $du = \frac{1}{3} x^{-2/3} dx$. Substituting gives $\int_{1/2}^{1} \frac{3u^2 du}{u^3 \sqrt{1+u^2}} = 3 \int_{1/2}^{1} \frac{du}{u\sqrt{1+u^2}} =$
$-3 \left(\text{csch}^{-1} x\right) \Big|_{1/2}^{1} = 3(\text{csch}^{-1}(1/2) - \text{csch}^{-1}(1)) = 3(\sinh^{-1}(2) - \sinh^{-1}(1)) = 3(\ln(2+\sqrt{5}) - \ln(1+\sqrt{2})) \approx 1.69.$

7.3.69 By symmetry, we compute $I = 2 \int_{0}^{\cosh^{-1}(17/15)} (17/15 - \cosh x) \, dx$ and divide by $2 \cosh^{-1}(17/15)$. The value of I is $2 \left((17/15)x - \sinh x\right) \Big|_{0}^{\cosh^{-1}(17/15)} = 2 \left((17/15) \cosh^{-1}(17/15) - \sinh(\cosh^{-1}(17/15))\right) = (34/15) \cosh^{-1}(17/15) - 16/15$. When we divide by $2 \cosh^{-1}(17/15)$ we obtain $\frac{17}{15} - \frac{8}{15 \cosh^{-1}(17/15)} \approx 0.09$.

7.3.71

a. Let a be the value that produces a sag of 10 feet. Note that $f(0) + \text{sag} = f(50)$, so $a + \text{sag} = a \cosh(50/a)$, and dividing by a yields $1 + \frac{\text{sag}}{a} = \cosh(50/a)$, which can be written as $\cosh(50/a) - 1 = 10/a$.

b. If $t = 10/a$, we have $\cosh(5t) - 1 = t$. Solving for t yields $t \approx 0.08$.

c. If $0.08 = 10/a$, then $a = 10/0.08 = 125$. The length of the power line is $2 \int_{0}^{50} \sqrt{1 + \sinh^2(x/125)} \, dx = 2 \int_{0}^{50} \cosh(x/125) \, dx = 250 \left(\sinh(x/125)\right) \Big|_{0}^{50} = 250 \sinh(50/125) = 250 \sinh(2/5) \approx 102.7$ feet.

7.3.73 We are seeking λ so that $7 = \sqrt{\frac{9.8\lambda}{2\pi}} \tanh\left(\frac{20\pi}{\lambda}\right)$. A computer algebra system finds the approximate root to be 32.81 meters.

7.3.75

a. $f'(x) = \text{sech}^2(x)$, so $f'(0) = 1$. The point of tangency is $(0, \tanh 0) = (0, 0)$, so the linearization is $L(x) = 0 + 1(x - 0) = x$.

b. When $d/\lambda < 0.05$, $2\pi d/\lambda$ is small. Because $\tanh x \approx x$ for small values of x, $\tanh(2\pi d/\lambda) \approx 2\pi d/\lambda$. Thus,
$$v = \sqrt{\frac{g\lambda}{2\pi}} \tanh\left(\frac{2\pi d}{\lambda}\right) \approx \sqrt{\frac{g\lambda}{2\pi} \cdot \frac{2\pi d}{\lambda}} = \sqrt{gd}.$$

c. $v = \sqrt{gd}$ is a function of depth only. As $d \to 0$, $v \to 0$ as well.

7.3.77

a. False. $\sinh(\ln 3)$ is a constant, so its derivative is zero.

b. False. $\frac{d}{dx} \cosh x = \sinh x$, not $-\sinh x$.

c. True. Note that $-\ln(\sqrt{2} - 1) = \ln\left(\frac{1}{\sqrt{2}-1}\right) = \ln\left(\frac{1}{\sqrt{2}-1} \cdot \frac{\sqrt{2}+1}{\sqrt{2}+1}\right) = \ln(\sqrt{2}+1)$.

d. False. $\int_0^1 \dfrac{dx}{2^2 - x^2} = \dfrac{1}{2}\tanh^{-1}(x/2)\Big|_0^1 = \dfrac{1}{2}\tanh^{-1}\left(\dfrac{1}{2}\right) \neq \dfrac{1}{2}\left(\coth^{-1}\dfrac{1}{2} - \coth^{-1} 0\right).$

7.3.79

a. $\cosh(0) = 1$

b. $\tanh(0) = 0$

c. Does not exist

d. $\operatorname{sech}(0) = 1$

e. $\coth(\ln 5) = \dfrac{5 + 1/5}{5 - 1/5} = \dfrac{13}{12}$

f. $\sinh(\ln 9) = \dfrac{9 - 1/9}{2} = \dfrac{40}{9}$

g. $\left(\dfrac{e + (1/e)}{2}\right)^2 = \left(\dfrac{e^2 + 1}{2e}\right)^2$

h. Does not exist

i. $\ln\left(17/8 + \sqrt{\dfrac{17^2 - 8^2}{8^2}}\right) = \ln 4$

j. $\sinh^{-1}\left(\dfrac{e^1 - e^{-1}}{2}\right) = 1$

7.3.81 $f'(x) = 2\sinh x \cosh x \cosh x + \sinh^2 x \sinh x = \sinh x(2\cosh^2 x + \sinh^2 x)$. Note that $\sinh x = 0$ for $x = 0$, while $2\cosh^2 x + \sinh^2 x > 0$ for all x. So the only critical point is $x = 0$.

7.3.83 $f'(x) = 2\tanh x \operatorname{sech}^2 x = \dfrac{2\sinh x}{\cosh^3 x}$.

$$f''(x) = \dfrac{2\cosh^4 x - 2(\sinh x)(3\cosh^2 x)(\sinh x)}{\cosh^6 x} = \dfrac{2\cosh^2 x - 6\sinh^2 x}{\cosh^4 x}.$$

This is zero for $\tanh^2 x = 1/3$, or $x = \tanh^{-1}(\pm\sqrt{1/3}) = \pm\tanh^{-1}(\sqrt{1/3})$.

7.3.85 This area would be the area under $\operatorname{sech} x$ between 0 and 1, minus the area of a quarter circle. Thus we have $\int_0^1 \operatorname{sech} x\, dx - \dfrac{\pi}{4} = \left(\tan^{-1}|\sinh x|\right)\Big|_0^1 - \dfrac{\pi}{4} = \tan^{-1}(\sinh(1)) - \dfrac{\pi}{4} \approx 0.08.$

7.3.87 Applying L'Hôpital's rule twice brings you back to the initial limit. If we write $\dfrac{\sinh x}{\cosh x} = \dfrac{(\sinh x)e^x}{(\cosh x)e^x} = \dfrac{e^{2x} - 1}{e^{2x} + 1}$, then applying L'Hôpital's rule once shows that the limit is 1.

7.3.89 $\lim\limits_{x \to 0} \dfrac{\tanh^{-1} x}{\tan(\pi x/2)} = \lim\limits_{x \to 0} \dfrac{\frac{1}{1-x^2}}{(\pi/2)\sec^2(\pi x/2)} = \dfrac{2}{\pi}\lim\limits_{x \to 0} \dfrac{\cos^2(\pi x/2)}{1 - x^2} = \dfrac{2}{\pi}.$

7.3.91 Let $y = (\tanh x)^x$. Then $\ln y = x\ln(\tanh x) = \dfrac{\ln \tanh x}{1/x}$. Note that $\lim\limits_{x \to 0}\dfrac{\ln \tanh x}{1/x} = \lim\limits_{x \to 0}\dfrac{1/(\sinh x \cosh x)}{-1/x^2} = \lim\limits_{x \to 0}\dfrac{-x^2}{\sinh x \cosh x} = \lim\limits_{x \to 0}\dfrac{-2x}{\sinh^2 x + \cosh^2 x} = 0.$
Thus, $\lim\limits_{x \to 0} y = \lim\limits_{x \to 0}(\tanh x)^x = e^0 = 1.$

7.3.93 Let A be the area under $3 - \cosh x$ between $\cosh^{-1}(-3)$ and $\cosh^{-1}(3)$. Then the volume of the kiln will be $6A$. By symmetry, $A = 2\int_0^{\cosh^{-1}(3)}(3 - \cosh x)\, dx = 2(3x - \sinh x)\Big|_0^{\cosh^{-1}(3)} = 6\cosh^{-1}(3) - 2\sinh(\cosh^{-1}(3)) \approx 4.92$. So the volume of the kiln is about $6 \cdot 4.92 \approx 29.5.$

7.3.95

a. $d(10) = \dfrac{75}{0.2}\ln\left(\cosh\left(\sqrt{\dfrac{(0.2)(9.8)}{75}} \cdot 10\right)\right) \approx 360.8.$

b. We need to solve $100 = \frac{75}{0.2} \ln\left(\cosh\left(\sqrt{\frac{(0.2)(9.8)}{75}} \cdot t\right)\right)$ for t. We have $t = \cosh^{-1}(e^{\frac{100 \cdot 0.2}{75}}) \cdot \sqrt{\frac{75}{(0.2)(9.8)}} \approx 4.72$ seconds. The first 200 feet take $t = \cosh^{-1}(e^{\frac{200 \cdot 0.2}{75}}) \cdot \sqrt{\frac{75}{(0.2)(9.8)}} \approx 6.97$ seconds, which means that the second 100 feet took about $6.97 - 4.72 = 2.25$ seconds. The average velocity over the first 100 feet is $\frac{100}{4.72} \approx 21.2$ meters per second, while over the second 100 feet is $\frac{100}{2.25} \approx 44.5$ meters per second.

7.3.97

a. $\sqrt{\frac{mg}{k}} \lim_{t \to \infty} \tanh\left(\sqrt{\frac{kg}{m}} t\right) = \sqrt{\frac{mg}{k}} \cdot 1 = \sqrt{\frac{mg}{k}}$.

b. $\sqrt{\frac{(75)(9.8)}{0.2}} \approx 60.6$ meters per second.

c. We seek t so that $v(t) = \sqrt{\frac{mg}{k}} \tanh\left(\sqrt{\frac{kg}{m}} t\right) = 0.95 \sqrt{\frac{mg}{k}}$.

So $\tanh\left(\sqrt{\frac{kg}{m}} t\right) = 0.95$, and $t = \tanh^{-1}(0.95) \sqrt{\frac{m}{kg}}$.

d. It takes $\tanh^{-1}(0.95) \sqrt{\frac{75}{(0.2)(9.8)}} \approx 11.33$ seconds to achieve 95 percent of terminal velocity. It takes $d(11.33) \approx 436.5$ feet to achieve this. So the cliff should be at least $436.5 + 300 = 736.5$ feet high.

7.3.99 For $y = A \sinh kx$, we have $y' = kA \cosh kx$ and $y'' = k^2 A \sinh kx$. So $y'' - k^2 y = k^2 A \sinh kx - k^2 A \sinh kx = 0$.

For $y = B \cosh kx$, we have $y' = kB \sinh kx$ and $y'' = k^2 B \cosh kx$. So $y'' - k^2 y = k^2 B \cosh kx - k^2 B \cosh kx = 0$.

7.3.101

$$\sinh(\cosh^{-1}(x)) = \sinh(\ln(x + \sqrt{x^2 - 1})) = \frac{e^{\ln(x + \sqrt{x^2 - 1})} - \frac{1}{e^{\ln(x + \sqrt{x^2 - 1})}}}{2}$$

$$= \frac{1}{2}\left(x + \sqrt{x^2 - 1} - \frac{1}{x + \sqrt{x^2 - 1}}\right).$$

Multiplying the last term by $\frac{x - \sqrt{x^2 - 1}}{x - \sqrt{x^2 - 1}}$ gives

$$\frac{1}{2}\left((x + \sqrt{x^2 - 1}) - (x - \sqrt{x^2 - 1})\right) = \sqrt{x^2 - 1}.$$

7.3.103 $\cosh(x + y) = \frac{e^{x+y} + e^{-(x+y)}}{2} = \frac{2e^{x+y} + 2e^{-(x+y)}}{4}$. This can be written as

$$\frac{1}{4}(e^{x+y} + e^{x-y} + e^{y-x} + e^{-(x+y)} + e^{x+y} - e^{x-y} - e^{y-x} + e^{-(x+y)})$$

$$= \left(\frac{e^x + e^{-x}}{2}\right)\left(\frac{e^y + e^{-y}}{2}\right) + \left(\frac{e^x - e^{-x}}{2}\right)\left(\frac{e^y - e^{-y}}{2}\right)$$

$$= \cosh x \cosh y + \sinh x \sinh y$$

7.3.105 First suppose $x \geq 0$. Then $\cosh^{-1}(\cosh x) = \ln(\cosh x + \sqrt{1 + \cosh^2 x}) = \ln(\cosh x + \sqrt{\sinh^2 x}) = \ln(\cosh x + \sinh x) = \ln e^x = x$.

Now suppose $x < 0$. Then $\cosh^{-1}(\cosh x) = \ln(\cosh x + \sqrt{1 + \cosh^2 x}) = \ln(\cosh x + \sqrt{\sinh^2 x}) = \ln(\cosh x - \sinh x) = \ln e^{-x} = -x$.

Thus $\cosh^{-1}(\cosh x) = |x|$.

7.3.107

a. $\int \operatorname{sech} x \, dx = \int \dfrac{\cosh x}{1 + \sinh^2 x} \, dx$. Let $u = \sinh x$ so that $du = \cosh x \, dx$. Then we have $\int \dfrac{1}{1 + u^2} \, du = \tan^{-1}(u) + C = \tan^{-1}(\sinh x) + C$.

b. Note that $\operatorname{sech} x = \dfrac{\operatorname{sech}^2 x}{\sqrt{\operatorname{sech}^2 x}} = \dfrac{\operatorname{sech}^2 x}{\sqrt{1 - \tanh^2 x}}$. So $\int \operatorname{sech} x \, dx = \int \dfrac{\operatorname{sech}^2 x}{\sqrt{1 - \tanh^2 x}} \, dx$. Let $u = \tanh x$ so that $du = \operatorname{sech}^2 x \, dx$. Substituting gives $\int \dfrac{1}{\sqrt{1 - u^2}} \, du = \sin^{-1}(u) + C = \sin^{-1}(\tanh x) + C$.

c. $\dfrac{d}{dx}(2 \tan^{-1}(e^x) + C) = 2 \left(\dfrac{1}{1 + e^{2x}} \right) e^x + 0 = \dfrac{2 e^x}{1 + e^{2x}} \cdot \dfrac{e^{-x}}{e^{-x}} = \dfrac{2}{e^{-x} + e^x} = \dfrac{1}{\cosh x} = \operatorname{sech} x.$

7.3.109 $L = \int_{\ln 2}^{\ln 8} \sqrt{1 + \operatorname{csch}^2 x} \, dx = \int_{\ln 2}^{\ln 8} \sqrt{\coth^2 x} \, dx = \int_{\ln 2}^{\ln 8} \coth x \, dx = \int_{\ln 2}^{\ln 8} \dfrac{\cosh x}{\sinh x} \, dx$. Let $u = \sinh x$ so that $du = \cosh x \, dx$. Substituting gives $\int_{3/4}^{63/16} \dfrac{1}{u} \, du = (\ln u) \Big|_{3/4}^{63/16} = \ln(63/16) - \ln(3/4) = \ln(21/4) \approx 1.66$.

7.3.111 Suppose $r > 0$ and $0 < x < 1$. If $u = x^r$, then $du = r x^{r-1} dx$. Substituting gives $\int \dfrac{du}{ru\sqrt{1 - u^2}} = -\dfrac{1}{r}(\operatorname{sech}^{-1}(u)) + C = -\dfrac{1}{r}(\operatorname{sech}^{-1}(x^r)) + C$.

Chapter Seven Review

1

a. False. For example, when t goes from 2 to 3, the value of y goes from 3 to 4, which is not doubling.

b. False. It grows from A to $A \cdot e^{.1} \approx 1.10517 A$, so the growth is about 10.517%.

c. False. For example, if $x = e$ and $y = 1$, this equation would imply that $\ln(e \cdot 1) = (\ln e)(\ln 1)$, which is not true because $1 \neq 0$.

d. True. $\sinh(\ln x) = \dfrac{e^{\ln x} - e^{-\ln x}}{2} = \dfrac{x - 1/x}{2} \cdot \dfrac{x}{x} = \dfrac{x^2 - 1}{2x}$.

3 Let $u = \ln x$ so that $du = \dfrac{1}{x} dx$. Then we have $\int_2^8 \dfrac{1}{u} du = (\ln u) \Big|_2^8 = \ln 4$.

5 Let $u = x^2 + 8x + 25$ so that $du = (2x + 8) dx = 2(x + 4) dx$. Then we have $\dfrac{1}{2} \int \dfrac{1}{u} du = \dfrac{1}{2} \ln |u| + C = \dfrac{1}{2} \ln(x^2 + 8x + 25) + C$.

7 $\int \dfrac{dx}{\sqrt{x^2 - 9}} = \cosh^{-1}(x/3) + C = \ln(x + \sqrt{x^2 - 9}) + C$.

9 Let $u = x^3$ so that $du = 3x^2\, dx$. Substituting gives
$$\frac{1}{3}\int_0^1 \frac{du}{3^2 - u^2} = \frac{1}{9}\left(\tanh^{-1}(u/3)\right)\Big|_0^1 = \frac{1}{9}\tanh^{-1}(1/3) \approx 0.0385.$$

11 We can write $g(x)$ as $g(x) = e^{(3x^2+1)\ln x}$, so
$$g'(x) = e^{(3x^2+1)\ln x}\left(6x\ln x + (3x^2+1)\frac{1}{x}\right) = x^{3x^2+1}\left(6x\ln x + 3x + \frac{1}{x}\right).$$

13 $f'(t) = \sinh t \sinh t + \cosh t \cosh t = \sinh^2 t + \cosh^2 t = \cosh 2t.$

15 $f'(x) = 2\cosh(3x-1)\cdot \sinh(3x-1)\cdot 3 = 6\cosh(3x-1)\sinh(3x-1) = 3\sinh(2(3x-1)) = 3\sinh(6x-2).$

17 $f'(x) = \dfrac{-\sin x}{1-\cos^2 x} = -\dfrac{\sin x}{\sin^2 x} = -\dfrac{1}{\sin x} = -\csc x.$

19 $f'(x) = -\dfrac{1}{\frac{1}{x^2}\sqrt{1-\frac{1}{x^4}}}\cdot \dfrac{-2}{x^3} = \dfrac{2}{x\sqrt{1-\frac{1}{x^4}}} = \dfrac{2x}{\sqrt{x^4-1}}.$

21 Because the half-life is 5.5, $k = \dfrac{\ln(1/2)}{5.5}$, and $y(t) = 75e^{\frac{\ln(1/2)}{5.5}t}$. We are seeking t so that $30 = 75e^{\frac{\ln(1/2)}{5.5}t}$, or $\dfrac{30}{75} = e^{\frac{\ln(1/2)}{5.5}t}$, so $\ln\left(\dfrac{30}{75}\right) = \dfrac{-\ln 2}{5.5}t$, so $t = \dfrac{5.5\ln(30/75)}{-\ln 2} \approx 7.3$ hours.

23

a. Because the doubling time is 2, $k = \dfrac{\ln 2}{2}$, so $y(t) = 29{,}000e^{\frac{\ln 2}{2}t}.$

b. $y(21) = 29{,}000e^{\frac{21\ln 2}{2}} \approx 41{,}996{,}486$ transistors. This closely approximates the actual number of transistors.

25 Growth is modeled by $p(t) = 150{,}000e^{kt}$ where $k = \ln(1.04)$. The population reaches $1{,}000{,}000$ when $1{,}000{,}000 = 150{,}000e^{\ln(1.04)t}$, or when $\ln(20/3) = \ln(1.04)t$, so when $t = \dfrac{\ln(20/3)}{\ln(1.04)} \approx 48.37$ years.

27 The domain of f is the set of all real numbers. Note that $\lim\limits_{x\to\infty} f(x) = \infty$ and $\lim\limits_{x\to-\infty} f(x) = 0$.

$f'(x) = e^x(2x-1) + (x^2-x)e^x = e^x(x^2+x-1)$, which is 0 for $x = \dfrac{-1\pm\sqrt{5}}{2}$. Note that $f'(x) > 0$ on the interval $(-\infty, (-1-\sqrt{5})/2)$ and on $((-1+\sqrt{5})/2, \infty)$, so f is increasing on those intervals, while $f'(x) < 0$ (and so f is decreasing) on $((-1-\sqrt{5})/2, (-1+\sqrt{5})/2)$. There is a local maximum at $(-1-\sqrt{5})/2$ and a local minimum at $(-1+\sqrt{5})/2$.

Note that $f''(x) = e^x(2x+1) + (x^2+x-1)e^x = e^x(x^2+3x)$, which is 0 for $x = 0$ and $x = -3$. $f''(x) > 0$ on $(-\infty, -3)$ and on $(0, \infty)$, so f is concave up on those intervals, while $f''(x) < 0$ on $(-3, 0)$, so f is concave down there. There are inflection points at $x = -3$ and at $x = 0$.

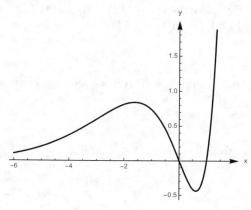

29

a.

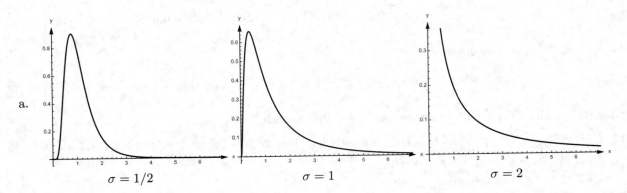

$\sigma = 1/2$ $\sigma = 1$ $\sigma = 2$

It appears that $\lim_{x \to 0} f(x) = 0$.

b. Let $x = e^y$, so that $y = \ln x$ and as $x \to 0$ we have $y \to -\infty$. Then $\lim_{x \to 0} f(x) = \lim_{y \to -\infty} \dfrac{e^{-y^2/2\sigma^2}}{\sigma\sqrt{2\pi}e^y} = \dfrac{1}{\sigma\sqrt{2\pi}} \lim_{y \to -\infty} \dfrac{1}{e^y \cdot e^{y^2/2\sigma^2}} = 0$.

c. Write f as $f(x) = \dfrac{1}{\sigma\sqrt{2\pi}} \left(\dfrac{1}{xe^{(\ln^2 x)/2\sigma^2}} \right)$. Then

$$f'(x) = \dfrac{-1}{\sigma\sqrt{2\pi}} \left(\dfrac{1}{x^2 e^{(\ln^2 x)/\sigma^2}} \right) \left(e^{(\ln^2 x)/2\sigma^2} + xe^{(\ln^2 x)/2\sigma^2} \cdot \dfrac{1}{\sigma^2} \ln x \cdot \dfrac{1}{x} \right) =$$

$$\dfrac{-1}{\sigma\sqrt{2\pi}} \left(\dfrac{1}{x^2} \right) \left(1 + \dfrac{\ln x}{\sigma^2} \right).$$

This quantity is 0 only when $1 + \dfrac{\ln x}{\sigma^2} = 0$, which occurs for $x^* = e^{-\sigma^2}$, and this critical number yields a maximum.

d. $f(e^{-\sigma^2}) = \dfrac{1}{\sigma\sqrt{2\pi}} \left(\dfrac{1}{(e^{-\sigma^2} e^{(\ln^2(e^{-\sigma^2}))/2\sigma^2})} \right) = \dfrac{1}{\sigma\sqrt{2\pi}} \left(\dfrac{1}{e^{-\sigma^2} e^{\sigma^2/2}} \right) = \dfrac{e^{\sigma^2/2}}{\sigma\sqrt{2\pi}}$.

e. Let $g(\sigma) = \dfrac{e^{\sigma^2/2}}{\sigma\sqrt{2\pi}}$. Then $g'(\sigma) = \dfrac{e^{\frac{\sigma^2}{2}}(\sigma^2 - 1)}{\sqrt{2\pi}\sigma^2}$, so there is a critical point at $\sigma = 1$. Note that g is decreasing on $(0, 1)$ and increasing on $(1, \infty)$, so g has a minimum at $\sigma = 1$.

31 $f'(\ln 3) = \sinh(\ln 3) = 4/3$. So the linearization is $L(x) = 5/3 + (4/3)(x - \ln 3)$. Using this approximation, we have $\cosh(1) \approx L(1) = (5/3) + (4/3)(1 - \ln 3) \approx 1.535$.

33

a. Because $\dfrac{d}{dx} \cosh x = \sinh x$ and $\dfrac{d}{dx} \sinh x = \cosh x$, we see that every nth derivative of $\cosh x$ where n is even is equal to $\cosh x$. So $\dfrac{d^6}{dx^6} \cosh x = \cosh x$.

b. $\dfrac{d}{dx} x \operatorname{sech} x = \operatorname{sech} x - x \operatorname{sech} x \tanh x = \operatorname{sech} x (1 - x \tanh x)$.

Chapter 8

Integration Techniques

8.1 Basic Approaches

8.1.1 Let $u = 4 - 7x$. Then $du = -7\,dx$ and we obtain $-\dfrac{1}{7}\displaystyle\int u^{-6}\,du$.

8.1.3 $\sin^2 x = \dfrac{1 - \cos 2x}{2}$.

8.1.5 Complete the square in the denominator to get $\displaystyle\int \dfrac{10}{(x-2)^2 + 1}\,dx$.

8.1.7 Let $u = 3 - 5x$ so that $du = -5\,dx$. Substituting gives
$$-\dfrac{1}{5}\int u^{-4}\,du = \dfrac{1}{15}u^{-3} + C = \dfrac{1}{15(3-5x)^3} + C.$$

8.1.9 Let $u = 2x - \pi/4$ so that $du = 2\,dx$. Substituting gives
$$\dfrac{1}{2}\int_{-\pi/4}^{\pi/2} \sin u\,du = \dfrac{1}{2}(-\cos u)\bigg|_{-\pi/4}^{\pi/2} = \dfrac{1}{2}\left(0 + \sqrt{2}/2\right) = \dfrac{\sqrt{2}}{4}.$$

8.1.11 Let $u = \ln(2x)$ so that $du = \dfrac{dx}{x}$. Substituting gives
$$\int u\,du = u^2/2 + C = \dfrac{1}{2}\ln^2 2x + C.$$

8.1.13 Let $u = e^x + 1$ so that $du = e^x\,dx$ Substituting gives
$$\int \dfrac{1}{u}\,du = \ln|u| + C = \ln(e^x + 1) + C.$$

8.1.15 Let $u = \ln x^2 = 2\ln x$. Then $du = \dfrac{2}{x}\,dx$. Substituting gives
$$\dfrac{1}{2}\int_0^4 u^2\,du = (u^3/6)\bigg|_0^4 = 32/3.$$

8.1.17 Let $u = s - 1$, so that $du = ds$ and $s = u + 1$. Note that $1 - 1 = 0$ and $2 - 1 = 1$, so the new limits of integration are 0 and 1. Substituting gives
$$\int_0^1 u^9(u+1)\,du = \int_0^1 (u^{10} + u^9)\,du = \left(\dfrac{u^{11}}{11} + \dfrac{u^{10}}{10}\right)\bigg|_0^1 = \dfrac{1}{11} + \dfrac{1}{10} = \dfrac{21}{110}.$$

8.1.19 Let $u = \ln w - 1$. Then $du = \dfrac{dw}{w}$ and $\ln w = u + 1$. Substituting gives

$$\int u^7(u+1)\, du = \int (u^8 + u^7)\, du = \frac{u^9}{9} + \frac{u^8}{8} + C = \frac{(\ln w - 1)^9}{9} + \frac{(\ln w - 1)^8}{8} + C.$$

8.1.21 $\displaystyle\int \frac{x}{x^2+4}\, dx + 2\int \frac{1}{x^2+4}\, dx = \frac{1}{2}\ln(x^2+4) + \tan^{-1}(x/2) + C.$

8.1.23 Let $u = 3e^x + 4$ so that $du = 3e^x\, dx$. Substituting gives

$$\frac{1}{3}\int \csc u\, du = -\frac{1}{3}\ln|\csc u + \cot u| + C = -\frac{1}{3}\ln|\csc(3e^x+4) + \cot(3e^x+4)| + C.$$

8.1.25 We may rewrite the integrand as $\displaystyle\int_0^{\pi/4}\left(\frac{\sec\theta}{\sec\theta\csc\theta} + \frac{\csc\theta}{\sec\theta\csc\theta}\right) d\theta$. This can then be simplified as

$$\int_0^{\pi/4}(\sin\theta + \cos\theta)\, d\theta = (-\cos\theta + \sin\theta)\Big|_0^{\pi/4} = \left(-\frac{\sqrt{2}}{2} + \frac{\sqrt{2}}{2}\right) - (-1 - 0) = 1.$$

8.1.27
$$\int \frac{2}{\sqrt{1-x^2}}\, dx - 3\int \frac{x}{\sqrt{1-x^2}}\, dx = 2\sin^{-1} x - 3\int \frac{x}{\sqrt{1-x^2}}\, dx.$$

Let $u = 1 - x^2$ so that $du = -2x\, dx$. Substituting gives

$$2\sin^{-1} x + \frac{3}{2}\int u^{-1/2}\, du = 2\sin^{-1} x + 3\sqrt{u} + C = 2\sin^{-1} x + 3\sqrt{1-x^2} + C.$$

8.1.29 $\displaystyle\int_{\pi/4}^{\pi/2} \sqrt{1+\cot^2 x}\, dx = \int_{\pi/4}^{\pi/2} \csc x\, dx = -\ln|\csc x + \cot x|\Big|_{\pi/4}^{\pi/2} = -\ln|1+0| + \ln|\sqrt{2}+1| = \ln(\sqrt{2}+1).$

8.1.31 Note that by completing the square, we have $x^2 - 2x + 10 = (x^2 - 2x + 1) + 9 = (x-1)^2 + 9$. So

$$\int \frac{dx}{(x-1)^2 + 3^2} = \frac{1}{3}\tan^{-1}\left(\frac{x-1}{3}\right) + C.$$

8.1.33 We reduce the integrand by long division to $x + 1 + \dfrac{2x+1}{x^2+x+2}$. Then we have

$$\int \left(x + 1 + \frac{2x+1}{x^2+x+2}\right) dx = \frac{x^2}{2} + x + \int \frac{2x+1}{x^2+x+1}\, dx.$$

For the remaining integral, we let $u = x^2 + x + 1$ so that $du = (2x+1)\, dx$. The last integral is therefore equal to $\ln|u| + C = \ln|x^2+x+1| + C$. So our final result is

$$\frac{x^2}{2} + x + \ln|x^2+x+1| + C. = \frac{x^2}{2} + x + \ln(x^2+x+1) + C.$$

8.1.35 By long division, we can write the integrand as $t^2 + t + \dfrac{1}{t^2+1}$. Then we have

$$\int_0^1 \left(t^2 + t + \frac{1}{t^2+1}\right) dt = \left(\frac{t^3}{3} + \frac{t^2}{2} + \tan^{-1} t\right)\Big|_0^1 = \left(\frac{1}{3} + \frac{1}{2} + \frac{\pi}{4}\right) - 0 = \frac{3\pi + 10}{12}.$$

8.1.37 Note that $27 - 6\theta - \theta^2 = -(\theta^2 + 6\theta + 9 - 36) = -((\theta+3)^2 - 36) = 36 - (\theta+3)^2$. Thus our integral is

$$\int \frac{d\theta}{\sqrt{36 - (\theta+3)^2}} = \sin^{-1}((\theta+3)/6) + C.$$

8.1.39

$$\int \frac{1}{1+\sin\theta} \cdot \frac{1-\sin\theta}{1-\sin\theta}\,d\theta = \int \frac{1-\sin\theta}{1-\sin^2\theta}\,d\theta = \int \frac{1-\sin\theta}{\cos^2\theta}\,d\theta$$

$$= \int \sec^2\theta\,d\theta - \int \frac{\sin\theta}{\cos^2\theta}\,d\theta = \tan\theta - \int \frac{\sin\theta}{\cos^2\theta}\,d\theta.$$

Let $u = \cos\theta$ so that $du = -\sin\theta\,d\theta$. Substituting gives

$$\tan\theta + \int u^{-2}\,du = \tan\theta - \frac{1}{u} + C = \tan\theta - \sec\theta + C.$$

8.1.41 $\int \frac{1}{\sec x - 1} \cdot \frac{\sec x + 1}{\sec x + 1}\,dx = \int \frac{\sec x + 1}{\sec^2 x - 1}\,dx = \int \frac{\sec x + 1}{\tan^2 x}\,dx = \int \frac{\sec x}{\tan^2 x}\,dx + \int \cot^2 x\,dx =$
$\int \cot x \csc x\,dx + \int \cot^2 x\,dx = -\csc x + \int (\csc^2 x - 1)\,dx = -\csc x - \cot x - x + C.$

8.1.43 Let $u = 1 + \sinh 3x$. Then $du = 3\cosh 3x\,dx$. Substituting gives

$$\frac{1}{3}\int \frac{1}{u}\,du = \frac{1}{3}\ln|u| + C = \frac{1}{3}\ln|1 + \sinh x| + C.$$

8.1.45 We rewrite the integral by multiplying the numerator and denominator of the integrand by e^x. We have

$$\int \frac{e^x}{e^x - 2e^{-x}}\,dx = \int \frac{e^x}{e^x - 2e^{-x}} \cdot \frac{e^x}{e^x}\,dx = \int \frac{e^{2x}}{e^{2x} - 2}\,dx.$$

Now let $u = e^{2x} - 2$ so that $du = 2e^{2x}\,dx$. Substituting gives

$$\frac{1}{2}\int \frac{du}{u} = \frac{1}{2}\ln|u| + C = \frac{1}{2}\ln|e^{2x} - 2| + C.$$

8.1.47 $\int \frac{dx}{x^{-1} + 1} = \int \frac{1}{x^{-1} + 1} \cdot \frac{x}{x}\,dx = \int \frac{x}{x+1}\,dx.$ Using long division, we have $\frac{x}{x+1} = 1 - \frac{1}{x+1}$. Then

$$\int \left(1 - \frac{1}{x+1}\right)dx = x - \ln|x+1| + C.$$

8.1.49 Let $u = 9 + \sqrt{t+1}$. Then $du = \frac{dt}{2\sqrt{t+1}}$, or $dt = 2(u-9)\,du$. Our integral is

$$\int \sqrt{9 + \sqrt{t+1}}\,dt = \int \sqrt{u}(2(u-9))\,du = 2\int (u^{3/2} - 9u^{1/2})\,du = 2\left(\frac{2}{5}u^{5/2} - 6u^{3/2}\right) + C$$

$$= \frac{4}{5}u^{3/2}(u - 15) + C = \frac{4}{5}(9 + \sqrt{t+1})^{3/2}(9 + \sqrt{t+1} - 15) + C$$

$$= \frac{4}{5}(9 + \sqrt{t+1})^{3/2}(\sqrt{t+1} - 6) + C.$$

8.1.51 $\int_{-1}^{0} \frac{x}{x^2 + 2x + 2}\,dx = \int_{-1}^{0} \frac{x}{(x+1)^2 + 1}\,dx.$ Let $u = x + 1$ so that $du = dx$. Substituting gives

$$\int_0^1 \frac{u-1}{u^2+1}\,du = \int_0^1 \frac{u}{u^2+1}\,du - \int_0^1 \frac{1}{u^2+1}\,du$$

$$= \left((1/2)\ln(u^2+1) - \tan^{-1}(u)\right)\Big|_0^1 = (1/2)\ln 2 - \frac{\pi}{4} = \frac{1}{4}(\ln 4 - \pi).$$

8.1.53 Let $u = e^x + 1$ so that $du = e^x\,dx$. Substituting gives

$$\int \sec u\,du = \ln|\sec u + \tan u| + C = \ln|\sec(e^x + 1) + \tan(e^x + 1)| + C.$$

8.1.55 Using the identity $\sin 2x = 2\sin x \cos x$, we have $2\int \sin^2 x \cos x\, dx$. Let $u = \sin x$ so that $du = \cos x\, dx$. We have $2\int u^2\, du = 2u^3/3 + C = 2(\sin^3 x)/3 + C$.

8.1.57 Rewrite the integral as $\int \frac{1}{\sqrt{x}} \cdot \frac{1}{1+(\sqrt{x})^2}\, dx$ and let $u = \sqrt{x}$. Then $du = \frac{1}{2\sqrt{x}}\, dx$, and substituting gives
$$2\int \frac{1}{1+u^2}\, du = 2\tan^{-1} u + C = 2\tan^{-1}\sqrt{x} + C.$$

8.1.59 Note that $x^2 + 6x + 13 = (x^2 + 6x + 9) + 4 = (x+3)^2 + 4$. Also note that we can write the numerator $x - 2 = x + 3 - 5 = \frac{1}{2}(2x+6) - 5$. We have
$$\int \frac{\frac{1}{2}(2x+6)}{x^2 + 6x + 13}\, dx - \int \frac{5}{(x+3)^2 + 4}\, dx.$$

For the first integral, let $u = x^2 + 6x + 13$ so that $du = (2x+6)\, dx$. We have (for just the first integral)
$$\frac{1}{2}\int \frac{1}{u}\, du = \frac{1}{2}\ln|u| + C = \frac{1}{2}\ln(x^2 + 6x + 13) + C.$$

The second integrand has antiderivative equal to $\frac{5}{2}\tan^{-1}((x+3)/2)$, so the original integral is equal to
$$\frac{1}{2}\ln(x^2 + 6x + 13) - \frac{5}{2}\tan^{-1}((x+3)/2) + C.$$

8.1.61 Let $u = e^x$ so that $du = e^x\, dx$. Substituting gives
$$\int \frac{1}{u^2 + 2u + 1}\, du = \int (u+1)^{-2}\, du = -\frac{1}{u+1} + C = -\frac{1}{e^x + 1} + C.$$

8.1.63 The denominator factors as $(x+1)^2$.
$$\int_1^3 \frac{2}{(x+1)^2}\, dx = -2\left(\frac{1}{x+1}\right)\Big|_1^3 = -2\left(\frac{1}{4} - \frac{1}{2}\right) = \frac{1}{2}.$$

8.1.65

a. False. This seem to use the untrue "identity" that $\frac{a}{b+c} = \frac{a}{b} + \frac{a}{c}$.

b. False. The degree of the numerator is already less than the degree of the denominator, so long division won't help.

c. False. This is false because $\frac{d}{dx}\ln|\sin x + 1| + C \neq \frac{1}{\sin x + 1}$. The substitution $u = \sin x + 1$ can't be carried out because $du = \cos x\, dx$ can't be accounted for.

d. False. In fact, $\int e^{-x}\, dx = -e^{-x} + C \neq \ln e^x + C$.

8.1.67
$$\int \csc x\, dx = \int (\csc x)\left(\frac{\csc x + \cot x}{\csc x + \cot x}\right)dx = \int \frac{\csc^2 x + \csc x \cot x}{\csc x + \cot x}\, dx.$$

Let $u = \csc x + \cot x$ so that $du = -\csc^2 x - \csc x \cot x\, dx$. Substituting then yields
$$-\int \frac{1}{u}\, du = -\ln|u| + C = -\ln|\csc x + \cot x| + C.$$

8.1. Basic Approaches

8.1.69

a. If $u = \tan x$ then $du = \sec^2 x\, dx$. Substituting gives $\int u\, du = \dfrac{u^2}{2} + C = \dfrac{\tan^2 x}{2} + C$.

b. If $u = \sec x$, then $du = \sec x \tan x\, dx$. Substituting gives $\int u\, du = \dfrac{u^2}{2} + C = \dfrac{\sec^2 x}{2} + C$.

c. The seemingly different answers are the same, since $\dfrac{\tan^2 x}{2}$ and $\dfrac{\sec^2 x}{2}$ differ by a constant. In fact, $\dfrac{\tan^2 x}{2} - \dfrac{\sec^2 x}{2} = -\dfrac{1}{2}$.

8.1.71

a. Let $u = x + 1$ so that $du = dx$. Note that $x = u - 1$, so that $x^2 = (u-1)^2$. Substituting gives

$$\int \frac{u^2 - 2u + 1}{u}\, du = \int (u - 2 + (1/u))\, du = u^2/2 - 2u + \ln|u| + C = (x+1)^2/2 - 2(x+1) + \ln|x+1| + C.$$

b. By long division, $\dfrac{x^2}{x+1} = x - 1 + \dfrac{1}{x+1}$. Thus,

$$\int \frac{x^2}{x+1}\, dx = \int \left(x - 1 + \frac{1}{x+1}\right) dx = x^2/2 - x + \ln|x+1| + C.$$

c. The seemingly different answers are the same, because they differ by a constant. In fact,

$$\frac{(x+1)^2}{2} - 2(x+1) + \ln|x+1| - \left(\frac{x^2}{2} - x + \ln|x+1|\right) = -\frac{3}{2}.$$

8.1.73

$$A = \int_2^4 \frac{x^2 - 1}{x^3 - 3x}\, dx.$$

Let $u = x^3 - 3x$ so that $du = 3x^2 - 3\, dx$. Substituting gives

$$A = \frac{1}{3}\int_2^{52} \frac{1}{u}\, du = \frac{1}{3} \ln u \Big|_2^{52} = \frac{1}{3}(\ln 52 - \ln 2) = \frac{\ln 26}{3}.$$

8.1.75

a. $V = \pi \displaystyle\int_0^2 (x^2 + 1)\, dx = \pi\left(x^3/3 + x\right)\Big|_0^2 = \pi(8/3 + 2) = \dfrac{14\pi}{3}$.

b. $V = 2\pi \displaystyle\int_0^2 x\sqrt{x^2 + 1}\, dx$. Let $u = x^2 + 1$ so that $du = 2x\, dx$. Substituting gives

$$\pi \int_1^5 u^{1/2}\, du = \frac{2\pi}{3}\left(u^{3/2}\right)\Big|_1^5 = \frac{2\pi}{3}(5\sqrt{5} - 1).$$

8.1.77

$$A = 2\pi \int_0^1 \sqrt{x+1}\sqrt{1 + \frac{1}{4(x+1)}}\, dx = 2\pi \int_0^1 \sqrt{x + 5/4}\, dx$$

$$= 2\pi \left((2/3)(x + 5/4)^{3/2}\right)\Big|_0^1 = \frac{4\pi}{3}(27/8 - 5\sqrt{5}/8) = \frac{9\pi}{2} - \frac{5\sqrt{5}\pi}{6}.$$

8.1.79 $L = \int_0^1 \sqrt{1 + \dfrac{25x^{1/2}}{16}}\, dx$. Let $u^2 = 1 + \dfrac{25x^{1/2}}{16}$. Then $2u\, du = \dfrac{25}{32\sqrt{x}}\, dx$. Note that $\sqrt{x} = \dfrac{16}{25}(u^2 - 1)$, and that $dx = \dfrac{64\sqrt{x}}{25} u\, du = \dfrac{1024}{625}(u^3 - u)\, du$. Substituting gives

$$L = \int_1^{\sqrt{41/16}} \frac{1024}{625}(u^4 - u^2)\, du = \frac{1024}{625}\left(u^5/5 - u^3/3\right)\Big|_1^{\sqrt{41/16}}$$

$$= \frac{1024}{625}((\sqrt{41/16})^5/5 - (\sqrt{41/16})^3/3 - (1/5 - 1/3)) = \frac{1024}{625}\left(\frac{2}{15} + \frac{1763\sqrt{41}}{15360}\right)$$

$$= \frac{2048 + 1763\sqrt{41}}{9375} \approx 1.423.$$

8.2 Integration by Parts

8.2.1 It is based on the product rule. In fact, the rule can be obtained by writing down the product rule, then integrating both sides and rearranging the terms in the result.

8.2.3 $u = \ln x$ so $du = \dfrac{dx}{x}$, and $dv = x\, dx$ so $v = \dfrac{x^2}{2}$. Then the integration by parts formula gives

$$\frac{x^2 \ln x}{2} - \int \frac{x^2}{2x}\, dx = \frac{x^2 \ln x}{2} - \int \frac{x}{2}\, dx = \frac{x^2 \ln x}{2} - \frac{x^2}{4} + C = \frac{x^2(2\ln x - 1)}{4} + C.$$

8.2.5 Those for which the choice for dv is easily integrated and when the resulting new integral is no more difficult than the original.

8.2.7 Let $u = \tan x + 2$, so that $du = \sec^2 x\, dx$. Then

$$\int \sec^2 x \ln(\tan x + 2)\, dx = \int \ln u\, du = u \ln u - u + C = (\tan x + 2)\ln(\tan x + 2) - \tan x + C.$$

8.2.9 Let $u = x$ and $dv = \cos 5x\, dx$. Then $du = dx$ and $v = \dfrac{1}{5}\sin 5x$. Then

$$\int x \cos 5x\, dx = \frac{1}{5} x \sin 5x - \frac{1}{5}\int \sin 5x\, dx = \frac{1}{5} x \sin 5x + \frac{1}{25}\cos 5x + C.$$

8.2.11 Let $u = t$ and $dv = e^{6t}\, dt$. Then $du = dt$ and $v = \dfrac{1}{6}\cdot e^{6t}$. Then

$$\int t e^{6t}\, dt = \frac{1}{6} t e^{6t} - \frac{1}{6}\int e^{6t}\, dt = \frac{1}{6} t e^{6t} - \frac{1}{36} e^{6t} + C.$$

8.2.13 Let $u = \ln 10x$ and $dv = x\, dx$. Then $du = \dfrac{1}{x}\, dx$ and $v = \dfrac{x^2}{2}$. Then

$$\int x \ln 10x\, dx = \frac{x^2}{2} \ln 10x - \frac{1}{2}\int x\, dx = \frac{x^2}{2}\ln 10x - \frac{x^2}{4} + C = \frac{x^2}{4}(2\ln 10x - 1) + C.$$

8.2.15 Let $u = 2w + 4$ and $dv = \cos 2w\, dw$. Then $du = 2\, dw$ and $v = \dfrac{1}{2}\sin 2w$. The integration by parts formula gives

$$(w + 2)\sin 2w - \int \sin 2w\, dw = (w + 2)\sin 2w + \frac{1}{2}\cos 2w + C.$$

8.2.17 Let $u = x$ and $dv = 3^x\, dx$. Then $du = dx$ and $v = \dfrac{3^x}{\ln 3}$. Then we have

$$\frac{x3^x}{\ln 3} - \frac{1}{\ln 3}\int 3^x\, dx = \frac{x3^x}{\ln 3} - \frac{3^x}{\ln^2 3} + C = \frac{3^x}{\ln 3}\left(x - \frac{1}{\ln 3}\right) + C.$$

8.2.19 Let $u = \ln x$ and $dv = x^{-10}\, dx$. Then $du = \dfrac{1}{x}\, dx$ and $v = -\dfrac{1}{9}x^{-9}$. Then

$$\int \frac{\ln x}{x^{10}}\, dx = -\frac{1}{9x^9}\ln x + \frac{1}{9}\int x^{-10}\, dx = -\frac{1}{9x^9}\ln x + -\frac{1}{81x^9} + C.$$

8.2.21
$$\int x \sin x \cos x\, dx = \frac{1}{2}\int x \cdot (2\sin x \cos x)\, dx = \frac{1}{2}\int x \sin 2x\, dx.$$

Now using the result of problem 10, we have

$$\int x \sin x \cos x\, dx = \frac{1}{2}\cdot\left(-\frac{1}{2}x\cos 2x + \frac{1}{4}\sin 2x\right) + C = -\frac{1}{4}x\cos 2x + \frac{1}{8}\sin 2x + C.$$

8.2.23 Let $u = x^2$ and $dv = \sin 2x\, dx$. Then $du = 2x\, dx$ and $v = -\frac{1}{2}\cos 2x$. Then we have

$$\int x^2 \sin 2x\, dx = -\frac{1}{2}x^2\cos 2x + \int x\cos 2x\, dx.$$

Now we consider computing this last term $\int x\cos 2x\, dx$ as a new problem. Let $u = x$ and $dv = \cos 2x\, dx$. Then $du = dx$ and $v = \dfrac{1}{2}\sin 2x$. So

$$\int x\cos 2x\, dx = \frac{1}{2}x\sin 2x - \frac{1}{2}\int \sin 2x\, dx = \frac{1}{2}x\sin 2x + \frac{1}{4}\cos 2x + C.$$

Combining these results we have

$$\int x^2 \sin 2x\, dx = -\frac{1}{2}x^2\cos 2x + \frac{1}{2}x\sin 2x + \frac{1}{4}\cos 2x + C.$$

8.2.25 Let $u = t^2$ and $dv = e^{-t}\, dt$. Then $du = 2t\, dt$ and $v = -e^{-t}$. We have

$$\int t^2 e^{-t}\, dt = -t^2 e^{-t} + 2\int t e^{-t}\, dt.$$

To compute this last integral, we let $u = t$ and $dv = e^{-t}\, dt$. Then

$$\int t e^{-t}\, dt = -t e^{-t} + \int e^{-t}\, dt = -t e^{-t} - e^{-t} + C.$$

Putting these results together, we obtain

$$\int t^2 e^{-t}\, dt = -t^2 e^{-t} + 2(-t e^{-t} - e^{-t}) + C = -e^{-t}(t^2 + 2t + 2) + C.$$

8.2.27 Let $u = \cos x$ and $dv = e^x\, dx$. Then $du = -\sin x\, dx$ and $v = e^x$. We have

$$\int e^x \cos x\, dx = e^x \cos x + \int e^x \sin x\, dx.$$

Now in order to compute the integral which comprises this last term, we let $u = \sin x$ and $dv = e^x\, dx$. Then $du = \cos x\, dx$ and $v = e^x$. Thus,

$$\int e^x \sin x\, dx = e^x \sin x - \int e^x \cos x\, dx.$$

Putting these results together gives

$$\int e^x \cos x\, dx = e^x \cos x + e^x \sin x - \int e^x \cos x\, dx$$

$$2\int e^x \cos x\, dx = e^x(\cos x + \sin x) + C$$

$$\int e^x \cos x\, dx = \frac{e^x}{2}(\cos x + \sin x) + C.$$

8.2.29 Let $u = \sin 4x$ and $dv = e^{-x}\, dx$. Then $du = 4\cos 4x\, dx$ and $v = -e^{-x}$. We have

$$\int e^{-x}\sin 4x\, dx = -e^{-x}\sin 4x + 4\int e^{-x}\cos 4x\, dx.$$

Now in order to compute the integral which comprises this last term, we let $u = \cos 4x$ and $dv = e^{-x}\, dx$. Then $du = -4\sin 4x\, dx$ and $v = -e^{-x}$. Thus,

$$\int e^{-x}\cos 4x\, dx = -e^{-x}\cos 4x - 4\int e^{-x}\sin 4x\, dx.$$

Putting these results together gives

$$\int e^{-x}\sin 4x\, dx = -e^{-x}\sin 4x - 4e^{-x}\cos 4x - 16\int e^{-x}\sin 4x\, dx$$

$$17\int e^{-x}\sin 4x\, dx = -e^{-x}\sin 4x - 4e^{-x}\cos 4x + C$$

$$\int e^{-x}\sin 4x\, dx = -\frac{e^{-x}}{17}(\sin 4x + 4\cos 4x) + C.$$

8.2.31 Let $u = e^{2x}$ and $dv = e^x \sin e^x\, dx$. Then $du = 2e^{2x}\, dx$ and $v = -\cos e^x$. Then we have

$$\int e^{3x} \sin e^x\, dx = -e^{2x}\cos e^x + 2\int e^{2x}\cos e^x\, dx.$$

To compute the last integral, we let $u = e^x$ and $dv = e^x \cos e^x\, dx$. Then $du = e^x\, dx$ and $v = \sin e^x$. Then the last integral is

$$2\int e^{2x}\cos e^x\, dx = 2e^x \sin e^x - 2\int e^x \sin e^x\, dx = 2e^x \sin e^x + 2\cos e^x + C.$$

Combining these results gives a final answer of

$$\int e^{3x}\sin e^x\, dx = -e^{2x}\cos e^x + 2e^x \sin e^x + 2\cos e^x + C.$$

8.2.33 Let $u = x$ and $dv = \sin x\, dx$. Then $du = dx$ and $v = -\cos x$. Then

$$\int_0^\pi x\sin x\, dx = -x\cos x\Big|_0^\pi + \int_0^\pi \cos x\, dx = \pi - 0 + \sin x\Big|_0^\pi = \pi - 0 + 0 - 0 = \pi.$$

8.2.35 Let $u = x$ and $dv = \cos 2x\, dx$. Then $du = dx$ and $v = \frac{1}{2}\sin 2x$. Then

$$\int_0^{\pi/2} x\cos 2x\, dx = \frac{1}{2}x\sin 2x\Big|_0^{\pi/2} - \frac{1}{2}\int_0^{\pi/2}\sin 2x\, dx = 0 - \left(\frac{1}{2}\cdot\frac{(-\cos 2x)}{2}\right)\Big|_0^{\pi/2} = -\frac{1}{4} - \frac{1}{4} = -\frac{1}{2}.$$

8.2.37 Let $u = \ln x$ and $dv = x^2\, dx$. Then $du = \frac{1}{x}\, dx$ and $v = \frac{x^3}{3}$. Then

$$\int_1^{e^2} x^2 \ln x\, dx = \frac{1}{3}x^3 \ln x\Big|_1^{e^2} - \frac{1}{3}\int_1^{e^2} x^2\, dx = \frac{2}{3}e^6 - \frac{1}{9}x^3\Big|_1^{e^2} = \frac{2}{3}e^6 - \frac{1}{9}(e^6 - 1) = \frac{5}{9}e^6 + \frac{1}{9}.$$

8.2. Integration by Parts

8.2.39 By problem 20, $\int \sin^{-1} y \, dy = y \sin^{-1} y + \sqrt{1-y^2}$. Thus,

$$\int_0^1 \sin^{-1} y \, dy = \left(y \sin^{-1} y + \sqrt{1-y^2}\right)\bigg|_0^1 = \left(\frac{\pi}{2} + 0\right) - (0+1) = \frac{\pi}{2} - 1 = \frac{\pi-2}{2}.$$

8.2.41

a. Let $u = \tan^{-1} x$ and $dv = dx$. Then $du = \dfrac{dx}{1+x^2}$ and $v = x$. Then

$$\int \tan^{-1} x \, dx = x \tan^{-1} x - \int \frac{x}{1+x^2} \, dx = x \tan^{-1} x - \frac{1}{2} \ln(1+x^2) + C.$$

b. Let $u = x^2$. Then $du = 2x \, dx$. Then

$$\int x \tan^{-1} x^2 \, dx = \int \frac{1}{2} \tan^{-1} u \, du = \frac{1}{2} u \tan^{-1} u - \frac{1}{4} \ln(1+u^2) + C = \frac{1}{2} x^2 \tan^{-1} x^2 - \frac{1}{4} \ln(1+x^4) + C.$$

8.2.43 Using shells, we have $\dfrac{V}{2\pi} = \displaystyle\int_0^{\ln 2} xe^{-x} \, dx$. Let $u = x$ and $dv = e^{-x} \, dx$, so that $du = dx$ and $v = -e^{-x}$. Then

$$\frac{V}{2\pi} = -xe^{-x}\bigg|_0^{\ln 2} + \int_0^{\ln 2} e^{-x} \, dx = -\frac{1}{2}\ln 2 - e^{-x}\bigg|_0^{\ln 2} = -\frac{\ln 2}{2} - \left(\frac{1}{2} - 1\right) = \frac{1}{2}(1 - \ln 2).$$

Thus $V = \pi(1 - \ln 2)$.

8.2.45 We have $V = \displaystyle\int_1^e \pi \ln x \, dx = \pi(x \ln x - x)\bigg|_1^e = \pi(e-e) - \pi(0-1) = \pi$.

8.2.47 Using disks, we have $\dfrac{V}{\pi} = \displaystyle\int_1^{e^2} x^2 \ln^2 x \, dx$. By problem 38, we have $\int x^2 \ln^2 x \, dx = \dfrac{1}{3} x^3 \ln^2 x - \dfrac{2}{9} x^3 \ln x + \dfrac{2}{27} x^3 + C$. Thus,

$$\frac{V}{\pi} = \left(\frac{1}{3} x^3 \ln^2 x - \frac{2}{9} x^3 \ln x + \frac{2}{27} x^3\right)\bigg|_1^{e^2} = \left(\frac{4}{3} e^6 - \frac{4}{9} e^6 + \frac{2}{27} e^6\right) - \left(\frac{2}{27}\right) = \frac{26}{27} e^6 - \frac{2}{27}.$$

Thus, $V = \dfrac{\pi}{27}\left(26 e^6 - 2\right)$.

8.2.49

a. False. For example, suppose $u = x$ and $dv = x \, dx$. Then $\int uv' \, dx = \int x^2 \, dx = \dfrac{x^3}{3} + C$, but

$$\int u \, dx \int v' \, dx = \left(\int x \, dx\right)^2 = \left(\frac{x^2}{2} + C\right)^2.$$

b. True. This is one way to write the integration by parts formula.

c. True. This is the integration by parts formula with the roles of u and v reversed.

8.2.51 Let $u = x^n$ and $dv = \cos ax \, dx$. Then $du = n x^{n-1} \, dx$ and $v = \dfrac{\sin ax}{a}$. Then

$$\int x^n \cos ax \, dx = \frac{x^n \sin ax}{a} - \frac{n}{a} \int x^{n-1} \sin ax \, dx.$$

8.2.53 Let $u = \ln^n x$ and $dv = dx$. Then $du = \dfrac{n \ln^{n-1}(x)}{x}\, dx$ and $v = x$. Then
$$\int \ln^n(x)\, dx = x \ln^n x - n \int \ln^{n-1}(x)\, dx.$$

8.2.55
$$\begin{aligned}\int x^2 \cos 5x\, dx &= \dfrac{x^2 \sin 5x}{5} - \dfrac{2}{5} \int x \sin 5x\, dx \\ &= \dfrac{x^2 \sin 5x}{5} - \dfrac{2}{5}\left(-\dfrac{x \cos 5x}{5} + \dfrac{1}{5}\int \cos 5x\, dx\right) \\ &= \dfrac{1}{5}\left(x^2 \sin 5x + \dfrac{2}{5}x\cos 5x - \dfrac{2}{25}\sin 5x\right) + C.\end{aligned}$$

8.2.57
$$\begin{aligned}\int_1^e \ln^3 x\, dx &= x \ln^3 x \Big|_1^e - 3\int_1^e \ln^2 x\, dx \\ &= e - 3\left(x \ln^2 x\right)\Big|_1^e + 6 \int_1^e \ln x\, dx \\ &= e - 3e + 6(x \ln x - x)\Big|_1^e \\ &= e - 3e + 6((e-e) - (0-1)) = 6 - 2e.\end{aligned}$$

8.2.59

a. Let $u = x$ and $dv = \dfrac{dx}{\sqrt{x+1}}$. Then $du = dx$ and $v = 2\sqrt{x+1}$. Then
$$\int \dfrac{x}{\sqrt{x+1}}\, dx = 2x\sqrt{x+1} - \int 2\sqrt{x+1}\, dx = 2x\sqrt{x+1} - \dfrac{4}{3}(x+1)^{3/2} + C = \dfrac{2}{3}\sqrt{x+1}(x-2) + C.$$

b. Let $u = x + 1$. Then $du = dx$ and $x = u - 1$. Substituting gives
$$\int \dfrac{x}{\sqrt{x+1}}\, dx = \int \dfrac{u-1}{\sqrt{u}}\, du = \int (u^{1/2} - u^{-1/2})\, du = \dfrac{2}{3}u^{3/2} - 2u^{1/2} + C = \dfrac{2}{3}(x+1)^{3/2} - 2(x+1)^{1/2} + C.$$

c. The answer to b can be written as the answer to a:
$$\dfrac{2}{3}(x+1)^{3/2} - 2(x+1)^{1/2} + C = \dfrac{2}{3}(x+1)^{1/2}(x+1-3) + C = \dfrac{2}{3}\sqrt{x+1}(x-2) + C.$$

8.2.61 Using the change of base formula, we have $\int \log_b x\, dx = \int \dfrac{\ln x}{\ln b}\, dx = \dfrac{1}{\ln b}(x \ln x - x) + C.$

8.2.63 Let $z = \sqrt{x}$, so that $dz = \dfrac{1}{2\sqrt{x}}\, dx$. Substituting yields $2\int \dfrac{\sqrt{x}\cos \sqrt{x}}{2\sqrt{x}}\, dx = 2\int z \cos z\, dz.$ Now let $u = z$ and $dv = \cos z\, dz$. then $du = dz$ and $v = \sin z$. Then by Integration by Parts, we have
$$2\int z \cos z\, dz = 2\left(z \sin z - \int \sin z\, dz\right) = 2z \sin z + 2 \cos z + C.$$
Thus, the original given integral is equal to $2(\sqrt{x}\sin \sqrt{x} + \cos \sqrt{x}) + C.$

8.2.65 Let $u = x$ and $dv = f''(x)\, dx$. Then $du = dx$ and $v = f'(x)$. We have
$$\int_a^b x f''(x)\, dx = x f'(x)\Big|_a^b - \int_a^b f'(x)\, dx = (0 - 0) - f(x)\Big|_a^b = -(f(b) - f(a)) = f(a) - f(b).$$

8.2.67 By the Fundamental Theorem, $f'(x) = \sqrt{\ln^2 x - 1}$. So the arc length is $\int_e^{e^3} \sqrt{1 + (f'(x))^2}\, dx = \int_e^{e^3} \ln x\, dx = (x \ln x - x)\Big|_e^{e^3} = 3e^3 - e^3 - (e - e) = 2e^3$.

8.2.69 Let V_1 be the volume generated when R is revolved about the x-axis, and V_2 the volume generated when R is revolved about the y-axis.

Using disks, we have

$$\frac{V_1}{\pi} = \int_0^\pi \sin^2 x\, dx = \frac{1}{2}\int_0^\pi (1 - \cos 2x)\, dx = \frac{1}{2}\left(\pi - \frac{1}{2}\int_0^\pi 2\cos 2x\, dx\right) = \frac{1}{2}\left(\pi - \frac{1}{2}\int_0^{2\pi} \cos u\, du\right) = \frac{\pi}{2},$$

where the ordinary substitution $u = 2x$ was made. So $V_1 = \dfrac{\pi^2}{2}$.

Using shells to compute V_2, we have $\dfrac{V_2}{2\pi} = \int_0^\pi x \sin x\, dx$. Letting $u = x$ and $dv = \sin x\, dx$, we have $du = dx$ and $v = -\cos x$. Thus, $\dfrac{V_2}{2\pi} = -x\cos x\Big|_0^\pi + \int_0^\pi \cos x\, dx = \pi$. Thus, $V_2 = 2\pi^2$, and $V_2 > V_1$.

8.2.71 Using shells, we have $\dfrac{V}{2\pi} = \int_0^{\pi/2} x \cos x\, dx$. Let $u = x$ and $dv = \cos x\, dx$, so that $du = dx$ and $v = \sin x$. We have $\dfrac{V}{2\pi} = x\sin x\Big|_0^{\pi/2} - \int_0^{\pi/2} \sin x\, dx = \dfrac{\pi}{2} - 1$.
Thus, $V = \pi(\pi - 2)$.

8.2.73 Let $u = \sin bx$ and $dv = e^{ax}\, dx$ so that $du = b\cos bx\, dx$ and $v = \dfrac{e^{ax}}{a}$. Then $\int e^{ax} \sin bx\, dx = \dfrac{e^{ax}}{a}\sin bx - \dfrac{b}{a}\int e^{ax}\cos bx\, dx$. Now let $u = \cos bx$ and $dv = e^{ax}$, so that $du = -b\sin bx$ and $v = \dfrac{e^{ax}}{a}$. Then $\int e^{ax}\cos bx\, dx = \dfrac{e^{ax}}{a}\cos bx + \dfrac{b}{a}\int e^{ax}\sin bx\, dx$. Putting these together yields

$$\int e^{ax}\sin bx\, dx = \dfrac{e^{ax}}{a}\sin bx - \dfrac{b}{a}\left(\dfrac{e^{ax}}{a}\cos bx + \dfrac{b}{a}\int e^{ax}\sin bx\, dx\right).$$

Multiplying through by a^2 and combining like terms yields

$$(a^2 + b^2)\int e^{ax}\sin bx\, dx = e^{ax}(a\sin bx - b\cos bx) + C,$$

so $\int e^{ax}\sin bx\, dx = \dfrac{e^{ax}}{a^2 + b^2}(a\sin bx - b\cos bx) + C$ as desired.

Now, returning to the integral mentioned above, and incorporating this last result, we have

$$\int e^{ax}\cos bx\, dx = \dfrac{e^{ax}}{a}\cos bx + \dfrac{b}{a}\cdot e^{ax}\cdot\dfrac{a\sin bx - b\cos bx}{a^2 + b^2} + C = \dfrac{e^{ax}}{a}\cdot\dfrac{(a^2+b^2)\cos bx + b(a\sin bx - b\cos bx)}{a^2 + b^2} + C$$

$$= e^{ax}\cdot\dfrac{a\cos bx + b\sin bx}{a^2 + b^2} + C.$$

8.2.75

a. We have $s(t) = 0$ when $\sin t = 0$, which occurs for $t = k\pi$, where k is an integer.

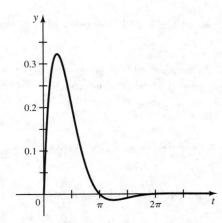

b. This is given by $\dfrac{1}{\pi}\displaystyle\int_0^\pi e^{-t}\sin t\,dt$. Using the previous problem, this is equal to

$$\frac{1}{\pi}e^{-t}\cdot\frac{-\sin t-\cos t}{2}\bigg|_0^\pi = \frac{1}{2\pi}(e^{-\pi}+1).$$

c. This is given by $\dfrac{1}{\pi}\displaystyle\int_{n\pi}^{(n+1)\pi} e^{-t}\sin t\,dt$. Using the previous problem, this is equal to

$$\frac{1}{\pi}e^{-t}\cdot\frac{-\sin t-\cos t}{2}\bigg|_{n\pi}^{(n+1)\pi} = \frac{1}{2\pi}\left(e^{-(n+1)\pi}(-\sin(n+1)\pi-\cos(n+1)\pi)-e^{-n\pi}(-\sin n\pi-\cos n\pi)\right)$$

$$=\frac{1}{2\pi}\left(-e^{-(n+1)\pi}\cos(n+1)\pi+e^{-n\pi}\cos n\pi\right) = \frac{e^{-n\pi}}{2\pi}\left(\cos n\pi-e^{-\pi}\cos(n+1)\pi\right)$$

$$=\frac{e^{-n\pi}}{2\pi}\left((-1)^n-e^{-\pi}(-1)^{n+1}\right) = (-1)^n\frac{e^{-n\pi}}{2\pi}\left(1+e^{-\pi}\right).$$

d. Each a_n is $e^{-\pi}$ times a_{n-1}.

8.2.77

a. If $u=f(x)$ then $du=f'(x)\,dx$, and if $dv=g(x)\,dx$, then $v=G_1(x)$. The integration by parts formula gives

$$\int f(x)g(x)\,dx = f(x)G_1(x) - \int f'(x)G_1(x)\,dx.$$

b. Letting $u=f'(x)$ yields $du=f''(x)\,dx$ and letting $dv=G_1(x)\,dx$ yields $v=G_2(x)$. Then continuing from part (a) we have

$$\int f(x)g(x)\,dx = f(x)G_1(x) - \left(f'(x)G_2(x) - \int f''(x)G_2(x)\,dx\right)$$

$$= f(x)G_1(x) - f'(x)G_2(x) + \int f''(x)G_2(x)\,dx.$$

Note that we end up with three terms which consist of products as indicated by the arrows, with the last product inside an integral sign. The signs alternate as indicated on the chart.

c. For the trailing integral from part (b), we let $u = f''(x)$ so that $du = f'''(x)\,dx$, and let $dv = G_2(x)\,dx$ so that $v = G_3(x)$. Then we have

$$\int f(x)g(x)\,dx = f(x)G_1(x) - f'(x)G_2(x) + \int f''(x)G_2(x)\,dx$$
$$= f(x)G_1(x) - f'(x)G_2(x) + f''(x)G_3(x) - \int f'''(x)G_3(x)\,dx.$$

f and its derivatives		g and its integrals
$f(x)$	+	$g(x)$
$f'(x)$	−	$G_1(x)$
$f''(x)$	+	$G_2(x)$
$f'''(x)$	−	$G_3(x)$

d. The table is as follows:

f and its derivatives		g and its integrals
x^2	+	$e^{0.5x}$
$2x$	−	$2e^{0.5x}$
2	+	$4e^{0.5x}$
0	−	$8e^{0.5x}$

So
$$\int x^2 e^{0.5x}\,dx = 2x^2 e^{0.5x} - 8xe^{0.5x} + 16e^{0.5x} + C.$$

Note that the last integral is $\int 0 \cdot 8e^{0.5x}\,dx = \int 0\,dx = C.$

e. We have:

f and its derivatives		g and its integrals
x^3	+	$\cos x$
$3x^2$	−	$\sin x$
$6x$	+	$-\cos x$
6	−	$-\sin x$
0	+	$\cos x$

So $\int x^3 \cos x\,dx = x^3 \sin x + 3x^2 \cos x - 6x \sin x - 6\cos x + C.$

f. $\dfrac{d^k}{dx^k}(p_n(x)) = 0$ for $k \geq n+1$, so the process will end.

8.2.79

a. $\int e^x \cos x\,dx = e^x \sin x + e^x \cos x - \int e^x \cos x\,dx.$

b. Adding $\int e^x \cos x\,dx$ to both sides of the previous equation gives $2\int e^x \cos x\,dx = e^x \sin x + e^x \cos x$, so
$$\int e^x \cos x\,dx = \frac{e^x \sin x + e^x \cos x}{2} + C.$$

c. The chart looks like this:

f and its derivatives		g and its integrals
e^{-2x}	$+$	$\sin 3x$
$-2e^{-2x}$	$-$	$-\frac{1}{3}\cos 3x$
$4e^{-2x}$	$+$	$-\frac{1}{9}\sin 3x$

So we have
$$\int e^{-2x}\sin 3x = -\frac{1}{3}e^{-2x}\cos 3x - \frac{2}{9}e^{-2x}\sin 3x - \frac{4}{9}e^{-2x}\sin 3x.$$

Adding $\frac{4}{9}\int e^{-2x}\sin 3x\,dx$ to both sides gives
$$\frac{13}{9}\int e^{-2x}\sin 3x\,dx = -\frac{1}{3}e^{-2x}\cos 3x - \frac{2}{9}e^{-2x}\sin 3x + D,$$

where D is an arbitrary constant so
$$\int e^{-2x}\sin 3x\,dx = -\frac{3}{13}e^{-2x}\cos 3x - \frac{2}{13}e^{-2x}\sin 3x + C.$$

8.2.81

a. To compute I_1, we let $u = -x^2$, so that $du = -2x\,dx$. Then an ordinary substitution yields
$$-\frac{1}{2}\int e^u\,du = -\frac{1}{2}e^u + C = -\frac{1}{2}e^{-x^2} + C.$$

b. To compute I_3, we let $u = x^2$ and $dv = xe^{-x^2}\,dx$, so that $du = 2x\,dx$ and $v = -\frac{1}{2}e^{-x^2}$ (by part (a)). Then $I_3 = -\frac{1}{2}x^2 e^{-x^2} + \int xe^{-x^2}\,dx = -\frac{1}{2}x^2 e^{-x^2} - \frac{1}{2}e^{-x^2} + C = -\frac{1}{2}e^{-x^2}(x^2+1) + C$.

c. To compute I_5, we let $u = x^4$ and $dv = xe^{-x^2}\,dx$. Then $du = 4x^3\,dx$ and $v = -\frac{1}{2}e^{-x^2}$. Then $I_5 = -\frac{1}{2}x^4 e^{-x^2} + 2\int x^3 e^{-x^2}\,dx = -\frac{1}{2}x^4 e^{-x^2} + 2I_3 = -\frac{1}{2}e^{-x^2}(x^4 + 2x^2 + 2)$.

d. Suppose that n is odd and that it is true that $I_n = -\frac{1}{2}e^{-x^2} p_{n-1}(x)$ where $p_{n-1}(x)$ is an even polynomial of degree $n-1$. We will show that I_{n+2} also has this property, so the result will follow by induction. Let $u = x^{n+1}$ and let $dv = xe^{-x^2}\,dx$, so that $du = (n+1)x^n\,dx$ and $v = -\frac{1}{2}e^{-x^2}$. Then $I_{n+2} = -\frac{1}{2}x^{n+1}e^{-x^2} + \frac{n+1}{2}I_n = -\frac{1}{2}e^{-x^2}(x^{n+1} + \frac{n+1}{2}p_{n-1}(x)) + C = -\frac{1}{2}e^{-x^2}p_{n+1}(x) + C$. Note that if $p_{n-1}(x)$ is an even polynomial of degree $n-1$ then $\frac{n+1}{2}p_{n-1}(x) + x^{n+1}$ is an even polynomial of degree $n+1$.

e. Note that $I_2 = -\frac{1}{2}xe^{-x^2} + \frac{1}{2}I_0$. (Using a similar technique to that above). Now if I_2 were expressible in terms of elementary functions, then I_0 would be as well, but we are given that it isn't. Similarly, we can express I_{2k} in terms of I_{2k-2} using Integration by Parts, and if any of these were expressible in terms of elementary functions, then the even numbered one below it would be. So the inability to express I_0 that way implies the inability to express I_2 that way, which implies the inability to express I_4 that way, and so on.

8.3 Trigonometric Integrals

8.3.1 The half-angle identities for sine and cosine:
$$\sin^2 x = \frac{1-\cos 2x}{2} \quad \text{and} \quad \cos^2 x = \frac{1+\cos 2x}{2},$$
which are conjugates.

8.3.3 To integrate $\sin^3 x$, write $\sin^3 x = \sin x \sin^2 x = \sin x(1-\cos^2 x)$, and let $u = \cos x$ so that $du = -\sin x\, dx$.

8.3.5 A reduction formula is a recursive formula involving integrals. Using it, one can rewrite an integral of a certain type in a simpler form – which can then perhaps be evaluated or further reduced.

8.3.7 One would compute this integral by letting $u = \tan x$, so that $du = \sec^2 x\, dx$. This substitution leads to the integral $\int u^{10}\, du$, which can easily be evaluated.

8.3.9 $\int \cos^3 x\, dx = \int \cos x(1-\sin^2 x)\, dx$ Let $u = \sin x$. Then $du = \cos x\, dx$. Substituting gives
$$\int (1-u^2)\, du = u - \frac{u^3}{3} + C = \sin x - \frac{\sin^3 x}{3} + C.$$

8.3.11 $\int \sin^2 3x\, dx = \frac{1}{2}\int (1-\cos 6x)\, dx = \frac{1}{2}\left(x - \frac{1}{6}\sin 6x\right) + C = \frac{x}{2} - \frac{1}{12}\sin 6x + C.$

8.3.13 $\int \sin^5 x\, dx = \int (\sin^2 x)^2 \sin x\, dx = \int (1-\cos^2 x)^2 \sin x\, dx$. Let $u = \cos x$ so that $du = -\sin x\, dx$. Substituting yields
$$-\int (1-u^2)^2\, du = \int (-u^4 + 2u^2 - 1)\, du = \frac{-u^5}{5} + \frac{2u^3}{3} - u + C = \frac{-\cos^5 x}{5} + \frac{2\cos^3 x}{3} - \cos x + C.$$

8.3.15 $\int \sin^3 x \cos^2 x\, dx = \int \sin x(1-\cos^2 x)(\cos^2 x)\, dx$. Let $u = \cos x$ so that $du = -\sin x\, dx$. Substituting gives
$$\int (u^4 - u^2)\, du = u^5/5 - u^3/3 + C = (\cos^5 x)/5 - (\cos^3 x)/3 + C.$$

8.3.17 $\int \cos^3 x \sqrt{\sin x}\, dx = \int \cos x(1-\sin^2 x)\sqrt{\sin x}\, dx$. Let $u = \sin x$ so that $du = \cos x\, dx$. Substituting gives
$$\int (1-u^2)u^{1/2}\, du = \int (u^{1/2} - u^{5/2})\, du = 2u^{3/2}/3 - 2u^{7/2}/7 + C = 2(\sin^{3/2} x)/3 - 2(\sin^{7/2} x)/7 + C.$$

8.3.19 $\int_0^{\pi/3} \sin^5 x \cos^{-2} x\, dx = \int_0^{\pi/3} (\sin x)\left(\frac{(1-\cos^2 x)^2}{\cos^2 x}\right) dx.$ Let $u = \cos x$ so that $du = -\sin x\, dx$. Substituting yields
$$-\int_1^{1/2} \frac{(1-u^2)^2}{u^2}\, du = \int_1^{1/2} (-u^{-2} + 2 - u^2)\, du = \left(\frac{1}{u} + 2u - \frac{u^3}{3}\right)\Big|_1^{1/2}$$
$$= \left(2 + 1 - \frac{1}{24} - \left(1 + 2 - \frac{1}{3}\right)\right) = \frac{7}{24}.$$

8.3.21 $\int_0^{\pi/2} \cos^3 x \sqrt{\sin^3 x}\, dx = \int_0^{\pi/2} (1-\sin^2 x)\sin^{3/2} x \cos x\, dx$. Let $u = \sin x$ so that $du = \cos x\, dx$. Substituting gives
$$\int_0^1 (1-u^2)u^{3/2}\, du = \int_0^1 (u^{3/2} - u^{7/2})\, du = \left(\frac{2}{5}u^{5/2} - \frac{2}{9}u^{9/2}\right)\bigg|_0^1 = \frac{2}{5} - \frac{2}{9} = \frac{8}{45}.$$

8.3.23
$$\int \sin^2 x \cos^2 x\, dx = \int \left(\frac{1-\cos 2x}{2}\right)\left(\frac{1+\cos 2x}{2}\right) dx = \frac{1}{4}\int 1 - \cos^2 2x\, dx$$
$$= \frac{1}{4}\int \left(1 - \frac{1+\cos 4x}{2}\right) dx = \frac{1}{4}\int \left(\frac{1}{2} - \frac{1}{2}\cos 4x\right) dx = \frac{1}{4}\left(\frac{x}{2} - \frac{\sin 4x}{8}\right) + C.$$

8.3.25 $\int \sin^2 x \cos^4 x\, dx = \int \left(\frac{1-\cos 2x}{2}\right)\left(\frac{1+\cos 2x}{2}\right)^2 dx = \frac{1}{8}\int (1-\cos^2 2x)(1+\cos 2x)\, dx = \frac{1}{8}\int 1 + \cos 2x - \cos^2 2x - \cos^3 2x\, dx = \frac{1}{8}\int 1 + \cos 2x - \frac{1+\cos 4x}{2} - \cos^3 2x\, dx = \frac{1}{8}\int \frac{1}{2} + \cos 2x - \frac{1}{2}\cos 4x\, dx - \frac{1}{8}\int (1-\sin^2 2x)\cos 2x\, dx = \frac{x}{16} + \frac{\sin 2x}{16} - \frac{\sin 4x}{64} - \frac{1}{8}\int (1-\sin^2 2x)\cos 2x\, dx$. To compute this last integral, we let $u = \sin 2x$, so that $du = 2\cos 2x\, dx$. Then
$$\int (1-\sin^2 2x)\cos 2x\, dx = \frac{1}{2}\int (1-u^2)\, du = \frac{1}{2}\left(u - \frac{u^3}{3}\right) + C = \frac{1}{2}\left(\sin 2x - \frac{\sin^3 2x}{3}\right) + C.$$

Thus, our original given integral is equal to
$$\frac{x}{16} + \frac{\sin 2x}{16} - \frac{\sin 4x}{64} - \frac{1}{8}\left(\frac{1}{2}\left(\sin 2x - \frac{\sin^3 2x}{3}\right)\right) + C = \frac{x}{16} - \frac{\sin 4x}{64} + \frac{\sin^3 2x}{48} + C.$$

8.3.27 $\int \tan^2 x\, dx = \int (\sec^2 x - 1)\, dx = \tan x - x + C.$

8.3.29 $\int \cot^4 x\, dx = \int \cot^2 x(\csc^2 x - 1)\, dx = \int (\cot^2 x \csc^2 x - (\csc^2 x - 1))\, dx = \int \cot^2 x \csc^2 x\, dx + \cot x + x$. Let $u = \cot x$ so that $du = -\csc^2 x\, dx$. Substituting gives
$$-\int u^2\, du + \cot x + x = -u^3/3 + \cot x + x + C = -(\cot^3 x)/3 + \cot x + x + C.$$

8.3.31
$$\int 20\tan^6 x\, dx = 20\int (\tan^4 x)(\sec^2 x - 1)\, dx = 20\int ((\tan^4 x)\sec^2 x - (\tan^2 x)(\sec^2 x - 1))\, dx$$
$$= 20\int (\tan^4 x \sec^2 x - \tan^2 x \sec^2 x + \sec^2 x - 1)\, dx.$$

Let $u = \tan x$ so that $du = \sec^2 x\, dx$. We have
$$20\left(\int (u^4 - u^2)\, du + \tan x - x\right) + C = 4u^5 - \frac{20u^3}{3} + 20\tan x - 20x + C$$
$$= 4\tan^5 x - \frac{20\tan^3 x}{3} + 20\tan x - 20x + C.$$

8.3.33 Let $u = \tan x$ so that $du = \sec^2 x\, dx$. Substituting gives $\int 10u^9\, du = u^{10} + C = \tan^{10} x + C.$

8.3.35 Let $u = \sec x$ so that $du = \sec x \tan x\, dx$. Substituting gives
$$\int u^2\, du = u^3/3 + C = (\sec^3 x)/3 + C.$$

8.3.37 Let $u = \ln \theta$ so that $du = \frac{1}{\theta}\, d\theta$. Substituting yields $\int \sec^4 u\, du = \int (\sec^2 u)(1 + \tan^2 u)\, du$. Let $w = \tan u$ so that $dw = \sec^2 u\, du$. Substituting again gives
$$\int (1 + w^2)\, dw = w + \frac{w^3}{3} + C = \tan(\ln(\theta)) + \frac{\tan^3(\ln(\theta))}{3} + C.$$

8.3.39
$$\int_{-\pi/3}^{\pi/3} \sqrt{\sec^2 \theta - 1}\, d\theta = 2\int_0^{\pi/3} \sqrt{\sec^2 \theta - 1}\, d\theta = 2\int_0^{\pi/3} \tan \theta\, d\theta$$
$$= -2 \ln |\cos \theta|\Big|_0^{\pi/3} = -2\ln(1/2) + 2\ln(1) = 2\ln 2.$$

8.3.41 $\int_0^{\pi/4} \sec^7 x \sin x\, dx = \int_0^{\pi/4} \sec^6 x \tan x\, dx = \int_0^{\pi/4} \sec^5 x \sec x \tan x\, dx$. Let $u = \sec x$ so that $du = \sec x \tan x\, dx$. Substituting gives
$$\int_1^{\sqrt{2}} u^5\, du = \frac{u^6}{6}\Big|_1^{\sqrt{2}} = \frac{8}{6} - \frac{1}{6} = \frac{7}{6}.$$

8.3.43
$$\int \tan^3 4x\, dx = \int (\tan 4x)(\sec^2 4x - 1)\, dx = \int (\tan 4x) \sec^2 4x\, dx - \int \tan 4x\, dx$$
$$= \int (\tan 4x) \sec^2 4x\, dx + \frac{\ln |\cos 4x|}{4} + C.$$
Let $u = \tan 4x$ so that $du = 4\sec^2 4x\, dx$. Substituting gives
$$\frac{1}{4}\int u\, du + \frac{\ln |\cos 4x|}{4} + C = \frac{u^2}{8} + \frac{\ln |\cos 4x|}{4} + C = \frac{\tan^2 4x}{8} + \frac{\ln |\cos 4x|}{4} + C.$$

8.3.45 Let $u = \tan x$ so that $du = \sec^2 x\, dx$. Then
$$\int \sec^2 x \tan^{1/2} x\, dx = \int u^{1/2}\, du = \frac{2}{3}u^{3/2} + C = \frac{2}{3}\tan^{3/2} x + C.$$

8.3.47 $\int \frac{\csc^4 x}{\cot^2 x}\, dx = \int (\csc^2 x)\left(\frac{\cot^2 x + 1}{\cot^2 x}\right) dx$. Let $u = \cot x$ so that $du = -\csc^2 x\, dx$. Substituting gives
$$-\int \frac{u^2 + 1}{u^2}\, du = \int -1 - u^{-2}\, du = -u + \frac{1}{u} + C = -\cot x + \tan x + C.$$

8.3.49 Let $u = \cot 5w$. Then $du = -5\csc^2 5w$. Substituting gives
$$-\frac{1}{5}\int_1^0 u^4\, du = -\frac{1}{25} u^5\Big|_1^0 = -\frac{1}{25}(0 - 1) = \frac{1}{25}.$$

8.3.51 $\int (\csc^2 x + \csc^4 x)\, dx = \int (1 + \csc^2 x) \csc^2 x\, dx$. Using the identity $\csc^2 x = 1 + \cot^2 x$, we can write this integral as $\int (2 + \cot^2 x) \csc^2 x\, dx$. Substituting $u = \cot x$ so that $du = -\csc^2 x\, dx$ gives
$$-\int (2 + u^2)\, du = -2u - \frac{u^3}{3} + C = -2\cot x - \frac{\cot^3 x}{3} + C.$$

8.3.53 $\int_0^{\pi/4} \sec^4 \theta \, d\theta = \int_0^{\pi/4} (\sec^2 \theta)(1 + \tan^2 \theta) \, d\theta$. Let $u = \tan \theta$ so that $du = \sec^2 \theta \, d\theta$. Note that when $\theta = 0$ we have $u = 0$ and when $\theta = \dfrac{\pi}{4}$ we have $u = 1$. So the original integral is equal to

$$\int_0^1 (1 + u^2) \, du = \left(u + \frac{u^3}{3}\right)\bigg|_0^1 = 1 + \frac{1}{3} = \frac{4}{3}.$$

8.3.55 $\int_{\pi/6}^{\pi/3} \cot^3 \theta \, d\theta = \int_{\pi/6}^{\pi/3} (\cot \theta)(\csc^2 \theta - 1) \, d\theta = \int_{\pi/6}^{\pi/3} \cot \theta \csc^2 \theta \, d\theta - \int_{\pi/6}^{\pi/3} \frac{\cos \theta}{\sin \theta} \, d\theta$. For the first integral, let $u = \cot \theta$ so that $du = -\csc^2 \theta \, d\theta$. For the second integral, let $w = \sin \theta$ so that $dw = \cos \theta \, d\theta$. Substituting gives

$$-\int_{\sqrt{3}}^{1/\sqrt{3}} u \, du - \int_{1/2}^{\sqrt{3}/2} \frac{1}{w} \, dw = -\frac{u^2}{2}\bigg|_{\sqrt{3}}^{1/\sqrt{3}} - \ln w \bigg|_{1/2}^{\sqrt{3}/2}$$

$$= -\frac{1}{2}\left(\frac{1}{3} - 3\right) - \left(\ln \sqrt{3} - \ln 2 - \ln 1 + \ln 2\right) = \frac{4}{3} - \frac{\ln 3}{2}.$$

8.3.57 $\int_0^\pi (1 - \cos 2x)^{3/2} \, dx = \int_0^\pi (2 \sin^2 x)^{3/2} \, dx = 2\sqrt{2} \int_0^\pi \sin^3 x \, dx = 2\sqrt{2} \int_0^\pi (\sin x)(1 - \cos^2 x) \, dx$. Let $u = \cos x$ so that $du = -\sin x \, dx$. Substituting yields

$$-2\sqrt{2} \int_1^{-1} (1 - u^2) \, du = 2\sqrt{2} \int_{-1}^1 (1 - u^2) \, du = 4\sqrt{2} \int_0^1 (1 - u^2) \, du = 4\sqrt{2} \left(u - \frac{u^3}{3}\right)\bigg|_0^1 = \frac{8\sqrt{2}}{3}.$$

8.3.59 $\int_0^{\pi/2} \sqrt{1 - \cos 2x} \, dx = \sqrt{2} \int_0^{\pi/2} \sin x \, dx = -\sqrt{2} \cos x \bigg|_0^{\pi/2} = \sqrt{2}.$

8.3.61 $\int_0^{\pi/4} (1 + \cos 4x)^{3/2} \, dx = \int_0^{\pi/4} (2 \cos^2 2x)^{3/2} \, dx = 2\sqrt{2} \int_0^{\pi/4} \cos^3 2x \, dx = 2\sqrt{2} \int_0^{\pi/4} (\cos 2x)(1 - \sin^2 2x) \, dx$. Let $u = \sin 2x$ so that $du = 2 \cos 2x \, dx$. Substituting gives

$$\sqrt{2} \int_0^1 (1 - u^2) \, du = \sqrt{2} \left(u - \frac{u^3}{3}\right)\bigg|_0^1 = \frac{2\sqrt{2}}{3}.$$

8.3.63

a. True. We have $\int_0^\pi \cos^{2m+1} x \, dx = \int_0^\pi (\cos^2 x)^m \cos x \, dx = \int_0^\pi (1 - \sin^2 x)^m \cos x \, dx$. Let $u = \sin x$ so that $du = \cos x \, dx$. Substituting yields $\int_0^0 (1 - u^2)^m \, du = 0$.

b. False. For example, suppose $m = 1$. Then $\int_0^\pi \sin x \, dx = -\cos x \bigg|_0^\pi = -(-1 - 1) = 2 \neq 0$.

8.3.65 $V = \int_0^{\pi/2} \pi \sin^4 x \cos^3 x \, dx = \pi \int_0^{\pi/2} \sin^4 x (1 - \sin^2 x) \cos x \, dx = \int_0^{\pi/2} (\sin^4 x - \sin^6 x) \cos x \, dx$. Let $u = \sin x$. Then $du = \cos x \, dx$ and substitution gives

$$\pi \int_0^1 (u^4 - u^6) \, du = \pi \left(\frac{u^5}{5} - \frac{u^7}{7}\right)\bigg|_0^1 = \pi \left(\frac{1}{5} - \frac{1}{7}\right) = \frac{2\pi}{35}.$$

8.3.67 $\int \sin 3x \cos 7x \, dx = \dfrac{1}{2}\left(\int \sin(-4x) \, dx + \int \sin 10x \, dx\right) = \dfrac{1}{2}\left(\dfrac{\cos(-4x)}{4} - \dfrac{\cos 10x}{10}\right) + C = \dfrac{\cos 4x}{8} - \dfrac{\cos 10x}{20} + C.$

8.3.69 $\int \sin 3x \sin 2x \, dx = \frac{1}{2}\left(\int \cos x \, dx - \int \cos 5x \, dx\right) = \frac{\sin x}{2} - \frac{\sin 5x}{10} + C.$

8.3.71

a. $\int_0^\pi \sin mx \sin nx \, dx = \frac{1}{2}\left(\int_0^\pi \cos(m-n)x \, dx - \int_0^\pi \cos(m+n)x \, dx\right) =$
$\frac{1}{2}\left(\frac{1}{m-n}\int_0^{(m-n)\pi} \cos u \, du - \frac{1}{m+n}\int_0^{(m+n)\pi} \cos v \, dv\right)$ where $u = (m-n)x$ and $v = (m+n)x$. But this yields $\frac{1}{2}\left(\frac{1}{m-n}\sin u\Big|_0^{(m-n)\pi} - \frac{1}{m+n}\sin v\Big|_0^{(m+n)\pi}\right) = \frac{1}{2}(0-0) = 0.$

b. $\int_0^\pi \cos mx \cos nx \, dx = \frac{1}{2}\left(\int_0^\pi \cos(m-n)x \, dx + \int_0^\pi \cos(m+n)x \, dx\right) = 0$ by the previous part of this problem,

c. $\int_0^\pi \sin mx \cos nx \, dx = \frac{1}{2}\left(\int_0^\pi \sin(m-n)x \, dx + \int_0^\pi \sin(m+n)x \, dx\right) =$
$\frac{1}{2}\left(\frac{1}{m-n}\int_0^{(m-n)\pi} \sin u \, du + \frac{1}{m+n}\int_0^{(m+n)\pi} \sin v \, dv\right)$ where $u = (m-n)x$ and $v = (m+n)x$. This quantity is equal to $\frac{-1}{2}\left(\frac{1}{m-n}\cos u\Big|_0^{(m-n)\pi} + \frac{1}{m+n}\cos v\Big|_0^{(m+n)\pi}\right) =$
$\frac{-1}{2}\left(\frac{1}{m-n}(\cos(m-n)\pi - 1) + \frac{1}{m+n}(\cos(m+n)\pi - 1)\right) =$
$\begin{cases} 0 & \text{if } m \text{ and } n \text{ are both even or both odd;} \\ \frac{1}{m-n} + \frac{1}{m+n} = \frac{2m}{m^2-n^2} & \text{otherwise.} \end{cases}$

8.3.73 For $n \neq 1$, $\int \tan^n x \, dx = \int (\tan^{n-2} x)(\sec^2 x - 1) \, dx = \int \tan^{n-2} x \sec^2 x \, dx - \int \tan^{n-2} x \, dx$. Let $u = \tan x$ so that $du = \sec^2 x \, dx$. Then substituting in the first of these last two integrals yields

$\int u^{n-2} \, du - \int \tan^{n-2} x \, dx = \frac{u^{n-1}}{n-1} - \int \tan^{n-2} x \, dx = \frac{\tan^{n-1} x}{n-1} - \int \tan^{n-2} x \, dx.$

Thus $\int_0^{\pi/4} \tan^3 x \, dx = \frac{\tan^2 x}{2}\Big|_0^{\pi/4} - \int_0^{\pi/4} \tan x \, dx = \frac{1}{2} + \ln|\cos x|\Big|_0^{\pi/4} = \frac{1}{2} - \frac{\ln 2}{2}.$

8.3.75

a. $\int_0^\pi \sin^2 x \, dx = \frac{1}{2}\int_0^\pi (1 - \cos 2x) \, dx =$
$\frac{1}{2}\left(x - \frac{\sin 2x}{2}\right)\Big|_0^\pi = \frac{\pi}{2}.$
$\int_0^\pi \sin^2 2x \, dx = \frac{1}{2}\int_0^\pi (1 - \cos 4x) \, dx =$
$\frac{1}{2}\left(x - \frac{\sin 4x}{4}\right)\Big|_0^\pi = \frac{\pi}{2}.$

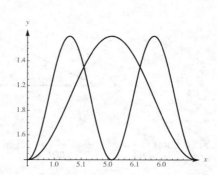

$$\int_0^\pi \sin^2 3x\, dx = \frac{1}{2}\int_0^\pi (1 - \cos 6x)\, dx =$$
$$\frac{1}{2}\left(x - \frac{\sin 6x}{6}\right)\Big|_0^\pi = \frac{\pi}{2}.$$

b. $\int_0^\pi \sin^2 4x\, dx = \frac{1}{2}\int_0^\pi (1 - \cos 8x)\, dx =$
$$\frac{1}{2}\left(x - \frac{\sin 8x}{8}\right)\Big|_0^\pi = \frac{\pi}{2}.$$

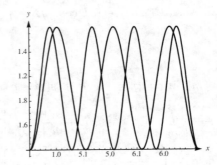

c. $\int_0^\pi \sin^2 nx\, dx = \frac{1}{2}\int_0^\pi (1 - \cos 2nx)\, dx = \frac{1}{2}\left(x - \frac{\sin 2nx}{2n}\right)\Big|_0^\pi = \frac{\pi}{2}.$

d. Yes. $\int_0^\pi \cos^2 nx\, dx = \frac{1}{2}\int_0^\pi (1 + \cos 2nx)\, dx = \frac{1}{2}\left(x + \frac{\sin 2nx}{2n}\right)\Big|_0^\pi = \frac{\pi}{2}.$

e. Claim: The corresponding integrals are all equal to $\frac{3\pi}{8}$. Proof:

$$\int_0^\pi \sin^4 nx\, dx = \int_0^\pi \left(\frac{1 - \cos 2nx}{2}\right)^2 dx$$
$$= \int_0^\pi \frac{1 - 2\cos 2nx + \cos^2 2nx}{4}\, dx = \int_0^\pi \frac{1}{4}\, dx - \frac{1}{2}\int_0^\pi \cos 2nx\, dx + \frac{1}{4}\int_0^\pi \cos^2 2nx\, dx$$
$$= \frac{\pi}{4} - \frac{1}{2}\left(\frac{\sin 2nx}{2n}\right)\Big|_0^\pi + \frac{1}{4}\cdot\frac{\pi}{2} = \frac{\pi}{4} - 0 + \frac{\pi}{8} = \frac{3\pi}{8}.$$

8.4 Trigonometric Substitutions

8.4.1 This would suggest $x = 3\sec\theta$, because then $\sqrt{x^2 - 9} = 3\sqrt{\sec^2\theta - 1} = 3\sqrt{\tan^2\theta} = 3\tan\theta$, for $\theta \in [0, \pi/2)$.

8.4.3 This would suggest $x = 10\sin\theta$, because then $\sqrt{100 - x^2} = 10\sqrt{1 - \sin^2\theta} = 10\sqrt{\cos^2\theta} = 10\cos\theta$, for $|\theta| \leq \frac{\pi}{2}$.

8.4.5

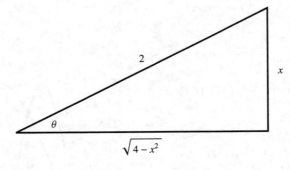

If $x = 2\sin\theta$ then $\frac{x^2}{4} = \sin^2\theta = \frac{1}{\csc^2\theta}$. Then $\cot^2\theta = \csc^2\theta - 1 = \frac{4}{x^2} - 1 = \frac{4 - x^2}{x^2}$. So $\cot\theta = \frac{\sqrt{4 - x^2}}{x}$ for $0 < |\theta| \leq \frac{\pi}{2}$.

8.4. Trigonometric Substitutions

8.4.7 Let $x = 5\sin\theta$, so that $dx = 5\cos\theta\, d\theta$. Note that $\sqrt{25-x^2} = 5\cos\theta$. Then $\displaystyle\int_0^{5/2} \frac{1}{\sqrt{25-x^2}}\, dx = \int_0^{\pi/6} \frac{5\cos\theta}{5\cos\theta}\, d\theta = \frac{\pi}{6}$.

Checking without using a trigonometric substitution:
$$\int_0^{5/2} \frac{dx}{\sqrt{25-x^2}} = \arcsin\left(\frac{x}{5}\right)\Big|_0^{5/2} = \left(\frac{\pi}{6} - 0\right) = \frac{\pi}{6}.$$

8.4.9 Let $x = 10\sin\theta$ so that $dx = 10\cos\theta\, d\theta$. Note that $\sqrt{100-x^2} = 10\cos\theta$. Then $\displaystyle\int_5^{5\sqrt{3}} \sqrt{100-x^2}\, dx =$
$$100\int_{\pi/6}^{\pi/3} \cos^2\theta\, d\theta = 50\int_{\pi/6}^{\pi/3}(1+\cos 2\theta)\, d\theta = 50\left(\theta + \frac{\sin 2\theta}{2}\right)\Big|_{\pi/6}^{\pi/3} = 50\left(\frac{\pi}{3} + \frac{\sqrt{3}}{4} - \left(\frac{\pi}{6} + \frac{\sqrt{3}}{4}\right)\right) = \frac{25\pi}{3}.$$

8.4.11 Let $x = \sin\theta$ so that $dx = \cos\theta\, d\theta$. Note that $\sqrt{1-x^2} = \sqrt{1-\sin^2\theta} = \cos\theta$. Substituting gives
$$\int_{\pi/6}^{\pi/3} \sin^2\theta\, d\theta = \int_{\pi/6}^{\pi/3} \frac{1-\cos 2\theta}{2}\, d\theta = \frac{\theta}{2}\Big|_{\pi/6}^{\pi/3} - \frac{\sin 2\theta}{4}\Big|_{\pi/6}^{\pi/3} = \frac{\pi}{12}.$$

8.4.13 Let $x = 4\sin\theta$ so that $dx = 4\cos\theta\, d\theta$. Note that $\sqrt{16-x^2} = 4\cos\theta$. Thus,
$$\int \frac{1}{\sqrt{16-x^2}}\, dx = \int \frac{4\cos\theta}{4\cos\theta}\, d\theta = \theta + C = \sin^{-1}\left(\frac{x}{4}\right) + C.$$

8.4.15 Let $x = 3\tan\theta$ so that $dx = 3\sec^2\theta\, d\theta$. Note that $\sqrt{x^2+9} = \sqrt{9(\tan^2\theta+1)} = 3\sec\theta$. Substituting gives
$$\int \frac{1}{9}\cot\theta\csc\theta\, d\theta = -\frac{1}{9}\csc\theta + C = -\frac{1}{9}\csc(\tan^{-1}(x/3)) + C = -\frac{\sqrt{x^2+9}}{9x} + C.$$

8.4.17 Let $x = 2\tan\theta$ so that $dx = 2\sec^2\theta\, d\theta$. Note that $x^2 + 4 = 4\tan^2\theta + 4 = 4(\tan^2\theta + 1) = 4\sec^2\theta$. Then
$$\int_0^1 \frac{x^2}{x^2+4}\, dx = \int_0^{\pi/4} \frac{4\tan^2\theta}{4\sec^2\theta} 2\sec^2\theta\, d\theta = 2\int_0^{\pi/4} \tan^2\theta\, d\theta$$
$$= 2\int_0^{\pi/4} (\sec^2\theta - 1)\, d\theta = 2(\tan\theta - \theta)\Big|_0^{\pi/4} = 2\left(1 - \frac{\pi}{4}\right) = 2 - \frac{\pi}{2}.$$

8.4.19 Let $x = 9\sec\theta$ with $\theta \in (0, \pi/2)$. Then $dx = 9\sec\theta\tan\theta\, d\theta$ and $\sqrt{x^2-81} = 9\tan\theta$. Then
$$\int \frac{1}{\sqrt{x^2-81}}\, dx = \int \frac{9\sec\theta\tan\theta}{9\tan\theta}\, d\theta = \int \sec\theta\, d\theta = \ln|\sec\theta + \tan\theta| + C = \ln\left|\frac{x}{9} + \frac{\sqrt{x^2-81}}{9}\right| + C.$$
Note that because $x > 9$, the absolute value signs are unnecessary, and the final result can be written as $\ln(\sqrt{x^2-81} + x) + C$.

8.4.21 Let $x = 8\sin\theta$ so that $dx = 8\cos\theta\, d\theta$ and $\sqrt{64-x^2} = 8\cos\theta$. Then,
$$\int \sqrt{64-x^2}\, dx = \int 64\cos^2\theta\, d\theta = 32\int (1+\cos 2\theta)\, d\theta = 32\theta + 16\sin 2\theta + C$$
$$= 32\theta + 32\sin\theta\cos\theta + C = 32\sin^{-1}\left(\frac{x}{8}\right) + \frac{x\sqrt{64-x^2}}{2} + C.$$

8.4.23 Let $x = 5\sin\theta$ so that $dx = 5\cos\theta\, d\theta$. Note that $\sqrt{25-x^2} = \sqrt{25-25\sin^2\theta} = 5\cos\theta$. Substituting gives
$$\int \frac{5\cos\theta}{125\cos^3\theta}\, d\theta = \frac{1}{25}\int \sec^2\theta\, d\theta = \frac{1}{25}\tan\theta + C = \frac{x}{25\sqrt{25-x^2}} + C.$$

8.4.25 Let $x = 3\sin\theta$ so that $dx = 3\cos\theta\, d\theta$ and $\sqrt{9-x^2} = 3\cos\theta$. Then

$$\int \frac{\sqrt{9-x^2}}{x}\, dx = \int \frac{3\cos\theta \cdot 3\cos\theta}{3\sin\theta}\, d\theta = 3\int \frac{1-\sin^2\theta}{\sin\theta}\, d\theta$$

$$= 3\left(\int \csc\theta\, d\theta - \int \sin\theta\, d\theta\right)\, d\theta = 3\left(-\ln|\csc\theta + \cot\theta| + \cos\theta\right)$$

$$= -3\ln\left|\frac{3}{x} + \frac{\sqrt{9-x^2}}{x}\right| + \sqrt{9-x^2} + C.$$

8.4.27 Let $x = \frac{1}{3}\tan\theta$ so that $dx = \frac{1}{3}\sec^2\theta\, d\theta$. Note that $\sqrt{9x^2+1} = \sec\theta$. Thus $\int_0^{1/3} \frac{1}{(9x^2+1)^{3/2}}\, dx = \int_0^{\pi/4} \frac{\frac{1}{3}\sec^2\theta}{\sec^3\theta}\, d\theta = \frac{1}{3}\int_0^{\pi/4} \cos\theta\, d\theta = \frac{1}{3}\sin\theta\bigg|_0^{\pi/4} = \frac{\sqrt{2}}{6}.$

8.4.29 Let $x = 2\tan\theta$. Then $dx = 2\sec^2\theta\, d\theta$ and $4 + x^2 = 4(\sec^2\theta)$. Then

$$\int \frac{dx}{(4+x^2)^2} = \int \frac{2\sec^2\theta}{2^4 \sec^4\theta}\, d\theta = \frac{1}{8}\int \cos^2\theta\, d\theta = \frac{1}{16}\int (1+\cos 2\theta)\, d\theta$$

$$= \frac{1}{16}\left(\theta + \frac{\sin 2\theta}{2}\right) + C = \frac{1}{16}(\theta + \sin\theta\cos\theta) + C = \frac{1}{16}\left(\tan^{-1}\frac{x}{2} + \frac{2x}{x^2+4}\right) + C.$$

8.4.31 Let $x = 4\sin\theta$ so that $dx = 4\cos\theta\, d\theta$. Note that $\sqrt{16-x^2} = 4\cos\theta$. Then $\int \frac{x^2}{\sqrt{16-x^2}}\, dx = \int \frac{16\sin^2\theta \cdot 4\cos\theta}{4\cos\theta}\, d\theta = 16\int \sin^2\theta\, d\theta = 8\int (1-\cos 2\theta)\, d\theta = 8\left(\theta - \frac{\sin 2\theta}{2}\right) + C = 8\theta - 8\sin\theta\cos\theta + C = 8\sin^{-1}\left(\frac{x}{4}\right) - \frac{x\sqrt{16-x^2}}{2} + C.$

8.4.33 Let $x = 3\sec\theta$ where $\theta \in (0, \pi/2)$. Then $dx = 3\sec\theta\tan\theta$ and $\sqrt{x^2-9} = 3\tan\theta$. Thus we have $\int \frac{\sqrt{x^2-9}}{x}\, dx = \int \frac{3\sec\theta\tan\theta \cdot 3\tan\theta}{3\sec\theta}\, d\theta = 3\int \tan^2\theta\, d\theta = 3\int \sec^2\theta - 1\, d\theta = 3(\tan\theta - \theta) + C = \sqrt{x^2-9} - 3\tan^{-1}\left(\frac{\sqrt{x^2-9}}{3}\right) + C = \sqrt{x^2-9} - 3\sec^{-1}(x/3) + C.$

8.4.35 Let $x = \sec\theta$ where $\theta \in (0, \pi/2)$. Then $dx = \sec\theta\tan\theta\, d\theta$ and $\sqrt{x^2-1} = \tan\theta$. Then

$$\int \frac{1}{x(x^2-1)^{3/2}}\, dx = \int \frac{\sec\theta\tan\theta}{\sec\theta\tan^3\theta}\, d\theta = \int \cot^2\theta\, d\theta$$

$$= \int \csc^2\theta - 1\, d\theta = -\cot\theta - \theta + C = -\frac{1}{\sqrt{x^2-1}} - \sec^{-1}x + C.$$

8.4.37 Let $x = \tan\theta$ so that $dx = \sec^2\theta\, d\theta$ and $\sqrt{1+x^2} = \sec\theta$. Substituting gives

$$\int_{\pi/6}^{\pi/4} \cot\theta\csc\theta\, d\theta = (-\csc\theta)\bigg|_{\pi/6}^{\pi/4} = -(\sqrt{2} - 2) = 2 - \sqrt{2}.$$

8.4.39 Let $x = 10\sin\theta$ so that $dx = 10\cos\theta\, d\theta$. Note that $\sqrt{100-x^2} = 10\cos\theta$. Thus,

$$\int \frac{x^2}{(100-x^2)^{3/2}}\, dx = \int \frac{100\sin^2\theta \cdot 10\cos\theta}{1000\cos^3\theta}\, d\theta = \int \tan^2\theta\, d\theta$$

$$= \int (\sec^2\theta - 1)\, d\theta = \tan\theta - \theta + C = \frac{x}{\sqrt{100-x^2}} - \sin^{-1}(x/10) + C.$$

8.4. Trigonometric Substitutions

8.4.41 Let $x = \dfrac{\tan\theta}{2}$ so that $dx = \dfrac{\sec^2\theta}{2}\,d\theta$ and $\sqrt{1+4x^2} = \sec\theta$. Then

$$\int \frac{1}{(1+4x^2)^{3/2}}\,dx = \frac{1}{2}\int \frac{\sec^2\theta}{\sec^3\theta}\,d\theta = \frac{1}{2}\int \cos\theta\,d\theta = \frac{\sin\theta}{2} + C = \frac{x}{\sqrt{1+4x^2}} + C.$$

8.4.43 Let $x = 4\tan\theta$ so that $dx = 4\sec^2\theta\,d\theta$. Note that $\sqrt{x^2+16} = 4\sec\theta$. Thus, $\displaystyle\int_0^{4/\sqrt{3}} \frac{1}{\sqrt{x^2+16}}\,dx =$
$\displaystyle\int_0^{\pi/6} \frac{4\sec^2\theta}{4\sec\theta}\,d\theta = \int_0^{\pi/6} \sec\theta\,d\theta = \ln|\sec\theta + \tan\theta|\Big|_0^{\pi/6} = \ln\left(\frac{2}{\sqrt{3}} + \frac{1}{\sqrt{3}}\right) - \ln 1 = \ln\frac{3}{\sqrt{3}} = \ln 3 - \frac{1}{2}\ln 3 = \frac{1}{2}\ln 3.$

8.4.45 Let $x = 9\sin\theta$ so that $dx = 9\cos\theta\,d\theta$. Note that $81 - x^2 = 81\cos^2\theta$. Thus,

$$\int \frac{x^3}{(81-x^2)^2}\,dx = \int \frac{9^3 \sin^3\theta \cdot 9\cos\theta}{9^4 \cos^4\theta}\,d\theta = \int \tan^3\theta\,d\theta$$

$$= \int (\tan\theta)(\sec^2\theta - 1)\,d\theta = \int \sec^2\theta\tan\theta\,d\theta - \int \tan\theta\,d\theta$$

$$= \frac{\sec^2\theta}{2} + \ln|\cos\theta| + C = \frac{81}{2(81-x^2)} + \ln\left|\frac{\sqrt{81-x^2}}{9}\right| + C.$$

This can be written as $\dfrac{81}{2(81-x^2)} + \ln\sqrt{81-x^2} + C$.

8.4.47 Let $x = 2\sec\theta$ so that $dx = 2\sec\theta\tan\theta\,d\theta$ and $x^2 - 4 = 4\tan^2\theta$. Thus,
$\displaystyle\int_{4/\sqrt{3}}^4 \frac{1}{x^2(x^2-4)}\,dx = \int_{\pi/6}^{\pi/3} \frac{2\sec\theta\tan\theta}{4\sec^2\theta \cdot 4\tan^2\theta}\,d\theta = \frac{1}{8}\int_{\pi/6}^{\pi/3} \frac{\cos^2\theta}{\sin\theta}\,d\theta = \frac{1}{8}\int_{\pi/6}^{\pi/3} \frac{1-\sin^2\theta}{\sin\theta}\,d\theta =$
$\dfrac{1}{8}\displaystyle\int_{\pi/6}^{\pi/3} \csc\theta - \sin\theta\,d\theta = \frac{1}{8}\left(-\ln|\csc\theta + \cot\theta| + \cos\theta\right)\Big|_{\pi/6}^{\pi/3} = \frac{1}{8}\left(-\ln(\sqrt{3}(2-\sqrt{3})) + \frac{1-\sqrt{3}}{2}\right) =$
$\dfrac{1}{16}\left(1 - \sqrt{3} - \ln(21 - 12\sqrt{3})\right)$.

8.4.49 Let $x = \tan\theta$ so that $dx = \sec^2\theta\,d\theta$. Note that $\sqrt{x^2+1} = \sqrt{\tan^2\theta + 1} = \sqrt{\sec^2\theta} = \sec\theta$. Substituting gives $\displaystyle\int_0^{\pi/6} \sec^3\theta\,d\theta$. Recall from section 8.2 number 48 that $\displaystyle\int \sec^3\theta\,d\theta = \frac{1}{2}\sec\theta\tan\theta + \frac{1}{2}\ln|\sec\theta + \tan\theta|$. Thus the original integral is equal to $\left(\dfrac{1}{2}\sec\theta\tan\theta + \dfrac{1}{2}\ln|\sec\theta + \tan\theta|\right)\Big|_0^{\pi/6} = \dfrac{1 \cdot 2 \cdot 1}{2 \cdot \sqrt{3} \cdot \sqrt{3}} + \dfrac{1}{2}\ln\left(\dfrac{2}{\sqrt{3}} + \dfrac{1}{\sqrt{3}}\right) = \dfrac{1}{3} + \dfrac{\ln 3}{4}.$

8.4.51 Let $x = 2\tan\theta$ so that $dx = 2\sec^2\theta\,d\theta$. Note that $\sqrt{4+x^2} = 2\sec\theta$. Then $\displaystyle\int \frac{x^2}{\sqrt{4+x^2}}\,dx =$
$\displaystyle\int \frac{4\tan^2\theta \cdot 2\sec^2\theta}{2\sec\theta}\,d\theta = 4\int \tan^2\theta \sec\theta\,d\theta = 4\int (\sec^2\theta - 1)\sec\theta\,d\theta = 4\left(\int \sec^3\theta\,d\theta - \int \sec\theta\,d\theta\right) =$
$4\left(\dfrac{1}{2}\left(\sec\theta\tan\theta + \int \sec\theta\,d\theta\right) - \int \sec\theta\,d\theta\right) = 2\sec\theta\tan\theta - 2\int \sec\theta\,d\theta = 2\sec\theta\tan\theta - 2\ln|\sec\theta + \tan\theta| + C = \dfrac{x\sqrt{4+x^2}}{2} - 2\ln\left|\dfrac{\sqrt{4+x^2}}{2} + \dfrac{x}{2}\right| + C$. This can be written as $\dfrac{x\sqrt{4+x^2}}{2} - 2\ln(x + \sqrt{4+x^2}) + C.$

8.4.53 Let $x = \dfrac{5}{3}\sec\theta$ where $\theta \in [0, \pi/2)$. Then $dx = \dfrac{5}{3}\sec\theta\tan\theta\,d\theta$ and $\sqrt{9x^2 - 25} = 5\tan\theta$. Thus,
$\displaystyle\int \frac{\sqrt{9x^2-25}}{x^3}\,dx = \int \frac{5\tan\theta \cdot \frac{5}{3} \cdot \sec\theta\tan\theta}{\frac{125}{27}\sec^3\theta}\,d\theta = \frac{9}{5}\int \frac{\tan^2\theta}{\sec^2\theta}\,d\theta = \frac{9}{5}\int \frac{\sec^2\theta - 1}{\sec^2\theta}\,d\theta = \frac{9}{5}\int 1 - \cos^2\theta\,d\theta =$
$\dfrac{9}{5}\displaystyle\int \sin^2\theta\,d\theta = \frac{9}{10}\int (1 - \cos 2\theta)\,d\theta = \frac{9\theta}{10} - \frac{9\sin 2\theta}{20} + C = \frac{9\theta}{10} - \frac{9\sin\theta\cos\theta}{10} = \frac{9\cos^{-1}(5/3x)}{10} - \frac{\sqrt{9x^2-25}}{2x^2} + C.$

8.4.55 Let $x = 10\sec\theta$ where $\theta \in (0, \pi/2)$. Then $dx = 10\sec\theta\tan\theta\,d\theta$ and $\sqrt{x^2 - 100} = 10\tan\theta$. Thus,

$$\int \frac{1}{x^3\sqrt{x^2-100}}\,dx = \int \frac{10\sec\theta\tan\theta}{10^3\sec^3\theta \cdot 10\tan\theta}\,d\theta = \frac{1}{1000}\int \cos^2\theta\,d\theta$$

$$= \frac{1}{2000}\int (1 + \cos 2\theta)\,d\theta = \frac{\theta}{2000} + \frac{\sin 2\theta}{4000} + C = \frac{\theta}{2000} + \frac{\sin\theta\cos\theta}{2000} + C$$

$$= \frac{\sec^{-1}(x/10)}{2000} + \frac{\sqrt{x^2-100}}{200x^2} + C.$$

8.4.57

a. False. In fact, we would have $\csc\theta = \dfrac{\sqrt{x^2+16}}{x}$.

b. True. Almost every number in the interval $[1, 2]$ is not in the domain of $\sqrt{1-x^2}$, so this integral isn't defined.

c. False. It does represent a finite real number, because $\sqrt{x^2-1}$ is continuous on the interval $[1, 2]$.

d. False. It can be so evaluated. The integral is equivalent to $\displaystyle\int \frac{1}{(x+2)^2+5}\,dx$, and this can be evaluated by the substitution $x + 2 = \sqrt{5}\tan\theta$.

8.4.59

a. Recall that the area of a circular sector subtended by an angle θ is given by $\dfrac{\theta r^2}{2}$. So the area of the cap is this area minus the area of the isosceles triangle with two sides of length r and angle between them θ. So

$$A_{\text{cap}} = A_{\text{sector}} - A_{\text{triangle}} = \frac{\theta r^2}{2} - \frac{r^2\sin\theta}{2} = \frac{r^2}{2}(\theta - \sin\theta).$$

b. For a cap we have $0 \le \theta \le \pi$ so $0 \le \theta/2 \le \pi/2$. By symmetry, $\dfrac{A_{\text{cap}}}{2} = \displaystyle\int_{r\cos\theta/2}^{r} \sqrt{r^2-x^2}\,dx$. Let $x = r\cos\alpha/2$ so that $dx = -\frac{r}{2}\sin\alpha/2\,d\alpha$. Then we have

$$\frac{A_{\text{cap}}}{2} = \int_\theta^0 r\sin(\alpha/2) \cdot -\frac{r}{2}\sin(\alpha/2)\,d\alpha$$

$$= \frac{r^2}{2}\int_0^\theta \sin^2(\alpha/2)\,d\alpha = \frac{r^2}{4}\int_0^\theta (1-\cos\alpha)\,d\alpha = \frac{r^2}{4}(\alpha - \sin\alpha)\Big|_0^\theta = \frac{r^2}{4}(\theta - \sin\theta).$$

Thus $A_{\text{cap}} = \dfrac{r^2}{2}(\theta - \sin\theta)$.

8.4.61 $\displaystyle\int \frac{1}{\sqrt{3-2x-x^2}}\,dx = \int \frac{1}{\sqrt{4-(x+1)^2}}\,dx = \int \frac{1}{\sqrt{4-u^2}}\,du$ where $u = x+1$. Then let $u = 2\sin\theta$ so that $du = 2\cos\theta\,d\theta$. We have $\displaystyle\int \frac{2\cos\theta}{2\cos\theta}\,d\theta = \theta + C = \sin^{-1}\left(\frac{x+1}{2}\right) + C$.

8.4.63 Note that the given integral can be written $\displaystyle\int \frac{1}{(x+3)^2+9}\,dx = \int \frac{1}{u^2+9}\,du$ where $u = x+3$. Now let $u = 3\tan\theta$ so that $du = 3\sec^2\theta\,d\theta$ and $u^2 + 9 = 9\sec^2\theta$. Thus we have

$$\int \frac{3\sec^2\theta}{9\sec^2\theta}\,d\theta = \frac{\theta}{3} + C = \frac{\tan^{-1}((x+3)/3)}{3} + C.$$

8.4.65 $\int_{1/2}^{(\sqrt{2}+3)/2\sqrt{2}} \frac{1}{8x^2 - 8x + 11} \, dx = \int_{1/2}^{(\sqrt{2}+3)/2\sqrt{2}} \frac{1}{8(x-1/2)^2 + 9} \, dx$. Let $u = x - 1/2$, so that our integral becomes $\int_0^{3/2\sqrt{2}} \frac{1}{8u^2 + 9} \, du$. Now let $u = \frac{3}{\sqrt{8}} \tan \theta$ so that $du = \frac{3}{\sqrt{8}} \sec^2 \theta \, d\theta$. Substituting gives

$$\int_0^{\pi/4} \frac{\frac{3}{\sqrt{8}} \sec^2 \theta}{9 \sec^2 \theta} \, d\theta = \frac{1}{6\sqrt{2}} \int d\theta = \frac{1}{6\sqrt{2}} \theta \Big|_0^{\pi/4} = \frac{\pi\sqrt{2}}{48}.$$

8.4.67 Note that the given integral can be written as $\int \frac{(x-4)^2}{(25 - (x-4)^2)^{3/2}} \, dx$. Let $u = x - 4$, and note that we have $\int \frac{u^2}{(25 - u^2)^{3/2}} \, du$. Now let $u = 5 \sin \theta$ so that $du = 5 \cos \theta \, d\theta$, and note that $\sqrt{25 - u^2} = 5 \cos \theta$. Thus we have

$$\int \frac{25 \sin^2 \theta \cdot 5 \cos \theta}{5^3 \cos^3 \theta} \, d\theta = \int \tan^2 \theta \, d\theta = \int \sec^2 \theta - 1 \, d\theta = \tan \theta - \theta + C = \frac{x-4}{\sqrt{25 - (x-4)^2}} - \sin^{-1}\left(\frac{x-4}{5}\right) + C.$$

8.4.69 $\int_{2+\sqrt{2}}^4 \frac{1}{\sqrt{(x-1)(x-3)}} \, dx = \int_{2+\sqrt{2}}^4 \frac{1}{\sqrt{(x-2)^2 - 1}} \, dx = \int_{\sqrt{2}}^2 \frac{1}{\sqrt{u^2 - 1}} \, du$, where $u = x - 2$. Now let $u = \sec \theta$, so that $du = \sec \theta \tan \theta \, d\theta$. Then $\int_{\pi/4}^{\pi/3} \frac{\sec \theta \tan \theta}{\tan \theta} \, d\theta = \int_{\pi/4}^{\pi/3} \sec \theta \, d\theta = \ln(\sec \theta + \tan \theta)\Big|_{\pi/4}^{\pi/3} = \ln(2 + \sqrt{3}) - \ln(\sqrt{2} + 1) = \ln\left(\frac{2+\sqrt{3}}{\sqrt{2}+1}\right) = \ln((2+\sqrt{3})(\sqrt{2} - 1))$.

8.4.71

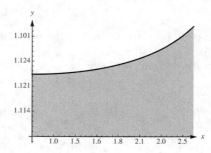

The area is given by $\int_0^{3/2} \frac{1}{(9-x^2)^2} \, dx$. Let $x = 3 \sin \theta$ so that $dx = 3 \cos \theta \, d\theta$. Then we have

$$\int_0^{\pi/6} \frac{3 \cos \theta}{9^2 \cos^4 \theta} \, d\theta = \frac{1}{27} \int_0^{\pi/6} \sec^3 \theta \, d\theta = \frac{1}{54} \left(\sec \theta \tan \theta + \ln|\sec \theta + \tan \theta| \right) \Big|_0^{\pi/6}$$
$$= \frac{1}{54} \left(\frac{2}{\sqrt{3}} \cdot \frac{1}{\sqrt{3}} + \ln\left(\frac{2}{\sqrt{3}} + \frac{1}{\sqrt{3}}\right) \right) = \frac{1}{81} + \frac{\ln(\sqrt{3})}{54}.$$

8.4.73

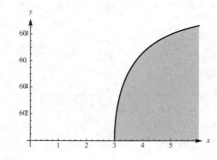

Let $x = 3\sec\theta$ for $x \in (0, \pi/2)$. Then $dx = 3\sec\theta\tan\theta\,d\theta$, and note that $|\tan\theta| = \tan\theta$. We have

$$\int_3^6 \frac{\sqrt{x^2-9}}{x}\,dx = \int_0^{\pi/3} \frac{3\sec\theta\tan\theta(3\tan\theta)}{3\sec\theta}\,d\theta =$$

8.4.75

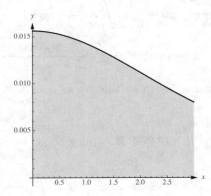

The area bounded by the curve and the axis on $[0, 3]$ is given by $\int_0^3 \frac{1}{(16+x^2)^{3/2}}\,dx$. Let $x = 4\tan\theta$ so that $dx = 4\sec^2\theta\,d\theta$. Substituting yields

$$\int_0^{\tan^{-1}(3/4)} \frac{4\sec^2\theta}{4^3\sec^3\theta}\,d\theta = \frac{1}{16}\int_0^{\tan^{-1}(3/4)} \cos\theta\,d\theta = \frac{\sin\theta}{16}\Big|_0^{\tan^{-1}(3/4)} = \frac{3}{80}.$$

8.4.77 Because $y = ax^2$, we have $1 + \left(\dfrac{dy}{dx}\right)^2 = 1 + 4a^2x^2$, so the arc length is given by $\int_0^{10} \sqrt{1+4a^2x^2}\,dx$. Let $x = \dfrac{1}{2a}\tan\theta$ so that $dx = \dfrac{1}{2a}\sec^2\theta\,d\theta$. Then we have

$$\frac{1}{2a}\int_0^{\tan^{-1}(20a)} \sec^2\theta\sec\theta\,d\theta = \frac{1}{2a}\int_0^{\tan^{-1}(20a)} \sec^3\theta\,d\theta = \frac{1}{4a}\left(\sec\theta\tan\theta + \ln|\sec\theta + \tan\theta|\right)\Big|_0^{\tan^{-1}(20a)}$$
$$= \frac{1}{4a}\left(\sqrt{1+400a^2}(20a) + \ln(\sqrt{1+400a^2} + 20a)\right).$$

8.4.79 If $1 < x = \sec\theta$ then we need $\theta \in (0, \pi/2)$. Alternatively, if $-1 > x = \sec\theta$, then we need $\theta \in (\pi/2, \pi)$. So, in the former, $\displaystyle\int \frac{1}{x\sqrt{x^2-1}}\,dx = \int \frac{\sec\theta\tan\theta}{\sec\theta\tan\theta}\,d\theta = \theta + C = \sec^{-1}x + C$. In the latter case, $\displaystyle\int \frac{1}{x\sqrt{x^2-1}}\,dx = \int \frac{\sec\theta}{\sec\theta(-\tan\theta)}\,d\theta = -\theta + C = -\sec^{-1}x + C$.

8.4.81

a. $E_x(a) = \dfrac{kQa}{2L}\displaystyle\int_{-L}^L \frac{dy}{(a^2+y^2)^{3/2}}$. Let $y = a\tan\theta$ so that $dy = a\sec^2\theta\,d\theta$. Then note that

$$\int_{-L}^L \frac{dy}{(a^2+y^2)^{3/2}} = 2\int_0^L \frac{dy}{(a^2+y^2)^{3/2}} = 2\int_0^{\tan^{-1}(L/a)} \frac{a\sec^2\theta}{a^3\sec^3\theta}\,d\theta = \frac{2}{a^2}\int_0^{\tan^{-1}(L/a)} \cos\theta\,d\theta$$
$$= \frac{2}{a^2}\sin\theta\Big|_0^{\tan^{-1}(L/a)} = \frac{2}{a^2}\cdot\frac{L}{\sqrt{a^2+L^2}} = \frac{2L}{a^2\sqrt{a^2+L^2}}.$$

Thus, $E_x(a) = \dfrac{kQ}{a\sqrt{a^2+L^2}}$.

b. Set $\rho = Q/(2L)$. Then because $\lim\limits_{L\to\infty} \dfrac{2L}{a^2\sqrt{a^2+L^2}} = \dfrac{2}{a^2}$, we have

$$E_x(a,0) \approx \frac{kQa}{2L} \lim_{L\to\infty} \int_{-L}^{L} \frac{dy}{(a^2+y^2)^{3/2}} = \frac{kQa}{2L} \lim_{L\to\infty} \frac{2L}{a^2\sqrt{a^2+L^2}} = \frac{kQa}{2L}\left(\frac{2}{a^2}\right) = \frac{2kQ}{2aL} = \frac{2k\rho}{a}.$$

8.4.83

a. The area of sector OAB is given by the formula $\dfrac{\theta}{2}a^2$ where $\sin\theta = x/a$. The area of triangle OBC is $\dfrac{x}{2}\sqrt{a^2-x^2}$. Thus,

$$F(x) = \frac{a^2 \sin^{-1}(x/a)}{2} + \frac{x\sqrt{a^2-x^2}}{2}.$$

b. By the first fundamental theorem of calculus, $F(x)$ is an antiderivative of $\sqrt{a^2-x^2}$, because $F(x) = \int_0^x \sqrt{a^2-t^2}\,dt$. Thus, any other antiderivative differs from this by a constant, so

$$\int \sqrt{a^2-x^2}\,dx = \frac{a^2\sin^{-1}(x/a)}{2} + \frac{x\sqrt{a^2-x^2}}{2} + C.$$

8.4.85

a. Because $t \in [0,\pi]$ so that $\sin t \geq 0$, we have

$$\int_a^b \sqrt{\frac{1-\cos t}{g(\cos a - \cos t)}}\,dt = \int_a^b \sqrt{\frac{(1-\cos t)(1+\cos t)}{g(1+\cos t)(\cos a - \cos t)}}\,dt = \int_a^b \sin t \sqrt{\frac{1}{g(1+\cos t)(\cos a - \cos t)}}\,dt.$$

Let $u = \cos t$ so that $du = -\sin t\,dt$. Then the given integral is equal to

$$-\frac{1}{\sqrt{g}}\int_{\cos a}^{\cos b} \sqrt{\frac{1}{(1+u)(\cos a - u)}}\,du.$$

Now we complete the square:

$$(1+u)(\cos a - u) = \cos a + (\cos a - 1)u - u^2$$

$$= -\left(u^2 - (\cos a - 1)u + \left(\frac{\cos a - 1}{2}\right)^2 - \left(\frac{\cos a - 1}{2}\right)^2\right) + \cos a$$

$$= \cos a + \left(\frac{\cos a - 1}{2}\right)^2 - \left(u - \frac{\cos a - 1}{2}\right)^2 = \left(\frac{\cos a + 1}{2}\right)^2 - \left(u - \frac{\cos a - 1}{2}\right)^2.$$

Thus, setting $v = u - \dfrac{\cos a - 1}{2}$ we have that the original integral is equal to

$$-\frac{1}{\sqrt{g}}\int_{(\cos a+1)/2}^{\cos b - \frac{\cos a - 1}{2}} \frac{1}{\sqrt{k^2 - v^2}}\,dv \text{ where } k = \frac{(\cos a + 1)}{2}.$$

Now, $\displaystyle\int \frac{1}{\sqrt{k^2-v^2}}\,dv = \int \frac{k\cos\theta}{k\cos\theta}\,d\theta = \theta + C = \sin^{-1}(v/k) + C$ where $v = k\sin\theta$.

Therefore, the original integral is equal to

$$-\frac{1}{\sqrt{g}} \sin^{-1}\left(\frac{2v}{\cos a + 1}\right)\Big|_{(\cos a + 1)/2}^{\cos b - (\cos a - 1)/2} = \frac{1}{\sqrt{g}}\left(\sin^{-1}\left(\frac{\cos a + 1}{\cos a + 1}\right) - \sin^{-1}\left(\frac{2\cos b - \cos a + 1}{\cos a + 1}\right)\right)$$

$$= \frac{1}{\sqrt{g}}\left(\frac{\pi}{2} - \sin^{-1}\left(\frac{2\cos b - \cos a + 1}{\cos a + 1}\right)\right).$$

Copyright © 2019 Pearson Education, Inc.

b. Letting $b = \pi$, we have that the integral is equal to
$$\frac{1}{\sqrt{g}}\left(\frac{\pi}{2} - \sin^{-1}\left(\frac{-2 - \cos a + 1}{\cos a + 1}\right)\right)$$
$$= \frac{1}{\sqrt{g}}\left(\frac{\pi}{2} - \sin^{-1}(-1)\right) = \frac{1}{\sqrt{g}}\left(\frac{\pi}{2} - \left(-\frac{\pi}{2}\right)\right) = \frac{\pi}{\sqrt{g}}.$$

8.5 Partial Fractions

8.5.1 Proper rational functions can be integrated using partial fraction decomposition.

8.5.3

a. $\dfrac{A}{x-3}$.

b. $\dfrac{A_1}{x-4}, \dfrac{A_2}{(x-4)^2}, \dfrac{A_3}{(x-4)^3}$.

c. $\dfrac{Ax+B}{x^2+2x+6}$.

8.5.5 $\dfrac{4x}{(x-4)(x-5)} = \dfrac{A}{x-4} + \dfrac{B}{x-5}$.

8.5.7 $\dfrac{x+3}{(x-5)^2} = \dfrac{A}{x-5} + \dfrac{B}{(x-5)^2}$.

8.5.9 $\dfrac{4}{x(x+1)(x-1)(x+2)(x-2)} = \dfrac{A}{x} + \dfrac{B}{x+1} + \dfrac{C}{x-1} + \dfrac{D}{x+2} + \dfrac{E}{x-2}$.

8.5.11 $\dfrac{1}{x(x^2+1)} = \dfrac{A}{x} + \dfrac{Bx+C}{x^2+1}$.

8.5.13 $\dfrac{x^4 + 12x^2}{(x-2)^2(x^2+x+2)^2} = \dfrac{A}{x-2} + \dfrac{B}{(x-2)^2} + \dfrac{Cx+D}{x^2+x+2} + \dfrac{Ex+F}{(x^2+x+2)^2}$.

8.5.15 $\dfrac{x}{(x-2)^2(x+2)^2(x^2+4)^2} = \dfrac{A}{x-2} + \dfrac{B}{(x-2)^2} + \dfrac{C}{x+2} + \dfrac{D}{(x+2)^2} + \dfrac{Ex+F}{x^2+4} + \dfrac{Gx+H}{(x^2+4)^2}$.

8.5.17 $\dfrac{5x-7}{x^2-3x+2} = \dfrac{5x-7}{(x-1)(x-2)} = \dfrac{A}{x-1} + \dfrac{B}{x-2}$. Thus, $A(x-2) + B(x-1) = 5x - 7$. Equating coefficients gives $A + B = 5$ and $-2A - B = -7$. Solving this system yields $A = 2$, $B = 3$. Thus,
$$\frac{5x-7}{x^2-3x+2} = \frac{2}{x-1} + \frac{3}{x-2}.$$

8.5.19 $\dfrac{6}{x^2-2x-8} = \dfrac{6}{(x-4)(x+2)} = \dfrac{A}{x-4} + \dfrac{B}{x+2}$. Thus, $6 = A(x+2) + B(x-4)$. Equating coefficients gives $A + B = 0$ and $2A - 4B = 6$. Solving this system yields $A = 1$ and $B = -1$. Thus,
$$\frac{6}{x^2-2x-8} = \frac{1}{x-4} - \frac{1}{x+2}.$$

8.5.21 By long division, $\dfrac{2x^2+5x+6}{x^2+3x+2} = 2 + \dfrac{-x+2}{(x+1)(x+2)}$. We use a partial fraction decomposition on $\dfrac{-x+2}{(x+1)(x+2)} = \dfrac{A}{x+1} + \dfrac{B}{x+2}$. Therefore $A(x+2) + B(x+1) = -x + 2$, so $A + B = -1$ and $2A + B = 2$. So $A = 3$ and $B = -4$. Thus
$$\frac{2x^2+5x+6}{x^2+3x+2} = 2 + \frac{3}{x+1} - \frac{4}{x+2}.$$

8.5. Partial Fractions

8.5.23 If we write $\dfrac{3}{(x-1)(x+2)} = \dfrac{A}{x-1} + \dfrac{B}{x+2}$, we have $3 = A(x+2) + B(x-1)$. Letting $x = -2$ yields $B = -1$ and letting $x = 1$ yields $A = 1$. Thus, the original integral is equal to

$$\int \left(\dfrac{1}{x-1} - \dfrac{1}{x+2} \right) dx = \ln|x-1| - \ln|x+2| + C = \ln\left|\dfrac{x-1}{x+2}\right| + C.$$

8.5.25 If we write $\dfrac{6}{x^2-1} = \dfrac{6}{(x-1)(x+1)} = \dfrac{A}{x-1} + \dfrac{B}{x+1}$, then we have $6 = A(x+1) + B(x-1)$. Letting $x = -1$ yields $B = -3$ and letting $x = 1$ yields $A = 3$. Thus, the original integral is equal to

$$\int \left(\dfrac{3}{x-1} - \dfrac{3}{x+1} \right) dx = 3\left(\ln|x-1| - \ln|x+1| \right) + C = 3\ln\left|\dfrac{x-1}{x+1}\right| + C.$$

8.5.27 $\dfrac{8x-5}{(x-1)(3x-2)} = \dfrac{A}{x-1} + \dfrac{B}{3x-2}$. Thus $8x - 5 = A(3x-2) + B(x-1) = (3A+B)x - 2A - B$. So $3A + B = 8$ and $-2A - B = -5$. Solving gives $A = 3$ and $B = -1$. We have

$$\int \dfrac{8x-5}{3x^2 - 5x + 2} dx = \int \left(\dfrac{3}{x-1} - \dfrac{1}{3x-2} \right) dx = 3\ln|x-1| - \dfrac{1}{3}\ln|3x-2| + C.$$

8.5.29 If we write $\dfrac{5x}{x^2 - x - 6} = \dfrac{A}{x-3} + \dfrac{B}{x+2}$, then we have $5x = A(x+2) + B(x-3)$. Letting $x = -2$ yields $B = 2$ and letting $x = 3$ yields $A = 3$. Thus the original integral is equal to

$$\int_{-1}^{2} \left(\dfrac{3}{x-3} + \dfrac{2}{x+2} \right) dx = (3\ln|x-3| + 2\ln|x+2|) \Big|_{-1}^{2} = \ln(16) - \ln(64) = -\ln 4.$$

8.5.31 Let $\dfrac{6x^2}{x^4 - 5x^2 + 4} = \dfrac{6x^2}{(x-2)(x+2)(x-1)(x+1)} = \dfrac{A}{x-2} + \dfrac{B}{x+2} + \dfrac{C}{x-1} + \dfrac{D}{x+1}$. Then $6x^2 = A(x+2)(x-1)(x+1) + B(x-2)(x-1)(x+1) + C(x-2)(x+2)(x+1) + D(x-2)(x+2)(x-1)$.

Letting $x = 2$ gives $A = 2$, letting $x = -2$ gives $B = -2$, letting $x = 1$ gives $C = -1$, and letting $x = -1$ gives $D = 1$. Thus, the original integral is equal to

$$\int \left(\dfrac{2}{x-2} - \dfrac{2}{x+2} - \dfrac{1}{x-1} + \dfrac{1}{x+1} \right) dx = \ln\left|\dfrac{(x-2)^2(x+1)}{(x+2)^2(x-1)}\right| + C.$$

8.5.33 After performing long division, we have that the original integrand is equal to $3 + \dfrac{13x-12}{(x-1)(x-2)}$, and if we write $\dfrac{13x-12}{(x-1)(x-2)} = \dfrac{A}{x-1} + \dfrac{B}{x-2}$, then $13x - 12 = A(x-2) + B(x-1)$. Letting $x = 1$ yields $A = -1$ and letting $x = 2$ yields $B = 14$. Thus the original integral is equal to

$$3x - \int \dfrac{1}{x-1} dx + 14 \int \dfrac{1}{x-2} dx = 3x - \ln|x-1| + 14\ln|x-2| + C.$$

8.5.35 Let $\dfrac{x^2 + 12x - 4}{x(x-2)(x+2)} = \dfrac{A}{x} + \dfrac{B}{x-2} + \dfrac{C}{x+2}$. Then $x^2 + 12x - 4 = A(x-2)(x+2) + Bx(x+2) + Cx(x-2)$. Letting $x = 0$ gives $A = 1$, letting $x = 2$ gives $B = 3$, and letting $x = -2$ gives $C = -3$. Thus, the original integral is equal to $\displaystyle\int \left(\dfrac{1}{x} + \dfrac{3}{x-2} - \dfrac{3}{x+2} \right) dx = \ln\left|\dfrac{x(x-2)^3}{(x+2)^3}\right| + C.$

8.5.37 If we write $\dfrac{1}{x^4 - 10x^2 + 9} = \dfrac{1}{(x-1)(x+1)(x-3)(x+3)} = \dfrac{A}{x-1} + \dfrac{B}{x+1} + \dfrac{C}{x-3} + \dfrac{D}{x+3}$ then $1 = A(x+1)(x-3)(x+3) + B(x-1)(x-3)(x+3) + C(x-1)(x+1)(x+3) + D(x-1)(x+1)(x-3)$.

Letting $x = -1$ yields $B = 1/16$. Letting $x = 3$ yields $C = 1/48$. Letting $x = -3$ yields $D = -1/48$, and letting $x = 1$ yields $A = -1/16$. Thus the original integral is equal to

$$\int \left(-\frac{1/16}{x-1} + \frac{1/16}{x+1} + \frac{1/48}{x-3} - \frac{1/48}{x+3}\right) dx$$

$$= -\frac{1}{16}\ln|x-1| + \frac{1}{16}\ln|x+1| + \frac{1}{48}\ln|x-3| - \frac{1}{48}\ln|x+3| + C$$

$$= \ln\left|\frac{(x+1)^3(x-3)}{(x-1)^3(x+3)}\right|^{1/48} + C.$$

8.5.39 If we write $\dfrac{81}{x^3 - 9x^2} = \dfrac{A}{x} + \dfrac{B}{x^2} + \dfrac{C}{x-9}$, then $81 = Ax(x-9) + B(x-9) + C(x^2)$. Letting $x = 0$ yields $B = -9$. Letting $x = 9$ yields $C = 1$. If we let $x = 10$, then we have $81 = 10A + B + 100C = 10A - 9 + 100$, so $A = -1$. Thus, the original integral is equal to $\int \left(-\dfrac{1}{x} - \dfrac{9}{x^2} + \dfrac{1}{x-9}\right) dx = \ln\left|\dfrac{(x-9)}{x}\right| + \dfrac{9}{x} + C.$

8.5.41 If we write $\dfrac{x}{(x+3)^2} = \dfrac{A}{x+3} + \dfrac{B}{(x+3)^2}$, then we have $x = A(x+3) + B$. Letting $x = -3$ yields $B = -3$, and then letting $x = -2$ yields $A = 1$. Thus the original integral is equal to

$$\int_{-1}^{1} \left(\frac{1}{x+3} - \frac{3}{(x+3)^2}\right) dx = \left(\ln|x+3| + \frac{3}{x+3}\right)\Big|_{-1}^{1} = \ln 4 + \frac{3}{4} - \left(\ln 2 + \frac{3}{2}\right) = \ln 2 - \frac{3}{4}.$$

8.5.43 If we write $\dfrac{2}{x^3 + x^2} = \dfrac{A}{x} + \dfrac{B}{x^2} + \dfrac{C}{x+1}$, then $2 = Ax(x+1) + B(x+1) + Cx^2$. Letting $x = 0$ yields $B = 2$, and letting $x = -1$ yields $C = 2$. Then letting $x = 1$ yields $A = -2$. So the original integral is equal to

$$\int \left(-\frac{2}{x} + \frac{2}{x^2} + \frac{2}{x+1}\right) dx = 2\left(\ln|x+1| - \ln|x|\right) - \frac{2}{x} + C.$$

8.5.45 If we write $\dfrac{x-5}{x^2(x+1)} = \dfrac{A}{x} + \dfrac{B}{x^2} + \dfrac{C}{x+1}$, then we have $x - 5 = Ax(x+1) + B(x+1) + Cx^2$. Letting $x = 0$ yields $B = -5$, and letting $x = -1$ yields $C = -6$. Then letting $x = 1$ yields $-4 = 2A - 10 - 6$, so $A = 6$. The original integral is thus equal to

$$\int \left(\frac{6}{x} - \frac{5}{x^2} - \frac{6}{x+1}\right) dx = 6\left(\ln|x| - \ln|x+1|\right) + \frac{5}{x} + C.$$

8.5.47 By long division, we can write the integrand as $x + \dfrac{2x}{(x-5)^2}$. We write

$$\frac{2x}{(x-5)^2} = \frac{A}{x-5} + \frac{B}{(x-5)^2},$$

so that $A(x-5) + B = 2x$. Then $A = 2$ and $B = 10$. We have

$$\int \frac{x^3 - 10x^2 + 27x}{x^2 - 10x + 25} dx = \int \left(x + \frac{2}{x-5} + \frac{10}{(x-5)^2}\right) dx = \frac{x^2}{2} + 2\ln|x-5| - \frac{10}{x-5} + C.$$

8.5.49 Let $\dfrac{x^2 - 4}{x^3 - 2x^2 + x} = \dfrac{A}{x} + \dfrac{B}{x-1} + \dfrac{C}{(x-1)^2}$. Then $x^2 - 4 = A(x-1)^2 + Bx(x-1) + Cx$. Letting $x = 0$ gives $A = -4$. Letting $x = 1$ gives $C = -3$. Letting $x = 2$ gives $0 = -4 + 2B - 6$, so $B = 5$. The given integral is thus equal to

$$\int \left(-\frac{4}{x} + \frac{5}{x-1} - \frac{3}{(x-1)^2}\right) dx = \ln\left|\frac{(x-1)^5}{x^4}\right| + \frac{3}{x-1} + C.$$

8.5. Partial Fractions

8.5.51 Let $\dfrac{x^2+x+2}{(x+1)(x^2+1)} = \dfrac{Ax+B}{x^2+1} + \dfrac{C}{x+1}$. Then $x^2+x+2 = (Ax+B)(x+1) + C(x^2+1)$. Letting $x=-1$ gives $C=1$. Letting $x=0$ gives $2 = B+1$, so $B=1$. Letting $x=1$ gives $4 = 2A+2+2$, so $A=0$. The original integral is therefore equal to

$$\int \left(\frac{1}{x^2+1} + \frac{1}{x+1}\right) dx = \tan^{-1}(x) + \ln|x+1| + C.$$

8.5.53 Let $\dfrac{2x^2+5x+5}{(x+1)(x^2+2x+2)} = \dfrac{Ax+B}{x^2+2x+2} + \dfrac{C}{x+1}$. Then $2x^2+5x+5 = (Ax+B)(x+1) + C(x^2+2x+2)$. Letting $x=-1$ gives $C=2$. Letting $x=0$ gives $5 = B+4$, so $B=1$. Letting $x=1$ gives $12 = 2A+2+2(5)$, so $A=0$. The original integral is therefore equal to

$$\int \left(\frac{1}{x^2+2x+2} + \frac{2}{x+1}\right) dx = \int \left(\frac{1}{(x+1)^2+1} + \frac{2}{x+1}\right) dx = \tan^{-1}(x+1) + \ln((x+1)^2) + C.$$

8.5.55 If we write $\dfrac{20x}{(x-1)(x^2+4x+5)} = \dfrac{A}{x-1} + \dfrac{Bx+C}{x^2+4x+5}$, then $20x = A(x^2+4x+5) + (Bx+C)(x-1)$. Letting $x=1$ yields $A=2$. Letting $x=0$ yields $0 = 10-C$, so $C=10$. Letting $x=2$ yields $40 = 34+2B+10$, so $B=-2$. The original integral is thus

$$\int \left(\frac{2}{x-1} - \frac{2x-10}{x^2+4x+5}\right) dx = \int \left(\frac{2}{x-1} - \frac{(2x+4)-14}{x^2+4x+5}\right) dx$$

$$= \int \left(\frac{2}{x-1} - \frac{(2x+4)}{x^2+4x+5} + \frac{14}{(x+2)^2+1}\right) dx$$

$$= \ln\left|\frac{(x-1)^2}{x^2+4x+5}\right| + 14\tan^{-1}(x+2) + C.$$

8.5.57 $\dfrac{x^3+5x}{(x^2+3)^2} = \dfrac{Ax+B}{x^2+3} + \dfrac{Cx+D}{(x^2+3)^2}$. Then

$$x^3+5x = (Ax+B)(x^2+3) + Cx+D = Ax^3 + Bx^2 + (3A+C)x + 3B+D.$$

Then $A=1$, $B=0$, $3A+C=5$, and $3B+D=0$. So $C=2$ and $D=0$. We have

$$\int \frac{x^3+5x}{(x^2+3)^2} dx = \int \left(\frac{x}{x^2+3} + \frac{2x}{(x^2+3)^2}\right) dx = \frac{1}{2}\ln|x^2+3| - \frac{1}{x^2+3} + C.$$

8.5.59 $\dfrac{x^3+6x^2+12x+6}{(x^2+6x+10)^2} = \dfrac{Ax+B}{x^2+6x+10} + \dfrac{Cx+D}{(x^2+6x+10)^2}$. Then

$$x^3+6x^2+12x+6 = (Ax+B)(x^2+6x+10) + Cx+D = Ax^3 + (B+6A)x^2 + (10A+6B+C)x + 10B+D.$$

Then $A=1$, $B=0$, $C=2$, and $D=6$. The given integral is then

$$\int \left(\frac{x}{x^2+6x+10} + \frac{2x+6}{(x^2+6x+10)^2}\right) dx = \int \frac{x}{(x+3)^2+1} dx - \frac{1}{x^2+6x+10} + C.$$

Note that

$$\int \frac{x}{(x+3)^2+1} dx = \int \left(\frac{x+3}{(x+3)^2+1} - \frac{3}{(x+3)^2+1}\right) dx = \frac{1}{2}\ln(x^2+6x+10) - 3\tan^{-1}(x+3) + C.$$

So the given integral is equal to

$$\frac{1}{2}\ln(x^2+6x+10) - 3\tan^{-1}(x+3) - \frac{1}{x^2+6x+10} + C.$$

8.5.61 If we write $\dfrac{2}{x(x^2+1)^2} = \dfrac{A}{x} + \dfrac{Bx+C}{x^2+1} + \dfrac{Dx+E}{(x^2+1)^2}$, then
$$2 = A(x^2+1)^2 + (Bx+C)x(x^2+1) + (Dx+E)x.$$

Letting $x = 0$ yields $A = 2$. Expanding the right-hand side yields $2 = (2+B)x^4 + Cx^3 + (4+B+D)x^2 + (C+E)x + 2$. Equating coefficients gives us the equations $2+B = 0$, $C = 0$, $4+B+D = 0$, and $C+E = 0$, from which we can deduce that $B = -2$, $C = 0$, $D = -2$ and $E = 0$. The original integral is thus equal to

$$\int \left(\frac{2}{x} - \frac{2x}{x^2+1} - \frac{2x}{(x^2+1)^2}\right) dx = 2\ln|x| - \ln|x^2+1| + \frac{1}{x^2+1} + C.$$

8.5.63 $\dfrac{9x^2+x+21}{(3x^2+7)^2} = \dfrac{Ax+B}{3x^2+7} + \dfrac{Cx+D}{(3x^2+7)^2}$. Then

$$9x^2 + x + 21 = (Ax+B)(3x^2+7) + Cx+D = 3Ax^3 + 3Bx^2 + (7A+C)x + 7B+D.$$

So $A = 0$, $B = 3$, $C = 1$, and $D = 0$. We have

$$\int \frac{9x^2+x+21}{(3x^2+7)^2} dx = \int \left(\frac{3}{3x^2+7} + \frac{x}{(3x^2+7)^2}\right) dx = \int \frac{1}{x^2+(7/3)} dx + \int \frac{x}{(3x^2+7)^2} dx$$
$$= \sqrt{\frac{3}{7}} \tan^{-1}\left(\sqrt{\frac{3}{7}}x\right) - \frac{1}{6(3x^2+7)} + C.$$

8.5.65

a. False. Because the given integrand is improper, the first step would be to use long division to write the integrand as the sum of a polynomial and a proper rational function.

b. False. This is easy to evaluate via the substitution $u = 3x^2 + x$.

c. False. The discriminant of the denominator is $b^2 - 4ac = 169 - 168 = 1 > 0$, so the denominator factors into linear factors of the real numbers.

d. True. The discriminant of the denominator is $b^2 - 4ac = 169 - 172 = -3 < 0$, so the given quadratic expression is irreducible.

8.5.67 We are seeking $\displaystyle\int_{-2}^{2} \frac{10}{x^2-2x-24} dx = \int_{-2}^{2} \frac{10}{(x-6)(x+4)} dx$. If we write

$$\frac{10}{x^2-2x-24} = \frac{10}{(x-6)(x+4)} = \frac{A}{x-6} + \frac{B}{x+4},$$

then $10 = A(x+4) + B(x-6)$. Letting $x = -4$ gives $B = -1$ and letting $x = 6$ gives $A = 1$. Thus the area in question is given by

$$-\int_{-2}^{2} \left(-\frac{1}{x+4} + \frac{1}{x-6}\right) dx = \left.(\ln(x+4) - \ln|x-6|)\right|_{-2}^{2} = \ln 6 - \ln 4 - (\ln 2 - \ln 8) = \ln 6.$$

8.5.69 Using disks, we have $\dfrac{V}{\pi} = \displaystyle\int_0^4 \frac{x^2}{(x+1)^2} dx$. Now we can perform long division to rewrite this integral as $\displaystyle\int_0^4 \left(1 - \frac{2x+2}{x^2+2x+1} + \frac{1}{(x+1)^2}\right) dx$. Thus we have $V = \pi\left(\left.x - \ln|x^2+2x+1| - \frac{1}{x+1}\right|_0^4\right) = \pi\left(4 - \ln(25) - \frac{1}{5} - (-1)\right) = \pi\left(\frac{24}{5} - \ln(25)\right)$.

8.5. Partial Fractions

8.5.71 Using disks, we have $\dfrac{V}{\pi} = \displaystyle\int_1^2 \dfrac{1}{x(3-x)}\,dx$. If we write

$$\dfrac{1}{x(3-x)} = \dfrac{A}{x} + \dfrac{B}{3-x},$$

then we have $1 = A(3-x) + Bx$. Letting $x = 0$ yields $A = 1/3$ and letting $x = 3$ yields $B = 1/3$. Thus we have

$$V = \dfrac{\pi}{3}\int_1^2 \left(\dfrac{1}{x} - \dfrac{1}{x-3}\right)dx = \dfrac{\pi}{3}\left(\ln x - \ln|x-3|\,\Big|_1^2\right) = \dfrac{\pi}{3}\cdot 2\ln 2.$$

8.5.73 If we write $\dfrac{1}{x^2-1} = \dfrac{A}{x-1} + \dfrac{B}{x+1}$, then $1 = A(x+1) + B(x-1)$, so $A = 1/2$ and $B = -1/2$. Thus we have

$$\dfrac{1}{2}\int \dfrac{1}{x-1} - \dfrac{1}{x+1}\,dx = \dfrac{1}{2}(\ln|x-1| - \ln|x+1|) + C.$$

Now let $x = \sec\theta$, so that $dx = \sec\theta\tan\theta\,d\theta$. Then the original integral is equal to $\displaystyle\int \csc\theta\,d\theta =$
$-\ln|\csc\theta + \cot\theta| + C = -\ln\left|\dfrac{x}{\sqrt{x^2-1}} + \dfrac{1}{\sqrt{x^2-1}}\right| + C = -\ln\left(\dfrac{|x+1|}{\sqrt{x^2-1}}\right) + C = -\ln\left(\sqrt{\left|\dfrac{x+1}{x-1}\right|}\right) + C =$
$\ln\left(\sqrt{\left|\dfrac{x-1}{x+1}\right|}\right) + C$. The two answers are equivalent.

8.5.75 $\dfrac{x^4 + 3x^2 + 1}{x(x^2+1)^2(x^2+x+4)^2} = \dfrac{A}{x} + \dfrac{Bx+C}{x^2+1} + \dfrac{Dx+E}{(x^2+1)^2} + \dfrac{Fx+G}{x^2+x+4} + \dfrac{Hx+I}{(x^2+x+4)^2}$. Using a computer algebra system, we find that $A = 1/16$, $B = -1/100$, $C = -1/10$, $D = 2/25$, $E = 3/50$, $F = -21/400$, $G = 19/400$, $H = -1/5$, and $I = -1/20$. Thus

$$\dfrac{x^4+3x^2+1}{x(x^2+1)^2(x^2+x+4)^2} = \dfrac{1}{16x} - \dfrac{x+10}{100(x^2+1)} + \dfrac{4x+3}{50(x^2+1)^2} + \dfrac{-21x+19}{400(x^2+x+4)} - \dfrac{4x+1}{20(x^2+x+4)^2}.$$

8.5.77 Let $u = e^x$ so that $du = e^x\,dx$. Then the original integral is equal to $\displaystyle\int \dfrac{1}{(u-1)(u+2)}\,du$. If we write $\dfrac{1}{(u-1)(u+2)} = \dfrac{A}{u-1} + \dfrac{B}{u+2}$, then $1 = A(u+2) + B(u-1)$. Letting $u = -2$ yields $B = -1/3$ and letting $u = 1$ yields $A = 1/3$. Thus we have

$$\int\left(\dfrac{1/3}{u-1} - \dfrac{1/3}{u+2}\right)du = \dfrac{1}{3}(\ln|u-1| - \ln|u+2|) + C = \dfrac{1}{3}\ln\left|\dfrac{e^x-1}{e^x+2}\right| + C.$$

8.5.79 This can be written as

$$\int \dfrac{dt}{\cos t(1+\sin t)} = \int \dfrac{\cos t}{\cos^2 t(1+\sin t)}\,dt = \int \dfrac{\cos t}{(1-\sin^2 t)(1+\sin t)}\,dt.$$

Now let $u = \sin t$ so that $du = \cos t\,dt$. Substituting gives $\displaystyle\int \dfrac{du}{(1-u^2)(1+u)} = \int \dfrac{du}{(1+u)^2(1-u)}$. If we write $\dfrac{1}{(1+u)^2(1-u)} = \dfrac{A}{1+u} + \dfrac{B}{(1+u)^2} + \dfrac{C}{1-u}$ then $1 = A(1+u)(1-u) + B(1-u) + C(1+u)^2$. Letting $u = -1$ yields $B = 1/2$. Letting $u = 1$ yields $C = 1/4$. Letting $u = 0$ then yields $A = 1/4$. Then our integral is equal to

$$\int\left(\dfrac{1/4}{1+u} + \dfrac{1/2}{(1+u)^2} + \dfrac{1/4}{1-u}\right)du = \dfrac{1}{4}\left(\dfrac{-2}{1+u} + \ln\left|\dfrac{1+u}{1-u}\right|\right) + C = \dfrac{1}{4}\left(\dfrac{-2}{1+\sin t} + \ln\left(\dfrac{1+\sin t}{1-\sin t}\right)\right) + C.$$

8.5.81 Let $u = e^x$. Then $du = e^x\,dx$. We have $\displaystyle\int \frac{e^{3x} + e^{2x} + e^x}{(e^{2x}+1)^2}\,dx = \int \frac{u^2+u+1}{(u^2+1)^2}\,du$.

Let $\dfrac{u^2+u+1}{(u^2+1)^2} = \dfrac{Au+B}{u^2+1} + \dfrac{Cu+D}{(u^2+1)^2}$. Then $u^2+u+1 = (Au+B)(u^2+1) + Cu+D = Au^3 + Bu^2 + (A+C)u + B + D$. Then $A = 0$, $B = 1$, $C = 1$, and $D = 0$. So we have

$$\int \frac{u^2+u+1}{(u^2+1)^2}\,du = \int \left(\frac{1}{u^2+1} + \frac{u}{(u^2+1)^2}\right)du = \tan^{-1} u - \frac{1}{2(u^2+1)} + C = \tan^{-1} e^x - \frac{1}{2(e^{2x}+1)} + C.$$

8.5.83 Let $u = e^x$, so that $du = e^x\,dx$. Then $\displaystyle\int \frac{1}{1+e^x} \cdot \frac{e^x}{e^x}\,dx = \int \frac{1}{u(1+u)}\,du$. If we write $\dfrac{1}{u(1+u)} = \dfrac{A}{u} + \dfrac{B}{1+u}$, then $1 = A(1+u) + Bu$. Letting $u = 0$ yields $A = 1$ and letting $u = -1$ yields $B = -1$, so the original integral is equal to $\displaystyle\int \left(\frac{1}{u} - \frac{1}{1+u}\right)du = \ln|u| - \ln|1+u| + C = x - \ln(1+e^x) + C$.

8.5.85

a. $\sec x = \dfrac{1}{\cos x} \cdot \dfrac{\cos x}{\cos x} = \dfrac{\cos x}{\cos^2 x} = \dfrac{\cos x}{1 - \sin^2 x}$.

b. $\displaystyle\int \sec x\,dx = \int \frac{\cos x}{1-\sin^2 x}\,dx$. Let $u = \sin x$ so that $du = \cos x\,dx$. Then our integral becomes $\displaystyle\int \frac{du}{1-u^2}$. If we write $\dfrac{1}{1-u^2} = \dfrac{1}{(1-u)(1+u)} = \dfrac{A}{1-u} + \dfrac{B}{1+u}$, we have $1 = A(1+u) + B(1-u)$, so $A = \tfrac{1}{2}$ and $B = \tfrac{1}{2}$. We have

$$\int \frac{du}{1-u^2} = \frac{1}{2}\int\left(\frac{1}{1-u} + \frac{1}{1+u}\right)du = \frac{1}{2}(\ln|1+u| - \ln|1-u|) + C = \frac{1}{2}\ln\left|\frac{1+\sin x}{1-\sin x}\right| + C.$$

8.5.87 If $x = 2\tan^{-1} u$, then $dx = \dfrac{2}{1+u^2}\,du$.

If $u = \tan(x/2)$, then $u^2 + 1 = \tan^2(x/2) + 1 = \sec^2(x/2) = \dfrac{1}{\cos^2(x/2)}$. Thus, $\dfrac{1}{u^2+1} = \cos^2(x/2) = \dfrac{1+\cos x}{2}$, so $\cos x = \dfrac{2}{u^2+1} - 1 = \dfrac{1-u^2}{u^2+1} = \dfrac{1-u^2}{1+u^2}$.

Also, $\sin x = 2\sin(x/2)\cos(x/2) = 2\tan(x/2)\cos^2(x/2) = \dfrac{2u}{1+u^2}$.

8.5.89 Using the substitution $x = 2\tan^{-1} u$ yields

$$\int \frac{1}{1 - \frac{1-u^2}{1+u^2}} \cdot \frac{2}{1+u^2}\,du = \int u^{-2}\,du = -\frac{1}{u} + C = -\cot(x/2) + C.$$

8.5.91 Using the substitution $\theta = 2\tan^{-1} u$ yields $\displaystyle\int_0^1 \frac{1}{\frac{1-u^2}{1+u^2} + \frac{2u}{1+u^2}} \cdot \frac{2}{1+u^2}\,du = \int_0^1 \frac{2}{1+2u-u^2}\,du = -2\int_0^1 \frac{1}{u^2-2u+1-2}\,du = -2\int_0^1 \frac{1}{(u-1)^2-2}\,du$. If we factor the denominator as the difference of squares we have $(u-1)^2 - 2 = (u-1-\sqrt{2})(u-1+\sqrt{2})$, and using a partial fractions decomposition yields

$$\frac{1}{\sqrt{2}}\int_0^1 \left(\frac{1}{u+\sqrt{2}-1} - \frac{1}{u-\sqrt{2}-1}\right)du = \frac{1}{\sqrt{2}}\ln\left|\frac{u+\sqrt{2}-1}{u-\sqrt{2}-1}\right|\bigg|_0^1 = \frac{1}{\sqrt{2}}\ln\left(\frac{\sqrt{2}+1}{\sqrt{2}-1}\right).$$

8.5.93 $s_A(t) = \displaystyle\int v_A(t)\,dt = 88\int \frac{t}{t+1}\,dt = 88\int\left(1 - \frac{1}{1+t}\right)dt = 88t - 88\ln(t+1) + C$. Because $s_A(0) = 0$, we see that $C = 0$, so $s_A(t) = 88t - 88\ln(1+t)$.

$$s_B(t) = \int v_B(t)\,dt = 88\int \frac{t^2}{(t+1)^2}\,dt = 88\int\left(1 - \frac{2t+2}{t^2+2t+1} + \frac{1}{(t+1)^2}\right)dt = 88(t - \ln(t^2+2t+1) - \frac{1}{t+1}) + D.$$ Because $s_B(0) = 0$, we see that $D = 88$, so $s_B(t) = 88(t - \ln(t^2+2t+1) - \frac{1}{t+1} + 1)$.

$$s_C(t) = \int v_C(t)\,dt = 88\int \frac{t^2}{t^2+1}\,dt = 88\int\left(1 - \frac{1}{t^2+1}\right)dt = 88(t - \tan^{-1}(t)) + E.$$ Because $s_C(0) = 0$, we have that $E = 0$. Thus $s_C(t) = 88t - 88\tan^{-1}(t)$.

a. $s_A(1) = 88(1 - \ln(2)) \approx 27$. $s_B(1) = 88(1 - \ln 4 - (1/2) + 1) \approx 88(3/2 - \ln 4) \approx 10$. $s_C(1) = 88(1 - \tan^{-1}(1)) \approx 18.9$. So car A travels farthest.

b. $s_A(5) = 88(5 - \ln 6) \approx 282$. $s_B(5) = 88(5 - \ln 36 - (1/6) + 1) \approx 198$. $s_C(5) = 88(5 - \tan^{-1}(5)) \approx 319$. So car C travels farthest.

c. See the development above.

d. Ultimately car C gains the lead. This can be seen by the fact that car C's velocity function is greater than that of the other cars.

8.5.95 First note that the numerator of the given integrand can be written as $x^8 - 4x^7 + 6x^6 - 4x^5 + x^4$, and this quantity when divided by $x^2 + 1$ yields $x^6 - 4x^5 + 5x^4 - 4x^2 + 4 - \dfrac{4}{1+x^2}$. Thus the given integral is equal to $\left(\dfrac{x^7}{7} - \dfrac{2x^6}{3} + x^5 - \dfrac{4x^3}{3} + 4x\right)\Big|_0^1 - (4\tan^{-1}(x))\Big|_0^1 = \dfrac{1}{7} - \dfrac{2}{3} + 1 - \dfrac{4}{3} + 4 - \pi = \dfrac{22}{7} - \pi$. Because the given integrand is positive on the interval $(0,1)$, we know that this integral is positive. Thus,

$$0 < \int_0^1 \frac{x^4(1-x)^4}{1+x^2}\,dx = \frac{22}{7} - \pi.$$

Adding π to both sides of this inequality yields $\pi < \dfrac{22}{7}$.

8.6 Integration Strategies

8.6.1 Because the integrand is a product of a polynomial and a trigonometric function, it makes sense to integrate by parts.

8.6.3 The presence of $64 - x^2$ in the integrand suggests the trigonometric substitution $x = 8\sin\theta$.

8.6.5 The method of partial fractions is appropriate because the integrand is a proper rational function.

8.6.7 Let $u = \cos\theta$, which implies that $du = -\sin\theta\,d\theta$, or $-du = \sin\theta\,d\theta$. The lower limit of integration becomes $u = 1$ and the upper limit becomes $u = 0$. So we have

$$\int_0^{\pi/2} \frac{\sin\theta}{1+\cos^2\theta}\,d\theta = -\int_1^0 \frac{1}{1+u^2} = \int_0^1 \frac{1}{1+u^2} = \tan^{-1}u\Big|_0^1 = \frac{\pi}{4}.$$

8.6.9 After completing the square of $8x - x^2$, we make a substitution of $u = x - 4$. The lower limit of integration becomes $u = 0$ and the upper limit of integration becomes $u = 2$. Integrating, we have

$$\int_4^6 \frac{dx}{\sqrt{8x-x^2}} = \int_4^6 \frac{dx}{\sqrt{16-(x-4)^2}} = \int_0^2 \frac{dx}{\sqrt{16-u^2}} = \sin^{-1}\frac{u}{4}\Big|_0^2 = \frac{\pi}{6}.$$

8.6.11 The integral is evaluated after expanding the integrand.

$$\int_0^{\pi/4} (\sec x - \cos x)^2 \, dx = \int_0^{\pi/4} (\sec^2 x - 2\underbrace{\sec x \cos x}_{1} + \cos^2 x) \, dx \qquad \text{Expand integrand.}$$

$$= \int_0^{\pi/4} \left(\sec^2 x - 2 + \frac{1 + \cos 2x}{2} \right) dx \qquad \cos^2 x = \frac{1 + \cos 2x}{2}$$

$$= \int_0^{\pi/4} \left(\sec^2 x - \frac{3}{2} + \frac{1}{2} \cos 2x \right) dx \qquad \text{Simplify integrand.}$$

$$= \left(\tan x - \frac{3}{2} x + \frac{1}{4} \sin 2x \right) \Big|_0^{\pi/4} = \frac{5}{4} - \frac{3\pi}{8} \qquad \text{Evaluate.}$$

8.6.13 Let $u = e^x$, which implies that $du = e^x dx$ and $dx = \frac{du}{e^x} = \frac{du}{u}$. The integrand is then expressed in terms of u:

$$\int \frac{dx}{e^x \sqrt{1 - e^{2x}}} = \int \frac{du}{u^2 \sqrt{1 - u^2}}.$$

The expression $1 - u^2$ suggests the trigonometric substitution $u = \sin\theta$, where $-\frac{\pi}{2} < \theta < \frac{\pi}{2}$. It follows that $du = \cos\theta \, d\theta$ and

$$\sqrt{1 - u^2} = \sqrt{1 - \sin^2\theta} = \sqrt{\cos^2\theta} = |\cos\theta| = \cos\theta.$$

Using the equation $u = \sin\theta$, we create a reference triangle that helps us evaluate the integral.

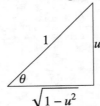

Summarizing what we've done so far and then evaluating the integral, we have

$$\int \frac{dx}{e^x \sqrt{1 - e^{2x}}} = \int \frac{du}{u^2 \sqrt{1 - u^2}} \qquad u\text{-substitution}$$

$$= \int \frac{\cos\theta}{\sin^2\theta \cos\theta} \, d\theta \qquad \text{Trigonometric substitution}$$

$$= \int \csc^2\theta \, d\theta \qquad \text{Simplify.}$$

$$= -\cot\theta + C \qquad \text{Integrate.}$$

$$= -\frac{\sqrt{1 - u^2}}{u} + C \qquad \text{Use the reference triangle.}$$

$$= -\frac{\sqrt{1 - e^{2x}}}{e^x} + C. \qquad \text{Replace } u \text{ with } e^x.$$

8.6.15 We let $u = \sqrt{x}$, which implies that $du = \frac{dx}{2\sqrt{x}}$ or $2du = \frac{dx}{\sqrt{x}}$. The lower limit of integration is $u = 1$ and the upper limit is $u = 2$. Therefore

$$\int_1^4 \frac{2^{\sqrt{x}}}{\sqrt{x}} \, dx = 2 \int_1^2 2^u \, du = \frac{2 \cdot 2^u}{\ln 2} \Big|_1^2 = \frac{4}{\ln 2}.$$

8.6. Integration Strategies

8.6.17 Let's evaluate the corresponding indefinite integral first. Let $z = w^2$ which implies that $dz = 2w\, dw$, or $\dfrac{dz}{2} = w\, dw$. We then have

$$\int w^3 e^{w^2}\, dw = \frac{1}{2}\int ze^z\, dz.$$

For the integral on the right, use integration by parts with the following choices.

$$u = z \qquad dv = e^z\, dz$$

$$du = dz \qquad v = e^z$$

Integrating by parts, we have

$$\int w^3 e^{w^2}\, dw = \frac{1}{2}\int ze^z\, dz = \frac{1}{2}\left(ze^z - \int e^z\, dz\right) = \frac{e^z}{2}(z-1) + C$$

After replacing z with w^2, we evaluate the definite integral:

$$\int_1^2 w^3 e^{w^2}\, dx = \left(\frac{1}{2}e^{w^2}(w^2-1)\right)\bigg|_1^2 = \frac{3e^4}{2}.$$

8.6.19 The integral has an odd power of $\sin x$, so we start by splitting off a factor of $\sin x$.

$$\begin{aligned}
\int_0^{\pi/2} \sin^7 x\, dx &= \int_0^{\pi/2} (\sin^2 x)^3 \sin x\, dx & &\text{Split off } \sin x.\\
&= \int_0^{\pi/2} (1-\cos^2 x)^3 \sin x\, dx & &\sin^2 x = 1-\cos^2 x\\
&= \int_1^0 -(1-u^2)^3\, du & &\text{Let } u = \cos x;\, du = -\sin x\, dx.\\
& & &x=0 \Rightarrow u=1;\, x=\frac{\pi}{2} \Rightarrow u=0\\
&= \int_0^1 (-u^6 + 3u^4 - 3u^2 + 1)\, du & &\text{Expand.}\\
&= \left(-\frac{u^7}{7} + \frac{3}{5}u^5 - u^3 + u\right)\bigg|_0^1 & &\text{Integrate.}\\
&= \frac{16}{35}. & &\text{Evaluate.}
\end{aligned}$$

8.6.21 We use integration by parts with the following choices.

$$u = \ln 3x \qquad dv = x^9\, dx$$

$$du = \frac{1}{x}dx \qquad v = \frac{x^{10}}{10}$$

Integrating by parts, we have

$$\int x^9 \ln 3x\, dx = \frac{x^{10}}{10}\ln 3x - \frac{1}{10}\int x^9\, dx = \frac{x^{10}}{10}\ln 3x - \frac{x^{10}}{100} + C.$$

8.6.23 Letting $u = \cos x$ and $du = -\sin x \, dx$, or $-du = \sin x \, dx$, we have

$$\int \frac{\sin x}{\cos^2 x + \cos x} \, dx = -\int \frac{du}{u^2 + u} = -\int \frac{du}{u(u+1)}.$$

To evaluate the integral on the right, we first find the constants A and B in the partial fraction decomposition

$$\frac{1}{u(u+1)} = \frac{A}{u} + \frac{B}{u+1}.$$

Multiplying both sides of this equation by $u(u+1)$, we have

$$1 = A(u+1) + Bu = (A+B)u + A.$$

It follows that $A = 1$ and $B = -1$. Using these constants in the partial fraction decomposition, integration is straightforward.

$$\int \frac{\sin x}{\cos^2 x + \cos x} \, dx = -\int \frac{du}{u(u+1)} \qquad u = \cos x; \, du = -\sin x \, dx$$
$$= -\int \left(\frac{1}{u} - \frac{1}{u+1}\right) du \qquad \text{Partial fractions}$$
$$= -(\ln|u| - \ln|u+1|) + C \qquad \text{Integrate.}$$
$$= \ln(\cos x + 1) - \ln|\cos x| + C. \qquad \text{Replace } u \text{ with } \cos x.$$

8.6.25 One approach for evaluating this integral is to use a trigonometric substitution. The expression $1 - x^2$ suggests the substitution $x = \sin \theta$ with $-\pi/2 < \theta < \pi/2$. With this substitution, $dx = \cos \theta \, d\theta$ and

$$\sqrt{1 - x^2} = \sqrt{1 - \sin^2 \theta} = \sqrt{\cos^2 \theta} = |\cos \theta| = \cos \theta.$$

Substituting these values into the integral and simplifying, we have

$$\int \frac{dx}{x\sqrt{1-x^2}} = \int \csc \theta \, d\theta = -\ln|\csc \theta + \cot \theta| + C.$$

Given that $\sin \theta = x$, a geometric description of the relationship between x and θ is given by the following reference triangle.

With the help of the reference triangle, we continue:

$$\int \frac{dx}{x\sqrt{1-x^2}} = -\ln|\csc \theta + \cot \theta| + C = -\ln\left|\frac{1}{x} + \frac{\sqrt{1-x^2}}{x}\right| + C = \ln\left|\frac{x}{1 + \sqrt{1-x^2}}\right| + C.$$

8.6.27 Because we have an even power of $\sin \frac{x}{2}$, we use the half-angle formula for $\sin^2 \frac{x}{2}$ to rewrite the integrand:

8.6. Integration Strategies

$$\int \sin^4 \frac{x}{2}\, dx = \int \underbrace{\left(\frac{1-\cos x}{2}\right)^2}_{\sin^2 \frac{x}{2}} dx = \frac{1}{4}\int \left(1 - 2\cos x + \cos^2 x\right) dx.$$

Using the half-angle formula for $\cos^2 x$, the evaluation may be completed:

$$\int \sin^4 \frac{x}{2}\, dx = \frac{1}{4}\int \left(1 - 2\cos x + \underbrace{\frac{1+\cos 2x}{2}}_{\cos^2 x}\right) dx = \frac{1}{4}\int \left(\frac{3}{2} - 2\cos x + \frac{\cos 2x}{2}\right) dx$$

$$= \frac{3x}{8} - \frac{1}{2}\sin x + \frac{1}{16}\sin 2x + C.$$

8.6.29 Before integrating, let's write $\cot x$ in terms of $\sin x$ and $\cos x$ and then simplify the integrand by multiplying the numerator and denominator by $\sin x$.

$$\frac{2\cos x + \overbrace{\frac{\cos x}{\sin x}}^{\cot x}}{1+\sin x} = \frac{2\sin x \cos x + \cos x}{\sin x + \sin^2 x}.$$

The integral is evaluated with the help of a u-substitution.

$$\int \frac{2\cos x + \cot x}{1+\sin x}\, dx = \int \frac{2\sin x \cos x + \cos x}{\sin x + \sin^2 x}\, dx \quad \text{Write in terms of } \sin x \text{ and } \cos x; \text{ simplify.}$$

$$= \int \frac{du}{u} \quad \text{Let } u = \sin x + \sin^2 x;\, du = (\cos x + 2\sin x \cos x)\, dx.$$

$$= \ln|u| + C \quad \text{Integrate.}$$

$$= \ln|\sin x + \sin^2 x| + C. \quad \text{Substitute } \sin x + \sin^2 x \text{ for } u.$$

8.6.31 Removing a factor of 3 from the square root, we have

$$\int \sqrt{36-9x^2}\, dx = \int \sqrt{9(4-x^2)}\, dx = 3\int \sqrt{4-x^2}\, dx.$$

The expression $4-x^2$ suggests the substitution $x = 2\sin\theta$, for $\frac{\pi}{2} \le \theta \le \frac{\pi}{2}$. It follows that $dx = 2\cos\theta\, d\theta$ and

$$\sqrt{4-x^2} = \sqrt{4-4\sin^2\theta} = \sqrt{4(1-\sin^2\theta)} = 2|\cos\theta| = 2\cos\theta.$$

Making these substitutions, we have

$$\int \sqrt{36-9x^2}\, dx = 3\int \sqrt{4-x^2}\, dx \quad \text{Factor.}$$

$$= 12\int \cos^2\theta\, d\theta \quad \text{Trigonometric substitution}$$

$$= 6\int (1+\cos 2\theta)\, d\theta \quad \cos^2\theta = \frac{1+\cos 2\theta}{2}$$

$$= 6\theta + 3\sin 2\theta + C \quad \text{Integrate.}$$

$$= 6\theta + 6\sin\theta\cos\theta + C. \quad \sin 2\theta = 2\sin\theta\cos\theta$$

From the equation $\sin\theta = \frac{x}{2}$ we have $\theta = \sin^{-1}\frac{x}{2}$. We use the following reference triangle to observe that $\cos\theta = \frac{\sqrt{4-x^2}}{2}$.

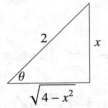

Therefore

$$\int \sqrt{36 - 9x^2}\,dx = 6\underbrace{\sin^{-1}\frac{x}{2}}_{\theta} + 6\underbrace{\left(\frac{x}{2}\right)}_{\sin\theta}\underbrace{\left(\frac{\sqrt{4-x^2}}{2}\right)}_{\cos\theta} + C = 6\sin^{-1}\frac{x}{2} + \frac{3}{2}x\sqrt{4-x^2} + C.$$

8.6.33 Letting $u = e^x$, we have $e^{2x} = (e^x)^2 = u^2$ and $du = e^x dx$. Provided that $a \neq 0$,

$$\int \frac{e^x}{a^2 + e^{2x}}\,dx = \int \frac{du}{a^2 + u^2} = \frac{1}{a}\tan^{-1}\frac{u}{a} + C = \frac{1}{a}\tan^{-1}\frac{e^x}{a} + C.$$

8.6.35 Let $u = \tan\theta$ which implies that $du = \sec^2\theta\,d\theta$. The lower limit of integration becomes $u = \tan 0 = 0$ and the upper limit becomes $u = \tan\pi/4 = 1$. Changing variables, we have

$$\int_0^{\pi/4}(\tan^2\theta + \tan\theta + 1)\sec^2\theta\,d\theta = \int_0^1 (u^2 + u + 1)\,du = \left(\frac{u^3}{3} + \frac{u^2}{2} + u\right)\Big|_0^1 = \frac{1}{3} + \frac{1}{2} + 1 = \frac{11}{6}$$

8.6.37 We start by multiplying the integrand by $\frac{1 + \sin 2x}{1 + \sin 2x}$.

$$\int_0^{\pi/6} \frac{1}{1 - \sin 2x}\cdot\frac{1 + \sin 2x}{1 + \sin 2x}\,dx = \int_0^{\pi/6}\frac{1 + \sin 2x}{1 - \sin^2 2x}\,dx \qquad \text{Simplify integrand.}$$

$$= \int_0^{\pi/6}\frac{1 + \sin 2x}{\cos^2 2x}\,dx \qquad 1 - \sin^2 2x = \cos^2 2x.$$

$$= \int_0^{\pi/6}\left(\sec^2 2x + \sec 2x\tan 2x\right)dx \qquad \text{Simplify integrand.}$$

$$= \frac{1}{2}\left(\tan 2x + \sec 2x\right)\Big|_0^{\pi/6} \qquad \text{Integrate.}$$

$$= \frac{1}{2}(\sqrt{3} + 2 - 1) = \frac{\sqrt{3} + 1}{2}. \qquad \text{Evaluate.}$$

8.6.39 We integrate by parts with the following variable choices.

$$u = \ln(\sin x) \qquad dv = \sin x\,dx$$

$$du = \frac{\cos x}{\sin x}\,dx \qquad v = -\cos x$$

It follows that

8.6. Integration Strategies

$$\int \sin x \ln(\sin x)\,dx = -\cos x \ln(\sin x) + \int \frac{\cos^2 x}{\sin x}\,dx \qquad \text{Integration by parts}$$

$$= -\cos x \ln(\sin x) + \int \frac{1-\sin^2 x}{\sin x}\,dx \qquad \cos^2 x = 1 - \sin^2 x$$

$$= -\cos x \ln(\sin x) + \int \left(\csc x - \sin x\right) dx \qquad \text{Simplify integrand.}$$

$$= -\cos x \ln(\sin x) - \ln|\csc x + \cot x| + \cos x + C. \qquad \text{Evaluate.}$$

8.6.41 Split $\csc^4 x$ and use the identity $\csc^2 x = 1 + \cot^2 x$:

$$\int \cot^{3/2} x \csc^4 x\,dx = \int \cot^{3/2} x \csc^2 x \csc^2 x\,dx \qquad \csc^4 x = \csc^2 x \csc^2 x$$

$$= \int \cot^{3/2} x(1 + \cot^2 x)\csc^2 x\,dx \qquad \csc^2 x = 1 + \cot^2 x$$

$$= -\int u^{3/2}(1 + u^2)\,du \qquad u = \cot x,\ du = -\csc^2 x\,dx$$

$$= -\int \left(u^{3/2} + u^{7/2}\right) du \qquad \text{Expand integrand.}$$

$$= -\frac{2}{5}u^{5/2} - \frac{2}{9}u^{9/2} + C \qquad \text{Evaluate.}$$

$$= -\frac{2}{5}\cot^{5/2} x - \frac{2}{9}\cot^{9/2} x + C. \qquad \text{Replace } u \text{ with } \cot x.$$

8.6.43 The integral $\int \dfrac{x^9}{\sqrt{1-x^{20}}}\,dx$ can be expressed as $\int \dfrac{x^9}{\sqrt{1-\left(x^{10}\right)^2}}\,dx$, so we let $u = x^{10}$ which implies that $du = 10x^9\,dx$, or $x^9\,dx = \dfrac{du}{10}$. Therefore

$$\int \frac{x^9}{\sqrt{1-\left(x^{10}\right)^2}}\,dx = \frac{1}{10}\int \frac{du}{\sqrt{1-u^2}} = \frac{\sin^{-1} u}{10} + C = \frac{\sin^{-1} x^{10}}{10} + C.$$

8.6.45 Letting $u = 1 + e^x$, it follows that $du = e^x\,dx$, $e^x = u - 1$, and $dx = \dfrac{du}{e^x} = \dfrac{du}{u-1}$. The lower limit of integration is $u = 2$ and the upper limit is $u = 3$. We have

$$\int_0^{\ln 2} \frac{dx}{(1+e^x)^2} = \int_2^3 \frac{du}{u^2(u-1)}.$$

Partial fractions is used to evaluate the integral on the right. An appropriate form for the partial fraction decomposition of the integrand is

$$\frac{1}{u^2(u-1)} = \frac{A}{u} + \frac{B}{u^2} + \frac{C}{u-1}.$$

Multiplying both sides by $u^2(u-1)$ leads to

$$1 = Au(u-1) + B(u-1) + Cu^2 = (A+C)u^2 + (B-A)u - B.$$

Equating coefficients of equal powers of u, we find that $A = -1$, $B = -1$, and $C = 1$. Therefore

$$\int_0^{\ln 2} \frac{dx}{(1+e^x)^2} = \int_2^3 \frac{du}{u^2(u-1)} = \int_2^3 \left(-\frac{1}{u} - \frac{1}{u^2} + \frac{1}{u-1}\right) du = \left.\left(-\ln|u| + \frac{1}{u} + \ln|u-1|\right)\right|_2^3$$

$$= \ln\frac{2}{3} - \frac{1}{6} + \ln 2 = \ln\frac{4}{3} - \frac{1}{6}.$$

8.6.47 Apply long division to obtain
$$\frac{2x^3 + x^2 - 2x - 4}{x^2 - x - 2} = 2x + 3 + \frac{5x + 2}{x^2 - x - 2}.$$
Factor $x^2 - x - 2$ and then find the partial fraction decomposition of $\frac{5x+2}{x^2-x-2}$.
$$\frac{5x + 2}{(x-2)(x+1)} = \frac{A}{x-2} + \frac{B}{x+1}$$
Multiply both sides by $(x-2)(x+1)$.
$$5x + 2 = A(x+1) + B(x-2).$$
Note that $x = 2$ implies that $A = 4$ and $x = -1$ implies that $B = 1$. Therefore
$$\int \left(\frac{2x^3 + x^2 - 2x - 4}{x^2 - x - 2}\right) dx = \int \left(2x + 3 + \frac{4}{x-2} + \frac{1}{x+1}\right) dx$$
$$= x^2 + 3x + 4\ln|x-2| + \ln|x+1| + C.$$

8.6.49 We start by splitting off factors of $\tan x$ and $\sec x$ and then we replace $\tan^2 x$ with $\sec^2 x - 1$:

$$\int \tan^3 x \sec^9 x\, dx = \int \tan^2 x \sec^8 x \sec x \tan x\, dx \qquad \text{Split off } \tan x \text{ and } \sec x.$$
$$= \int (\sec^2 x - 1) \sec^8 x \sec x \tan x\, dx \qquad \tan^2 x = \sec^2 x - 1$$
$$= \int (\sec^{10} x - \sec^8 x) \sec x \tan x\, dx \qquad \text{Expand integrand.}$$
$$= \int (u^{10} - u^8)\, du \qquad u = \sec x,\ du = \sec x \tan x\, dx$$
$$= \frac{u^{11}}{11} - \frac{u^9}{9} + C \qquad \text{Integrate.}$$
$$= \frac{\sec^{11} x}{11} - \frac{\sec^9 x}{9} + C. \qquad \text{Replace } u \text{ with } \sec x.$$

8.6.51 Rewrite $\sec^{7/4} x$ as $\sec^{3/4} x \sec x$ and let $u = \sec x$. Then $du = \sec x \tan$ and after making this substitution, the lower limit of integration is $u = \sec 0 = 1$ and the upper limit is $u = \sec\frac{\pi}{3} = 2$. Integrating, we have
$$\int_0^{\pi/3} \tan x \sec^{7/4} x\, dx = \int_0^{\pi/3} \sec^{3/4} x \sec x \tan x\, dx = \int_1^2 u^{3/4}\, du = \frac{4}{7} u^{7/4}\Big|_1^2 = \frac{4}{7}(2^{7/4} - 1).$$

8.6.53 We begin with the substitution $u = e^x$, $du = e^x\, dx$, and then split apart the integrand.
$$\int e^x \cot^3 e^x\, dx = \int \cot^3 u\, du = \int \cot^2 u \cot u\, du = \int (\csc^2 u - 1) \cot u\, du = \int \csc^2 u \cot u\, du - \int \cot u\, du$$

Using the substitution $v = \cot u$ in the first integral with $dv = -\csc^2 u\, du$ or $-dv = \csc^2 u\, du$, we find that
$$\int \csc^2 u \cot u\, du = -\int v\, dv = -\frac{v^2}{2} + C = -\frac{\cot^2 u}{2} + C.$$

8.6. Integration Strategies

Therefore

$$\int e^x \cot^3 e^x \, dx = \int \csc^2 u \cot u \, du - \int \cot u \, du = -\frac{\cot^2 u}{2} - \ln|\sin u| + C = -\frac{\cot^2 e^x}{2} - \ln|\sin e^x| + C.$$

8.6.55 The denominator factors as $x^3 + x = x(x^2 + 1)$ and therefore the form of the partial fraction decomposition is

$$\frac{3x^2 + 3x + 1}{x(x^2 + 1)} = \frac{A}{x} + \frac{Bx + C}{x^2 + 1}.$$

Multiplying both sides by $x(x^2 + 1)$ leads to

$$3x^2 + 3x + 1 = A(x^2 + 1) + (Bx + C)x = (A + B)x^2 + Cx + A.$$

By equating coefficients of equal powers of x, we find that $A = 1$, $B = 2$, and $C = 3$. The original integral can be written as

$$\int \frac{3x^2 + 3x + 1}{x^3 + x} \, dx = \int \frac{dx}{x} + \int \frac{2x + 3}{x^2 + 1} \, dx = \int \frac{dx}{x} + \int \frac{2x}{x^2 + 1} \, dx + \int \frac{3}{x^2 + 1} \, dx.$$

Using a substitution if $u = x^2 + 1$ in the second integral with $du = 2x \, dx$, we find that $\int \frac{2x}{x^2+1} \, dx = \ln(x^2 + 1) + C$. Completing the integration process we have

$$\int \frac{3x^2 + 3x + 1}{x^3 + x} \, dx = \int \frac{dx}{x} + \int \frac{2x}{x^2 + 1} \, dx + \int \frac{3}{x^2 + 1} \, dx = \ln|x| + \ln(x^2 + 1) + 3 \tan^{-1} x + C.$$

8.6.57 Some observations are in order before we start integrating. Notice that

$$\int \sin \sqrt{x} \, dx = \int \sqrt{x} \, \frac{\sin \sqrt{x}}{\sqrt{x}} \, dx.$$

Also observe that $\int \frac{\sin \sqrt{x}}{\sqrt{x}} \, dx = -2 \cos \sqrt{x} + C$ because $\frac{d}{dx} \left(-2 \cos \sqrt{x} \right) = \frac{\sin \sqrt{x}}{\sqrt{x}}$. These facts will help us evaluate the integral using integration by parts with the following variable choices.

$$u = \sqrt{x} \qquad dv = \frac{\sin \sqrt{x}}{\sqrt{x}} \, dx$$
$$du = \frac{dx}{2\sqrt{x}} \qquad v = -2 \cos \sqrt{x}$$

Integrating by parts, we have

$$\int \sin \sqrt{x} \, dx = -2\sqrt{x} \cos \sqrt{x} + \int \frac{\cos \sqrt{x}}{\sqrt{x}} \, dx$$
$$= -2\sqrt{x} \cos \sqrt{x} + 2 \sin \sqrt{x} + C.$$

8.6.59 Factor the denominator and make the trigonometric substitution $x = \tan \theta$, where $-\pi/2 < \theta < \pi/2$. Then $dx = \sec^2 \theta \, d\theta$ and

$$\int \frac{dx}{x^4 + x^2} = \int \frac{dx}{x^2(x^2 + 1)} = \int \frac{\sec^2 \theta}{\tan^2 \theta \underbrace{(\tan^2 \theta + 1)}_{\sec^2 \theta}} \, d\theta = \int \cot^2 \theta \, d\theta = \int (\csc^2 \theta - 1) d\theta = -\cot \theta - \theta + C.$$

Because $\tan \theta = x$, we use the following reference triangle and the substitution $\theta = \tan^{-1} x$ to complete the integration.

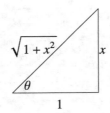

$$\int \frac{dx}{x^4+x^2} = -\cot\theta - \theta + C = -\frac{1}{x} - \tan^{-1} x + C.$$

8.6.61 Letting $u = \sin^{-1} x$ we have $x = \sin u$ and $dx = \cos u\, du$. In this case, $x = 0 \Rightarrow u = \sin^{-1} 0 = 0$ and $x = \frac{\sqrt{2}}{2} \Rightarrow u = \sin^{-1}\left(\frac{\sqrt{2}}{2}\right) = \frac{\pi}{4}$. So the integral becomes

$$\int_0^{\sqrt{2}/2} e^{\sin^{-1} x}\, dx = \int_0^{\pi/4} e^u \cos u\, du.$$

Using integration by parts twice (similar to the solution in the previous problem), it can be shown that

$$\int e^u \cos u\, du = \frac{e^u}{2}\bigl(\cos u + \sin u\bigr) + C.$$

Now we summarize what we've done so far and then complete the exercise:

$$\int_0^{\sqrt{2}/2} e^{\sin^{-1} x}\, dx = \int_0^{\pi/4} e^u \cos u\, du \qquad \text{u-substitution}$$

$$= \frac{e^u}{2}\bigl(\cos u + \sin u\bigr)\bigg|_0^{\pi/4} \qquad \text{Integraton by parts (twice)}$$

$$= \frac{e^{\pi/4}}{2}\left(\frac{\sqrt{2}}{2} + \frac{\sqrt{2}}{2}\right) - \frac{1}{2} \qquad \text{Evaluate.}$$

$$= \frac{\sqrt{2}\, e^{\pi/4} - 1}{2}. \qquad \text{Simplify.}$$

8.6.63 We apply integration by parts with the following choices for our variables.

$$u = \ln x \qquad dv = x^a\, dx$$
$$du = \tfrac{1}{x} dx \qquad v = \frac{x^{a+1}}{a+1}$$

The integration process is straightforward:

$$\int x^a \ln x\, dx = \frac{x^{a+1} \ln x}{a+1} - \int \frac{x^a}{a+1}\, dx \qquad \text{Integration by parts}$$

$$= \frac{x^{a+1} \ln x}{a+1} - \frac{x^{a+1}}{(a+1)^2} + C \qquad \text{Evaluate.}$$

$$= \frac{x^{a+1}}{a+1}\left(\ln x - \frac{1}{a+1}\right) + C. \qquad \text{Factor.}$$

8.6.65 By factoring the denominator, we can apply the formula $\int \frac{dx}{\sqrt{a^2-x^2}} = \sin^{-1} \frac{x}{a} + C$:

$$\int_0^{1/6} \frac{dx}{\sqrt{1-9x^2}} = \int_0^{1/6} \frac{dx}{\sqrt{9(\frac{1}{9}-x^2)}} = \frac{1}{3}\int_0^{1/6} \frac{dx}{\sqrt{\left(\frac{1}{3}\right)^2 - x^2}} = \frac{1}{3}\sin^{-1}(3x)\bigg|_0^{1/6} = \frac{\pi}{18}.$$

8.6.67 $\sqrt{1-9x^2} = \sqrt{1-(3x)^2}$ suggests the change of variables $3x = \sin\theta$, $-\frac{\pi}{2} < \theta < \frac{\pi}{2}$. This substitution implies that $3\,dx = \cos\theta\,d\theta$ and

$$\sqrt{1-9x^2} = \sqrt{1-(3x)^2} = \sqrt{1-\sin^2\theta} = \sqrt{\cos^2\theta} = |\cos\theta| = \cos\theta.$$

Using these substitutions and the following reference triangle, the integral is evaluated as follows.

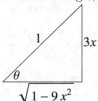

$$\int \frac{x^2}{\sqrt{1-(3x)^2}}\,dx = \frac{1}{27}\int \frac{\sin^2\theta}{\cos\theta}\cdot\cos\theta\,d\theta \qquad \text{Trigonometric substitution}$$

$$= \frac{1}{27}\int \sin^2\theta\,d\theta \qquad \text{Simplify.}$$

$$= \frac{1}{54}\int (1-\cos 2\theta)\,d\theta \qquad \sin^2\theta = \frac{1-\cos 2\theta}{2}$$

$$= \frac{1}{54}\left(\theta - \frac{1}{2}\sin 2\theta\right) + C \qquad \text{Evaluate.}$$

$$= \frac{1}{54}(\theta - \sin\theta\cos\theta) + C \qquad \frac{1}{2}\sin 2\theta = \sin\theta\cos\theta$$

$$= \frac{1}{54}\left(\sin^{-1} 3x - 3x\sqrt{1-9x^2}\right) + C \qquad \text{Reference triangle}$$

8.6.69 The presence of the expression $1-x^2$ in the integrand suggests a trigonometric substitution of $x = \sin\theta$, where $-\frac{\pi}{2} < \theta < \frac{\pi}{2}$. For this choice, $dx = \cos\theta\,d\theta$, $1-x^2 = 1-\sin^2\theta = \cos^2\theta$ and $\sqrt{1-x^2} = |\cos\theta| = \cos\theta$. This trigonometric substitution gets the integration process started.

$$\int \frac{dx}{1-x^2+\sqrt{1-x^2}} = \int \frac{\cos\theta}{\cos^2\theta + \cos\theta}\,d\theta \qquad \text{Trigonometric substitution}$$

$$= \int \frac{d\theta}{\cos\theta + 1} \qquad \text{Simplify.}$$

$$= \int \frac{1}{\cos\theta+1}\cdot\frac{1-\cos\theta}{1-\cos\theta}\,d\theta \qquad \text{Multiply by } \frac{1-\cos\theta}{1-\cos\theta}.$$

$$= \int \frac{1-\cos\theta}{\sin^2\theta}\,d\theta \qquad \text{Simplify.}$$

$$= \int (\csc^2\theta - \csc\theta\cot\theta)\,d\theta \qquad \frac{1-\cos\theta}{\sin^2\theta} = \frac{1}{\sin^2\theta} - \frac{\cos\theta}{\sin^2\theta} = \csc^2\theta - \csc\theta\cot\theta$$

$$= -\cot\theta + \csc\theta + C \qquad \text{Evaluate.}$$

Given that $\sin\theta = x$, a geometric description of the relationship between x and θ is given by the following reference triangle.

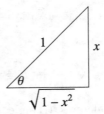

Therefore

$$\int \frac{dx}{1-x^2+\sqrt{1-x^2}} = -\cot\theta + \csc\theta = -\frac{\sqrt{1-x^2}}{x} + \frac{1}{x} + C = \frac{1-\sqrt{1-x^2}}{x} + C.$$

8.6.71 We start by multiplying the numerator and the denominator of the integrand by $(1-\cos x)$:

$$\frac{1-\cos x}{1+\cos x} \cdot \frac{1-\cos x}{1-\cos x} = \frac{1-2\cos x + \cos^2 x}{1-\cos^2 x} \qquad \text{Multiply by } \frac{1-\cos x}{1-\cos x}.$$

$$= \frac{1-2\cos x + \cos^2 x}{\sin^2 x} \qquad \text{Trigonometric Identity.}$$

$$= \csc^2 x - 2\cot x \csc x + \cot^2 x \qquad \text{Simplify.}$$

$$= \csc^2 x - 2\cot x \csc x + \csc^2 x - 1 \qquad \text{Replace } \cot^2 x \text{ with } \csc^2 x - 1.$$

$$= 2\csc^2 x - 2\cot x \csc x - 1 \qquad \text{Simplify.}$$

With the transformed integrand, the integral is easier to evaluate:

$$\int \frac{1-\cos x}{1+\cos x}\,dx = \int (2\csc^2 x - 2\cot x \csc x - 1)\,dx = -2\cot x + 2\csc x - x + C.$$

8.6.73 Letting $u = 1 + \sqrt{x}$ it follows that $du = \frac{dx}{2\sqrt{x}}$. Rearranging these equations, we have $\sqrt{x} = u - 1$ and $dx = 2\sqrt{x}\,du = 2(u-1)du$. After the substitution is made, the lower limit of integration is $u = 4$ and the upper limit is $u = 5$. Therefore

$$\int_9^{16} \sqrt{1+\sqrt{x}}\,dx = 2\int_4^5 \sqrt{u}(u-1)\,du$$

$$= 2\int_4^5 (u^{3/2} - u^{1/2})\,du$$

$$= 2\left(\frac{2}{5}u^{5/2} - \frac{2}{3}u^{3/2}\right)\bigg|_4^5$$

$$= 4u^{3/2}\left(\frac{u}{5} - \frac{1}{3}\right)\bigg|_4^5$$

$$= \frac{4u^{3/2}}{15}(3u-5)\bigg|_4^5$$

$$= \frac{40\sqrt{5}}{3} - \frac{224}{15}.$$

8.6.75 Letting $u = \tan^{-1}\sqrt{x}$, it follows that $du = \frac{1}{1+(\sqrt{x})^2} \cdot \frac{1}{2\sqrt{x}}dx = \frac{dx}{2(\sqrt{x}+x^{3/2})}$. Then $2du = \frac{dx}{\sqrt{x}+x^{3/2}}$. The lower limit of integration is $u = \tan^{-1} 1 = \frac{\pi}{4}$ and the upper limit is $u = \tan^{-1}\sqrt{3} = \frac{\pi}{3}$. Therefore

8.6. Integration Strategies

$$\int_1^3 \frac{\tan^{-1}\sqrt{x}}{x^{1/2}+x^{3/2}}\,dx = \int_{\pi/4}^{\pi/3} 2u\,du = u^2\Big|_{\pi/4}^{\pi/3} = \frac{7\pi^2}{144}.$$

8.6.77 We use integration by parts with the following choices for the variables:

$$u = \cos^{-1} x \qquad dv = dx$$

$$du = -\frac{dx}{\sqrt{1-x^2}} \qquad v = x$$

Now we apply integration by parts followed by integration by substitution.

$$\int \cos^{-1} x\,dx = x\cos^{-1} x + \int \frac{x}{\sqrt{1-x^2}}\,dx \qquad \text{Integration by parts}$$
$$= x\cos^{-1} x - \frac{1}{2}\int u^{-1/2}\,du \qquad u = 1-x^2;\ du = -2x\,dx$$
$$= x\cos^{-1} x - u^{1/2} + C \qquad \text{Integrate.}$$
$$= x\cos^{-1} x - \sqrt{1-x^2} + C. \qquad \text{Replace } u \text{ with } 1-x^2.$$

8.6.79 We apply integration by parts with the following choices for the variables.

$$u = \sin^{-1} x \qquad dv = \frac{dx}{x^2}$$

$$du = \frac{dx}{\sqrt{1-x^2}} \qquad v = -\frac{1}{x}$$

Using these substitutions with integration by parts, along with the answer to Exercise 25, we evaluate the integral.

$$\int \frac{\sin^{-1} x}{x^2}\,dx = -\frac{\sin^{-1} x}{x} + \int \frac{dx}{x\sqrt{1-x^2}} = -\frac{\sin^{-1} x}{x} + \ln\left|\frac{x}{1+\sqrt{1-x^2}}\right| + C.$$

8.6.81 Factoring the denominator, we have

$$x^5 + 2x^3 + x = x(x^4 + 2x^2 + 1) = x(x^2+1)^2.$$

So the partial fraction decomposition of the integrand has the form

$$\frac{x^4 + 2x^3 + 5x^2 + 2x + 1}{x(x^2+1)^2} = \frac{A}{x} + \frac{Bx+C}{x^2+1} + \frac{Dx+E}{(x^2+1)^2}.$$

Multiplying by $x(x^2+1)^2$, we have

$$x^4 + 2x^3 + 5x^2 + 2x + 1 = A(x^2+1)^2 + (Bx+C)x(x^2+1) + (Dx+E)x.$$

To find the constants, we expand the right hand side of equation and group equal powers of x:

$$x^4 + 2x^3 + 5x^2 + 2x + 1 = (A+B)x^4 + Cx^3 + (2A+B+D)x^2 + (C+E)x + A.$$

Comparing coefficients of equal powers, we see that $A = 1$, $B = 0$, $C = 2$, $D = 3$, and $E = 0$. Using these values, the integral is evaluated with the help of a u-substitution in the third integral.

$$\int \frac{x^4 + 2x^3 + 5x^2 + 2x + 1}{x(x^2+1)^2} dx = \int \frac{dx}{x} + \int \frac{2}{x^2+1} dx + \int \frac{3x}{(x^2+1)^2} dx \quad \text{Partial fractions.}$$

$$= \int \frac{dx}{x} + \int \frac{2}{x^2+1} dx + \frac{3}{2} \int \frac{du}{u^2} \quad u = x^2+1, \; \frac{1}{2}du = x\, dx$$

$$= \ln|x| + 2\tan^{-1} x - \frac{3}{2u} + C \quad \text{Evaluate.}$$

$$= \ln|x| + 2\tan^{-1} x - \frac{3}{2(x^2+1)} + C. \quad \text{Replace } u \text{ with } x^2+1.$$

8.6.83 We simplify the integrand first using the substitution $u = e^x$, so that $du = e^x dx$ and

$$\int e^x \sin^{998}(e^x) \cos^3(e^x)\, dx = \int \sin^{998} u \cos^3 u\, du.$$

We then continue the integration process, by first splitting $\cos^3 u$.

$$\int e^x \sin^{998}(e^x) \cos^3(e^x)\, dx = \int \sin^{998} u \cos^3 u\, du \quad u = e^x,\; du = e^x dx$$

$$= \int \sin^{998} u \cos^2 u \cos u\, du \quad \text{Split } \cos^3 u.$$

$$= \int \sin^{998} u (1 - \sin^2 u) \cos u\, du \quad \cos^2 x = 1 - \sin^2 x$$

$$= \int (\sin^{998} u - \sin^{1000} u) \cos u\, du \quad \text{Expand.}$$

$$= \int (w^{998} - w^{1000})\, dw \quad w = \sin u,\; dw = \cos u\, du$$

$$= \frac{w^{999}}{999} - \frac{w^{1001}}{1000} + C \quad \text{Integrate.}$$

$$= \frac{\sin^{999} u}{999} - \frac{\sin^{1001} u}{1001} + C \quad \text{Replace } w \text{ with } \sin u.$$

$$= \frac{\sin^{999}(e^x)}{999} - \frac{\sin^{1001}(e^x)}{1001} + C \quad \text{Replace } u \text{ with } e^x.$$

8.6.85

a. True. For example, the method of partial fractions or a trigonometric substitution can be used to evaluate the integral.

b. True. See Example 4.

c. False. It is easiest to use the substitution $u = \tan 3x$.

d. False. The integral $\int \frac{u^2}{u-1} du$ would imply that $du = dx$, which is not the case here.

8.6.87 The surface area is given by $S = 2\pi \int_0^{\pi/2} \sin x \sqrt{1 + \cos^2 x}\, dx$. Let $u = \cos x$ so that $du = -\sin x\, dx$. Then $S = -2\pi \int_1^0 \sqrt{1+u^2}\, du = 2\pi \int_0^1 \sqrt{1+u^2}\, du$. Now let $u = \tan\theta$ with $du = \sec^2\theta\, d\theta$. Substituting gives

$$2\pi \int_0^{\pi/4} \sec^3\theta\, d\theta = \pi (\sec\theta \tan\theta + \ln|\sec\theta + \tan\theta|) \Big|_0^{\pi/4} = \pi\left(\sqrt{2} + \ln(1+\sqrt{2})\right).$$

8.6. Integration Strategies

The last evaluated integral came from the reduction formula for $\int \sec^3 \theta \, d\theta$ together with the integral formula for $\int \sec \theta \, d\theta$.

8.6.89 We use the Disk method.

$$V = \int_0^{\pi/4} \pi \frac{dx}{(1-\sin x)^2} = \pi \int_0^{\pi/4} \frac{(1+\sin x)^2}{(1-\sin x)^2(1+\sin x)^2} \, dx = \pi \int_0^{\pi/4} \frac{1 + 2\sin x + \sin^2 x}{(1-\sin^2 x)^2} \, dx.$$

$$= \pi \int_0^{\pi/4} \frac{1 + 2\sin x + \sin^2 x}{\cos^4 x} \, dx$$

$$= \pi \int_0^{\pi/4} \sec^4 x \, dx + \pi \int_0^{\pi/4} 2 \tan x \sec^3 x \, dx + \pi \int_0^{\pi/4} \tan^2 x \sec^2 x \, dx.$$

For the first integral in the sum, write the integrand as $\sec^4 x = \sec^2 x \sec^2 x = (1 + \tan^2 x) \sec^2 x$. Then let $u = \tan x$ so that $du = \sec^2 x \, dx$. The first integral is equal to

$$\pi \int_0^1 (1 + u^2) \, du = \pi \left(u + \frac{u^3}{3} \right) \Big|_0^1 = \pi \left(1 + \frac{1}{3} \right) = \frac{4\pi}{3}.$$

For the second integral in the above sum, let $u = \sec x$ so that $du = \sec x \tan x \, dx$. We have

$$\pi \int_1^{\sqrt{2}} 2u^2 \, du = \frac{2\pi u^3}{3} \Big|_1^{\sqrt{2}} = \frac{4\sqrt{2}\pi}{3} - \frac{2\pi}{3}.$$

For the third integral in the sum, let $u = \tan x$ so that $du = \sec^2 x \, dx$. Then that last integral is equal to

$$\pi \int_0^1 u^2 \, du = \frac{\pi u^3}{3} \Big|_0^1 = \frac{\pi}{3}.$$

Adding these three results together gives

$$V = \pi \left(\frac{4}{3} + \frac{4\sqrt{2}}{3} - \frac{2}{3} + \frac{1}{3} \right) = \pi \left(\frac{4\sqrt{2}}{3} + 1 \right) = \frac{\pi}{3} \left(4\sqrt{2} + 3 \right).$$

8.6.91 The work is given by

$$W = \int_0^{\pi/3} 1000(9.8)\pi \sec^2 y \left(\frac{\pi}{3} - y \right) dy = 9800\pi \left(\int_0^{\pi/3} \frac{\pi}{3} \sec^2 y \, dy - \int_0^{\pi/3} y \sec^2 y \, dy \right).$$

For the second integral, we integrate by parts. Let $u = y$ and $dv = \sec^2 y \, dy$. Then $du = dy$ and $v = \tan y$. Then

$$\int \sec^2 y \, dy = y \tan y - \int \tan y \, dy = y \tan y + \ln |\cos y| + C.$$

Then

$$W = 9800\pi \left(\frac{\pi}{3} \tan y \Big|_0^{\pi/3} - (y \tan y + \ln |\cos y|) \Big|_0^{\pi/3} \right) = 9800\pi \left(\frac{\pi}{3} \sqrt{3} - \left(\frac{\pi}{3} \sqrt{3} + \ln \frac{1}{2} \right) \right) = 9800\pi \ln 2.$$

8.6.93 Start with a substitution of $u = e^x$. Then $du = e^x \, dx$, which implies that $dx = \frac{du}{e^x} = \frac{du}{u}$. Making this substitution, we have

$$\int \frac{68}{e^{2x} + 2e^x + 17} \, dx = \int \frac{68}{u(u^2 + 2u + 17)} \, du.$$

The denominator has one simple linear factor u and one irreducible quadratic factor $u^2 + 2u + 17$. Therefore the partial fraction decomposition has the form

$$\frac{68}{u(u^2 + 2u + 17)} = \frac{A}{u} + \frac{Bu + C}{u^2 + 2u + 17}.$$

Multiplying both sides by $u(u^2 + 2u + 17)$, we have

$$68 = A(u^2 + 2u + 17) + (Bu + C)u = (A + B)u^2 + (2A + C)u + 17A.$$

By equating coefficients of equal powers of x, we find that $A = 4$, $B = -4$, and $C = -8$. Using these values, we have

$$\int \frac{68}{u(u^2 + 2u + 17)}\,du = \int \left(\frac{4}{u} + \frac{-4u - 8}{u^2 + 2u + 17}\right)du = \int \frac{4}{u}\,du - \int \frac{4u + 8}{u^2 + 2u + 17}\,du.$$

The first integral on the right equals $4 \ln|u| + C$. For the second integral on the right, we use a substitution of $w = u^2 + 2u + 17$, which implies that $dw = (2u + 2)du$. We write the numerator in the second integral as $4u + 8 = 2(2u + 2) + 4$ and split the integral:

$$\int \frac{68}{u(u^2 + 2u + 17)}\,du = \int \frac{4}{u}\,du - 2 \underbrace{\int \frac{2u + 2}{u^2 + 2u + 17}\,du}_{w = u^2 + 2u + 17} - 4 \underbrace{\int \frac{du}{u^2 + 2u + 17}}_{(u+1)^2 + 16}$$

$$= 4 \ln|u| - 2 \ln|u^2 + 2u + 17| - \tan^{-1}\left(\frac{u + 1}{4}\right) + C.$$

Replacing u with e^x and then simplifying, we have

$$\int \frac{68}{e^{2x} + 2e^x + 17}\,dx = 4 \ln e^x - 2 \ln|e^{2x} + 2e^x + 17| - \tan^{-1}\left(\frac{e^x + 1}{4}\right) + C$$

$$= 4x - 2 \ln(e^{2x} + 2e^x + 17) - \tan^{-1}\left(\frac{e^x + 1}{4}\right) + C.$$

8.6.95 Let $u = \tan x$. Then $du = \sec^2 x\,dx$, and

$$dx = \frac{du}{\sec^2 x} = \frac{du}{\tan^2 x + 1} = \frac{du}{u^2 + 1}.$$

With this change of variable, the integral becomes

$$\int \frac{dx}{1 - \tan^2 x} = \int \frac{du}{(1 - u^2)(u^2 + 1)} = \int \frac{du}{(1 + u)(1 - u)(u^2 + 1)}.$$

The form of the partial fraction decomposition of the integrand on the right is

$$\frac{1}{(1 + u)(1 - u)(u^2 + 1)} = \frac{A}{1 + u} + \frac{B}{1 - u} + \frac{Cu + D}{u^2 + 1}.$$

Multiplying both sides by $(1 + u)(1 - u)(u^2 + 1)$ and then regrouping terms, we have

$$1 = A(1 - u)(u^2 + 1) + B(1 + u)(u^2 + 1) + (Cu + D)(1 - u^2)$$
$$= (-A + B - C)u^3 + (A + B - D)u^2 + (-A + B + C)u + A + B + D.$$

By comparing coefficients of equal powers, we obtain the system of equations

8.6. Integration Strategies

$$-A+B-C=0, A+B-D=0, -A+B+C=0, A+B+D=1.$$

You can verify that the values satisfying this system are $A=1/4$, $B=1/4$, $C=0$, $D=1/2$. The partial fraction decomposition makes the integration process straightforward:

$$\int \frac{dx}{1-\tan^2 x} = \int \frac{du}{(1-u^2)(u^2+1)} \qquad \text{u-substitution}$$

$$= \frac{1}{4}\int \frac{du}{1+u} + \frac{1}{4}\int \frac{du}{1-u} + \frac{1}{2}\int \frac{du}{u^2+1} \qquad \text{Partial fractions}$$

$$= \frac{1}{4}\ln|u+1| - \frac{1}{4}\ln|u-1| + \frac{1}{2}\tan^{-1} u + C \qquad \text{Integrate.}$$

$$= \frac{1}{4}\ln|\tan x+1| - \frac{1}{4}\ln|\tan x-1| + \frac{1}{2}\tan^{-1}(\tan x) + C \qquad \text{Replace u with $\tan x$.}$$

$$= \frac{1}{4}\ln|\tan x+1| - \frac{1}{4}\ln|\tan x-1| + \frac{x}{2} + C \qquad \text{Simplify.}$$

8.6.97 We integrate by parts with the following variable assignments.

$$u = \tan^{-1}\sqrt[3]{x} \qquad dv = dx$$

$$du = \frac{dx}{3x^{2/3}(1+x^{2/3})} \qquad v = x$$

Integrating by parts, we have

$$\int \tan^{-1}\sqrt[3]{x}\, dx = x\tan^{-1}\sqrt[3]{x} - \int \frac{x^{1/3}}{3(1+x^{2/3})}\, dx.$$

For the integral on the right, we make a substitution of $v = x^{1/3}$ which implies that $dv = \frac{dx}{3x^{2/3}}$, and $dx = 3v^2\, dv$. With this change of variable, we have

$$\int \frac{x^{1/3}}{3(1+x^{2/3})}\, dx = \int \frac{v}{3(1+v^2)} 3v^2\, dv = \int \frac{v^3}{1+v^2}\, dv.$$

It follows by long division that $\frac{v^3}{1+v^2} = v - \frac{v}{1+v^2}$. Letting $w = 1+v^2$, it follows that $dw = 2v\, dv$ and

$$\int \frac{v^3}{1+v^2}\, dv = \int v\, dv - \int \frac{v}{1+v^2}\, dv$$

$$= \int v\, dv - \frac{1}{2}\int \frac{dw}{w}$$

$$= \frac{v^2}{2} - \frac{1}{2}\ln|w| + C$$

$$= \frac{v^2}{2} - \frac{1}{2}\ln(v^2+1) + C.$$

Summarizing and then completing our work, we have

$$\int \tan^{-1} \sqrt[3]{x}\, d = x\tan^{-1}\sqrt[3]{x} - \int \frac{x^{1/3}}{3(1+x^{2/3})}\,dx \qquad \text{Integration by parts}$$

$$= x\tan^{-1}\sqrt[3]{x} - \int \frac{v^3}{1+v^2}\,dv \qquad v = x^{1/3},\ dv = \frac{dx}{3x^{2/3}}$$

$$= x\tan^{-1}\sqrt[3]{x} - \left(\int v\,dv - \int \frac{v}{1+v^2}\,dv\right) \qquad \frac{v^3}{1+v^2} = v - \frac{v}{1+v^2}$$

$$= x\tan^{-1}\sqrt[3]{x} - \left(\frac{v^2}{2} - \frac{1}{2}\ln(1+v^2)\right) + C \qquad \text{Integrate.}$$

$$= x\tan^{-1}\sqrt[3]{x} - \frac{x^{2/3}}{2} + \frac{1}{2}\ln(1+x^{2/3}) + C. \qquad \text{Replace } v \text{ with } x^{1/3};\ \text{simplify.}$$

8.6.99 The surface area is given by

$$S = \int_0^{\pi/4} 2\pi \tan x\sqrt{1+\sec^4 x}\,dx = 2\pi\int_0^{\pi/4} \frac{\sqrt{1+\sec^4 x}\,\tan x\sec x}{\sec x}\,dx.$$

Let $u = \sec x$ so that $du = \sec x\tan x\,dx$. Substituting gives $2\pi \int_1^{\sqrt{2}} \frac{\sqrt{1+u^4}}{u}\,du$. Now let $u^2 = \tan\theta$ so that $2u\,du = \sec^2\theta\,d\theta$. Substituting gives

$$2\pi \int_{\pi/4}^{\tan^{-1} 2} \frac{\sec\theta}{\sqrt{\tan\theta}} \cdot \frac{\sec^2\theta}{2\sqrt{\tan\theta}}\,d\theta = \pi \int_{\pi/4}^{\tan^{-1} 2} \frac{\sec^3\theta}{\tan\theta}\cdot\frac{\tan\theta}{\tan\theta}\,d\theta = \pi \int_{\pi/4}^{\tan^{-1} 2} \frac{\sec^2\theta\sec\theta\tan\theta}{\sec^2\theta - 1}\,d\theta.$$

Now let $w = \sec\theta$ so that $dw = \sec\theta\tan\theta\,d\theta$. Then

$$S = \pi\int_{\sqrt{2}}^{\sqrt{5}} \frac{w^2}{w^2 - 1}\,dw.$$

Using long division and then a partial fraction decomposition, we can write this integrand as

$$\frac{w^2}{w^2-1} = 1 - \frac{1/2}{w+1} + \frac{1/2}{w-1}.$$

Therefore

$$S = \pi \int_{\sqrt{2}}^{\sqrt{5}} \left(1 - \frac{1}{2(w+1)} + \frac{1}{2(w-1)}\right)\,dw = \pi\left(w - \frac{1}{2}\ln(w+1) + \frac{1}{2}\ln(w-1)\right)\Big|_{\sqrt{2}}^{\sqrt{5}}$$

$$= \pi\left(\sqrt{5} - \sqrt{2} + \frac{1}{2}\ln\frac{\sqrt{5}-1}{\sqrt{5}+1} - \frac{1}{2}\ln\frac{\sqrt{2}-1}{\sqrt{2}+1}\right) \approx 3.83908.$$

8.7 Other Methods of Integration

8.7.1 The power rule, substitution, integration by parts, and partial fraction decomposition are examples of analytical methods.

8.7.3 The computer algebra system may use a different algorithm than whoever prepared the table, so the results may look different – however, they should differ by a constant.

8.7.5 Let $u = e^x$. Then $du = e^x\,dx$. Substituting yields

$$\int \cos^3 u\,du = -\frac{1}{3}\sin^3 u + \sin u + C = -\frac{1}{3}\sin^3(e^x) + \sin(e^x) + C.$$

8.7. Other Methods of Integration

8.7.7 Using table entry 17, we have $\int \cos^{-1}(x)\,dx = x\cos^{-1}x - \sqrt{1-x^2} + C$.

8.7.9 Using table entry 77 for $\int \frac{1}{\sqrt{x^2+a^2}}\,dx$, we see that $\int \frac{1}{\sqrt{x^2+16}}\,dx = \ln(x+\sqrt{x^2+16}) + C$.

8.7.11 Using table entry 90 for $\int \frac{x}{ax+b}\,dx$, we see that $\int \frac{3u}{2u+7}\,du = \frac{3u}{2} - \frac{21}{4}\ln|2u+7| + C$.

8.7.13 Using table entry 47, we have $\int \frac{dx}{1-\cos 4x} = -\frac{1}{4}\cot(2x) + C$.

8.7.15 Using table entry 94 for $\int \frac{x}{\sqrt{ax+b}}\,dx$, we have that $\int \frac{x\,dx}{\sqrt{4x+1}} = \frac{1}{24}(4x-2)\sqrt{4x+1} + C = \frac{1}{12}(2x-1)\sqrt{4x+1} + C$.

8.7.17 Using table entry 69 for $\int \frac{1}{\sqrt{x^2-a^2}}\,dx$ we have $\int \frac{1}{\sqrt{9x^2-100}}\,du = \frac{1}{3}\int \frac{1}{\sqrt{u^2-100}}\,du$ where $u = 3x$. This is then equal to $\frac{1}{3}\ln|u+\sqrt{u^2-100}| + C = \frac{1}{3}\ln|3x+\sqrt{9x^2-100}| + C$. This can be written as $\frac{1}{3}\ln|x+\sqrt{x^2-(100/9)}| + C$.

8.7.19 $\int \frac{e^x}{\sqrt{e^{2x}+4}}\,dx = \int \frac{1}{\sqrt{u^2+4}}\,du$ where $u = e^x$. Then using table entry 77, we have $\ln(u+\sqrt{u^2+4}) + C = \ln(e^x + \sqrt{e^{2x}+4}) + C$.

8.7.21 $\int \frac{\cos x}{\sin^2 x + 2\sin x}\,dx = \int \frac{1}{u^2+2u}\,du = \int \frac{1}{(u+1)^2-1}\,du$ where $u = \sin x$. Then using table entry 74, we have
$$-\frac{1}{2}\ln\left|\frac{u+2}{u}\right| + C = -\frac{1}{2}\ln\left|\frac{\sin x + 2}{\sin x}\right| + C.$$

8.7.23 Let $u = \ln x$, so that $du = \frac{1}{x}dx$. Substituting yields $\int u\sin^{-1}(u)\,du = \frac{2u^2-1}{4}\sin^{-1}(u) + \frac{u\sqrt{1-u^2}}{4} + C = \frac{2\ln^2 x - 1}{4}\sin^{-1}(\ln x) + \frac{\ln x\sqrt{1-\ln^2 x}}{4} + C$, where we used table entry 105.

8.7.25 Using table entry 84 for $\int \frac{1}{(a^2+x^2)^{3/2}}\,dx$, we have $\int \frac{1}{(16+9x^2)^{3/2}}\,dx = \frac{1}{3}\int \frac{1}{(16+u^2)^{3/2}}\,du = \frac{1}{3}\cdot \frac{3x}{16\sqrt{16+9x^2}} + C = \frac{x}{16\sqrt{16+9x^2}} + C$.

8.7.27 Using table entry 62 for $\int \frac{1}{x\sqrt{a^2-x^2}}\,dx$, we have $\int \frac{1}{x\sqrt{(12)^2-x^2}}\,dx = \frac{-1}{12}\ln\left|\frac{12+\sqrt{144-x^2}}{x}\right| + C$.

8.7.29 Using table entry 103 for $\int \ln^n x\,dx$ we have that $\int \ln^2 x\,dx = x\ln^2 x - 2\int \ln x\,dx = x\ln^2 x - 2(x\ln x - x) + C = 2x + x\ln^2 x - 2x\ln x + C$.

8.7.31 Note that $\sqrt{x^2+10x} = \sqrt{x^2+10x+25-25} = \sqrt{(x+5)^2-5^2}$. Thus, the given integral is equal to $\int \sqrt{u^2-5^2}\,du$ where $u = x+5$. Using table entry 68 we have that this is equal to $\frac{u}{2}\sqrt{u^2-25} - \frac{25}{2}\ln|u+\sqrt{u^2-25}| + C = \frac{x+5}{2}\sqrt{(x+5)^2-25} - \frac{25}{2}\ln|x+5+\sqrt{(x+5)^2-25}| + C$.

8.7.33 $\int \dfrac{1}{x^2+2x+10}\,dx = \int \dfrac{1}{(x+1)^2+9}\,dx$. Using table entry 14 this is equal to $\dfrac{1}{3}\tan^{-1}\left(\dfrac{x+1}{3}\right)+C$.

8.7.35 $\int \dfrac{1}{x(x^{10}+1)}\,dx = \int \dfrac{10x^9}{10x^{10}(x^{10}+1)}\,dx$. Let $u=x^{10}+1$ so that $du=10x^9\,dx$. Substituting yields $\int \dfrac{1}{10}\cdot\dfrac{1}{(u-1)u}\,du$. Using table entry 96 for $\int \dfrac{dx}{x(ax+b)}$ we have

$$\dfrac{1}{10}\ln\left|\dfrac{u-1}{u}\right|+C = \dfrac{1}{10}\ln\left|\dfrac{x^{10}}{x^{10}+1}\right|+C.$$

8.7.37 $\int \dfrac{1}{\sqrt{x^2-6x}}\,dx = \int \dfrac{1}{\sqrt{(x-3)^2-9}}\,dx$. Using table entry 69, this is equal to $\ln|x-3+\sqrt{(x-3)^2-9}|+C$. This can also be written as $2\ln(\sqrt{x-6}+\sqrt{x})+C$.

8.7.39 Let $u=x^3$, so that $du=3x^2\,dx$. Substituting yields

$$\dfrac{1}{3}\int \dfrac{\tan^{-1}(u)}{u^2}\,du = -\dfrac{1}{3}\left(\dfrac{1}{u}\tan^{-1}(u)-\int \dfrac{1}{u(1+u^2)}\,du\right) = -\dfrac{1}{3}\left(\dfrac{\tan^{-1}(x^3)}{x^3}\right)+\dfrac{1}{3}\int \dfrac{1}{u(1+u^2)}\,du,$$

where we used table entry 18. Now let $w=u^2+1$ so that $dw=2u\,du$. This last integral is thus equal to

$$\dfrac{1}{3}\int \dfrac{2u\,du}{2u^2(1+u^2)} = \dfrac{1}{6}\int \dfrac{dw}{(w-1)w} = -\dfrac{1}{6}\ln\left|\dfrac{w}{w-1}\right|+C = -\dfrac{1}{6}\ln\left|\dfrac{u^2+1}{u^2}\right|+C = -\dfrac{1}{6}\ln\left|\dfrac{x^6+1}{x^6}\right|+C,$$

where we used table entry 96. Thus the original integral is equal to $-\dfrac{1}{3}\left(\dfrac{\tan^{-1}(x^3)}{x^3}\right)-\dfrac{1}{6}\ln\left|\dfrac{x^6+1}{x^6}\right|+C$. This can be written as $-\dfrac{\tan^{-1}(x^3)}{3x^3}+\ln\left|\dfrac{x}{(x^6+1)^{1/6}}\right|+C$.

8.7.41 The integral which gives the length of the curve is $\dfrac{1}{2}\int_0^8 \sqrt{4+x^2}\,dx$. Using table entry 76, we have

$$\dfrac{1}{2}\left(\dfrac{x}{2}\sqrt{4+x^2}+\dfrac{4}{2}\ln(x+\sqrt{4+x^2})\right)\bigg|_0^8 = 4\sqrt{17}+\ln(8+2\sqrt{17})-\ln 2 = 4\sqrt{17}+\ln(4+\sqrt{17})\approx 18.59.$$

8.7.43 The integral which gives the length of the curve is $\int_0^{\ln 2}\sqrt{1+e^{2x}}\,dx = \int_1^2 \dfrac{\sqrt{1+u^2}}{u}\,du$ where $u=e^x$. This is equal to

$$\left(\sqrt{1+u^2}-\ln\left|\dfrac{1+\sqrt{1+u^2}}{u}\right|\right)\bigg|_1^2 = \sqrt{5}-\sqrt{2}+\ln(1+\sqrt{2})-\ln\left(\dfrac{1+\sqrt{5}}{2}\right)\approx 1.22.$$

Note that we used table entry 83.

8.7.45 Using the method of shells, we have $\dfrac{V}{2\pi} = \int_0^{12}\dfrac{x}{\sqrt{x+4}}\,dx = \dfrac{2}{3}(x-8)\sqrt{x+4}\bigg|_0^{12} = \dfrac{32}{3}+\dfrac{32}{3} = \dfrac{64}{3}$. Thus $V=\dfrac{128\pi}{3}$. We used table entry 94 to compute the integral.

8.7.47 Using the method of disks, we have $\dfrac{V}{\pi} = \int_0^{\pi/2}\sin^2 y\,dy = \dfrac{y}{2}-\dfrac{\sin 2y}{4}\bigg|_0^{\pi/2} = \dfrac{\pi}{4}$. Thus $V=\dfrac{\pi^2}{4}$. We used table entry 31 to compute the antiderivative.

8.7.49 $\int \dfrac{x}{\sqrt{2x+3}}\,dx = \dfrac{1}{3}(x-3)\sqrt{3+2x}+C$.

8.7.51 $\int \tan^2 3x\, dx = \frac{1}{3} \tan 3x - x + C.$

8.7.53 $\int_0^4 (9+x^2)^{3/2}\, dx = \frac{1540 + 243 \ln 3}{8} \approx 225.870.$

8.7.55 $\int \frac{(x^2-a^2)^{3/2}}{x}\, dx = \frac{1}{3}(x^2-a^2)^{3/2} - a^2\sqrt{x^2-a^2} + a^3 \cos^{-1}(a/x) + C.$

8.7.57 $\int_0^{\pi/2} \frac{1}{1+\tan^2 t}\, dt = \frac{\pi}{4} \approx 0.785.$

8.7.59 $\int (a^2-x^2)^{3/2}\, dx = \frac{-x}{8}(2x^2-5a^2)\sqrt{a^2-x^2} + \frac{3a^4}{8}\sin^{-1}(x/a) + C.$

8.7.61 $\int_0^1 \ln x \ln(1+x)\, dx = 2 - \frac{\pi^2}{12} - \ln 4 \approx -0.209.$

8.7.63 $\int_1^2 \frac{x^{19}}{\sqrt{x^4-1}}\, dx = \frac{27{,}456\sqrt{15}}{7} \approx 15{,}190.9.$

8.7.65 $\int x^3 e^{2x}\, dx = \frac{1}{2}x^3 e^{2x} - \frac{3}{2}\int x^2 e^{2x}\, dx = \frac{1}{2}x^3 e^{2x} - \frac{3}{2}\left(\frac{1}{2}x^2 e^{2x} - \int x e^{2x}\, dx\right) = \frac{1}{2}x^3 e^{2x} - \frac{3}{4}x^2 e^{2x} + \frac{3}{2}\left(\frac{1}{2}x e^{2x} - \frac{1}{2}\int e^{2x}\, dx\right) = \frac{1}{2}x^3 e^{2x} - \frac{3}{4}x^2 e^{2x} + \frac{3}{4}x e^{2x} - \frac{3}{8}e^{2x} + C.$

8.7.67 Let $u = 3y$. Then $\int \tan^4 3y\, dy = \frac{1}{3}\int \tan^4 u\, du = \frac{1}{3}\left(\frac{1}{3}\tan^3 u - \int \tan^2 u\, du\right) = \frac{1}{9}\tan^3 u - \frac{1}{3}\left(\tan u - \int du\right) = \frac{1}{9}\tan^3 3y - \frac{1}{3}\tan 3y + y + C.$

8.7.69 By inspection, the curves intersect at $x=1$ where they both have value $\sqrt{2}$. The area we are seeking is given by $\int_0^1 \left(\frac{2}{\sqrt{\sqrt{x}+1}} - \sqrt{\sqrt{x}+x}\right) dx$. Using a computer algebra system, we find that
$$\int_0^1 \left(\frac{2}{\sqrt{\sqrt{x}+1}} - \sqrt{\sqrt{x}+x}\right) dx = \frac{1}{24}\left(128 - 78\sqrt{2} - 3\ln(3 + 2\sqrt{2})\right).$$

8.7.71 Let $u = ax + b$, so that $du = a\, dx$. Then we have
$$\frac{1}{a}\int \frac{ax+b-b}{ax+b}\, dx = \frac{1}{a}\int \left(1 - \frac{b}{ax+b}\right) dx = \frac{1}{a}\left(x - \frac{b}{a}\int \frac{1}{u}\, du\right)$$
$$= \frac{1}{a}\left(x - \frac{b}{a}\ln|u|\right) + C = \frac{1}{a}\left(x - \frac{b}{a}\ln|ax+b|\right) + C.$$

8.7.73 Let $u = ax + b$, so that $du = a\, dx$. Then we have
$$\frac{1}{a}\int \frac{u-b}{a}u^n\, du = \frac{1}{a^2}\int (u^{n+1} - bu^n)\, du = \frac{1}{a^2}\left(\frac{u^{n+2}}{n+2} - \frac{bu^{n+1}}{n+1}\right) + C$$
$$= \frac{1}{a^2}\left(\frac{(ax+b)^{n+2}}{n+2} - \frac{b(ax+b)^{n+1}}{n+1}\right) + C.$$

8.7.75

a. True. This is because these are equal for $x > 1$.

b. True. Note that $\dfrac{1}{9} = 0.\overline{11}$.

8.7.77 The two answers differ by a constant, namely the constant one. This can be seen as follows:

$$\dfrac{2\sin(x/2)}{\cos(x/2) + \sin(x/2)} = \dfrac{2\tan(x/2)}{1 + \tan(x/2)} = 2\dfrac{\sin x}{1 + \cos x} \cdot \dfrac{1}{1 + \frac{\sin x}{1+\cos x}}$$

$$= \dfrac{2\sin x}{1 + \cos x + \sin x} = \dfrac{2\sin x}{1 + \cos x + \sin x} \cdot \dfrac{1 - \cos x - \sin x}{1 - \cos x - \sin x}$$

$$= \dfrac{2\sin x(1 - \cos x - \sin x)}{-2\sin x \cos x} = \dfrac{\sin x + \cos x - 1}{\cos x} = \dfrac{\sin x - 1}{\cos x} + 1.$$

8.7.79 Let $u = 2x$. Then

$$\int x \sin^{-1} 2x \, dx = \dfrac{1}{4} \int u \sin^{-1} u \, du = \dfrac{2u^2 - 1}{16} \sin^{-1} u + \dfrac{u\sqrt{1 - u^2}}{16} + C = \dfrac{8x^2 - 1}{16} \sin^{-1} 2x + \dfrac{x\sqrt{1 - 4x^2}}{8} + C.$$

8.7.81 $\displaystyle\int x^{-2} \tan^{-1} x \, dx = -(1/x) \tan^{-1} x + \int \dfrac{1}{x(1 + x^2)} \, dx = \dfrac{-\tan^{-1} x}{x} + \ln\left(\dfrac{|x|}{\sqrt{1 + x^2}}\right) + C.$

8.7.83

a. Note that $\displaystyle\int_0^b f(x) \, dx = \int_0^{b/2} f(x) \, dx + \int_{b/2}^b f(x) \, dx$. For the second integral, let $u = b - x$. Then $du = -dx$. The second integral is equal to $-\displaystyle\int_{b-(b/2)}^0 f(b - u) \, du = \int_0^{b/2} f(b - u) \, du$. Because this can be written as $\displaystyle\int_0^{b/2} f(b - x) \, dx$, we have $\displaystyle\int_0^b f(x) \, dx = \int_0^{b/2} (f(x) + f(b - x)) \, dx$.

b.

$$\int_0^{\pi/4} \ln(1 + \tan x) \, dx = \int_0^{\pi/8} (\ln(1 + \tan x) + \ln(1 + \tan(\pi/4 - x))) \, dx$$

$$= \int_0^{\pi/8} \left(\ln(1 + \tan x) + \ln\left(1 + \dfrac{\tan \pi/4 - \tan x}{1 + \tan \pi/4 \tan x}\right)\right) dx$$

$$= \int_0^{\pi/8} \left(\ln(1 + \tan x) + \ln\left(1 + \dfrac{1 - \tan x}{1 + \tan x}\right)\right) dx$$

$$= \int_0^{\pi/8} \left(\ln(1 + \tan x) + \ln\left(\dfrac{2}{1 + \tan x}\right)\right) = \int_0^{\pi/8} \ln(2) \, dx = \dfrac{\pi \ln 2}{8}.$$

8.7.85

a. Using a computer algebra system, we have

θ_0	T	Relative Error
0.1	6.27927	0.000623603
0.2	6.26762	0.0024778
0.3	6.24854	0.0051388
0.4	6.22253	0.00965413
0.5	6.19021	0.0147967
0.6	6.15236	0.0208215
0.7	6.10979	0.0275963
0.8	6.06338	0.0349831
0.9	6.01399	0.0428433
1.0	5.96247	0.0510427

b. All of these values are within 10 percent of 2π.

8.7.87

a. The result holds.

b. First note that $\int_0^{\pi/2} \cos^n x\, dx = -\int_{\pi/2}^0 \cos^n(\pi/2 - \theta)\, d\theta = \int_0^{\pi/2} \sin^n \theta\, d\theta$. So we only need to show the result for $\int_0^{\pi/2} \sin^{10} x\, dx$. Repeatedly using the reduction formula we have

$$\int_0^{\pi/2} \sin^{10} x\, dx = \frac{9}{10}\int_0^{\pi/2} \sin^8 x\, dx = \frac{9 \cdot 7}{10 \cdot 8}\int_0^{\pi/2} \sin^6 x\, dx$$

$$= \frac{9 \cdot 7 \cdot 5}{10 \cdot 8 \cdot 6}\int_0^{\pi/2} \sin^4 x\, dx = \frac{9 \cdot 7 \cdot 5 \cdot 3}{10 \cdot 8 \cdot 6 \cdot 4}\int_0^{\pi/2} \sin^2 x\, dx$$

$$= \frac{9 \cdot 7 \cdot 5 \cdot 3}{10 \cdot 8 \cdot 6 \cdot 4 \cdot 2}\int_0^{\pi/2} dx = \frac{63\pi}{2^9}.$$

c. The values decrease as n increases.

8.8 Numerical Integration

8.8.1 $\Delta x = \frac{18 - 4}{28} = \frac{1}{2}$.

8.8.3 The Trapezoidal Rule approximates the definite integral by using a trapezoid over each subinterval rather than a rectangle.

8.8.5 $M(4) = 2(2) + 5(2) + 6(2) + 8(2) = 42$.

8.8.7 $S(4) = (1 + 4(4) + 2(7) + 4(5) + 5)\frac{2}{3} = \frac{112}{3} \approx 37.33$.

8.8.9 The endpoints of the subintervals are $-1, 1, 3, 5, 7,$ and 9. The trapezoidal rule uses the value of f at each of these endpoints.

8.8.11 The absolute error is $|\pi - 3.14| \approx 0.0015926536$. The relative error is $\frac{|\pi - 3.14|}{\pi} \approx 5 \times 10^{-4}$.

8.8.13 The absolute error is $|e - 2.72| \approx 0.0017181715$. The relative error is $\frac{|e - 2.72|}{e} \approx 6.32 \times 10^{-4}$.

8.8.15
For $n = 1$, we have $f(6) \cdot 8 = 72 \cdot 8 = 576$.
For $n = 2$ we have $f(4) \cdot 4 + f(8) \cdot 4 = 32 \cdot 4 + 128 \cdot 4 = 640$.
For $n = 4$, we have $f(3) \cdot 2 + f(5) \cdot 2 + f(7) \cdot 2 + f(9) \cdot 2 = 18 \cdot 2 + 50 \cdot 2 + 98 \cdot 2 + 162 \cdot 2 = 656$.

8.8.17 We have

$$\frac{1}{6}(\sin(\pi/12) + \sin(\pi/4) + \sin(5\pi/12) + \sin(7\pi/12) + \sin(3\pi/4) + \sin(11\pi/12)) \approx 0.6439505509.$$

8.8.19 For $n = 2$ we have $T(2) = \frac{4}{2}(f(2) + 2f(6) + f(10)) = 2(8 + 2(72) + 200) = 704$.

Using the results of number 15: For $n = 4$, note that $T(4) = \frac{T(2) + M(2)}{2} = \frac{704 + 640}{2} = 672$.

Using the results of number 15: For $n = 8$ we have that $T(8) = \frac{T(4) + M(4)}{2} = \frac{672 + 656}{2} = 664$.

8.8.21 We have

$$T(6) = \frac{1}{12}\left(\sin 0 + 2\sin\frac{\pi}{6} + 2\sin\frac{\pi}{3} + 2\sin\frac{\pi}{2} + 2\sin\frac{2\pi}{3} + 2\sin\frac{5\pi}{6} + \sin\pi\right)$$

$$= \frac{1}{6}\left(\frac{1}{2} + \frac{\sqrt{3}}{2} + 1 + \frac{\sqrt{3}}{2} + \frac{1}{2}\right) = \frac{1}{6}\left(2 + \sqrt{3}\right).$$

8.8.23 $S(4) = \dfrac{\pi}{12}\left(\sqrt{\sin 0} + 4\sqrt{\sin\dfrac{\pi}{4}} + 2\sqrt{\sin\dfrac{\pi}{2}} + 4\sqrt{\sin\dfrac{3\pi}{4}} + \sqrt{\sin\pi}\right) \approx 2.28477.$

$$S(8) \approx \frac{\pi}{24}\left(\sqrt{\sin 0} + 4\sqrt{\sin\frac{\pi}{8}} + 2\sqrt{\sin\frac{\pi}{4}} + 4\sqrt{\sin\frac{3\pi}{8}} + \right.$$

$$\left. 2\sqrt{\sin\frac{\pi}{2}} + 4\sqrt{\sin\frac{5\pi}{8}} . + 2\sqrt{\sin\frac{3\pi}{4}} + 4\sqrt{\sin\frac{7\pi}{8}} + \sqrt{\sin\pi}\right) \approx 2.35646.$$

8.8.25

$$S(10) = \frac{1}{6}\left(e^{-4} + 4e^{-2.25} + 2e^{-1} + 4e^{-0.25} + 2e^0 + 4e^{-0.25} + 2e^{-1}\right.$$
$$\left. + 4e^{-2.25} + 2e^{-4} + 4e^{-6.25} + e^{-9}\right) \approx 1.7680.$$

8.8.27 The width of each subinterval is $1/25$, so

$$M(25) = \frac{1}{25}(\sin\pi/50 + \sin 3\pi/50 + \sin 5\pi/50 + \cdots + \sin 49\pi/50) \approx 0.63704.$$

Because $\int_0^1 \sin\pi x\, dx = \dfrac{2}{\pi}$, the absolute error is $|2/\pi - M(25)| \approx 4.19 \times 10^{-4}$ and the relative error is this number divided by $2/\pi$ which is approximately 6.58×10^{-4}. The Trapezoidal Rule yields approximately 0.63578, with a relative error of ≈ 0.00132.

8.8.29

n	$M(n)$	Absolute Error	$T(n)$	Absolute Error
4	99	1	102	2
8	99.75	0.250	100.5	0.5
16	99.9375	0.0625	100.125	0.125
32	99.984375	0.0156	100.03125	0.03125

8.8.31

n	$M(n)$	Absolute Error	$T(n)$	Absolute Error
4	1.50968181	9.7×10^{-3}	1.48067370	1.9×10^{-2}
8	1.50241228	2.4×10^{-3}	1.49517776	4.8×10^{-3}
16	1.50060256	6.0×10^{-4}	1.49879502	1.2×10^{-3}
32	1.50015061	1.5×10^{-4}	1.49969879	3.0×10^{-4}

8.8.33 Because the given function has odd symmetry about the midpoint of the interval $[0, \pi]$, the midpoint rule calculates to be zero for all even values of n, as does the trapezoidal rule.

8.8.35 $T(4) = 3\,(9.1 + 2(12) + 2(26) + 2(46) + 53) = 690.3$ million ft^3.
$S(4) = 2(9.1 + 4(12) + 2(26) + 4(46) + 53) = 692.2$ million ft^3.

8.8.37 Answers may vary.

$$\overline{T} = \frac{1}{12}\int_0^{12} T(t)\,dt \approx \frac{1}{12}\text{Trapezoid}(12)$$
$$\approx \frac{1}{24}(47 + 2(50 + 46 + 45 + 48 + 52 + 54 + 61 + 62 + 63 + 63 + 59) + 55) = 54.5.$$

8.8.39 Answers may vary.

$$\overline{T} = \frac{1}{12}\int_0^{12} T(t)\,dt \approx \frac{1}{12}\text{Trapezoid}(12)$$
$$\approx \frac{1}{24}(35 + 2(34 + 34 + 36 + 36 + 37 + 37 + 36 + 35 + 35 + 34 + 33) + 32)) \approx 35.0.$$

8.8.41

a. Left Riemann sum: $\frac{1}{120}(70 \cdot 20 + 130 \cdot 25 + 200 \cdot 15 + 239 \cdot 30 + 311 \cdot 20 + 355 \cdot 10) = 204.917$.
Right Riemann sum: $\frac{1}{120}(130 \cdot 20 + 200 \cdot 25 + 239 \cdot 15 + 311 \cdot 30 + 355 \cdot 20 + 375 \cdot 10) = 261.375$.
Trapezoidal rule:

$$\frac{(70+130)\cdot 20}{2\cdot 120} + \frac{(130+200)\cdot 25}{2\cdot 120} + \frac{(200+239)\cdot 15}{2\cdot 120} + \frac{(239+311)\cdot 30}{2\cdot 120}$$
$$+ \frac{(311+355)\cdot 20}{2\cdot 120} + \frac{(355+375)\cdot 10}{2\cdot 120} = 233.146$$

These are approximations to the average temperature of the curling iron on the interval $[0, 120]$.

b. Because the function is increasing, the left Riemann sum is an underestimate and the right Riemann sum is an overestimate. The Trapezoidal rule appears to be a slight underestimate. The followings is a plot of the points with straight line segments connecting them. The shape suggests that the actual function is concave down on the interval, so the trapezoidal rule will be an underestimate.

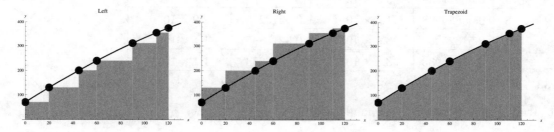

c. The change in temperature over the time interval is

$$\int_0^{120} T'(t)\,dt = T(120) - T(0) = 375 - 70 = 305 \text{ degrees}.$$

8.8.43

a. The net change in elevation is $\int_0^5 v(t)\,dt$, which can be approximated by the trapezoidal rule to give

$$\frac{(0+100)\cdot 1}{2} + \frac{(100+120)\cdot 0.5}{2} + \frac{(120+150)\cdot 1.5}{2}$$
$$+ \frac{(150+110)\cdot 0.5}{2} + \frac{(110+90)\cdot 0.5}{2} + \frac{(90+80)\cdot 1}{2} = 507.5.$$

If we add the net change to the original elevation of 5400 feet, we have an elevation of approximately 5907.5 feet.

b. The right Riemann sum is

$$100 \cdot 1 + 120 \cdot 0.5 + 150 \cdot 1.5 + 110 \cdot 0.5 + 90 \cdot 0.5 + 80 \cdot 1 = 565.$$

If we add the net change to the original elevation of 5400 feet, we have an elevation of approximately 5965 feet.

c. The elevation can be estimated by

$$5400 + \int_0^5 g(t)\, dt = 5400 + \int_0^5 (3.49t^3 - 43.21t^2 + 142.43t - 1.75)\, dt \approx 5916.52,$$

so the elevation of the balloon is about 5917 feet.

8.8.45

n	$T(n)$	Absolute Error	$S(n)$	Absolute Error
25	3.19623162	—	—	—
50	3.19495398	4.3×10^{-4}	3.19452809	4.5×10^{-8}

8.8.47

n	$T(n)$	Absolute Error	$S(n)$	Absolute Error
50	1.00008509	—	—	—
100	1.00002127	2.1×10^{-5}	1.00000000	4.6×10^{-9}

8.8.49

n	$T(n)$	Absolute Error	$S(n)$	Absolute Error
4	1820	284	—	—
8	1607.75	71.8	1537	1
16	1553.9844	18	1536.0625	6.3×10^{-2}
32	1540.4990	4.5	1536.0039	3.9×10^{-3}

8.8.51

n	$T(n)$	Absolute Error	$S(n)$	Absolute Error
4	0.46911538	5.3×10^{-2}	—	—
8	0.50826998	1.3×10^{-2}	0.52132152	2.9×10^{-4}
16	0.51825968	3.4×10^{-3}	0.52158957	1.7×10^{-5}
32	0.52076933	8.4×10^{-4}	0.52160588	1.1×10^{-6}

8.8.53

a. True. Note that $\Delta x = \dfrac{b-a}{18}$, and $E_S(18) \leq \dfrac{1 \cdot (b-a)}{10 \cdot 18} \cdot (\Delta x)^4 = \dfrac{(\Delta x)^5}{10}$.

b. False. Because $E_M(n) \leq \dfrac{k(b-a)^3}{24n^2}$, we have $E_M(3n) \leq \dfrac{k(b-a)^3}{24(9n^2)}$, so the error decreases by a factor of about 9.

c. True. Because $E_T(n) \leq \dfrac{k(b-a)^3}{12n^2}$, we have $E_T(4n) \leq \dfrac{k(b-a)^3}{24(16n^2)}$, so the error decreases by a factor of about 16.

8.8.55 $\displaystyle\int_0^{\pi/2} \cos^9 x\, dx = \dfrac{128}{315}.$

n	$M(n)$	Absolute Error	$T(n)$	Absolute Error
4	0.40635058	1.4×10^{-6}	0.40634783	1.4×10^{-6}
8	0.40634921	7.6×10^{-10}	0.40634921	7.6×10^{-9}
16	0.40634921	6.6×10^{-13}	0.40634921	6.6×10^{-13}
32	0.40634921	8.9×10^{-16}	0.40634921	7.8×10^{-16}

8.8.57 $\int_0^\pi \ln(5 + 3\cos x)x\, dx = \pi \ln(9/2)$.

n	$M(n)$	Absolute Error	$T(n)$	Absolute Error
4	4.72531820	1.2×10^{-4}	4.72507878	1.2×10^{-4}
8	4.72519851	9.1×10^{-9}	4.72519849	9.1×10^{-9}
16	4.72519850	0.	4.72519850	8.9×10^{-16}
32	4.72519850	0.	4.72519850	8.9×10^{-19}

8.8.59 $\int_0^\pi \dfrac{\cos x}{(5/4) - \cos x}\, dx = \dfrac{2\pi}{3} \approx 2.094395102$.

n	$S(n)$	Absolute Error
4	1.916439963	0.178
8	2.080919302	0.0135
16	2.094341841	5.3×10^{-5}

8.8.61 $\int_0^\pi \sin 6x \cos 3x\, dx = \dfrac{4}{9} = 0.\overline{4}$.

n	$S(n)$	Absolute Error
8	0.0305049084	0.475
16	0.4540112289	0.0096
32	0.44487	0.00087

8.8.63 We are computing $\dfrac{1}{3\sqrt{2\pi}} \int_{66}^{72} e^{-(x-69)^2/18}\, dx$. If we use Simpson's rule we obtain

n	$S(n)$
4	0.683
8	0.683

So about 68.3%.

8.8.65

a. For even n we have $S(n) = \dfrac{20}{n}(365)(f(a) + 4f(x_1) + 4f(x_2) + \ldots + 4f(x_{n-1}) + f(b))$. For $n = 6$ since there are 6 decades between 1940 and 2000, we have $S(6) \approx 160,000$ millions of barrels produced.

b. Following part (a) with $n = 6$ we have $S(n) \approx 68,000$ millions of barrels imported.

8.8.67

a. $M(50) \approx 35.43456$.

b. $f'(x) = \dfrac{1}{2}(x^3 + 1)^{-1/2} 3x^2 = \dfrac{3x^2}{2(x^3 + 1)^{1/2}}$. Then $f''(x) = \dfrac{2(x^3 + 1)^{1/2}(6x) - 3x^2(x^3 + 1)^{-1/2}(3x^2)}{4(x^3 + 1)}$.

Multiplying the numerator and denominator by $(x^3 + 1)^{1/2}$ yields

$$\dfrac{12x(x^3 + 1) - 9x^4}{4(x^3 + 1)^{3/2}} = \dfrac{3x^4 + 12x}{4(x^3 + 1)^{3/2}} = \dfrac{3x(x^3 + 4)}{4(x^3 + 1)^{3/2}}.$$

c. Because f'' is decreasing on $[1, 6]$, $|f''(x)| \leq f''(1) = \dfrac{15}{8\sqrt{2}}$.

d. Note that $\frac{15}{8\sqrt{2}} < 1.33$. We have $E_M \leq \frac{k(b-a)}{24}(\Delta x)^2 = \frac{1.33(5)}{24}\left(\frac{1}{10}\right)^2 \leq 0.0028$.

8.8.69

a. $T(40) = \frac{1}{80}\left(\sin 1 + 2\sum_{i=1}^{39} \sin e^{i/40} + \sin e\right) \approx 0.8748$.

b. $f(x) = \sin e^x$, so $f'(x) = e^x \cos e^x$. $f''(x) = -e^{2x}\sin e^x + e^x \cos e^x = e^x(\cos e^x - e^x \sin e^x)$.

c. $|f''(x)| = |e^x| \cdot |\cos e^x - e^x \sin e^x| \leq |e^x|(|\cos e^x| + |e^x \sin e^x|) \leq e(1+e) < 10.11$. However, the graph of the absolute value of $f''(x)$ reveals that is is actually bounded by 5.75 on the interval $[0,1]$.

d. Because $E_T(n) \leq k(b-a)^3/12n^2 = 6/(12(40^2)) = 0.0003125$, $T(40)$ is accurate to at least 3 decimal places.

8.8.71

a. $S(20) \approx 0.97775$.

b. The error is bounded by $\frac{f''''(x)}{180}\left(\frac{1}{20}\right)^4 \leq \frac{1}{180 \cdot 20^4} \leq 3.5 \times 10^{-8}$.

8.8.73 $\int_0^{2\pi} \sqrt{a^2 \cos^2 t + b^2 \sin^2 t}\, dt$, $a = 4$, $b = 8$.

n	$S(n)$
4	41.88790205
8	39.05860599

8.8.75

a. The exact value is $\int_0^4 x^3\, dx = \left.\frac{x^4}{4}\right|_0^4 = 64$. Simpson's rule gives $S(2) = (0 + 4 \cdot 8 + 64)\frac{2}{3} = 64$. The values match exactly.

b. $S(4) = (0 + 4 \cdot 1 + 2 \cdot 8 + 4 \cdot 27 + 64)\frac{1}{3} = 64$. This also matches the exact value exactly.

c. The 4th derivative of x^3 is 0, so the value of K in the theorem can be taken to be 0, so the error is 0.

d. Any polynomial of degree 3 or less has 4th derivative equal to 0, so the value of K in the theorem can be taken to be 0, so the error is 0.

8.8.77 The trapezoidal rule will be an overestimate in this case. This is because of the fact that if the function is above the axis and concave up on the given interval, then each trapezoid on each subinterval lies over the area under the curve for that corresponding subinterval.

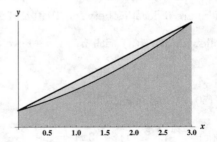

8.8.79 Using the given formula, we have

$$S(2n) = \frac{2M(n) + T(n)}{3}.$$

So $S(20) = \frac{1}{3}(2M(10) + T(10)) \approx 1.000001$.

8.9 Improper Integrals

8.9.1 The interval of integration is infinite or the integrand is unbounded on the interval of integration.

8.9.3 $\lim_{b\to\infty} \int_2^b \frac{dx}{x^{1/5}}$. We have

$$\lim_{b\to\infty} \frac{5}{4} \cdot x^{4/5} \Big|_2^b = \lim_{b\to\infty} \left(\frac{5}{4} \cdot b^{4/5} - \frac{5}{4} \cdot 2^{4/5}\right) = \infty.$$

The integral diverges.

8.9.5 This is $\int_{-\infty}^{\infty} f(x)\,dx$.

8.9.7 $\int_3^\infty x^{-2}\,dx = \lim_{b\to\infty} \int_3^b x^{-2}\,dx = \lim_{b\to\infty} \left(-\frac{1}{x}\right)\Big|_3^b = \lim_{b\to\infty} \left(\frac{1}{3} - \frac{1}{b}\right) = \frac{1}{3}.$

8.9.9 $\int_2^\infty \frac{dx}{\sqrt{x}} = \lim_{b\to\infty} \int_2^b \frac{dx}{\sqrt{x}} = \lim_{b\to\infty} 2\sqrt{x}\Big|_2^b = \lim_{b\to\infty} 2(\sqrt{b} - \sqrt{2}) = \infty$, so the integral diverges.

8.9.11 $\int_0^\infty e^{-ax}\,dx = \lim_{b\to\infty} \int_0^b e^{-ax}\,dx = \lim_{b\to\infty} \frac{e^{-ax}}{-a}\Big|_0^b = \lim_{b\to\infty} \left(-\frac{1}{ae^{ab}} + \frac{1}{a}\right) = \frac{1}{a}.$

8.9.13 $\int_0^\infty \cos x\,dx = \lim_{b\to\infty} \int_0^b \cos x\,dx = \lim_{b\to\infty} \sin x\Big|_0^b = \lim_{b\to\infty} \sin b$, which does not exist so the integral diverges.

8.9.15 The given integral is equal to $\lim_{b\to-\infty} \int_b^0 \frac{dx}{x^2+100} + \lim_{b\to\infty} \int_0^b \frac{dx}{x^2+100}$, if both limits exist.

$$\lim_{b\to-\infty} \int_b^0 \frac{dx}{x^2+100} = \lim_{b\to-\infty} \frac{1}{10}\tan^{-1}\frac{x}{10}\Big|_b^0 = \frac{1}{10}\left(0 + \frac{\pi}{2}\right) = \frac{\pi}{20}.$$

$$\lim_{b\to\infty} \int_0^b \frac{dx}{x^2+100} = \lim_{b\to\infty} \frac{1}{10}\tan^{-1}\frac{x}{10}\Big|_0^b = \frac{1}{10}\left(\frac{\pi}{2} - 0\right) = \frac{\pi}{20}.$$

So the given integral has value $\frac{\pi}{20} + \frac{\pi}{20} = \frac{\pi}{10}$.

8.9.17 $\lim_{b\to\infty} \int_7^b \frac{dx}{(x+1)^{1/3}} = \lim_{b\to\infty} \frac{3}{2}(x+1)^{2/3}\Big|_7^b = \lim_{b\to\infty} \left(\frac{3}{2}(b+1)^{2/3} - \frac{3}{2}7^{2/3}\right) = \infty.$
The integral diverges.

8.9.19 Let $u = x^3 + x$ so that $du = (3x^2+1)\,dx$ Substituting gives

$$\int_2^\infty \frac{1}{u}\,du = \lim_{b\to\infty} \int_2^b \frac{1}{u}\,du = \lim_{b\to\infty} \ln u\Big|_2^b = \lim_{b\to\infty}(\ln b - \ln 2) = \infty.$$

The given integral diverges.

8.9.21 $\int_2^\infty \frac{\cos(\pi/x)}{x^2}\,dx = \lim_{b\to\infty} \int_2^b \frac{\cos(\pi/x)}{x^2}\,dx = \lim_{b\to\infty} \left(-\frac{1}{\pi}\sin(\pi/x)\right)\Big|_2^b = \lim_{b\to\infty} \frac{1}{\pi}(1 - \sin(\pi/b)) = \frac{1}{\pi}.$

8.9.23 $\int_0^\infty \dfrac{e^u}{e^{2u}+1}\,du = \lim\limits_{b\to\infty}\int_0^b \dfrac{e^u}{e^{2u}+1}\,du$. Let $u = e^u$ so that $du = e^u\,du$. After substitution we have
$\lim\limits_{b\to\infty}\int_1^{e^b} \dfrac{1}{u^2+1}\,du = \lim\limits_{b\to\infty}\left(\tan^{-1}(u)\right)\Big|_1^{e^b} = \lim\limits_{b\to\infty}(\tan^{-1}(e^b) - \tan^{-1}(1)) = \pi/2 - \pi/4 = \pi/4.$

8.9.25 $\int_{-\infty}^\infty \dfrac{e^{3x}}{1+e^{6x}}\,dx = \lim\limits_{a\to-\infty}\int_a^0 \dfrac{e^{3x}}{1+e^{6x}}\,dx + \lim\limits_{b\to\infty}\int_0^b \dfrac{e^{3x}}{1+e^{6x}}\,dx$. Let $u = e^{3x}$ so that $du = 3e^{3x}$. Note that
$$\int \dfrac{e^{3x}}{1+e^{6x}}\,dx = \dfrac{1}{3}\int \dfrac{du}{1+u^2} = \dfrac{1}{3}\tan^{-1} u + C = \dfrac{1}{3}\tan^{-1} e^{3x} + C.$$
So we have
$\dfrac{1}{3}\lim\limits_{a\to-\infty}(\tan^{-1} 1 - \tan^{-1} e^{3a}) + \dfrac{1}{3}\lim\limits_{b\to\infty}(\tan^{-1} e^{3b} - \tan^{-1} 1) = \dfrac{1}{3}\left(\dfrac{\pi}{4} - 0\right) + \dfrac{1}{3}\left(\dfrac{\pi}{2} - \dfrac{\pi}{4}\right) = \dfrac{\pi}{6}.$

8.9.27
$$\int_{-\infty}^\infty xe^{-x^2}\,dx = \lim\limits_{b\to-\infty}\int_b^0 xe^{-x^2} + \lim\limits_{b\to\infty}\int_0^b xe^{-x^2}$$
$$= \lim\limits_{b\to-\infty}\left(-\dfrac{1}{2}e^{-x^2}\right)\Big|_b^0 + \lim\limits_{b\to\infty}\left(-\dfrac{1}{2}e^{-x^2}\right)\Big|_0^b$$
$$= \lim\limits_{b\to-\infty}\left(-\dfrac{1}{2} + \dfrac{1}{2}e^{-b^2}\right) + \lim\limits_{b\to\infty}\left(-\dfrac{1}{2}e^{-b^2} + \dfrac{1}{2}\right) = -\dfrac{1}{2} + \dfrac{1}{2} = 0.$$

8.9.29 By symmetry, the given integral is equal to $2\int_0^\infty \dfrac{(\tan^{-1} t)^2}{t^2+1}\,dt$ if the integral exists. Let $u = \tan^{-1} t$ so that $du = \dfrac{dt}{t^2+1}$. We have
$$2\lim\limits_{b\to\infty}\int_0^{\tan^{-1} b} u^2\,du = 2\lim\limits_{b\to\infty}\dfrac{u^3}{3}\Big|_0^{\tan^{-1} b} = 2\lim\limits_{b\to\infty}\dfrac{(\tan^{-1} b)^3}{3} = \dfrac{2(\pi/2)^3}{3} = \dfrac{\pi^3}{12}.$$

8.9.31 $\int_1^\infty \dfrac{1}{v(v+1)}\,dv = \lim\limits_{b\to\infty}\int_1^b\left(\dfrac{1}{v} - \dfrac{1}{v+1}\right)dv = \lim\limits_{b\to\infty}\ln\left|\dfrac{v}{v+1}\right|\Big|_1^b = \lim\limits_{b\to\infty}\ln\left|\dfrac{b}{b+1}\right| - \ln(1/2) = 0 - \ln(1/2) = \ln 2.$

8.9.33 $\int_2^\infty \dfrac{dy}{y\ln y} = \lim\limits_{b\to\infty}\int_2^b \dfrac{dy}{y\ln y} = \lim\limits_{b\to\infty}(\ln(\ln y))\Big|_2^b = \lim\limits_{b\to\infty}(\ln(\ln b) - \ln(\ln 2)) = \infty$, so the integral diverges.

8.9.35
$$\int_{-\infty}^0 \dfrac{dx}{\sqrt[3]{2-x}} = \lim\limits_{b\to-\infty}\int_b^0 (2-x)^{-1/3}\,dx = \lim\limits_{b\to-\infty}\left(-\dfrac{3}{2}(2-x)^{2/3}\right)\Big|_b^0$$
$$= \lim\limits_{b\to-\infty}\left(\dfrac{3}{2}\left(-2^{2/3} + (2-b)^{2/3}\right)\right) = \infty,$$
so the integral diverges.

8.9.37 $\int_0^8 \dfrac{dx}{\sqrt[3]{x}} = \lim\limits_{c\to 0^+}\int_c^8 x^{-1/3}\,dx = \lim\limits_{c\to 0^+}\left(\dfrac{3}{2}x^{2/3}\right)\Big|_c^8 = \dfrac{3}{2}\lim\limits_{c\to 0^+}(4 - c^{2/3}) = 6.$

8.9.39 $\int_0^{\pi/2} \tan\theta\,d\theta = \lim\limits_{c\to\pi/2^-}\int_0^c \tan\theta\,d\theta = \lim\limits_{c\to\pi/2^-}(\ln\sec\theta)\Big|_0^c = \lim\limits_{c\to\pi/2^-}\ln\sec c = \infty$, so the integral diverges.

8.9. Improper Integrals

8.9.41 $\lim_{b \to (\pi/2)^-} \int_0^b \sec x \tan x \, dx = \lim_{b \to (\pi/2)^-} (\sec x) \Big|_0^b = \lim_{b \to (\pi/2)^-} (\sec b - 1) = \infty$. The given integral diverges.

8.9.43 Note that $\int \dfrac{e^{\sqrt{x}}}{\sqrt{x}} \, dx = 2e^{\sqrt{x}} + C$. Thus,

$$\lim_{c \to 0^+} \int_c^1 \frac{e^{\sqrt{x}}}{\sqrt{x}} \, dx = \lim_{c \to 0^+} \left(2e^{\sqrt{x}}\right) \Big|_c^1 = \lim_{c \to 0^+} \left(2e - 2e^{\sqrt{c}}\right) = 2e - 2.$$

8.9.45 $\int_0^1 \dfrac{x^3}{x^4 - 1} \, dx = \lim_{c \to 1^-} \int_0^c \dfrac{x^3}{x^4 - 1} \, dx = \lim_{c \to 1^-} \left(\dfrac{1}{4} \ln|x^4 - 1|\right) \Big|_0^c = \dfrac{1}{4} \lim_{c \to 1^-} \ln|c^4 - 1| = -\infty$, so the integral diverges.

8.9.47

$$\int_0^{10} \frac{dx}{\sqrt[4]{10 - x}} = \lim_{c \to 10^-} \int_0^c (10 - x)^{-1/4} \, dx = \lim_{c \to 10^-} \left(-\frac{4}{3}(10 - x)^{3/4}\right) \Big|_0^c$$

$$= \frac{4}{3} \lim_{c \to 10^-} \left(10^{3/4} - (10 - c)^{3/4}\right) = \frac{4}{3} 10^{3/4}.$$

8.9.49 The given integrand isn't defined for $x = 1$, so we write the given integral as

$$\int_0^1 \frac{dx}{(x - 1)^2} + \int_1^2 \frac{dx}{(x - 1)^2}.$$

We will show that the given integral diverges by showing that $\int_0^1 \dfrac{dx}{(x - 1)^2}$ diverges. We have

$$\lim_{b \to 1^-} \int_0^b \frac{dx}{(x - 1)^2} = \lim_{b \to 1^-} -\frac{1}{x - 1} \Big|_0^b = \lim_{b \to 1^-} \left(-\frac{1}{b - 1} + 1\right) = \infty.$$

8.9.51 By symmetry,

$$\int_{-2}^2 \frac{dp}{\sqrt{4 - p^2}} = 2 \int_0^2 \frac{dp}{\sqrt{4 - p^2}} = 2 \lim_{c \to 2^-} \int_0^c \frac{dp}{\sqrt{4 - p^2}}$$

$$= 2 \lim_{c \to 2^-} \left(\sin^{-1}(p/2)\right) \Big|_0^c = 2 \lim_{c \to 2^-} \left(\sin^{-1}(c/2) - \sin^{-1} 0\right)$$

$$= 2(\sin^{-1} 1 - 0) = \pi.$$

8.9.53 Integration by parts gives $\int \ln x \, dx = x \ln x - x + C$, so

$$\int_0^1 \ln x \, dx = \lim_{c \to 0^+} \int_c^1 \ln x \, dx = \lim_{c \to 0^+} (x \ln x - x) \Big|_c^1 = \lim_{c \to 0^+} (-1 - (c \ln c - c)) = -1,$$

because $\lim_{c \to 0^+} c \ln c = 0$.

8.9.55 Write the given integral as $\lim_{a \to 0^+} \int_a^{\ln 2} \dfrac{e^x}{\sqrt{e^{2x} - 1}} \, dx$ and then let $u = e^x$ so that $du = e^x \, dx$. Substituting gives $\lim_{a \to 0^+} \int_{e^a}^2 \dfrac{du}{\sqrt{u^2 - 1}}$. Then we let $u = \sec \theta$ with $du = \sec \theta \tan \theta \, d\theta$. Substituting again gives

$$\lim_{a \to 0^+} \int_{\sec^{-1} e^a}^{\pi/3} \sec \theta \, d\theta = \lim_{a \to 0^+} \ln|\sec \theta + \tan \theta| \Big|_{\sec^{-1} e^a}^{\pi/3}$$

$$= \lim_{a \to 0^+} \left(\ln(2 + \sqrt{3}) - \ln\left(e^a + \sqrt{e^{2a} - 1}\right)\right) = \ln(2 + \sqrt{3}).$$

8.9.57 The function $e^{-|x|}$ is even, so $\int_{-\infty}^{\infty} e^{-|x|}\, dx = 2\int_{0}^{\infty} e^{-x}\, dx$. We have

$$2\lim_{b\to\infty}\int_0^b e^{-x}\,dx = 2\lim_{b\to\infty} -e^{-x}\Big|_0^b = 2\lim_{b\to\infty}(-e^{-b}+1) = 2.$$

8.9.59 We have the relation $B = I\int_0^\infty e^{-rt}\,dt = \dfrac{I}{r}$, using the result that $\int_0^\infty e^{-ax}\,dx = \dfrac{1}{a}$ for $a>0$. Therefore $B = \dfrac{5000}{0.12} = \$41{,}666.67$.

8.9.61 As in Example 6, we have

$$\text{AUC}_i = \int_0^\infty C_i(t)\,dt = 250\int_0^\infty e^{-0.08t}\,dt = \frac{250}{0.08} = 3125$$

and

$$\text{AUC}_o = \int_0^\infty C_0(t)\,dt = 200\int_0^\infty (e^{-0.08t} - e^{-1.8t})\,dt = 200\left(\frac{1}{0.08} - \frac{1}{1.8}\right) = \frac{21{,}500}{9} \approx 2389,$$

(here we use the fact that $\int_0^\infty e^{-ax}\,dx = \dfrac{1}{a}$ for $a > 0$). Therefore the bioavailability of the drug is

$$F = \frac{\text{AUC}_o}{\text{AUC}_i} = \frac{21{,}500}{9\cdot 3125} \approx 0.764.$$

8.9.63 Evaluate the improper integral

$$\int_0^\infty te^{-at}\,dt = \lim_{b\to\infty}\int_0^b te^{-at}\,dt = \lim_{b\to\infty}\left(-\frac{e^{-at}(at+1)}{a^2}\right)\Big|_0^b = \frac{1}{a^2}\lim_{b\to\infty}(1 - e^{-ab}(ab+1)) = \frac{1}{a^2},$$

provided $a>0$. Therefore $0.00005\int_0^\infty te^{-0.00005t}\,dt = \dfrac{0.00005}{0.00005^2} = 20{,}000$ hrs.

8.9.65 Using the result from Example 2, we see that the volume is given by $V = \pi\int_1^\infty x^{-4}\,dx = \dfrac{\pi}{4-1} = \dfrac{\pi}{3}$.

8.9.67 Using the result from Example 2, we see that the volume is given by

$$V = \pi\int_1^\infty \left(\frac{1}{x^2} + \frac{1}{x^3}\right)dx = \frac{\pi}{2-1} + \frac{\pi}{3-1} = \frac{3\pi}{2}.$$

8.9.69 $V = \pi\int_2^\infty \dfrac{dx}{x(\ln x)^2} = \pi\lim_{b\to\infty}\int_2^b \dfrac{dx}{x(\ln x)^2} = \pi\lim_{b\to\infty}\left(-\dfrac{1}{\ln x}\right)\Big|_2^b = \pi\lim_{b\to\infty}\left(\dfrac{1}{\ln 2} - \dfrac{1}{\ln b}\right) = \dfrac{\pi}{\ln 2}.$

8.9.71 Using disks, we have

$$V = \pi\int_1^2 (x-1)^{-1/2}\,dx = \pi\lim_{c\to 1^+}\int_c^2 (x-1)^{-1/2}\,dx = \pi\lim_{c\to 1^+}\left(2(x-1)^{1/2}\right)\Big|_c^2 = 2\pi\lim_{c\to 1^+}(1 - \sqrt{c-1}) = 2\pi.$$

8.9.73 $V = \pi\int_0^{\pi/2} \tan^2 x\,dx = \lim_{a\to\frac{\pi}{2}^-}\pi\int_0^a \tan^2 x\,dx = \lim_{a\to\frac{\pi}{2}^-}\pi\int_0^a (\sec^2 x - 1)\,dx = \lim_{a\to\frac{\pi}{2}^-}\pi(\tan x - x)\Big|_0^a = \lim_{a\to\frac{\pi}{2}^-}\pi(\tan a - a) = \infty$, so the volume doesn't exist.

8.9. Improper Integrals

8.9.75 Using shells, we have

$$V = 2\pi \int_0^4 x(4-x)^{-1/3}\,dx = 2\pi \int_0^4 (4-u)u^{-1/3}\,du = 2\pi \int_0^4 (4u^{-1/3} - u^{2/3})\,du$$

via the substitution $u = 4 - x$. Therefore

$$V = 2\pi \lim_{c \to 0+} \int_c^4 (4u^{-1/3} - u^{2/3})\,du = 2\pi \lim_{c \to 0+} \left(6u^{2/3} - \frac{3}{5}u^{5/3}\right)\bigg|_c^4 = \frac{72 \cdot 2^{1/3} \pi}{5}.$$

8.9.77 $x^3 + 1 > x^3$ for $x \geq 1$, so $\dfrac{1}{x^3+1} < \dfrac{1}{x^3}$ for $x \geq 1$. Because $\displaystyle\int_1^\infty \frac{1}{x^3}\,dx$ converges, $\displaystyle\int_1^\infty \frac{1}{x^3+1}\,dx$ converges as well by the Comparison Test. Note that the fact that $\displaystyle\int_1^\infty \frac{1}{x^3}\,dx$ converges to $\dfrac{1}{2}$ follows from Example 2, Case 1.

8.9.79 For $x \geq 3$, $0 < \ln x \leq x$, so $0 < \dfrac{1}{x} \leq \dfrac{1}{\ln x}$. Note that

$$\int_3^\infty \frac{dx}{x} = \lim_{b \to \infty} \int_3^b \frac{dx}{x} = \lim_{b \to \infty} \ln x \bigg|_3^b = \lim_{b \to \infty} (\ln b - \ln 3) = \infty.$$

Using the Comparison Test, because $\displaystyle\int_3^\infty \frac{dx}{x}$ diverges, $\displaystyle\int_3^\infty \frac{dx}{\ln x}$ diverges as well.

8.9.81 $0 \leq \sin^2 x \leq 1$, so $0 \leq \dfrac{\sin^2 x}{x^2} \leq \dfrac{1}{x^2}$ for $x \geq 1$. By Example 2, $\displaystyle\int_1^\infty \frac{dx}{x^2}$ converges. Therefore $\displaystyle\int_1^\infty \frac{\sin^2 x}{x^2}\,dx$ converges as well by the Comparison Test.

8.9.83 $2 + \cos x \geq 1$, so $\dfrac{2 + \cos x}{\sqrt{x}} \geq \dfrac{1}{\sqrt{x}} \geq 0$. By Example 2, $\displaystyle\int_1^\infty \frac{dx}{\sqrt{x}}$ diverges, so $\displaystyle\int_1^\infty \frac{2 + \cos x}{\sqrt{x}}\,dx$ diverges by the Comparison Test.

8.9.85 $\sqrt{x^{1/3} + x} \geq \sqrt{x^{1/3}} = x^{1/6}$. So for $0 \leq x \leq 1$, $0 < \dfrac{1}{\sqrt{x^{1/3}+x}} \leq \dfrac{1}{x^{1/6}}$. We will show that $\displaystyle\int_0^1 \frac{dx}{x^{1/6}}$ converges and conclude that $\displaystyle\int_0^1 \frac{1}{\sqrt{x^{1/3}+x}}\,dx$ converges by the Comparison Test. We have

$$\lim_{a \to 0+} \int_a^1 \frac{dx}{x^{1/6}} = \lim_{a \to 0+} \frac{6}{5}x^{5/6}\bigg|_a^1 = \lim_{a \to 0+} \frac{6}{5}\left(1 - a^{5/6}\right) = \frac{6}{5}.$$

8.9.87

a. True. The area under the curve $y = f(x)$ from 0 to ∞ is less than the area under $y = g(x)$ on this interval, which by assumption is finite.

b. False. For example, take $f(x) = 1$; then $\displaystyle\int_0^\infty f(x)\,dx = \infty$.

c. False. For example, take $p = 1/2$ and $q = 1$.

d. True. The area under the curve $y = x^{-q}$ from 1 to ∞ is less than the area under $y = x^{-p}$ on this interval, which by assumption is finite.

e. True. Using the result in Example 2, we see that this integral exists if and only if $3p + 2 > 1$, which is equivalent to $p > -1/3$.

8.9.89 The region R has area $A = \int_0^\infty e^{-bx}\,dx - \int_0^\infty e^{-ax}\,dx = \dfrac{1}{b} - \dfrac{1}{a}$.

8.9.91

a. We have
$$A(a,b) = \int_b^\infty e^{-ax}\,dx = \lim_{c\to\infty} \int_b^c e^{-ax}\,dx = \lim_{c\to\infty}\left(-\frac{1}{a}e^{-ax}\right)\bigg|_b^c = \frac{1}{a}\lim_{c\to\infty}(e^{-ab}-e^{-ac}) = \frac{e^{-ab}}{a}.$$

b. Solving $e^{-ab} = 2a$ for b gives $b = g(a) = -\dfrac{1}{a}\ln 2a$.

c. The function g has $g'(x) = \dfrac{1}{x^2}\ln 2x - \dfrac{1}{x^2} = \dfrac{\ln 2x - 1}{x^2}$, so g has a critical point at $x = e/2$, and the first derivative test shows that g takes a minimum at this point. Hence $b^* = g(e/2) = -\dfrac{2}{e}$.

8.9.93 In the following, we will make the substitution $u = \sqrt{x-1}$ so that $du = \dfrac{dx}{2\sqrt{x-1}}$ and $x = u^2 + 1$.

$$\int_1^\infty \frac{dx}{x\sqrt{x-1}} = \int_1^2 \frac{dx}{x\sqrt{x-1}} + \int_2^\infty \frac{dx}{x\sqrt{x-1}}$$

$$= \lim_{a\to 1^+}\int_a^2 \frac{dx}{x\sqrt{x-1}} + \lim_{b\to\infty}\int_2^b \frac{dx}{x\sqrt{x-1}}$$

$$= 2\lim_{a\to 1^+}\int_{\sqrt{a-1}}^1 \frac{du}{u^2+1} + 2\lim_{b\to\infty}\int_1^{\sqrt{b-1}} \frac{du}{u^2+1}$$

$$= 2\lim_{a\to 1^+}\tan^{-1} u\bigg|_{\sqrt{a-1}}^1 + 2\lim_{b\to\infty}\tan^{-1} u\bigg|_1^{\sqrt{b-1}}$$

$$= 2\lim_{a\to 1^+}\left(\tan^{-1} 1 - \tan^{-1}\sqrt{a-1}\right) + 2\lim_{b\to\infty}\left(\tan^{-1}\sqrt{b-1} - \tan^{-1} 1\right)$$

$$= 2\left(\frac{\pi}{4} - 0\right) + 2\left(\frac{\pi}{2} - \frac{\pi}{4}\right) = \pi.$$

8.9.95 These integrals cannot be evaluated by finding an antiderivative of their integrands. However, if we make the substitution $u = \pi/2 - x$ we find that $\int_0^{\pi/2} \ln\sin x\,dx = -\int_{\pi/2}^0 \ln\cos u\,du = \int_0^{\pi/2} \ln\cos x\,dx$. We also have

$$\int_{\pi/2}^\pi \ln\sin x\,dx = \int_0^{\pi/2} \ln\sin(x+\pi/2)\,dx = \int_0^{\pi/2} \ln\cos x\,dx,$$

and therefore

$$\int_0^{\pi/2} \ln\sin x\,dx = \frac{1}{2}\int_0^\pi \ln\sin x\,dx = \int_0^{\pi/2} \ln\sin 2y\,dy = \int_0^{\pi/2} (\ln\sin y + \ln\cos y + \ln 2)\,dy.$$

This implies $\int_0^{\pi/2} \ln\cos x\,dx = -\int_0^{\pi/2} \ln 2\,dx = -\dfrac{\pi\ln 2}{2}$.

8.9.97 This integral cannot be evaluated by finding an antiderivative of the integrand; however the result may be verified by numerical approximation.

8.9.99

a. We have $\int_0^\infty e^{-ax}\cos bx\,dx = \lim_{c\to\infty}\int_0^c e^{-ax}\cos bx\,dx = \lim_{c\to\infty}\left(\dfrac{e^{-ax}(b\sin bx - a\cos bx)}{a^2+b^2}\right)\bigg|_0^c = \lim_{c\to\infty}\dfrac{a + e^{-ac}(b\sin bc - a\cos bc)}{a^2+b^2} = \dfrac{a}{a^2+b^2}$.

8.9. Improper Integrals

b. We have $\int_0^\infty e^{-ax} \sin bx \, dx = \lim_{c \to \infty} \int_0^c e^{-ax} \sin bx \, dx = \lim_{c \to \infty} \left(-\frac{e^{-ax}(a \sin bx + b \cos bx)}{a^2 + b^2} \right) \Big|_0^c =$
$\lim_{c \to \infty} \frac{b - e^{-ac}(a \sin bc + b \cos bc)}{a^2 + b^2} = \frac{b}{a^2 + b^2}.$

8.9.101

a. Integrate by parts with $u = \sqrt{x} \ln x$ and $v = -1/(1+x)$:

$$\int_0^\infty \frac{\sqrt{x} \ln x}{(1+x)^2} \, dx = \frac{1}{2} \int_0^\infty \frac{\ln x + 2}{\sqrt{x}(x+1)} \, dx = \frac{1}{2} \int_0^\infty \frac{\ln x}{\sqrt{x}(x+1)} \, dx + \int_0^\infty \frac{dx}{\sqrt{x}(x+1)}.$$

(The integration by parts is legitimate for this improper integral because the product uv has limit 0 as $x \to \infty$ and as $x \to 0^+$).

b. Let $y = 1/x$. Then $dy = -\frac{1}{x^2} \, dx$.

c. We have

$$\int_0^1 \frac{\ln x}{\sqrt{x}(x+1)} \, dx = \int_\infty^1 \frac{-\ln y}{\frac{1}{\sqrt{y}}\left(\frac{1}{y}+1\right)} \left(-\frac{dy}{y^2} \right) = -\int_1^\infty \frac{\ln y}{\sqrt{y}(1+y)} \, dy,$$

and hence $\int_0^\infty \frac{\ln x}{\sqrt{x}(x+1)} \, dx = 0$.

d. The change of variables $z = \sqrt{x}$ gives $\int_0^\infty \frac{dx}{\sqrt{x}(x+1)} = 2 \int_0^\infty \frac{dz}{z^2+1} = \pi$.

8.9.103 The Laplace transform of $f(t) = e^{at}$ is given by $F(s) = \int_0^\infty e^{-st} e^{at} \, dt = \int_0^\infty e^{-(s-a)t} \, dt = \frac{1}{s-a}$, using the formula $\int_0^\infty e^{-cx} \, dx = \frac{1}{c}$ for $c > 0$.

8.9.105 The Laplace transform of $f(t) = \sin at$ is given by $F(s) = \int_0^\infty e^{-st} \sin at \, dt = \frac{a}{s^2 + a^2}$ (this formula is derived in the solution to problem 99 b).

8.9.107

a. Make the substitution $x = y + 2$; then

$$\int_1^3 \frac{dx}{\sqrt{(x-1)(3-x)}} = \int_{-1}^1 \frac{dy}{\sqrt{1-y^2}} = 2 \int_0^1 \frac{dy}{\sqrt{1-y^2}} = 2 \lim_{c \to 1^-} (\sin^{-1} x) \Big|_0^c = \pi.$$

b. The substitution $y = e^x$ gives

$$\int_1^\infty \frac{dx}{e^{x+1} + e^{3-x}} = \frac{1}{e} \int_1^\infty \frac{e^x \, dx}{e^{2x} + e^2} = \frac{1}{e} \int_e^\infty \frac{dy}{y^2 + e^2} = \frac{1}{e} \lim_{b \to \infty} \left(\frac{1}{e} \tan^{-1} \left(\frac{y}{e}\right) \right) \Big|_e^b = \frac{\pi}{4e^2}.$$

8.9.109

a. We have $W = GMm \int_R^\infty x^{-2} \, dx = GMm \lim_{b \to \infty} \left(-\frac{1}{x} \right) \Big|_R^b = \frac{GMm}{R} \approx 6.279 \times 10^7$ m J.

b. Solve $\frac{1}{2} v_e^2 = 6.279 \times 10^7$ to obtain $v_e \approx 11.207$ km/s.

c. We need $\frac{GM}{R} \geq \frac{1}{2} c^2 \iff R \leq \frac{2GM}{c^2} \approx 9$ mm.

8.9.111

a. Repeatedly applying this reduction formula gives $\Gamma(p+1) = p(p-1) \cdots 2 \cdot 1 \cdot \int_0^\infty e^{-x}\, dx = p! \cdot 1 = p!$.

b. We have $\Gamma\left(\dfrac{1}{2}\right) = \int_0^\infty x^{-1/2} e^{-x}\, dx = 2 \int_0^\infty e^{-u^2}\, du = \sqrt{\pi}$.

Chapter Eight Review

1

a. True. Two applications of integration by parts are needed to reduce to $\int x^2 e^{2x}\, dx$.

b. False. This integral can be done using a trigonometric substitution.

c. False. Both are correct, since $-\cos^2 x = \sin^2 x - 1$, so $\sin^2 x$ and $-\cos^2 x$ differ by a constant.

d. True. Recall that $2 \sin x \cos x = \sin 2x$.

e. False. Use long division to write the integrand as the sum of a polynomial and a proper rational function.

3 Let $u = x + 4$ so that $du = dx$. Note that $x = u - 4$. Substituting gives

$$\int \frac{3(u-4)}{\sqrt{u}}\, du = 3 \int (u^{1/2} - 4u^{-1/2})\, du = 3((2/3)u^{3/2} - 8u^{1/2}) + C$$
$$= 2(x+4)^{3/2} - 24\sqrt{x+4} + C = 2\sqrt{x+4}(x+4-12) + C = 2\sqrt{x+4}(x-8) + C.$$

5 Let $u = x$ and $dv = (1/2)(x+2)^{-1/2}\, dx$. Then $du = dx$ and $v = (x+2)^{1/2}$. We have

$$\int \frac{x}{2\sqrt{x+2}}\, dx = x\sqrt{x+2} - \int \sqrt{x+2}\, dx = x\sqrt{x+2} - \frac{2}{3}(x+2)^{3/2} + C$$
$$= \frac{1}{3}\sqrt{x+2}(3x - 2x - 4) + C = \frac{1}{3}\sqrt{x+2}(x - 4) + C.$$

7
$$\int_{-2}^{1} \frac{3}{(x+2)^2 + 9}\, dx = \left(\tan^{-1}((x+2)/3)\right)\bigg|_{-2}^{1} = \pi/4 - 0 = \pi/4.$$

9 $\int_0^{\pi/4} \cos^5 2x \sin^2 2x\, dx = \int_0^{\pi/4} \cos(2x)(1 - \sin^2 2x)^2 \sin^2 2x\, dx.$ Let $u = \sin 2x$ so that $du = 2\cos 2x\, dx$. Substituting gives

$$\frac{1}{2} \int_0^1 (1 - u^2)^2 u^2\, du = \frac{1}{2} \int_0^1 (u^2 - 2u^4 + u^6)\, du = \frac{1}{2}\left(u^3/3 - 2u^5/5 + u^7/7\right)\bigg|_0^1 = \frac{1}{2}(1/3 - 2/5 + 1/7) - 0 = \frac{4}{105}.$$

11 Let $u = \sqrt{t-1}$. Then $u^2 + 1 = t$, and $2u\, du = dt$. Substituting gives

$$\int \frac{u^2}{u^2+1}\, du = \int \left(1 - \frac{1}{u^2+1}\right) du = u - \tan^{-1} u + C = \sqrt{t-1} - \tan^{-1}\sqrt{t-1} + C.$$

13 We integrate by parts. Let $u = e^{3x}$ and $dv = \sin 6x\, dx$. Then $du = 3e^{3x}\, dx$ and $v = -\dfrac{1}{6}\cos 6x$. Our integral is then equal to

$$I = \int_0^\pi e^{3x} \sin 6x\, dx = -\frac{1}{6} e^{3x} \cos 6x \bigg|_0^\pi + \frac{1}{2} \int_0^\pi e^{3x} \cos 6x\, dx. = -\frac{1}{6}e^{3\pi} + \frac{1}{6} + \frac{1}{2} \int_0^\pi e^{3x} \cos 6x\, dx.$$

We integrate by parts again, letting $u = e^{3x}$ and $dv = \cos 6x\, dx$. Then $du = 3e^{3x}\, dx$ and $v = \frac{1}{6}\sin 6x$. We have

$$I = -\frac{1}{6}e^{3\pi} + \frac{1}{6} + \frac{1}{2}\left(\frac{1}{6}e^{3x}\sin 6x\bigg|_0^\pi - \frac{1}{2}\int_0^\pi e^{3x}\sin 6x\, dx\right)$$

$$= -\frac{1}{6}e^{3\pi} + \frac{1}{6} - \frac{1}{4}I.$$

So

$$\frac{5}{4}I = \frac{1}{6}\left(1 - e^{3\pi}\right),$$

and therefore

$$I = \frac{2}{15}\left(1 - e^{3\pi}\right).$$

15 We simplify the integrand via long division and then partial fractions. We have

$$\frac{3x^5 + 48x^3 + 3x^2 + 16}{x^3 + 16x} = 3x^2 + \frac{3x^2 + 16}{x(x^2 + 16)} = 3x^2 + \frac{1}{x} + \frac{2x}{x^2 + 16}.$$

Therefore our integral is equal to

$$\int_1^2 \left(3x^2 + \frac{1}{x} + \frac{2x}{x^2 + 16}\right) dx = \left(x^3 + \ln|x| + \ln|x^2 + 16|\right)\bigg|_1^2 = (8 + \ln 2 + \ln 20) - (1 + \ln 17) = 7 + \ln 40 - \ln 17.$$

17 Write $\dfrac{2x^2 + 7x + 4}{x^3 + 2x^2 + 2x} = \dfrac{Ax + B}{x^2 + 2x + 2} + \dfrac{C}{x}$. Then $2x^2 + 7x + 4 = Ax^2 + Bx + C(x^2 + 2x + 2)$. Letting $x = 0$ gives $C = 2$. Letting $x = 1$ gives $13 = A + B + 10$, so $A + B = 3$. Letting $x = -1$ gives $-1 = A - B + 2$, so $A - B = -3$. Solving this system of linear equations gives $A = 0$ and $B = 3$. Thus our integral is equal to

$$\int \left(\frac{3}{x^2 + 2x + 2} + \frac{2}{x}\right) dx = \int \left(\frac{3}{(x+1)^2 + 1} + \frac{2}{x}\right) dx = 3\tan^{-1}(x+1) + 2\ln|x| + C.$$

19 If we write

$$\frac{x^2 + 4x + 7}{(x+3)^3} = \frac{A}{x+3} + \frac{B}{(x+3)^2} + \frac{C}{(x+3)^3}$$

and solve, we obtain $A = 1$, $B = -2$, and $C = 4$. We then have

$$\int \frac{1}{x+3}\, dx - \int \frac{2}{(x+3)^2}\, dx + \int \frac{4}{(x+3)^3}\, dx = \ln|x+3| + \frac{2}{x+3} - \frac{2}{(x+3)^2} + C.$$

21 Let $x = \sec\theta$ so that $dx = \sec\theta\tan\theta\, d\theta$ and $\sqrt{x^2 - 1} = \tan\theta$. Substituting gives

$$\int_{\pi/4}^{\pi/3} \tan^2\theta\, d\theta = \int_{\pi/4}^{\pi/3} (\sec^2\theta - 1)\, d\theta = (\tan\theta - \theta)\bigg|_{\pi/4}^{\pi/3} = \sqrt{3} - \pi/3 - (1 - \pi/4) = \sqrt{3} - 1 - \frac{\pi}{12}.$$

23 $\int \tan^4 t \sec^2 t\, dt$. Let $u = \tan t$ so that $du = \sec^2 t\, dt$. Substituting gives

$$\int u^4\, du = u^5/5 + C = \frac{1}{5}\tan^5 t + C.$$

25 Complete the square in the denominator of the integrand as follows:

$$4x^2 + 12x + 10 = 4(x^2 + 3x + 9/4 - 9/4) + 10 = (4x^2 + 12x + 9) - 9 + 10 = (2x + 3)^2 + 1.$$

Then we have

$$\int_{-3/2}^{-1} \frac{dx}{(2x+3)^2 + 1}.$$

. Let $u = 2x + 3$ and $du = 2\, dx$. Substituting gives

$$\frac{1}{2}\int_0^1 \frac{du}{u^2 + 1} = \frac{1}{2}\tan^{-1} u\bigg|_0^1 = \frac{\pi}{8}.$$

27 Note that
$$w^2 + 2w - 8 = w^2 + 2w + 1 - 9 = (w+1)^2 - 9.$$

Let $u = w + 1$ with $du = dw$. Substituting gives $\int \dfrac{1}{u^2\sqrt{u^2-9}}\, du$. Now let $u = 3\sec\theta$ with $du = 3\sec\theta\tan\theta\, d\theta$. Now we have

$$\int \frac{1}{9\sec^2\theta\sqrt{9\sec^2\theta - 9}} 3\sec\theta\tan\theta\, d\theta = \int \frac{3\tan\theta}{27\sec\theta\tan\theta}\, d\theta$$
$$= \frac{1}{9}\int \cos\theta\, d\theta = \frac{1}{9}\sin\theta + C = \frac{1}{9}\cdot\frac{\sqrt{u^2-9}}{u} + C$$
$$= \frac{1}{9}\cdot\frac{\sqrt{(w+1)^2-9}}{w+1} + C = \frac{\sqrt{w^2+2w-8}}{9(w+1)} + C.$$

29 The integral can be written as $\int \cot^4 x \csc^2 x\, dx$. Let $u = \cot x$ so that $du = -\csc^2 x\, dx$. Substituting gives

$$-\int u^4\, du = -\frac{u^5}{5} + C = -\frac{\cot^5 x}{5} + C.$$

31 Let $u = x$ and $dv = \sinh x\, dx$. Then $du = dx$ and $v = \cosh x$. We have

$$\int x\sinh x\, dx = x\cosh x - \int \cosh x\, dx = x\cosh x - \sinh x + C.$$

33 $\int \tan^2 3\theta \sec^2 3\theta(\sec 3\theta \tan 3\theta)\, d\theta = \int (\sec^2 3\theta - 1)\sec^2 3\theta(\sec 3\theta \tan 3\theta)\, d\theta.$
Let $u = \sec 3\theta$ so that $du = 3\sec 3\theta \tan 3\theta\, d\theta$. Substituting gives

$$\frac{1}{3}\int (u^2 - 1)u^2\, du = \frac{1}{3}\int (u^4 - u^2)\, du = \frac{u^5}{15} - \frac{u^3}{9} + C = \frac{1}{15}\sec^5 3\theta - \frac{1}{9}\sec^3 3\theta + C.$$

35 Let $x = 2\tan\theta$ so that $dx = 2\sec^2\theta\, d\theta$ and $\sqrt{4x^2 + 16} = 2\sqrt{x^2 + 4} = 4\sec\theta$. Substituting gives

$$\int \frac{8\tan^3\theta}{4\sec\theta} 2\sec^2\theta\, d\theta = 4\int \tan\theta \sec\theta(\sec^2\theta - 1)\, d\theta.$$

Let $u = \sec\theta$ so that $du = \sec\theta\tan\theta\, d\theta$. Substituting again gives

$$4\int (u^2 - 1)\, du = \frac{4}{3}u^3 - 4u + C = \frac{4}{3}\sec^3\theta - 4\sec\theta + C$$
$$= \frac{4}{3}(\sec(\tan^{-1}(x/2)))^3 - 4\sec\tan^{-1}(x/2) + C = \frac{1}{6}\sqrt{x^2+4}(x^2+4) - 2\sqrt{x^2+4} + C$$
$$= \frac{1}{6}\sqrt{x^2+4}(x^2+4-12) + C = \frac{1}{6}\sqrt{x^2+4}(x^2-8) + C.$$

37 Write $\dfrac{3x^3 + 4x^2 + 6x}{(x+1)^2(x^2+4)} = \dfrac{Ax+B}{x^2+4} + \dfrac{C}{x+1} + \dfrac{D}{(x+1)^2}$. Then $3x^3 + 4x^2 + 6x = (Ax+B)(x+1)^2 + C(x+1)(x^2+4) + D(x^2+4)$. Letting $x = -1$ gives $D = -1$. Letting $x = 0$ gives $0 = B + 4C - 4$, so $B + 4C = 4$. Letting $x = 1$ gives $13 = 4A + 4B + 10C - 5$, so $9 = 2A + 2B + 5C$. Letting $x = 2$ gives $52 = 18A + 9B + 24C - 8$, so $60 = 18A + 9B + 24C$, so $20 = 6A + 3B + 8C$. Solving the system of linear equations gives $A = 2$, $B = 0$, and $C = 1$. Our integral is thus equal to

$$\int \left(\frac{2x}{x^2+4} + \frac{1}{x+1} - \frac{1}{(x+1)^2}\right) dx = \ln\left|(x^2+4)(x+1)\right| + \frac{1}{x+1} + C.$$

39 $\int \dfrac{1}{2+e^t}\,dt = \int \dfrac{e^t}{(e^t+2)e^t}\,dt$. Let $u=e^t$, so that $du = e^t\,dt$. Then we have $\int \dfrac{1}{(u+2)u}\,du$. If we write
$$\dfrac{1}{(u+2)u} = \dfrac{A}{u} + \dfrac{B}{u+2},$$
then $A(u+2) + Bu = 1$, and we find that $A = 1/2$ and $B = -1/2$. Then we integrate
$$\int \left(\dfrac{1/2}{u} - \dfrac{1/2}{u+2}\right) du = \dfrac{1}{2}\left(\ln|u| - \ln|u+2|\right) + C = \dfrac{1}{2}\left(t - \ln(2+e^t)\right) + C.$$

41
$$\int \dfrac{1}{1+\cos 4\theta} \cdot \dfrac{1-\cos 4\theta}{1-\cos 4\theta}\,d\theta = \int \dfrac{1-\cos 4\theta}{\sin^2 4\theta}\,d\theta = \int (\csc^2 4\theta - \csc 4\theta \cot 4\theta)\,d\theta = \dfrac{1}{4}(-\cot 4\theta + \csc 4\theta) + C.$$

43 Two applications of integration by parts gives
$$\int e^x \sin x\,dx = e^x \sin x - \int e^x \cos x\,dx = e^x \sin x - e^x \cos x - \int e^x \sin x\,dx.$$
Thus,
$$2\int e^x \sin x\,dx = e^x \sin x - e^x \cos x,$$
so
$$\int e^x \sin x\,dx = \dfrac{1}{2}(e^x \sin x - e^x \cos x) + C.$$

45 We decompose the integrand via partial fractions. If we write $\dfrac{2x^3 + 5x^2 + 13x + 9}{x^2(x^2+4x+9)} = \dfrac{A}{x} + \dfrac{B}{x^2} + \dfrac{Cx+D}{x^2+4x+9}$, we can solve and obtain $A = B = C = 1$ and $D = 0$. Then we have
$$\int \dfrac{1}{x}\,dx + \int \dfrac{1}{x^2}\,dx + \int \dfrac{x}{x^2+4x+9}\,dx = \ln|x| - \dfrac{1}{x} + \int \dfrac{x}{(x+2)^2+5}\,dx = \ln|x| - \dfrac{1}{x} + \int \dfrac{u-2}{u^2+5}\,du,$$
where $u = x+2$ so that $du = dx$. Then we have
$$\ln|x| - \dfrac{1}{x} + \int \dfrac{u}{u^2+5}\,du - \int \dfrac{2}{u^2+5}\,du = \ln|x| - \dfrac{1}{x} + \dfrac{1}{2}\ln|u^2+5| - \dfrac{2}{\sqrt{5}}\tan^{-1}\dfrac{u}{\sqrt{5}} + C$$
$$= \ln|x| - \dfrac{1}{x} + \dfrac{1}{2}\ln|x^2+4x+9| - \dfrac{2}{\sqrt{5}}\tan^{-1}\left(\dfrac{x+2}{\sqrt{5}}\right) + C.$$

47 Let $u = 4\theta$, Then
$$\int \cos^2 4\theta\,d\theta = \dfrac{1}{4}\int \cos^2 u\,du = \dfrac{1}{4}\int \dfrac{1+\cos 2u}{2}\,du = \dfrac{u}{8} + \dfrac{\sin 2u}{16} + C = \dfrac{\theta}{2} + \dfrac{\sin 8\theta}{16} + C.$$

49 Let $u = \sec 2z$. Then $du = 2\sec 2z \tan 2z\,dz$ and
$$\int \sec^{49} 2z \tan 2z\,dz = \dfrac{1}{2}\int u^{48}\,du = \dfrac{1}{2}\cdot \dfrac{u^{49}}{49} + C = \dfrac{\sec^{49} 2z}{98} + C.$$

51 Let $u = \cos 4\theta$ so that $du = -4\sin\theta\,d\theta$. Then
$$\int_0^{\pi/4} \sin^5 4\theta\,d\theta = \int_0^{\pi/4} \sin^4 4\theta \cdot \sin 4\theta\,d\theta = -\dfrac{1}{4}\int_1^{-1}(1-u^2)^2\,du = \dfrac{1}{4}\int_{-1}^{1}(1-u^2)^2\,du$$
$$= \dfrac{1}{4}\int_{-1}^{1}(1 - 2u^2 + u^4)\,du = \dfrac{1}{4}\left(u - \dfrac{2}{3}u^3 + \dfrac{1}{5}u^5\right)\Big|_{-1}^{1}$$
$$= \dfrac{1}{4}\left(1 - \dfrac{2}{3} + \dfrac{1}{5} - \left(-1 + \dfrac{2}{3} - \dfrac{1}{5}\right)\right) = \dfrac{4}{15}.$$

53 If we let $u^6 = x$, then $6u^5\, du = dx$. Substituting yields

$$\int \frac{6u^5}{u^3+u^2}\, du = 6\int \frac{u^3}{u+1}\, du = 6\int \left(u^2 - u + 1 - \frac{1}{u+1}\right) du$$
$$= 2u^3 - 3u^2 + 6u - 6\ln|u+1| + C = 2\sqrt{x} - 3\sqrt[3]{x} + 6\sqrt[6]{x} - 6\ln(\sqrt[6]{x}+1) + C.$$

55 We may write the given integral as $\dfrac{1}{\sqrt{2}}\displaystyle\int \dfrac{dy}{y^2\sqrt{9-y^2}}$. Let $y = 3\sin\theta$ so that $dy = 3\cos\theta\, d\theta$. Then

$$\frac{1}{\sqrt{2}}\int \frac{dy}{y^2\sqrt{9-y^2}} = \frac{1}{\sqrt{2}}\int \frac{3\cos\theta\, d\theta}{9\sin^2\theta \cdot 3\cos\theta} = \frac{1}{9\sqrt{2}}\int \csc^2\theta\, d\theta = -\frac{1}{9\sqrt{2}}\cot\theta + C.$$

Now use $\sin\theta = y/3$, $\cos\theta = (1/3)\sqrt{9-y^2}$ to obtain

$$\frac{1}{\sqrt{2}}\int \frac{dy}{y^2\sqrt{9-y^2}} = -\frac{1}{9\sqrt{2}\,y}\sqrt{9-y^2} + C.$$

57 Let $x = (3/2)\tan\theta$ so that $dx = (3/2)\sec^2\theta\, d\theta$, and note that $(3/2)\tan(\pi/6) = \sqrt{3}/2$. Then

$$\int_0^{\sqrt{3}/2} \frac{4}{9+4x^2}\, dx = \int_0^{\pi/6} \frac{4}{9\sec^2\theta} \cdot \frac{3}{2}\sec^2\theta\, d\theta = \frac{2}{3}\int_0^{\pi/6} d\theta = \frac{\pi}{9}.$$

59 Let $u = \cosh x$ so that $du = \sinh x\, dx$. Then

$$\int \frac{\sinh x}{\cosh^2 x}\, dx = \int u^{-2}\, du = \frac{-1}{u} + C = \frac{-1}{\cosh x} + C = -\operatorname{sech} x + C.$$

61 Let $u = \sinh x$ so that $du = \cosh x\, dx$. Substituting gives

$$\int_0^{\sqrt{3}} \frac{1}{\sqrt{4-u^2}}\, du = \left(\sin^{-1}(u/2)\right)\Big|_0^{\sqrt{3}} = \frac{\pi}{3}.$$

63 Using the method of partial fractions, we express $\dfrac{1}{x^2-2x-15} = \dfrac{1}{(x-5)(x+3)} = \dfrac{A}{x-5} + \dfrac{B}{x+3}$. Clearing denominators gives $1 = A(x+3) + B(x-5)$ and comparing coefficients gives $A+B = 0$, $3A-5B = 1$ which has solution $A = 1/8$, $B = -1/8$. Hence

$$\int \frac{dx}{x^2-2x-15} = \frac{1}{8}\int \left(\frac{1}{x-5} - \frac{1}{x+3}\right) dx = \frac{1}{8}\ln\left|\frac{x-5}{x+3}\right| + C.$$

65 Using the method of partial fractions, we express $\dfrac{1}{(y+1)(y^2+1)} = \dfrac{A}{y+1} + \dfrac{By+C}{y^2+1}$. Clearing denominators gives $1 = A(y^2+1) + (By+C)(y+1)$ and comparing coefficients gives $A+B = 0$, $B+C = 0$ and $A+C = 1$, which has solution $A = 1/2$, $B = -1/2$, $C = 1/2$. Hence

$$\int_0^1 \frac{dy}{(y+1)(y^2+1)} = \frac{1}{2}\int_0^1 \left(\frac{1}{y+1} + \frac{1-y}{y^2+1}\right) dy = \frac{1}{2}\left(\ln(y+1) + \tan^{-1} y - \frac{1}{2}\ln(y^2+1)\right)\Big|_0^1 = \frac{1}{4}\ln 2 + \frac{\pi}{8}.$$

67 First consider $\displaystyle\int_0^1 \frac{dx}{\sqrt[3]{|x-1|}} = \lim_{a\to 1^-}\int \frac{dx}{\sqrt[3]{1-x}} = \lim_{a\to 1^-} -\frac{3}{2}(1-x)^{2/3}\Big|_0^a = \lim_{a\to 1^-} -\frac{3}{2}\left((1-a)^{2/3} - 1\right) = \frac{3}{2}$.

Now consider $\displaystyle\int_1^2 \frac{dx}{\sqrt[3]{|x-1|}} = \lim_{b\to 1^+}\int_b^2 \frac{dx}{\sqrt[3]{x-1}} = \lim_{b\to 1^+} \frac{3}{2}(x-1)^{2/3}\Big|_b^2 = \lim_{b\to 1^+} \frac{3}{2}\left(1 - (b-1)^{2/3}\right) = \frac{3}{2}$.

Thus, $\displaystyle\int_0^2 \frac{dx}{\sqrt[3]{|x-1|}} = \frac{3}{2} + \frac{3}{2} = 3$.

69 Factor $x^2 - x - 2 = (x-2)(x+1)$ and use the method of partial fractions: $\dfrac{1}{(x-2)(x+1)} = \dfrac{A}{x-2} + \dfrac{B}{x+1}$. Clearing denominators gives $1 = A(x+1) + B(x-2)$ and equating coefficients gives $A + B = 0$, $A - 2B = 1$, so $A = 1/3$, $B = -1/3$. Therefore

$$\int \frac{dx}{x^2 - x - 2} = \frac{1}{3}\int \left(\frac{1}{x-2} - \frac{1}{x+1}\right)dx = \frac{1}{3}\ln\left|\frac{x-2}{x+1}\right| + C.$$

71 As a preliminary step, observe that

$$\frac{2x^2 - 4x}{x^2 - 4} = \frac{2x(x-2)}{(x-2)(x+2)} = \frac{2x}{x+2} = \frac{2(x+2-2)}{x+2} = 2 - \frac{4}{x+2}.$$

Hence

$$\int \frac{2x^2 - 4x}{x^2 - 4}\,dx = 2x - 4\ln|x+2| + C = 2(x - 2\ln|x+2|) + C.$$

73 Make the preliminary substitution $x = e^{2t}$, $dx = 2e^{2t}$. Then we have

$$\int \frac{e^{2t}}{(1+e^{4t})^{3/2}}\,dt = \frac{1}{2}\int \frac{dx}{(1+x^2)^{3/2}}.$$

Now let $x = \tan\theta$, $dx = \sec^2\theta\,d\theta$ and

$$\frac{1}{2}\int \frac{dx}{(1+x^2)^{3/2}} = \frac{1}{2}\int \frac{\sec^2\theta}{\sec^3\theta}\,d\theta = \frac{1}{2}\int \cos\theta\,d\theta = \frac{1}{2}\sin\theta + C.$$

Now

$$\sin\theta = \frac{\tan\theta}{\sec\theta} = \frac{x}{\sqrt{1+x^2}} = \frac{e^{2t}}{\sqrt{1+e^{4t}}}.$$

Therefore

$$\int \frac{e^{2t}}{(1+e^{4t})^{3/2}}\,dt = \frac{1}{2}\frac{e^{2t}}{\sqrt{1+e^{4t}}} + C.$$

75

a. Let $u = e^x$. Then $du = e^x\,dx$ and we have

$$\int \sec u \tan u\,du = \sec u + C = \sec e^x + C.$$

b. We integrate by parts with $u = e^x$ and $dv = e^x \sec e^x \tan e^x\,dx$. Then $du = e^x\,dx$ and $v = \sec e^x$ (by part a). Then we have

$$e^x \sec e^x - \int e^x \sec e^x\,dx = e^x \sec e^x - \ln|\sec e^x + \tan e^x| + C.$$

77 Using the table of integrals entry 77, we find that

$$\int \frac{dx}{x\sqrt{ax-b}} = \frac{2}{\sqrt{b}}\tan^{-1}\left(\sqrt{\frac{ax-b}{b}}\right) + C.$$

Therefore

$$\int \frac{dx}{x\sqrt{4x-6}} = \frac{\sqrt{6}}{3}\tan^{-1}\left(\sqrt{\frac{2x-3}{3}}\right) + C.$$

79 Using the table of integrals reduction formula (number 45) twice, we have

$$\int \sec^5 x \, dx = \frac{\sec^3 x \tan x}{4} + \frac{3}{4} \int \sec^3 x \, dx$$

$$= \frac{\sec^3 x \tan x}{4} + \frac{3}{4} \left(\frac{\sec x \tan x}{2} + \frac{1}{2} \int \sec x \, dx \right)$$

$$= \frac{\sec^3 x \tan x}{4} + \frac{3}{4} \left(\frac{\sec x \tan x}{2} + \frac{1}{2} (\ln|\sec x + \tan x|) \right) + C$$

$$= \frac{\sec^3 x \tan x}{4} + \frac{3 \sec x \tan x}{8} + \frac{3}{8} (\ln|\sec x + \tan x| + C.$$

81 Let $u = 1 + \sin x$ so that $du = \cos x \, dx$. We have

$$\int_1^2 \ln^3 u \, du = \left(u \ln^3 u - 3u \ln^2 u + 6u \ln u - 6u \right) \Big|_1^2$$

$$= 2 \ln^3 2 - 6 \ln^2 2 + 12 \ln 2 - 12 - (0 - 6) = 2(\ln^3 2 - 3 \ln^2 2 + 6 \ln 2 - 3).$$

83 First evaluate

$$\int_0^b xe^{-x} \, dx = -e^{-x}(x+1) \Big|_0^b = 1 - (b+1)e^{-b}.$$

Then

$$\int_0^\infty xe^{-x} \, dx = \lim_{b \to \infty} \left(1 - (b+1)e^{-b} \right) = 1.$$

85 First take $0 < c < 3$ and evaluate

$$\int_0^c \frac{dx}{\sqrt{9-x^2}} = \sin^{-1}\left(\frac{x}{3}\right) \Big|_0^c = \sin^{-1}\left(\frac{c}{3}\right).$$

Then

$$\int_0^3 \frac{dx}{\sqrt{9-x^2}} = \lim_{c \to 3^-} \sin^{-1}\left(\frac{c}{3}\right) = \frac{\pi}{2}.$$

87

$$\int_1^\infty \frac{dx}{x(x-1)^{1/3}} \, dx = \int_1^2 \frac{dx}{x(x-1)^{1/3}} \, dx + \int_2^\infty \frac{dx}{x(x-1)^{1/3}} \, dx$$

$$= \lim_{a \to 1^+} \int_a^2 \frac{dx}{x(x-1)^{1/3}} \, dx + \lim_{b \to \infty} \int_2^b \frac{dx}{x(x-1)^{1/3}} \, dx$$

$$= \lim_{a \to 1^-} \left(-\log\left(\sqrt[3]{x-1}+1\right) + \frac{1}{2} \log\left((x-1)^{2/3} - \sqrt[3]{x-1}+1\right) + \sqrt{3} \tan^{-1}\left(\frac{2\sqrt[3]{x-1}-1}{\sqrt{3}}\right) \right) \Big|_a^2$$

$$+ \lim_{b \to \infty} \left(-\log\left(\sqrt[3]{x-1}+1\right) + \frac{1}{2} \log\left((x-1)^{2/3} - \sqrt[3]{x-1}+1\right) + \sqrt{3} \tan^{-1}\left(\frac{2\sqrt[3]{x-1}-1}{\sqrt{3}}\right) \right) \Big|_2^b$$

$$= \frac{\pi}{\sqrt{3}} - \ln 2 + \frac{\pi}{\sqrt{3}} + \ln 2 = \frac{2\pi}{\sqrt{3}}.$$

89 $x^5 + x^4 + x^3 + 1 > x^5$ for $x \geq 1$, so $0 < \frac{1}{x^5 + x^4 + x^3 + 1} < \frac{1}{x^5}$. Because $\int_1^\infty \frac{1}{x^5} \, dx$ converges, so does the given integral by the Comparison Test.

91 $\frac{x^3}{\sqrt{x^7-1}} \cdot \frac{\frac{1}{x^3}}{\frac{1}{\sqrt{x^6}}} = \frac{1}{\sqrt{x - \frac{1}{x^6}}} > \frac{1}{\sqrt{x}}$ for $x \geq 3$. Then because $\int_3^\infty \frac{dx}{\sqrt{x}}$ diverges, so does $\int_3^\infty \frac{x^3}{\sqrt{x^7-1}} \, dx$ by the Comparison Test.

93 Using a Computer Algebra System, we find that
$$\int_{-1}^{1} e^{-2x^2}\, dx \approx 1.196.$$

95 $\Delta x = \dfrac{8-0}{4} = 2$. $M(4) = 2(4+4+6+8) = 44$. $T(4) = \dfrac{2}{2}(3 + 2(5) + 2(7) + 2(5) + 5) = 42$.
$S(4) = \dfrac{2}{3}(3 + 4(5) + 2(7) + 4(5) + 5) = \dfrac{124}{3}$.

97 $\Delta x = \dfrac{1}{40}$. $M(40) \approx 0.398236$, $T(40) \approx 0.398771$, and $S(40) \approx 0.398416$.

99

n	$T_2(n)$	$T_4(n)$	$T_8(n)$
4	0.880619	0.886319	1.036632
8	0.881704	0.886227	0.886319
16	0.881986	0.886227	0.886227
32	0.882058	0.886227	0.886227

Based on these results, we conclude that $\int_0^\infty e^{-x^2}\, dx \approx 0.886227$.

101 The volume generated by revolving around the x-axis is
$$V_x = \pi \int_0^\pi \sin^2 x\, dx = \pi \left(\frac{x}{2} - \frac{\sin 2x}{4}\right)\Big|_0^\pi = \frac{\pi^2}{2},$$
and the volume generated by revolving around the y-axis is
$$V_y = 2\pi \int_0^\pi x \sin x\, dx = 2\pi(\sin x - x \cos x)\Big|_0^\pi = 2\pi^2,$$
so the greater volume is obtained by revolving around the y-axis.

103 The volume is
$$V = \pi \int_1^e (\ln x)^2\, dx = \pi x((\ln x)^2 - 2\ln x + 2)\Big|_1^e = \pi(e-2).$$

105 The volume is
$$V = 2\pi \int_1^e (x-1)\ln x\, dx = \frac{\pi}{2} x\left(2(x-2)\ln x - x + 4\right)\Big|_1^e = \frac{\pi}{2}\left(e^2 - 3\right).$$

107

a. Observe that
$$\int_{1/2}^b \ln x\, dx = (x \ln x - x)\Big|_{1/2}^b \approx b \ln b - b + 0.847;$$
solve $b \ln b - b + 0.847 = 0$ numerically to obtain $b \approx 1.603$.

b. Similarly, we have
$$\int_{1/3}^b \ln x\, dx = (x \ln x - x)\Big|_{1/3}^b \approx b \ln b - b + 0.700;$$
solve $b \ln b - b + 0.700 = 0$ numerically to obtain $b \approx 1.870$.

c. In general, the pair (a, b) must satisfy the equation

$$\int_a^b \ln x \, dx = (b \ln b - b) - (a \ln a - a) = 0,$$

which gives $b \ln b - b = a \ln a - a$.

d. As a increases there is less negative area to the right of $x = 1$, so $b = g(a)$ is a decreasing function of a.

109 The average velocity is

$$\bar{v} = \frac{1}{\pi} \int_0^\pi 10 \sin 3t \, dt = -\frac{10}{3\pi} \cos 3t \bigg|_0^\pi = \frac{20}{3\pi}.$$

111 The number of cars is given by the integral

$$\int_0^4 800 t e^{-t/2} \, dt = -1600(t + 2) e^{-t/2} \bigg|_0^4 = 3200(1 - 3e^{-2}) \approx 1901.$$

113

a. Using integration by parts, we find that

$$I(p) = \int_1^e \frac{\ln x}{x^p} \, dx = -\frac{x^{1-p}}{(p-1)^2} \left((p-1) \ln x + 1\right) \bigg|_1^e = \frac{1}{(p-1)^2} \left(1 - p e^{1-p}\right)$$

for $p \neq 1$, and using the substitution $u = \ln x$ gives

$$I(1) = \int_1^e \frac{\ln x}{x} \, dx = \frac{(\ln x)^2}{2} \bigg|_1^e = \frac{1}{2}.$$

b. We have

$$\lim_{p \to \infty} I(p) = \lim_{p \to \infty} \frac{1}{(p-1)^2} (1 - p e^{1-p}) = \lim_{p \to \infty} \left(\frac{1}{(p-1)^2} - \frac{pe}{(p-1)^2} e^{-p} \right) = 0,$$

and

$$\lim_{p \to -\infty} I(p) = \lim_{p \to -\infty} \left(\frac{1}{(p-1)^2} - \frac{pe}{(p-1)^2} e^{-p} \right) = \infty,$$

since e^{-p} grows much faster than $(p-1)^2/p$ as $p \to -\infty$.

c. By inspection we see that $I(0) = 1$.

115 Use a calculator program for, say, Simpson's rule with $n = 100$, but replace the limits with 0.00000001 and 0.99999999 to avoid errors coming from trying to evaluate the function at $x = 0$ or $x = 1$; we obtain

$$\int_0^1 \frac{x^2 - x}{\ln x} \, dx \approx 0.4054651.$$

117 Numerically approximate the integral to obtain

$$\int_0^1 \frac{\sin^{-1} x}{x} \, dx \approx 1.0889;$$

therefore $n = \pi \ln 2 / 1.0889 = 2$. (When approximating the integral, replace the lower limit 0 with a small positive number like 0.00001 to avoid an error from evaluating the integrand at $x = 0$.)

119

a. Using integration by parts or a CAS, we find that the volume is $V_1(a) = \pi \int_1^a (\ln x)^2 \, dx = \pi x((\ln x)^2 - 2\ln x + 2)\Big|_1^a = \pi[(a\ln^2 a - 2a\ln a + 2(a-1)]$.

b. Using integration by parts or a CAS, we find that the volume is $V_2(a) = 2\pi \int_1^a x \ln x \, dx = \frac{\pi}{2} x^2 (2\ln x - 1)\Big|_1^a = \frac{\pi}{2}(2a^2 \ln a - a^2 + 1)$.

c. As shown by the graph, $V_2(a) > V_1(a)$ for all $a > 1$.

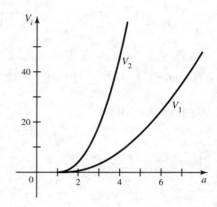

121 We have
$$V_1 = \pi \int_0^b e^{-2ax} \, dx = -\frac{\pi}{2a} e^{-2ax} \Big|_0^b = \frac{\pi}{2a}\left(1 - e^{-2ab}\right),$$
and
$$V_2 = \pi \int_b^\infty e^{-2ax} \, dx = \lim_{c \to \infty}\left(-\frac{\pi}{2a} e^{-2ax}\Big|_b^c\right) = \frac{\pi}{2a} e^{-2ab}.$$
Equating and solving gives $e^{-2ab} = 1/2$, which is equivalent to $ab = (1/2)\ln 2$.

123
$$A = \int_0^{\pi/4}(\sec x - \tan x) \, dx = (\ln|\sec x + \tan x| + \ln|\cos x|)\Big|_0^{\pi/4}$$
$$= \ln\left(\sqrt{2}+1\right) + \ln\left(\frac{\sqrt{2}}{2}\right) - (0+0) = \ln\left(1+\frac{\sqrt{2}}{2}\right).$$

125

a. From the reduction formula in Section 8.3 we have
$$\int_0^\pi \sin^m x \, dx = \frac{-\sin^{m-1} x \cos x}{m}\Big|_0^\pi + \frac{m-1}{m}\int_0^\pi \sin^{m-2} \, dx = \frac{m-1}{m}\int_0^\pi \sin^{m-2} \, dx.$$

b. We proceed by strong induction on n. For $n = 1$, we have
$$\int_0^\pi \sin^2 x \, dx = \int_0^\pi \left(\frac{1}{2} - \frac{\sin 2x}{2}\right) dx = \left(\frac{1}{2}x + \frac{\cos 2x}{4}\right)\Big|_0^\pi = \frac{\pi}{2}.$$
Assume the result holds for $k < n$. Then by the reduction formula, we have
$$\int_0^\pi \sin^{2n} x \, dx = \frac{2n-1}{2n}\int_0^\pi \sin^{2n-2} x \, dx = \frac{2n-1}{2n} \cdot \frac{2n-3}{2n-2} \cdots \frac{1}{2}\pi.$$

c. For $n = 1$ we have $\int_0^\pi \sin 3x\, dx = \int_0^\pi \sin x(1 - \cos^2 x)\, dx = \int_0^\pi \sin x\, dx - \int_0^\pi \sin x \cos^2 x\, dx. = -\cos x \Big|_0^\pi - \int_0^\pi \sin x \cos^2 x\, dx = 2 - \int_0^\pi \sin x \cos^2 x\, dx.$ Let $u = \cos x$. Then $du = -\sin x\, dx$. Then we have

$$2 + \int_1^{-1} u^2\, du = 2 + \frac{u^3}{3}\Big|_1^{-1} = 2 - \frac{2}{3} = \frac{4}{3}.$$

If the result holds for integers less than n, then by the reduction formula we have

$$\int_0^\pi \sin^{2n+1} x\, dx = \frac{-\sin^{2n}(-\cos x)}{2n+1}\Big|_0^\pi + \frac{2n}{2n+1}\int_0^\pi \sin^{2n-1} x\, dx = \frac{2n}{2n+1} \cdot \frac{2n-2}{2n-1} \cdots \frac{2}{3} \cdot 2.$$

d. Because $0 \leq \sin x \leq 1$, if we multiply through by $\sin^m x$ we have $0 \leq \sin^{m+1} x \leq \sin^m x$ for $0 \leq x \leq \pi$. Therefore $\int_0^\pi 0\, dx \leq \int_0^\pi \sin^{m+1} x\, dx \leq \int_0^\pi \sin^m x\, dx$. So $0 \leq \int_0^\pi \sin^{m+1} x\, dx \leq \int_0^\pi \sin^m x\, dx$.

e. By part d we have

$$\int_0^\pi \sin^{2n+1} x\, dx \leq \int_0^\pi \sin^{2n} x\, dx \leq \int_0^\pi \sin^{2n-1} x\, dx.$$

Therefore

$$\frac{2n}{2n+1} \cdot \frac{2n-2}{2n-1} \cdots \frac{2}{3} \cdot 2 \leq \frac{2n-1}{2n} \cdot \frac{2n-3}{2n-2} \cdots \frac{3}{4} \cdot \frac{1}{2} \cdot \pi \leq \frac{2n-2}{2n-1} \cdot \frac{2n-4}{2n-3} \cdots \frac{2}{3} \cdot 2$$

for any integer $n \geq 2$.

f. Let $A_n = \frac{2n}{2n+1} \cdot \frac{2n-2}{2n-1} \cdots \frac{2}{3} \cdot 2$ and let $B_n = \frac{2n-1}{2n} \cdot \frac{2n-3}{2n-2} \cdots \frac{3}{4}$ and let $C_n = \frac{2n-2}{2n-1} \cdot \frac{2n-4}{2n-3} \cdots \frac{2}{3} \cdot 2$. The inequality in part e can be written as

$$A_n \leq B_n\left(\frac{\pi}{2}\right) \leq C_n.$$

From this it follows that $\frac{A_n}{B_n} \leq \frac{\pi}{2}$ and that $\frac{\pi}{2} \leq \frac{C_n}{B_n}$. The inequality $\frac{A_n}{B_n} \leq \frac{\pi}{2}$ can be written as

$$\frac{(2n)^2}{2n+1} \cdot \frac{(2n-2)^2}{(2n-1)^2} \cdots \frac{2^2}{3^2} \leq \frac{\pi}{2},$$

and the inequality $\frac{\pi}{2} \leq \frac{C_n}{B_n}$ can be written as

$$\frac{\pi}{2} \leq 2n \cdot \frac{(2n-2)^2}{(2n-1)^2} \cdot \frac{(2n-4)^2}{(2n-3)^2} \cdots \frac{2^2}{3^2},$$

so

$$\frac{2n}{2n+1} \cdot \frac{\pi}{2} \leq \frac{(2n)^2}{2n+1} \cdot \frac{(2n-2)^2}{(2n-1)^2} \cdots \frac{2^2}{3^2}.$$

Putting these together gives the desired result.

g. Because $\lim_{n \to \infty} \frac{2n}{2n+1} \cdot \frac{\pi}{2} = \frac{\pi}{2}$, the result follows immediately from the Squeeze Theorem.

Chapter 9

Differential Equations

9.1 Basic Ideas

9.1.1

   a This differential equation is order 1, so there is 1 arbitrary constant.

   b. It is linear.

9.1.3 Yes. Note that $y'''(t) = 0$ and $y'(t) = 2$.

9.1.5 The portion of the domain of $y = 2\sec t$ that contains the point $(\pi, -2)$ is the interval $\left(\dfrac{\pi}{2}, \dfrac{3\pi}{2}\right)$.

9.1.7 Yes, it is a solution. Note that $y'(t) = -5Ce^{-5t}$, so $y'(t) + 5y(t) = 0$.

9.1.9 Yes, it is a solution. $y'(t) = 4C_1 \cos 4t - 4C_2 \sin 4t$, so $y''(t) = -16C_1 \sin 4t - 16C_2 \cos 4t$, so $y''(t) + 16y(t) = 0$.

9.1.11 $u'(t) = Ce^{1/(4t^4)}\left(-\dfrac{4}{4}\right)t^{-5} = -\dfrac{u(t)}{t^5}$. Thus $u'(t) + \dfrac{u(t)}{t^5} = -\dfrac{u(t)}{t^5} + \dfrac{u(t)}{t^5} = 0$.

9.1.13
$$g'(x) = -2C_1 e^{-2x} + C_2 e^{-2x} + -2C_2 x e^{-2x},$$
so
$$g''(x) = 4C_1 e^{-2x} - 2C_2 e^{-2x} + -2C_2 e^{-2x} + 4C_2 x e^{-2x} = 4C_1 e^{-2x} - 4C_2 e^{-2x} + 4C_2 x e^{-2x}.$$
Thus,
$$g''(x) + 4g'(x) + 4g(x) = 4C_1 e^{-2x} - 4C_2 e^{-2x} + 4C_2 x e^{-2x} + 4(-2C_1 e^{-2x} + C_2 e^{-2x} + -2C_2 x e^{-2x}) + 4(C_1 e^{-2x} + C_2 x e^{-2x} + 2) = 8.$$

9.1.15 $u'(t) = 5C_1 t^4 - 4C_2 t^{-5} - 3t^2$, so $u''(t) = 20C_1 t^3 + 20C_2 t^{-6} - 6t$. Thus,
$$t^2 u''(t) - 20u(t) = 20C_1 t^5 + 20C_2 t^{-4} - 6t^3 - 20\left(C_1 t^5 + C_2 t^{-4} - t^3\right) = 14t^3.$$

9.1.17 Yes, it is a solution. $y'(t) = 32e^{2t}$, so $y'(t) - 2y(t) = 32e^{2t} - (32e^{2t} - 20) = 20$. Also, $y(0) = 16 - 10 = 6$.

9.1.19 Yes, it is a solution. $y'(t) = 9 \sin 3t$, so $y''(t) = 27 \cos 3t$. Thus, $y''(t) + 9y(t) = 27 \cos 3t - 27 \cos 3t = 0$. Also, $y'(0) = 0$ and $y(0) = -3$.

9.1.21 $y(t) = \displaystyle\int (3 + e^{-2t})\,dt = 3t - \dfrac{1}{2}e^{-2t} + C.$

435

9.1.23 $y(x) = \int (4\tan 2x - 3\cos x)\,dx = -2\ln|\cos 2x| - 3\sin x + C = 2\ln|\sec 2x| - 3\sin x + C$.

9.1.25 $y'(t) = \int (60t^4 - 4 + 12t^{-3})\,dt = 12t^5 - 4t - 6t^{-2} + C$. $y(t) = \int (12t^5 - 4t - 6t^{-2} + C)\,dt = 2t^6 - 2t^2 + 6t^{-1} + C_1 t + C_2$.

9.1.27 $u'(x) = \int (55x^9 + 36x^7 - 21x^5 + 10x^{-3})\,dx = 5.5x^{10} + \frac{9}{2}x^8 - \frac{7}{2}x^6 - 5x^{-2} + C_1$.
$u(x) = \int (5.5x^{10} + \frac{9}{2}x^8 - \frac{7}{2}x^6 - 5x^{-2} + C)\,dx = \frac{1}{2}x^{11} + \frac{1}{2}x^9 - \frac{1}{2}x^7 + 5x^{-1} + C_1 x + C_2$.

9.1.29 $u(x) = \int \frac{2x}{x^2+4}\,dx - \int \frac{2}{x^2+4}\,dx = \ln(x^2+4) - \tan^{-1}(x/2) + C$.

9.1.31 $y'(x) = \int \frac{x}{(1-x^2)^{3/2}}\,dx$. Let $u = 1 - x^2$, so that $du = -2x\,dx$. Substituting gives
$y'(x) = -\frac{1}{2}\int u^{-3/2}\,du = u^{-1/2} + C_1 = \frac{1}{\sqrt{1-x^2}} + C_1\,dx$. $y(x) = \int \left(\frac{1}{\sqrt{1-x^2}} + C_1\right)dx = \sin^{-1}(x) + C_1 x + C_2$.

9.1.33 $y(t) = \int (1 + e^t)\,dt = t + e^t + C$. Because $y(0) = 4 = 1 + C$, we have $C = 3$. Thus, $y(t) = t + e^t + 3$.

9.1.35 $y(x) = \int (3x^2 - 3x^{-4})\,dx = x^3 + x^{-3} + C$. Because $y(1) = 0 = 1 + 1 + C$, we have $C = -2$. So $y(x) = x^3 + x^{-3} - 2$, $x > 0$.

9.1.37 $y'(t) = \int (12t - 20t^3)\,dt = 6t^2 - 5t^4 + C_1$. Because $y'(0) = 0 = 0 + C_1$, we have $C_1 = 0$. $y(t) = \int (6t^2 - 5t^4)\,dt = 2t^3 - t^5 + C_2$. Because $y(0) = 1 = 0 - 0 + C_2$, we have $C_2 = 1$. Thus, $y(t) = 2t^3 - t^5 + 1$.

9.1.39 Using the result of number 40 below, we have $y'(t) = te^t - e^t + C_1$, and because $y'(0) = 1 = 0 - 1 + C_1$, we have $C_1 = 2$. Thus $y'(t) = te^t - e^t + 2$. $y(t) = \int y'(t)\,dt = \int (te^t - e^t + 2)\,dt = te^t - e^t - e^t + 2t + C_2 = te^t - 2e^t + 2t + C_2$. Because $y(0) = 0 = 0 - 2 + 0 + C_2$, we have $C_2 = 2$. Thus, $y(t) = te^t - 2e^t + 2t + 2$.

9.1.41 $u(x) = \int \left(\frac{1}{x^2+4^2} - 4\right)dx = \frac{1}{4}\tan^{-1}(x/4) - 4x + C$. Because $u(0) = 2 = 0 - 0 + C$, we have $C = 2$. Thus, $u(x) = \frac{1}{4}\tan^{-1}(x/4) - 4x + 2$.

9.1.43

a. $v(t) = -9.8t + 29.4$. $s(t) = -4.9t^2 + 29.4t + 30$. The domain of each function is approximately $0 \leq t \leq 6.89$.

b. The object reaches its high point when $-9.8t + 29.4 = 0$, or $t = \frac{29.4}{9.8} = 3$. At that time its position is $s(3) \approx 74.1$ meters.

9.1.45 We have $p(t) = (1500 - 20H)e^{.05t} + 20H$. The amount of resource is increasing when $1500 - 20H > 0$, which occurs for $H < 75$. The amount of resource is constant when $1500 - 20H = 0$, which occurs for $H = 75$. If $H = 100$, the resource is zero when $(1500 - 2000)e^{.05t} + 2000 = 0$, which occurs for $t = 20\ln 4 \approx 28$.

9.1.47 The height function is given by

$$h(t) = \left(\sqrt{1.96} - \frac{0.3\sqrt{2 \cdot 9.8}}{1.5} \cdot \frac{t}{2}\right)^2 \approx (1.4 - 0.44t)^2 \text{ for } 0 \leq t \leq 3.16.$$

The tank is empty when $h(t) = 0$, which occurs after about 3.16 seconds.

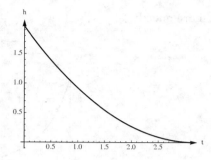

9.1.49

a. False. That is a specific solution. The general solution is $t + C$.

b. False. It is second order, but is not linear.

c. True. First find the general solution, and then find the specific solution which satisfies the initial condition.

9.1.51

a. $y'(t) = C_1 \cos t - C_2 \sin t$, so $y''(t) = -C_1 \sin t - C_2 \cos t$. Thus, $y''(t) + y(t) = 0$.

b. $y'(t) = 2C_2 \cos 2t - 2C_2 \sin 2t$, so $y''(t) = -4C_2 \sin 2t - 4C_2 \cos 2t$. Thus, $y''(t) + 4y(t) = 0$.

c. A general solution appears to be $y(t) = C_1 \sin kt + C_2 \cos kt$. Then $y'(t) = kC_1 \cos kt - kC_2 \sin kt$, so $y''(t) = -k^2 C_1 \sin kt - k^2 C_2 \cos kt$. And then $y''(t) + k^2 y(t) = 0$.

9.1.53

a. Let $p(t) = \dfrac{K}{1 + Ce^{-rt}}$. Note that $1 - \dfrac{P}{K} = 1 - \dfrac{1}{1 + Ce^{-rt}} = \dfrac{Ce^{-rt}}{1 + Ce^{-rt}}$. We have
$$p'(t) = \dfrac{KCre^{-rt}}{(1 + Ce^{-rt})^2} = r \cdot \dfrac{K}{1 + Ce^{-rt}} \cdot \dfrac{Ce^{-rt}}{1 + Ce^{-rt}} = rp\left(1 - \dfrac{p}{K}\right).$$

b. If $p(0) = 50 = \dfrac{K}{1 + C}$, then $50 + 50C = K$, so $C = \dfrac{K - 50}{50}$.

c. We have $p(t) = \dfrac{300}{1 + 5e^{-0.1t}}$.

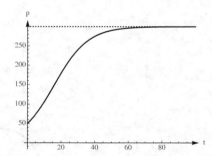

d. $\lim\limits_{t \to \infty} \dfrac{300}{1 + 5e^{-0.1t}} = \dfrac{300}{1 + 0} = 300$, which is consistent with the graph from part c.

9.1.55

a. If $y(t) = y_0 e^{-kt}$, then $y(0) = y_0$, and $y'(t) = -ky_0 e^{-kt}$, so $y'(t) = -ky(t)$.

b. Let $y(t) = \dfrac{y_0}{y_0 kt + 1}$. Then $y(0) = y_0$, and $y'(t) = \dfrac{-y_0^2 k}{(y_0 kt + 1)^2} = -k(y(t))^2$.

c. The first order reaction decays more quickly.

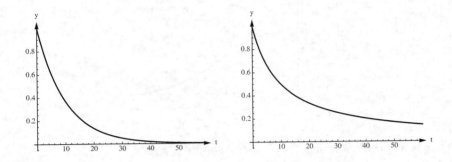

9.2 Direction Fields and Euler's Method

9.2.1 Choose a regular grid of points in the ty-plane, and for each point P, make a small line segment with slope $f(t, y)$.

9.2.3 $u_0 = y(3) = 1$. $u_1 = u_0 + f(3, 1)(.1) = 1 + .6 = 1.6$.

9.2.5

a. This matches with D.

b. This matches with B.

c. This matches with A.

d. This matches with C.

9.2.7

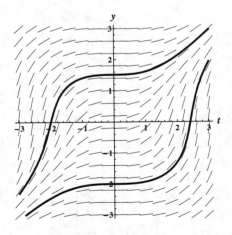

9.2.9

An initial condition of $y(0) = -1$ leads to a constant solution. For any other initial condition, the solutions are increasing over time.

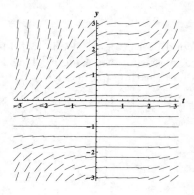

9.2.11

An initial condition of $y(0) = 1$ leads to a constant solution. Initial conditions $y(0) = A$ lead to solutions that are increasing over time if $A > 1$ and solutions that are decreasing over time if $A < 1$.

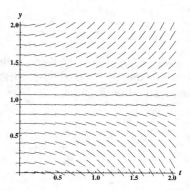

9.2.13

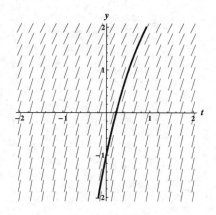

9.2.15

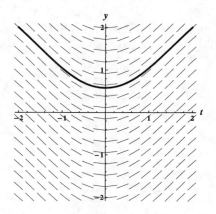

9.2.17

a. The solutions $y = 1$ and $y = -1$ are constant.

b. Solutions are increasing when both $y > 1$ and $y > -1$, (so for $y > 1$) and when both $y < 1$ and $y < -1$ (so for $y < -1$). In other words, for $|y| > 1$. Solutions are decreasing for $|y| < 1$.

c. Initial condition $y(0) = A$ leads to solutions that are increasing over time if $|A| > 1$ and decreasing over time if $|A| < 1$.

d.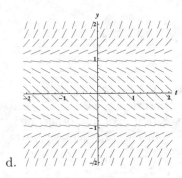

9.2.19

a. The solutions $y = \pi/2$ and $y = -\pi/2$ are constant.

b. Solutions are increasing when $|y| < \pi/2$ and decreasing when $\pi/2 < |y| < \pi$.

c. Initial condition $y(0) = a$ leads to solutions that are increasing over time if $|A| < \pi/2$ and decreasing over time if $|A| > \pi/2$.

d.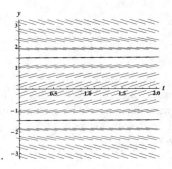

9.2.21

Equilibrium solutions are $P(t) = 0$ and $P(t) = 500$.

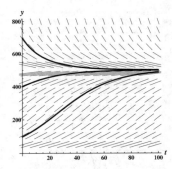

9.2.23

Equilibrium solutions are $P(t) = 0$ and $P(t) = 3200$.

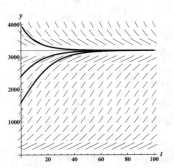

9.2. Direction Fields and Euler's Method

9.2.25 $u_0 = 2$. $u_1 = 2 + f(0,2)(0.5) = 2 + (4)(0.5) = 4$. $u_2 = 4 + f(0.5, 4)(0.5) = 4 + (8)(0.5) = 8$. So $y(0.5) \approx 4$ and $y(1) \approx 8$.

9.2.27 $u_0 = 1$. $u_1 = 1 + f(0,1)(0.1) = 1 + 0.1 = 1.1$. $u_2 = 1.1 + f(0.1, 1.1)(0.1) = 1.1 + (0.9)(0.1) = 1.19$. So $y(0.1) \approx 1.1$ and $y(0.2) \approx 1.19$.

9.2.29

a.

Δt	approximation of $y(0.2)$	approximation of $y(0.4)$
0.2	0.8	0.64
0.1	0.81	0.65610
0.05	0.81451	0.66342
0.025	0.81665	0.66692

b. $e^{-0.2} \approx 0.818731$ and $e^{-0.4} \approx 0.67032$.

Δt	error in approximation of $y(0.2)$	error in approximation of $y(0.4)$
0.2	0.018371	0.03032
0.1	0.00873	0.01422
0.05	0.00422	0.00690
0.025	0.00208	0.00340

c. The time step $\Delta t = 0.025$ has the smallest errors. A smaller time step generally produces more accurate results.

d. Halving the time steps results in approximately halving the error.

9.2.31

a.

Δt	approximation of $y(0.2)$	approximation of $y(0.4)$
0.2	3.2	3.36
0.1	3.19	3.3439
0.05	3.18549	3.33658
0.025	3.18335	3.33308

b. $4 - e^{-0.2} \approx 3.18127$ and $4 - e^{-0.4} \approx 3.32968$.

Δt	error in approximation of $y(0.2)$	error in approximation of $y(0.4)$
0.2	0.0187308	0.03032
0.1	0.00873075	0.01422
0.05	0.0042245	0.00689961
0.025	0.0020789	0.00339988

c. The time step $\Delta t = 0.025$ has the smallest errors. A smaller time step generally produces more accurate results.

d. Halving the time steps results in approximately halving the error.

9.2.33

a. The computations yield:

t_k	0	0.2	0.4	0.6	0.8	1	1.2	1.4	1.6	1.8	2
u_k	1	0.6	0.36	0.216	0.1296	0.07776	0.046656	0.0279936	0.0167962	0.0100777	0.00604662

So $y(2) \approx 0.00604662$.

b. $y(2) = e^{-4} \approx 0.0183156$, so the error is about $0.0183156 - 0.00604662 = 0.012269$.

c. The computations yield:

t_k	0	0.1	0.2	0.3	0.4	0.5	0.6	0.7	0.8	0.9	1
u_k	1	0.8	0.64	0.512	0.4096	0.32768	0.262144	0.209715	0.167772	0.134218	0.107374

t_k	1.1	1.2	1.3	1.4	1.5	1.6	1.7	1.8	1.9	2
u_k	0.0859	0.0687	0.0550	0.0440	0.0352	0.02815	0.02252	0.01801	0.01441	0.01153

So $y(2) \approx 0.01153$ The error is about $0.0183156 - 0.01153 = 0.0067856$.

d. The error with twice as many steps is about half the other error.

9.2.35

a. After many calculations, we arrive at $y(4) \approx 3.05765$.

b. $y(4) = 3 + 5e^{-4} \approx 3.09158$, so the error is about $3.09158 - 3.05765 = .03393$.

c. After many calculations, we arrive at $y(4) = 3.0739$. The error is about $3.09158 - 3.0739 = 0.01768$.

d. The error with twice as many steps is about half the other error.

9.2.37

a. True.

b. False. It allows you to compute approximations.

9.2.39

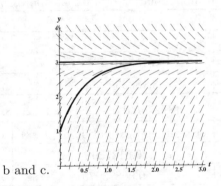

a. $y = 3$ is an equilibrium solution, because $6 - 2(3) = 0$.

b and c.

9.2.41

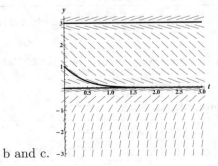

a. Solve $y(y-3) = 0$ to get equilibrium solutions $y = 0$ and $y = 3$.

b and c.

9.2.43

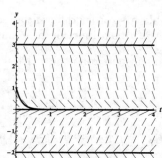

a. The equilibrium solutions are $y=0$, $y=-2$ and $y=3$.

b and c.

9.2.45

a. $\Delta t = \dfrac{b-a}{N}$.

b. Recall that $u_0 = A$ and $t_0 = a$. So $u_1 = A + f(a, A)\left(\dfrac{b-a}{N}\right)$.

c. $u_{k+1} = u_k + f(t_k, u_k)\left(\dfrac{b-a}{N}\right)$, where $t_k = a + k\left(\dfrac{b-a}{N}\right)$ for $k = 0, 1, 2, \ldots, N-1$.

9.2.47

a.

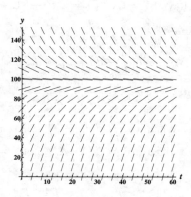

b. The solutions are increasing for $A < 98$ and decreasing for $A > 98$.

c. The equilibrium solution is $v(t) = 98$.

9.2.49

a. We have $u_0 = y(0) = 1$, and $u_{k+1} = u_k + f(t_k, u_k)h = u_k + au_k h = u_k(1+ah)$ for $k = 0, 1, 2 \ldots$.

b. Suppose $u_k = (1+ah)^k$. Then $u_{k+1} = u_k(1+ah) = (1+ah)^k(1+ah) = (1+ah)^{k+1}$, $k = 0, 1, 2, \ldots$.

c. $\lim\limits_{h \to 0} u_k = \lim\limits_{h \to 0}(1+ah)^k = \lim\limits_{h \to 0}(1+ah)^{t_k/h} = \lim\limits_{h \to 0}\left((1+ah)^{1/h}\right)^{t_k} = (e^a)^{t_k} = e^{at_k} = y(t_k)$.

9.3 Separable Differential Equations

9.3.1 A separable first-order differential equation is one that can be written in the form $g(y)y'(t) = h(t)$, where the factor $g(y)$ is a function of y and $h(t)$ is a function of t.

9.3.3 No, this equation cannot be written in the required form.

9.3.5 We have $y'(t) = t^3$, so $\int \frac{dy}{dt} dt = \int t^3 dt$, so $y = \frac{t^4}{4} + C$.

9.3.7 $y\frac{dy}{dt} = 3t^2$, so $\int y\, dy = \int 3t^2\, dt$. Thus, $\frac{y^2}{2} = t^3 + C$, and thus $y = \pm\sqrt{2t^3 + C}$.

9.3.9 We have $\int e^{-y/2}\, dy = \int \sin t\, dt$, and so $-2e^{-y/2} = -\cos t + C$. Thus, $y = -2\ln\left(\frac{1}{2}\cos t + C\right)$.

9.3.11 $\frac{1}{y^2}\frac{dy}{dx} = \frac{1}{x^2}$, so $\int \frac{1}{y^2}\frac{dy}{dx}\, dx = \int \frac{1}{x^2}\, dx$, so $-\frac{1}{y} = -\frac{1}{x} + C = \frac{Cx - 1}{x}$. Thus, $y = \frac{x}{1 - Cx}$. If we replace the arbitrary constant C by its opposite, this can be written as $y = \frac{x}{1 + Cx}$.

9.3.13 $-\frac{2}{y^3}\frac{dy}{dt} = \sin t$, so $\int \frac{-2}{y^3}\frac{dy}{dt}\, dt = \int \sin t\, dt$. Thus, $\frac{1}{y^2} = -\cos t + C$. Solving for y gives $y = \pm\frac{1}{\sqrt{C - \cos t}}$.

9.3.15 $e^u u'(x) = e^{2x}$, so $\int e^u\, du = \int e^{2x}\, dx$, and $e^u = \frac{1}{2}e^{2x} + C$. Thus, $u = \ln\left(\frac{e^{2x}}{2} + C\right)$.

9.3.17 This is separable, and is already written in the desired form. We have $\int 2y\, dy = \int 3t^2\, dt$, so $y^2 = t^3 + C$. Because $y(0) = 9$, we have $81 = C$, so $y = \sqrt{t^3 + 81}$.

9.3.19 This equation is not separable.

9.3.21 The equation is separable. We have $\int \frac{1}{y}\, dy = \int e^t\, dt$, so $\ln|y| = e^t + C$, and $|y| = e^{e^t}e^C = Ae^{e^t}$. Then $|y(0)| = 1 = Ae$, so $A = \frac{1}{e}$ and $y = -\frac{1}{e}e^{e^t} = -e^{-1}e^{e^t} = -e^{e^t - 1}$.

9.3.23 This equation is separable. We have $\int e^y\, dy = \int e^x\, dx$, and thus $e^y = e^x + C$. Therefore, $y = \ln(e^x + C)$. Substituting $y(0) = \ln 3$ gives $\ln 3 = \ln(1 + C)$, so $C = 2$ and the solution to this initial value problem is $y = \ln(e^x + 2)$.

9.3.25 This equation is separable. We have $\int e^y\, dy = \int \frac{\ln^3 t}{t}\, dt$. For the integral on the right-hand side, let $u = \ln t$ so that $du = \frac{1}{t}\, dt$. Then $\int \frac{\ln^3 t}{t}\, dt = \int u^3\, du = \frac{u^4}{4} + C = \frac{\ln^4 t}{4} + C$. Thus we have

$$e^y = \frac{\ln^4 t}{4} + C.$$

Because $y(1) = 0$, we have $1 = 0 + C$, so $y = \ln\left(\frac{\ln^4 t}{4} + 1\right)$.

9.3.27 The equation is separable. We have $\int 2y\, dy = \int \sec^2 t\, dt$, so $y^2 = \tan t + C$. Because $y(\pi/4) = 1$ we have $C = 0$, and $y = \sqrt{\tan t}$, Note that the original differential equation is undefined for $y = 0$, which corresponds to $t = 0$ and $t = \pi$, so the solution is $y = \sqrt{\tan t}$ for $0 < t < \pi/2$.

9.3. Separable Differential Equations

9.3.29 The equation is separable.
$yy'(t) = t$, so $\int y\, dy = \int t\, dt$, so $\frac{y^2}{2} = \frac{t^2}{2} + C$. Because $y(1) = 2$, we have $2 = 1/2 + C$, so $C = 3/2$. So $y^2 = t^2 + 3$, and $y = \sqrt{t^2 + 3}$.

9.3.31 This is separable, and can be written as $y'(t) = \frac{1}{t}$. Thus, $\int y'(t)\, dt = \int \frac{dt}{t} = \ln t + C$, so $y(t) = \ln t + C$. Because $y(1) = 2 = 0 + C$, we have $C = 2$. Thus, $y(t) = \ln t + 2$.

9.3.33 We may separate variables and write $\int (y^2 - 1)\, dy = \int 2t^2\, dt$. Then $\frac{y^3}{3} - y = \frac{2t^3}{3} + C$. Because $y(0) = 0$, it follows that $C = 0$. So we have $y^3 - 3y = 2t^3$. From the original differential equation, we see that $y'(t)$ is undefined for $y = \pm 1$. Using the solution, $y = 1$ for $2t^3 = -2$, or $t = -1$, and likewise $y = -1$ implies that $t = 1$. So the solution is $y^3 - 3y = 2t^3$ for $-1 < t < 1$, and the graph is the portion of the curve passing through the origin.

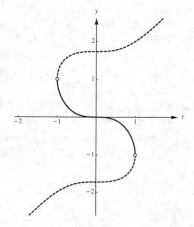

9.3.35 $(\sin u)u'(x) = \cos(x/2)$, so $\int \sin u\, du = \int \cos(x/2)\, dx$. We have $-\cos u = 2\sin(x/2) + C$. Because $u(\pi) = \pi/2$, we have $-\cos \pi/2 = 2\sin(\pi/2) + C$, so $0 = 2(1) + C$, so $C = -2$. Thus, $-\cos u = 2\sin(x/2) - 2$ describes the solution. Note that the original differential equation is undefined for $u = 0$ which corresponds to $x = \pi/3$ and $x = 5\pi/3$. So the solution is

$$-\cos u = 2\sin\left(\frac{x}{2}\right) - 2, \text{ for } \frac{\pi}{3} < x < \frac{5\pi}{3}.$$

Not that the non-dotted curve contains the point $(\pi, \pi/2)$.

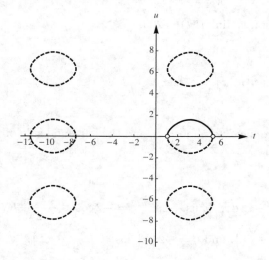

9.3.37 $\sqrt{y+4}\,y'(x) = \sqrt{x+1}$, so $\int \sqrt{y+4}\,dy = \int \sqrt{x+1}\,dx$, so $\frac{2}{3}(y+4)^{3/2} = \frac{2}{3}(x+1)^{3/2} + C$. Because $y(3) = 5$, we have $\frac{2}{3}(27) = \frac{16}{3} + C$, so $C = \frac{38}{3}$. Thus, $\frac{2}{3}(y+4)^{3/2} = \frac{2}{3}(x+1)^{3/2} + \frac{38}{3}$, so $(y+4)^{3/2} = (x+1)^{3/2} + 19$

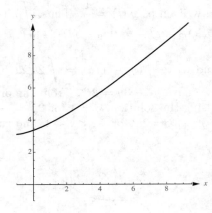

9.3.39

a. This equation is separable, so we have $\int \frac{200}{P(200-P)}\,dP = \int 0.08\,dt$, so $\int \left(\frac{1}{P} + \frac{1}{200-P}\right) dP = 0.08t + C$. Therefore, $\ln\left|\frac{P}{200-P}\right| = 0.08t + C$. Substituting $P(0) = 50$ gives $-\ln 3 = C$, and solving for P gives $P(t) = \frac{200}{3e^{-0.08t} + 1}$.

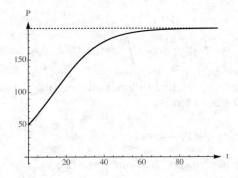

b. The steady-state population is $\lim_{t \to \infty} P(t) = 200$.

9.3.41

a. True. It can be written as $u^7 u'(x) = x^{-2}$.

b. False.

c. True. When separated, we have $ye^y y'(x) = x$, and the left-hand side of the equation can be integrated by parts.

9.3.43

a. $\int e^{-y/2}\,dy = \int (4x \sin x^2 - x)\,dx$. Thus, $-2e^{-y/2} = -2\cos x^2 - \frac{x^2}{2} + K$. Then $e^{-y/2} = \cos x^2 + \frac{x^2}{4} + C$, so $-\frac{y}{2} = \ln\left(\cos x^2 + \frac{x^2}{4} + C\right)$, and $y = -2\ln\left(\cos x^2 + \frac{x^2}{4} + C\right)$.

b. When $y(0) = 0$ we have $0 = -2\ln(1 + C)$, so $C = 0$. When $y(0) = \ln(1/4)$, we have $\ln(1/4) = -2\ln(1 + C)$, so $\ln 2 = \ln(1 + C)$, so $C = 1$. When $y(\sqrt{\pi/2}) = 0$, we have $0 = -2\ln(0 + \pi/8 + C)$, so $C = 1 - \dfrac{\pi}{8}$.

c.

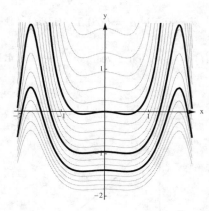

9.3.45 Differentiating implicitly gives $2x + 2yy' = 0$, so $y' = -\dfrac{x}{y}$. We are thus seeking curves so that $\dfrac{dy}{dx} = \dfrac{y}{x}$. We have $\dfrac{dy}{y} = \dfrac{dx}{x}$, so $\ln|y| = \ln|x| + C$ so $y = e^C |x| = kx$. So the family of curves we are seeking is the collection of curves $y = kx$.

9.3.47

a. We have $mv'(t) = mg - kv^2$, so $v'(t) = g - av^2$ with $a = k/m$.

b. We solve $av^2 = g$ to obtain the terminal velocity $\tilde{v} = \sqrt{g/a} = \sqrt{gm/k}$.

c. This equation is separable, so we have $\displaystyle\int \dfrac{1}{g - av^2} dv = \int dt$, so $-\dfrac{1}{a} \displaystyle\int \dfrac{1}{v^2 - \tilde{v}^2} dv = t + D$.

Thus, $-\dfrac{1}{2a\tilde{v}} \displaystyle\int \left(\dfrac{1}{v - \tilde{v}} - \dfrac{1}{v + \tilde{v}} \right) dv = t + D$, and $-\dfrac{1}{2a\tilde{v}} \ln\left| \dfrac{v - \tilde{v}}{v + \tilde{v}} \right| = t + D$, hence $\dfrac{v - \tilde{v}}{v + \tilde{v}} = Ce^{-2a\tilde{v}t}$

The initial condition $v(0) = 0$ gives $C = -1$, and solving for v gives $v = \dfrac{1 - e^{-2a\tilde{v}t}}{1 + e^{-2a\tilde{v}t}} \tilde{v}$. This can be written as $v = (\sqrt{g/a}) \dfrac{e^{2\sqrt{ag}t} - 1}{e^{2\sqrt{ag}t} + 1}$.

d. We have $a = 0.1$, $\tilde{v} \approx 9.90$ m/s

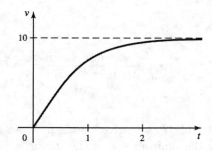

9.3.49

a. The equation $h' = -k\sqrt{h}$ is separable, so we have $\int \dfrac{dh}{\sqrt{h}} = \int -k\,dt$, so $2\sqrt{h} = -kt + C$. The initial condition $h(0) = H$ gives $C = 2\sqrt{H}$, so the solution is $h = \left(\sqrt{H} - \dfrac{kt}{2}\right)^2$.

b. The solution for $k = 0.1$ and $H = 0.5$ is $h = \left(\sqrt{0.5} - 0.05t\right)^2$, $0 \le t \le 14.1$.

c. The tank is drained when $h(t) = 0$, which gives $t = \dfrac{\sqrt{0.5}}{0.05} \approx 14.1$ s.

d.

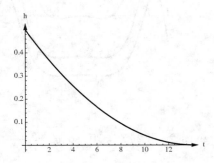

9.3.51

a. The growth rate is positive when $0 < M < 4$. The function $R(M)$ has derivative
$$R'(M) = -r\left(\ln\left(\frac{M}{K}\right) + M \cdot \frac{K}{M} \cdot \frac{1}{K}\right) = -r\left(\ln\left(\frac{M}{K}\right) + 1\right),$$
which is 0 when $M/4 = 1/e$ or $M = 4/e$. We also observe that $\lim\limits_{M \to 0^+} R(M) = 0$ and $R(4) = 0$, so $R(M)$ takes its maximum at the critical point $M = 4/e$.

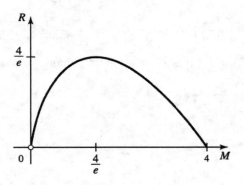

b. The equation is separable, so we have $\int \dfrac{dM}{M(\ln M - \ln K)} = -\int r\,dt$, so $\ln|\ln M - \ln K| = -rt + D$, and thus $\ln\left(\dfrac{M}{K}\right) = Ce^{-rt}$. Therefore $M = Ke^{Ce^{-rt}}$.

The conditions $r = 1$, $K = 4$ and $M_0 = 1$ give $C = -\ln 4$ and $M = 4e^{(-\ln 4)e^{-t}} = 4^{1-e^{-t}}$. Observe that $\lim\limits_{t \to \infty} M(t) = 4e^0 = 4$, so the limiting size of the tumor is 4. In general, the limiting size of the tumor is $\lim\limits_{t \to \infty} Ke^{Ce^{-rt}} = K$, because $r > 0$.

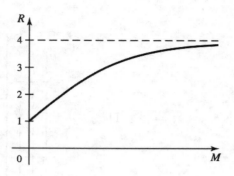

9.3.53

a. $\frac{1}{y^2}y'(t) = 1$, so $\int \frac{dy}{y^2} = \int 1\, dt$. Thus $-\frac{1}{y} = t + C$. Because $y_0 = 1$ we have $-1 = C$. Thus, $y = \frac{1}{1-t}$ for $t < 1$.

b. $\frac{1}{y^3}y'(t) = 1$, so $\int \frac{dy}{y^3} = \int 1\, dt$. Thus $-\frac{1}{2y^2} = t + C$. Because $y_0 = \frac{1}{\sqrt{2}}$ we have $-1 = C$. Thus, $\frac{1}{2y^2} = 1 - t$ and $y = \frac{1}{\sqrt{2}\sqrt{1-t}}$ for $t < 1$.

c. $\frac{1}{y^{n+1}}y'(t) = 1$, so $\int \frac{dy}{y^{n+1}} = \int 1\, dt$. Thus $-\frac{1}{ny^n} = t + C$. Because $y_0 = n^{-1/n}$ we have $-1 = C$. Thus, $\frac{1}{ny^n} = 1 - t$, and $ny^n = \frac{1}{1-t}$. Thus, $y = \frac{1}{(n(1-t))^{1/n}}$ for $t < 1$.

We have $\lim_{t \to 1^-} \frac{1}{(n(1-t))^{1/n}} = \infty$.

9.4 Special First-Order Linear Differential Equations

9.4.1 Because $y(0) = 4$, we have $4 = C - 13$, so $C = 17$. Thus, the solution is $y = 17e^{-10t} - 13$.

9.4.3 The general solution is $y = Ce^{-4t} - \left(\frac{6}{-4}\right) = Ce^{-4t} + \frac{3}{2}$.

9.4.5 Because $k = 3$ and $b = -4$, we have $y = Ce^{3t} + \frac{4}{3}$.

9.4.7 Because $k = -2$ and $b = -4$, we have $y = Ce^{-2x} - 2$.

9.4.9 Because $k = -12$ and $b = 15$, we have $u = Ce^{-12t} + \frac{5}{4}$.

9.4.11 Because $k = 3$ and $b = -6$, we have $y = Ce^{3t} + 2$. Because $y(0) = 9$, we have $9 = C + 2$, so $C = 7$. Thus, the solution is $y = 7e^{3t} + 2$.

9.4.13 Because $k = 2$ and $b = 8$, we have $y = Ce^{2t} - 4$. Because $y(0) = 0$, we have $0 = C - 4$, so $C = 4$. Thus, the solution is $y = 4e^{2t} - 4$.

9.4.15 Because $k = 3$ and $b = 12$, we have $y = Ce^{3t} - 4$. Because $y(1) = 4$, we have $4 = Ce^3 - 4$, so $C = 8e^{-3}$. Thus, the solution is $y = 8e^{3t-3} - 4$.

9.4.17

The equilibrium solution is $y = \frac{18}{12} = \frac{3}{2}$. The solution is unstable.

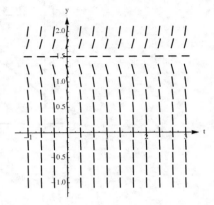

9.4.19

The equilibrium solution is $y = -3$. The solution is stable.

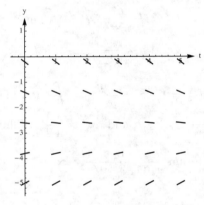

9.4.21

The equilibrium solution is $u = -3$. The solution is stable.

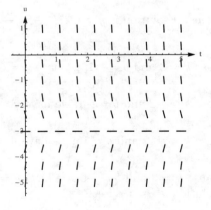

9.4.23 Because $k = 0.005$ and $b = -500$, we have $B = Ce^{0.005t} + \frac{500}{0.005} = Ce^{0.005t} + 100000$. Because $B(0) = 50,000$, we have $50000 = C + 100000$, so $C = -50000$. Thus, $B = -50000e^{0.005t} + 100000$.

The balance is zero when $t = \ln(2)/0.005 \approx 139$ months.

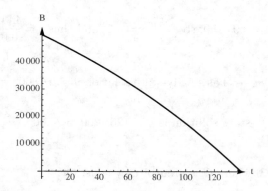

9.4.25 Because $k = 0.0075$ and $b = -1500$, we have $B = Ce^{0.0075t} + \dfrac{1500}{0.0075} = Ce^{0.0075t} + 200000$. Because $B(0) = 100,000$, we have $100000 = C + 200000$, so $C = -100000$. Thus, $B = -100000e^{0.0075t} + 200000$. The balance is zero when $t = \ln(2)/0.0075 \approx 93$ months.

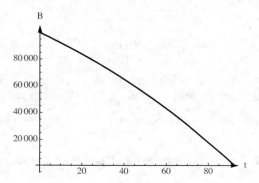

9.4.27 We know that the solution has the form $T(t) = (90-25)e^{-kt} + 25 = 65e^{-kt} + 25$. Because $T(1) = 85$, we have $65e^{-k} + 25 = 85$. Thus, $k = -\ln(60/65) \approx 0.08$. Thus, $T(t) = 65e^{-0.08t} + 25$. We have $T(t) = 30$ when $65e^{-0.08t} + 25 = 30$, or when $e^{-0.08t} = \dfrac{1}{13}$. This occurs when $t = \ln(13)/0.08 \approx 32$ minutes after the coffee is first poured.

9.4.29 $T(t) = (5-20)e^{-kt} + 20 = -15e^{-kt} + 20$. Because $T(1) = 7$, we have $-15e^{-k} + 20 = 7$, so $k = \ln(15/13) \approx 0.143$. Thus $T(t) = -15e^{-0.143t} + 20$. The milk will reach 18 degrees when $-15e^{-0.143t} + 20 = 18$, which occurs when $e^{-0.143t} = 2/15$. Thus, when $t \approx \dfrac{\ln(2/15)}{-0.143} \approx 14$ minutes.

9.4.31

a. The general solution is $y(t) = Ce^{-0.02t} + 150$; substitute $y(0) = 0$ to obtain $C + 150 = 0$, so $C = -150$; hence $y(t) = 150(1 - e^{-0.02t})$.

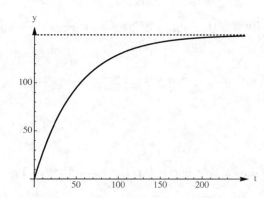

b. The steady-state level is $\lim\limits_{t \to \infty} 150(1 - e^{-0.02t}) = 150$ mg.

c. We have $150(1 - e^{-0.02t}) = 0.9 \cdot 150$, so $e^{-0.02t} = 0.1$, and thus $t = \dfrac{\ln 10}{0.02} \approx 115.1$ hours.

9.4.33

a. The equation $y'(t) = 0.008y - h$ has steady-state solution $y = h/0.008 = 125h$, so we solve $y_0 = 2000 = 125h$ to obtain $h = 16$.

b. If $h = 200$, then the steady-state solution is $y = 125 \cdot 200 = 25{,}000$.

9.4.35

a. False. It is $y(t) = Ce^{2t} + 9$ where C is an arbitrary constant.

b. True. Note that if $y(t) = 0$, then $y'(t) = 0$, not $k(0) - b = -b$ as dictated by the equation.

c. False. It is not separable.

d. False. It approaches it asymptotically.

9.4.37

a. Note that $k = 0.03$ and $b = -600$. Thus, $B(t) = Ce^{0.03t} + 20{,}000$. If $B_0 = B(0)$ is the amount borrowed, then $B_0 = C + 20{,}000$, so $C = B_0 - 20{,}000 = 20{,}000$.

Thus, $B(t) = 20{,}000 e^{0.03t} + 20{,}000$. This is an increasing function because its derivative is positive. It is occurring because the amount paid monthly is less than the monthly accruing interest.

b. Because $20000 \cdot 0.03 = 600$, the amount borrowed should be less than 20000 if the balance is to be decreasing.

c. The maximum amount that can be borrowed and not have the unpaid balance increase is B_0 when $rB_0 - m = 0$, or $B_0 = \dfrac{m}{r}$.

9.4.39 $\displaystyle\int (ty'(t) + y(t))\,dt = \int (1 + t)\,dt$, so $ty(t) = t + \dfrac{t^2}{2} + C$ for $t > 0$. Thus, $y(t) = 1 + \dfrac{t}{2} + \dfrac{C}{t}$. Because $y(1) = 4$, we have $4 = 1 + 1/2 + C$, so $C = 5/2$. Thus, $y(t) = 1 + \dfrac{t}{2} + \dfrac{5}{2t}$ for $t > 0$.

9.4.41 $\displaystyle\int (e^{-t}y'(t) - e^{-t}y(t))\,dt = \int e^{2t}\,dt$, so $e^{-t}y(t) = e^{2t}/2 + C$. Thus, $y(t) = e^{3t}/2 + Ce^t$. Because $y(0) = 4$, we have $4 = 1/2 + C$, so $C = 7/2$. The solution is therefore $y(t) = e^{3t}/2 + 7e^t/2$.

9.4.43

a. Let $v = y^{1-p}$. Then $v'(t) = (1-p)y^{-p}y'(t)$, so $y'(t) = \dfrac{y^p}{1-p}v'(t)$.

b. Given $y'(t) + ay = by^p$, we have $\dfrac{y^p}{1-p}v'(t) + ay = by^p$, so $\dfrac{1}{1-p}v'(t) + ay^{1-p} = b$, so
$$v'(t) = (p-1)av(t) + b(1-p).$$
Then $v(t) = Ce^{(p-1)at} - \dfrac{b(1-p)}{(p-1)a} = Ce^{(p-1)at} + \dfrac{b}{a}$. Therefore,
$$y(t) = \left(Ce^{(p-1)at} + \dfrac{b}{a}\right)^{1/(1-p)}.$$

9.4.45 Let $p(t) = \exp\left(\displaystyle\int 1/t\,dt\right) = e^{\ln t} = t$. The original differential equation can be written as $t(y'(t) + (1/t)y(t)) = 0$, or $ty'(t) + y(t) = 0$. Integrating both sides with respect to t gives $ty(t) = C$, so $y(t) = \dfrac{C}{t}$. Because $y(1) = 6$, we have $C = 6$, and $y = \dfrac{6}{t}$ for $t > 0$.

9.4.47 Let $p(t) = \exp\left(\int \frac{2t}{t^2+1}\,dt\right) = \exp(\ln(t^2+1)) = t^2+1$. The original differential equation can be written as $(t^2+1)y'(t) + (2t)y(t) = (t^2+1)(1+3t^2)$. Integrating both sides with respect to t gives

$$(t^2+1)y(t) = \int (3t^4 + 4t^2 + 1)\,dt = 3t^5/5 + 4t^3/3 + t + C.$$

Because $y(1) = 4$, we have $8 = 3/5 + 4/3 + 1 + C$, so $C = \frac{76}{15}$. Thus,

$$y(t) = \frac{3t^5/5 + 4t^3/3 + t + \frac{76}{15}}{t^2+1} = \frac{9t^5 + 20t^3 + 15t + 76}{15(t^2+1)}.$$

9.5 Modeling with Differential Equations

9.5.1 The growth rate function specifies the rate of growth of the population. If the growth rate function is positive, then the population is increasing, while the population is decreasing when the growth rate function is negative.

9.5.3 If the growth rate function is positive and decreasing, then the population is increasing. Whether or not the population is increasing is completely determined by whether the growth rate function is positive or negative.

9.5.5 Is is a linear, first-order differential equation.

9.5.7 The solution curves in the FH-plane are closed curves that circulate around the equilibrium point.

9.5.9

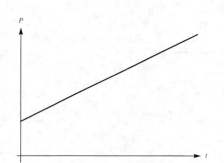

9.5.11

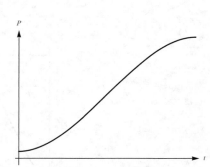

9.5.13

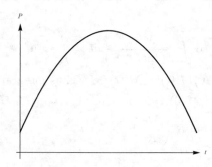

9.5.15 We must solve $P'(t) = 0.2P\left(1 - \frac{P}{300}\right)$. We have $\int \frac{300P'(t)}{P(300-P)}\,dt = \int 0.2\,dt$, which can be written as $\int \left(\frac{P'(t)}{P(t)} + \frac{P'(t)}{300 - P(t)}\right) dt = \int 0.2\,dt$. Thus, $\ln\left(\frac{P(t)}{300 - P(t)}\right) = 0.2t + C$. Taking the exponential of both sides and reciprocating gives $\dfrac{300 - P(t)}{P(t)} = Ae^{-0.2t}$. Because $P(0) = 50$, we have $A = 5$. Thus $\dfrac{300}{P(t)} = 5e^{-0.2t} + 1$, so $P(t) = \dfrac{300}{5e^{-0.2t} + 1}$ for $t \geq 0$.

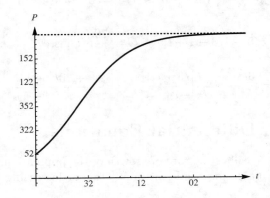

9.5.17 We must solve $P'(t) = rP\left(1 - \dfrac{P}{2000}\right)$. We have $\displaystyle\int \dfrac{2000 P'(t)}{P(2000 - P)}\, dt = \int r\, dt$, which can be written as $\displaystyle\int \left(\dfrac{P'(t)}{P(t)} + \dfrac{P'(t)}{2000 - P(t)}\right) dt = \int r\, dt$. Thus, $\ln\left(\dfrac{P(t)}{2000 - P(t)}\right) = rt + C$. Taking the exponential of both sides and reciprocating gives $\dfrac{2000 - P(t)}{P(t)} = Ae^{-rt}$. Because $P(0) = 200$, we have $A = 9$. Thus $\dfrac{2000}{P(t)} = 9e^{-rt} + 1$, so $P(t) = \dfrac{2000}{9e^{-rt} + 1}$. Now because $P(1) = 600$, we have $r = \ln(27/7)$. So $P(t) = \dfrac{2000}{9e^{\ln(7/27)t} + 1}$ for $t \geq 0$.

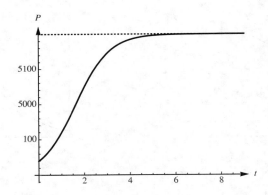

9.5.19 We have $\dfrac{M'(t)}{M \ln(M/K)} = -r$, so $\displaystyle\int \dfrac{M'(t)}{M \ln(M/K)}\, dt = \int -r\, dt$. Note that $\dfrac{d}{dt} \ln(M/K) = (K/M)(1/K) M'(t) = M'(t)/M$. Thus integrating gives $\ln|\ln(M/K)| = -rt + C$, and thus $\ln(M/K) = Ae^{-rt}$. Because $M(0) = M_0$, we have $A = \ln(M_0/K)$. Thus $M(t) = K(\exp(\ln(M_0/K)e^{-rt})) = K\left(\dfrac{M_0}{K}\right)^{e^{-rt}}$ for $t \geq 0$.

9.5.21 With $r = 0.05$, $K = 1200$, and $M_0 = 90$, we have

$$M(t) = 1200 \left(\dfrac{3}{40}\right)^{e^{-0.05t}} \quad \text{for } t \geq 0.$$

9.5. Modeling with Differential Equations

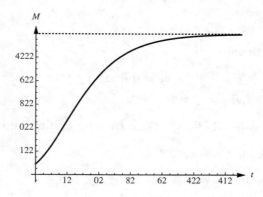

9.5.23

a. The initial mass of copper sulfate is $m_0 = 0$. We have $m'(t) = -\dfrac{4}{500}m(t) + 20 \cdot 4 = -\dfrac{1}{125}m(t) + 80$.

b. This is an equation of the form $m'(t) = km + b$, so the solution is $m(t) = Ce^{kt} - \dfrac{b}{k} = Ce^{-0.008t} + 10{,}000$. Because $m(0) = 0$, we have $C = -10{,}000$. Thus, $m(t) = -10{,}000e^{-0.008t} + 10{,}000$ for $t \geq 0$.

9.5.25

a. The initial mass of the sugar is $m_0 = 2000 \cdot 40 = 80{,}000$. We have $m'(t) = -\dfrac{10}{2000}m(t) + 10 \cdot 10 = -\dfrac{1}{200}m(t) + 100$.

b. This is an equation of the form $m'(t) = km + b$, so the solution is $m(t) = Ce^{kt} - \dfrac{b}{k} = Ce^{-0.005t} + 20{,}000$. Because $m(0) = 80000$, we have $C = 60{,}000$. Thus, $m(t) = 60{,}000e^{-0.005t} + 20{,}000$ for $t \geq 0$.

9.5.27

a. x is the predator and y is the prey.

b. $-3x + 6xy = 0$ when $x = 0$ or $y = 1/2$. $y - 4xy = 0$ when $y = 0$ or $x = 1/4$.
   The desired lines are $x = 1/4$ and $y = 1/2$.

c. The equilibrium points are where both equations are zero simultaneously, so they are $(0,0)$ and $(1/4, 1/2)$.

d. Note that $x'(t) = 3x(2y - 1)$ and $y'(t) = y(1 - 4x)$.

   - For $0 < x < 1/4$ and $0 < y < 1/2$, we have $x' < 0$ and $y' > 0$.
   - For $0 < x < 1/4$ and $y > 1/2$, we have $x' > 0$ and $y' > 0$.
   - For $x > 1/4$ and $0 < y < 1/2$, we have $x' < 0$ and $y' < 0$.
   - For $x > 1/4$ and $y > 1/2$, we have $x' > 0$ and $y' < 0$.

e. The direction of the solution is clockwise.

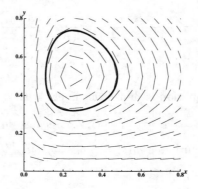

9.5.29

a. x is the predator and y is the prey.

b. $-3x + xy = 0$ when $x = 0$ or $y = 3$. $2y - xy = 0$ when $y = 0$ or $x = 2$. The desired lines are $x = 2$ and $y = 3$.

c. The equilibrium points are where both equations are zero simultaneously, so they are $(0,0)$ and $(2,3)$.

d. Note that $x'(t) = x(y-3)$ and $y'(t) = y(2-x)$.
- For $0 < x < 2$ and $0 < y < 3$, we have $x' < 0$ and $y' > 0$.
- For $0 < x < 2$ and $y > 3$, we have $x' > 0$ and $y' > 0$.
- For $x > 2$ and $0 < y < 3$, we have $x' < 0$ and $y' < 0$.
- For $x > 2$ and $y > 3$, we have $x' > 0$ and $y' < 0$.

e. The direction of the solution is clockwise.

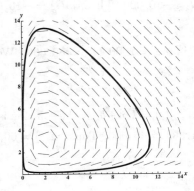

9.5.31

a. True. The growth rate function is the derivative, so where it is positive, the population is increasing.

b. True. In the limit, the solution in the tank is the same as the solution being poured in.

c. True. In the absence of predators, we assume that the prey population increases exponentially.

9.5.33 We must solve $P'(t) = rP\left(1 - \dfrac{P}{K}\right)$. We have $\displaystyle\int \dfrac{KP'(t)}{P(K-P)}\,dt = \int r\,dt$, which can be written as $\displaystyle\int \left(\dfrac{P'(t)}{P(t)} + \dfrac{P'(t)}{K-P(t)}\right) dt = \int r\,dt$. Thus, $\ln\left(\dfrac{P(t)}{K-P(t)}\right) = rt + C$. Taking the exponential of both sides and reciprocating gives $\dfrac{K-P(t)}{P(t)} = Ae^{-rt}$. Because $P(0) = P_0$, we have $A = \dfrac{K-P_0}{P_0}$. Thus $\dfrac{K}{P(t)} = \left(\dfrac{K-P_0}{P_0}\right)e^{-rt} + 1$, so $P(t) = \dfrac{K}{\frac{K-P_0}{P_0}e^{-rt} + 1} = \dfrac{KP_0}{(K-P_0)e^{-rt} + P_0}$.

9.5.35

a. Note that this differential equation is first-order linear with $k = -\dfrac{R}{V}$ and $b = C_i R$, so the solution is $m(t) = Ce^{(-R/V)t} - \dfrac{C_i R}{-R/V} = Ce^{(-R/V)t} + C_i V$. Because $m(0) = m_0$, we have $m_0 = C + C_i V$, so $C = m_0 - C_i V$. Thus, the solution is $m(t) = (m_0 - C_i V)e^{-Rt/V} + C_i V$.

b. $m(0) = (m_0 - C_i V) + C_i V = m_0$.

c. Note that $\lim_{t \to \infty} e^{-Rt/V} = 0$, so $\lim_{t \to \infty} m(t) = C_i V$. In the limit, the solution in the tank is the incoming solution, so the amount of material in the tank is the amount per unit volume times the volume of the tank.

d. Increasing R causes the graph to approach the asymptote more quickly.

9.5.37

a. We can write the equation as $I'(t) = -\dfrac{1}{RC} I(t)$. This is a first-order linear equation with $k = -\dfrac{1}{RC}$ and $b = 0$. Thus, the solution is $I(t) = Ce^{kt} - \dfrac{b}{k} = Ae^{-t/RC}$. Because $I(0) = \dfrac{V}{R}$, we have $A = \dfrac{V}{R}$. Thus, $I(t) = \dfrac{V}{R} e^{-t/RC}$. The current decays exponentially with decay constant $-\dfrac{1}{RC}$.

b. We can write the equation as $Q'(t) = -\dfrac{1}{RC} Q + \dfrac{V}{R}$. This is a first-order linear equation with $k = -\dfrac{1}{RC}$ and $b = \dfrac{V}{R}$. Thus, the solution is $Q(t) = Ce^{kt} - \dfrac{b}{k} = Ae^{-t/RC} - \dfrac{(V/R)}{(-1/RC)} = Ae^{-t/RC} + VC$. Because $Q(0) = 0$, we have $A = -VC$. Thus,
$$Q(t) = (-VC)\, e^{-t/RC} + VC.$$

In the long run, the charge has limit VC.

9.5.39

a. We have $\dfrac{dy}{dx} = \dfrac{y'(t)}{x'(t)} = \dfrac{cy - dxy}{-ax + bxy} = \dfrac{y(c - dx)}{x(by - a)}$. Thus we can write $\dfrac{by - a}{y} \dfrac{dy}{dx} = \dfrac{c - dx}{x}$, or
$$\left(b - \dfrac{a}{y}\right) \dfrac{dy}{dx} = \dfrac{c}{x} - d.$$

Thus, the equation is separable.

b. We have $\displaystyle\int \left(b - \dfrac{a}{y}\right) \dfrac{dy}{dx}\, dx = \int \left(\dfrac{c}{x} - d\right) dx$, so $by - a \ln y = c \ln x - dx + K$, so $dx + by = c \ln x + a \ln y + K$. Taking the exponential function of both sides gives $e^{dx+by} = Cx^c y^a$ for an arbitrary constant C.

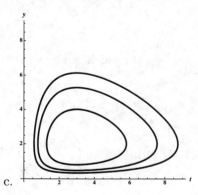

c.

Chapter Nine Review

1

a. False. It is a first-order, linear differential equation, but it isn't separable.

b. False. It is a first-order, separable differential equation, but it isn't linear.

c. **True.** Note that $y' = 1 - t^{-2}$, so $ty' = t - t^{-1}$. Thus, $ty' + y = t - t^{-1} + (t + t^{-1}) = 2t$. Also, $y(1) = 2$.

d. **True.**

e. **False.** In general, Euler's method gives approximate solutions.

3 $y'(t) = -2y + 6$, so $\displaystyle\int \frac{y'(t)}{-2y+6}\,dt = \int 1\,dt$, and therefore $-\dfrac{1}{2}\ln|-2y+6| = t + K$. It follows that $|-2y+6| = Ae^{-2t}$. We can write $-2y + 6 = \pm Ae^{-2t}$, so $y(t) = Ce^{-2t} + 3$.

5 $y'(t) = 2ty$, so $\displaystyle\int \frac{y'(t)}{y}\,dt = \int 2t\,dt$, and therefore, $\ln|y| = t^2 + K$. It follows that $y = Ce^{t^2}$.

7 $\dfrac{y'(t)}{y} = \dfrac{1}{t^2+1}$. Integrating both sides with respect to t gives $\ln|y| = \tan^{-1}(t) + K$, so $y = Ce^{\tan^{-1}(t)}$.

9 $\displaystyle\int \frac{y'(t)}{y^2+1}\,dt = \int (2t+1)\,dt$, so $\tan^{-1}(y) = t^2 + t + C$, and thus $y = \tan(t^2 + t + C)$.

11 $\displaystyle\int y'(t)\,dt = \int (2t + \cos t)\,dt$, so $y(t) = t^2 + \sin t + C$. Because $y(0) = 1$, we have $1 = 0 + 0 + C$, so $C = 1$. Thus, $y(t) = t^2 + \sin t + 1$.

13 $\displaystyle\int \frac{Q'(t)}{Q-8}\,dt = \int 1\,dt$, so $\ln|Q-8| = t + K$. Thus, $Q - 8 = Ce^t$. Because $Q(1) = 0$, we have $-8 = Ce$, so $C = -\dfrac{8}{e}$. Thus, $Q = -8e^{t-1} + 8$.

15 $u^{-1/3}u'(t) = t^{-1/3}$. Thus $\displaystyle\int u^{-1/3}u'(t)\,dt = \int t^{-1/3}\,dt$. We have $\dfrac{3}{2}u^{2/3} = \dfrac{3}{2}t^{2/3} + C$. Because $u(1) = 8$, we have $6 = \dfrac{3}{2} + C$, so $C = \dfrac{9}{2}$. Thus, $u^{2/3} = t^{2/3} + 3$, and $u = (t^{2/3} + 3)^{3/2}$ for $t > 0$.

17 $\dfrac{s'(t)}{s} = \dfrac{1}{t(t^2+1)}$, so $\displaystyle\int \frac{ds}{s} = \int \frac{1}{t(t^2+1)}\,dt$, which can be written as $\displaystyle\int \frac{ds}{s} = \int \left(\frac{1}{t} - \frac{t}{t^2+1}\right)dt$. Then we have

$$\ln|s| = \ln|t| - \frac{1}{2}\ln(1+t^2) + C,$$

Because $s(1) = 1$, we have $C = \ln\sqrt{2}$. Then $\ln|s| = \ln\left(\dfrac{\sqrt{2}|t|}{\sqrt{1+t^2}}\right)$, and

$$|s| = \frac{\sqrt{2}|t|}{\sqrt{1+t^2}}.$$

Because $s(1) = 1$, we have

$$s = \frac{\sqrt{2}\,t}{\sqrt{1+t^2}}.$$

19

a.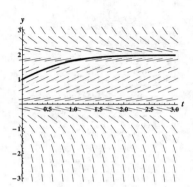

b.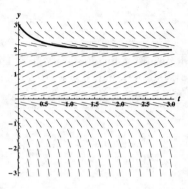

c. The solutions are increasing when $0 < A < 2$.

d. The solutions are decreasing when $A < 0$ or when $A > 2$.

e. The equilibrium solutions are $y = 0$ and $y = 2$.

21

a. $u_0 = 1$. $u_1 = 1 + f(0,1)(0.1) = 1 + 1/20 = 1.05$. Also, $u_2 = 1.05 + f(0.1, 1.05)(0.1) = 1.05 + 0.047619 = 1.09762$. Thus, $y(0.1) \approx 1.05$ and $y(0.2) \approx 1.09762$.

b. $u_0 = 1$. $u_1 = 1 + f(0,1)(0.05) = 1 + 0.025 = 1.025$. Also, $u_2 = 1.025 + f(0.05, 1.025)(0.05) = 1.04939$. $u_3 = 1.04939 + f(0.1, 1.04939)(.05) = 1.07321$. $u_4 = 1.07321 + f(0.15, 1.07321)(.05) = 1.0961$. Thus, $y(0.1) \approx 1.04939$ and $y(0.2) \approx 1.09651$.

c. For part a, we have $\dfrac{|1.09762 - \sqrt{1.2}|}{\sqrt{1.2}} = 0.00198539$.

For part b, we have $\dfrac{|109651 - \sqrt{1.2}|}{\sqrt{1.2}} = 0.000972103$. Part b is more accurate.

23

The equilibrium solutions are $y = -3$ (unstable) and $y = 0$ (stable), and $y = 5$ (unstable).

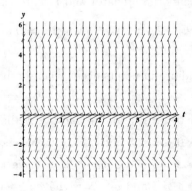

25

The equilibrium solutions are $y = -1$ (unstable) and $y = 0$ (stable), and $y = 2$ (unstable).

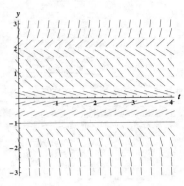

27

a. The logistic equation has solution $P(t) = \dfrac{KP_0}{(K-P_0)e^{-rt} + P_0} = \dfrac{32000}{1580e^{-rt} + 20}$. We are seeking r so that $80 = \dfrac{32000}{1580e^{-20r} + 20}$. Solving for r gives $r = \dfrac{\ln(79/19)}{20} \approx 0.0713$.

b. The solution is $P(t) = \dfrac{32000}{1580e^{-0.0713t} + 20} = \dfrac{1600}{79e^{-0.0713t} + 1}$ for $t \geq 0$.

c. We are seeking t so that $800 = \dfrac{1600}{79e^{-0.0713t} + 1}$. Solving for t gives $t \approx 61$ hours.

29

a. The initial mass of the sugar is $m_0 = 0$. We have $m'(t) = -\dfrac{0.5}{100}m(t) + 100 \cdot 0.5 = -\dfrac{1}{200}m(t) + 10$. This is an equation of the form $m'(t) = km + b$, so the solution is $m(t) = Ce^{kt} - \dfrac{b}{k} = Ce^{-0.005t} + 2000$. Because $m(0) = 0$, we have $C = -2000$. Thus, $m(t) = -2000e^{-0.005t} + 2000$.

b. The steady-state mass is $\lim\limits_{t \to \infty} m(t) = 2000$.

c. The mass reaches $0.95(2000)$ when $0.05 = e^{-0.005t}$, or $t = \ln(0.05)/(-0.005) \approx 599$ minutes.

31

a. x represents the predator and y represents the prey.

b. $x'(t) = 2x(-2 + y)$ is zero when $x = 0$ and when $y = 2$. $y'(t) = y(5 - x)$ is zero when $y = 0$ and when $x = 5$.

c. The equilibrium points occur when both equations are zero simultaneously, which occurs at $(0, 0)$ and $(5, 2)$.

d.
– For $0 < x < 5$ and $0 < y < 2$, we have $x' < 0$ and $y' > 0$.

– For $0 < x < 5$ and $y > 2$, we have $x' > 0$ and $y' > 0$.

– For $x > 5$ and $0 < y < 2$, we have $x' < 0$ and $y' < 0$.

– For $x > 5$ and $y > 2$, we have $x' > 0$ and $y' < 0$.

e. The solution evolves in the clockwise direction.

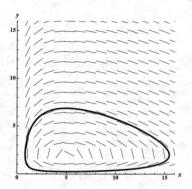

33

a. If $y(t) = t^p$ is a solution, then $p(p-1)t^p + 2pt^p - 12t^p = 0$, so $p^2 - p + 2p - 12 = 0$, so $p^2 + p - 12 = (p-3)(p+4) = 0$. Thus, $p_1 = 3$ and $p_2 = -4$.

b. If $y(t) = C_1 t^3 + C_2 t^{-4}$, and if $y(1) = 0$, then $C_1 + C_2 = 0$. Note that $y'(t) = 3C_1 t^2 - 4C_2 t^{-5}$, so if $y'(1) = 7$, we have $7 = 3c_1 - 4C_2$. Solving the system of linear equations in C_1 and C_2 gives $C_1 = 1$ and $C_2 = -1$. Thus the solution we seek is $y(t) = t^3 - t^{-4}$, $t > 0$.

Chapter 10

Sequences and Infinite Series

10.1 An Overview

10.1.1 A *sequence* is an ordered list of numbers $a_1, a_2, a_3, \ldots$, often written $\{a_1, a_2, \ldots\}$ or $\{a_n\}$. For example, the natural numbers $\{1, 2, 3, \ldots\}$ are a sequence where $a_n = n$ for every n.

10.1.3 $a_1 = 1$ (given); $a_2 = 1 \cdot a_1 = 1$; $a_3 = 2 \cdot a_2 = 2$; $a_4 = 3 \cdot a_3 = 6$; $a_5 = 4 \cdot a_4 = 24$.

10.1.5 $a_n = (-1)^{n+1} n$ for $n = 1, 2, 3, \ldots$. Or we could write $a_n = (-1)^n (n+1)$ for $n = 0, 1, 2, \ldots$. Other answer are possible.

10.1.7 The sequence has limit e.

10.1.9 $S_1 = \sum_{k=1}^{1} k^2 = 1$; $S_2 = \sum_{k=1}^{2} k^2 = 1 + 4 = 5$; $S_3 = \sum_{k=1}^{3} k^2 = 1 + 4 + 9 = 14$; $S_4 = \sum_{k=1}^{4} k^2 = 1 + 4 + 9 + 16 = 30$.

10.1.11 One possible answer is $\sum_{k=1}^{\infty} 10$.

10.1.13 $a_1 = \dfrac{1}{10}$; $a_2 = \dfrac{1}{100}$; $a_3 = \dfrac{1}{1000}$; $a_4 = \dfrac{1}{10000}$.

10.1.15 $a_1 = -\dfrac{1}{2}$, $a_2 = \dfrac{1}{2^2} = \dfrac{1}{4}$, $a_3 = -\dfrac{2}{2^3} = -\dfrac{1}{8}$, $a_4 = \dfrac{1}{2^4} = \dfrac{1}{16}$.

10.1.17 $a_1 = \dfrac{2^2}{2+1} = \dfrac{4}{3}$, $a_2 = \dfrac{2^3}{2^2+1} = \dfrac{8}{5}$, $a_3 = \dfrac{2^4}{2^3+1} = \dfrac{16}{9}$, $a_4 = \dfrac{2^5}{2^4+1} = \dfrac{32}{17}$.

10.1.19 $a_1 = 1 + \sin(\pi/2) = 2$; $a_2 = 1 + \sin(2\pi/2) = 1 + \sin(\pi) = 1$; $a_3 = 1 + \sin(3\pi/2) = 0$; $a_4 = 1 + \sin(4\pi/2) = 1 + \sin(2\pi) = 1$.

10.1.21 $a_1 = 2$, $a_2 = 2(2) = 4$, $a_3 = 2(4) = 8$, $a_4 = 2(8) = 16$.

10.1.23 $a_1 = 10$ (given); $a_2 = 3 \cdot a_1 - 12 = 30 - 12 = 18$; $a_3 = 3 \cdot a_2 - 12 = 54 - 12 = 42$; $a_4 = 3 \cdot a_3 - 12 = 126 - 12 = 114$.

10.1.25 $a_0 = 1$(given); $a_2 = \dfrac{1}{1+1} = \dfrac{1}{2}$; $a_3 = \dfrac{1}{1+\frac{1}{2}} = \dfrac{2}{3}$; $a_4 = \dfrac{1}{1+\frac{2}{3}} = \dfrac{3}{5}$.

10.1.27

  a. $\dfrac{1}{32}, \dfrac{1}{64}$.

463

b. $a_1 = 1; a_{n+1} = \dfrac{a_n}{2}$.

c. $a_n = \dfrac{1}{2^{n-1}}$.

10.1.29

a. 32, 64.

b. $a_1 = 1; a_{n+1} = 2a_n$.

c. $a_n = 2^{n-1}$.

10.1.31

a. 243, 729.

b. $a_1 = 1; a_{n+1} = 3a_n$.

c. $a_n = 3^{n-1}$.

10.1.33

a. $-5, 5$.

b. $a_1 = -5, a_{n+1} = -a_n$.

c. $a_n = (-1)^n 5$.

10.1.35 $a_1 = 9, a_2 = 99, a_3 = 999, a_4 = 9999$. This sequence diverges, because the terms get larger without bound.

10.1.37 $a_1 = \dfrac{1}{10}, a_2 = \dfrac{1}{100}, a_3 = \dfrac{1}{1000}, a_4 = \dfrac{1}{10,000}$. This sequence converges to zero.

10.1.39 $a_1 = 3 + \cos \pi = 2.$ $a_2 = 3 + \cos 2\pi = 4.$ $a_3 = 3 + \cos 3\pi = 2.$ $a_4 = 3 + \cos 4\pi = 4.$ This sequence diverges.

10.1.41 $a_1 = 1 + 1 = 2, a_2 = 1 + 1 = 2, a_3 = 2, a_4 = 2$. This constant sequence converges to 2.

10.1.43 $a_0 = 100, a_1 = 0.5 \cdot 100 + 50 = 100, a_2 = 0.5 \cdot 100 + 50 = 100, a_3 = 0.5 \cdot 100 + 50 = 100, a_4 = 0.5 \cdot 100 + 50 = 100$. This constant sequence converges to 100.

10.1.45

n	a_n
1	0.83333
2	0.96154
3	0.99206
4	0.99840
5	0.99968
6	0.99994
7	0.99999
8	0.99999
9	0.99999
10	0.99999

This sequence appears to converge to 1.

10.1. An Overview

10.1.47

n	1	2	3	4	5	6	7	8	9	10
a_n	2	6	12	20	30	42	56	72	90	110

This sequence appears to diverge.

10.1.49

a. 2.5, 2.25, 2.125, 2.0625.

b. The limit appears to be 2.

10.1.51

n	1	2	3	4	5	6	7	8	9	10
a_n	3	3.5000	3.7500	3.8750	3.9375	3.9688	3.9844	3.9922	3.9961	3.9980

The limit appears to be 4.

10.1.53

n	0	1	2	3	4	5	6	7	8	9
a_n	1	5	21	85	341	1365	5461	21,845	87,381	349,525

This sequence appears to diverge.

10.1.55

n	a_n
1	8
2	4.41421356
3	4.05050149
4	4.00629289
5	4.00078630
6	4.00009828
7	4.00001229
8	4.00000154
9	4.00000019
10	4.00000002

The limit appears to be 4.

10.1.57

a. 20, 10, 5, 2.5.

b. $h_n = 20(0.5)^n$.

10.1.59

a. 30, 7.5, 1.875, 0.46875.

b. $h_n = 30(0.25)^n$.

10.1.61 $S_1 = 0.3$, $S_2 = 0.33$, $S_3 = 0.333$, $S_4 = 0.3333$. It appears that the infinite series has a value of $0.3333\ldots = \dfrac{1}{3}$.

10.1.63 $S_1 = 4$, $S_2 = 4.9$, $S_3 = 4.99$, $S_4 = 4.999$. The infinite series has a value of $4.999\cdots = 5$.

10.1.65 $S_1 = \dfrac{6}{10} = 0.6$, $S_2 = \dfrac{66}{100} = 0.66$, $S_3 = \dfrac{666}{1000} = 0.666$, $S_4 = \dfrac{6666}{10000} = 0.6666$. The limit is $\dfrac{2}{3}$.

10.1.67

a. $S_1 = \dfrac{2}{3}$, $S_2 = \dfrac{4}{5}$, $S_3 = \dfrac{6}{7}$, $S_4 = \dfrac{8}{9}$.

b. It appears that $S_n = \dfrac{2n}{2n+1}$. $S_5 = \dfrac{10}{11}$, $S_6 = \dfrac{12}{13}$, $S_7 = \dfrac{14}{15}$, $S_8 = \dfrac{16}{17}$.

c. The series has a value of 1 (the partial sums converge to 1).

10.1.69

a. $S_1 = 9$, $S_2 = 9.9$, $S_3 = 9.99$, $S_4 = 9.999$.

b. $S_n = 10 - (0.1)^{n-1}$. $S_5 = 9.9999$, $S_6 = 9.99999$, $S_7 = 9.999999$, $S_8 = 9.9999999$.

c. The partial sums converge to 10, which is the value of the series.

10.1.71

a. True. For example, $S_2 = 1 + 2 = 3$, and $S_4 = a_1 + a_2 + a_3 + a_4 = 1 + 2 + 3 + 4 = 10$.

b. False. For example, $\dfrac{1}{2}, \dfrac{3}{4}, \dfrac{7}{8}, \ldots$ where $a_n = 1 - \dfrac{1}{2^n}$ converges to 1, but each term is greater than the previous one.

c. True. In order for the partial sums to converge, they must get closer and closer together. In order for this to happen, the difference between successive partial sums, which is just the value of a_n, must approach zero.

10.1.73

a. $M_0 = 20$, $M_1 = 20 \cdot 0.5 = 10$, $M_2 = 20 \cdot 0.5^2 = 5$, $M_3 = 20 \cdot 0.5^3 = 2.5$, $M_4 = 20 \cdot 0.5^4 = 1.25$

b. $M_n = 20 \cdot 0.5^n$.

c. The initial mass is $M_0 = 20$. We are given that 50% of the mass is gone after each decade, so that $M_{n+1} = 0.5 \cdot M_n$, $n \geq 0$.

d. The amount of material goes to 0.

10.1.75

a. $d_0 = 200$, $d_1 = 200 \cdot .95 = 190$, $d_2 = 200 \cdot .95^2 = 180.5$, $d_3 = 200 \cdot .95^3 = 171.475$, $d_4 = 200 \cdot .95^4 = 162.90125$.

b. $d_n = 200(0.95)^n$, $n \geq 0$.

c. We are given $d_0 = 200$; because 5% of the drug is washed out every hour, that means that 95% of the preceding amount is left every hour, so that $d_{n+1} = 0.95 \cdot d_n$.

d. The sequence converges to 0.

10.1.77 Using the work from the previous problem:

a. Here $h_0 = 20$, $r = 0.75$, so $S_0 = 40$, $S_1 = 40 + 40 \cdot 0.75 = 70$, $S_2 = S_1 + 40 \cdot (0.75)^2 = 92.5$, $S_3 = S_2 + 40 \cdot (0.75)^3 = 109.375$, $S_4 = S_3 + 40 \cdot (0.75)^4 = 122.03125$

10.2. Sequences

b.

n	0	1	2	3	4	5
a_n	40	70	92.5	109.375	122.0313	131.5234
n	6	7	8	9	10	11
a_n	138.6426	143.9819	147.9865	150.9898	153.2424	154.9318
n	12	13	14	15	16	17
a_n	156.1988	157.1491	157.8618	158.3964	158.7973	159.0980
n	18	19	20	21	22	23
a_n	159.3235	159.4926	159.6195	159.715	159.786	159.839

The sequence converges to 160.

10.1.79

n	a_n
0	0.800
1	0.697
2	0.767
3	0.720
4	0.752
5	0.730
6	0.745
7	0.735
8	0.742
9	0.737
10	0.740
11	0.738
12	0.740
13	0.739
14	0.739

The limit appears to be 0.739.

10.2 Sequences

10.2.1 There are many examples; one is $a_n = \frac{1}{n}$. This sequence is nonincreasing (in fact, it is decreasing) and has a limit of 0.

10.2.3 There are many examples; one is $a_n = \frac{1}{n}$. This sequence is nonincreasing (in fact, it is decreasing), is bounded above by 1 and below by 0, and has a limit of 0.

10.2.5 $\{r^n\}$ converges for $-1 < r \leq 1$. It diverges for all other values of r (see Theorem 10.3).

10.2.7 Because $1.00001 > 1$, the sequence diverges; because $1.00001 > 0$, the divergence is monotone.

10.2.9 Because $|-0.7| < 1$, the sequence converges to 0; because $-0.7 < 0$, it does not do so monotonically. The sequence converges by oscillation.

10.2.11 $\{e^{n/100}\}$ grows faster than $\{n^{100}\}$ as $n \to \infty$.

10.2.13 Divide numerator and denominator by n^4 to get $\lim\limits_{n\to\infty} \dfrac{1/n}{1+\frac{1}{n^4}} = 0$.

10.2.15 Divide numerator and denominator by n^3 to get $\lim\limits_{n\to\infty} \dfrac{3-n^{-3}}{2+n^{-3}} = \dfrac{3}{2}$.

10.2.17 Because $\lim\limits_{n\to\infty} \dfrac{10n}{10n+4} = 1$, and because the inverse tangent function is continuous, the given sequence has limit $\tan^{-1} 1 = \pi/4$.

10.2.19 $\lim\limits_{n\to\infty} (1+\cos(1/n)) = 1+\cos(0) = 2$.

10.2.21 $\ln(\sin(1/n)) + \ln n = \ln(n\sin(1/n)) = \ln\left(\dfrac{\sin(1/n)}{1/n}\right)$. As $n\to\infty$, $\sin(1/n)/(1/n) \to 1$, so the limit of the original sequence is $\ln 1 = 0$.

10.2.23 Divide numerator by $n^2 = \sqrt{n^4}$ and the denominator by n^2 to obtain
$$\lim_{n\to\infty} \frac{\sqrt{4+3/n^3}}{8+1/n^2} = \frac{\sqrt{4+0}}{8+0} = \frac{2}{8} = \frac{1}{4}.$$

10.2.25 The sequence is equivalent to $\left\{\dfrac{2\ln n}{\ln 3 + \ln n}\right\}$. Divide numerator and denominator by $\ln n$ to obtain
$$\lim_{n\to\infty} \frac{2}{(\ln 3/\ln n)+1} = \frac{2}{1} = 2.$$

10.2.27 This is the sequence
$$\frac{2^{n+1}}{3^n} = 2\cdot\left(\frac{2}{3}\right)^n;$$
because $0 < \dfrac{2}{3} < 1$, the sequence converges monotonically to zero.

10.2.29 Because $|0.5| < 1$ and $|0.75| < 1$, the sequence converges to $0 + 3\cdot 0 = 0$.

10.2.31 Divide numerator and denominator by 3^n to get $\lim\limits_{n\to\infty} \dfrac{3+(1/3^{n-1})}{1} = 3$.

10.2.33 Because $\dfrac{(n+1)!}{n!} = n+1$, this sequence diverges.

10.2.35 $\lim\limits_{n\to\infty} \tan^{-1} n = \dfrac{\pi}{2}$.

10.2.37 Multiply by $\dfrac{\sqrt{n^2+1}+n}{\sqrt{n^2+1}+n}$ to obtain
$$\lim_{n\to\infty}\left(\sqrt{n^2+1}-n\right) = \lim_{n\to\infty} \frac{\left(\sqrt{n^2+1}-n\right)\left(\sqrt{n^2+1}+n\right)}{\sqrt{n^2+1}+n} = \lim_{n\to\infty} \frac{1}{\sqrt{n^2+1}+n} = 0.$$

10.2.39 Find the limit of the logarithm of the expression, which is $n\ln\left(1+\dfrac{2}{n}\right)$. Using L'Hôpital's rule:
$$\lim_{n\to\infty} n\ln\left(1+\frac{2}{n}\right) = \lim_{n\to\infty} \frac{\ln\left(1+\frac{2}{n}\right)}{1/n} = \lim_{n\to\infty} \frac{\frac{1}{1+(2/n)}\left(-\frac{2}{n^2}\right)}{-1/n^2} = \lim_{n\to\infty} \frac{2}{1+(2/n)} = 2.$$

Thus the limit of the original expression is e^2.

10.2.41 Let $y = \sqrt[n]{e^{3n+4}} = e^{\frac{3n+4}{n}}$. Then $\ln y = \frac{3n+4}{n}$. Then $\lim\limits_{n\to\infty} \ln y = \lim\limits_{n\to\infty} \frac{3n+4}{n} = 3$. So $\lim\limits_{n\to\infty} \sqrt[n]{e^{3n+4}} = e^3$.

10.2.43 Take the logarithm of the expression and use L'Hôpital's rule:

$$\lim_{n\to\infty} \frac{n}{2} \ln\left(1 + \frac{1}{2n}\right) = \lim_{n\to\infty} \frac{\ln(1+(1/2n))}{2/n} = \lim_{n\to\infty} \frac{\frac{1}{1+(1/2n)} \cdot \frac{-1}{2n^2}}{-2/n^2} = \lim_{n\to\infty} \frac{1}{4(1+(1/2n))} = \frac{1}{4}.$$

Thus the original limit is $e^{1/4}$.

10.2.45 Using L'Hôpital's rule: $\lim\limits_{n\to\infty} \dfrac{n}{e^n + 3n} = \lim\limits_{n\to\infty} \dfrac{1}{e^n + 3} = 0$.

10.2.47 Taking logs, we have $\lim\limits_{n\to\infty} \dfrac{1}{n} \ln(1/n) = \lim\limits_{n\to\infty} -\dfrac{\ln n}{n} = \lim\limits_{n\to\infty} -\dfrac{1}{n} = 0$ by L'Hôpital's rule. Thus the original sequence has limit $e^0 = 1$.

10.2.49 Except for a finite number of terms, this sequence is just $a_n = ne^{-n}$, so it has the same limit as this sequence. Note that $\lim\limits_{n\to\infty} \dfrac{n}{e^n} = \lim\limits_{n\to\infty} \dfrac{1}{e^n} = 0$, by L'Hôpital's rule.

10.2.51 $\lim\limits_{n\to\infty} n\sin(6/n) = \lim\limits_{n\to\infty} \dfrac{\sin(6/n)}{1/n} = \lim\limits_{n\to\infty} \dfrac{\frac{-6\cos(6/n)}{n^2}}{(-1/n^2)} = \lim\limits_{n\to\infty} 6\cos(6/n) = 6 \cdot \cos 0 = 6$.

10.2.53 When n is an integer, $\sin\left(\dfrac{n\pi}{2}\right)$ oscillates between the values ± 1 and 0, so this sequence does not converge.

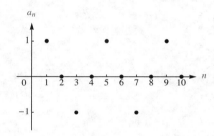

10.2.55 Because
$$-\frac{1}{2^n} \leq \frac{(-1)^n}{2^n} \leq \frac{1}{2^n},$$
and because $\lim\limits_{n\to\infty} -\dfrac{1}{2^n} = \lim\limits_{n\to\infty} \dfrac{1}{2^n} = 0$, the given sequence has limit zero by the Squeeze Theorem.

10.2.57 Ignoring the factor of $(-1)^n$ for the moment, we see, taking logs, that $\lim\limits_{n\to\infty} \dfrac{\ln n}{n} = 0$, so that $\lim\limits_{n\to\infty} \sqrt[n]{n} = e^0 = 1$. Taking the sign into account, the odd terms converge to -1 while the even terms converge to 1. Thus the sequence does not converge.

10.2.59 This sequence diverges. To see this, call the given sequence a_n, and assume it converges to limit L. Then because the sequence $b_n = \dfrac{n}{n+1}$ converges to 1, the sequence $c_n = \dfrac{a_n}{b_n}$ would converge to L as well. But $c_n = \sin^3 \dfrac{\pi n}{2}$ doesn't converge (because it is $1, -1, 1, -1 \cdots$), so the given sequence doesn't converge either.

10.2.61 This is the sequence $\dfrac{\cos n}{e^n}$; the numerator is bounded in absolute value by 1 and the denominator increases without bound, so the limit is zero.

10.2.63 Because $\lim_{n\to\infty} \tan^{-1} n = \frac{\pi}{2}$, $\lim_{n\to\infty} \frac{\tan^{-1} n}{n} = 0$.

10.2.65 Because $-1 \leq \cos n \leq 1$, we have $-\frac{1}{n} \leq \frac{\cos n}{n} \leq \frac{1}{n}$. Because both $-\frac{1}{n}$ and $\frac{1}{n}$ have limit 0 as $n \to \infty$, the given sequence does as well.

10.2.67 Because $-1 \leq \sin n \leq 1$ for all n, the given sequence satisfies $-\frac{1}{2^n} \leq \frac{\sin n}{2^n} \leq \frac{1}{2^n}$, and because both $\pm\frac{1}{2^n} \to 0$ as $n \to \infty$, the given sequence converges to zero as well by the Squeeze Theorem.

10.2.69 Evaluate the limit of each term separately: $\lim_{n\to\infty} \frac{75^{n-1}}{99^n} = \frac{1}{99} \lim_{n\to\infty} \left(\frac{75}{99}\right)^{n-1} = 0$, while $-\frac{5^n}{8^n} \leq \frac{5^n \sin n}{8^n} \leq \frac{5^n}{8^n}$, so by the Squeeze Theorem, this second term converges to 0 as well. Thus the sum of the terms converges to zero.

10.2.71

a. After the nth dose is given, the amount of drug in the bloodstream is $d_n = 0.5 \cdot d_{n-1} + 80$, because the half-life is one day. The initial condition is $d_1 = 80$.

b. The limit of this sequence is 160 mg.

c. Let $L = \lim_{n\to\infty} d_n$. Then from the recurrence relation, we have $d_n = 0.5 \cdot d_{n-1} + 80$, and thus $\lim_{n\to\infty} d_n = 0.5 \cdot \lim_{n\to\infty} d_{n-1} + 80$, so $L = 0.5 \cdot L + 80$, and therefore $L = 160$.

10.2.73

a.
$$B_0 = 0$$
$$B_1 = 1.0075 \cdot B_0 + \$100 = \$100$$
$$B_2 = 1.0075 \cdot B_1 + \$100 = \$200.75$$
$$B_3 = 1.0075 \cdot B_2 + \$100 = \$302.26$$
$$B_4 = 1.0075 \cdot B_3 + \$100 = \$404.52$$
$$B_5 = 1.0075 \cdot B_4 + \$100 = \$507.56$$

b. $B_n = 1.0075 \cdot B_{n-1} + \$100$.

c. Using a calculator or computer program, $B_n > \$5,000$ during the 43rd month.

10.2.75 Because $n! \ll n^n$ by Theorem 10.6, we have $\lim_{n\to\infty} \frac{n!}{n^n} = 0$.

10.2.77 Theorem 10.6 indicates that $\ln^q n \ll n^p$, so $\ln^{20} n \ll n^{10}$, so $\lim_{n\to\infty} \frac{n^{10}}{\ln^{20} n} = \infty$.

10.2.79 By Theorem 10.6, $n^p \ll b^n$, so $n^{1000} \ll 2^n$, and thus $\lim_{n\to\infty} \frac{n^{1000}}{2^n} = 0$.

10.2.81 Dividing the numerator and denominator by 6^n gives $a_n = \frac{1 + (1/2)^n}{1 + (n^{100}/6^n)}$. By Theorem 10.6, $n^{100} \ll 6^n$. Thus $\lim_{n\to\infty} a_n = \frac{1+0}{1+0} = 1$.

10.2.83

a. True. See Theorem 10.2 part 4.

b. False. For example, if $a_n = 1/n$ and $b_n = e^n$, then $\lim\limits_{n\to\infty} a_n b_n = \infty$.

c. True. The definition of the limit of a sequence involves only the behavior of the nth term of a sequence as n gets large (see the Definition of Limit of a Sequence). Thus suppose a_n, b_n differ in only finitely many terms, and that M is large enough so that $a_n = b_n$ for $n > M$. Suppose a_n has limit L. Then for $\varepsilon > 0$, if N is such that $|a_n - L| < \varepsilon$ for $n > N$, first increase N if required so that $N > M$ as well. Then we also have $|b_n - L| < \varepsilon$ for $n > N$. Thus a_n and b_n have the same limit. A similar argument applies if a_n has no limit.

d. True. Note that a_n converges to zero. Intuitively, the nonzero terms of b_n are those of a_n, which converge to zero. More formally, given ϵ, choose N_1 such that for $n > N_1$, $a_n < \epsilon$. Let $N = 2N_1 + 1$. Then for $n > N$, consider b_n. If n is even, then $b_n = 0$ so certainly $b_n < \epsilon$. If n is odd, then $b_n = a_{(n-1)/2}$, and $(n-1)/2 > ((2N_1 + 1) - 1)/2 = N_1$ so that $a_{(n-1)/2} < \epsilon$. Thus b_n converges to zero as well.

e. False. If $\{a_n\}$ happens to converge to zero, the statement is true. But consider for example $a_n = 2 + \dfrac{1}{n}$. Then $\lim\limits_{n\to\infty} a_n = 2$, but $(-1)^n a_n$ does not converge (it oscillates between positive and negative values increasingly close to ± 2).

f. True. Suppose $\{0.000001 a_n\}$ converged to L, and let $\epsilon > 0$ be given. Choose N such that for $n > N$, $|0.000001 a_n - L| < \epsilon \cdot 0.000001$. Dividing through by 0.000001, we get that for $n > N$, $|a_n - 1000000 L| < \epsilon$, so that a_n converges as well (to $1000000 L$).

10.2.85

a. The first three terms of the sequence are 0.3, 0.42, and 0.4872. Because we are given that the sequence is monotonic, the sequence is nondecreasing.

b. Let its limit be L, then $\lim\limits_{n\to\infty} a_{n+1} = \lim\limits_{n\to\infty}(2a_n(1-a_n)) = 2(\lim\limits_{n\to\infty} a_n)(1 - \lim\limits_{n\to\infty} a_n)$, so $L = 2L(1-L) = 2L - 2L^2$, and thus $2L^2 - L = 0$, so $L = 0, \dfrac{1}{2}$. Thus the limit appears to be either 0 or 1/2; with the given initial condition, doing a few iterations by hand (as above) confirms that the sequence converges to 1/2.

10.2.87

a. The first 3 terms of the sequence are 3, 2.236068, and 2.058171. Because we are given that the sequence is monotonic, it is nonincreasing.

b. Let its limit be L. Then $\lim\limits_{n\to\infty} a_{n+1} = \sqrt{2 + \lim\limits_{n\to\infty} a_n}$, so $L = \sqrt{2 + L}$. Thus we have $L^2 = 2 + L$, so $L^2 - L - 2 = 0$, and thus $L = -1, 2$. A square root can never be negative, so this sequence must converge to 2.

10.2.89

a. $d_{n+1} = 0.4 d_n + 75$; $d_1 = 75$.

b. By examining the first few terms of the sequence, it appears to be increasing and bounded above by 125. To confirm this, observe that $d_2 = 105$ and so $d_2 > d_1$. Assume that $d_{k+1} > d_k$, for some positive integer k. Then $d_{k+2} = 0.4 d_{k+1} + 75 > 0.4 d_k + 75 = d_{k+1}$. Therefore $d_{k+2} > d_{k+1}$. So by mathematical induction, it follows that $\{d_n\}$ is an increasing sequence. Because $\{d_n\}$ is an increasing sequence and $d_1 = 75$, the sequence is bounded below by 75. To show that the sequence is indeed bounded above by 125, first observe that $d_1 = 75 < 125$. Now assume that $d_k < 125$, for some positive integer k for $k \geq 1$. Then $d_{k+1} = 0.4 d_k + 75 \leq 0.4(125) + 75 = 125$, and therefore $d_{k+1} < 125$. So by mathematical induction, the sequence is bounded above by 125. Because $\{d_n\}$ is bounded below and above, it is bounded. So by Theorem 10.5, it converges.

c. Assume that the sequence $\{d_n\}$ converges to L. Taking limits of the recurrence relation, we have $\lim_{n\to\infty} d_n = 0.4 \lim_{n\to\infty} d_n + 75 = 0.4L + 75$. So the limit is $L = 75/0.6 = 125$. In the long run there will be approximately 125 mg of medication in the blood

10.2.91 The approximate first few values of this sequence are:

n	0	1	2	3	4	5	6
c_n	0.7071	0.6325	0.6136	0.6088	0.6076	0.6074	0.6073

The value of the constant appears to be around 0.607.

10.2.93

a. We begin by showing that $a_{n+1} > a_n$, for $n \geq 1$. This inequality holds for $n = 0$ because $a_1 = 7 > a_0$. Assume that $a_{n+1} > a_n$ for a nonnegative integer $n = k$. Then $a_{k+2} = (1/3)a_{k+1} + 6 > (1/3)a_k + 6 = a_{k+1}$. Therefore $a_{n+1} > a_n$ holds for $n = k+1$. So by mathematical induction, we have established that $a_{n+1} > a_n$, for $n \geq 1$, which implies that $\{a_n\}$ is a monotonic sequence. Next, we establish that $a_n < 9$, for $n \geq 1$. This result holds for $n = 0$ because $a_0 = 3 < 9$. Now assume that $a_k < 9$ for a nonnegative integer k. Then $a_{k+1} = (1/3)a_k + 6 < (1/3)(9) + 6 = 9$. So by mathematical induction, we've shown that $a_n < 9$, for $n \geq 1$. Because $\{a_n\}$ is an increasing sequence and $a_0 = 3$, it follows that $3 \leq a_n < 9$. Therefore $\{a_n\}$ is bounded. Because $\{a_n\}$ is monotonic and bounded, it converges.

b. Because $\{a_n\}_{n=0}^{\infty}$ is bounded above and is increasing, it converges. Assume that L is the limit of the sequence. Taking the limit of both sides of the recurrence relation gives $L = \frac{1}{3}L + 6$, or $3L = L + 18$, so $L = 9$.

10.2.95

a. $a_0 = \sqrt{2} \approx 1.41421$, $a_1 = \sqrt{2 + \sqrt{2}} \approx 1.84776$, $a_2 = \sqrt{2 + \sqrt{2 + \sqrt{2}}} \approx 1.96157$,
$a_3 = \sqrt{2 + \sqrt{2 + \sqrt{2 + \sqrt{2}}}} \approx 1.99037$, and $a_4 = \sqrt{2 + \sqrt{2 + \sqrt{2 + \sqrt{2 + \sqrt{2}}}}} \approx 1.99759$.

b. We have $a_1 = \sqrt{2 + \sqrt{2}}$, so it is the case that $a_1 > a_0$. Now assume that $a_k > a_{k-1}$ for some $k \geq 1$. Then $a_{k+1} = \sqrt{2 + a_k} > \sqrt{2 + a_{k-1}} = a_k$. So we have $a_{k+1} > a_k$. So by mathematical induction, it follows that a_n is increasing. Because the first term in the sequence is $\sqrt{2}$ and the sequence is increasing, the sequence is bounded below by $\sqrt{2}$. It appears that all of the terms in the sequence are less than 2 and we now use mathematical induction to prove this. We have $a_0 = \sqrt{2} < 2$. Suppose $a_k < 2$ for some $k \geq 1$. Then $a_{k+1} = \sqrt{2 + a_k} < \sqrt{2 + 2} = 2$. So it follows that $\sqrt{2} < a_n < 2$ for $n \geq 0$.

c. Because $\{a_n\}_{n=0}^{\infty}$ is bounded and increasing, it converges. Assume that L is the limit of the sequence. Taking the limit of both sides of the recurrence relation gives $L = \sqrt{2 + L}$, so $L^2 = 2 + L$, or $L^2 - L - 2 = (L-2)(L+1) = 0$. So $L = 2$ or $L = -1$ and because terms of the sequence are all positive, the limit is 2.

10.2.97

a. $f_0 = f_1 = 1, f_2 = 2, f_3 = 3, f_4 = 5, f_5 = 8, f_6 = 13, f_7 = 21, f_8 = 34, f_9 = 55, f_{10} = 89$.

b. The sequence is clearly not bounded.

c. $\frac{f_{10}}{f_9} \approx 1.61818$

10.2. Sequences

d. We use induction. Note that

$$\frac{1}{\sqrt{5}}\left(\varphi + \frac{1}{\varphi}\right) = \frac{1}{\sqrt{5}}\left(\frac{1+\sqrt{5}}{2} + \frac{2}{1+\sqrt{5}}\right) = \frac{1}{\sqrt{5}}\left(\frac{1+2\sqrt{5}+5+4}{2(1+\sqrt{5})}\right) = 1 = f_1.$$

Also note that

$$\frac{1}{\sqrt{5}}\left(\varphi^2 - \frac{1}{\varphi^2}\right) = \frac{1}{\sqrt{5}}\left(\frac{3+\sqrt{5}}{2} - \frac{2}{3+\sqrt{5}}\right) = \frac{1}{\sqrt{5}}\left(\frac{9+6\sqrt{5}+5-4}{2(3+\sqrt{5})}\right) = 1 = f_2.$$

Now note that

$$f_{n-1} + f_{n-2} = \frac{1}{\sqrt{5}}(\varphi^{n-1} - (-1)^{n-1}\varphi^{1-n} + \varphi^{n-2} - (-1)^{n-2}\varphi^{2-n})$$

$$= \frac{1}{\sqrt{5}}((\varphi^{n-1} + \varphi^{n-2}) - (-1)^n(\varphi^{2-n} - \varphi^{1-n})).$$

Now, note that $\varphi - 1 = \frac{1}{\varphi}$, so that

$$\varphi^{n-1} + \varphi^{n-2} = \varphi^{n-1}\left(1 + \frac{1}{\varphi}\right) = \varphi^{n-1} \cdot \varphi = \varphi^n$$

and

$$\varphi^{2-n} - \varphi^{1-n} = \varphi^{-n}(\varphi^2 - \varphi) = \varphi^{-n}(\varphi(\varphi - 1)) = \varphi^{-n}.$$

Making these substitutions, we get

$$f_n = f_{n-1} + f_{n-2} = \frac{1}{\sqrt{5}}(\varphi^n - (-1)^n\varphi^{-n})$$

10.2.99

a. Define a_n as given in the problem statement. Then we can *define* the value of the continued fraction to be $\lim_{n\to\infty} a_n$.

b. $a_0 = 1$, $a_1 = 1 + \frac{1}{a_0} = 2$, $a_2 = 1 + \frac{1}{a_1} = \frac{3}{2} = 1.5$, $a_3 = 1 + \frac{1}{a_2} = \frac{5}{3} \approx 1.667$, $a_4 = 1 + \frac{1}{a_3} = \frac{8}{5} = 1.6$, $a_5 = 1 + \frac{1}{a_4} = \frac{13}{8} = 1.625$.

c. From the list above, the values of the sequence alternately decrease and increase, so we would expect that the limit is somewhere between 1.6 and 1.625.

d. Assume that the limit is equal to L. Then from $a_{n+1} = 1 + \frac{1}{a_n}$, we have $\lim_{n\to\infty} a_{n+1} = 1 + \frac{1}{\lim_{n\to\infty} a_n}$, so $L = 1 + \frac{1}{L}$, and thus $L^2 - L - 1 = 0$. Therefore, $L = \frac{1 \pm \sqrt{5}}{2}$, and because L is clearly positive, it must be equal to $\frac{1+\sqrt{5}}{2} \approx 1.618$.

e. Here $a_0 = a$ and $a_{n+1} = a + \frac{b}{a_n}$. Assuming that $\lim_{n\to\infty} a_n = L$ we have $L = a + \frac{b}{L}$, so $L^2 = aL + b$, and thus $L^2 - aL - b = 0$. Therefore, $L = \frac{a \pm \sqrt{a^2 + 4b}}{2}$, and because $L > 0$ we have $L = \frac{a + \sqrt{a^2 + 4b}}{2}$.

10.2.101 Let $\varepsilon > 0$ be given and let N be an integer with $N > \frac{1}{\varepsilon}$. Then if $n > N$, we have

$$\left|\frac{1}{n} - 0\right| = \frac{1}{n} < \frac{1}{N} < \varepsilon.$$

Copyright © 2019 Pearson Education, Inc.

10.2.103 Let $\varepsilon > 0$ be given. We wish to find N such that for $n > N$, $\left|\frac{3n^2}{4n^2+1} - \frac{3}{4}\right| = \left|\frac{-3}{4(4n^2+1)}\right| = \frac{3}{4(4n^2+1)} < \varepsilon$. But this means that $3 < 4\varepsilon(4n^2+1)$, or $16\varepsilon n^2 + (4\varepsilon - 3) > 0$. Solving the quadratic, we get $n > \frac{1}{4}\sqrt{\frac{3}{\varepsilon} - 4}$, provided $\varepsilon < 3/4$. So let $N = \frac{1}{4}\sqrt{\frac{3}{\varepsilon}}$ if $\epsilon < 3/4$ and let $N = 1$ otherwise.

10.2.105 Let $\varepsilon > 0$ be given. We wish to find N such that for $n > N$, $\left|\frac{cn}{bn+1} - \frac{c}{b}\right| = \left|\frac{-c}{b(bn+1)}\right| = \frac{c}{b(bn+1)} < \varepsilon$. But this means that $\varepsilon b^2 n + (b\varepsilon - c) > 0$, so that $N > \frac{c}{b^2\varepsilon}$ will work.

10.2.107 First note that for $a = 1$ we already know that $\{n^n\}$ grows fast than $\{n!\}$. So if $a > 1$, then $n^{an} \geq n^n$, so that $\{n^{an}\}$ grows faster than $\{n!\}$ for $a > 1$ as well. To settle the case $a < 1$, recall Stirling's formula which states that for large values of n,

$$n! \sim \sqrt{2\pi n}\, n^n e^{-n}.$$

Thus

$$\lim_{n\to\infty} \frac{n!}{n^{an}} = \lim_{n\to\infty} \frac{\sqrt{2\pi n}\, n^n e^{-n}}{n^{an}}$$
$$= \sqrt{2\pi} \lim_{n\to\infty} n^{\frac{1}{2} + (1-a)n} e^{-n}$$
$$\geq \sqrt{2\pi} \lim_{n\to\infty} n^{(1-a)n} e^{-n}$$
$$= \sqrt{2\pi} \lim_{n\to\infty} e^{(1-a)n \ln n} e^{-n}$$
$$= \sqrt{2\pi} \lim_{n\to\infty} e^{((1-a)\ln n - 1)n}.$$

If $a < 1$ then $(1-a)\ln n - 1 > 0$ for large values of n because $1 - a > 0$, so that this limit is infinite. Hence $\{n!\}$ grows faster than $\{n^{an}\}$ exactly when $a < 1$.

10.2.109 $\{(n-2)^2 + 6(n-2) - 9\}_{n=3}^{\infty} = \{n^2 + 2n - 17\}_{n=3}^{\infty}$.

10.2.111 For $b = 2$, $2^3 > 3!$ but $16 = 2^4 < 4! = 24$, so the crossover point is $n = 4$. For e, $e^5 \approx 148.41 > 5! = 120$ while $e^6 \approx 403.4 < 6! = 720$, so the crossover point is $n = 6$. For 10, $24! \approx 6.2 \times 10^{23} < 10^{24}$, while $25! \approx 1.55 \times 10^{25} > 10^{25}$, so the crossover point is $n = 25$.

10.3 Infinite Series

10.3.1 The ratio is the common ratio between successive terms in the sum.

10.3.3 No. For example, the geometric series with $a_n = 3 \cdot 2^n$ does not have a finite sum.

10.3.5

a. $a = \frac{2}{3}$; $r = \frac{1}{5}$.

b. $a = \frac{1}{27}$; $r = -\frac{1}{3}$.

10.3.7 $S_n = \left(\frac{1}{4} - \frac{1}{5}\right) + \left(\frac{1}{5} - \frac{1}{6}\right) + \ldots + \left(\frac{1}{n+3} - \frac{1}{n+4}\right) = \frac{1}{4} - \frac{1}{n+4}$. $S_{36} = \frac{1}{4} - \frac{1}{40} = \frac{10-1}{40} = \frac{9}{40}$.

10.3.9 $S = 1 \cdot \frac{1 - 3^9}{1 - 3} = \frac{19682}{2} = 9841$.

10.3. Infinite Series

10.3.11 $S = 1 \cdot \dfrac{1 - (4/25)^{21}}{1 - 4/25} = \dfrac{25^{21} - 4^{21}}{25^{21} - 4 \cdot 25^{20}} \approx 1.1905.$

10.3.13 $S = 1 \cdot \dfrac{1 - (-3/4)^{10}}{1 + 3/4} = \dfrac{4^{10} - 3^{10}}{4^{10} + 3 \cdot 4^9} = \dfrac{141361}{262144} \approx 0.5392.$

10.3.15 $\dfrac{1}{4} \sum_{k=0}^{6} \dfrac{1}{3^k} = \dfrac{1}{4}\left(\dfrac{1 - (1/3)^7}{1 - (1/3)}\right) = \dfrac{1093}{2916}.$

10.3.17 $A_{60} = 250 + 250(1.002) + \cdots + 250(1.002)^{59} = 250\left(\dfrac{1 - 1.002^{60}}{1 - 1.002}\right) = 15,920.22.$

10.3.19

a. The nth partial sum is $S_n = a\left(\dfrac{1 - r^n}{1 - r}\right) = \dfrac{1 - (-2/7)^n}{1 - (-2/7)} = \dfrac{7}{9}\left(1 - \left(-\dfrac{2}{7}\right)^n\right)$. So the sum of the series is $S = \lim_{n \to \infty} S_n = \dfrac{7}{9}(1 - 0) = \dfrac{7}{9}.$

b. $S = \sum_{k=0}^{\infty} \left(-\dfrac{2}{7}\right)^k = \dfrac{1}{1 - (-2/7)} = \dfrac{7}{9}.$

10.3.21 $\dfrac{1}{1 - 1/4} = \dfrac{4}{3}.$

10.3.23 $\dfrac{1}{1 + 9/10} = \dfrac{10}{19}.$

10.3.25 $\dfrac{1}{1 - 0.9} = 10.$

10.3.27 Divergent, because $r > 1$.

10.3.29 $\dfrac{e^{-2}}{1 - e^{-2}} = \dfrac{1}{e^2 - 1}.$

10.3.31 $\dfrac{2^{-3}}{1 - 2^{-3}} = \dfrac{1}{7}.$

10.3.33 $\dfrac{1/625}{1 - 1/5} = \dfrac{1}{500}.$

10.3.35 $3 \cdot \dfrac{1}{1 + 1/\pi} = \dfrac{3\pi}{\pi + 1}.$

10.3.37 $\dfrac{1}{1 - e/\pi} = \dfrac{\pi}{\pi - e}.$

10.3.39 $\dfrac{0.15^2}{1.15} = \dfrac{9}{460} \approx 0.0196.$

10.3.41 $\dfrac{4/12}{1 - (1/2)} = \dfrac{4/12}{11/12} = \dfrac{4}{11}.$

10.3.43 $A_5 = 200 + 200(0.25) + \cdots + 200(0.25)^4 = 200\left(\dfrac{1 - 0.25^5}{1 - 0.25}\right) = 266.406.$

$A_{10} = 200 + 200(0.25) + \cdots + 200(0.25)^9 = 200\left(\dfrac{1 - 0.25^{10}}{1 - 0.25}\right) = 266.666.$

$A_{30} = 200 + 200(0.25) + \cdots + 200(0.25)^{29} = 200\left(\dfrac{1 - 0.25^{30}}{1 - 0.25}\right) = 266.667.$

$\lim_{n \to \infty} 200\left(\dfrac{1}{1 - 0.25}\right) = 266.\overline{66}$ mg.

10.3.45 $200 + 200(0.5) + 200(0.5)^2 + \cdots = \dfrac{200}{1 - 0.5} = 400$ mg.

10.3.47 $0.\overline{3} = 0.333\ldots = \sum\limits_{k=1}^{\infty} 3(0.1)^k = \dfrac{0.3}{1 - 0.1} = \dfrac{0.3}{0.9} = \dfrac{1}{3}$.

10.3.49 $0.\overline{037} = 0.037037037\ldots = \sum\limits_{k=1}^{\infty} 37(0.001)^k = \dfrac{0.037}{1 - 0.001} = \dfrac{0.037}{0.999} = \dfrac{1}{27}$.

10.3.51 $0.\overline{456} = 0.456456456\ldots = \sum\limits_{k=0}^{\infty} 0.456 \cdot 10^{-3k} = \dfrac{0.456}{1 - 1/1000} = \dfrac{456}{999} = \dfrac{152}{333}$.

10.3.53 $0.00\overline{952} = 0.00952952\ldots = \sum\limits_{k=0}^{\infty} 0.00952 \cdot 10^{-3k} = \dfrac{0.00952}{1 - 1/1000} = \dfrac{9.52}{999} = \dfrac{952}{99900} = \dfrac{238}{24975}$.

10.3.55 The second part of each term cancels with the first part of the succeeding term, so $S_n = \dfrac{1}{1+1} - \dfrac{1}{n+2} = \dfrac{n}{2n+4}$, and $\lim\limits_{n\to\infty} \dfrac{n}{2n+4} = \dfrac{1}{2}$.

10.3.57 $\dfrac{1}{(k+6)(k+7)} = \dfrac{1}{k+6} - \dfrac{1}{k+7}$, so the series given is the same as $\sum\limits_{k=1}^{\infty}\left(\dfrac{1}{k+6} - \dfrac{1}{k+7}\right)$. In that series, the second part of each term cancels with the first part of the succeeding term, so $S_n = \dfrac{1}{1+6} - \dfrac{1}{n+7}$. Thus $\lim\limits_{n\to\infty} S_n = \dfrac{1}{7}$.

10.3.59 Note that $\dfrac{4}{(4k-3)(4k+1)} = \dfrac{1}{4k-3} - \dfrac{1}{4k+1}$.

Thus the given series is the same as $\sum\limits_{k=3}^{\infty}\left(\dfrac{1}{4k-3} - \dfrac{1}{4k+1}\right) = \sum\limits_{k=1}^{\infty}\left(\dfrac{1}{4k+5} - \dfrac{1}{4k+9}\right)$. In this last series, the second part of each term cancels with the first part of the succeeding term, so we have $S_n = \dfrac{1}{9} - \dfrac{1}{4n+9}$, and thus $\lim\limits_{n\to\infty} S_n = \dfrac{1}{9}$.

10.3.61 $\ln\left(\dfrac{k+1}{k}\right) = \ln(k+1) - \ln k$, so the series given is the same as $\sum\limits_{k=1}^{\infty}(\ln(k+1) - \ln k)$, in which the first part of each term cancels with the second part of the next term, so we have $S_n = \ln(n+1) - \ln 1 = \ln(n+1)$, and thus the series diverges.

10.3.63 $\dfrac{1}{(k+p)(k+p+1)} = \dfrac{1}{k+p} - \dfrac{1}{k+p+1}$, so that $\sum\limits_{k=1}^{\infty} \dfrac{1}{(k+p)(k+p+1)} = \sum\limits_{k=1}^{\infty}\left(\dfrac{1}{k+p} - \dfrac{1}{k+p+1}\right)$ and this series telescopes to give $S_n = \dfrac{1}{p+1} - \dfrac{1}{n+p+1} = \dfrac{n}{n(p+1) + (p+1)^2}$ so that $\lim\limits_{n\to\infty} S_n = \dfrac{1}{p+1}$.

10.3.65 Let $a_n = \dfrac{1}{\sqrt{n+1}} - \dfrac{1}{\sqrt{n+3}}$. Then the second term of a_n cancels with the first term of a_{n+2}, so the series telescopes and $S_n = \dfrac{1}{\sqrt{2}} + \dfrac{1}{\sqrt{3}} - \dfrac{1}{\sqrt{n-1+3}} - \dfrac{1}{\sqrt{n+3}}$ and thus the sum of the series is the limit of S_n, which is $\dfrac{1}{\sqrt{2}} + \dfrac{1}{\sqrt{3}}$.

10.3. Infinite Series

10.3.67 $\dfrac{3}{k^2+5k+4} = \dfrac{3}{(k+1)(k+4)} = \dfrac{1}{k+1} - \dfrac{1}{k+4}$. So our series can be written as

$$\sum_{k=1}^{\infty}\left(\dfrac{1}{k+1} - \dfrac{1}{k+4}\right).$$

Starting with S_4, the second part of each term cancels the first part of the term that is 3 terms ahead, so

$$S_n = \dfrac{1}{2} + \dfrac{1}{3} + \dfrac{1}{4} - \dfrac{1}{n+2} - \dfrac{1}{n+3} - \dfrac{1}{n+4},$$

and $\lim\limits_{n\to\infty} S_n = \dfrac{1}{2} + \dfrac{1}{3} + \dfrac{1}{4} = \dfrac{13}{12}$.

10.3.69 This series telescopes to give

$$S_n = \tan^{-1}(n+1) - \tan^{-1}(1) = \tan^{-1}(n+1) - \dfrac{\pi}{4}.$$

Then because $\lim\limits_{n\to\infty}\tan^{-1}(n+1) = \dfrac{\pi}{2}$, the sum of the series is equal to $\lim\limits_{n\to\infty} S_n = \dfrac{\pi}{4}$.

10.3.71

a. Because the first part of each term cancels the second part of the previous term, the nth partial sum telescopes to be $S_n = \dfrac{4}{3} - \dfrac{4}{3^{n+1}}$. Thus, the sum of the series is $\lim\limits_{n\to\infty} S_n = \dfrac{4}{3}$.

b. $\displaystyle\sum_{k=1}^{\infty}\left(\dfrac{4}{3^k} - \dfrac{4}{3^{k+1}}\right) = \sum_{k=1}^{\infty}\dfrac{4}{3^k} - \sum_{k=1}^{\infty}\dfrac{4}{3^{k+1}} = \dfrac{4/3}{1-1/3} - \dfrac{4/9}{1-1/3} = 2 - \dfrac{2}{3} = \dfrac{4}{3}$.

10.3.73 $16k^2 + 8k - 3 = (4k+3)(4k-1)$, so $\dfrac{1}{16k^2+8k-3} = \dfrac{1}{(4k+3)(4k-1)} = \dfrac{1}{4}\left(\dfrac{1}{4k-1} - \dfrac{1}{4k+3}\right)$. Thus the series given is equal to $\dfrac{1}{4}\displaystyle\sum_{k=0}^{\infty}\left(\dfrac{1}{4k-1} - \dfrac{1}{4k+3}\right)$. This series telescopes, so $S_n = \dfrac{1}{4}\left(-1 - \dfrac{1}{4n-1}\right)$, so the sum of the series is equal to $\lim\limits_{n\to\infty} S_n = -\dfrac{1}{4}$.

10.3.75 $\displaystyle\sum_{k=0}^{\infty}\left(\dfrac{1}{4}\right)^k 5^{3-k} = 5^3 \sum_{k=0}^{\infty}\left(\dfrac{1}{20}\right)^k = 5^3 \cdot \dfrac{1}{1-1/20} = \dfrac{5^3 \cdot 20}{19} = \dfrac{2500}{19}$.

10.3.77 This can be written as $\dfrac{1}{3}\displaystyle\sum_{k=1}^{\infty}\left(-\dfrac{2}{3}\right)^k$. This is a geometric series with ratio $r = -\dfrac{2}{3}$ so the sum is

$$\dfrac{1}{3} \cdot \dfrac{-2/3}{1-(-2/3)} = \dfrac{1}{3}\cdot\left(-\dfrac{2}{5}\right) = -\dfrac{2}{15}.$$

10.3.79 Note that

$$\dfrac{\ln((k+1)k^{-1})}{(\ln k)\ln(k+1)} = \dfrac{\ln(k+1)}{(\ln k)\ln(k+1)} - \dfrac{\ln k}{(\ln k)\ln(k+1)} = \dfrac{1}{\ln k} - \dfrac{1}{\ln(k+1)}.$$

In the partial sum S_n, the first part of each term cancels the second part of the preceding term, so we have $S_n = \dfrac{1}{\ln 2} - \dfrac{1}{\ln(n+2)}$. Thus we have $\lim\limits_{n\to\infty} S_n = \dfrac{1}{\ln 2}$.

10.3.81 $\displaystyle\sum_{k=0}^{\infty}\left(3\left(\dfrac{2}{5}\right)^k - 2\left(\dfrac{5}{7}\right)^k\right) = 3\sum_{k=0}^{\infty}\left(\dfrac{2}{5}\right)^k - 2\sum_{k=0}^{\infty}\left(\dfrac{5}{7}\right)^k = 3\left(\dfrac{1}{3/5}\right) - 2\left(\dfrac{1}{2/7}\right) = 5 - 7 = -2$.

10.3.83 $\sum_{k=1}^{\infty}\left(\frac{1}{3}\left(\frac{5}{6}\right)^k + \frac{3}{5}\left(\frac{7}{9}\right)^k\right) = \frac{1}{3}\sum_{k=1}^{\infty}\left(\frac{5}{6}\right)^k + \frac{3}{5}\sum_{k=1}^{\infty}\left(\frac{7}{9}\right)^k = \frac{1}{3}\left(\frac{5/6}{1/6}\right) + \frac{3}{5}\left(\frac{7/9}{2/9}\right) = \frac{5}{3} + \frac{21}{10} = \frac{113}{30}.$

10.3.85 $\sum_{k=1}^{\infty}\left(\left(\frac{1}{6}\right)^k + \left(\frac{1}{3}\right)^{k-1}\right) = \sum_{k=1}^{\infty}\left(\frac{1}{6}\right)^k + \sum_{k=1}^{\infty}\left(\frac{1}{3}\right)^{k-1} = \frac{1/6}{5/6} + \frac{1}{2/3} = \frac{17}{10}.$

10.3.87

a. True. The two series differ by a finite amount ($\sum_{k=1}^{9} a_k$), so if one converges, so does the other.

b. True. The same argument applies as in part (a).

c. False. If $\sum a_k$ converges, then $a_k \to 0$ as $k \to \infty$, so that $a_k + 0.0001 \to 0.0001$ as $k \to \infty$, so that $\sum(a_k + 0.0001)$ cannot converge.

d. False. Suppose $p = -1.0001$. Then $\sum p^k$ diverges but $p + .0001 = -0.9991$ so that $\sum(p + .0001)^k$ converges.

e. True. $\left(\frac{\pi}{e}\right)^{-k} = \left(\frac{e}{\pi}\right)^k$; because $e < \pi$, this is a geometric series with ratio less than 1.

f. False. For example, let $0 < a < 1$ and $b > 1$.

g. True. Suppose $a > \frac{1}{2}$. Then we want $a = \sum_{k=0}^{\infty} r^k = \frac{1}{1-r}$. Solving for r gives $r = 1 - \frac{1}{a}$. Because $a > 0$ we have $r < 1$; because $a > \frac{1}{2}$ we have $r > 1 - \frac{1}{1/2} = -1$. Thus $|r| < 1$ so that $\sum_{k=0}^{\infty} r^k$ converges, and it converges to a.

10.3.89

a. $0.00110011\ldots = \frac{1}{2^3} + \frac{1}{2^4} + \frac{1}{2^7} + \frac{1}{2^8} + \frac{1}{2^{11}} + \frac{1}{2^{12}} + \cdots = \frac{3}{2^4} + \frac{3}{2^8} + \frac{3}{2^{12}} + \cdots = \frac{3/2^4}{1 - 1/2^4} = \frac{3}{15} = \frac{1}{5}.$

b. $0.0011001100110011 = \frac{3}{2^4} + \frac{3}{2^8} + \frac{3}{2^{12}} + \frac{3}{2^{16}} \approx 0.19999695.$

10.3.91 We seek r so that $\frac{r^6}{1-r^2} = 10$, so $r^6 = 10 - 10r^2$, so $r^6 + 10r^2 - 10 = 0$. Using a computer algebra system, we obtain $r \approx 0.96$.

10.3.93

$$\begin{aligned}
A_n &= A_{n-1}(1+r) - M \\
&= (A_{n-2}(1+r) - M)(1+r) - M \\
&= (A_{n-3}(1+r) - M)(1+r)^2 - (M + M(1+r)) \\
&= A_{n-3}(1+r)^3 - (M + M(1+r) + M(1+r)^2) = \cdots \\
&= A_0(1+r)^n - (M + M(1+r) + M(1+r)^2 + \cdots + M(1+r)^{n+1}) \\
&= A_0(1+r)^n - M\left(\frac{1 - (1+r)^n}{1 - (1+r)}\right) \\
&= A_0(1+r)^n - \frac{M}{r}\left((1+r)^n - 1\right)
\end{aligned}$$

10.3.95 $180000(1.005)^n - \dfrac{1000}{0.005}((1.005)^n - 1) = 0$, so $-20000(1.005)^n + 200000 = 0$. Then $(1.005)^n = 10$, so $n = \dfrac{\ln 10}{\ln 1.005} \approx 461.67$. Rounding up we have 462 months.

10.3.97 $\displaystyle\sum_{k=1}^{\infty} ca_k = \lim_{n\to\infty}\sum_{k=1}^{n} ca_k = \lim_{n\to\infty} c\sum_{k=1}^{n} a_k = c\lim_{n\to\infty}\sum_{k=1}^{n} a_k$, so that one sum diverges if and only if the other one does.

10.3.99 At the nth stage, there are 2^{n-1} triangles of area $A_n = \dfrac{1}{8}A_{n-1} = \dfrac{1}{8^{n-1}}A_1$, so the total area of the triangles formed at the nth stage is $\dfrac{2^{n-1}}{8^{n-1}}A_1 = \left(\dfrac{1}{4}\right)^{n-1} A_1$. Thus the total area under the parabola is

$$\sum_{n=1}^{\infty} \left(\frac{1}{4}\right)^{n-1} A_1 = A_1 \sum_{n=1}^{\infty} \left(\frac{1}{4}\right)^{n-1} = A_1 \frac{1}{1 - 1/4} = \frac{4}{3}A_1.$$

10.3.101

a. I_{n+1} is obtained by I_n by dividing each edge into three equal parts, removing the middle part, and adding two parts equal to it. Thus 3 equal parts turn into 4, so $L_{n+1} = \dfrac{4}{3}L_n$. This is a geometric sequence with a ratio greater than 1, so the nth term grows without bound.

b. As the result of part (a), I_n has $3 \cdot 4^n$ sides of length $\dfrac{1}{3^n}$; each of those sides turns into an added triangle in I_{n+1} of side length 3^{-n-1}. Thus the added area in I_{n+1} consists of $3 \cdot 4^n$ equilateral triangles with side 3^{-n-1}. The area of an equilateral triangle with side x is $\dfrac{x^2\sqrt{3}}{4}$. Thus $A_{n+1} = A_n + 3 \cdot 4^n \cdot \dfrac{3^{-2n-2}\sqrt{3}}{4} = A_n + \dfrac{\sqrt{3}}{12}\cdot\left(\dfrac{4}{9}\right)^n$, and $A_0 = \dfrac{\sqrt{3}}{4}$. Thus $A_{n+1} = A_0 + \displaystyle\sum_{i=0}^{n} \dfrac{\sqrt{3}}{12}\cdot\left(\dfrac{4}{9}\right)^i$, so that

$$A_\infty = A_0 + \frac{\sqrt{3}}{12}\sum_{i=0}^{\infty}\left(\frac{4}{9}\right)^i = \frac{\sqrt{3}}{4} + \frac{\sqrt{3}}{12}\cdot\frac{1}{1-4/9} = \frac{\sqrt{3}}{4}\left(1 + \frac{3}{5}\right) = \frac{2}{5}\sqrt{3}.$$

10.3.103 $|S - S_n| = \left|\displaystyle\sum_{i=n}^{\infty} r^k\right| = \left|\dfrac{r^n}{1-r}\right|$ because the latter sum is simply a geometric series with first term r^n and ratio r.

10.3.105

a. Solve $\left|\dfrac{(-0.8)^n}{1.8}\right| = \dfrac{0.8^n}{1.8} < 10^{-6}$ for n to get $n = 60$.

b. Solve $\dfrac{0.2^n}{0.8} < 10^{-6}$ for n to get $n = 9$.

10.3.107

a. Solve $\dfrac{1/\pi^n}{1 - 1/\pi} < 10^{-6}$ for n to get $n = 13$.

b. Solve $\dfrac{1/e^n}{1 - 1/e} < 10^{-6}$ for n to get $n = 15$.

10.3.109

a. $f(x) = \sum_{k=0}^{\infty}(-1)^k x^k = \dfrac{1}{1+x}$; because f is a geometric series, $f(x)$ exists only when the ratio, $-x$, is such that $|-x| = |x| < 1$. Then $f(0) = 1$, $f(0.2) = \dfrac{1}{1.2} = \dfrac{5}{6}$, $f(0.5) = \dfrac{1}{1+.05} = \dfrac{2}{3}$. Neither $f(1)$ nor $f(1.5)$ exists.

b. The domain of f is $\{x : |x| < 1\}$.

10.3.111 $f(x)$ is a geometric series with ratio $\dfrac{1}{1+x}$; thus $f(x)$ converges when $\left|\dfrac{1}{1+x}\right| < 1$. For $x > -1$, $\left|\dfrac{1}{1+x}\right| = \dfrac{1}{1+x}$ and $\dfrac{1}{1+x} < 1$ when $1 < 1+x$, $x > 0$. For $x < -1$, $\left|\dfrac{1}{1+x}\right| = \dfrac{1}{-1-x}$, and this is less than 1 when $1 < -1-x$, i.e. $x < -2$. So $f(x)$ converges for $x > 0$ and for $x < -2$. When $f(x)$ converges, its value is $\dfrac{1}{1-\frac{1}{1+x}} = \dfrac{1+x}{x}$, so $f(x) = 3$ when $1+x = 3x$, $x = \dfrac{1}{2}$.

10.4 The Divergence and Integral Tests

10.4.1 The series diverges.

10.4.3 We can be sure that $\lim\limits_{k\to\infty} a_k = 0$.

10.4.5 For the same values of p as in the previous problem – it converges for $p > 1$, and diverges for all other values of p.

10.4.7 The remainder of an infinite series is the error in approximating a convergent infinite series by a finite number of terms.

10.4.9 $a_k = \dfrac{k}{2k+1}$ and $\lim\limits_{k\to\infty} a_k = \dfrac{1}{2}$, so the series diverges.

10.4.11 $a_k = \dfrac{1}{1000+k}$ and $\lim\limits_{k\to\infty} a_k = 0$, so the divergence test is inconclusive.

10.4.13 $a_k = \dfrac{k}{\ln k}$ and $\lim\limits_{k\to\infty} a_k = \infty$, so the series diverges.

10.4.15 $a_k = \dfrac{\sqrt{k}}{\ln^{10} k}$ and $\lim\limits_{k\to\infty} a_k = \infty$, so the series diverges.

10.4.17 Let $f(x) = \dfrac{1}{\sqrt{x+8}}$. $f(x)$ is obviously continuous and decreasing for $x \geq 1$. Because

$$\int_1^{\infty} \dfrac{1}{\sqrt{x+8}}\,dx = \lim_{b\to\infty}\int_1^b \dfrac{1}{\sqrt{x+8}}\,dx = \lim_{b\to\infty} 2\sqrt{x+8}\Big|_1^b = \lim_{b\to\infty}(2\sqrt{b+8} - 6) = \infty,$$

the series diverges.

10.4.19 The function $f(x) = \dfrac{1}{\sqrt[3]{5x+3}}$ is positive, decreasing, and continuous for $x \geq 1$. The relevant integral is

$$\int_1^{\infty} \dfrac{dx}{\sqrt[3]{5x+3}} = \lim_{b\to\infty} \dfrac{3}{10}(5x+3)^{2/3}\Big|_1^b = \dfrac{3}{10}\lim_{b\to\infty}\left((5b+3)^{2/3} - 4\right) = \infty.$$

Because the integral diverges, the given series also diverges by the Integral Test.

10.4. The Divergence and Integral Tests

10.4.21 Let $f(x) = x \cdot e^{-2x^2}$. This function is continuous for $x \geq 1$. Its derivative is $e^{-2x^2}(1 - 4x^2) < 0$ for $x \geq 1$, so $f(x)$ is decreasing. The relevant integral is

$$\int_1^\infty x \cdot e^{-2x^2}\, dx = -\frac{1}{4} \lim_{b \to \infty} e^{-2x^2}\Big|_1^b = -\frac{1}{4}\lim_{b \to \infty}\left(e^{-2b^2} - e^{-2}\right) = \frac{1}{4e^2}.$$

Because the integral converges, the given series also converges by the Integral Test.

10.4.23 $\displaystyle\sum_{k=1}^\infty \frac{1}{k^{1/5}}$ is a p-series with $p = \frac{1}{5} < 1$. The series diverges by the p-series test.

10.4.25 We will use the Integral Test. $f(x) = \dfrac{x^3}{e^{x^4}}$ is continuous and positive, and $f'(x) = \dfrac{x^2(3 - 4x^4)}{e^{x^4}} < 0$ for $x > 1$. Thus f is decreasing for $x > 1$. The relevant integral is

$$\int_1^\infty \frac{x^3}{e^{x^4}}\, dx = \lim_{b \to \infty}\left(-\frac{e^{-x^4}}{4}\right)\Big|_1^b = \frac{1}{4e}.$$

Because the integral converges, the given series also converges by the Integral Test.

10.4.27 $a_k = k^{1/k}$. In order to compute $\lim_{k \to \infty} a_k$, we let $y_k = \ln a_k = \dfrac{\ln k}{k}$. By Theorem 10.6, (or by L'Hôpital's rule) $\lim_{k \to \infty} y_k = 0$, so $\lim_{k \to \infty} a_k = e^0 = 1$. The given series thus diverges by the Divergence Test.

10.4.29 This is a p-series with $p = 10$, so this series converges by the p-series test.

10.4.31 $\displaystyle\sum_{k=3}^\infty \frac{1}{(k - 2)^4} = \sum_{k=1}^\infty \frac{1}{k^4}$, which is a p-series with $p = 4$, thus convergent by the p-series test.

10.4.33 Let $f(x) = \dfrac{x}{e^x}$. $f(x)$ is clearly continuous for $x > 1$, and its derivative, $f'(x) = \dfrac{e^x - xe^x}{e^{2x}} = (1 - x)\dfrac{e^x}{e^{2x}}$, is negative for $x > 1$ so that $f(x)$ is decreasing. Because $\displaystyle\int_1^\infty f(x)\, dx = 2e^{-1}$, the series converges by the Integral Test.

10.4.35 Because $\displaystyle\lim_{k \to \infty}\left(\frac{k}{k + 10}\right)^k = \lim_{k \to \infty}\frac{1}{(1 + (10/k))^k} = \frac{1}{e^{10}} \neq 0$, the series diverges by the Divergence Test.

10.4.37 $\displaystyle\sum_{k=1}^\infty \frac{1}{\sqrt[3]{k}} = \sum_{k=1}^\infty \frac{1}{k^{1/3}}$ is a p-series with $p = 1/3$, thus divergent by the p-series test.

10.4.39

a. $S \approx S_2 = 1 + \dfrac{1}{2^7} = 1.0078125$.

b. $R_2 < \displaystyle\int_2^\infty \frac{dx}{x^7} = \lim_{b \to \infty}\int_2^b x^{-7}\, dx = \lim_{b \to \infty} -\frac{1}{6x^6}\Big|_2^b = \lim_{b \to \infty} -\left(\frac{1}{6b^6} - \frac{1}{384}\right) \approx 0.0026042$.

c.

$$L_2 = S_2 + \int_3^\infty \frac{dx}{x^7} = S_2 + \lim_{b \to \infty}\int_3^b x^{-7}\, dx$$

$$= S_2 + \lim_{b \to \infty} -\frac{1}{6x^6}\Big|_3^b = S_2 + \lim_{b \to \infty}-\left(\frac{1}{6b^6} - \frac{1}{4374}\right)$$

$$\approx 1.0078125 + 0.000228624 \approx 1.0080411.$$

$$U_2 = S_2 + \int_2^\infty \frac{dx}{x^7} = S_2 + \lim_{b\to\infty} \int_2^b x^{-7}\, dx$$
$$= S_2 + \lim_{b\to\infty} -\frac{1}{6x^6}\Big|_2^b = S_2 + \lim_{b\to\infty} -\left(\frac{1}{6b^6} - \frac{1}{384}\right)$$
$$\approx 1.0078125 + 0.0026042 \approx 1.0104167.$$

So
$$1.0080411 \leq S \leq 1.0104167.$$

10.4.41

a. The remainder R_n is bounded by $\int_n^\infty \frac{1}{x^6}\, dx = \frac{1}{5n^5}$.

b. We solve $\frac{1}{5n^5} < 10^{-3}$ to get $n = 3$.

c. $L_n = S_n + \int_{n+1}^\infty \frac{1}{x^6}\, dx = S_n + \frac{1}{5(n+1)^5}$, and $U_n = S_n + \int_n^\infty \frac{1}{x^6}\, dx = S_n + \frac{1}{5n^5}$.

d. $S_{10} \approx 1.017341512$, so $L_{10} \approx 1.017341512 + \frac{1}{5\cdot 11^5} \approx 1.017342754$, and $U_{10} \approx 1.017341512 + \frac{1}{5\cdot 10^5} \approx 1.017343512$.

10.4.43

a. The remainder R_n is bounded by $\int_n^\infty \frac{1}{3^x}\, dx = \frac{1}{3^n \ln(3)}$.

b. We solve $\frac{1}{3^n \ln(3)} < 10^{-3}$ to obtain $n = 7$.

c. $L_n = S_n + \int_{n+1}^\infty \frac{1}{3^x}\, dx = S_n + \frac{1}{3^{n+1} \ln(3)}$, and $U_n = S_n + \int_n^\infty \frac{1}{3^x}\, dx = S_n + \frac{1}{3^n \ln(3)}$.

d. $S_{10} \approx 0.4999915325$, so $L_{10} \approx 0.4999915325 + \frac{1}{3^{11} \ln 3} \approx 0.4999966708$, and $U_{10} \approx 0.4999915325 + \frac{1}{3^{10} \ln 3} \approx 0.5000069475$.

10.4.45 We first try to determine what n will suffice. We need $R_n < \frac{1}{10^4}$. We have $R_n = \int_n^\infty \frac{dx}{x^7} = \frac{1}{6n^6}$. Setting this less than $\frac{1}{10^4}$ and solving for n we have $n > \left(\frac{10^4}{6}\right)^{1/6} \approx 3.44$, so $n = 4$ should suffice. We have
$$S_4 = 1 + \frac{1}{2^7} + \frac{1}{3^7} + \frac{1}{4^7} \approx 1.0083.$$

10.4.47

a. False. This is a geometric series.

b. True. If we reindex the series, we obtain $\sum_{k=1}^\infty \frac{1}{\sqrt{k}}$, which is a p-series with $p = \frac{1}{2}$.

c. False. Both series converge, but not necessarily to the same value.

d. True. The partial sums are an increasing sequence.

10.4. The Divergence and Integral Tests

e. False. Let $p = 1.0005$; then $-p + .001 = -(p - .001) = -0.9995$, so that $\sum k^{-p}$ converges (p-series) but $\sum k^{-p+.001}$ diverges.

f. False. Let $a_k = \dfrac{1}{k}$, the harmonic series.

10.4.49 This is a geometric series with $r = \dfrac{1}{e}$ and $\left|\dfrac{1}{e}\right| < 1$, so the series converges the Geometric Series Test.

10.4.51 Converges by the Integral Test because $\displaystyle\int_1^\infty \dfrac{1}{(3x+1)(3x+4)}\,dx = \int_1^\infty \dfrac{1}{3(3x+1)} - \dfrac{1}{3(3x+4)}\,dx =$
$\displaystyle\lim_{b\to\infty} \int_1^b \left(\dfrac{1}{3(3x+1)} - \dfrac{1}{3(3x+4)}\right) dx = \lim_{b\to\infty} \dfrac{1}{9}\left(\ln\left(\dfrac{3x+1}{3x+4}\right)\right)\bigg|_1^b = \lim_{b\to\infty} = -\dfrac{1}{9}\cdot \ln(4/7) \approx 0.06217 < \infty.$

10.4.53 Diverges by the Divergence Test because $\displaystyle\lim_{k\to\infty} a_k = \lim_{k\to\infty} \dfrac{k}{\sqrt{k^2+1}} = 1 \neq 0$.

10.4.55 We use the Integral Test. The function $f(x) = \dfrac{4}{x\sqrt{\ln x}}$ is positive, decreasing, and continuous for $x \geq 3$. The relevant integral is

$$\lim_{b\to\infty}\int_3^b \dfrac{4}{x\sqrt{\ln x}}\,dx = 8\lim_{b\to\infty}\sqrt{\ln x}\bigg|_3^b = 8\lim_{b\to\infty}(\sqrt{\ln b}-\sqrt{\ln 3}) = \infty.$$

The given series therefore diverges by the Integral Test.

10.4.57 The series can be written as $\displaystyle\sum_{k=1}^\infty \left(\dfrac{6}{5}\right)^k$, which is a divergent geometric series because $r = \dfrac{6}{5} > 1$.

10.4.59 Reindexing gives $\displaystyle\sum_{k=4}^\infty \dfrac{3}{(k-3)^4} = 3\sum_{k=1}^\infty \dfrac{1}{k^4}$, which is convergent because $\displaystyle\sum_{k=1}^\infty \dfrac{1}{k^4}$ is a convergent p-series with $p = 4$ by the p-series test.

10.4.61 This series can be written as $2\displaystyle\sum_{k=4}^\infty \dfrac{1}{k^2}$, so it converges, because $\displaystyle\sum_{k=1}^\infty \dfrac{1}{k^2}$ is a convergent p-series with $p = 2$.

10.4.63 This can be written in the form $9\displaystyle\sum_{k=1}^\infty \left(\dfrac{3}{5}\right)^k$ so it is convergent because $\displaystyle\sum_{k=1}^\infty \left(\dfrac{3}{5}\right)^k$ is a convergent geometric series. In fact, its value is $9\left(\dfrac{3/5}{1-3/5}\right) = 9\left(\dfrac{3}{2}\right) = \dfrac{27}{2}$.

10.4.65

a. Note that $\displaystyle\int \dfrac{1}{x\ln x(\ln\ln x)^p}\,dx = \dfrac{1}{1-p}(\ln\ln x)^{1-p}$, and thus the improper integral with bounds n and ∞ exists only if $p > 1$ because $\ln\ln x > 0$ for $x > e$. So this series converges for $p > 1$.

b. For large values of z, clearly $\sqrt{z} > \ln z$, so that $z > (\ln z)^2$. Write $z = \ln x$; then for large x, $\ln x > (\ln\ln x)^2$; multiplying both sides by $x\ln x$ we have that $x\ln^2 x > x\ln x(\ln\ln x)^2$, so that the first series converges faster because the terms get smaller faster.

10.4.67 To approximate the sequence for $\zeta(x)$, note that the remainder R_n after n terms is bounded by

$$\int_n^\infty \frac{1}{z^x}\,dz = \frac{1}{x-1}n^{1-x}.$$

For $x = 3$, if we wish to approximate the value to within 10^{-3}, we must solve $\frac{1}{2}n^{-2} < 10^{-3}$, so that $n = 23$, and $\sum_{k=1}^{23} \frac{1}{k^3} \approx 1.201151926$. The true value is ≈ 1.202056903.

For $x = 5$, if we wish to approximate the value to within 10^{-3}, we must solve $\frac{1}{4}n^{-4} < 10^{-3}$, so that $n = 4$, and $\sum_{k=1}^{4} \frac{1}{k^5} \approx 1.036341789$. The true value is ≈ 1.036927755.

10.4.69 $\sum_{k=1}^\infty \frac{1}{k^2} = \sum_{k=1}^\infty \frac{1}{(2k)^2} + \sum_{k=1}^\infty \frac{1}{(2k-1)^2}$, splitting the series into even and odd terms. But $\sum_{k=1}^\infty \frac{1}{(2k)^2} = \frac{1}{4}\sum_{k=1}^\infty \frac{1}{k^2}$. Thus $\frac{\pi^2}{6} = \frac{1}{4} \cdot \frac{\pi^2}{6} + \sum_{k=1}^\infty \frac{1}{(2k-1)^2}$, so that the sum in question is $\frac{3\pi^2}{24} = \frac{\pi^2}{8}$.

10.4.71 Let $S_n = \sum_{k=1}^n \frac{1}{\sqrt{k}}$. Then this looks like a left Riemann sum for the function $y = \frac{1}{\sqrt{x}}$ on $[1, n+1]$. Because each rectangle lies above the curve itself, we see that S_n is bounded below by the integral of $\frac{1}{\sqrt{x}}$ on $[1, n+1]$. Now,

$$\int_1^{n+1} \frac{1}{\sqrt{x}}\,dx = \int_1^{n+1} x^{-1/2}\,dx = 2\sqrt{x}\Big|_1^{n+1} = 2\sqrt{n+1} - 2.$$

This integral diverges as $n \to \infty$, so the series does as well by the bound above.

10.4.73

a. $x_1 = \sum_{k=2}^{2} \frac{1}{k} = \frac{1}{2}$, $x_2 = \sum_{k=3}^{4} \frac{1}{k} = \frac{1}{3} + \frac{1}{4} = \frac{7}{12}$, $x_3 = \sum_{k=4}^{6} \frac{1}{k} = \frac{1}{4} + \frac{1}{5} + \frac{1}{6} = \frac{37}{60}$.

b. x_n has n terms. Each term is bounded below by $\frac{1}{2n}$ and bounded above by $\frac{1}{n+1}$. Thus $x_n \geq n \cdot \frac{1}{2n} = \frac{1}{2}$, and $x_n \leq n \cdot \frac{1}{n+1} < n \cdot \frac{1}{n} = 1$.

c. The right Riemann sum for $\int_1^2 \frac{dx}{x}$ using n subintervals has n rectangles of width $\frac{1}{n}$; the right edges of those rectangles are at $1 + \frac{i}{n} = \frac{n+i}{n}$ for $i = 1, 2, \ldots, n$. The height of such a rectangle is the value of $\frac{1}{x}$ at the right endpoint, which is $\frac{n}{n+i}$. Thus the area of the rectangle is $\frac{1}{n} \cdot \frac{n}{n+i} = \frac{1}{n+i}$. Adding up over all the rectangles gives x_n.

d. The limit $\lim_{n \to \infty} x_n$ is the limit of the right Riemann sum as the width of the rectangles approaches zero. This is precisely $\int_1^2 \frac{dx}{x} = \ln x \Big|_1^2 = \ln 2$.

10.4.75

a. Dividing both sides of the recurrence equation by f_n gives $\dfrac{f_{n+1}}{f_n} = 1 + \dfrac{f_{n-1}}{f_n}$. Let the limit of the ratio of successive terms be L. Taking the limit of the previous equation gives $L = 1 + \dfrac{1}{L}$. Thus $L^2 = L + 1$, so $L^2 - L - 1 = 0$. The quadratic formula gives $L = \dfrac{1 \pm \sqrt{1 - 4 \cdot (-1)}}{2}$, but we know that all the terms are positive, so we must have $L = \dfrac{1 + \sqrt{5}}{2} = \phi \approx 1.618$.

b. Write the recurrence in the form $f_{n-1} = f_{n+1} - f_n$ and divide both sides by f_{n+1}. Then we have $\dfrac{f_{n-1}}{f_{n+1}} = 1 - \dfrac{f_n}{f_{n+1}}$. Taking the limit gives $1 - \dfrac{1}{\phi}$ on the right-hand side.

c. Consider the harmonic series with the given groupings, and compare it with the sum of $\dfrac{f_{k-1}}{f_{k+1}}$ as shown. The first three terms match exactly. The sum of the next two are $\dfrac{1}{4} + \dfrac{1}{5} > \dfrac{1}{5} + \dfrac{1}{5} = \dfrac{2}{5}$. The sum of the next three are $\dfrac{1}{6} + \dfrac{1}{7} + \dfrac{1}{8} > \dfrac{1}{8} + \dfrac{1}{8} + \dfrac{1}{8} = \dfrac{3}{8}$. The sum of the next five are $\dfrac{1}{9} + \cdots + \dfrac{1}{13} > 5 \cdot \dfrac{1}{13} = \dfrac{5}{13}$. Thus the harmonic series is bounded below by the series $\displaystyle\sum_{k=1}^{\infty} \dfrac{f_{k-1}}{f_{k+1}}$.

d. The result above implies that the harmonic series diverges, because the series $\displaystyle\sum_{k=1}^{\infty} \dfrac{f_{k-1}}{f_{k+1}}$ diverges, since its general term has limit $1 - \dfrac{1}{\phi} \neq 0$.

10.5 Comparison Tests

10.5.1 Given a series of positive terms $\sum a_k$ that you suspect converges, find a series $\sum b_k$ that you know converges, for which $\lim\limits_{k \to \infty} \dfrac{a_k}{b_k} = L$ where $L \geq 0$ is a finite number. If you are successful, you will have shown that the series $\sum a_k$ converges.

Given a series of positive terms $\sum a_k$ that you suspect diverges, find a series $\sum b_k$ that you know diverges, for which $\lim\limits_{k \to \infty} \dfrac{a_k}{b_k} = L$ where $L > 0$ (including the case $L = \infty$). If you are successful, you will have shown that $\sum a_k$ diverges.

10.5.3 $\displaystyle\sum_{k=1}^{\infty} \dfrac{1}{k^2}$.

10.5.5 $\displaystyle\sum_{k=1}^{\infty} \left(\dfrac{2}{3}\right)^k$.

10.5.7 $\displaystyle\sum_{k=1}^{\infty} \dfrac{1}{k}$.

10.5.9 $\dfrac{1}{k^2 + 4} < \dfrac{1}{k^2}$, and $\displaystyle\sum_{k=1}^{\infty} \dfrac{1}{k^2}$ converges, so $\displaystyle\sum_{k=1}^{\infty} \dfrac{1}{k^2 + 4}$ converges as well, by the Comparison Test.

10.5.11 Use the Limit Comparison Test with $\left\{\dfrac{1}{k}\right\}$. The ratio of the terms of the two series is $\dfrac{k^3 - k}{k^3 + 4}$ which has limit 1 as $k \to \infty$. Because the comparison series diverges, the given series does as well.

10.5.13 Because $\dfrac{\sqrt{k}}{k^2 + 3} < \dfrac{\sqrt{k}}{k^2} = \dfrac{1}{k^{3/2}}$ and $\displaystyle\sum_{k=1}^{\infty} \dfrac{1}{k^{3/2}}$ is a convergent p-series, $\displaystyle\sum_{k=1}^{\infty} \dfrac{\sqrt{k}}{k^2 + 3}$ converges by the Comparison Test.

10.5.15 We use the Limit Comparison Test with the convergent series $\displaystyle\sum_{k=1}^{\infty} \left(\dfrac{4}{5}\right)^k$. We have

$$\lim_{k \to \infty} \dfrac{\dfrac{4^k}{5^k - 3}}{\left(\dfrac{4}{5}\right)^k} = \lim_{k \to \infty} \dfrac{5^k}{5^k - 3} = \lim_{k \to \infty} \dfrac{1}{1 - \dfrac{3}{5^k}} = 1.$$

Because $0 < 1 < \infty$, we have that the given series is convergent by the Limit Comparison Test.

10.5.17 For all k, $\dfrac{1}{k^{3/2} + 1} < \dfrac{1}{k^{3/2}}$. The series whose terms are $\dfrac{1}{k^{3/2}}$ is a p-series which converges, so the given series converges as well by the Comparison Test.

10.5.19 Note that $\dfrac{1 + \cos^2 k}{k - 3} > \dfrac{1}{k - 3} > \dfrac{1}{k}$ for $k \geq 4$. Because $\displaystyle\sum_{k=1}^{\infty} \dfrac{1}{k}$ is the divergent harmonic series, we have that $\displaystyle\sum_{k=4}^{\infty} \dfrac{1 + \cos^2 k}{k - 3}$ diverges as well by the Comparison Test.

10.5.21 We use the Limit Comparison Test with the divergent harmonic series $\displaystyle\sum_{k=1}^{\infty} \dfrac{1}{k}$. We have

$$\lim_{k \to \infty} \dfrac{\dfrac{21k^5 + 3k^3 + 28k^2 + 4}{8k^6 + 2k^3 - 1}}{\dfrac{1}{k}} = \lim_{k \to \infty} \dfrac{21k^6 + 3k^4 + 28k^3 + 4k}{8k^6 + 2k^3 - 1} = \dfrac{21}{8}.$$

Because $0 < \dfrac{21}{8} < \infty$, the given series diverges by the Limit Comparison Test.

10.5.23 $\sin(1/k) > 0$ for $k \geq 1$, so we can apply the Comparison Test with $\displaystyle\sum_{k=1}^{\infty} \dfrac{1}{k^2}$. $\sin(1/k) < 1$, so $\dfrac{\sin(1/k)}{k^2} < \dfrac{1}{k^2}$. Because the comparison series converges, the given series converges as well by the Comparison Test.

10.5.25 Use the Limit Comparison Test with $\displaystyle\sum_{k=1}^{\infty} \dfrac{1}{k}$. The ratio of the terms of the two series is $\dfrac{k}{2k - \sqrt{k}} = \dfrac{1}{2 - 1/\sqrt{k}}$, which has limit $1/2$ as $k \to \infty$. Because the comparison series diverges, the given series does as well by the Limit Comparison Test.

10.5.27 Note that $2 + (-1)^k \leq 3$ for $k \geq 1$. It follows that $\dfrac{2 + (-1)^k}{k^2} \leq \dfrac{3}{k^2}$ for $k \geq 1$. Because $\displaystyle\sum_{k=1}^{\infty} \dfrac{3}{k^2} = 3\displaystyle\sum_{k=1}^{\infty} \dfrac{1}{k^2}$ and $\displaystyle\sum_{k=1}^{\infty} \dfrac{1}{k^2}$ is a convergent p-series, $\displaystyle\sum_{k=1}^{\infty} \dfrac{3}{k^2}$ converges. It follows that the given series converges by the Comparison Test.

10.5. Comparison Tests

10.5.29 Use the Limit Comparison Test with $\dfrac{k^{2/3}}{k^{3/2}}$. The ratio of corresponding terms of the two series is

$$\frac{\sqrt[3]{k^2+1}}{\sqrt{k^3+1}} \cdot \frac{k^{3/2}}{k^{2/3}} = \frac{\sqrt[3]{k^2+1}}{\sqrt[3]{k^2}} \cdot \frac{\sqrt{k^3}}{\sqrt{k^3+1}},$$

which has limit 1 as $k \to \infty$. The comparison series is the series whose terms are $k^{2/3 - 3/2} = k^{-5/6}$, which is a p-series with $p < 1$, so it, and the given series, both diverge.

10.5.31 We use the Limit Comparison Test with the divergent series $\sum_{k=1}^{\infty} \dfrac{1}{\sqrt{k}}$. We have

$$L = \lim_{k \to \infty} \frac{\frac{20}{\sqrt[3]{k} + \sqrt{k}}}{\frac{1}{\sqrt{k}}} = \lim_{k \to \infty} \frac{20 k^{1/2}}{k^{1/3} + k^{1/2}} = \lim_{k \to \infty} \frac{20}{(1/k^{1/6}) + 1} = 20.$$

Because $0 < L < \infty$, we have that the given series diverges by the Limit Comparison Test.

10.5.33 We use the Limit Comparison Test with the divergent series $\sum_{k=1}^{\infty} \dfrac{1}{k}$. We have

$$L = \lim_{k \to \infty} \frac{\frac{1}{2^{\ln k}}}{\frac{1}{k}} = \cdot \lim_{k \to \infty} \frac{k}{2^{\ln k}} = \lim_{k \to \infty} \frac{e^{\ln k}}{2^{\ln k}} = \lim_{k \to \infty} \left(\frac{e}{2}\right)^{\ln k} = \infty.$$

Because $L = \infty$ and $\sum_{k=1}^{\infty} \dfrac{1}{k}$ diverges, then $\sum_{k=1}^{\infty} \dfrac{1}{2^{\ln k}}$ diverges by part 3 of the Limit Comparison Test.

10.5.35 We use the Limit Comparison Test with the convergent $p$series $\sum_{k=2}^{\infty} \dfrac{1}{k^{5/4}}$. We have

$$L = \lim_{k \to \infty} \frac{\frac{1}{4^{\ln k}}}{\frac{1}{k^{5/4}}} = \lim_{k \to \infty} \frac{k^{5/4}}{4^{\ln k}} = \lim_{k \to \infty} \frac{e^{\frac{5}{4} \ln k}}{4^{\ln k}} = \lim_{k \to \infty} \left(\frac{e^{5/4}}{4}\right)^{\ln k} = 0.$$

The last limit follows from the fact that $\dfrac{e^{5/4}}{4} \approx 0.87 < 1$. Because $L = 0$ and $\sum_{k=1}^{\infty} \dfrac{1}{k^{5/4}}$ converges, the given series converges by the Limit Comparison Test.

10.5.37

a. False. For example, let $\{a_k\}$ be all zeros, and $\{b_k\}$ be all 1's.

b. True. This is a result of the Comparison Test.

c. True. Both of these statements follow from the Comparison Test.

d. True. The limit of the ratio is 1 in this case, so the two series both converge.

10.5.39 Method 1: We use the Comparison Test with the convergent p-series $\sum_{k=1}^{\infty} \dfrac{1}{k^2}$. Note that $\dfrac{1}{k^2 + 2k + 1} \leq \dfrac{1}{k^2}$. Thus the given series converges by the Comparison Test.

Method 2: We rewrite the summand. We have $\sum_{k=1}^{\infty} \frac{1}{k^2+2k+1} = \sum_{k=1}^{\infty} \frac{1}{(k+1)^2} = \sum_{k=2}^{\infty} \frac{1}{k^2}$. Because $\sum_{k=1}^{\infty} \frac{1}{k^2}$ is a convergent series, $\sum_{k=2}^{\infty} \frac{1}{k^2}$ converges as well.

10.5.41 Use the Divergence Test: $\lim_{k\to\infty} a_k = \lim_{k\to\infty} \left(1 + \frac{2}{k}\right)^k = e^2 \neq 0$, so the given series diverges.

10.5.43 Use the Divergence Test: $\lim_{k\to\infty} \frac{k^2+2k+1}{3k^2+1} = \frac{1}{3} \neq 0$, so the given series diverges.

10.5.45 Use the Limit Comparison Test with the harmonic series. Note that $\lim_{k\to\infty} \frac{\frac{1}{\ln k}}{\frac{1}{k}} = \lim_{k\to\infty} \frac{k}{\ln k} = \infty$, and because the harmonic series diverges, the given series does as well.

10.5.47 Use the Limit Comparison Test with the series whose kth term is $\frac{1}{k^{3/2}}$. Note that $\lim_{k\to\infty} \frac{\frac{1}{\sqrt{k^3-k+1}}}{\frac{\sqrt{k^3}}{1}} = \lim_{k\to\infty} \sqrt{\frac{k^3}{k^3-k+1}} = \sqrt{1} = 1$, and the series $\sum_{k=1}^{\infty} \frac{1}{k^{3/2}}$ converges because it is a p-series with $p = \frac{3}{2}$. Thus, the given series also converges.

10.5.49 Use the Comparison Test. Note that $\frac{1}{k} + 2^{-k} > \frac{1}{k}$ for $k \geq 1$. Because the harmonic series $\sum_{k=1}^{\infty} \frac{1}{k}$ diverges, so does the given series.

10.5.51 Use the Limit Comparison Test with the convergent p-series $\sum_{k=1}^{\infty} \frac{1}{k^3}$. The ratio of corresponding terms is $\frac{k^{11}}{k^{11}+3}$, which has limit 1 as $k \to \infty$. Therefore the given series converges by the Limit Comparison Test.

10.5.53 This is a p-series with exponent greater than 1, so it converges.

10.5.55 $\ln\left(\frac{k+2}{k+1}\right) = \ln(k+2) - \ln(k+1)$, so this series telescopes. We get $\sum_{k=1}^{n} \ln\left(\frac{k+2}{k+1}\right) = \ln(n+2) - \ln 2$. Because $\lim_{n\to\infty} (\ln(n+2) - \ln 2) = \infty$, the sequence of partial sums diverges, so the given series is divergent.

10.5.57 For $k > 7$, $\ln k > 2$ so note that $\frac{1}{k^{\ln k}} < \frac{1}{k^2}$. Because $\sum_{k=1}^{\infty} \frac{1}{k^2}$ converges, the given series converges as well by the Comparison Test.

10.5.59 Use the Limit Comparison Test with the harmonic series. $\frac{\tan(1/k)}{1/k}$ has limit 1 as $k \to \infty$ because $\lim_{x\to 0} \frac{\tan x}{x} = 1$. Thus the original series diverges.

10.5.61 Note that $\frac{1}{(2k+1)\cdot(2k+3)} = \frac{1}{2}\left(\frac{1}{2k+1} - \frac{1}{2k+3}\right)$. Thus this series telescopes.

$$\sum_{k=0}^{n} \frac{1}{(2k+1)(2k+3)} = \frac{1}{2}\sum_{k=0}^{n}\left(\frac{1}{2k+1} - \frac{1}{2k+3}\right) = \frac{1}{2}\left(-\frac{1}{2n+3} + 1\right),$$

so the given series converges to $1/2$, because that is the limit of the sequence of partial sums.

10.5.63 Use the Limit Comparison Test: $\lim_{k\to\infty} \frac{a_k^2}{a_k} = \lim_{k\to\infty} a_k = 0$, because $\sum a_k$ converges. By the Limit Comparison Test part 2, the series $\sum a_k^2$ must converge as well.

10.5.65 To prove case (2), assume $L = 0$ and that $\sum b_k$ converges. Because $L = 0$, for every $\varepsilon > 0$, there is some N such that for all $n > N$, $|\frac{a_k}{b_k}| < \varepsilon$. Take $\varepsilon = 1$; this then says that there is some N such that for all $n > N$, $0 < a_k < b_k$. By the Comparison Test, because $\sum b_k$ converges, so does $\sum a_k$. To prove case (3), because $L = \infty$, then $\lim_{k\to\infty} \frac{b_k}{a_k} = 0$, so by the argument above, we have $0 < b_k < a_k$ for sufficient large k. But $\sum b_k$ diverges, so by the Comparison Test, $\sum a_k$ does as well.

10.5.67 The sum on the left is simply the left Riemann sum over n equal intervals between 0 and 1 for $f(x) = x^p$. The limit of the sum is thus $\int_0^1 x^p dx = \frac{1}{p+1} x^{p+1} \Big|_0^1 = \frac{1}{p+1}$, because p is positive.

10.6 Alternating Series

10.6.1 Because $S_{n+1} - S_n = a_{n+1}$ alternates signs.

10.6.3 Because
$$S = S_{2n+1} + (a_{2n} - a_{2n+1}) + (a_{2n+2} - a_{2n+3}) + \cdots$$
and each term of the form $a_{2k} - a_{2k+1} > 0$, so that $S_{2n+1} < S$. Also
$$S = S_{2n} + (-a_{2n+1} + a_{2n+2}) + (-a_{2n+3} + a_{2n+4}) + \cdots$$
and each term of the form $-a_{2k+1} + a_{2k+2} < 0$, so that $S < S_{2n}$. Thus the sum of the series is trapped between the odd partial sums and the even partial sums.

10.6.5 The remainder is less than the first neglected term because
$$S - S_n = (-1)^{n+1}(a_{n+1} + (-a_{n+2} + a_{n+3}) + \cdots)$$
so that the sum of the series *after* the first disregarded term has the opposite sign from the first disregarded term.

10.6.7 No. If the terms are positive, then the absolute value of each term is the term itself, so convergence and absolute convergence would mean the same thing in this context.

10.6.9 Yes. For example, $\sum \frac{(-1)^k}{k^3}$ converges absolutely and thus not conditionally (see the definition).

10.6.11 The terms of the series decrease in magnitude, and $\lim_{k\to\infty} \frac{1}{2k+1} = 0$, so the given series converges.

10.6.13 $\lim_{k\to\infty} \frac{k}{3k+2} = \frac{1}{3} \neq 0$, so the given series diverges.

10.6.15 The terms of the series decrease in magnitude, and $\lim_{k\to\infty} \frac{1}{k^3} = 0$, so the given series converges.

10.6.17 The terms of the series decrease in magnitude, and $\lim_{k\to\infty} \frac{k^2}{k^3+1} = \lim_{k\to\infty} \frac{1/k}{1+1/k^3} = 0$, so the given series converges. In order to see that the terms decrease in magnitude, consider $f(x) = \frac{x^2}{x^3+1}$. Then $f'(x) = \frac{x(2-x^3)}{(1+x^3)^2}$, which is negative for $x \geq 2$.

10.6.19 $\lim\limits_{k\to\infty} \frac{k^2-1}{k^2+3} = 1$, so the terms of the series do not tend to zero and thus the given series diverges.

10.6.21 $\lim\limits_{k\to\infty} \left(1 + \frac{1}{k}\right) = 1$, so the given series diverges.

10.6.23 Because $\lim\limits_{k\to\infty} \frac{k^{11}+2k^5+1}{4k(k^{10}+1)} = \frac{1}{4}$, $\lim\limits_{k\to\infty} \left((-1)^{k+1} \frac{k^{11}+2k^5+1}{4k(k^{10}+1)}\right) \neq 0$. So the given series diverges by the Divergence Test.

10.6.25 $\lim\limits_{k\to\infty} k^{1/k} = 1$ (for example, take logs and apply L'Hôpital's rule), so the given series diverges by the Divergence Test.

10.6.27 $\frac{1}{\sqrt{k^2+4}}$ is decreasing and tends to zero as $k \to \infty$, so the given series converges.

10.6.29 $S_4 = -\frac{1}{1} + \frac{1}{16} - \frac{1}{81} + \frac{1}{256} \approx -0.945939$. $|S - S_4| \leq |a_5| = \frac{1}{625} \approx 0.0016$.

10.6.31 $S_5 = 1 - \frac{1}{3} + \frac{1}{21} - \frac{1}{91} + \frac{1}{273} \approx 0.70696$. $|S - S_5| \leq |a_5| = \frac{1}{25+625+1} \approx 0.001536$.

10.6.33 We want $\frac{1}{n+1} < 10^{-4}$, or $n+1 > 10^4$, so $n = 10^4$.

10.6.35 The series starts with $k = 0$, so we want $\frac{1}{2n+1} < 10^{-4}$, or $2n + 1 > 10^4$, $n = 5000$.

10.6.37 We want $\frac{1}{(n+1)^4} < 10^{-4}$, or $(n+1)^4 > 10^4$, so $n = 10$.

10.6.39 To figure out how many terms we need to sum, we must find n such that $\frac{1}{(n+1)^5} < 10^{-3}$, so that $(n+1)^5 > 1000$; this occurs first for $n = 3$. Thus $-\frac{1}{1} + \frac{1}{2^5} - \frac{1}{3^5} \approx -0.972865$.

10.6.41 To figure out how many terms we need to sum, we must find n so that $\frac{n}{n^2+1} < 10^{-3}$, so that $\frac{n^2+1}{n} = n + (1/n) > 1000$. This occurs first for $n = 1000$. We have $\sum\limits_{k=1}^{1000} \frac{(-1)^k k}{k^2+1} \approx -0.269111$.

10.6.43 To figure how many terms we need to sum, we must find n such that $\frac{1}{(n+1)^{n+1}} < 10^{-3}$, or $(n+1)^{n+1} > 1000$, so $n = 4$ ($5^5 = 3125$). Thus the approximation is $\sum\limits_{k=1}^{4} \frac{(-1)^n}{n^n} \approx -.7831307870$.

10.6.45 The series of absolute values is a p-series with $p = 2/3$, so it diverges. The given alternating series does converge, though, by the Alternating Series Test. Thus, the given series is conditionally convergent.

10.6.47 Note that the absolute value of the kth term is $\frac{1}{(k+1)!}$, and also note that for $k \geq 1$, $(k+1)! > k^2$, and thus $\frac{1}{(k+1)!} < \frac{1}{k^2}$. Now because $\sum\limits_{k=1}^{\infty} \frac{1}{k^2}$ is a convergent p-series, the given series converges absolutely by the Comparison Test.

10.6.49 This series of all positive terms is convergent, because it is a geometric series with $r = \frac{3}{4}$. Therefore it converges absolutely.

10.6. Alternating Series

10.6.51 The series of absolute values is $\sum_{k=1}^{\infty} \frac{|\cos(k)|}{k^3}$, which converges by the Comparison Test because $\frac{|\cos(k)|}{k^3} \leq \frac{1}{k^3}$. Thus the series converges absolutely.

10.6.53 The absolute value of the kth term of this series has limit $\pi/2$ as $k \to \infty$, so the given series is divergent by the Divergence Test.

10.6.55 Note that the absolute value of the kth term is $\frac{1}{2\sqrt{k}-1}$. We will use the Limit Comparison Test with the divergent p-series $\sum_{k=1}^{\infty} \frac{1}{\sqrt{k}}$. We have

$$\lim_{k \to \infty} \frac{\frac{1}{2\sqrt{k}-1}}{\frac{1}{\sqrt{k}}} = \lim_{k \to \infty} \frac{\sqrt{k}}{2\sqrt{k}-1} = \frac{1}{2}.$$

Therefore the given series is not absolutely convergent. However, it is the case that $\left\{\frac{1}{2\sqrt{k}-1}\right\}$ is decreasing and has limit 0, so by the Alternating Series Test, the given series is convergent. Therefore, the given series is conditionally convergent.

10.6.57 The series of absolute values is $\sum_{k=1}^{\infty} \frac{k}{2k+1}$, but $\lim_{k \to \infty} \frac{k}{2k+1} = \frac{1}{2}$, so by the Divergence Test, this series diverges. The original series does not converge conditionally, either, because $\lim_{k \to \infty} a_k = \frac{1}{2} \neq 0$.

10.6.59 The series of absolute values is $\sum_{k=1}^{\infty} \frac{\tan^{-1}(k)}{k^3}$, which converges by the Comparison Test because $\frac{\tan^{-1}(k)}{k^3} < \frac{\pi}{2} \cdot \frac{1}{k^3}$, and $\sum_{k=1}^{\infty} \frac{\pi}{2} \cdot \frac{1}{k^3}$ converges because it is a constant multiple of a convergent p-series. So the original series converges absolutely.

10.6.61 The absolute value of the kth term is $\frac{k^2+1}{k^3-1}$. We use the Limit Comparison Test with the divergent harmonic series. We have

$$L = \lim_{k \to \infty} \frac{\frac{k^2+1}{k^3-1}}{\frac{1}{k}} = \lim_{k \to \infty} \frac{k^3+k}{k^3-1} = 1.$$

The given series therefore does not converge absolutely. However, the given series does converge by the Alternating Series Test. Note that if $f(x) = \frac{x^2+1}{x^3-1}$, then $f'(x) = \frac{-(2x+3x^2+x^4)}{(x^3-1)^2}$ which is less than 0 for $x \geq 1$. Therefore the absolute value of the sequence of terms of the original series is decreasing and has limit 0, so the given series is conditionally convergent.

10.6.63 For $k \geq 3$,

$$\frac{k!}{k^k} = \frac{1 \cdot 2 \cdot 3 \cdots k}{k \cdot k \cdot k \cdots k} = \frac{1}{k} \cdot \frac{2}{k} \cdot \frac{3}{k} \cdots \frac{k-1}{k} \cdot \frac{k}{k} \leq \frac{1}{k} \cdot \frac{2}{k} \cdot \frac{k}{k} \cdot \frac{k}{k} \cdots \frac{k}{k} = \frac{2}{k^2}.$$

Because the series $\sum_{k=3}^{\infty} \frac{2}{k^2}$ is a convergent series (it is a constant multiple of a p-series), it follows that the given series converges absolutely by the Comparison Test.

10.6.65

a. False. For example, consider the alternating harmonic series.

b. True. This is part of Theorem 10.21.

c. True. This statement is simply saying that a convergent series converges.

d. True. This is part of Theorem 10.21.

e. False. Let $a_k = \dfrac{1}{k}$.

f. True. Use the Comparison Test: $\lim\limits_{k\to\infty} \dfrac{a_k^2}{a_k} = \lim\limits_{k\to\infty} a_k = 0$ because $\sum a_k$ converges, so $\sum a_k^2$ and $\sum a_k$ converge or diverge together. Because the latter converges, so does the former.

g. True, by definition. If $\sum |a_k|$ converged, the original series would converge absolutely, not conditionally.

10.6.67 $\displaystyle\sum_{k=1}^{\infty} \dfrac{1}{k^2} - \sum_{k=1}^{\infty} \dfrac{(-1)^{k+1}}{k^2} = 2\sum_{k=1}^{\infty} \dfrac{1}{(2k)^2} = 2 \cdot \dfrac{1}{4} \sum_{k=1}^{\infty} \dfrac{1}{k^2}$, and thus $\displaystyle\sum_{k=1}^{\infty} \dfrac{(-1)^{k+1}}{k^2} = \dfrac{\pi^2}{6} - \dfrac{1}{2} \cdot \dfrac{\pi^2}{6} = \dfrac{\pi^2}{12}$.

10.6.69 Both series diverge, so comparisons of their values are not meaningful.

10.6.71

a. The sequences $\{a_{2n}\} = \left\{\dfrac{1}{n}\right\}$ and $\{a_{2n-1}\} = \left\{\dfrac{1}{n^2}\right\}$ both converge to 0, so $\{a_n\}$ converges to 0.

b.
$$S_{2n} = \left(\dfrac{1}{1^2} - \dfrac{1}{1}\right) + \left(\dfrac{1}{2^2} - \dfrac{1}{2}\right) + \left(\dfrac{1}{3^2} - \dfrac{1}{3}\right) + \cdots + \left(\dfrac{1}{n^2} - \dfrac{1}{n}\right) = \sum_{k=1}^{\infty} \left(\dfrac{1}{k^2} - \dfrac{1}{k}\right).$$

c. Recall that $\displaystyle\sum_{k=1}^{\infty} \dfrac{1}{k}$ diverges while $\displaystyle\sum_{k=1}^{\infty} \dfrac{1}{k^2}$ diverges. Therefore, $S_{2n} = \displaystyle\sum_{k=1}^{\infty}\left(\dfrac{1}{k^2} - \dfrac{1}{k}\right)$ diverges. So $\{S_n\}$ diverges, which means that the given alternating series diverges. This does not violate the Alternating Series Test because the sequence $\{a_n\}$ is not a nonincreasing sequence.

10.7 The Ratio and Root Tests

10.7.1 Given a series $\sum a_k$ of terms, compute $\lim\limits_{k\to\infty} \left|\dfrac{a_{k+1}}{a_k}\right|$ and call it r. If $r < 1$, the given series converges absolutely. If $r > 1$ (including $r = \infty$), the given series diverges. If $r = 1$, the test is inconclusive.

10.7.3 $\dfrac{1000!}{998!} = \dfrac{1000(999)(998!)}{998!} = 1000(999) = 999{,}000$.

10.7.5 $\dfrac{k!}{(k+2)!} = \dfrac{k!}{(k+2)(k+1)(k!)} = \dfrac{1}{(k+2)(k+1)}$.

10.7.7 The Ratio Test.

10.7.9 The absolute value of the ratio between successive terms is $\left|\dfrac{a_{k+1}}{a_k}\right| = \dfrac{1}{(k+1)!} \cdot \dfrac{(k)!}{1} = \dfrac{1}{k+1}$, which goes to zero as $k \to \infty$, so the given series converges absolutely by the Ratio Test.

10.7.11 The kth root of the absolute value of the kth term is $\dfrac{10k^3 + k}{9k^3 + k + 1}$. The limit of this as $k \to \infty$ is $\dfrac{10}{9} > 1$, so the given series diverges by the Root Test.

10.7. The Ratio and Root Tests

10.7.13 The absolute value of the ratio between successive terms is $\left|\frac{a_{k+1}}{a_k}\right| = \frac{(k+1)^2}{4(k+1)} \cdot \frac{4^k}{(k)^2} = \frac{1}{4}\left(\frac{k+1}{k}\right)^2$. The limit is $1/4$ as $k \to \infty$, so the given series converges absolutely by the Ratio Test.

10.7.15 The absolute value of the ratio between successive terms is $\left|\frac{a_{k+1}}{a_k}\right| = \frac{(k+1)e^{-(k+1)}}{(k)e^{-(k)}} = \frac{k+1}{(k)e}$. The limit of this ratio as $k \to \infty$ is $1/e < 1$, so the given series converges absolutely by the Ratio Test.

10.7.17 The absolute value of the ratio between successive terms is $\left|\frac{a_{k+1}}{a_k}\right| = \frac{k^2 7^{k+1}}{(k+1)^2 7^k} = \frac{7k^2}{(k+1)^2}$ which has limit $7 > 1$ as $k \to \infty$. Therefore the given series diverges by the Ratio Test.

10.7.19 The absolute value of the ratio between successive terms is

$$\frac{2^{k+1}}{(k+1)^{99}} \cdot \frac{(k)^{99}}{2^k} = 2\left(\frac{k}{k+1}\right)^{99};$$

the limit as $k \to \infty$ is 2, so the given series diverges by the Ratio Test.

10.7.21 The absolute value of the ratio between successive terms is

$$\frac{((k+1)!)^2}{(2(k+1))!} \cdot \frac{(2k)!}{((k)!)^2} = \frac{(k+1)^2}{(2k+2)(2k+1)};$$

the limit as $k \to \infty$ is $1/4$, so the given series converges absolutely by the Ratio Test.

10.7.23 The kth root of the absolute value of the kth term is $\left(\frac{k}{k+1}\right)^{2k}$. The limit of this as $k \to \infty$ is $e^{-2} < 1$, so the given series converges absolutely by the Root Test.

10.7.25 The kth root of the absolute value of the kth term is $\frac{3k^2 + 4k}{2k^2 + 1}$ which has limit $\frac{3}{2} > 1$ as $k \to \infty$, so the given series diverges by the Root Test.

10.7.27 The kth root of the absolute value of the kth term is $\frac{1}{k}$. The limit of this as $k \to \infty$ is 0, so the given series converges absolutely by the Root Test.

10.7.29 We use the Ratio Test. The limit of the absolute values of the ratio of the terms is

$$L = \lim_{k \to \infty} \frac{k!k!((k+1)^{2k+2}}{(k+1)!(k+1)!k^{2k}} = \lim_{k \to \infty} \frac{(k+1)^{2k+2}}{(k+1)(k+1)k^{2k}} = \lim_{k \to \infty} \frac{(k+1)^{2k}}{k^{2k}} = \lim_{k \to \infty} \left(\left(1 + \frac{1}{k}\right)^k\right)^2 = e^2 > 1.$$

So the given series diverges by the Ratio Test.

10.7.31

a. False. For example, when $n = 1$ we have $1!1! = 1$ but $(2 \cdot 1)! = 2! = 2$.

b. True. $\frac{(2n)!}{(2n-1)!} = \frac{2n(2n-1)!}{(2n-1)!} = 2n$.

c. True. By the Root Test, $\sum a_k$ converges, so therefore $\sum 10 a_k$ converges.

d. True. In this case, the limit of the ratio of terms is 1.

10.7.33 Note that $\sum_{k=1}^{\infty}\left((-1)^k \frac{2^{k+1}}{9^{k-1}}\right) = \sum_{k=1}^{\infty} 18\left(-\frac{2}{9}\right)^k$ is a geometric series with $r = -\frac{2}{9}$. Because $|r| < 1$, the given series converges absolutely.

10.7.35 We use the Ratio Test. We have

$$\lim_{k \to \infty} \frac{\frac{(2k+3)!}{((k+1)!)^2}}{\frac{(2k+1)!}{(k!)^2}} = \lim_{k \to \infty} \frac{(2k+3)!(k!)^2}{(2k+1)!((k+1)!)^2} = \lim_{k \to \infty} \frac{(2k+3)(2k+2)}{(k+1)^2} = 4 > 1.$$

Therefore the given series diverges by the Ratio Test.

10.7.37 Use the Ratio Test: the ratio of successive terms is $\frac{(k+1)^{100}}{(k+2)!} \cdot \frac{(k+1)!}{k^{100}} = \left(\frac{k+1}{k}\right)^{100} \cdot \frac{1}{k+2}$. This has limit $1^{100} \cdot 0 = 0$ as $k \to \infty$, so the given series converges absolutely by the Ratio Test.

10.7.39 We use the Limit Comparison Test with the divergent series $\sum_{k=1}^{\infty} \frac{1}{k}$ on the corresponding series with all positive terms to obtain

$$L = \lim_{k \to \infty} \frac{\frac{k^3}{\sqrt{k^8+1}}}{\frac{1}{k}} = \lim_{k \to \infty} \frac{k^4}{\sqrt{k^8+1}} = \lim_{k \to \infty} \frac{1}{\sqrt{1+1/k^8}} = 1.$$

Therefore the given series is not absolutely convergent. To see that it is convergent (and therefore conditionally convergent), note that $\lim_{k \to \infty} \frac{k^3}{\sqrt{k^8+1}} = 0$, and that the sequence is decreasing, so by the Alternating Series Test, the series is convergent. To see that the sequence decreases, note that if $f(x) = \frac{x^3}{\sqrt{x^8+1}}$, that $f'(x) = \frac{x^2(3-x^8)}{(x^8+1)^{3/2}}$ which is less than 0 for $x \geq 2$. This shows that the the sequence $|a_k|$ is decreasing, at least for $k \geq 2$.

10.7.41 Use the Root Test. The kth root of the absolute value of the kth term is $(k^{1/k} - 1)^2$, which has limit 0 as $k \to \infty$, so the given series converges absolutely by the Root Test.

10.7.43 Use the Ratio Test. $\left|\frac{a_{k+1}}{a_k}\right| = \frac{2^{k+1}(k+1)!}{(k+1)^{k+1}} \cdot \frac{(k)^k}{2^k(k)!} = 2\left(\frac{k}{k+1}\right)^k$, which has limit $\frac{2}{e}$ as $k \to \infty$, so the given series converges absolutely.

10.7.45 We have $|a_k| = \frac{1}{k^{0.99}}$, and $\sum_{k=1}^{\infty} \frac{1}{k^{0.99}}$ is a p-series with $p = 0.99$. Because $p < 1$, the series $\sum |a_k|$ diverges. However, because the sequence $\left\{\frac{1}{k^{0.99}}\right\}$ decreases to 0, the given series converges by the Alternating Series Test. Therefore, the given series is conditionally convergent.

10.7.47 This series is $\sum_{k=1}^{\infty} \frac{k^2}{k!}$. By the Ratio Test, $\left|\frac{a_{k+1}}{a_k}\right| = \frac{(k+1)^2}{(k+1)!} \cdot \frac{k!}{k^2} = \frac{1}{k+1}\left(\frac{k+1}{k}\right)^2$, which has limit 0 as $k \to \infty$, so the given series converges absolutely.

10.7.49 For the corresponding series of all positive terms, we use the Limit Comparison Test with the divergent series $\sum_{k=1}^{\infty} \frac{1}{k^{3/4}}$. We have

$$L = \lim_{k \to \infty} \frac{\frac{1}{\sqrt{k^{3/2}+k}}}{\frac{1}{k^{3/4}}} = \lim_{k \to \infty} \sqrt{\frac{k^{3/2}}{k^{3/2}+k}} = 1.$$

Therefore the given series does not converge absolutely. However, because $\left\{\dfrac{1}{\sqrt{k^{3/2}+k}}\right\}$ is decreasing to 0, the given series is convergent by the Alternating Series Test. Thus, the given series is conditionally convergent.

10.7.51 For $p \leq 1$ and $k > e$, $\dfrac{\ln k}{k^p} > \dfrac{1}{k^p}$. The series $\sum_{k=1}^{\infty} \dfrac{1}{k^p}$ diverges, so the given series diverges. For $p > 1$, let $q < p - 1$; then for sufficiently large k, $\ln k < k^q$, so that by the Comparison Test, $\dfrac{\ln k}{k^p} < \dfrac{k^q}{k^p} = \dfrac{1}{k^{p-q}}$. But $p - q > 1$, so that $\sum_{k=1}^{\infty} \dfrac{1}{k^{p-q}}$ is a convergent p-series. Thus the original series is convergent precisely when $p > 1$.

10.7.53 For $p \leq 1$, $\dfrac{(\ln k)^p}{k^p} > \dfrac{1}{k^p}$, and $\sum_{k=1}^{\infty} \dfrac{1}{k^p}$ diverges for $p \leq 1$, so the original series diverges. For $p > 1$, let $q < p - 1$; then for sufficiently large k, $(\ln k)^p < k^q$. Note that $\dfrac{(\ln k)^p}{k^p} < \dfrac{k^q}{k^p} = \dfrac{1}{k^{p-q}}$. But $p - q > 1$, so $\sum_{k=1}^{\infty} \dfrac{1}{k^{p-q}}$ converges, so the given series converges. Thus, the given series converges exactly for $p > 1$.

10.7.55 Use the Ratio Test:
$$\frac{a_{k+1}}{a_k} = \frac{(k+1)p^{k+1}}{k+2} \cdot \frac{k+1}{kp^k} = p \cdot \frac{k^2 + 2k + 1}{k^2 + 2k},$$
and this expression has limit p as $k \to \infty$. Thus the series converges for $p > 1$.

10.7.57 $\lim\limits_{k \to \infty} a_k = \lim\limits_{k \to \infty} \left(1 - \dfrac{p}{k}\right)^k = e^{-p} \neq 0$, so this sequence diverges for all p by the Divergence Test.

10.7.59 $\left|\dfrac{a_{k+1}}{a_k}\right| = \left|\dfrac{x^{k+1}}{x^k}\right| = |x|$. This has limit $|x|$ as $k \to \infty$, so the series converges absolutely for $|x| < 1$. It clearly does not converge for $x = \pm 1$.

10.7.61 $\left|\dfrac{a_{k+1}}{a_k}\right| = \left|\dfrac{x^{k+1}}{(k+1)^2} \cdot \dfrac{k^2}{x^k}\right| = |x| \left(\dfrac{k}{k+1}\right)^2$, which has limit $|x|$ as $k \to \infty$. Thus the series converges for $|x| < 1$. When $x = 1$, the series is $\sum_{k=1}^{\infty} \dfrac{1}{k^2}$, which converges, and for $x = -1$, the series is $\sum_{k=1}^{\infty} \dfrac{(-1)^k}{k^2}$ which converges. Thus the original series converges for $|x| \leq 1$.

10.7.63 $\left|\dfrac{a_{k+1}}{a_k}\right| = \left|\dfrac{x^{k+1}}{2^{k+1}} \cdot \dfrac{2^k}{x^k}\right| = \dfrac{|x|}{2}$, which has limit $\dfrac{|x|}{2}$ as $k \to \infty$. Thus the series converges for $|x| < 2$. For $x = \pm 2$, it is obviously divergent. So the series converges for $-2 < x < 2$.

10.8 Choosing a Convergence Test

10.8.1 The presence of the power k makes the Root Test a good candidate for determining convergence.

10.8.3 The Divergence Test is appropriate because $\lim\limits_{k \to \infty} \dfrac{2k^2}{k^2 - k - 2} \neq 0$.

10.8.5 Reindex the sum $\sum_{k=10}^{\infty} \dfrac{1}{(k-9)^5}$ to obtain $\sum_{k=1}^{\infty} \dfrac{1}{k^5}$ and apply the p-series Test. Another approach is to use the Limit Comparison Test with the comparison series $\sum_{k=10}^{\infty} \dfrac{1}{k^5}$.

10.8.7 Use the Comparison Test or Limit Comparison Test with the comparison series $\sum_{k=10}^{\infty} \frac{1}{k^2}$.

10.8.9 The series is an alternating series and the function $f(x) = \dfrac{1}{\sqrt{2^x + \ln x}}$ is nonincreasing for $x \geq 1$, so the Alternating Series Test is an appropriate test to apply to this series.

10.8.11 Taking the limit of the kth-term of the series, we have $\lim_{k\to\infty} \dfrac{2k^4 + k}{4k^4 - 8k} = \dfrac{1}{2}$. The terms of the series $\sum_{k=1}^{\infty} \dfrac{2k^4 + k}{4k^4 - 8k}$ do not approach 0, so the series diverges by the Divergence Test.

10.8.13 We apply the Limit Comparison Test using the series $\sum_{k=3}^{\infty} \dfrac{1}{k}$ as the comparison series. This comparison series diverges because it is the tail of the divergent harmonic series $\sum_{k=1}^{\infty} \dfrac{1}{k}$. Calculating L, with the help of l'Hôpital's Rule, we have

$$L = \lim_{k\to\infty} \frac{5/(2+\ln k)}{1/k} = \lim_{k\to\infty} \frac{5k}{2+\ln k} = \lim_{k\to\infty} \frac{5}{1/k} = \lim_{k\to\infty} 5k = \infty.$$

By case (3) of the Limit Comparison Test, the series $\sum_{k=3}^{\infty} \dfrac{5}{2+\ln k}$ diverges because the comparison series $\sum_{k=3}^{\infty} \dfrac{1}{k}$ diverges.

10.8.15 The presence of $k!$ in the sum suggests that the Ratio Test might be an appropriate approach here.

$$r = \lim_{k\to\infty} \left| \frac{(-7)^{k+1}/(k+1)!}{(-7)^k/k!} \right| = \lim_{k\to\infty} \frac{7^{k+1}}{(k+1)!} \cdot \frac{k!}{7^k} = \lim_{k\to\infty} \frac{7}{k+1} = 0$$

Because $r < 1$, the series $\sum_{k=1}^{\infty} \dfrac{(-7)^k}{k!}$ converges by the Ratio Test.

10.8.17 The limit of the kth-term of the series is $\lim_{k\to\infty} \dfrac{(-k)^3}{3k^3 + 2} = -\dfrac{1}{3}$. Therefore $\sum_{k=1}^{\infty} \dfrac{(-k)^3}{3k^3 + 2}$ diverges by the Divergence Test because $\lim_{k\to\infty} \dfrac{(-k)^3}{3k^3 + 2} \neq 0$.

10.8.19 Writing out the first few terms of the series, we have

$$\sum_{k=0}^{\infty} \frac{3^{k+4}}{5^{k-2}} = \frac{3^4}{5^{-2}} + \frac{3^5}{5^{-1}} + \frac{3^6}{5^0} + \frac{3^7}{5^1} + \cdots.$$

This is a geometric series with $r = 3/5$. Because $r < 1$, the series converges.

10.8.21 The Alternating Series can be used here to establish convergence of the series. Another approach is to first establish that the positive-term series $\sum_{k=1}^{\infty} \dfrac{k}{k^3 + 1}$ converges. Notice that

$$\frac{k}{k^3+1} < \frac{k}{k^3} = \frac{1}{k^2}.$$

It follows that $\sum_{k=1}^{\infty} \frac{k}{k^3+1}$ converges by the Comparison Test because $\sum_{k=1}^{\infty} \frac{1}{k^2}$ is a convergent p-series. This implies that the alternating series $\sum_{k=1}^{\infty} \frac{(-1)^k k}{k^3+1}$ is absolutely convergent. Therefore $\sum_{k=1}^{\infty} \frac{(-1)^k k}{k^3+1}$ converges because every absolutely convergent series converges.

10.8.23 Apply the Ratio Test.

$$r = \lim_{k \to \infty} \left| \frac{(k+1)^5/5^{k+1}}{k^5/5^k} \right| = \lim_{k \to \infty} \frac{(k+1)^5}{5^{k+1}} \cdot \frac{5^k}{k^5} = \lim_{k \to \infty} \frac{1}{5}\left(\frac{k+1}{k}\right)^5 = \frac{1}{5}$$

Because $r < 1$, the series $\sum_{k=1}^{\infty} \frac{k^5}{5^k}$ converges by the Ratio Test.

10.8.25 The terms of the series are positive and decreasing. We apply the Integral Test, with the help of the substitution $u = \sqrt{x}$.

$$\int_1^{\infty} \frac{dx}{\sqrt{x} e^{\sqrt{x}}} = \lim_{b \to \infty} \int_1^b \frac{dx}{\sqrt{x} e^{\sqrt{x}}} = \lim_{b \to \infty} \int_1^{\sqrt{b}} \frac{2}{e^u} du = \lim_{b \to \infty} \left(-\frac{2}{e^u}\right)\bigg|_1^{\sqrt{b}} = \lim_{b \to \infty} \left(\frac{2}{e} - \frac{2}{e^{\sqrt{b}}}\right) = \frac{2}{e}$$

Because the integral converges, it follows from the Integral Test that the series converges.

10.8.27 Adding 3 to the inequality $\cos 5k \leq 1$, and then dividing by k^3, we obtain

$$\frac{3 + \cos 5k}{k^3} \leq \frac{4}{k^3}.$$

Because $\sum_{k=1}^{\infty} \frac{1}{k^3}$ is a convergent p-series, $\sum_{k=1}^{\infty} \frac{4}{k^3}$ converges. So by the Comparison Test, $\sum_{k=1}^{\infty} \frac{3 + \cos 5k}{k^3}$ converges.

10.8.29 We first examine the series $\sum_{k=1}^{\infty} \frac{10^k}{k^{10}}$ using the Ratio Test.

$$r = \lim_{k \to \infty} \left| \frac{a_{k+1}}{a_k} \right| = \lim_{k \to \infty} \frac{10^{k+1}/(k+1)^{10}}{10^k/k^{10}} = \lim_{k \to \infty} \frac{10^{k+1}}{(k+1)^{10}} \cdot \frac{k^{10}}{10^k} = 10 \lim_{k \to \infty} \left(\frac{k}{k+1}\right)^{10} = 10.$$

Because $r > 1$, the series $\sum_{k=1}^{\infty} \frac{10^k}{k^{10}}$ diverges by the Ratio Test. Because $\frac{10^k + 1}{k^{10}} > \frac{10^k}{k^{10}}$ and $\sum_{k=1}^{\infty} \frac{10^k}{k^{10}}$ diverges, the series $\sum_{k=1}^{\infty} \frac{10^k + 1}{k^{10}}$ diverges by the Comparison Test.

10.8.31 Note that $\frac{1}{j^2+4} < \frac{1}{j^2}$. Because $\sum_{j=1}^{\infty} \frac{1}{j^2}$ is a convergent p-series, it follows that $\sum_{j=1}^{\infty} \frac{1}{j^2+4}$ converges by the Comparison Test. Therefore $\sum_{j=1}^{\infty} \frac{5}{j^2+4}$ converges.

10.8.33 Using the comparison series $\sum_{k=3}^{\infty} \frac{1}{k}$, we calculate the limit L using L'Hôpital's Rule.

$$L = \lim_{k \to \infty} \frac{1/(k^{1/3} \ln k)}{1/k} = \lim_{k \to \infty} \frac{1}{k^{1/3} \ln k} \cdot \frac{k}{1} = \lim_{k \to \infty} \frac{k^{2/3}}{\ln k} = \lim_{k \to \infty} \frac{\frac{2}{3} k^{-1/3}}{\frac{1}{k}} = \lim_{k \to \infty} \frac{2}{3} k^{2/3} = \infty$$

The series $\sum_{k=3}^{\infty} \frac{1}{k}$ diverges because it is the tail of the divergent harmonic series. Because $L = \infty$, it follows from case (3) of the Limit Comparison Test that $\sum_{k=3}^{\infty} \frac{1}{k^{1/3} \ln k}$ diverges.

10.8.35 We apply the Root Test.

$$\rho = \lim_{k \to \infty} \left| \frac{2^k 3^k}{k^k} \right|^{1/k} = \lim_{k \to \infty} \frac{6}{k} = 0$$

Because $\rho < 1$, the series $\sum_{k=1}^{\infty} \frac{2^k 3^k}{k^k}$ converges by the Root Test.

10.8.37 We apply the Root Test.

$$\rho = \lim_{k \to \infty} \left| (-1)^k \left(\frac{5k}{3k+7} \right)^k \right|^{1/k} = \lim_{k \to \infty} \frac{5k}{3k+7} = \frac{5}{3}$$

Because $\rho > 1$, the series $\sum_{k=1}^{\infty} \left(\frac{5k}{3k+7} \right)^k$ diverges by the Root Test.

10.8.39 Applying the Ratio Test, we have

$$r = \lim_{k \to \infty} \left| \frac{\frac{5^{k+1}((k+1)!)^2}{(2k+2)!}}{\frac{5^k (k!)^2}{(2k)!}} \right| = \lim_{k \to \infty} \frac{5^k 5 (k+1)!(k+1)!}{(2k+2)(2k+1)(2k)!} \cdot \frac{(2k)!}{5^k (k!)(k!)} = \lim_{k \to \infty} \frac{5(k+1)}{2(2k+1)} = \frac{5}{4}.$$

Because $r > 1$, the series $\sum_{k=1}^{\infty} \frac{5^k (k!)^2}{(2k)!}$ diverges by the Ratio Test.

10.8.41 We apply the Ratio Test.

$$r = \lim_{k \to \infty} \left| \frac{2^{k+1}/(3^{k+1} - 2^{k+1})}{2^k/(3^k - 2^k)} \right| = \lim_{k \to \infty} \frac{2(3^k - 2^k)}{3^{k+1} - 2^{k+1}} = \lim_{k \to \infty} \frac{2(3^k/3^k - 2^k/3^k)}{3^{k+1}/3^k - 2^{k+1}/3^k} = \lim_{k \to \infty} \frac{2(1 - (2/3)^k)}{3 - 2(2/3)^k} = \frac{2}{3}$$

Because $r < 1$, it follows that $\sum_{k=1}^{\infty} \frac{2^k}{3^k - 2^k}$ converges.

10.8.43 The values of $\cos \frac{(3k-1)\pi}{3}$ oscillate between $-1/2$ and $1/2$, which implies that $\lim_{k \to \infty} \cos \frac{(3k-1)\pi}{3} \neq 0$. So by the Divergence Test, the series $\sum_{k=1}^{\infty} \cos \frac{(3k-1)\pi}{3}$ diverges.

10.8.45 We will use the Integral Test, and we first show that the terms of the series $\sum_{k=1}^{\infty} \frac{k^4}{e^{k^5}}$ are decreasing by demonstrating that the positive and continuous function $f(x) = \frac{x^4}{e^{x^5}}$ is decreasing. In this case,

$$f'(x) = \frac{e^{x^5} 4x^3 - x^4 e^{x^5} 5x^4}{e^{2x^5}} = \frac{x^3 e^{x^5}(4 - 5x^5)}{e^{2x^5}} = \frac{x^3(4 - 5x^5)}{e^{x^5}}.$$

Therefore $f'(x) < 0$ for $x \geq 1$ which implies that the terms of the series $\sum_{k=1}^{\infty} \frac{k^4}{e^{k^5}}$ are decreasing. Next, we integrate with the help of the substitution $u = x^5$, which implies that $du = 5x^4 dx$.

$$\int_1^{\infty} \frac{x^4}{e^{x^5}} dx = \lim_{b \to \infty} \int_1^b \frac{x^4}{e^{x^5}} dx$$

$$= \lim_{b \to \infty} \frac{1}{5} \int_1^{b^5} \frac{1}{e^u} dx$$

$$= \lim_{b \to \infty} \frac{1}{5} \left(-\frac{1}{e^u} \right) \Big|_1^{b^5}$$

$$= \lim_{b \to \infty} \frac{1}{5} \left(\frac{1}{e} - \frac{1}{e^{b^5}} \right) = \frac{1}{5e}$$

Because the integral converges, it follows by the Integral Test that the series $\sum_{k=1}^{\infty} \frac{k^4}{e^{k^5}}$ converges.

10.8.47 We apply the Ratio Test.

$$r = \lim_{k \to \infty} \left| \frac{\frac{(4k+4)!}{((k+1)!)^4}}{\frac{(4k)!}{(k!)^4}} \right|$$

$$= \lim_{k \to \infty} \frac{(4k+4)!}{((k+1)!)^4} \cdot \frac{(k!)^4}{(4k)!}$$

$$= \lim_{k \to \infty} \frac{(4k+4)(4k+3)(4k+2)(4k+1)(4k)!}{((k+1)k!)^4} \cdot \frac{(k!)^4}{(4k)!}$$

$$= \lim_{k \to \infty} \frac{(4k+4)(4k+3)(4k+2)(4k+1)}{(k+1)^4} = 4^4$$

Because $r > 1$, it follows from the Ratio Test that the series $\sum_{k=1}^{\infty} \frac{(4k)!}{(k!)^4}$ diverges.

10.8.49 We apply the Comparison Test using the comparison series $\sum_{k=1}^{\infty} \frac{1}{k^{6/5}}$. Note that

$$\frac{\sqrt[5]{k}}{\sqrt[5]{k^7 + 1}} < \frac{\sqrt[5]{k}}{\sqrt[5]{k^7}} = \frac{1}{k^{6/5}}.$$

Because $\sum_{k=1}^{\infty} \frac{1}{k^{6/5}}$ is a convergent p-series, $\sum_{k=1}^{\infty} \frac{\sqrt[5]{k}}{\sqrt[5]{k^7 + 1}}$ converges by the Comparison Test.

10.8.51 The limit of the kth-term of the series is

$$\lim_{k\to\infty} \frac{7^k + 11^k}{11^k} = \lim_{k\to\infty}\left(\frac{7^k}{11^k} + \frac{11^k}{11^k}\right) = \lim_{k\to\infty}\left(\left(\frac{7}{11}\right)^k + 1\right) = 1$$

Because $\displaystyle\lim_{k\to\infty} \frac{7^k + 11^k}{11^k} \neq 0$, it follows that $\displaystyle\sum_{k=1}^{\infty} \frac{7^k + 11^k}{11^k}$ diverges by the Divergence Test.

10.8.53 We apply the Limit Comparison Test using the convergent p-series $\displaystyle\sum_{k=1}^{\infty} \frac{1}{k^9}$ as the comparison series. Observe that

$$L = \lim_{k\to\infty} \frac{\sin\frac{1}{k^9}}{\frac{1}{k^9}} = 1.$$

Because $0 < L < \infty$, we conclude that the series $\displaystyle\sum_{k=1}^{\infty} \sin\frac{1}{k^9}$ converges by the Limit Comparison Test.

10.8.55 Because $\displaystyle\lim_{k\to\infty} \cos\frac{1}{k^9} = 1 \neq 0$, $\displaystyle\sum_{k=1}^{\infty} \cos\frac{1}{k^9}$ diverges by the Divergence Test.

10.8.57 We note that

$$\sum_{k=1}^{\infty} 5^{1-2k} = \sum_{k=1}^{\infty} 5 \cdot 5^{-2k} = \sum_{k=1}^{\infty} \frac{5}{25^k}.$$

So $\displaystyle\sum_{k=1}^{\infty} 5^{1-2k}$ is a geometric series with $r = \frac{1}{25}$. Because $|r| < 1$, the series converges.

10.8.59 We first establish that $\displaystyle\sum_{k=1}^{\infty} \frac{k!}{k^k}$ is a convergent series using the Ratio Test. Taking the limit of successive terms of $\displaystyle\sum_{k=1}^{\infty} \frac{k!}{k^k}$, we have

$$r = \lim_{k\to\infty} \left|\frac{\frac{(k+1)!}{(k+1)^{k+1}}}{\frac{k!}{k^k}}\right| = \lim_{k\to\infty} \frac{(k+1)k!}{(k+1)^k(k+1)} \cdot \frac{k^k}{k!} = \lim_{k\to\infty} \frac{k^k}{(k+1)^k} = \lim_{k\to\infty} \frac{1}{\left(1+\frac{1}{k}\right)^k} = \frac{1}{e},$$

where the last equality follows from the fact that $\displaystyle\lim_{k\to\infty}(1+a/k)^k = e^a$ for any real number a (see the solution to Exercise 18). Because $r < 1$, the series $\displaystyle\sum_{k=1}^{\infty} \frac{k!}{k^k}$ converges by the Ratio Test. It is also the case that $\dfrac{k!}{k^k + 3} < \dfrac{k!}{k^k}$ which implies that $\displaystyle\sum_{k=1}^{\infty} \frac{k!}{k^k + 3}$ converges by the Comparison Test.

10.8.61 We apply the Limit Comparison Test by using the divergent harmonic series $\displaystyle\sum_{k=1}^{\infty} \frac{1}{k}$ as our comparison series. With the help of l'Hôpital's Rule, we calculate L.

$$L = \lim_{k\to\infty} \frac{\frac{1}{\ln(e^k+1)}}{\frac{1}{k}} = \lim_{k\to\infty} \frac{k}{\ln(e^k+1)} = \lim_{k\to\infty} \frac{1}{\frac{e^k}{e^k+1}} = \lim_{k\to\infty}\left(1 + \frac{1}{e^k}\right) = 1$$

Because $0 < L < \infty$, $\displaystyle\sum_{k=1}^{\infty} \frac{1}{\ln(e^k+1)}$ diverges by the Limit Comparison Test.

10.8. Choosing a Convergence Test

10.8.63 In the solution to Exercise 18, we showed that $\lim_{k\to\infty}(1+a/k)^k = e^a$. We use this result in the following limit calculation. Applying the Root Test, we have

$$\rho = \lim_{k\to\infty} \sqrt[k]{\left|\left(\frac{k+a}{k}\right)^{k^2}\right|} = \lim_{k\to\infty}\left(1+\frac{a}{k}\right)^k = e^a.$$

Because $a > 0$, $\rho = e^a > 1$. So the series diverges by the Root Test.

10.8.65 The nth-partial sum of $\sum_{k=1}^{\infty}\left(\cos\frac{1}{k} - \cos\frac{1}{k+1}\right)$ is

$$S_n = \left(\cos 1 - \cos\frac{1}{2}\right) + \left(\cos\frac{1}{2} - \cos\frac{1}{3}\right) + \cdots + \left(\cos\frac{1}{n} - \cos\frac{1}{n+1}\right) = \cos 1 - \cos\frac{1}{n+1}.$$

It follows that $\lim_{n\to\infty} S_n = \cos 1 - \cos 0 = \cos 1 - 1$. Because the sequence of partial sums converges, the series $\sum_{k=1}^{\infty}\left(\cos\frac{1}{k} - \cos\frac{1}{k+1}\right)$ converges.

10.8.67 Apply the Limit Comparison Test using the convergent geometric series $\sum_{j=1}^{\infty}\frac{1}{2^j}$ as the comparison series.

$$L = \lim_{j\to\infty} \frac{\frac{\cot^{-1}\frac{1}{j}}{2^j}}{\frac{1}{2^j}} = \lim_{j\to\infty} \cot^{-1}\frac{1}{j} = \frac{\pi}{2}$$

Because $0 < L < \infty$, the series $\sum_{j=1}^{\infty}\frac{\cot^{-1}\frac{1}{j}}{2^j}$ converges by the Limit Comparison Test.

10.8.69 Note that

$$\lim_{k\to\infty}\left(1+\frac{1}{2k}\right)^k = e^{1/2}.$$

Because the limit does not equal 0, the series $\sum_{k=1}^{\infty}\left(1+\frac{1}{2k}\right)^k$ diverges by the Divergence Test.

10.8.71 We apply the Limit Comparison Test with the comparison series $\sum_{k=1}^{\infty}\frac{1}{k^{5/4}}$.

$$L = \lim_{k\to\infty}\frac{\ln^2 k/k^{3/2}}{1/k^{5/4}} = \lim_{k\to\infty}\frac{\ln^2 k}{k^{1/4}} = \lim_{k\to\infty}\frac{2(\ln k)/k}{\frac{1}{4}k^{-3/4}} = 8\lim_{k\to\infty}\frac{\ln k/k}{k^{-3/4}} = 8\lim_{k\to\infty}\frac{\ln k}{k^{1/4}} = 32\lim_{k\to\infty}\frac{1}{k^{1/4}} = 0.$$

Because the series $\sum_{k=1}^{\infty}\frac{1}{k^{5/4}}$ converges, it follows from case (2) of the Limit Comparison Test that $\sum_{k=1}^{\infty}\frac{\ln^2 k}{k^{3/2}}$ converges.

10.8.73 Use the Ratio Test.

$$r = \lim_{k\to\infty}\left|\frac{(k+1)^2 \cdot 1.0001^{-k-1}}{k^2 \cdot 1.0001^{-k}}\right| = \frac{1}{1.001}$$

Because $r < 1$, the series $\sum_{k=0}^{\infty} k^2 \cdot 1.001^{-k}$ converges by the Ratio Test.

10.8.75 The terms of the series decrease in magnitude, for $k \geq 1$. Furthermore $\lim_{k\to\infty} \frac{1}{k^k} = 0$. Therefore, the series converges by the Alternating Series Test.

10.8.77 The series diverges by the Divergence Test because

$$\lim_{k\to\infty} \frac{3k}{\sqrt[4]{k^4+3}} = \lim_{k\to\infty} \frac{3k/k}{(\sqrt[4]{k^4+3})/k} = \lim_{k\to\infty} \frac{3}{\sqrt[4]{k^4/k^4 + 3/k^4}} = \lim_{k\to\infty} \frac{3}{\sqrt[4]{1+3/k^4}} \neq 0.$$

10.8.79 Apply the Limit Comparison Test using the divergent p-series $\sum_{k=1}^{\infty} \frac{1}{\sqrt{k}}$ as the comparison series.

$$L = \lim_{k\to\infty} \frac{\tan^{-1}\frac{1}{\sqrt{k}}}{\frac{1}{\sqrt{k}}} = \lim_{k\to\infty} \frac{\frac{1}{1+(1/\sqrt{k})^2}\left(-\frac{1}{2}k^{-3/2}\right)}{-\frac{1}{2}k^{-3/2}} = \lim_{k\to\infty} \frac{1}{1+(1/\sqrt{k})^2} = 1$$

Because $0 < L < \infty$, $\sum_{k=1}^{\infty} \tan^{-1}\frac{1}{\sqrt{k}}$ diverges by the Limit Comparison Test.

10.8.81 The nth-partial sum of the series $\sum_{k=1}^{\infty} \left(\frac{1}{\sqrt{k+2}} - \frac{1}{\sqrt{k}}\right)$ is

$$S_n = \left(\frac{1}{\sqrt{3}} - 1\right) + \left(\frac{1}{2} - \frac{1}{\sqrt{2}}\right) + \left(\frac{1}{\sqrt{5}} - \frac{1}{\sqrt{3}}\right) + \left(\frac{1}{\sqrt{6}} - \frac{1}{2}\right) + \cdots + \left(\frac{1}{\sqrt{n+1}} - \frac{1}{\sqrt{n-1}}\right) + \left(\frac{1}{\sqrt{n+2}} - \frac{1}{\sqrt{n}}\right).$$

For $n \geq 4$, S_n simplifies to $S_n = -1 - \frac{1}{\sqrt{2}} + \frac{1}{\sqrt{n+1}} + \frac{1}{\sqrt{n+2}}$. So the limit of the sequence of partial sums is $\lim_{n\to\infty} S_n = -1 - \frac{1}{\sqrt{2}}$. Therefore the series $\sum_{k=1}^{\infty} \left(\frac{1}{\sqrt{k+2}} - \frac{1}{\sqrt{k}}\right)$ converges (to this limit).

10.8.83 We apply the Integral Test using the function $f(x) = \frac{1}{x \ln^{10} x}$. This function is continuous, positive, and decreasing for $x \geq 2$. Integrating and using the substitution $u = \ln x$, we have

$$\int_2^{\infty} \frac{dx}{x \ln^{10} x} = \lim_{b\to\infty} \int_2^b \frac{dx}{x \ln^{10} x}$$
$$= \lim_{b\to\infty} \int_{\ln 2}^{\ln b} \frac{du}{u^{10}}$$
$$= \lim_{b\to\infty} \left(-\frac{1}{9u^9}\right)\bigg|_{\ln 2}^{\ln b}$$
$$= \lim_{b\to\infty} \left(-\frac{1}{9\ln^9 b} + \frac{1}{9\ln^9 2}\right) = \frac{1}{9\ln^9 2}$$

The integral converges which implies that the series $\sum_{j=2}^{\infty} \frac{1}{j \ln^{10} j}$ converges by the Integral Test.

10.8.85 The series $\frac{1}{2\cdot 3} + \frac{1}{4\cdot 5} + \frac{1}{6\cdot 7} + \frac{1}{8\cdot 9} + \cdots$ is expressed in summation notation as $\sum_{k=1}^{\infty} \frac{1}{2k(2k+1)}$.

We apply the Limit Comparison Test using the convergent p-series $\sum_{k=1}^{\infty} \frac{1}{k^2}$.

$$L = \lim_{k\to\infty} \frac{1/(2k(2k+1))}{1/k^2} = \lim_{k\to\infty} \frac{k^2}{4k^2 + 2k} = \frac{1}{4}$$

10.8. Choosing a Convergence Test

So we have $0 < L < \infty$ and it follows that $\sum_{k=1}^{\infty} \dfrac{1}{2k(2k+1)}$ converges by the Limit Comparison Test.

10.8.87

a. False. For example, the Limit Comparison Series can be used to show that $\sum_{k=2}^{\infty} \dfrac{1}{k^2-1}$ converges using the comparison series $\sum_{k=2}^{\infty} \dfrac{1}{k^2}$, but this comparison series will not work with the Comparison Test.

b. True.

c. True.

d. False. The Alternating Series Test can only be used to show that a series converges.

10.8.89 Using the sum $1 + 2 + \cdots + k = k(k+1)/2$ from Chapter 5, we have

$$\sum_{k=1}^{\infty} \frac{k}{1+2+\cdots+k} = \sum_{k=1}^{\infty} \frac{2k}{k(k+1)} = \sum_{k=1}^{\infty} \frac{2}{k+1}.$$

We apply the Limit Comparison Test using the divergent harmonic series $\sum_{k=1}^{\infty} \dfrac{1}{k}$ as the comparison series.

$$L = \lim_{k \to \infty} \frac{2/(k+1)}{1/k} = \lim_{k \to \infty} \frac{2k}{k+1} = 2$$

Because $0 < L < \infty$, $\sum_{k=1}^{\infty} \dfrac{k}{1+2+\cdots+k}$ diverges by the Limit Comparison Test.

10.8.91 The series $\sum_{k=0}^{\infty} \ln\left(\dfrac{2k+1}{2k+4}\right)$ consists of negative terms because $(2k+1)/(2k+4) < 1$. Notice that $\sum_{k=0}^{\infty} \ln\left(\dfrac{2k+1}{2k+4}\right) = -\sum_{k=0}^{\infty} \ln\left(\dfrac{2k+4}{2k+1}\right)$ where $\sum_{k=0}^{\infty} \ln\left(\dfrac{2k+4}{2k+1}\right)$ consists of positive terms. So we show that the series $\sum_{k=0}^{\infty} \ln\left(\dfrac{2k+1}{2k+4}\right)$ diverges by showing that $\sum_{k=0}^{\infty} \ln\left(\dfrac{2k+4}{2k+1}\right)$ diverges. To do this, we apply the Limit Comparison test using the divergent comparison series $\sum_{k=0}^{\infty} \dfrac{1}{k}$.

$$L = \lim_{k \to \infty} \frac{\ln\left((2k+4)/(2k+1)\right)}{1/k} \qquad \text{Limit Comparison Test}$$

$$= \lim_{k \to \infty} \frac{\frac{2}{2k+4} - \frac{2}{2k+1}}{-1/k^2} \qquad \text{L'Hôpital's Rule}$$

$$= \lim_{k \to \infty} \frac{2(2k+1) - 2(2k+4)}{(2k+1)(2k+4)} \cdot \frac{-k^2}{1} \qquad \text{Common denominators.}$$

$$= \lim_{k \to \infty} \frac{6k^2}{4k^2 + 10k + 4} = \frac{3}{2} \qquad \text{Evaluate.}$$

Because $0 < L < \infty$, the series $\sum_{k=0}^{\infty} \ln\left(\dfrac{2k+4}{2k+1}\right)$ diverges and so $\sum_{k=0}^{\infty} \ln\left(\dfrac{2k+1}{2k+4}\right)$ diverges.

10.8.93 We apply the Limit Comparison Test with the comparison series $\sum_{k=1}^{\infty} \frac{1}{k}$:

$$\lim_{k \to \infty} \frac{1/(2^{\ln k} + 2)}{1/k} = \lim_{k \to \infty} \frac{e^{\ln k}}{2^{\ln k} + 2} = \lim_{k \to \infty} \frac{(e/2)^{\ln k}}{1 + 2^{(1-\ln k)}} = \frac{\lim_{k \to \infty} \left(\frac{e}{2}\right)^{\ln k}}{1 - 0} = \infty.$$

Because $L = \infty$ and $\sum_{k=1}^{\infty} \frac{1}{k}$ diverges (because it is the harmonic series), it follows by part 3 of the Limit Comparison Test that $\sum_{k=1}^{\infty} \frac{1}{2^{\ln k} + 2}$ diverges.

Chapter Ten Review

1

a. False. Let $a_n = 1 - \frac{1}{n}$. This sequence has limit 1.

b. False. The terms of a sequence tending to zero is necessary but not sufficient for convergence of the series.

c. True. This is the definition of convergence of a series.

d. False. If a series converges absolutely, the definition says that it does not converge conditionally.

e. True. It has limit 1 as $n \to \infty$.

f. False. The subsequence of the even terms has limit 1 and the subsequence of odd terms has limit -1, so the sequence does not have a limit.

g. False. It diverges by the Divergence Test because $\lim_{k \to \infty} \frac{k^2}{k^2 + 1} = 1 \neq 0$.

h. True. The given series converges by the Limit Comparison Test with the series $\sum_{k=1}^{\infty} \frac{1}{k^2}$, and thus its sequence of partial sums converges.

3 $\sum_{k=0}^{9} (0.2)^k = \frac{1 - (0.2)^{10}}{1 - 0.2} \approx 1.25$, and $\sum_{k=2}^{9} (0.2)^k = (0.2)^2 \frac{1 - (0.2)^8}{1 - 0.2} \approx 0.0499999$.

5 Because the series converges, we must have $\lim_{k \to \infty} a_k = 0$. Because it converges to 8, the partial sums converge to 8, so that $\lim_{k \to \infty} S_k = 8$.

7 Consider the constant sequence with $a_k = 1$ for all k. The sequence $\{a_k\}$ converges to 1, but the corresponding series $\sum a_k$ diverges by the divergence test.

9

a. For $|x| < 1$, $\lim_{k \to \infty} x^k = 0$, so this limit is zero.

b. This is a geometric series with ratio $-4/5$, so the sum is $\frac{1}{1 + 4/5} = \frac{5}{9}$.

11

a. This sequence converges because $\lim_{k\to\infty} \frac{k}{k+1} = \lim_{k\to\infty} \frac{1}{1+\frac{1}{k}} = \frac{1}{1+0} = 1$.

b. Because the sequence of terms has limit 1 (which means its limit isn't zero) this series diverges by the Divergence Test.

13 Because $\lim_{n\to\infty} \frac{3n^3+4n}{6n^3+5} = \frac{1}{2} \neq 0$, the sequence $\left\{(-1)^n \frac{3n^3+4n}{6n^3+5}\right\}$ oscillates and doesn't converge.

15 $\lim_{n\to\infty} \frac{2^n + 5^{n+1}}{5^n} = \lim_{n\to\infty} \left(\frac{2}{5}\right)^n + 5 = 0 + 5 = 5$.

17 $\lim_{n\to\infty} \frac{8^n}{n!} = 0$ because exponentials grow more slowly than factorials.

19 Because $-1 \leq \sin n \leq 1$, we have

$$\frac{-(3n^2+2n+1)}{4n^3+n} \leq \frac{(3n^2+2n+1)\sin n}{4n^3+n} \leq \frac{(3n^2+2n+1)}{4n^3+n}.$$

Because $\lim_{n\to\infty} \frac{-(3n^2+2n+1)}{4n^3+n} = 0 = \lim_{n\to\infty} \frac{(3n^2+2n+1)}{4n^3+n}$, we must have the the original sequence has limit 0.

21 Take logs, and then evaluate $\lim_{n\to\infty} \frac{1}{\ln n} \ln(1/n) = \lim_{n\to\infty} (-1) = -1$, so the original limit is e^{-1}.

23 $a_n = (-1/0.9)^n = (-10/9)^n$. The terms grow without bound so the sequence does not converge.

25

a. 80, 48, 32, 24, 20

b. Assume that $L = \lim_{n\to\infty} a_n$. Then $L = \frac{L}{2} + 8$, so $2L = L + 16$, so $L = 16$.

27 $\sum_{k=1}^{\infty} 3(1.001)^k = 3\sum_{k=1}^{\infty} (1.001)^k$. This is a geometric series with ratio greater than 1, so it diverges.

29 Because $\sum_{k=1}^{\infty} \left(\frac{1}{3}\right)^k$ converges, but $\sum_{k=0}^{\infty} \left(\frac{4}{3}\right)^k$ diverges, the given series diverges.

31 Because $\ln\left(\frac{2k+1}{2k-1}\right) = \ln(2k+1) - \ln(2k-1)$, this series telescopes. The formula for S_n is $S_n = (\ln 3 - \ln 1) + (\ln 5 - \ln 3) + \cdots + (\ln(2n+1) - \ln(2n-1)) = \ln(2n+1)$. The given series therefore diverges because $\lim_{n\to\infty} S_n = \infty$.

33 This series telescopes, and the formula for S_n is

$$S_n = (\tan^{-1} 2 - \tan^{-1} 0) + (\tan^{-1} 3 - \tan^{-1} 1) + (\tan^{-1} 4 - \tan^{-1} 2) + \cdots + (\tan^{-1}(n+2) - \tan^{-1} n)$$
$$= \tan^{-1}(n+2) + \tan^{-1}(n+1) - \tan^{-1} 1 - \tan^{-1} 0.$$

So

$$\lim_{n\to\infty} S_n = \frac{\pi}{2} + \frac{\pi}{2} - \frac{\pi}{4} = \frac{3\pi}{4}.$$

35 Note that $\dfrac{9}{(3k-2)(3k+1)} = \dfrac{3}{3k-2} - \dfrac{3}{3k+1}$. This series telescopes.

$$S_n = \left(3 - \dfrac{3}{4}\right) + \left(\dfrac{3}{4} - \dfrac{3}{7}\right) + \cdots + \left(\dfrac{3}{3n-2} - \dfrac{3}{3n+1}\right) = 3 - \dfrac{3}{3n+1},$$

so that $\lim_{n \to \infty} S_n = 3$, which is the value of the series.

37 $\displaystyle\sum_{k=1}^{\infty} \dfrac{2^k}{3^{k+2}} = \dfrac{1}{9} \sum_{k=1}^{\infty} \left(\dfrac{2}{3}\right)^k = \dfrac{1}{9} \cdot \dfrac{2/3}{1 - 2/3} = \dfrac{2}{9}.$

39 $0.314141414\ldots = 0.3 + (0.014 + 0.00014 + 0.00000014 + \cdots) = \dfrac{3}{10} + \dfrac{0.014}{1 - 0.01} = \dfrac{3}{10} + \dfrac{14}{990} = \dfrac{311}{990}.$

41 Note that $d_2 = 100 + \dfrac{1}{2}(100)$, $d_3 = 100 + \dfrac{1}{2}\left(100 + \dfrac{1}{2}(100)\right) = 100 + \dfrac{1}{2}(100) + \dfrac{1}{2^2}(100)$, and

$$d_n = 100 + \dfrac{1}{2}(100) + \cdots + \dfrac{1}{2^{n-1}}(100).$$

The steady state is $\lim_{n \to \infty} d_n$, which is the sum of the geometric series which is given by $\dfrac{100}{1 - (1/2)} = 200$ mg. So the steady-state amount of medication in the blood is 200 mg.

43 The series can be written $\displaystyle\sum_{k=1}^{\infty} \dfrac{1}{k^{2/3}}$, which is a p-series with $p = 2/3 < 1$, so this series diverges.

45 The kth term does not have limit 0, so the given series diverges by the Divergence Test. Note that

$$\lim_{k \to \infty} \dfrac{2k^4 + k^3 + 1}{3k^4 + 4} = \dfrac{2}{3} \neq 0.$$

47 We have $\dfrac{7 + \sin k}{k^2} \leq \dfrac{8}{k^2}$ and $\displaystyle\sum_{k=1}^{\infty} \dfrac{8}{k^2}$ converges (it is a p-series with $p > 1$), so by the Comparison Test, the given series converges.

49 Note that $\sqrt{9k^{12} + 2} \geq \sqrt{9k^{12}} \geq \sqrt{k^{12}} = k^6$ for $k \geq 1$. It follows that

$$\dfrac{k^4}{\sqrt{9k^{12} + 2}} \leq \dfrac{k^4}{k^6} = \dfrac{1}{k^2}.$$

Because $\displaystyle\sum_{k=1}^{\infty} \dfrac{1}{k^2}$ converges, the given series converges by the Comparison Test.

51 This is a geometric series with ratio $2/e < 1$, so the given series converges.

53 We use the Root Test. The kth root of the absolute value of the kth term is $\dfrac{k}{2k+3}$ which has limit $\dfrac{1}{2}$ as $k \to \infty$, so the given series converges.

55 Applying the Ratio Test:

$$\lim_{k \to \infty} \dfrac{a_{k+1}}{a_k} = \lim_{k \to \infty} \dfrac{\frac{(k+1)!}{e^{k+1}(k+1)^{k+1}}}{\frac{k!}{e^k k^k}} = \lim_{k \to \infty} \dfrac{(k+1)k^k}{e(k+1)^{k+1}} = \lim_{k \to \infty} \dfrac{1}{e} \cdot \left(\dfrac{k}{k+1}\right)^k = \dfrac{1}{e^2},$$

so the given series converges.

57 The series can be written as $2\sum_{k=1}^{\infty} \frac{5^k}{4^k}$, and because $\sum_{k=1}^{\infty} \left(\frac{5}{4}\right)^k$ is a divergent geometric series, the given series is divergent.

59 Use the Ratio Test. The ratio of successive terms is $\frac{2 \cdot 4^{j+1}}{(2j+3)!} \cdot \frac{(2j+1)!}{2 \cdot 4^j} = \frac{4}{(2j+3)(2j+2)}$. This has limit 0 as $j \to \infty$, so the given series converges.

61 We use the Integral Test. Let $f(x) = \frac{\ln x}{x^{3/2}}$. Then $f'(x) = \frac{2 - 3\ln x}{2x^{5/2}}$, which is less than zero for $x \geq 2$. So f is decreasing, positive and continuous for $x \geq 3$. Using Integration by Parts, it can be shown that

$$\int f(x)\,dx = -\frac{4}{\sqrt{x}} - \frac{2\ln x}{\sqrt{x}},$$

so $\lim_{b \to \infty} \int_3^b f(x)\,dx = \frac{4}{\sqrt{3}} + \frac{2\ln 3}{\sqrt{3}}$. Because this integral converges, the given series converges by the Integral Test.

63 Use the Comparison Test: $\frac{3}{2 + e^k} < \frac{3}{e^k}$, but $\sum_{k=1}^{\infty} \frac{3}{e^k}$ converges because it is a geometric series with ratio $\frac{1}{e} < 1$. Thus the original series converges as well.

65 $a_k = \frac{k^{1/k}}{k^3} = \frac{1}{k^{3-1/k}}$. For $k \geq 2$, then, $a_k < \frac{1}{k^2}$. Because $\sum_{k=1}^{\infty} \frac{1}{k^2}$ converges, the given series also converges, by the Comparison Test.

67 Use the Ratio Test: $\frac{a_{k+1}}{a_k} = \frac{(k+1)^5}{e^{k+1}} \cdot \frac{e^k}{k^5} = \frac{1}{e} \cdot \left(\frac{k+1}{k}\right)^5$, which has limit $1/e < 1$ as $k \to \infty$. Thus the given series converges.

69 We use the Limit Comparison Test with $\sum_{k=1}^{\infty} \frac{1}{k^2}$. We have

$$L = \lim_{k \to \infty} \frac{1 - \cos \frac{1}{k}}{\frac{1}{k^2}} = \lim_{k \to \infty} \frac{\left(\sin \frac{1}{k}\right) \frac{1}{k^2}}{\frac{2}{k^3}} = \frac{1}{2} \lim_{k \to \infty} \frac{\sin \frac{1}{k}}{\frac{1}{k}} = \frac{1}{2}.$$

Because $0 < L < \infty$, and because $\sum_{k=1}^{\infty} \frac{1}{k^2}$ converges, we have that the given series converges.

71 We use the Limit Comparison Test with the convergent p-series $\sum_{k=1}^{\infty} \frac{1}{k^2}$. We have

$$L = \lim_{k \to \infty} \frac{\left(1 - \cos \frac{1}{k}\right)^2}{\frac{1}{k^2}} = \lim_{k \to \infty} \frac{2\left(1 - \cos \frac{1}{k}\right)\left(\sin \frac{1}{k}\right)\left(-\frac{1}{k^2}\right)}{-\frac{2}{k^3}}$$

$$= \lim_{k \to \infty} k\left(1 - \cos \frac{1}{k}\right)\left(\sin \frac{1}{k}\right) = \lim_{k \to \infty} \left(1 - \cos \frac{1}{k}\right) \lim_{k \to \infty} \frac{\sin(1/k)}{(1/k)} = 0 \cdot 1 = 0.$$

Because $L = 0$ and the comparison series converges, by the Limit Comparison Test, the given series converges.

73 Use the Limit Comparison Test with the harmonic series. Note that $\lim_{k\to\infty} \frac{\coth k}{k} \cdot \frac{k}{1} = \lim_{k\to\infty} \coth k = 1$. Because the harmonic series diverges, the given series does as well.

75 Use the Divergence Test. $\lim_{k\to\infty} \tanh k = \lim_{k\to\infty} \frac{e^k + e^{-k}}{e^k - e^{-k}} = 1 \neq 0$, so the given series diverges.

77 The corresponding series of all positive terms is the divergent p-series $\sum_{k=1}^{\infty} \frac{1}{k^{3/7}}$. However, the sequence $\left\{\frac{1}{k^{3/7}}\right\}$ is nonincreasing and has limit 0, so the given alternating series is convergent by the Alternating Series Test. Therefore, the given series is conditionally convergent.

79 $|a_k| = \frac{1}{k^2 - 1}$. Use the Limit Comparison Test with the convergent series $\sum_{k=1}^{\infty} \frac{1}{k^2}$. Because

$$\lim_{k\to\infty} \frac{\frac{1}{k^2-1}}{\frac{1}{k^2}} = \lim_{k\to\infty} \frac{k^2}{k^2 - 1} = 1,$$

the given series converges absolutely.

81 The sequence $\left\{\frac{k^2 + 4}{2k^2 + 1}\right\}$ has limit $\frac{1}{2}$, which implies that the sequence $\left\{(-1)^{k+1} \frac{k^2+4}{2k^2+1}\right\}$ oscillates and diverges. The given series therefore diverges, because by the Divergence Test, the kth term does not have limit 0.

83 Use the Root Test on the absolute values of the sequence of terms:

$$\lim_{k\to\infty} |\sqrt[k]{a_k}| = \lim_{k\to\infty} \sqrt[k]{ke^{-k}} = \lim_{k\to\infty} \frac{k^{1/k}}{e} = \frac{1}{e}.$$

Thus, the original series is absolutely convergent by the Root Test.

85 Use the Ratio Test on the absolute values of the sequence of terms: $\lim_{k\to\infty} \left|\frac{a_{k+1}}{a_k}\right| = \lim_{k\to\infty} \frac{10}{k+1} = 0$, so the series converges absolutely.

87 Because $k^2 \ll 2^k$, $\lim_{k\to\infty} \frac{-2 \cdot (-2)^k}{k^2} \neq 0$. The given series thus diverges by the Divergence Test. Alternatively, this result can be derived from either the Ratio Test or the Root Test.

89

a. $\sum_{k=1}^{5} \frac{1}{k^5} \approx 1.03666$.

b. $R_5 < \int_{5}^{\infty} \frac{dx}{x^5} = \lim_{b\to\infty} \int_{5}^{b} \frac{dx}{x^5} = \lim_{b\to\infty} -\frac{1}{4}\left(\frac{1}{b^4} - \frac{1}{5^4}\right) = \frac{1}{4 \cdot 5^4} = 0.0004$.

c. Note that $\int_{6}^{\infty} \frac{dx}{x^5} = \frac{1}{4 \cdot 6^4} \approx 0.00019$. So $L_5 = S_5 + \int_{6}^{\infty} \frac{dx}{x^5} \approx 1.03685$, and $U_5 = S_5 + \int_{5}^{\infty} \frac{dx}{x^5} \approx 1.03706$.

91 We have $R_n = \int_{n}^{\infty} \frac{dx}{(2x+5)^3} = \frac{1}{4(2n+5)^2}$. We want $\frac{1}{4(2n+5)^2} < \frac{1}{10^4}$, which leads to $n \geq 23$. Therefore $\sum_{k=1}^{\infty} \frac{1}{(2k+5)^3} \approx \sum_{k=1}^{23} \frac{1}{(2k+5)^3} \approx 0.0067$.

93 The maximum error is a_{n+1}, so we want $a_{n+1} = \dfrac{1}{(k+1)^4} < 10^{-8}$, or $(k+1)^4 > 10^8$, so $k = 100$.

95

a. Let T_n be the amount of additional tunnel dug during week n. Then $T_0 = 100$ and $T_n = .95 \cdot T_{n-1} = (.95)^n T_0 = 100(0.95)^n$, so the total distance dug in N weeks is

$$S_N = 100 \sum_{k=0}^{N-1} (0.95)^k = 100 \left(\frac{1 - (0.95)^N}{1 - 0.95} \right) = 2000(1 - 0.95^N).$$

Then $S_{10} \approx 802.5$ meters and $S_{20} \approx 1283.03$ meters.

b. The longest possible tunnel is $S_\infty = 100 \sum_{k=0}^{\infty} (0.95)^k = \dfrac{100}{1 - 0.95} = 2000$ meters.

97

a. The area of a circle of radius r is πr^2. For $r = 2^{1-n}$, this is $2^{2-2n}\pi$. There are 2^{n-1} circles on the nth page, so the total area of circles on the nth page is $2^{n-1} \cdot \pi 2^{2-2n} = 2^{1-n}\pi$.

b. The sum of the areas on all pages is $\displaystyle\sum_{k=1}^{\infty} 2^{1-k}\pi = 2\pi \sum_{k=1}^{\infty} 2^{-k} = 2\pi \cdot \dfrac{1/2}{1/2} = 2\pi$.

99

a. $T_1 = \dfrac{\sqrt{3}}{16}$ and $T_2 = \dfrac{7\sqrt{3}}{64}$.

b. At stage n, 3^{n-1} triangles of side length $1/2^n$ are removed. Each of those triangles has an area of $\dfrac{\sqrt{3}}{4 \cdot 4^n} = \dfrac{\sqrt{3}}{4^{n+1}}$, so a total of

$$3^{n-1} \cdot \frac{\sqrt{3}}{4^{n+1}} = \frac{\sqrt{3}}{16} \cdot \left(\frac{3}{4}\right)^{n-1}$$

is removed at each stage. Thus

$$T_n = \frac{\sqrt{3}}{16} \sum_{k=1}^{n} \left(\frac{3}{4}\right)^{k-1} = \frac{\sqrt{3}}{16} \sum_{k=0}^{n-1} \left(\frac{3}{4}\right)^k = \frac{\sqrt{3}}{4}\left(1 - \left(\frac{3}{4}\right)^n\right).$$

c. $\displaystyle\lim_{n\to\infty} T_n = \frac{\sqrt{3}}{4}$ because $\left(\dfrac{3}{4}\right)^n \to 0$ as $n \to \infty$.

d. The area of the triangle was originally $\dfrac{\sqrt{3}}{4}$, so none of the original area is left.

101 Ignoring the initial drop for the moment, the height after the nth bounce is $10p^n$, so the total time spent in that bounce is $2 \cdot \sqrt{2 \cdot 10p^n/g}$ seconds. The total time before the ball comes to rest (now including the time for the initial drop) is then $\sqrt{20/g} + \displaystyle\sum_{i=1}^{\infty} 2 \cdot \sqrt{2 \cdot 10p^n/g} = \sqrt{\dfrac{20}{g}} + 2\sqrt{\dfrac{20}{g}} \sum_{i=1}^{\infty} (\sqrt{p})^n = \sqrt{\dfrac{20}{g}} + 2\sqrt{\dfrac{20}{g}} \dfrac{\sqrt{p}}{1 - \sqrt{p}} = \sqrt{\dfrac{20}{g}}\left(1 + \dfrac{2\sqrt{p}}{1 - \sqrt{p}}\right) = \sqrt{\dfrac{20}{g}}\left(\dfrac{1 + \sqrt{p}}{1 - \sqrt{p}}\right)$ seconds.

Chapter 11

Power Series

11.1 Approximating Functions With Polynomials

11.1.1 Let the polynomial be $p(x)$. Then $p(0) = f(0)$, $p'(0) = f'(0)$, and $p''(0) = f''(0)$.

11.1.3 The approximations are $p_0(0.1) = 1$, $p_1(0.1) = 1 + \dfrac{0.1}{2} = 1.05$, and $p_2(0.1) = 1 + \dfrac{0.1}{2} - \dfrac{0.01}{8} = 1.04875$.

11.1.5 The third-order Taylor polynomial is $p_3(x) = f(0) + f'(0) + \dfrac{f''(0)}{2!}x^2 + \dfrac{f'''(0)}{3!}x^3 = 1 + x^2 + x^3$. $f(0.2) \approx 1 + (0.2)^2 + (0.3)^3 = 1.048$.

11.1.7 The third-order Taylor polynomial is

$$p_3(x) = f(2) + f'(2)(x-2) + \dfrac{f''(2)}{2!}(x-2)^2 + \dfrac{f'''(2)}{3!}(x-2)^3 = 1 + (x-2) + 2(x-2)^3$$

, so $f(1.9) \approx 1 - 0.1 + 2(-0.1)^3 = 0.898$.

11.1.9

a. Note that $f(1) = 8$, and $f'(x) = 12\sqrt{x}$, so $f'(1) = 12$. Thus, $p_1(x) = 8 + 12(x-1)$.

b. $f''(x) = 6/\sqrt{x}$, so $f''(1) = 6$. Thus $p_2(x) = 8 + 12(x-2) + 3(x-1)^2$.

c. $p_1(1.1) = 12(.1) + 8 = 9.2$. $p_2(1.1) = 3(.1)^2 + 12(.1) + 8 = 9.23$.

11.1.11

a. $f'(x) = -2e^{-2x}$, so $p_1(x) = f(0) + f'(0)(x) = 1 - 2x$.

b. $f''(x) = 4e^{-2x}$, so $p_2(x) = f(0) + f'(0)(x) + \dfrac{1}{2}f''(0)(x^2) = 1 - 2x + 2x^2$.

c. $p_1(0.2) = 0.6$, and $p_2(0.2) = 1 - 2(0.2) + 2(0.04) = 0.68$.

11.1.13

a. $f'(x) = -\dfrac{1}{(x+1)^2}$, so $p_1(x) = f(0) + f'(0)(x) = 1 - x$.

b. $f''(x) = \dfrac{2}{(x+1)^3}$, so $p_2(x) = f(0) + f'(0)(x) + \dfrac{1}{2}f''(0)x^2 = 1 - x + x^2$.

c. $p_1(0.05) = 0.95$, and $p_2(0.05) = 1 - 0.05 + 0.0025 = 0.9525$.

11.1.15

a. $f'(x) = (1/3)x^{-2/3}$, so $p_1(x) = f(8) + f'(8)(x-8) = 2 + \dfrac{1}{12}(x-8)$.

b. $f''(x) = (-2/9)x^{-5/3}$, so $p_2(x) = f(8) + f'(8)(x-8) + \dfrac{1}{2}f''(8)(x-8)^2 = 2 + \dfrac{1}{12}(x-8) - \dfrac{1}{288}(x-8)^2$.

c. $p_1(7.5) \approx 1.958333333$, $p_2(7.5) \approx 1.957465278$.

11.1.17 $f(0) = 1, f'(0) = -6\sin 0 = 0, f''(0) = -36\cos 0 = -36, f'''(0) = 6^3 \sin 0 = 0$, and $f^{(4)}(0) = 6^4 \cos 0 = 1296$, so that $p_1(x) = 1, p_2(x) = p_3(x) = 1 - 18x^2, p_4(x) = 1 - 18x^2 + 54x^4$.

11.1.19 $f(0) = 1, f'(0) = -3(1+0)^{-4} = -3, f''(0) = 12(1+0)^{-5} = 12, f'''(0) = -60(1+0)^{-6} = -60$, and $f^{(4)}(0) = 360(1+0)^{-7} = 360$, so that $p_1(x) = 1 - 3x, p_2(x) = 1 - 3x + 6x^2, p_3(x) = 1 - 3x + 6x^2 - 10x^3$, $p_4(x) = 1 - 3x + 6x^2 - 10x^3 + 15x^4$.

11.1.21 Note that $f(1) = 1, f'(1) = 3, f''(1) = 6$, and $f'''(1) = 6$. Thus, $p_1(x) = 1 + 3(x-1)$, and $p_2(x) = 1 + 3(x-1) + 3(x-1)^2$, and $p_3(x) = 1 + 3(x-1) + 3(x-1)^2 + (x-1)^3$.

11.1.23 $f(x) = \ln x$ so $f(e) = 1, f'(e) = \dfrac{1}{e}, f''(e) = -\dfrac{1}{e^2}$, and $f'''(e) = \dfrac{2}{e^3}$.
So,
$$p_3(x) = 1 + \dfrac{1}{e}(x-e) - \dfrac{1}{2e^2}(x-e)^2 + \dfrac{1}{3e^3}(x-e)^3.$$

11.1.25 $f(0) = 0, f'(0) = \dfrac{-1}{1-0} = -1, f''(0) = \dfrac{1}{(1-0)^2} = -1$, so that $p_1(x) = -x, p_2(x) = -x - \dfrac{1}{2}x^2$.

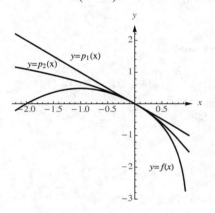

11.1.27

a. $p_1(x) = \dfrac{\sqrt{2}}{2} + \dfrac{\sqrt{2}}{2}\left(x - \dfrac{\pi}{4}\right), p_2(x) = \dfrac{\sqrt{2}}{2} + \dfrac{\sqrt{2}}{2}\left(x - \dfrac{\pi}{4}\right) - \dfrac{\sqrt{2}}{4}\left(x - \dfrac{\pi}{4}\right)^2$.

b.

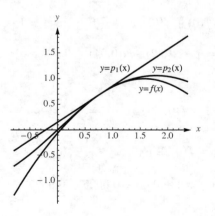

11.1. Approximating Functions With Polynomials

11.1.29

a. $p_2(0.05) = 1.0246875$.

b. The absolute error is $\sqrt{1.05} - p_2(1.05) \approx 1.024695077 - 1.0246875 = 7.58 \times 10^{-6}$.

11.1.31

a. $p_2(0.15) = 0.86125$.

b. The absolute error is $p_2(0.15) - e^{-0.15} \approx 0.86125 - 0.8607079764 \approx 5.42 \times 10^{-4}$.

11.1.33

a. The 3rd-order Taylor polynomial of $f(x) = e^{-x}$ centered at $a = 0$ is
$$p(x) = 1 + x + \frac{1}{2}x^2 + \frac{1}{6}x^3,$$
so the approximate value of $e^{0.12}$ is $p(0.12) = 1.127488$.

b. The absolute error is $\left|e^{0.12} - 1.127488\right| \approx 8.85 \times 10^{-6}$. (Answers may vary if intermediate calculations are rounded.)

11.1.35

a. The 3rd-order Taylor polynomial of $f(x) = \tan x$ centered at $a = 0$ is
$$p(x) = x + \frac{1}{3}x^3$$
so the approximate value of $\tan(-0.1)$ is $p(-0.1) \approx -0.10033333$.

b. The absolute error is $|\tan(-0.1) - (-0.10033333)| \approx 1.34 \times 10^{-6}$. (Answers may vary if intermediate calculations are rounded.)

11.1.37

a. The 3rd-order Taylor polynomial of $f(x) = \sqrt{x}$ centered at $a = 1$ is
$$p(x) = 1 + \frac{1}{2}(x-1) - \frac{1}{8}(x-1)^2 + \frac{1}{16}(x-1)^3,$$
so the approximate value of $\sqrt{1.06}$ is $p(1.06) \approx 1.0295635$.

b. The absolute error is $\left|\sqrt{1.06} - 1.0295635\right| \approx 4.86 \times 10^{-7}$. (Answers may vary if intermediate calculations are rounded.)

11.1.39

a. The 3rd-order Taylor polynomial of $f(x) = \sinh x$ centered at $a = 0$ is
$$p(x) = x + \frac{1}{6}x^3,$$
so the approximate value of $\sinh 0.5$ is $p(0.5) \approx 0.52083333$.

b. The absolute error is $|\sinh(0.5) - 0.52083333| \approx 2.62 \times 10^{-4}$. (Answers may vary if intermediate calculations are rounded.)

11.1.41 $R_n(x) = \dfrac{\sin^{(n+1)}(c)}{(n+1)!}x^{n+1}$ for some c between 0 and x.

11.1.43 $R_n(x) = \dfrac{(-1)^n e^{-c}}{(n+1)!} x^{n+1}$ for some c between 0 and x.

11.1.45 $R_n(x) = \dfrac{\sin^{(n+1)}(c)}{(n+1)!} \left(x - \dfrac{\pi}{2}\right)^{n+1}$ for some c between $\dfrac{\pi}{2}$ and x.

11.1.47 $f(x) = \sin x$, so $f^{(4)}(x) = \cos x$. Because $\cos x$ is bounded in magnitude by 1, the remainder is bounded by $|R_4(x)| \le \dfrac{0.3^5}{5!} \approx 2.03 \times 10^{-5}$.

11.1.49 $f(x) = e^x$, so $f^{(5)}(x) = e^x$. Because $e^{0.25}$ is bounded by 2, $|R_4(x)| \le 2 \cdot \dfrac{0.25^5}{5!} \approx 1.63 \times 10^{-5}$.

11.1.51 $f(x) = e^{-x}$, so $f^{(5)}(x) = -e^{-x}$. Because $f^{(5)}$ achieves its maximum magnitude in the range at $x = 0$, which has absolute value 1, $|R_4(x)| \le 1 \cdot \dfrac{0.5^5}{5!} \approx 2.6 \times 10^{-4}$.

11.1.53 Here $n = 3$ or 4, so use $n = 4$, and $M = 1$ because $f^{(5)}(x) = \cos x$, so that $R_4(x) \le \dfrac{(\pi/4)^5}{5!} \approx 2.49 \times 10^{-3}$.

11.1.55 $n = 2$ and $M = e^{1/2} < 2$, so $R_2(x) \le 2 \cdot \dfrac{(1/2)^3}{3!} \approx 4.17 \times 10^{-2}$.

11.1.57 $n = 2$; $f^{(3)}(x) = \dfrac{2}{(1+x)^3}$, which achieves its maximum at $x = -0.2$: $|f^{(3)}(x)| = \dfrac{2}{0.8^3} < 4$. Then $R_2(x) \le 4 \cdot \dfrac{0.2^3}{3!} \approx 5.4 \times 10^{-3}$.

11.1.59 Use the Taylor series for e^x at $x = 0$. The derivatives of e^x are e^x. On $[-0.5, 0]$, the maximum magnitude of any derivative is thus 1 and $x = 0$, so $|R_n(-0.5)| \le \dfrac{0.5^{n+1}}{(n+1)!}$, so for $R_n(-0.5) < 10^{-3}$ we need $n = 4$.

11.1.61 Use the Taylor series for $\cos x$ at $x = 0$. The magnitude of any derivative of $\cos x$ is bounded by 1, so $|R_n(-0.25)| \le \dfrac{0.25^{n+1}}{(n+1)!}$, so for $|R_n(-0.25)| < 10^{-3}$ we need $n = 3$.

11.1.63 Use the Taylor series for $f(x) = \sqrt{x}$ at $x = 1$. Then $|f^{(n+1)}(x)| = \dfrac{1 \cdot 3 \cdots (2n-1)}{2^{n+1}} x^{-(2n+1)/2}$, which achieves its maximum on $[1, 1.06]$ at $x = 1$. Then

$$|R_n(1.06)| \le \dfrac{1 \cdot 3 \cdots (2n-1)}{2^{n+1}} \cdot \dfrac{(1.06 - 1)^{n+1}}{(n+1)!},$$

and for $|R_n(0.06)| < 10^{-3}$ we need $n = 1$.

11.1.65

a. False. If $f(x) = e^{-2x}$, then $f^{(n)}(x) = (-1)^n 2^n e^{-2x}$, so that $f^{(n)}(0) \ne 0$ and all powers of x are present in the Taylor series.

b. True. The constant term of the Taylor series is $f(0) = 1$. Higher-order terms all involve derivatives of $f(x) = x^5 - 1$ evaluated at $x = 0$; clearly for $n < 5$, $f^{(n)}(0) = 0$, and for $n > 5$, the derivative itself vanishes. Only for $n = 5$, where $f^{(5)}(x) = 5!$, is the derivative nonzero, so the coefficient of x^5 in the Taylor series is $f^{(5)}(0)/5! = 1$ and the Taylor polynomial of order 10 is in fact $x^5 - 1$.

c. True. The odd derivatives of $\sqrt{1 + x^2}$ vanish at $x = 0$, while the even ones do not.

11.1. Approximating Functions With Polynomials

d. True. Because then $f''(a) = 0$.

11.1.67

a. This matches (C) because for $f(x) = (1+2x)^{1/2}$, $f''(x) = -(1+2x)^{-3/2}$ so $\dfrac{f''(0)}{2!} = \dfrac{-1}{2}$.

b. This matches (E) because for $f(x) = (1+2x)^{-1/2}$, $f''(x) = 3(1+2x)^{-5/2}$, so $\dfrac{f''(0)}{2!} = \dfrac{3}{2}$.

c. This matches (A) because $f^{(n)}(x) = 2^n e^{2x}$, so that $f^{(n)}(0) = 2^n$, which is (A)'s pattern.

d. This matches (D) because $f''(x) = 8(1+2x)^{-3}$ and $f''(0) = 8$, so that $f''(0)/2! = 4$.

e. This matches (B) because $f'(x) = -6(1+2x)^{-4}$ so that $f'(0) = -6$.

f. This matches (F) because $f^{(n)}(x) = (-2)^n e^{-2x}$, so $f^{(n)}(0) = (-2)^n$, which is (F)'s pattern.

11.1.69

a. $p_2(0.1) = 0.1$. The maximum error in the approximation is $1 \cdot \dfrac{0.1^3}{3!} \approx 1.67 \times 10^{-4}$.

b. $p_2(0.2) = 0.2$. The maximum error in the approximation is $1 \cdot \dfrac{0.2^3}{3!} \approx 1.33 \times 10^{-3}$.

11.1.71

a. $p_3(0.1) = 1 - .01/2 = 0.995$. The maximum error is $1 \cdot \dfrac{0.1^4}{4!} \approx 4.17 \times 10^{-6}$.

b. $p_3(0.2) = 1 - .04/2 = 0.98$. The maximum error is $1 \cdot \dfrac{0.2^4}{4!} \approx 6.67 \times 10^{-5}$.

11.1.73

a. $p_1(0.1) = 1.05$. Because $|f''(x)| = \dfrac{1}{4}(1+x)^{-3/2}$ has a maximum value of $1/4$ at $x = 0$, the maximum error is $\dfrac{1}{4} \cdot \dfrac{0.1^2}{2} = 1.25 \times 10^{-3}$.

b. $p_1(0.2) = 1.1$. The maximum error is $\dfrac{1}{4} \cdot \dfrac{0.2^2}{2} = 5 \times 10^{-3}$.

11.1.75

a. $p_1(0.1) = 1.1$. Because $f''(x) = e^x$ is less than 2 on $[0, 0.1]$, the maximum error is less than $2 \cdot \dfrac{0.1^2}{2!} = \dfrac{1}{100}$.

b. $p_1(0.2) = 1.2$. The maximum error is less than $2 \cdot \dfrac{0.2^2}{2!} = .04 = \dfrac{1}{25}$.

11.1.77

a.

	$|\sec x - p_2(x)|$	$|\sec x - p_4(x)|$
-0.2	3.39×10^{-4}	5.51×10^{-6}
-0.1	2.09×10^{-5}	8.51×10^{-8}
0.0	0	0
0.1	2.09×10^{-5}	8.51×10^{-8}
0.2	3.39×10^{-4}	5.51×10^{-6}

b. The errors are get smaller as $|x|$ decreases.

11.1.79

a.

	$|e^{-x} - p_1(x)|$	$|e^{-x} - p_2(x)|$
−0.2	2.14×10^{-2}	1.4×10^{-3}
−0.1	5.17×10^{-3}	1.71×10^{-4}
0.0	0	0
0.1	4.84×10^{-3}	1.63×10^{-4}
0.2	1.87×10^{-2}	1.27×10^{-3}

b. The errors get smaller as $|x|$ decreases.

11.1.81 The true value of $e^{0.35} \approx 1.419067549$. The 6th-order Taylor polynomial for e^x centered at $x = 0$ is

$$p_6(x) = 1 + x + \frac{x^2}{2} + \frac{x^3}{6} + \frac{x^4}{24} + \frac{x^5}{120} + \frac{x^6}{720}.$$

Evaluating the polynomials at $x = 0.35$ produces the following table:

n	$p_n(0.35)$	$|p_n(0.35) - e^{0.35}|$
1	1.350000000	6.91×10^{-2}
2	1.411250000	7.82×10^{-3}
3	1.418395833	6.72×10^{-4}
4	1.419021094	4.65×10^{-5}
5	1.419064862	2.69×10^{-6}
6	1.419067415	1.33×10^{-7}

The 6th-order Taylor polynomial for e^x centered at $x = \ln 2$ is

$$p_6(x) = 2 + 2(x - \ln 2) + (x - \ln 2)^2 + \frac{1}{3}(x - \ln 2)^3 + \frac{1}{12}(x - \ln 2)^4$$
$$+ \frac{1}{60}(x - \ln 2)^5 + \frac{1}{360}(x - \ln 2)^6.$$

Evaluating the polynomials at $x = 0.35$ produces the following table:

n	$p_n(0.35)$	$|p_n(0.35) - e^{0.35}|$
1	1.313705639	1.05×10^{-1}
2	1.431455626	1.24×10^{-2}
3	1.417987101	1.08×10^{-3}
4	1.419142523	7.50×10^{-5}
5	1.419063227	4.32×10^{-6}
6	1.419067762	2.13×10^{-7}

Comparing the tables shows that using the polynomial centered at $x = 0$ is more accurate for all n. To see why, consider the remainder. Let $f(x) = e^x$. By Theorem 9.2, the magnitude of the remainder when approximating $f(0.35)$ by the polynomial p_n centered at 0 is:

$$|R_n(0.35)| = \frac{|f^{(n+1)}(c)|}{(n+1)!}(0.35)^{n+1} = \frac{e^c}{(n+1)!}(0.35)^{n+1}$$

for some c with $0 < c < 0.35$ while the magnitude of the remainder when approximating $f(0.35)$ by the polynomial p_n centered at $\ln 2$ is:

$$|R_n(0.35)| = \frac{|f^{(n+1)}(c)|}{(n+1)!}|0.35 - \ln 2|^{n+1} = \frac{e^c}{(n+1)!}(\ln 2 - 0.35)^{n+1}$$

11.1. Approximating Functions With Polynomials

for some c with $0.35 < c < \ln 2$. Because $\ln 2 - 0.35 \approx 0.35$, the relative size of the magnitudes of the remainders is determined by e^c in each remainder. Because e^x is an increasing function, the remainder in using the polynomial centered at 0 will be less than the remainder in using the polynomial centered at $\ln 2$, and the former polynomial will be more accurate.

11.1.83

a. The slope of the tangent line to $f(x)$ at $x = a$ is by definition $f'(a)$; by the point-slope form for the equation of a line, we have $y - f(a) = f'(a)(x-a)$, or $y = f(a) + f'(a)(x-a)$.

b. The Taylor polynomial centered at a is $p_1(x) = f(a) + f'(a)(x-a)$, which is the tangent line at a.

11.1.85

a. We have

$$f(0) = f^{(4)}(0) = \sin 0 = 0 \qquad f(\pi) = f^{(4)}(\pi) = \sin \pi = 0$$
$$f'(0) = f^{(5)}(0) = \cos 0 = 1 \qquad f'(\pi) = f^{(5)}(0) = \cos \pi = -1$$
$$f''(0) = -\sin 0 = 0 \qquad f''(\pi) = -\sin \pi = 0$$
$$f'''(0) = -\cos 0 = -1 \qquad f'''(\pi) = -\cos \pi = 1.$$

Thus

$$p_5(x) = x - \frac{x^3}{3!} + \frac{x^5}{5!}$$
$$q_5(x) = -(x-\pi) + \frac{1}{3!}(x-\pi)^3 - \frac{1}{5!}(x-\pi)^5.$$

b. A plot of the three functions, with $\sin x$ the black solid line, $p_5(x)$ the dashed line, and $q_5(x)$ the dotted line is below.

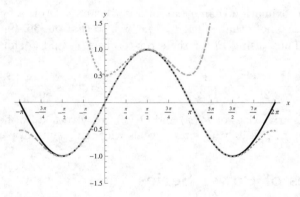

$p_5(x)$ and $\sin x$ are almost indistinguishable on $[-\pi/2, \pi/2]$, after which $p_5(x)$ diverges pretty quickly from $\sin x$. $q_5(x)$ is reasonably close to $\sin x$ over the entire range, but the two are almost indistinguishable on $[\pi/2, 3\pi/2]$. $p_5(x)$ is a better approximation than $q_5(x)$ on about $[-\pi, \pi/2)$, while $q_5(x)$ is better on about $(\pi/2, 2\pi]$.

c. Evaluating the errors gives

x	$|\sin x - p_5(x)|$	$|\sin x - q_5(x)|$
$\frac{\pi}{4}$	3.6×10^{-5}	7.4×10^{-2}
$\frac{\pi}{2}$	4.5×10^{-3}	4.5×10^{-3}
$\frac{3\pi}{4}$	7.4×10^{-2}	3.6×10^{-5}
$\frac{5\pi}{4}$	2.3	3.6×10^{-5}
$\frac{7\pi}{4}$	20.4	7.4×10^{-2}

d. $p_5(x)$ is a better approximation than $q_5(x)$ only at $x = \frac{\pi}{4}$, in accordance with part (b). The two are equal at $x = \frac{\pi}{2}$, after which $q_5(x)$ is a substantially better approximation than $p_5(x)$.

11.1.87

a. We have

$$f(36) = \sqrt{36} = 6 \qquad\qquad f(49) = \sqrt{49} = 7$$
$$f'(36) = \frac{1}{2} \cdot \frac{1}{\sqrt{36}} = \frac{1}{12} \qquad\qquad f'(49) = \frac{1}{2} \cdot \frac{1}{\sqrt{49}} = \frac{1}{14}.$$

Thus

$$p_1(x) = 6 + \frac{1}{12}(x - 36) \qquad\qquad q_1(x) = 7 + \frac{1}{14}(x - 49).$$

b. Evaluating the errors gives

x	$|\sqrt{x} - p_1(x)|$	$|\sqrt{x} - q_1(x)|$
37	5.7×10^{-4}	6.0×10^{-2}
39	5.0×10^{-3}	4.1×10^{-2}
41	1.4×10^{-2}	2.5×10^{-2}
43	2.6×10^{-2}	1.4×10^{-2}
45	4.2×10^{-2}	6.1×10^{-3}
47	6.1×10^{-2}	1.5×10^{-3}

c. $p_1(x)$ is a better approximation than $q_1(x)$ for $x \leq 41$, and $q_1(x)$ is a better approximation for $x \geq 43$. To see why this is true, note that $f''(x) = -\frac{1}{4}x^{-3/2}$, so that on $[36, 49]$ it is bounded in magnitude by $\frac{1}{4} \cdot 36^{-3/2} = \frac{1}{864}$. . Thus (using P_1 for the error term for p_1 and Q_1 for the error term for q_1)

$$P_1(x) \leq \frac{1}{864} \cdot \frac{|x - 36|^2}{2!} = \frac{1}{1728}(x - 36)^2, \qquad Q_1(x) \leq \frac{1}{864} \cdot \frac{|x - 49|^2}{2!} = \frac{1}{1728}(x - 49)^2.$$

It follows that the relative sizes of $P_1(x)$ and $Q_1(x)$ are governed by the distance of x from 36 and 49. Looking at the different possibilities for x reveals why the results in part (b) hold.

11.2 Properties of Power Series

11.2.1 $c_0 + c_1 x + c_2 x^2 + c_3 x^3$.

11.2.3 Generally the Ratio Test or Root Test is used.

11.2.5 The radius of convergence does not change, but the interval of convergence may change at the endpoints.

11.2.7 The series converges for $|x - 2| \leq 10$, so the radius of convergence is 10 and the interval is $[-8, 12]$.

11.2.9 Using the Root Test: $\lim_{k \to \infty} \sqrt[k]{|a_k|} = \lim_{k \to \infty} |2x| = |2x|$. So the radius of convergence is $\frac{1}{2}$. At $x = 1/2$ the series is $\sum_{k=0}^{\infty} 1$ which diverges, and at $x = -1/2$ the series is $\sum_{k=0}^{\infty} (-1)^k$ which also diverges. So the interval of convergence is $(-1/2, 1/2)$

11.2. Properties of Power Series

11.2.11 Using the Ratio Test: $\lim_{k\to\infty}\left|\frac{a_{k+1}}{a_k}\right| = \lim_{k\to\infty}\left|\frac{(k+1)^{k+1}x^{k+1}}{k^k x^k}\right| = \lim_{k\to\infty}(k+1)\left(\frac{k+1}{k}\right)^k |x| = \infty$ (for $x \neq 0$) because $\lim_{k\to\infty}\left(\frac{k+1}{k}\right)^k = e$. Thus, the radius of convergence is 0 and the series only converges at $x = 0$.

11.2.13 Using the Root Test: $\lim_{k\to\infty}\sqrt[k]{|a_k|} = \lim_{k\to\infty}\sin(1/k)|x| = \sin(0)|x| = 0$. Thus, the radius of convergence is ∞ and the interval of convergence is $(-\infty, \infty)$.

11.2.15 Using the Root Test: $\lim_{k\to\infty}\sqrt[k]{|a_k|} = \lim_{k\to\infty}\frac{|x|}{3} = \frac{|x|}{3}$, so the radius of convergence is 3. At $x = -3$, the series is $\sum(-1)^k$, which diverges. At $x = 3$, the series is $\sum 1$, which diverges. So the interval of convergence is $(-3, 3)$.

11.2.17 Using the Root Test: $\lim_{k\to\infty}\sqrt[k]{|a_k|} = \lim_{k\to\infty}\frac{|x|}{k} = 0$, so the radius of convergence is infinite and the interval of convergence is $(-\infty, \infty)$.

11.2.19 We use the Ratio Test. The limit of the absolute value of the ratio of successive terms is

$$\lim_{k\to\infty}|x|\frac{2^k + 1}{2^{k+1} + 1} = |x|\lim_{k\to\infty}\frac{1 + 1/2^k}{2 + 1/2^k} = \frac{|x|}{2},$$

so the radius of convergence is 2. For $x = 2$, we have the series $\sum_{k=1}^{\infty}\frac{2^k}{2^k + 1}$ which diverges by the Divergence Test because $\lim_{k\to\infty}\frac{2^k}{2^k + 1} = 1 \neq 0$. Similarly, the series for $x = -2$ fails the divergence test because the kth term does not have limit 0 as $k \to \infty$. So the interval of convergence is $(-2, 2)$.

11.2.21 The given series is $\sum_{k=1}^{\infty}\frac{(-1)^k x^{2k}}{k!}$. Note that the limit of the absolute value of successive terms is

$$\lim_{k\to\infty}\frac{x^2}{k+1} = 0,$$

which is less than 1 for all x. So the radius of convergence is ∞ and the interval of convergence is $(-\infty, \infty)$.

11.2.23 We use the Root Test. The limit of the kth root of the absolute value of the kth term is

$$\lim_{k\to\infty}\frac{|x-1|}{k^{1/k}} = |x-1|.$$

This is less than 1 for $0 < x < 2$. The radius of convergence is 1. For $x = 0$, we have -1 times the Harmonic Series (which diverges) and for $x = 2$ we have the Alternating Harmonic series which converges by the Alternating Series Test. Therefore, the interval of convergence is $(0, 2]$.

11.2.25 Using the Ratio Test:

$$\lim_{k\to\infty}\left|\frac{a_{k+1}}{a_k}\right| = \lim_{k\to\infty}|4x - 1|\frac{k^2 + 4}{k^2 + 2k + 5} = |4x - 1|.$$

This is less than 1 for $0 < x < \frac{1}{2}$. The radius of convergence is $\frac{1}{4}$. For $x = \frac{1}{2}$ we obtain the convergent series $\sum_{k=0}^{\infty}\frac{1}{k^2 + 4}$, which converges because $\frac{1}{k^2 + 4} < \frac{1}{k^2}$, and $\sum_{k=0}^{\infty}\frac{1}{k^2}$ is a convergent p-series. For $x = 0$, we obtain the alternating version of that same series, which is clearly absolutely convergent and therefore convergent. The interval of convergence is $\left[0, \frac{1}{2}\right]$.

11.2.27 We use the Root Test. The limit of the kth root of the absolute value of the kth term is
$$\lim_{k\to\infty} |2x-4|\frac{k^{10/k}}{10} = \frac{|2x-4|}{10}.$$
This is less than one for $-10 < 2x - 4 < 10$, or $-3 < x < 7$. The radius of convergence is therefore 5. For the endpoints, the series diverges by the Divergence Test. The interval of convergence is thus $(-3, 7)$.

11.2.29 Using the Ratio Test: $\lim_{k\to\infty} \left|\frac{(k+1)^2 x^{2k+2}}{(k+1)!} \cdot \frac{k!}{k^2 x^{2k}}\right| = \lim_{k\to\infty} \frac{k+1}{k^2} x^2 = 0$, so the radius of convergence is infinite, and the interval of convergence is $(-\infty, \infty)$.

11.2.31 Using the Ratio Test: $\lim_{k\to\infty} \left|\frac{a_{k+1}}{a_k}\right| = \left|\frac{x^{2k+3}}{3^k} \cdot \frac{3^{k-1}}{x^{2k+1}}\right| = \frac{x^2}{3}$ so that the radius of convergence is $\sqrt{3}$. At $x = \sqrt{3}$, the series is $\sum_{k=1}^{\infty} 3\sqrt{3}$, which diverges. At $x = -\sqrt{3}$, the series is $\sum_{k=1}^{\infty} (-3\sqrt{3})$, which also diverges, so the interval of convergence is $(-\sqrt{3}, \sqrt{3})$.

11.2.33 Using the Root Test: $\lim_{k\to\infty} \sqrt[k]{|a_k|} = \lim_{k\to\infty} \frac{(|x-1|)k}{k+1} = |x-1|$, so the series converges when $|x-1| < 1$, so for $0 < x < 2$. The radius of convergence is 1. At $x = 2$, the series diverges by the Divergence Test. At $x = 0$, the series diverges as well by the Divergence Test. Thus the interval of convergence is $(0, 2)$.

11.2.35 Using the Ratio Test:
$$\lim_{k\to\infty} \left|\frac{a_{k+1}}{a_k}\right| = \left|\frac{(k+1)^{20} x^{k+1}}{(2k+3)!} \cdot \frac{(2k+1)!}{x^k k^{20}}\right| = \lim_{k\to\infty} \left(\frac{k+1}{k}\right)^{20} \frac{|x|}{(2k+2)(2k+3)} = 0,$$
so the radius of convergence is infinite, and the interval of convergence is $(-\infty, \infty)$.

11.2.37 Using the Ratio Test: $\lim_{k\to\infty} \left|\frac{a_{k+1}}{a_k}\right| = \lim_{k\to\infty} \left|\frac{(k+1)! \, x^{k+1}}{(k+1)^{k+1}} \cdot \frac{k^k}{k! \, x^k}\right| = \lim_{k\to\infty} \left(\frac{k}{k+1}\right)^k |x| = \frac{1}{e}|x|$. The radius of convergence is therefore e.

11.2.39 Using the Root Test:
$$\lim_{k\to\infty} \sqrt[k]{|a_k|} = |x| \lim_{k\to\infty} \left(\frac{k}{k+4}\right)^k = |x| \lim_{k\to\infty} \frac{1}{\left(1+\frac{4}{k}\right)^k} = \frac{|x|}{e^4}.$$
This is less than 1 for $|x| < e^4$, so the radius of convergence is e^4.

11.2.41 $f(3x) = \frac{1}{1-3x} = \sum_{k=0}^{\infty} 3^k x^k$, which converges for $|x| < 1/3$, and diverges at the endpoints.

11.2.43 $h(x) = \frac{2x^3}{1-x} = \sum_{k=0}^{\infty} 2x^{k+3}$, which converges for $|x| < 1$ and is divergent at the endpoints.

11.2.45 $p(x) = \frac{4x^{12}}{1-x} = \sum_{k=0}^{\infty} 4x^{k+12} = 4\sum_{k=0}^{\infty} x^{k+12}$, which converges for $|x| < 1$. It is divergent at the endpoints.

11.2.47 $f(3x) = \ln(1-3x) = -\sum_{k=1}^{\infty} \frac{(3x)^k}{k} = -\sum_{k=1}^{\infty} \frac{3^k}{k} x^k$ for $-1 \leq 3x < 1$, or $-\frac{1}{3} \leq x < \frac{1}{3}$. The radius of convergence is $\frac{1}{3}$. The series diverges at $\frac{1}{3}$ (harmonic series), and converges at $-\frac{1}{3}$ (alternating harmonic series). The interval of convergence is $\left[-\frac{1}{3}, \frac{1}{3}\right)$.

11.2.49 $p(x) = 2x^6 \ln(1-x) = -2 \sum_{k=1}^{\infty} \frac{x^{k+6}}{k}$ for $-1 \leq x < 1$, so the radius of convergence is 1. The series diverges at 1 (harmonic series) but converges at -1 (alternating harmonic series), so the interval of convergence is $[-1, 1)$.

11.2.51 The power series for $f(x)$ is $\sum_{k=0}^{\infty}(2x)^k$, which is convergent for $-1 < 2x < 1$, so for $-1/2 < x < 1/2$. The power series for $g(x) = f'(x)$ is $\sum_{k=1}^{\infty} k(2x)^{k-1} \cdot 2 = 2\sum_{k=1}^{\infty} k(2x)^{k-1}$, also convergent on $|x| < 1/2$. At the endpoints the series diverges by the Divergence Test. The interval of convergence is $\left(-\frac{1}{2}, \frac{1}{2}\right)$.

11.2.53 $f(x) = \frac{1}{1+x} = \sum_{k=0}^{\infty}(-x)^k$ for $-1 < -x < 1$, or $-1 < x < 1$. So

$$g(x) = -\frac{1}{(1+x)^2} = \sum_{k=1}^{\infty} -k(-x)^{k-1} = \sum_{k=1}^{\infty}(-1)^k k x^{k-1},$$

for $-1 < x < 1$. At each endpoint $x = \pm 1$, the series diverges by the Divergence Test. So the interval of convergence is $(-1, 1)$.

11.2.55 The power series for $\frac{1}{1-3x}$ is $\sum_{k=0}^{\infty}(3x)^k$, which is convergent on $|x| < 1/3$. Because $g(x) = \ln(1-3x) = -3\int \frac{1}{1-3x}\, dx$ and because $g(0) = 0$, the power series for $g(x)$ is $-3\sum_{k=0}^{\infty} 3^k \frac{1}{k+1} x^{k+1} = -\sum_{k=1}^{\infty} \frac{3^k}{k} x^k$. The series converges at $x = -\frac{1}{3}$ (alternating harmonic) and diverges at $x = \frac{1}{3}$ (harmonic series). The interval of convergence is $\left[-\frac{1}{3}, \frac{1}{3}\right)$.

11.2.57 Recall that $\frac{d}{dx}\left(\frac{1}{1+x^2}\right) = -\frac{2x}{(1+x^2)^2}$, and by Example 3c, $\frac{1}{1+x^2} = 1 - x^2 + x^4 - x^6 + \cdots$. So

$$-\frac{2x}{(1+x^2)^2} = \frac{d}{dx}(1 - x^2 + x^4 - x^6 + \cdots) = -2x + 4x^3 - 6x^5 - \cdots = \sum_{k=1}^{\infty}(-1)^{k+1} 2k x^{2k-1},$$

for $|x| < 1$. Checking the endpoints for $x = \pm 1$ reveals that the series diverges by the Divergence Test.

11.2.59 Note that $f(x) = \frac{3}{3+x} = \frac{1}{1+(1/3)x}$. Let $g(x) = \frac{1}{1+x}$. The power series for $g(x)$ is $\sum_{k=0}^{\infty}(-1)^k x^k$, so the power series for $f(x) = g((1/3)x)$ is $\sum_{k=0}^{\infty}(-1)^k 3^{-k} x^k = \sum_{k=0}^{\infty}\left(\frac{-x}{3}\right)^k$. Using the Ratio Test:

$$\lim_{k \to \infty}\left|\frac{a_{k+1}}{a_k}\right| = \lim_{k \to \infty}\left|\frac{3^{-(k+1)}x^{k+1}}{3^{-k}x^k}\right| = \frac{|x|}{3},$$

so the radius of convergence is 3. The series diverges at both endpoints by the Divergence test. The interval of convergence is $(-3, 3)$.

11.2.61 Note that $f(x) = \ln\sqrt{4-x^2} = \frac{1}{2}\ln(4-x^2) = \frac{1}{2}\left(\ln 4 + \ln\left(1 - \frac{x^2}{4}\right)\right) = \ln 2 + \frac{1}{2}\ln\left(1 - \frac{x^2}{4}\right)$. Now, the power series for $g(x) = \ln(1-x)$ is $-\sum_{k=1}^{\infty}\frac{1}{k}x^k$, so the power series for $f(x)$ is $\ln 2 - \frac{1}{2}\sum_{k=1}^{\infty}\frac{1}{k} \cdot \frac{x^{2k}}{4^k} =$

$\ln 2 - \sum_{k=1}^{\infty} \frac{x^{2k}}{k2^{2k+1}}$. Now, $\lim_{k\to\infty} \left|\frac{a_{k+1}}{a_k}\right| = \lim_{k\to\infty} \left|\frac{x^{2k+2}}{(k+1)2^{2k+3}} \cdot \frac{k2^{2k+1}}{x^{2k}}\right| = \lim_{k\to\infty} \frac{k}{4(k+1)} x^2 = \frac{x^2}{4}$, so that the radius of convergence is 2. The series diverges at both endpoints, so its interval of convergence is $(-2, 2)$.

11.2.63

a. True. This power series is centered at $x = 3$, so its interval of convergence will be symmetric about 3.

b. True. Use the Root Test. At the endpoints, the series diverges by the Divergence Test.

c. True. Substitute x^2 for x in the series.

d. True. Because the power series is zero on the interval, all its derivatives are as well, which implies (differentiating the power series) that all the c_k are zero.

11.2.65 The power series for $f(x-a)$ is $\sum c_k(x-a)^k$. Then $\sum c_k(x-a)^k$ converges if and only if $|x-a| < R$, which happens if and only if $a - R < x < a + R$, so the radius of convergence is the same.

11.2.67 This is a geometric series with ratio $\sqrt{x} - 2$, so its sum is $\frac{1}{1-(\sqrt{x}-2)} = \frac{1}{3-\sqrt{x}}$, for $-1 < \sqrt{x} - 2 < 1$. Simplifying the inequality, we have $1 < \sqrt{x} < 3$, or $1 < x < 9$.

11.2.69 This is a geometric series with ratio e^{-x}, so its sum is $\frac{1}{1-e^{-x}}$, for $e^{-x} < 1$, or $e^x > 1$ or $x > 0$.

11.2.71 This is a geometric series with ratio $(x^2-1)/3$, so its sum is $\frac{1}{1 - \frac{x^2-1}{3}} = \frac{3}{3-(x^2-1)} = \frac{3}{4-x^2}$, convergent for $|x^2 - 1| < 3$, so that $-2 < x^2 < 4$ or $-2 < x < 2$. It diverges at both endpoints.

11.2.73 Substitute $-3x$ for x in the power series for e^x to get $e^{-3x} = \sum_{k=0}^{\infty} \frac{(-3x)^k}{k!} = \sum_{k=0}^{\infty} (-1)^k \frac{3^k}{k!} x^k$. The series converges for all x.

11.2.75 The power series for $x^m f(x)$ is $\sum c_k x^{k+m}$. The radius of convergence of this power series is determined by the limit
$$\lim_{k\to\infty} \left|\frac{c_{k+1} x^{k+1+m}}{c_k x^{k+m}}\right| = \lim_{k\to\infty} \left|\frac{c_{k+1} x^{k+1}}{c_k x^k}\right|,$$
and the right-hand side is the limit used to determine the radius of convergence for the power series for $f(x)$. Thus the two have the same radius of convergence.

11.2.77

a. $f(x)g(x) = c_0 d_0 + (c_0 d_1 + c_1 d_0)x + (c_0 d_2 + c_1 d_1 + c_2 d_0)x^2 + \dots$

b. The coefficient of x^n in $f(x)g(x)$ is $\sum_{i=0}^{n} c_i d_{n-i}$.

11.3 Taylor Series

11.3.1 The nth Taylor Polynomial is the nth sum of the corresponding Taylor Series.

11.3.3 $\sum_{k=0}^{\infty} \frac{(x-2)^k}{k!}$.

11.3. Taylor Series

11.3.5 Substitute x^2 for x in the Taylor series. By theorems proved in the previous section about power series, the interval of convergence does not change except perhaps at the endpoints of the interval.

11.3.7 It means that the limit of the remainder term is zero.

11.3.9

a. $f(x) = \dfrac{1}{x^2}$, so $f(1) = 1$. $f'(x) = -\dfrac{2}{x^3}$, so $f'(1) = -2$. $f''(x) = \dfrac{6}{x^4}$, so $f''(1) = 6$. $f'''(x) = -\dfrac{24}{x^5}$, so $f'''(1) = -24$. So the Taylor series is

$$1 - 2(x-1) + 6\frac{(x-1)^2}{2!} - 24\frac{(x-1)^3}{3!} + \cdots = 1 - 2(x-1) + 3(x-1)^2 - 4(x-1)^3 + \cdots.$$

b. $\displaystyle\sum_{k=0}^{\infty}(-1)^k(k+1)(x-1)^k$.

c. $r = \displaystyle\lim_{k\to\infty}\left|\dfrac{(-1)^{k+1}(k+2)(x-1)^{k+1}}{(-1)^k(k+1)(x-1)^k}\right| = |x-1|$. This is less than 1 for $-1 < x-1 < 1$, or $0 < x < 2$. The series diverges at the endpoints by the Divergence Test, so the interval of convergence is $(0, 2)$.

11.3.11

a. Note that $f(0) = 1$, $f'(0) = -1$, $f''(0) = 1$, and $f'''(0) = -1$. So the Maclaurin series is $1 - x + x^2/2 - x^3/6 + \cdots$.

b. $\displaystyle\sum_{k=0}^{\infty}(-1)^k\dfrac{x^k}{k!}$.

c. The series converges on $(-\infty, \infty)$, as can be seen from the Ratio Test.

11.3.13

a. Note that $f(0) = 2$, $f'(x) = \dfrac{6}{(1-x)^4}$, so $f'(0) = 6$, $f''(x) = \dfrac{24}{(1-x)^5}$, so $f''(0) = 24$, and $f'''(x) = \dfrac{120}{(1-x)^6}$, so $f'''(0) = 120$. Thus the Maclaurin series is

$$2 + 6x + 12x^2 + 20x^3 + \cdots.$$

b. $\displaystyle\sum_{k=0}^{\infty}(k+1)(k+2)x^k$.

c. Using the Ratio Test, we see that the series converges on $(-1, 1)$. It does not converge at the endpoints by the Divergence Test.

11.3.15

a. For $f(x) = \dfrac{1}{1+x^2}$, we have $f(0) = 1$, $f'(x) = -\dfrac{2x}{(1+x^2)^2}$ so $f'(0) = 0$. $f''(x) = \dfrac{2(3x^2-1)}{(1+x^2)^3}$, so $f''(0) = -2$. Similarly, $f'''(0) = 0$ and $f^{(4)} = 24$, $f^{(5)}(0) = 0$, and $f^{(6)}(0) = 6!$. Thus the series is $1 - x^2 + x^4 - x^6 + \cdots$.

b. $\displaystyle\sum_{k=0}^{\infty}(-1)^k x^{2k}$.

c. The absolute value of the ratio of consecutive terms is x^2, so by the Ratio Test, the radius of convergence is 1. The series diverges at the endpoints by the Divergence Test, so the interval of convergence is $(-1, 1)$.

11.3.17

a. Note that $f(0) = 1$, and that $f^{(n)}(0) = 2^n$. Thus, the series is given by $1 + 2x + \dfrac{4x^2}{2} + \dfrac{8x^3}{6} + \cdots$.

b. $\displaystyle\sum_{k=0}^{\infty} \dfrac{(2x)^k}{k!}$.

c. The absolute value of the ratio of consecutive terms is $\dfrac{2|x|}{n}$, which has limit 0 as $n \to \infty$. So by the Ratio Test, the interval of convergence is $(-\infty, \infty)$.

11.3.19

a. We have $f(0) = 0$, while $f'(x) = \dfrac{2}{x^2 + 4}$, so $f'(0) = \dfrac{1}{2}$. $f''(x) = -\dfrac{4x}{(x^2+4)^2}$, so $f''(0) = 0$. Continuing, we have $f'''(x) = \dfrac{4(3x^2 - 4)}{(x^2+4)^3}$, so $f'''(0) = -\dfrac{1}{4}$. In a similar manner, $f^{(4)}(0) = 0$, $f^{(5)}(0) = \dfrac{3}{4}$, $f^{(6)}(0) = 0$, and $f^{(7)}(0) = -\dfrac{45}{8}$. So the series is $\dfrac{x}{2} - \dfrac{x^3}{24} + \dfrac{x^5}{160} - \dfrac{x^7}{896} + \cdots$.

b. $\displaystyle\sum_{k=0}^{\infty} (-1)^k \dfrac{1}{(2k+1) \cdot 2^{2k+1}} x^{2k+1}$.

c. The ratio of consecutive terms has limit $\dfrac{x^2}{4}$, so by the Ratio Test, the radius of convergence is $|x| < 2$. Also, at the endpoints we have convergence by the Alternating Series Test, so the interval of convergence is $[-2, 2]$.

11.3.21

a. Note that $f(0) = 1$, $f'(0) = \ln 3$, $f''(0) = \ln^2 3$, $f''(0) = \ln^3 3$. So the first four terms of the desired series are $1 + (\ln 3)x + \dfrac{\ln^2 3}{2}x^2 + \dfrac{\ln^3 3}{6}x^3 + \cdots$.

b. $\displaystyle\sum_{k=0}^{\infty} \dfrac{(\ln^k 3) x^k}{k!}$.

c. The ratio of successive terms is $\dfrac{(\ln^{k+1} 3)x^{k+1}}{(k+1)!} \cdot \dfrac{k!}{(\ln^k 3)x^k} = \dfrac{\ln 3}{k+1}x$, and the limit as $k \to \infty$ of this quantity is 0, so the interval of convergence is $(-\infty, \infty)$.

11.3.23

a. Note that $f(0) = 1$, $f'(0) = 0$, $f''(0) = 9$, $f'''(0) = 0$, etc. The first terms of the series are $1 + 9x^2/2 + 81x^4/4! + 3^6 x^6/6! + \cdots$.

b. $\displaystyle\sum_{k=0}^{\infty} \dfrac{(3x)^{2k}}{(2k)!}$.

c. The absolute value of the ratio of successive terms is $\left|\dfrac{(3x)^{2k+2}}{(2k+2)!} \cdot \dfrac{(2k)!}{(3x)^{2k}}\right| = \dfrac{1}{(2k+2)(2k+1)} \cdot 9x^2$, which has limit 0 as $x \to \infty$. The interval of convergence is therefore $(-\infty, \infty)$.

11.3. Taylor Series

11.3.25

a. $f(3) = 0$, $f'(x) = \dfrac{1}{x-2}$, so $f'(3) = 1$. $f''(x) = -\dfrac{1}{(x-2)^2}$, so $f''(3) = -1$. $f'''(x) = \dfrac{2}{(x-2)^3}$, so $f'''(3) = 2$, and $f^{(4)}(x) = -\dfrac{6}{(x-2)^4}$, so $f^{(4)}(3) = -6$. The series is

$$(x-3) - \frac{1}{2}(x-3)^2 + \frac{1}{3}(x-3)^3 - \frac{1}{4}(x-3)^4 + \cdots.$$

b. $\displaystyle\sum_{k=1}^{\infty} \frac{(-1)^{k+1}(x-3)^k}{k}$.

c. The limit of the absolute value of the ratio of successive terms is $x - 3$. So the series converges for $-1 < x - 3 < 1$, or $2 < x < 4$. At the left endpoint it diverges (it is the opposite of the harmonic series) and at the right endpoint it converges (it is the alternating harmonic series), so the interval of convergence is $(2, 4]$.

11.3.27

a. Note that $f(\pi/2) = 1$, $f'(\pi/2) = \cos(\pi/2) = 0$, $f''(\pi/2) = -\sin(\pi/2) = -1$, $f'''(\pi/2) = -\cos(\pi/2) = 0$, and so on. Thus the series is given by $1 - \dfrac{1}{2}\left(x - \dfrac{\pi}{2}\right)^2 + \dfrac{1}{24}\left(x - \dfrac{\pi}{2}\right)^4 - \dfrac{1}{720}\left(x - \dfrac{\pi}{2}\right)^6 + \cdots$.

b. $\displaystyle\sum_{k=0}^{\infty}(-1)^k \frac{1}{(2k)!}\left(x - \frac{\pi}{2}\right)^{2k}$.

11.3.29

a. Note that $f^{(k)}(1) = (-1)^k \dfrac{k!}{1^{k+1}} = (-1)^k \cdot k!$. Thus the series is given by $1 - (x-1) + (x-1)^2 - (x-1)^3 + \cdots$.

b. $\displaystyle\sum_{k=0}^{\infty}(-1)^k (x-1)^k$.

11.3.31

a. Note that $f^{(k)}(3) = (-1)^{k-1} \dfrac{(k-1)!}{3^k}$. Thus the series is given by $\ln(3) + \dfrac{x-3}{3} - \dfrac{1}{18}(x-3)^2 + \dfrac{1}{81}(x-3)^3 + \cdots$.

b. $\ln 3 + \displaystyle\sum_{k=1}^{\infty}(-1)^{k+1} \frac{1}{k \cdot 3^k}(x-3)^k$.

11.3.33

a. Note that $f(1) = 2$, $f'(1) = 2\ln 2$, $f''(1) = 2\ln^2 2$, $f'''(1) = 2\ln^3 2$. The first terms of the series are $2 + (2\ln 2)(x-1) + (\ln^2 2)(x-1)^2 + \dfrac{(\ln^3 2)(x-1)^3}{3} + \cdots$.

b. $\displaystyle\sum_{k=0}^{\infty} \frac{2(x-1)^k \ln^k 2}{k!}$.

11.3.35 Because the Taylor series for $\ln(1+x)$ is $x - \dfrac{x^2}{2} + \dfrac{x^3}{3} - \dfrac{x^4}{4} + \cdots$, the first four terms of the Taylor series for $\ln(1+x^2)$ are $x^2 - \dfrac{x^4}{2} + \dfrac{x^6}{3} - \dfrac{x^8}{4} + \cdots$, obtained by substituting x^2 for x.

11.3.37 Because the Taylor series for $\frac{1}{1-x} = 1 + x + x^2 + x^3 + \cdots$, the first four terms of the Taylor series for $\frac{1}{1-2x}$ are $1 + 2x + 4x^2 + 8x^3 + \cdots$ obtained by substituting $2x$ for x.

11.3.39 The Taylor series for $e^x - 1$ is the Taylor series for e^x, less the constant term of 1, so it is $x + \frac{x^2}{2} + \frac{x^3}{3!} + \frac{x^4}{4!} + \cdots$. Thus, the first four terms of the Taylor series for $\frac{e^x-1}{x}$ are $1 + \frac{x}{2!} + \frac{x^2}{3!} + \frac{x^3}{4!} + \cdots$, obtained by dividing the terms of the first series by x.

11.3.41 Because the Taylor series for $(1+x)^{-1}$ is $1 - x + x^2 - x^3 + \cdots$, if we substitute x^4 for x, we obtain $1 - x^4 + x^8 - x^{12} + \cdots$.

11.3.43 The Taylor series for $\sinh x$ is $x + \frac{x^3}{6} + \frac{x^5}{120} + \frac{x^7}{5040} + \cdots$. Thus, the Taylor series for $\sinh x^2$ is $x^2 + \frac{x^6}{6} + \frac{x^{10}}{120} + \frac{x^{14}}{5040} + \cdots$ obtained by substituting x^2 for x.

11.3.45

a. The binomial coefficients are $\binom{-2}{0} = 1$, $\binom{-2}{1} = \frac{-2}{1!} = -2$, $\binom{-2}{2} = \frac{(-2)(-3)}{2!} = 3$, $\binom{-2}{3} = \frac{(-2)(-3)(-4)}{3!} = -4$.
Thus the first four terms of the series are $1 - 2x + 3x^2 - 4x^3 + \cdots$.

b. $1 - 2 \cdot 0.1 + 3 \cdot 0.01 - 4 \cdot 0.001 = 0.826$

11.3.47

a. The binomial coefficients are $\binom{1/4}{0} = 1$, $\binom{1/4}{1} = \frac{1/4}{1} = \frac{1}{4}$, $\binom{1/4}{2} = \frac{(1/4)(-3/4)}{2!} = -\frac{3}{32}$, $\binom{1/4}{3} = \frac{(1/4)(-3/4)(-7/4)}{3!} = \frac{7}{128}$, so the first four terms of the series are $1 + \frac{1}{4}x - \frac{3}{32}x^2 + \frac{7}{128}x^3 + \cdots$.

b. Substitute $x = 0.12$ to get approximately 1.029.

11.3.49

a. The binomial coefficients are $\binom{-2/3}{0} = 1$, $\binom{-2/3}{1} = -\frac{2}{3}$, $\binom{-2/3}{2} = \frac{(-2/3)(-5/3)}{2!} = \frac{5}{9}$, $\binom{-2/3}{3} = \frac{(-2/3)(-5/3)(-8/3)}{3!} = -\frac{40}{81}$, so the first four terms of the series are $1 - \frac{2}{3}x + \frac{5}{9}x^2 - \frac{40}{81}x^3 + \cdots$.

b. Substitute $x = 0.18$ to get 0.89512.

11.3.51 $\sqrt{1+x^2} = 1 + \frac{x^2}{2} - \frac{x^4}{8} + \frac{x^6}{16} - \cdots$. By the Ratio Test, the radius of convergence is 1. At the endpoints, the series obtained are convergent by the Alternating Series Test. Thus, the interval of convergence is $[-1, 1]$.

11.3.53 $\sqrt{9 - 9x} = 3\sqrt{1-x} = 3 - \frac{3}{2}x - \frac{3}{8}x^2 - \frac{3}{16}x^3 - \cdots$. The interval of convergence is $[-1, 1]$.

11.3.55 $\sqrt{a^2 + x^2} = a\sqrt{1 + \frac{x^2}{a^2}} = a + \frac{x^2}{2a} - \frac{x^4}{8a^3} + \frac{x^6}{16a^5} - \cdots$. The series converges when $\frac{x^2}{a^2}$ is less than 1 in magnitude, so the radius of convergence is a. The series given by the endpoints is convergent by the Alternating Series Test, so the interval of convergence is $[-a, a]$.

11.3.57 $(1+4x)^{-2} = 1 - 2(4x) + 3(4x)^2 - 4(4x)^3 + \cdots = 1 - 8x + 48x^2 - 256x^3 + \cdots$.

11.3.59 $\dfrac{1}{(4+x^2)^2} = (4+x^2)^{-2} = \dfrac{1}{16}(1+(x^2/4))^{-2} = \dfrac{1}{16}\left(1 - 2\cdot\dfrac{x^2}{4} + 3\cdot\dfrac{x^4}{16} - 4\cdot\dfrac{x^6}{64} + \cdots\right) = \dfrac{1}{16} - \dfrac{1}{32}x^2 + \dfrac{3}{256}x^4 - \dfrac{1}{256}x^6 + \cdots$

11.3.61 $(3+4x)^{-2} = \dfrac{1}{9}\left(1+\dfrac{4x}{3}\right)^{-2} = \dfrac{1}{9} - \dfrac{2}{9}\left(\dfrac{4x}{3}\right) + \dfrac{3}{9}\left(\dfrac{4x}{3}\right)^2 - \dfrac{4}{9}\left(\dfrac{4x}{3}\right)^3 + \cdots$.

11.3.63 The interval of convergence for the Taylor series for $f(x) = \sin x$ is $(-\infty, \infty)$. The remainder is $R_n(x) = \dfrac{f^{(n+1)}(c)}{(n+1)!}x^{n+1}$ for some c. Because $f^{(n+1)}(x)$ is $\pm \sin x$ or $\pm \cos x$, we have

$$\lim_{n\to\infty} |R_n(x)| \leq \lim_{n\to\infty} \dfrac{1}{(n+1)!}|x^{n+1}| = 0$$

for any x.

11.3.65 The interval of convergence for the Taylor series for e^{-x} is $(-\infty, \infty)$. The remainder is $R_n(x) = \dfrac{(-1)^{n+1}e^{-c}}{(n+1)!}x^{n+1}$ for some c. Thus $\lim_{n\to\infty} |R_n(x)| = 0$ for any x.

11.3.67

a. False. Not all of its derivatives are defined at zero - in fact, none of them are.

b. True. The derivatives of $\csc x$ involve positive powers of $\csc x$ and $\cot x$, both of which are defined at $\pi/2$, so that $\csc x$ has continuous derivatives at $\pi/2$.

c. False. For example, the Taylor series for $f(x^2)$ doesn't converge at $x = 1.9$, because the Taylor series for $f(x)$ doesn't converge at $1.9^2 = 3.61$.

d. False. The Taylor series centered at 1 involves derivatives of f evaluated at 1, not at 0.

e. True. The follows because the Taylor series must itself be an even function.

11.3.69

a. The relevant Taylor series are: $e^x = 1 + x + \dfrac{x^2}{2!} + \dfrac{x^3}{3!} + \dfrac{x^4}{4!} + \dfrac{x^5}{5!} + \dfrac{x^6}{6!} + \cdots$ and $e^{-x} = 1 - x + \dfrac{x^2}{2!} - \dfrac{x^3}{3!} + \dfrac{x^4}{4!} - \dfrac{x^5}{5!} + \dfrac{x^6}{6!} + \cdots$. Thus the first four terms of the resulting series are $\dfrac{1}{2}(e^x + e^{-x}) = 1 + \dfrac{x^2}{2!} + \dfrac{x^4}{4!} + \dfrac{x^6}{6!} + \cdots$.

b. Because each series converges (absolutely) on $(-\infty, \infty)$, so does their sum. The radius of convergence is ∞.

11.3.71

a. Use the binomial theorem. The binomial coefficients are $\binom{-2/3}{0} = 1$, $\binom{-2/3}{1} = -\dfrac{2}{3}$, $\binom{-2/3}{2} = \dfrac{(-2/3)(-5/3)}{2!} = \dfrac{5}{9}$, $\binom{-2/3}{3} = \dfrac{(-2/3)(-5/3)(-8/3)}{3!} = -\dfrac{40}{81}$ and then, substituting x^2 for x, we obtain $1 - \dfrac{2}{3}x^2 + \dfrac{5}{9}x^4 - \dfrac{40}{81}x^6 + \cdots$.

b. The radius of convergence is determined from $|x^2| < 1$, so it is 1.

11.3.73

a. From the binomial formula, the Taylor series for $(1-x)^p$ is $\sum \binom{p}{k}(-1)^k x^k$, so the Taylor series for $(1-x^2)^p$ is $\sum \binom{p}{k}(-1)^k x^{2k}$. Here $p = 1/2$, and the binomial coefficients are $\binom{1/2}{0} = 1$, $\binom{1/2}{1} = \frac{1/2}{1!} = \frac{1}{2}$, $\binom{1/2}{2} = \frac{(1/2)(-1/2)}{2!} = -\frac{1}{8}$, $\binom{1/2}{3} = \frac{(1/2)(-1/2)(-3/2)}{3!} = \frac{1}{16}$ so that
$$(1-x^2)^{1/2} = 1 - \frac{1}{2}x^2 - \frac{1}{8}x^4 - \frac{1}{16}x^6 + \cdots.$$

b. The radius of convergence is determined from $|x^2| < 1$, so it is 1.

11.3.75

a. $f(x) = (1+x^2)^{-2}$; using the binomial series and substituting x^2 for x we obtain $1 - 2x^2 + 3x^4 - 4x^6 + \cdots$.

b. The radius of convergence is determined from $|x^2| < 1$, so it is 1.

11.3.77 Because $f(64) = 4$, and $f'(x) = \frac{1}{3}x^{-2/3}$, $f'(64) = \frac{1}{48}$, $f''(x) = -\frac{2}{9}x^{-5/3}$, $f''(64) = -\frac{1}{4608}$, $f'''(x) = \frac{10}{27}x^{-8/3}$, and $f'''(64) = \frac{10}{1769472} = \frac{5}{884736}$, the first four terms of the Taylor series are $4 + \frac{1}{48}(x-64) - \frac{1}{4608 \cdot 2!}(x-64)^2 + \frac{5}{884736 \cdot 3!}(x-64)^3$. Evaluating at $x = 60$, we obtain 3.914870274.

11.3.79 Because $f(16) = 2$, and $f'(x) = \frac{1}{4}x^{-3/4}$, $f'(16) = \frac{1}{32}$, $f''(x) = -\frac{3}{16}x^{-7/4}$, $f''(16) = -\frac{3}{2048}$, $f'''(x) = \frac{21}{64}x^{-11/4}$, and $f'''(16) = \frac{21}{131072}$, the first four terms of the Taylor series are $2 + \frac{1}{32}(x-16) - \frac{3}{2048 \cdot 2!}(x-16)^2 + \frac{21}{131072 \cdot 3!}(x-16)^3$. Evaluating at $x = 13$, we get 1.898937225.

11.3.81 Evaluate the binomial coefficient
$$\binom{1/2}{k} = \frac{(1/2)(-1/2)(-3/2)\cdots(1/2-k+1)}{k!} = \frac{(1/2)(-1/2)\cdots((3-2k)/2)}{k!}$$
$$= (-1)^{k-1} 2^{-k} \frac{1 \cdot 3 \cdots (2k-3)}{k!} = (-1)^{k-1} 2^{-k} \frac{(2k-2)!}{2^{k-1} \cdot (k-1)! \cdot k!} = (-1)^{k-1} 2^{1-2k} \cdot \frac{1}{k}\binom{2k-2}{k-1}.$$

This is the coefficient of x^k in the Taylor series for $\sqrt{1+x}$. Substituting $4x$ for x, the Taylor series becomes
$$\sum_{k=0}^{\infty}(-1)^{k-1} 2^{1-2k} \cdot \frac{1}{k}\binom{2k-2}{k-1}(4x)^k = \sum_{k=0}^{\infty}(-1)^{k-1}\frac{2}{k}\binom{2k-2}{k-1}x^k.$$

If we can show that k divides $\binom{2k-2}{k-1}$, we will be done, for then the coefficient of x^k will be an integer. But
$$\binom{2k-2}{k-1} - \binom{2k-2}{k-2} = \frac{(2k-2)!}{(k-1)!(k-1)!} - \frac{(2k-2)!}{(k-2)!k!} = \frac{(2k-2)!}{(k-1)!(k-1)!} - \frac{(2k-2)!(k-1)}{(k-1)!(k-1)!k}$$
$$= \frac{k(2k-2)! - (k-1)(2k-2)!}{k(k-1)!(k-1)!} = \frac{1}{k}\frac{(2k-2)!}{(k-1)!(k-1)!} = \frac{1}{k}\binom{2k-2}{k-1}$$

and thus we have shown that k divides $\binom{2k-2}{k-1}$.

11.3. Taylor Series

11.3.83

a. The Maclaurin series for $\sin x$ is $x - \frac{1}{3!}x^3 + \frac{1}{5!}x^5 - \frac{1}{7!}x^7 + \cdots$. Squaring the first four terms yields

$$\left(x - \frac{1}{3!}x^3 + \frac{1}{5!}x^5 - \frac{1}{7!}x^7\right)^2$$

$$= x^2 - \frac{2}{3!}x^4 + \left(\frac{2}{5!} + \frac{1}{3!3!}\right)x^6 + \left(-2 \cdot \frac{1}{7!} - 2 \cdot \frac{1}{3!5!}\right)x^8$$

$$= x^2 - \frac{1}{3}x^4 + \frac{2}{45}x^6 - \frac{1}{315}x^8.$$

b. The Maclaurin series for $\cos x$ is $1 - \frac{1}{2}x^2 + \frac{1}{4!}x^4 - \frac{1}{6!}x^6 + \frac{1}{8!}x^8 - \cdots$. Substituting $2x$ for x in the Maclaurin series for $\cos x$ and then computing $(1 - \cos 2x)/2$, we obtain

$$(1 - (1 - \frac{1}{2}(2x)^2 + \frac{1}{4!}(2x)^4 - \frac{1}{6!}(2x)^6 + \frac{1}{8!}(2x)^8)/2$$

$$= (2x^2 - \frac{2}{3}x^4 + \frac{4}{45}x^6 - \frac{2}{315}x^8)/2$$

$$= x^2 - \frac{1}{3}x^4 + \frac{2}{45}x^6 - \frac{1}{315}x^8,$$

and the two are the same.

c. If $f(x) = \sin^2 x$, then $f(0) = 0$, $f'(x) = \sin 2x$, so $f'(0) = 0$. $f''(x) = 2\cos 2x$, so $f''(x) = 2$, $f'''(x) = -4\sin 2x$, so $f'''(0) = 0$. Note that from this point $f^{(n)}(0) = 0$ if n is odd and $f^{(n)}(0) = \pm 2^{n-1}$ if n is even, with the signs alternating for every other even n. Thus, the series for $\sin^2 x$ is

$$2x^2/2 - 8x^4/4! + 32x^6/6! - 128x^8/8! + \cdots = x^2 - \frac{1}{3}x^4 + \frac{2}{45}x^6 - \frac{1}{315}x^8 + \cdots.$$

11.3.85 There are many solutions. For example, first find a series that has $(-1, 1)$ as an interval of convergence, say $\dfrac{1}{1-x} = \sum_{k=0}^{\infty} x^k$. Then the series $\dfrac{1}{1-x/2} = \sum_{k=0}^{\infty} \left(\dfrac{x}{2}\right)^k$ has $(-2, 2)$ as its interval of convergence. Now shift the series up so that it is centered at 4. We have $\sum_{k=0}^{\infty} \left(\dfrac{x-4}{2}\right)^k$, which has interval of convergence $(2, 6)$.

11.3.87 Use the Taylor series for $\cos x$ centered at $\pi/4$:

$$\frac{\sqrt{2}}{2}\left(1 - (x - \pi/4) - \frac{1}{2}(x - \pi/4)^2 + \frac{1}{6}(x - \pi/4)^3 + \cdots\right).$$

The remainder after n terms (because the derivatives of $\cos x$ are bounded by 1 in magnitude) is

$$|R_n(x)| \leq \frac{1}{(n+1)!} \cdot \left(\frac{\pi}{4} - \frac{2\pi}{9}\right)^{n+1}.$$

Solving for $|R_n(x)| < 10^{-4}$, we obtain $n = 3$. Evaluating the first four terms (through $n = 3$) of the series we get 0.7660427050. The true value is ≈ 0.7660444431.

11.3.89

a. Use the Taylor series for $(125 + x)^{1/3}$ centered at $x = 0$. Using the first four terms and evaluating at $x = 3$ gives a result (5.03968) accurate to within 10^{-4}.

b. Use the Taylor series for $x^{1/3}$ centered at $x = 125$. Note that this gives the identical Taylor series except that the exponential terms are $(x - 125)^n$ rather than x^n. Thus we need terms up through $(x - 125)^3$, just as before, evaluated at $x = 128$, and we obtain the identical result.

c. Because the two Taylor series are the same except for the shifting, the results are equivalent.

11.3.91 Consider the remainder after the first term of the Taylor series. Taylor's Theorem indicates that
$$R_1(x) = \frac{f''(c)}{2}(x-a)^2 \quad \text{for some } c \text{ between } x \text{ and } a,$$
so that $f(x) = f(a) + f'(a)(x-a) + \frac{f''(c)}{2}(x-a)^2$. But $f'(a) = 0$, so that for every x in an interval containing a, there is a c between x and a such that $f(x) = f(a) + \frac{f''(c)}{2}(x-a)^2$.

  a. If $f''(x) > 0$ on the interval containing a, then for every x in that interval, we have $f(x) = f(a) + \frac{f''(c)}{2}(x-a)^2$ for some c between x and a. But $f''(c) > 0$ and $(x-a)^2 > 0$, so that $f(x) > f(a)$ and a is a local minimum.

  b. If $f''(x) < 0$ on the interval containing a, then for every x in that interval, we have $f(x) = f(a) + \frac{f''(c)}{2}(x-a)^2$ for some c between x and a. But $f''(c) < 0$ and $(x-a)^2 > 0$, so that $f(x) < f(a)$ and a is a local maximum.

11.4 Working with Taylor Series

11.4.1 Replace f and g by their Taylor series centered at a, and evaluate the limit.

11.4.3 Substitute -0.6 for x in the Taylor series for e^x centered at 0. Note that this series is an alternating series, so the error can easily be estimated by looking at the magnitude of the first neglected term.

11.4.5
$$\frac{d}{dx}\sinh x = \frac{d}{dx}\left(x + \frac{x^3}{3!} + \frac{x^5}{5!} + \cdots\right)$$
$$= 1 + \frac{3x^2}{3!} + \frac{5x^4}{5!} + \cdots = 1 + \frac{x^2}{2!} + \frac{x^4}{4!} + \cdots + \frac{x^{2k}}{(2k)!} + \cdots.$$

11.4.7 Because $e^x = 1 + x + \frac{x^2}{2!} + \frac{x^3}{3!} + \cdots$, we have $\frac{e^x - 1}{x} = 1 + \frac{x}{2!} + \cdots$, so $\lim_{x \to 0} \frac{e^x - 1}{x} = 1$.

11.4.9 Because $-\ln(1-x) = x + \frac{x^2}{2} + \frac{x^3}{3} + \frac{x^4}{4} + \frac{x^5}{5} + \cdots$, we have $\frac{-x - \ln(1-x)}{x^2} = \frac{1}{2} + \frac{x}{3} + \frac{x^2}{4} + \cdots$, so $\lim_{x \to 0} \frac{-x - \ln(1-x)}{x^2} = \frac{1}{2}$.

11.4.11 We compute that
$$\frac{e^x - e^{-x}}{x} = \frac{1}{x}\left(\left(1 + x + \frac{x^2}{2} + \frac{x^3}{6} + \cdots\right) - \left(1 - x + \frac{x^2}{2} - \frac{x^3}{6} + \cdots\right)\right)$$
$$= \frac{1}{x}\left(2x + \frac{x^3}{3} + \cdots\right) = 2 + \frac{x^2}{3} + \cdots$$
so the limit of $\frac{e^x - e^{-x}}{x}$ as $x \to 0$ is 2.

11.4.13 We compute that
$$\frac{2\cos 2x - 2 + 4x^2}{2x^4} = \frac{1}{2x^4}\left(2\left(1 - \frac{(2x)^2}{2} + \frac{(2x)^4}{24} - \frac{(2x)^6}{720} + \cdots\right) - 2 + 4x^2\right)$$
$$= \frac{1}{2x^4}\left(\frac{(2x)^4}{12} - \frac{(2x)^6}{360} + \cdots\right) = \frac{2}{3} - \frac{4x^2}{45} + \cdots$$
so the limit of $\frac{2\cos 2x - 2 + 4x^2}{2x^4}$ as $x \to 0$ is $\frac{2}{3}$.

Copyright © 2019 Pearson Education, Inc.

11.4. Working with Taylor Series

11.4.15 We have $\ln(1+x) = x - \frac{1}{2}x^2 + \frac{1}{3}x^3 - \frac{1}{4}x^4 + \cdots$, so that

$$\frac{\ln(1+x) - x + x^2/2}{x^3} = \frac{x^3/3 - x^4/4 + \cdots}{x^3} = \frac{1}{3} - \frac{x}{4} + \cdots$$

so that $\lim\limits_{x \to 0} \frac{\ln(1+x) - x + x^2/2}{x^3} = \frac{1}{3}$.

11.4.17 We compute that

$$\frac{3\tan^{-1}x - 3x + x^3}{x^5} = \frac{1}{x^5}\left(3\left(x - \frac{x^3}{3} + \frac{x^5}{5} - \frac{x^7}{7} + \cdots\right) - 3x + x^3\right)$$

$$= \frac{1}{x^5}\left(\frac{3x^5}{5} - \frac{3x^7}{7} + \cdots\right) = \frac{3}{5} - \frac{3x^2}{7} + \cdots$$

so the limit of $\frac{3\tan^{-1}x - 3x + x^3}{x^5}$ as $x \to 0$ is $\frac{3}{5}$.

11.4.19 The Taylor series for $\sin 2x$ centered at 0 is

$$\sin 2x = 2x - \frac{1}{3!}(2x)^3 + \frac{1}{5!}(2x)^5 - \frac{1}{7!}(2x)^7 + \cdots = 2x - \frac{4}{3}x^3 + \frac{4}{15}x^5 - \frac{8}{315}x^7 + \cdots.$$

Thus

$$\frac{12x - 8x^3 - 6\sin 2x}{x^5} = \frac{12 - 8x^3 - (12x - 8x^3 + \frac{8}{5}x^5 - \frac{16}{105}x^7 + \cdots)}{x^5}$$

$$= -\frac{8}{5} + \frac{16}{105}x^2 - \cdots,$$

so $\lim\limits_{x \to 0} \frac{12x - 8x^3 - 6\sin 2x}{x^5} = -\frac{8}{5}$.

11.4.21 The Taylor series for $\ln(x-1)$ centered at 2 is

$$\ln(x-1) = (x-2) - \frac{1}{2}(x-2)^2 + \cdots.$$

We compute that

$$\frac{x-2}{\ln(x-1)} = \frac{x-2}{(x-2) - \frac{1}{2}(x-2)^2 + \cdots} = \frac{1}{1 - \frac{1}{2}(x-2) + \cdots}$$

so the limit of $\frac{x-2}{\ln(x-1)}$ as $x \to 2$ is 1.

11.4.23 Computing Taylor series centers at 0 gives

$$e^{-2x} = 1 - 2x + \frac{1}{2!}(-2x)^2 + \frac{1}{3!}(-2x)^3 + \cdots = 1 - 2x + 2x^2 - \frac{4}{3}x^3 + \cdots$$

$$e^{-x/2} = 1 - \frac{x}{2} + \frac{1}{2!}\left(-\frac{x}{2}\right)^2 + \frac{1}{3!}\left(-\frac{x}{2}\right)^3 + \cdots = 1 - \frac{x}{2} + \frac{1}{8}x^2 - \frac{1}{48}x^3 + \cdots.$$

Thus

$$\frac{e^{-2x} - 4e^{-x/2} + 3}{2x^2} = \frac{1 - 2x + 2x^2 - \frac{4}{3}x^3 + \cdots - (4 - 2x + \frac{1}{2}x^2 - \frac{1}{12}x^3 + \cdots) + 3}{2x^2}$$

$$= \frac{\frac{3}{2}x^2 - \frac{5}{4}x^3 + \cdots}{2x^2}$$

$$= \frac{3}{4} - \frac{5}{8}x + \cdots$$

so $\lim\limits_{x \to 0} \frac{e^{-2x} - 4e^{-x/2} + 3}{2x^2} = \frac{3}{4}$.

11.4.25

a. $f'(x) = \dfrac{d}{dx}\left(\sum_{k=0}^{\infty} \dfrac{x^k}{k!}\right) = \sum_{k=1}^{\infty} k\dfrac{x^{k-1}}{k!} = \sum_{k=0}^{\infty} \dfrac{x^k}{k!} = f(x).$

b. $f'(x) = e^x$ as well.

c. The series converges on $(-\infty, \infty)$.

11.4.27

a. $f'(x) = \dfrac{d}{dx}(\ln(1+x)) = \dfrac{d}{dx}(\sum_{k=1}^{\infty}(-1)^{k+1}\dfrac{1}{k}x^k) = \sum_{k=1}^{\infty}(-1)^{k+1}x^{k-1} = \sum_{k=0}^{\infty}(-1)^k x^k.$

b. This is the power series for $\dfrac{1}{1+x}$.

c. The Taylor series for $\ln(1+x)$ converges on $(-1, 1)$, as does the Taylor series for $\dfrac{1}{1+x}$.

11.4.29

a.
$$f'(x) = \dfrac{d}{dx}(e^{-2x}) = \dfrac{d}{dx}(\sum_{k=0}^{\infty}\dfrac{(-2x)^k}{k!}) = \dfrac{d}{dx}(\sum_{k=0}^{\infty}(-2)^k\dfrac{x^k}{k!}) = -2\sum_{k=1}^{\infty}(-2)^{k-1}\dfrac{x^{k-1}}{(k-1)!} = -2\sum_{k=0}^{\infty}\dfrac{(-2x)^k}{k!}.$$

b. This is the Taylor series for $-2e^{-2x}$.

c. Because the Taylor series for e^{-2x} converges on $(-\infty, \infty)$, so does this one.

11.4.31

a. $\tan^{-1} x = x - \dfrac{x^3}{3} + \dfrac{x^5}{5} - \cdots$, so $\dfrac{d}{dx}\tan^{-1} x^2 = 1 - x^2 + x^4 - x^6 + \cdots$.

b. This is the series for $\dfrac{1}{1+x^2}$.

c. Because the series for $\tan^{-1} x$ has a radius of convergence of 1, this series does too. Checking the endpoints shows that the interval of convergence is $(-1, 1)$.

11.4.33

a. Because $y(0) = 2$, we have $0 = y'(0) - y(0) = y'(0) - 2$ so that $y'(0) = 2$. Differentiating the equation gives $y''(0) = y'(0)$, so that $y''(0) = 2$. Successive derivatives also have the value 2 at 0, so the Taylor series is $2\sum_{k=0}^{\infty}\dfrac{t^k}{k!}$.

b. $2\sum_{k=0}^{\infty}\dfrac{t^k}{k!} = 2e^t$.

11.4.35

a. $y(0) = 2$, so that $y'(0) = 16$. Differentiating, $y''(t) - 3y'(t) = 0$, so that $y''(0) = 48$, and in general $y^{(k)}(0) = 3y^{(k-1)}(0) = 3^{k-1} \cdot 16$. Thus the power series is $2 + \dfrac{16}{3}\sum_{k=1}^{\infty}\dfrac{(3t)^k}{k!} = 2 + \sum_{k=1}^{\infty}\dfrac{3^{k-1}16}{k!}t^k$.

b. $2 + \dfrac{16}{3}\sum_{k=1}^{\infty}\dfrac{(3t)^k}{k!} = 2 + \dfrac{16}{3}(e^{3t} - 1) = \dfrac{16}{3}e^{3t} - \dfrac{10}{3}$.

11.4. Working with Taylor Series

11.4.37 The Taylor series for e^{-x^2} is $\sum_{k=0}^{\infty}(-1)^k \frac{x^{2k}}{k!}$. Thus, the desired integral is

$$\int_0^{0.25} \sum_{k=0}^{\infty}(-1)^k \frac{x^{2k}}{k!}\, dx = \sum_{k=0}^{\infty}(-1)^k \frac{x^{2k+1}}{(2k+1)k!}\bigg|_0^{0.25} = \sum_{k=0}^{\infty}(-1)^k \frac{1}{(2k+1)k! 4^{2k+1}}.$$

Because this is an alternating series, to approximate it to within 10^{-4}, we must find n such that $a_{n+1} < 10^{-4}$, or $\frac{1}{(2n+3)(n+1)! \cdot 4^{2n+3}} < 10^{-4}$. This occurs for $n=1$, so $\sum_{k=0}^{1}(-1)^k \frac{1}{(2k+1) \cdot k! \cdot 4^{2k+1}} = \frac{1}{4} - \frac{1}{192} \approx 0.245$.

11.4.39 The Taylor series for $\cos 2x^2$ is $\sum_{k=0}^{\infty}(-1)^k \frac{(2x^2)^{2k}}{(2k)!} = \sum_{k=0}^{\infty}(-1)^k \frac{4^k x^{4k}}{(2k)!}$. Note that $\cos x$ is an even function, so we compute the integral from 0 to 0.35 and double it:

$$2\int_0^{0.35} \sum_{k=0}^{\infty}(-1)^k \frac{4^k x^{4k}}{(2k)!}\, dx = 2\left(\sum_{k=0}^{\infty}(-1)^k \frac{4^k x^{4k+1}}{(4k+1)(2k)!}\right)\bigg|_0^{0.35} = 2\left(\sum_{k=0}^{\infty}(-1)^k \frac{4^k (0.35)^{4k+1}}{(4k+1)(2k)!}\right).$$

Because this is an alternating series, to approximate it to within $\frac{1}{2} \cdot 10^{-4}$, we must find n such that $a_{n+1} < \frac{1}{2} \cdot 10^{-4}$, or $\frac{4^{n+1}(0.35)^{4n+5}}{(4n+3)(2n+2)!} < \frac{1}{2} \cdot 10^{-4}$. This occurs first for $n=1$, and we have $2\left(0.35 - \frac{4 \cdot (0.35)^5}{5 \cdot 2!}\right) \approx 0.696$.

11.4.41 $\tan^{-1} x = x - x^3/3 + x^5/5 - x^7/7 + x^9/9 - \cdots$, so $\int \tan^{-1} x\, dx = \int (x - x^3/3 + x^5/5 - x^7/7 + x^9/9 - \cdots)\, dx = C + \frac{x^2}{2} - \frac{x^4}{12} + \frac{x^6}{30} - \frac{x^8}{56} + \cdots$. Thus,

$$\int_0^{0.35} \tan^{-1} x\, dx = \frac{(0.35)^2}{2} - \frac{(0.35)^4}{12} + \frac{(0.35)^6}{30} - \frac{(0.35)^8}{56} + \cdots.$$

Note that this series is alternating, and $\frac{(0.35)^6}{30} < 10^{-4}$, so we add the first two terms to approximate the integral to the desired accuracy. Calculating gives approximately 0.060.

11.4.43 The Taylor series for $(1+x^6)^{-1/2}$ is $\sum_{k=0}^{\infty}\binom{-1/2}{k} x^{6k}$, so the desired integral is

$$\int_0^{0.5} \sum_{k=0}^{\infty}\binom{-1/2}{k} x^{6k}\, dx = \sum_{k=0}^{\infty}\frac{1}{6k+1}\binom{-1/2}{k} x^{6k+1}\bigg|_0^{0.5} = \sum_{k=0}^{\infty}\frac{1}{6k+1}\binom{-1/2}{k}(0.5)^{6k+1}.$$

This is an alternating series because the binomial coefficients alternate in sign, so to approximate it to within 10^{-4}, we must find n such that $a_{n+1} < 10^{-4}$, or $\left|\frac{1}{6n+7}\binom{-1/2}{n+1}(0.5)^{6n+7}\right| < 10^{-4}$. This occurs first for $n=1$, so we have $\binom{-1/2}{0} 0.5 + \frac{1}{7}\binom{-1/2}{1}(0.5)^7 \approx 0.499$.

11.4.45 Use the Taylor series for e^x at 0: $1 + \frac{2}{1!} + \frac{2^2}{2!} + \frac{2^3}{3!}$.

11.4.47 Use the Taylor series for $\cos x$ at 0: $1 - \frac{2^2}{2!} + \frac{2^4}{4!} - \frac{2^6}{6!}$.

11.4.49 Use the Taylor series for $\ln(1+x)$ evaluated at $x = 1/2$: $\frac{1}{2} - \frac{1}{2} \cdot \frac{1}{4} + \frac{1}{3} \cdot \frac{1}{8} - \frac{1}{4} \cdot \frac{1}{16}$.

11.4.51 The Taylor series for f centered at 0 is $\dfrac{-1+\sum_{k=0}^{\infty}\frac{x^k}{k!}}{x} = \dfrac{\sum_{k=1}^{\infty}\frac{x^k}{k!}}{x} = \sum_{k=1}^{\infty}\dfrac{x^{k-1}}{k!} = \sum_{k=0}^{\infty}\dfrac{x^k}{(k+1)!}$.

Evaluating both sides at $x=1$, we have $e-1 = \sum_{k=0}^{\infty}\dfrac{1}{(k+1)!}$.

11.4.53 The Maclaurin series for $\ln(1+x)$ is

$$x - \frac{1}{2}x^2 + \frac{1}{3}x^3 - \frac{1}{4}x^4 + \cdots = \sum_{k=1}^{\infty}(-1)^{k+1}\frac{x^k}{k}.$$

By the Ratio Test, $\lim_{k\to\infty}\left|\dfrac{a_{k+1}}{a_k}\right| = \lim_{k\to\infty}\left|\dfrac{x^{k+1}k}{x^k(k+1)}\right| = |x|$, so the radius of convergence is 1. The series diverges at -1 and converges at 1, so the interval of convergence is $(-1,1]$. Evaluating at 1 gives $\ln 2 = \sum_{k=1}^{\infty}(-1)^{k+1}\dfrac{1}{k} = 1 - \dfrac{1}{2} + \dfrac{1}{3} - \dfrac{1}{4} + \cdots$.

11.4.55 $\sum_{k=0}^{\infty}\dfrac{x^k}{2^k} = \sum_{k=0}^{\infty}\left(\dfrac{x}{2}\right)^k = \dfrac{1}{1-\frac{x}{2}} = \dfrac{2}{2-x}$.

11.4.57 $\sum_{k=0}^{\infty}(-1)^k\dfrac{x^{2k}}{4^k} = \sum_{k=0}^{\infty}\left(\dfrac{-x^2}{4}\right)^k = \dfrac{1}{1+\frac{x^2}{4}} = \dfrac{4}{4+x^2}$.

11.4.59 $\ln(1+x) = -\sum_{k=1}^{\infty}(-1)^k\dfrac{x^k}{k}$, so $\ln(1-x) = -\sum_{k=1}^{\infty}\dfrac{x^k}{k}$, and finally $-\ln(1-x) = \sum_{k=1}^{\infty}\dfrac{x^k}{k}$.

11.4.61

$$\sum_{k=1}^{\infty}(-1)^k\frac{kx^{k+1}}{3^k} = \sum_{k=1}^{\infty}(-1)^k\frac{k}{3^k}x^{k+1} = \sum_{k=1}^{\infty}k\left(-\frac{1}{3}\right)^k x^{k+1}$$

$$= x^2\sum_{k=1}^{\infty}\left(-\frac{1}{3}\right)^k kx^{k-1} = x^2\sum_{k=1}^{\infty}\left(-\frac{1}{3}\right)^k \frac{d}{dx}(x^k)$$

$$= x^2\frac{d}{dx}\left(\sum_{k=1}^{\infty}\left(-\frac{x}{3}\right)^k\right) = x^2\frac{d}{dx}\left(\frac{1}{1+\frac{x}{3}}\right) = -\frac{3x^2}{(x+3)^2}.$$

11.4.63 $\sum_{k=2}^{\infty}\dfrac{k(k-1)x^k}{3^k} = x^2\sum_{k=2}^{\infty}\dfrac{k(k-1)x^{k-2}}{3^k} = x^2\dfrac{d^2}{dx^2}\left(\sum_{k=2}^{\infty}\dfrac{x^k}{3^k}\right)$

$= x^2\dfrac{d^2}{dx^2}\left(\sum_{k=2}^{\infty}\left(\dfrac{x}{3}\right)^k\right) = x^2\dfrac{d^2}{dx^2}\left(\dfrac{x^2}{9}\cdot\dfrac{1}{1-\frac{x}{3}}\right) = x^2\dfrac{d^2}{dx^2}\left(\dfrac{x^2}{9-3x}\right) = x^2\dfrac{-6}{(x-3)^3} = \dfrac{-6x^2}{(x-3)^3}$.

11.4.65

a. False. This is because $\dfrac{1}{1-x}$ is not continuous at 1, which is in the interval of integration.

b. False. The Ratio Test shows that the radius of convergence for the Taylor series for $\tan^{-1}x$ centered at 0 is 1.

c. True. $\sum_{k=0}^{\infty}\dfrac{x^k}{k!} = e^x$. Substitute $x = \ln 2$.

11.4. Working with Taylor Series

11.4.67 The Taylor series for $\sin x$ centered at 0 is

$$\sin x = x - \frac{x^3}{6} + \frac{x^5}{120} - \cdots.$$

We compute that

$$\frac{\sin ax}{\sin bx} = \frac{ax - \frac{(ax)^3}{6} + \frac{(ax)^5}{120} - \cdots}{bx - \frac{(bx)^3}{6} + \frac{(bx)^5}{120} - \cdots}$$

$$= \frac{a - \frac{a^3 x^2}{6} + \frac{a^5 x^4}{120} - \cdots}{b - \frac{b^3 x^2}{6} + \frac{b^5 x^4}{120} - \cdots}$$

so the limit of $\frac{\sin ax}{\sin bx}$ as $x \to 0$ is $\frac{a}{b}$.

11.4.69 Compute instead the limit of the log of this expression, $\lim_{x \to 0} \frac{\ln(\sin x/x)}{x^2}$. If the Taylor expansion of $\ln(\sin x/x)$ is $\sum_{k=0}^{\infty} c_k x^k$, then $\lim_{x \to 0} \frac{\ln(\sin x/x)}{x^2} = \lim_{x \to 0} \sum_{k=0}^{\infty} c_k x^{k-2} = \lim_{x \to 0} c_0 x^{-2} + c_1 x^{-1} + c_2$, because the higher-order terms have positive powers of x and thus approach zero as x does. So compute the terms of the Taylor series of $\ln\left(\frac{\sin x}{x}\right)$ up through the quadratic term. The relevant Taylor series are: $\frac{\sin x}{x} = 1 - \frac{1}{6}x^2 + \frac{1}{120}x^4 - \cdots$, $\ln(1+x) = x - \frac{1}{2}x^2 + \frac{1}{3}x^3 - \cdots$ and we substitute the Taylor series for $\frac{\sin x}{x} - 1$ for x in the Taylor series for $\ln(1+x)$. Because the lowest power of x in the first Taylor series is 2, it follows that only the linear term in the series for $\ln(1+x)$ will give any powers of x that are at most quadratic. The only term that results is $-\frac{1}{6}x^2$. Thus $c_0 = c_1 = 0$ in the above, and $c_2 = -\frac{1}{6}$, so that $\lim_{x \to 0} \frac{\ln(\sin x/x)}{x^2} = -\frac{1}{6}$ and thus $\lim_{x \to 0} \left(\frac{\sin x}{x}\right)^{1/x^2} = e^{-1/6}$.

11.4.71 The Taylor series we need are $\cos x = 1 - \frac{1}{2}x^2 + \frac{1}{24}x^4 + \ldots$, and $e^t = 1 + t + \frac{1}{2!}t^2 + \frac{1}{3!}t^3 + \frac{1}{4!}t^4 + \ldots$. We are looking for powers of x^3 and x^4 that occur when the first series is substituted for t in the second series. Clearly there will be no odd powers of x, because $\cos x$ has only even powers. Thus the coefficient of x^3 is zero, so that $f^{(3)}(0) = 0$. The coefficient of x^4 comes from the expansion of $1 - \frac{1}{2}x^2 + \frac{1}{24}x^4$ in each term of e^t. Higher powers of x clearly cannot contribute to the coefficient of x^4. Thus consider $\left(1 - \frac{1}{2}x^2 + \frac{1}{24}x^4\right)^k$. The term $-\frac{1}{2}x^2$ generates $\binom{k}{2}$ terms of value $\frac{1}{4}x^4$ for $k \geq 2$, while the other term generates k terms of value $\frac{1}{24}x^4$ for $k \geq 1$. These terms all have to be divided by the $k!$ appearing in the series for e^t. So the total coefficient of x^4 is $\frac{1}{24}\sum_{k=1}^{\infty} \frac{k}{k!} + \frac{1}{4}\sum_{k=2}^{\infty}\binom{k}{2}\frac{1}{k!}, = \frac{1}{24}\sum_{k=1}^{\infty}\frac{1}{(k-1)!} + \frac{1}{4}\sum_{k=2}^{\infty}\frac{1}{2\cdot(k-2)!}, = \frac{1}{24}\sum_{k=0}^{\infty}\frac{1}{k!} + \frac{1}{8}\sum_{k=0}^{\infty}\frac{1}{k!},$ $= \frac{1}{24}e + \frac{1}{8}e = \frac{e}{6}.$ Thus $f^{(4)}(0) = \frac{e}{6} \cdot 4! = 4e$.

11.4.73 The Taylor series for $\sin t^2$ is $\sin t^2 = t^2 - \frac{1}{3!}t^6 + \frac{1}{5!}t^{10} - \ldots$, so that $\int_0^x \sin t^2 \, dt = \frac{1}{3}t^3 - \frac{1}{7 \cdot 3!}t^7 + \cdots \Big|_0^x = \frac{1}{3}x^3 - \frac{1}{7 \cdot 3!}x^7 + \ldots$. Thus $f^{(3)}(0) = \frac{3!}{3} = 2$ and $f^{(4)}(0) = 0$.

11.4.75 Consider the series $\sum_{k=1}^{\infty} x^k = \frac{x}{1-x}$. Differentiating both sides gives $\frac{1}{(1-x)^2} = \sum_{k=0}^{\infty} kx^{k-1} =$

$\frac{1}{x}\sum_{k=0}^{\infty} kx^k$ so that $\frac{x}{(1-x)^2} = \sum_{k=0}^{\infty} kx^k$. Evaluate both sides at $x = 1/2$ to see that the sum of the series is $\frac{1/2}{(1-1/2)^2} = 2$. Thus the expected number of tosses is 2.

11.4.77 We look first for a Taylor series for $(1 - k^2 \sin^2 \theta)^{-1/2}$. Because $(1 - k^2 x^2)^{-1/2} = (1 - (kx)^2)^{-1/2} = \sum_{i=0}^{\infty} \binom{-1/2}{i}(kx)^{2i}$, and $\sin \theta = \theta - \frac{1}{3!}\theta^3 + \frac{1}{5!}\theta^5 - \ldots$, substituting the second series into the first gives

$$\frac{1}{\sqrt{1 - k^2 \sin^2 \theta}} = 1 + \frac{1}{2}k^2\theta^2 + \left(-\frac{1}{6}k^2 + \frac{3}{8}k^4\right)\theta^4$$
$$+ \left(\frac{1}{45}k^2 - \frac{1}{4}k^4 + \frac{5}{16}k^6\right)\theta^6 + \left(\frac{-1}{630}k^2 + \frac{3}{40}k^4 - \frac{5}{16}k^6 + \frac{35}{128}k^8\right)\theta^8 + \ldots.$$

Integrating with respect to θ and evaluating at $\pi/2$ (the value of the antiderivative is 0 at 0) gives

$$\frac{1}{2}\pi + \frac{1}{48}k^2\pi^3 + \frac{1}{160}\left(-\frac{1}{6}k^2 + \frac{3}{8}k^4\right)\pi^5 + \frac{1}{896}\left(\frac{1}{45}k^2 - \frac{1}{4}k^4 + \frac{5}{16}k^6\right)\pi^7$$
$$+ \frac{1}{4608}\left(-\frac{1}{630}k^2 + \frac{3}{40}k^4 - \frac{5}{16}k^6 + \frac{35}{128}k^8\right)\pi^9.$$

Evaluating these terms for $k = 0.1$ gives $F(0.1) \approx 1.574749680$. The true value is approximately 1.574745562.

11.4.79

a. By the Fundamental Theorem, $S'(x) = \sin x^2$, $C'(x) = \cos x^2$.

b. The relevant Taylor series are $\sin t^2 = t^2 - \frac{1}{3!}t^6 + \frac{1}{5!}t^{10} - \frac{1}{7!}t^{14} + \ldots$, and $\cos t^2 = 1 - \frac{1}{2!}t^4 + \frac{1}{4!}t^8 - \frac{1}{6!}t^{12} + \ldots$. Integrating, we have $S(x) = \frac{1}{3}x^3 - \frac{1}{7 \cdot 3!}x^7 + \frac{1}{11 \cdot 5!}x^{11} - \frac{1}{15 \cdot 7!}x^{15} + \ldots$, and $C(x) = x - \frac{1}{5 \cdot 2!}x^5 + \frac{1}{9 \cdot 4!}x^9 - \frac{1}{13 \cdot 6!}x^{13} + \ldots$.

c. $S(0.05) \approx \frac{1}{3}(0.05)^3 - \frac{1}{42}(0.05)^7 + \frac{1}{1320}(0.05)^{11} - \frac{1}{75600}(0.05)^{15} \approx 4.166664807 \times 10^{-5}$. $C(-0.25) \approx (-0.25) - \frac{1}{10}(-0.25)^5 + \frac{1}{216}(-0.25)^9 - \frac{1}{9360}(-0.25)^{13} \approx -.2499023616$.

d. The series is alternating. Because $a_{n+1} = \frac{1}{(4n+7)(2n+3)!}(0.05)^{4n+7}$, and this is less than 10^{-4} for $n = 0$, only one term is required.

e. The series is alternating. Because $a_{n+1} = \frac{1}{(4n+5)(2n+2)!}(0.25)^{4n+5}$, and this is less than 10^{-6} for $n = 1$, two terms are required.

11.4.81

a. $J_0(x) = 1 - \frac{1}{4}x^2 + \frac{1}{16 \cdot 2!^2}x^4 - \frac{1}{2^6 \cdot 3!^2}x^6 + \ldots$.

b. Using the Ratio Test: $\left|\frac{a_{k+1}}{a_k}\right| = \frac{x^{2k+2}}{2^{2k+2}((k+1)!)^2} \cdot \frac{2^{2k}(k!)^2}{x^{2k}} = \frac{x^2}{4(k+1)^2}$, which has limit 0 as $k \to \infty$ for any x. Thus the radius of convergence is infinite and the interval of convergence is $(-\infty, \infty)$.

11.4. Working with Taylor Series

c. Starting only with terms up through x^8, we have $J_0(x) = 1 - \frac{1}{4}x^2 + \frac{1}{64}x^4 - \frac{1}{2304}x^6 + \frac{1}{147456}x^8 + \cdots$, $J_0'(x) = -\frac{1}{2}x + \frac{1}{16}x^3 - \frac{1}{384}x^5 + \frac{1}{18432}x^7 + \cdots$, $J_0''(x) = -\frac{1}{2} + \frac{3}{16}x^2 - \frac{5}{384}x^4 + \frac{7}{18432}x^6 + \cdots$ so that $x^2 J_0(x) = x^2 - \frac{1}{4}x^4 + \frac{1}{64}x^6 - \frac{1}{2304}x^8 + \frac{1}{147456}x^{10} + \cdots$, $xJ_0'(x) = -\frac{1}{2}x^2 + \frac{1}{16}x^4 - \frac{1}{384}x^6 + \frac{1}{18432}x^8 + \cdots$, $x^2 J_0''(x) = -\frac{1}{2}x^2 + \frac{3}{16}x^4 - \frac{5}{384}x^6 + \frac{7}{18432}x^8 + \cdots$, and $x^2 J_0''(x) + xJ_0'(x) + x^2 J_0(x) = 0$.

11.4.83

a. Because $f(a) = g(a) = 0$, we use the Taylor series for $f(x)$ and $g(x)$ centered at a to compute that

$$\lim_{x \to a} \frac{f(x)}{g(x)} = \lim_{x \to a} \frac{f(a) + f'(a)(x-a) + \frac{1}{2}f''(a)(x-a)^2 + \cdots}{g(a) + g'(a)(x-a) + \frac{1}{2}g''(a)(x-a)^2 + \cdots}$$

$$= \lim_{x \to a} \frac{f'(a)(x-a) + \frac{1}{2}f''(a)(x-a)^2 + \cdots}{g'(a)(x-a) + \frac{1}{2}g''(a)(x-a)^2 + \cdots}$$

$$= \lim_{x \to a} \frac{f'(a) + \frac{1}{2}f''(a)(x-a) + \cdots}{g'(a) + \frac{1}{2}g''(a)(x-a) + \cdots} = \frac{f'(a)}{g'(a)}.$$

Because $f'(x)$ and $g'(x)$ are assumed to be continuous at a and $g'(a) \neq 0$,

$$\frac{f'(a)}{g'(a)} = \lim_{x \to a} \frac{f'(x)}{g'(x)}$$

and we have that

$$\lim_{x \to a} \frac{f(x)}{g(x)} = \lim_{x \to a} \frac{f'(x)}{g'(x)}$$

which is one form of L'Hôpital's Rule.

b. Because $f(a) = g(a) = f'(a) = g'(a) = 0$, we use the Taylor series for $f(x)$ and $g(x)$ centered at a to compute that

$$\lim_{x \to a} \frac{f(x)}{g(x)} = \lim_{x \to a} \frac{f(a) + f'(a)(x-a) + \frac{1}{2}f''(a)(x-a)^2 + \frac{1}{6}f'''(a)(x-a)^3 + \cdots}{g(a) + g'(a)(x-a) + \frac{1}{2}g''(a)(x-a)^2 + \frac{1}{6}g'''(a)(x-a)^3 + \cdots}$$

$$= \lim_{x \to a} \frac{\frac{1}{2}f''(a)(x-a)^2 + \frac{1}{6}f'''(a)(x-a)^3 + \cdots}{\frac{1}{2}g''(a)(x-a)^2 + \frac{1}{6}g'''(a)(x-a)^3 + \cdots}$$

$$= \lim_{x \to a} \frac{\frac{1}{2}f''(a) + \frac{1}{6}f'''(a)(x-a) + \cdots}{\frac{1}{2}g''(a) + \frac{1}{6}g'''(a)(x-a) + \cdots} = \frac{f''(a)}{g''(a)}.$$

Because $f''(x)$ and $g''(x)$ are assumed to be continuous at a and $g''(a) \neq 0$,

$$\frac{f''(a)}{g''(a)} = \lim_{x \to a} \frac{f''(x)}{g''(x)}$$

and we have that

$$\lim_{x \to a} \frac{f(x)}{g(x)} = \lim_{x \to a} \frac{f''(x)}{g''(x)}$$

which is consistent with two applications of L'Hôpital's Rule.

Chapter Eleven Review

1

a. True. The approximations tend to get better as n increases in size, and also when the value being approximated is closer to the center of the series. Because 2.1 is closer to 2 than 2.2 is, and because $3 > 2$, we should have $|p_3(2.1) - f(2.1)| < |p_2(2.2) - f(2.2)|$.

b. False. The interval of convergence may or may not include the endpoints.

c. True. The interval of convergence is an interval centered at 0, and the endpoints may or may not be included.

d. True. Because $f(x)$ is a polynomial, all its derivatives vanish after a certain point (in this case, $f^{(12)}(x)$ is the last nonzero derivative).

e. True.

3 $p_2(x) = 1 - \dfrac{3}{2}x^2$.

5 $p_2(x) = 1 - (x-1) + \dfrac{3}{2}(x-1)^2$.

7 $p_2(x) = 1 - \dfrac{1}{2}(x-1)^2$.

9 $p_3(x) = \dfrac{5}{4} + \dfrac{3(x-\ln 2)}{4} + \dfrac{5(x-\ln 2)^2}{8} + \dfrac{(x-\ln 2)^3}{8}$.

11

a. $p_1(x) = 1 + x$, and $p_2(x) = 1 + x + \dfrac{x^2}{2}$.

b.

n	$p_n(-0.08)$	$\|p_n(-0.08) - e^{-0.08}\|$
1	0.92	3.1×10^{-3}
2	0.923	8.4×10^{-5}

13

a. $p_1(x) = \dfrac{\sqrt{2}}{2} + \dfrac{\sqrt{2}}{2}\left(x - \dfrac{\pi}{4}\right)$, and $p_2(x) = \dfrac{\sqrt{2}}{2} + \dfrac{\sqrt{2}}{2}\left(x - \dfrac{\pi}{4}\right) - \dfrac{\sqrt{2}}{4}\left(x - \dfrac{\pi}{4}\right)^2$.

b.

n	$p_n(\pi/5)$	$\|p_n(\pi/5) - \sin(\pi/5)\|$
1	0.596	8.2×10^{-3}
2	0.587	4.7×10^{-4}

15 The derivatives of $\sin x$ are bounded in magnitude by 1, so $|R_n(x)| \leq M\dfrac{|x|^{n+1}}{(n+1)!} \leq \dfrac{|x|^{n+1}}{(n+1)!}$. But $|x| < \pi$, so $|R_3(x)| \leq \dfrac{\pi^4}{24}$.

17 Using the Ratio Test, $\lim\limits_{k\to\infty}\left|\dfrac{a_{k+1}}{a_k}\right| = \lim\limits_{k\to\infty}\left|\dfrac{(k+1)^2 x^{k+1}}{(k+1)!} \cdot \dfrac{k!}{k^2 x^k}\right| = \lim\limits_{k\to\infty}\left(\dfrac{k+1}{k}\right)^2 \dfrac{|x|}{k+1} = 0$, so the radius of convergene is ∞ and the interval of convergence is $(-\infty, \infty)$.

19 Using the Ratio Test, $\lim_{k \to \infty} \dfrac{a_{k+1}}{a_k} = \lim_{k \to \infty} \left| \dfrac{(x+1)^{2k+2}}{(k+1)!} \cdot \dfrac{k!}{(x+1)^{2k}} \right| = \lim_{k \to \infty} \dfrac{1}{k+1}(x+1)^2 = 0$, so the radius of convergene is ∞ and the interval of convergence is $(-\infty, \infty)$.

21 By the Root Test, $\lim_{k \to \infty} \sqrt[k]{|a_k|} = \lim_{k \to \infty} \left(\dfrac{|x|}{9} \right)^3 = \dfrac{|x^3|}{729}$, so the series converges for $|x| < 9$. The radius of convergence is 9. The series given by letting $x = \pm 9$ are both divergent by the Divergence Test. Thus, $(-9, 9)$ is the interval of convergence.

23 By the Ratio Test, $\lim_{k \to \infty} \left| \dfrac{(x+2)^{k+1}}{2^{k+1}\ln(k+1)} \cdot \dfrac{2^k \ln k}{(x+2)^k} \right| = \lim_{k \to \infty} \dfrac{\ln k}{2\ln(k+1)}|x+2| = \dfrac{|x+2|}{2}$. The radius of convergence is thus 2, and a check of the endpoints gives the divergent series $\sum \dfrac{1}{\ln k}$ at $x = 0$ and the convergent alternating series $\sum \dfrac{(-1)^k}{\ln k}$ at $x = -4$. The interval of convergence is therefore $[-4, 0)$.

25
$$r = \lim_{k \to \infty} \left| \dfrac{\frac{(-1)^{k+1}(2x+1)^{k+1}}{(k+1)^2 3^{k+1}}}{\frac{(-1)^k(2x+1)^k}{k^2 3^k}} \right| = \lim_{k \to \infty} \dfrac{k^2 |2x+1|}{(k+1)^2 3} = \dfrac{|2x+1|}{3}.$$

This is less than 1 for $-3 < 2x+1 < 3$, so for $-2 < x < 1$. So the radius of convergence is $\dfrac{3}{2}$, and the interval of convergence is $[-2, 1]$, because at the endpoints the series converges absolutely as the corresponding series of all positive terms is a convergent p-series.

27
$$r = \lim_{k \to \infty} \left| \dfrac{\frac{(3k+3)!x^{k+1}}{((k+1)!)^3}}{\frac{(3k)!x^k}{(k!)^3}} \right| = \lim_{k \to \infty} \dfrac{(3k+3)(3k+2)(3k+1)|x|}{(k+1)^3} = 27|x|.$$

This is less than 1 for $|x| < \dfrac{1}{27}$. The radius of convergence is $R = \dfrac{1}{27}$.

29 The Maclaurin series for $f(x)$ is $\sum_{k=0}^{\infty} x^{2k}$. By the Root Test, this converges for $|x^2| < 1$, so $-1 < x < 1$. It diverges at both endpoints, so the interval of convergence is $(-1, 1)$.

31 The Maclaurin series for $f(x)$ is $\sum_{k=0}^{\infty}(-5x)^k = \sum_{k=0}^{\infty}(-5)^k x^k$. By the Root Test, this has radius of convergence $1/5$. Checking the endpoints, we obtain an interval of convergence of $(-1/5, 1/5)$.

33 Note that $\dfrac{1}{1-10x} = \sum_{k=0}^{\infty}(10x)^k$, so $\dfrac{1}{10} \cdot \dfrac{1}{1-10x} = \dfrac{1}{10}\sum_{k=0}^{\infty}(10x)^k$. Taking the derivative of $\dfrac{1}{10} \cdot \dfrac{1}{1-10x}$ gives $f(x)$. Thus, the Maclaurin series for $f(x)$ is $\dfrac{1}{10}\sum_{k=1}^{\infty} 10k(10x)^{k-1} = \sum_{k=1}^{\infty} k(10x)^{k-1}$. Using the Ratio Test, we see that the radius of convergence is $1/10$, and checking endpoints we obtain an interval of convergence of $(-1/10, 1/10)$.

35 The first three terms are $1 + 3x + \dfrac{9x^2}{2}$. The series is $\sum_{k=0}^{\infty} \dfrac{(3x)^k}{k!}$.

37 The first three terms are $-(x - \pi/2) + \dfrac{1}{6}(x - \pi/2)^3 - \dfrac{1}{120}(x - \pi/2)^5$. The series is
$$\sum_{k=0}^{\infty}(-1)^{k+1} \dfrac{1}{(2k+1)!}\left(x - \dfrac{\pi}{2}\right)^{2k+1}.$$

39 The first three terms are $4x - \frac{1}{3}(4x)^3 + \frac{1}{5}(4x)^5$. The series is $\sum_{k=0}^{\infty} (-1)^k \frac{(4x)^{2k+1}}{2k+1}$.

41 The nth derivative of $\cosh(2x-2)$ at $x=1$ is 0 if n is odd and is 2^n if n is even. The first 3 terms of the series are thus $1 + 2(x-1)^2 + \frac{2}{3}(x-1)^4$. The whole series can be written as $\sum_{k=0}^{\infty} \frac{2^{2k}}{(2k)!}(x-1)^{2k}$.

43 $f(x) = \binom{1/3}{0} + \binom{1/3}{1}x + \binom{1/3}{2}x^2 + \cdots = 1 + \frac{1}{3}x - \frac{1}{9}x^2 + \cdots$.

45 $f(x) = \binom{-3}{0} + \binom{-3}{1}\frac{x}{2} + \binom{-3}{2}\frac{x^2}{4} + \cdots = 1 - \frac{3}{2}x + \frac{3}{2}x^2 + \cdots$.

47 Note that $f(x) = \sinh x + \cosh x = e^x$. $R_n(x) = \frac{f^{(n+1)}(c)}{(n+1)!}x^{n+1} = \frac{e^c}{(n+1)!}x^{n+1}$ for some c. $\lim_{n \to \infty} \frac{e^c}{(n+1)!}x^{n+1} = 0$ for all x.

49 The Taylor series for $\cos x$ centered at 0 is
$$\cos x = 1 - \frac{x^2}{2} + \frac{x^4}{24} - \frac{x^6}{720} + \cdots.$$
We compute that
$$\begin{aligned}\frac{x^2/2 - 1 + \cos x}{x^4} &= \frac{1}{x^4}\left(x^2/2 - 1 + \left(1 - \frac{x^2}{2} + \frac{x^4}{24} - \frac{x^6}{720} + \cdots\right)\right) \\ &= \frac{1}{x^4}\left(\frac{x^4}{24} - \frac{x^6}{720} + \cdots\right) = \frac{1}{24} - \frac{x^2}{720} + \cdots\end{aligned}$$
so the limit of $\frac{x^2/2 - 1 + \cos x}{x^4}$ as $x \to 0$ is $\frac{1}{24}$.

51 The Taylor series for $\ln(x-3)$ centered at 4 is
$$\ln(x-3) = (x-4) - \frac{1}{2}(x-4)^2 + \frac{1}{3}(x-4)^3 - \cdots.$$
We compute that
$$\begin{aligned}\frac{\ln(x-3)}{x^2-16} &= \frac{1}{(x-4)(x+4)}\left((x-4) - \frac{1}{2}(x-4)^2 + \frac{1}{3}(x-4)^3 - \cdots\right) \\ &= \frac{1}{(x-4)(x+4)}\left((x-4)\left(1 - \frac{1}{2}(x-4) + \frac{1}{3}(x-4)^2 - \cdots\right)\right) \\ &= \frac{1}{x+4}\left(1 - \frac{1}{2}(x-4) + \frac{1}{3}(x-4)^2 - \cdots\right)\end{aligned}$$
so the limit of $\frac{\ln(x-3)}{x^2-16}$ as $x \to 4$ is $\frac{1}{8}$.

53 The Taylor series for $\sec x$ centered at 0 is
$$\sec x = 1 + \frac{x^2}{2} + \frac{5x^4}{24} + \frac{61x^6}{720} + \cdots$$
and the Taylor series for $\cos x$ centered at 0 is
$$\cos x = 1 - \frac{x^2}{2} + \frac{x^4}{24} - \frac{x^6}{720} + \cdots.$$

Chapter Eleven Review

We compute that

$$\frac{\sec x - \cos x - x^2}{x^4}$$

$$= \frac{1}{x^4}\left(\left(1 + \frac{x^2}{2} + \frac{5x^4}{24} + \frac{61x^6}{720} + \cdots\right) - \left(1 - \frac{x^2}{2} + \frac{x^4}{24} - \frac{x^6}{720} + \cdots\right) - x^2\right)$$

$$= \frac{1}{x^4}\left(\frac{x^4}{6} + \frac{31x^6}{360} + \cdots\right) = \frac{1}{6} + \frac{31x^2}{360} + \cdots$$

so the limit of $\dfrac{\sec x - \cos x - x^2}{x^4}$ as $x \to 0$ is $\dfrac{1}{6}$.

55 We have $e^{-x^2} = 1 - x^2 + \dfrac{x^4}{2} - \dfrac{x^6}{6} + \dfrac{x^8}{24} - \cdots$, so $\displaystyle\int e^{-x^2}\,dx = \int\left(1 - x^2 + \dfrac{x^4}{2} - \dfrac{x^6}{6} + \dfrac{x^8}{24} - \cdots\right)dx = C + x - \dfrac{x^3}{3} + \dfrac{x^5}{10} - \dfrac{x^7}{42} + \cdots$. Thus, $\displaystyle\int_0^{1/2} e^{-x^2}\,dx = (0.5) - \dfrac{(0.5)^3}{3} + \dfrac{(0.5)^5}{10} - \dfrac{(0.5)^7}{42} + \cdots$. Because $(0.5)^7/42 < 0.001$, we can calculate the approximation using the first three numbers shown, arriving at approximately 0.461.

57 $x\cos x = x - \dfrac{x^3}{2} + \dfrac{x^5}{24} - \dfrac{x^7}{720} + \cdots$, so $\displaystyle\int x\cos x\,dx = \int\left(x - \dfrac{x^3}{2} + \dfrac{x^5}{24} - \dfrac{x^7}{720} + \cdots\right)dx = C + \dfrac{x^2}{2} - \dfrac{x^4}{8} + \dfrac{x^6}{144} - \dfrac{x^8}{5760} + \dfrac{x^{10}}{403200} - \cdots$. Thus $\displaystyle\int_0^1 x\cos x\,dx = \dfrac{1}{2} - \dfrac{1}{8} + \dfrac{1}{144} - \dfrac{1}{5760} + \cdots$. Because $\dfrac{1}{5760} < 0.001$, we add the first three terms to approximate to the desired accuracy. Calculating gives $\displaystyle\int_0^1 x\cos x\,dx \approx 0.382$.

59 The series for $f(x) = \sqrt{x}$ centered at $a = 121$ is $11 + \dfrac{x-121}{22} - \dfrac{(x-121)^2}{10648} + \dfrac{(x-121)^3}{2576816} + \cdots$. Letting $x = 119$ gives $\sqrt{119} \approx 11 - \dfrac{1}{11} - \dfrac{1}{2 \cdot 11^3} - \dfrac{1}{2 \cdot 11^5}$.

61 $\tan^{-1} x = x - x^3/3 + x^5/5 - x^7/7 + x^9/9 + \cdots$, so $\tan^{-1}(-1/3) \approx \dfrac{-1}{3} + \dfrac{1}{3\cdot 3^3} - \dfrac{1}{5\cdot 3^5} + \dfrac{1}{7\cdot 3^7}$.

63 Because $y(0) = 4$, we have $y'(0) - 16 + 12 = 0$, so $y'(0) = 4$. Differentiating the equation $n-1$ times and evaluating at 0 we obtain $y^{(n)}(0) = 4y^{(n-1)}(0)$, so that $y^{(n)}(0) = 4^n$. The Taylor series for $y(x)$ is thus $y(x) = 4 + 4x + \dfrac{4^2 x^2}{2!} + \dfrac{4^3 x^3}{3!} + \cdots$, or $y(x) = 3 + e^{4x}$.

65

a. The Taylor series for $\ln(1+x)$ is $\displaystyle\sum_{k=1}^\infty (-1)^{k+1}\dfrac{x^k}{k}$. Evaluating at $x = 1$ gives $\ln 2 = \displaystyle\sum_{k=1}^\infty (-1)^{k+1}\dfrac{1}{k}$.

b. The Taylor series for $\ln(1-x)$ is $-\displaystyle\sum_{k=1}^\infty \dfrac{x^k}{k}$. Evaluating at $x = 1/2$ gives $\ln(1/2) = -\displaystyle\sum_{k=1}^\infty \dfrac{1}{k 2^k}$, so that $\ln 2 = \displaystyle\sum_{k=1}^\infty \dfrac{1}{k 2^k}$.

c. $f(x) = \ln\left(\dfrac{1+x}{1-x}\right) = \ln(1+x) - \ln(1-x)$. Using the two Taylor series above we have $f(x) = \displaystyle\sum_{k=1}^\infty (-1)^{k+1}\dfrac{x^k}{k} - \left(-\sum_{k=1}^\infty \dfrac{x^k}{k}\right) = \sum_{k=1}^\infty (1+(-1)^{k+1})\dfrac{x^k}{k} = 2\sum_{k=0}^\infty \dfrac{x^{2k+1}}{2k+1}$.

d. Because $\dfrac{1+x}{1-x} = 2$ when $x = \dfrac{1}{3}$, the resulting infinite series for $\ln 2$ is $2\displaystyle\sum_{k=0}^{\infty} \dfrac{1}{3^{2k+1}(2k+1)}$.

e. The first four terms of each series are: $1 - \dfrac{1}{2} + \dfrac{1}{3} - \dfrac{1}{4} \approx 0.5833333333$, $\dfrac{1}{2} + \dfrac{1}{8} + \dfrac{1}{24} + \dfrac{1}{64} \approx 0.6822916667$, $\dfrac{2}{3} + \dfrac{2}{81} + \dfrac{2}{1215} + \dfrac{2}{15309} \approx 0.6931347573$ The true value is $\ln 2 \approx 0.6931471806$. The third series converges the fastest, because it has 3^{k+1} in the denominator as opposed to 2^k, so its terms get small faster.

Chapter 12

Parametric and Polar Curves

12.1 Parametric Equations

12.1.1 Given an input value of t, the point $(x(t), y(t))$ can be plotted in the xy-plane, generating a curve.

12.1.3 Let $x = R\cos(\pi t/5)$ and $y = -R\sin(\pi t/5)$. Note that as t ranges from 0 to 10, $\pi t/5$ ranges from 0 to 2π. Because $x^2 + y^2 = R^2$, this curve represents a circle of radius R. Note also that for $t = 0$ the initial point is $(R, 0)$, and for small values of t the plotted points are in the third quadrant — so the curve is being traced with clockwise orientation.

12.1.5 Let $x = t^2$ and $y = t$ for $t \in (-\infty, \infty)$.

12.1.7 $\dfrac{d}{dx}\bigg|_{t=2} = \left(\dfrac{6t}{-6t^2}\right)\bigg|_{t=2} = \left(\dfrac{-1}{t}\right)\bigg|_{t=2} = -\dfrac{1}{2}$.

12.1.9 $x = t, y = t, 0 \le t \le 6$; $x = 2t, y = 2t, 0 \le t \le 3$; $x = 3t, y = 3t, 0 \le t \le 2$. (Answers may vary).

12.1.11

a.

t	x	y
-10	-20	-34
-5	-10	-19
0	0	-4
5	10	11
10	20	26

b.
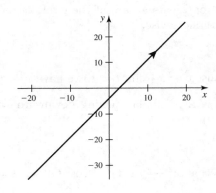

c. Solving $x = 2t$ for t yields $t = x/2$, so $y = 3t - 4 = 3x/2 - 4$.

d. The curve is the line segment from $(-20, -34)$ to $(20, 26)$.

12.1.13

a.

t	x	y
−5	11	−18
−3	9	−12
0	6	−3
3	3	6
5	1	12

b.

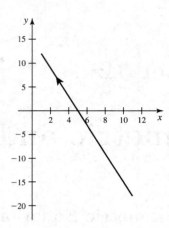

c. Solving $x = -t + 6$ for t yields $t = 6 - x$, so $y = 3t - 3 = 18 - 3x - 3 = 15 - 3x$.

d. The curve is the line segment from $(11, -18)$ to $(1, 12)$.

12.1.15

a. Because $t = x - 3$, we have $y = 1 - (x - 3) = 4 - x$.

b. This is a line segment starting at $(3, 1)$ and ending at $(4, 0)$.

12.1.17

a. Solving $x = \sqrt{t} + 4$ for t yields $t = (x - 4)^2$. Thus, $y = 3\sqrt{t} = 3(x - 4)$, where x ranges from 4 to 8. Note that all $t \geq 0$, $x > 0$, and $y > 0$.

b. The curve is the line segment from $(4, 0)$ to $(8, 12)$.

12.1.19

a. Note that $x^2 + y^2 = 9\cos^2 t + 9\sin^2 t = 9$.

b. This represents an arc of the circle of radius 3 centered at the origin from $(-3, 0)$ to $(3, 0)$ traversed counterclockwise.

12.1.21

a. Because $\cos^2 t + \sin^2 t = 1$, we have $x^2 + y = 1$, so $y = 1 - x^2$, $-1 \leq x \leq 1$.

b. This is a parabola opening downward with a vertex at $(0, 1)$, and starting at $(1, 0)$ and ending at $(-1, 0)$.

12.1.23

a. Note that $x^2 + (y - 1)^2 = \cos^2 t + \sin^2 t = 1$,

b. We have a circle of radius 1 centered at $(0, 1)$, traversed counterclockwise starting at $(1, 1)$.

12.1.25

a. Solving $x = r - 1$ for r yields $r = x + 1$. Thus, $y = r^3 = (x + 1)^3$, where $-5 \leq x \leq 3$.

b. The curve is the part of the standard cubic curve, shifted one unit to the left, from $(-5, -64)$ to $(3, 64)$.

12.1. Parametric Equations

12.1.27

a. Note that $x^2 + y^2 = 49\cos^2 2t + 49\sin^2 2t = 49$.

b. This represents an arc of the circle of radius 7 centered at the origin from $(-7,0)$ to $(-7,0)$ traversed counterclockwise. (So the whole circle is represented.)

12.1.29

a. $y = 1$, $-\infty < x < \infty$.

b. Because $y = 1$, this is a horizontal line with slope 0 and y-intercept 1.

12.1.31 Note that $x^2 + y^2 = 4\sin^2 8t + 4\cos^2 8t = 4$, so the curve is the circle $x^2 + y^2 = 4$.

12.1.33 Note that because $t = x$, we have $y = \sqrt{4 - t^2} = \sqrt{4 - x^2}$.

12.1.35 Because $\sec^2 t - 1 = \tan^2 t$, we have $y = x^2$.

12.1.37 Let $x = 4\cos t$ and $y = 4\sin t$ for $0 \leq t \leq 2\pi$. Then $x^2 + y^2 = 16\cos^2 t + 16\sin^2 t = 16$.

12.1.39 Let $x = \cos t + 2$ and $y = \sin t + 3$ for $0 \leq t \leq 2\pi$. Then $(x-2)^2 + (y-3)^2 = 1$, which is a circle with the desired center and radius and orientation.

12.1.41 Let $x = x_0 + at$ and $y = y_0 + bt$. Letting $(x_0, y_0) = (0, 0)$, and then finding a and b so that the curve is at the point Q when $t = 1$ yields $x = 2t$, $y = 8t$ for $0 \leq t \leq 1$.

12.1.43 Let $x = t$ and $y = 2t^2 - 4$, $-1 \leq t \leq 5$.

12.1.45 Let $x = 2$

12.1.47 Let $x = -2 + 4t$ and $y = 3 - 6t$, $0 \leq t \leq 1$, and $x = t + 1$, $y = 8t - 11$ for $1 \leq t \leq 2$.

12.1.49 Let $x = 1 + 2t$ and $y = 1 + 4t$, for $-\infty < t < \infty$. Note that $y = 2(1 + 2t) - 1$, so $y = 2x - 1$.

12.1.51 Let $x = t^2$ and $y = t$, for $0 \leq t < \infty$. Note that $x = t^2 = y^2$, and that the starting point is $(0, 0)$.

12.1.53 Let t be time in minutes, so $0 \leq t \leq 1.5$ Let $x = 400\cos(4\pi/3)t$ and $y = 400\sin(4\pi/3)t$. Then because $x^2 + y^2 = 400^2$, the path is a circle of radius 400. Note that the values of x and y are the same at $t = 0$ and $t = 1.5$, and that the circle is traversed counterclockwise.

12.1.55 Let t be time in seconds, so $0 \leq t \leq 24$ Let $x = 50\cos(\pi/12)t$ and $y = 50\sin(\pi/12)t$. Then because $x^2 + y^2 = 50^2$, the path is a circle of radius 50. Note that the values of x and y are the same at $t = 0$ and $t = 24$, and that the circle is traversed counterclockwise.

12.1.57

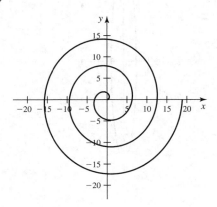

12.1.59

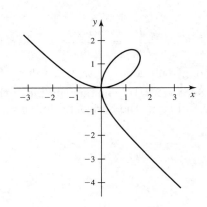

12.1.61

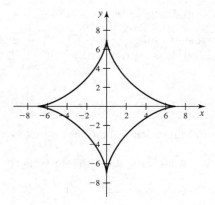

12.1.63 Let $x = 1 + \cos^2 t - \sin^2 t$ and $y = t$, for $-\infty < t < \infty$. Note that because $1 - \sin^2 t = \cos^2 t$, we have $x = 2\cos^2 t$, $y = t$.

12.1.65 The packet lands when $y = 0$, so when $-4.9t^2 + 4000 = 0$ for $t > 0$. This occurs when $t = \sqrt{\frac{4000}{4.9}} \approx 28.571$ seconds. At that time, $x \approx 100 \cdot 28.57 = 2857$ meters.

12.1.67

a. $\dfrac{dy}{dx} = \dfrac{dy/dt}{dx/dt} = -\dfrac{8}{4} = -2$ for all t. Because the curve is a line, the tangent line to the curve at the given point is the line itself.

b.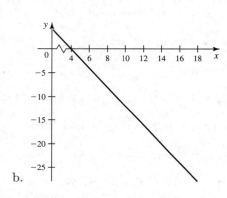

12.1.69

a. $\dfrac{dy}{dx} = \dfrac{dy/dt}{dx/dt} = \dfrac{8\cos t}{-\sin t} = -8\cot t$. At the given value of t, the value of $\dfrac{dy}{dx}$ is $-8\cot \pi/2 = 0$. The tangent line at the point $(0, 8)$ is thus the horizontal line $y = 8$.

b.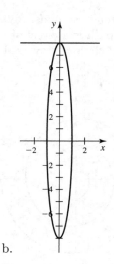

12.1. Parametric Equations

12.1.71

a. $\dfrac{dy}{dx} = \dfrac{dy/dt}{dx/dt} = \dfrac{1 + \frac{1}{t^2}}{1 - \frac{1}{t^2}} = \dfrac{t^2 + 1}{t^2 - 1}$. At the given value of t, the derivative doesn't exist, and the tangent line is the vertical line $x = 2$, tangent at the point $(2, 0)$.

b.

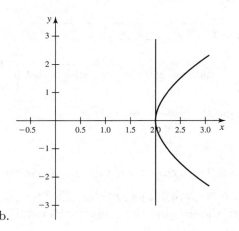

12.1.73 The point corresponding to $t = 2$ is $(3, 10)$. $\dfrac{dy}{dx} = \dfrac{dy/dt}{dx/dt} = \dfrac{3t^2 + 1}{2t}$, so the slope at $t = 2$ is $\dfrac{13}{4}$. The equation of the tangent line is therefore $y - 10 = \dfrac{13}{4}(x - 3)$, or $y = \dfrac{13}{4}x + \dfrac{1}{4}$.

12.1.75 The point corresponding to $t = \pi/4$ is
$$\left(\dfrac{4\sqrt{2} + \pi\sqrt{2}}{8}, \dfrac{4\sqrt{2} - \pi\sqrt{2}}{8} \right).$$
$\dfrac{dy}{dx} = \dfrac{dy/dt}{dx/dt} = \dfrac{\cos t - (\cos t - t \sin t)}{-\sin t + (\sin t + t \cos t)} = \tan t$. At $t = \pi/4$, we have a slope of 1. The equation of the tangent line is thus $y - \dfrac{4\sqrt{2} - \pi\sqrt{2}}{8} = 1 \left(x - \dfrac{4\sqrt{2} + \pi\sqrt{2}}{8} \right)$, or $y = x - \dfrac{\pi\sqrt{2}}{4}$.

12.1.77 $\dfrac{dy}{dx} = \dfrac{dy/dt}{dx/dt} = \dfrac{4\cos t}{-4\sin t} = -\cot t$. We seek t so that $\cot t = -1/2$, so $t = \cot^{-1}(-1/2)$. The corresponding points on the curve are $\left(-\dfrac{4\sqrt{5}}{5}, \dfrac{8\sqrt{5}}{5} \right)$ and $\left(\dfrac{4\sqrt{5}}{5}, -\dfrac{8\sqrt{5}}{5} \right)$.

12.1.79 $\dfrac{dy}{dx} = \dfrac{dy/dt}{dx/dt} = \dfrac{1 + (1/t^2)}{1 - (1/t^2)} = \dfrac{t^2 + 1}{t^2 - 1}$. We seek t so that $\dfrac{t^2 + 1}{t^2 - 1} = 1$, which never occurs. Thus, there are no points on this curve with slope 1.

12.1.81 The arc length is $\displaystyle\int_0^2 \sqrt{9 + 16}\, dt = 5t \Big|_0^2 = 10$.

12.1.83 The arc length is
$$\int_0^\pi \sqrt{(-\sin t - \cos t)^2 + (-\sin t + \cos t)^2}\, dt = \int_0^\pi \sqrt{2\sin^2 t + 2\cos^2 t}\, dt = \sqrt{2}\, t \Big|_0^\pi = \pi\sqrt{2}.$$

12.1.85 The arc length is
$$\int_0^1 \sqrt{(4t^3)^2 + (2t^5)^2}\, dt = \int_0^1 \sqrt{16t^6 + 4t^{10}}\, dt = \int_0^1 t^3 \sqrt{16 + 4t^4}\, dt = \dfrac{1}{16}\int_0^1 16t^3 \sqrt{16 + 4t^4}\, dt.$$
Let $u = 16 + t^4$ so that $du = 16t^3\, dt$. Then we have
$$\dfrac{1}{16}\int_{16}^{20} u^{1/2}\, du = \dfrac{2}{48} u^{3/2} \Big|_{16}^{20} = \dfrac{1}{3}\left(5\sqrt{5} - 8 \right).$$

12.1.87 The arc length is

$$\int_0^{\pi/4} \sqrt{((3\cos^2 2t)(-2\sin 2t))^2 + ((3\sin^2 2t)(2\cos 2t))^2}\, dt = \int_0^{\pi/4} \sqrt{36\cos^4 2t \sin^2 2t + 36\sin^4 2t \cos^2 2t}\, dt$$

$$= 6\int_0^{\pi/4} \cos 2t \sin 2t \sqrt{\cos^2 2t + \sin^2 2t}\, dt = 6\int_0^{\pi/4} \cos 2t \sin 2t\, dt = 3\int_0^{\pi/4} \sin 4t\, dt$$

$$= -\frac{3}{4}\cos 4t\Big|_0^{\pi/4} = -\frac{3}{4}(-1-1) = \frac{3}{2}.$$

12.1.89

a. False. This generates a circle in the counterclockwise direction.

b. True. Note that when t is increased by one, the value of $2\pi t$ is increased by 2π, which is the period of both the sine and the cosine functions.

c. False. This generates only the portion of the parabola in the first quadrant, omitting the portion in the second quadrant.

d. True. They describe the portion of the unit circle in the 4th and 1st quadrants.

e. True. This ellipse has vertical tangents at $t = 0$ and $t = \pi$.

12.1.91

The entire curve is traversed for $0 \leq t \leq 2\pi$.

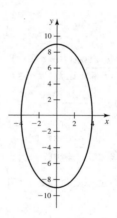

12.1.93

Let $x = 3\cos t$ and $y = \frac{3}{2}\sin t$ for $0 \leq t \leq 2\pi$. Then the major axis on the x-axis has length $2 \cdot 3 = 6$ and the minor axis on the y-axis has length $2 \cdot \frac{3}{2} = 3$. Note that $\left(\frac{x}{3}\right)^2 + \left(\frac{2y}{3}\right)^2 = \cos^2 t + \sin^2 t = 1$.

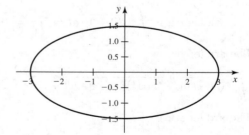

12.1.95

a. For $(1+s, 2s) = (1+2t, 3t)$, we must have $1+s = 1+2t$ and $2s = 3t$, so that $s = 2t$ and $2s = 3t$. The only solution to this pair of equations is $s = t = 0$, so these two lines intersect when $s = t = 0$, at the point $(1, 0)$.

12.1. Parametric Equations

b. For $(2+5s, 1+s) = (4+10t, 3+2t)$, we must have $2+5s = 4+10t$ and $1+s = 3+2t$, so that $s = 2+2t$ and $s = 2+2t$. This pair of equations has no solutions, so the lines are parallel.

c. For $(1+3s, 4+2s) = (4-3t, 6+4t)$, we must have $1+3s = 4-3t$ and $4+2s = 6+4t$, so that $s = 1-t$ and $s = 1+2t$. The only solution to this pair of equations is $s = 1$ and $t = 0$, so these two lines intersect for these values of s and t, at the point $(4, 6)$.

12.1.97

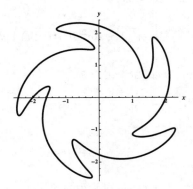

12.1.99

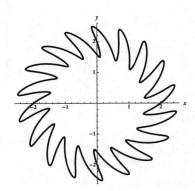

12.1.101

a. $\dfrac{dy}{dx} = \dfrac{dy/dt}{dx/dt} = \dfrac{2\cos t}{2\cos 2t}$. This is zero when $\cos t = 0$ but $\cos 2t \neq 0$, which occurs for $t = \pi/2$ and $t = 3\pi/2$. The corresponding points on the graph are $(0, 2)$ and $(0, -2)$.

b. Using the derivative obtained above, we seek points where $\cos 2t = 0$ but $\cos t \neq 0$. This occurs for $t = \pi/4, 3\pi/4, 5\pi/4$, and $7\pi/4$. The corresponding points on the curve are $(1, \sqrt{2})$, $(-1, \sqrt{2})$, $(-1, -\sqrt{2})$, and $(1, -\sqrt{2})$.

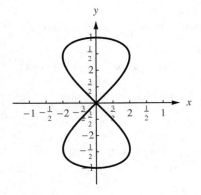

12.1.103 One arch is traced out by the parametric equations for $0 \leq t \leq 2\pi$. The base of the arch is in the interval $0 \leq x \leq 6\pi$. So the area is given by

$$\int_0^{6\pi} y\, dx = 9 \int_0^{2\pi} (1 - \cos t)(1 - \cos t)\, dt = 9 \int_0^{2\pi} (1 - 2\cos t + \cos^2 t)\, dt$$

$$= 9 \int_0^{2\pi} \left(1 - 2\cos t + \frac{1}{2} + \frac{1}{2}\cos 2t\right) dt = 9\left(\frac{3t}{2} - 2\sin t + \frac{1}{4}\sin 2t\right)\bigg|_0^{2\pi} = 27\pi.$$

12.1.105 The area is given by

$$4\int_{\pi/2}^{0} \sin^3 t(3\cos^2 t)(-\sin t)\,dt = 12\int_0^{\pi/2}\sin^4 t\cos^2 t\,dt = \frac{3}{2}\int_0^{\pi/2}(1-\cos 2t)^2(1+\cos 2t)\,dt$$

$$= \frac{3}{2}\int_0^{\pi/2}(1-\cos^2 2t)(1-\cos 2t)\,dt$$

$$= \frac{3}{2}\int_0^{\pi/2}(1-\cos 2t - \cos^2 2t + \cos^3 2t)\,dt$$

$$= \frac{3}{2}\int_0^{\pi/2}\left(1 - \cos 2t - \frac{1+\cos 4t}{2} + (1-\sin^2 2t)(\cos 2t)\right)dt$$

$$= \frac{3}{2}\int_0^{\pi/2}\left(\frac{1}{2} - \frac{\cos 4t}{2} - \sin^2 2t\cos 2t\right)dt$$

$$= \frac{3}{2}\left(\frac{t}{2} - \frac{\sin 4t}{8} - \frac{\sin^3 2t}{6}\right)\bigg|_0^{\pi/2} = \frac{3\pi}{8}.$$

12.1.107

a. Observe that $x^2 + (y-4)^2 = 9\cos^2 t + 9\sin^2 t = 9$. So the curve is a circle of radius 3 centered at $(0,4)$.

b. Because the circle lies above the x-axis, the corresponding solid of revolution is a torus. The surface area is

$$S = \int_0^{2\pi} 2\pi g(t)\sqrt{(f'(t))^2 + (g'(t))^2}\,dt = \int_0^{2\pi} 2\pi(3\sin t + 4)\sqrt{(-3\sin t)^2 + (3\cos t)^2}\,dt$$

$$= 6\pi\int_0^{2\pi}(3\sin t + 4)\,dt = 6\pi(-3\cos t + 4t)\bigg|_0^{2\pi} = 6\pi(8\pi) = 48\pi^2.$$

12.1.109 The area is given by

$$\int_0^{2\pi} 2\pi(1-\cos t)\sqrt{(1-\cos t)^2 + \sin^2 t}\,dt = \int_0^{2\pi} 2\pi(1-\cos t)\sqrt{2-2\cos t}\,dt$$

$$= 2\pi\sqrt{2}\int_0^{2\pi}(1-\cos t)^{3/2}\,dt = 2\pi\sqrt{2}\int_0^{2\pi}\left(2\sin^2 \frac{t}{2}\right)^{3/2}dt$$

$$= 8\pi\int_0^{2\pi}\sin^3\frac{t}{2}\,dt = 8\pi\int_0^{2\pi}\left(1-\cos^2\frac{t}{2}\right)\left(\sin\frac{t}{2}\right)dt.$$

Now let $u = \cos\frac{t}{2}$ so that $du = -\frac{1}{2}\sin\frac{t}{2}$. Substituting gives

$$-16\pi\int_1^{-1}(1-u^2)\,du = 16\pi\left(u - \frac{u^3}{3}\right)\bigg|_{-1}^{1}$$

$$= 16\pi\cdot\frac{4}{3} = \frac{64\pi}{3}.$$

12.1.111 The area is given by $\int_0^1 2\pi(e^{3t}+1)\sqrt{4e^{4t}+9e^{6t}}\,dt$ which can be evaluated with a CAS to give about 1445.9.

12.1.113

a.

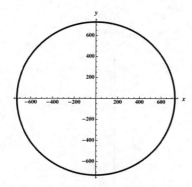 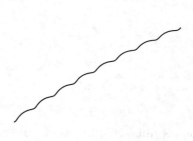

b.

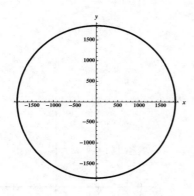 

12.2 Polar Coordinates

12.2.1 The coordinates $(2, \pi/6)$, $(2, -11\pi/6)$, and $(-2, 7\pi/6)$ all give rise to the same point. Also, the coordinates $(-3, -\pi/2)$, $(3, \pi/2)$ and $(-3, 3\pi/2)$ give rise to the same point.

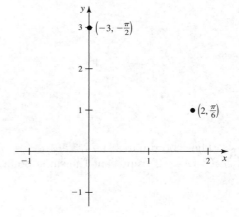

12.2.3 If a point has Cartesian coordinates (x, y) then $r^2 = x^2 + y^2$ and $\tan\theta = y/x$ for $x \neq 0$. If $x = 0$, then $\theta = \pi/2$ and $r = y$.

12.2.5 Because $x = r\cos\theta$, we have that the vertical line $x = 5$ has polar equation $r = 5\sec\theta$.

12.2.7 x-axis symmetry occurs if (r,θ) on the graph implies $(r,-\theta)$ is on the graph. y-axis symmetry occurs if (r,θ) on the graph implies $(r,\pi-\theta) = (-r,-\theta)$ is on the graph. Symmetry about the origin occurs if (r,θ) on the graph implies $(-r,\theta) = (r,\theta+\pi)$ is on the graph.

12.2.9 The coordinates $(2,\pi/4)$, $(-2,5\pi/4)$, and $(2,9\pi/4)$ represent the same point.

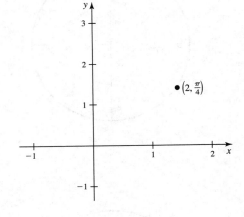

12.2.11 The coordinates $(-1,-\pi/3)$, $(1,2\pi/3)$ and $(1,-4\pi/3)$ represent the same point.

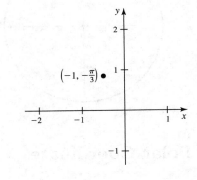

12.2.13 The coordinates $(-4,3\pi/2)$, $(4,\pi/2)$ and $(-4,-\pi/2)$ represent the same point.

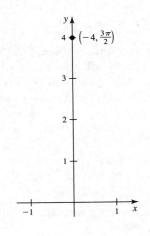

12.2.15

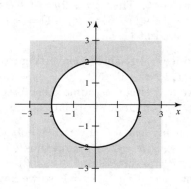

12.2.17

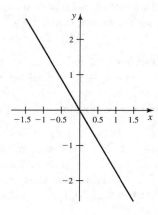

12.2.19

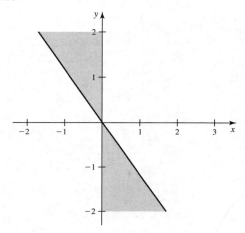

12.2.21

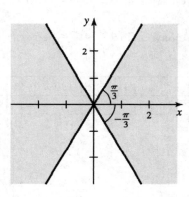

12.2.23 The angle (in radians) is $\frac{\pi}{2} - \frac{3\pi}{4}$, so the polar coordinates are $\left(100, \frac{\pi}{2} - \frac{3\pi}{4}\right) = \left(100, -\frac{\pi}{4}\right)$.

12.2.25 $x = 3\cos(\pi/4) = \frac{3\sqrt{2}}{2}$. $y = 3\sin(\pi/4) = \frac{3\sqrt{2}}{2}$.

12.2.27 $x = \cos(-\pi/3) = \frac{1}{2}$. $y = \sin(-\pi/3) = -\frac{\sqrt{3}}{2}$.

12.2.29 $x = -4\cos(3\pi/4) = 2\sqrt{2}$. $y = -4\sin(3\pi/4) = -2\sqrt{2}$.

12.2.31 $r^2 = x^2 + y^2 = 4 + 4 = 8$, so $r = \sqrt{8}$. $\tan\theta = 1$, so $\theta = \pi/4$, so $(2\sqrt{2}, \pi/4)$ is one representation of this point, and $(-2\sqrt{2}, -3\pi/4)$ is another.

12.2.33 $r^2 = x^2 + y^2 = 1 + 3 = 4$, so $r = \pm 2$. $\tan\theta = \sqrt{3}$, so $\theta = \pi/3, 4\pi/3$. $(2, \pi/3)$ is one representation of this point, and $(-2, -2\pi/3)$ is another.

12.2.35 $r^2 = 64$, so $r = \pm 8$. $\tan\theta = -\sqrt{3}$, so $\theta = -\pi/3, 2\pi/3$. One representation of the given point is $(8, 2\pi/3)$, and $(-8, -\pi/3)$ is another.

12.2.37 $x = -4$, so this is the vertical line through $(-4, 0)$.

12.2.39 Because $x^2 + y^2 = r^2 = 4$, this is a circle of radius 2 centered at the origin.

12.2.41 Note that $x^2 + y^2 = r^2 = 4\sin^2\theta + 8\sin\theta\cos\theta + 4\cos^2\theta = 4 + 8\sin\theta\cos\theta$. Also note that $x = r\cos\theta = 2\sin\theta\cos\theta + 2\cos^2\theta$ and $y = r\sin\theta = 2\sin^2\theta + 2\sin\theta\cos\theta$. Thus, $2x + 2y = 4 + 8\sin\theta\cos\theta$. If we combine these, we see that $x^2 + y^2 - (2x + 2y) = 0$. Thus $(x^2 - 2x + 1) + (y^2 - 2y + 1) = 2$, so we have the circle $(x - 1)^2 + (y - 1)^2 = 2$. This is a circle of radius $\sqrt{2}$ centered at $(1, 1)$.

12.2.43 Multiplying each side of the equation by r, we have $r^2 = 6r\cos\theta + 8r\sin\theta$. Converting to Cartesian coordinates, we have $x^2 + y^2 = 6x + 8y$. Completing the square leads to $x^2 - 6x + 9 + y^2 - 8y + 16 = 25$ and then factoring we arrive at $(x - 3)^2 + (y - 4)^2 = 5^2$. So the curve is a circle of radius 5 centered at $(3,4)$. This result can also be obtained by inspection using the summary box on circles in polar coordinates.

12.2.45 $r\cos\theta = \sin 2\theta = 2\sin\theta\cos\theta$. Note that if $\cos\theta = 0$, then r can be any real number, and the equation is satisfied. For $\cos\theta \neq 0$, we have $x = r\cos\theta = 2\sin\theta\cos\theta$, so $r = 2\sin\theta$, and thus $y = r\sin\theta = 2\sin^2\theta$. Thus $x^2 + y^2 - 2y = 4\sin^2\theta\cos^2\theta + 4\sin^2\theta\sin^2\theta - 4\sin^2\theta = 4\sin^2\theta(\sin^2\theta + \cos^2\theta) - 4\sin^2\theta = 4\sin^2\theta - 4\sin^2\theta = 0$. Note also that $x^2 + y^2 - 2y = 0$ is equivalent to $x^2 + (y - 1)^2 = 1$, so we have a circle of radius one centered at $(0, 1)$, as well as the line $x = 0$ which is the y-axis.

12.2.47 $r = 8\sin\theta$, so $r^2 = 8r\sin\theta$, so $x^2 + y^2 = 8y$. This can be written $x^2 + (y - 4)^2 = 16$, which represents a circle of radius 4 centered at $(0, 4)$.

12.2.49 We have $r\sin\theta = r^2\cos^2\theta$, so $r = \dfrac{\sin\theta}{\cos^2\theta} = \tan\theta\sec\theta$.

12.2.51 We have $r\sin\theta = \dfrac{1}{r\cos\theta}$, so $r^2 = \sec\theta\csc\theta$.

12.2.53

θ	0	$\pi/6$	$\pi/4$	$\pi/3$	$\pi/2$	$2\pi/3$	$3\pi/4$	$5\pi/6$	π
r	8	$4\sqrt{3}$	$4\sqrt{2}$	4	0	-4	$-4\sqrt{2}$	$-4\sqrt{3}$	-8

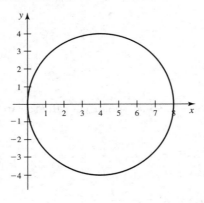

12.2.55 $r(\sin\theta - 2\cos\theta) = 0$ when $r = 0$ or when $\tan\theta = 2$, so the curve is a straight line through the origin of slope 2.

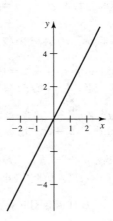

12.2. Polar Coordinates

12.2.57

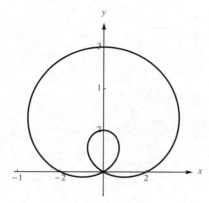

12.2.59

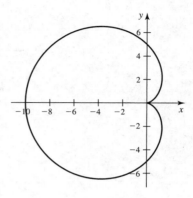

12.2.61

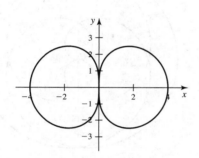

12.2.63

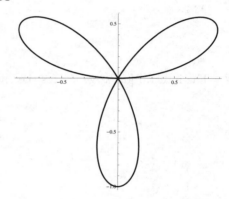

12.2.65 Points B, D, F, H, J and L have y-coordinate 0, so the graph is at the pole for each of these points. Points E, I, and M have maximal radius, so these correspond to the points at the tips of the outer loops. The points C, G and K correspond to the tips of the smaller loops. Point A corresponds to the polar point $(1, 0)$.

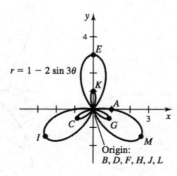

12.2.67 Points B, D, F, H, J, L, N and P are at the origin. C, G, K and O are on the ends of the long loops, while A, E, I and M are at the ends of the smaller loops.

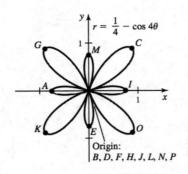

12.2.69 The interval $[0, 8\pi]$ generates the entire graph.

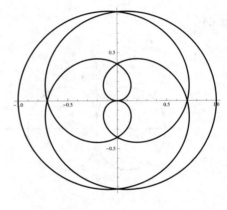

12.2.71 The interval $[0, 2\pi]$ generates the entire graph.

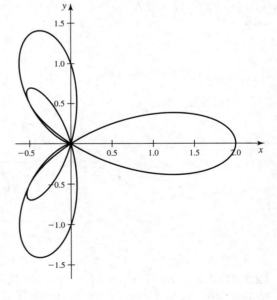

12.2. Polar Coordinates

12.2.73 The interval $[0, 5\pi]$ generates the entire graph.

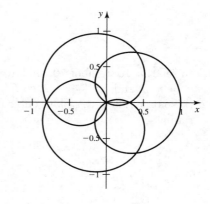

12.2.75 The interval $[0, 2\pi]$ generates the entire graph.

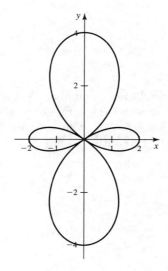

12.2.77

a. True. Note that $r^2 = 8$ and $\tan\theta = -1$.

b. True. Their intersection point (in Cartesian coordinates) is $(4, -2)$.

c. False. They intersect at the polar coordinates $(2, \pi/4)$ and $(2, 5\pi/4)$.

d. True. Note that for $\theta = \frac{3\pi}{2}$ we have $r = -3$. But the polar point $\left(-3, \frac{3\pi}{2}\right)$ is the same as the polar point $\left(3, \frac{\pi}{2}\right)$.

e. True. The first is the line $x = 2$ because $x = r\cos\theta = 2\sec\theta\cos\theta = 2$, and the second is $y = 3$ because $y = r\sin\theta = 3\csc\theta\sin\theta = 3$.

12.2.79 Consider the circle with center $C(r_0, \theta_0)$, and let A be the origin and $B(r, \theta)$ be a point on the circle not collinear with A and C. Note that the length of side BC is R, and that the angle CAB has measure $\theta - \theta_0$. Applying the law of cosines to triangle CAB yields the equation $R^2 = r^2 + r_0^2 - 2rr_0\cos(\theta - \theta_0)$, which is equivalent to the given equation.

12.2.81 In relation to number 79, we have $r_0 = 2$ and $\theta_0 = \pi/3$, and $R^2 - 4 = 12$, so $R^2 = 16$. Thus this is a circle with polar center $(2, \pi/3)$ and radius 4.

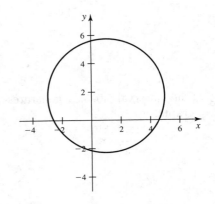

12.2.83 In relation to number 78, we have $a = 2$ and $b = 3$. So $R^2 - a^2 - b^2 = R^2 - 13 = 3$, and thus $R^2 = 16$. Thus we have a circle centered at $(2, 3)$ with radius 4.

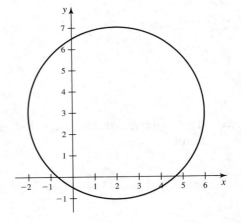

12.2.85

a. During this 30-minute period, the plane moves on a straight line from the point (using Cartesian coordintes) $(0, 150)$, to $(100 \cos \frac{\pi}{6}, 100 \sin \frac{\pi}{6}) = (50\sqrt{3}, 50)$. So the distance travelled is

$$\sqrt{(50\sqrt{3})^2 + 100^2} \approx 132.3 \text{ miles.}$$

b. The average velocity is $\dfrac{132.3}{1/2} = 264.6$ miles per hour.

12.2.87 Using problem 86b, this is the line with $r_0 = 3$ and $\theta_0 = \frac{\pi}{3}$. So it is the line through the polar point $(3, \pi/3)$ in the direction of angle $\pi/3 + \pi/2 = 5\pi/6$. The Cartesian equation is $y = -\frac{x}{\sqrt{3}} + 2\sqrt{3}$.

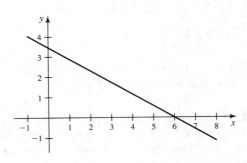

12.2.89 Using problem 86a, this is the line with $b = 3$ and $m = 4$, so $y = 4x + 3$.

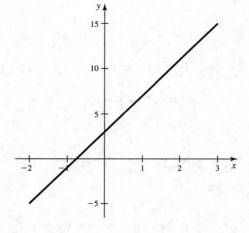

12.2.91

a. This matches (A), because we have $|a| = 1 = |b|$, and the graph is a cardioid.

b. This matches (C). This has an inner loop because $|a| = 1 < 2 = |b|$. Note that $r = 1$ when $\theta = 0$, so it can't be (D).

c. This matches (B). This has $|a| = 2 > 1 = |b|$, so it has an oval-like shape.

d. This matches (D). This has an inner loop because $|a| = 1 < 2 = |b|$. Note that $r = -1$ when $\theta = 0$, so this can't be (C).

e. This matches (E). Note that there is an inner loop because $|a| = 1 < 2 = |b|$, and that $r = 3$ when $\theta = \pi/2$.

f. This matches (F).

12.2.93

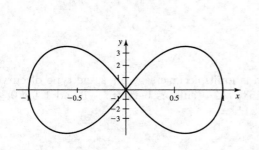

12.2.95

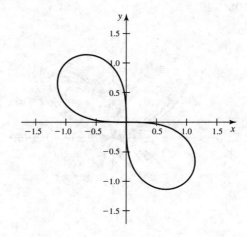

12.2.97

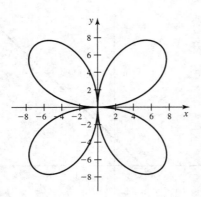

12.2.99

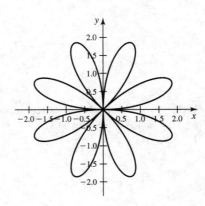

12.2.101 Note that $a\sin m\theta = 0$ for $\theta = \dfrac{k\pi}{m}$, $k = 1, 2, \ldots, 2m$. Thus the graph is back at the pole $r = 0$ for each of these values, and each of these gives rise to a distinct petal of the rose if m is odd. If m is even, then by symmetry, each petal for $k = 1, 2, \ldots \dfrac{m}{2}$ is equivalent to one for $k = \dfrac{m}{2} + 1, \dfrac{m}{2} + 2, \ldots, m$. (Note that this follows because the sine function is odd.) A similar result holds for the rose $r = a\cos\theta$.

12.2.103 For $a = 1$, the spiral winds outward counterclockwise. For $a = -1$, the spiral winds inward counterclockwise.

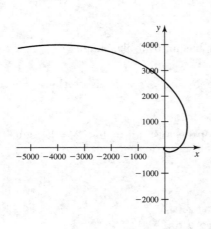

12.2.105

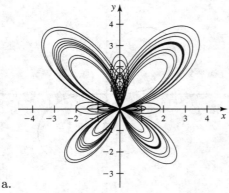

a.

b. It adds multiple layers of the same type of curve as $\sin^5\theta/12$ oscillates between -1 and 1 for $0 \leq \theta \leq 24\pi$.

12.2.107

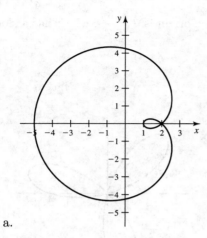

a.

b. $r = 3 - 4\cos \pi t$ is a limaçon, and $x - 2 = r\cos \pi t$ and $y = r\sin \pi t$ is a circle, and the composition of a limaçon and a circle is a limaçon.

12.2.109 Because $\sin(\theta/2) = \sin(\pi - \theta/2) = \sin((2\pi - \theta)/2)$, we have that the graph is symmetric with respect to the x-axis.

12.2.111 Note that $\cos(2\theta) = \cos^2 \theta - \sin^2 \theta$, so $r^2 = a^2(\cos^2 \theta - \sin^2 \theta)$, so $r^4 = a^2(r^2 \cos^2 \theta - r^2 \sin^2 \theta)$, so $(x^2 + y^2)^2 = a^2(x^2 - y^2)$.

12.3 Calculus in Polar Coordinates

12.3.1 Because $x = r\cos\theta$ and $y = r\sin\theta$, we have $x = f(\theta)\cos\theta$ and $y = f(\theta)\sin(\theta)$.

12.3.3 Because slope is given relative to the horizontal and vertical coordinates, it is given by $\dfrac{dy}{dx}$, not by $\dfrac{dr}{d\theta}$.

12.3.5 The slope is $\tan \dfrac{\pi}{3} = \sqrt{3}$.

12.3.7 The area is given by $\displaystyle\int_0^\pi \dfrac{1}{2}\theta\, d\theta = \dfrac{\theta^2}{4}\bigg|_0^\pi = \dfrac{\pi^2}{4}$.

12.3.9 Both curves pass through the origin, but for different values of θ.

12.3.11 $\dfrac{dy}{dx} = \dfrac{-\cos\theta\sin\theta + (1-\sin\theta)\cos\theta}{-\cos^2\theta - (1-\sin\theta)\sin\theta}$. At $(1/2, \pi/6)$, we have $\dfrac{dy}{dx} = \dfrac{0}{-1} = 0$.

12.3.13 $\dfrac{dy}{dx} = \dfrac{16\cos\theta\sin\theta}{-8\sin^2\theta + 8\cos^2\theta}$. At $(4, 5\pi/6)$ we have $\dfrac{dy}{dx} = \dfrac{-4\sqrt{3}}{4} = -\sqrt{3}$.

12.3.15 $\dfrac{dy}{dx} = \dfrac{-3\sin^2\theta + (6+3\cos\theta)\cos\theta}{-3\cos\theta\sin\theta - (6+3\cos\theta)\sin\theta}$. At both $(3, \pi)$ and $(9, 0)$, this doesn't exist.

12.3.17 $\dfrac{dy}{dx} = \dfrac{-8\sin(2\theta)\sin\theta + 4\cos(2\theta)\cos\theta}{-8\sin(2\theta)\cos\theta - 4\cos(2\theta)\sin\theta}$. The tips of the leaves occur at $\theta = 0, \pi/2, \pi$ and $3\pi/2$. At 0 and at π, we have that $\dfrac{dy}{dx}$ doesn't exist. At $\pi/2$ and $3\pi/2$ we have $\dfrac{dy}{dx} = 0$.

12.3.19 The slopes of the tangent lines are given by $\tan(\pi/4) = 1$ and $\tan(-\pi/4) = -1$.

12.3.21 $\cos\theta + \sin\theta = 0$ implies that $\cos\theta = -\sin\theta$, or $\tan\theta = -1$. So $\theta = \dfrac{3\pi}{4}$ or $\theta = \dfrac{7\pi}{4}$. Both of these represent the same line. So the line tangent to the curve at the origin is $\theta = \dfrac{3\pi}{4}$, which has a slope of -1.

12.3.23

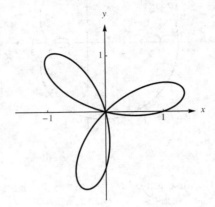

a. The curved is generated for $[0, \pi]$.

b. $\cos 3\theta + \sin 3\theta = 0$ implies that $\tan 3\theta = -1$, which means that $3\theta = \dfrac{3\pi}{4}, \dfrac{7\pi}{4}$, and $\dfrac{11\pi}{4}$. So $\theta = \dfrac{\pi}{4}, \dfrac{7\pi}{12}$, and $\dfrac{11\pi}{12}$. The slopes are $\tan\dfrac{\pi}{4} = 1$, $\tan\dfrac{7\pi}{12} \approx -3.73$, and $\tan\dfrac{11\pi}{12} \approx -0.27$.

12.3.25 Note that the curve is at the origin at $\pi/2$, so there is vertical tangent at $(0, \pi/2)$. Also, $\dfrac{dy}{dx} = \dfrac{-4\sin^2\theta + 4\cos^2\theta}{-8\sin\theta\cos\theta} = \dfrac{1 - 2\sin^2\theta}{\sin(2\theta)}$. Thus, there are horizontal tangents at $\pi/4$ and $3\pi/4$ (at the polar points $(2\sqrt{2}, \pi/4)$ and $(-2\sqrt{2}, 3\pi/4)$). There is also a vertical tangent where $\theta = 0$, at the point $(4, 0)$.

12.3.27 Using the double angle identities somewhat liberally:

$$\dfrac{dy}{dx} = \dfrac{2\cos(2\theta)\sin\theta + \sin(2\theta)\cos\theta}{2\cos(2\theta)\cos\theta - \sin(2\theta)\sin\theta} = \dfrac{\sin\theta(\cos 2\theta + \cos^2\theta)}{\cos\theta(\cos(2\theta) - \sin^2\theta)} = \dfrac{\sin\theta(3\cos^2\theta - 1)}{\cos\theta(1 - 3\sin^2\theta)} = \dfrac{\sin\theta(3\cos^2\theta - 1)}{\cos\theta(3\cos^2\theta - 2)}.$$

The numerator is 0 for $\theta = 0$ and for $\theta = \pm\cos^{-1}(\pm\sqrt{3}/3)$, so there are horizontal tangents at the corresponding points $(0, 0)$, $(0.943, 0.955)$, $(-0.943, 2.186)$, $(0.943, 4.097)$, and $(-0.943, 5.328)$. The denominator is 0 for $\theta = \pi/2$ and $3\pi/2$, and for $\theta = \pm\cos^{-1}(\pm\sqrt{6}/3)$, so there are vertical tangents at $(0, 0)$, $(0.943, 0.615)$, $(-0.943, 2.526)$, $(0.943, 3.757)$, and $(-0.943, 5.668)$.

12.3.29 Suppose $2\cos\theta = 1 + \cos\theta$. Then $\cos\theta = 1$, so this occurs for $\theta = 0$ and $\theta = 2\pi$. At those values, $r = 2$, so the curves intersect at the polar point $(2, 0)$. The curves also intersect when $r = 0$, which occurs for $\theta = \pi/2$ and $\theta = 3\pi/2$ for the first curve and $\theta = \pi$ for the second.

12.3.31 $2\sin 2\theta = 1$ implies that $\sin 2\theta = \dfrac{1}{2}$, so $2\theta = \dfrac{\pi}{6}, \dfrac{5\pi}{6}, \dfrac{13\pi}{6}$, and $\dfrac{17\pi}{6}$, and therefore $\theta = \dfrac{\pi}{12}, \dfrac{5\pi}{12}, \dfrac{13\pi}{12}$, and $\dfrac{17\pi}{12}$. Using symmetry, we also see that they intersect at $\theta = \dfrac{7\pi}{12}, \dfrac{11\pi}{12}, \dfrac{19\pi}{12}$, and $\dfrac{23\pi}{12}$. So the curves intersect at the points

$$\left(1, \dfrac{\pi}{12}\right), \left(1, \dfrac{5\pi}{12}\right), \left(1, \dfrac{7\pi}{12}\right), \left(1, \dfrac{11\pi}{12}\right), \left(1, \dfrac{13\pi}{12}\right), \left(1, \dfrac{17\pi}{12}\right), \left(1, \dfrac{19\pi}{12}\right), \left(1, \dfrac{23\pi}{12}\right).$$

12.3.33 $A = 2 \cdot \dfrac{1}{2}\displaystyle\int_0^{\pi/2} \cos\theta\, d\theta = \sin\theta\Big|_0^{\pi/2} = 1.$

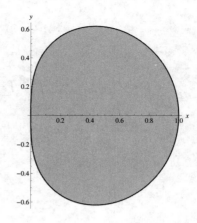

12.3.35 $A = \dfrac{1}{2}\displaystyle\int_0^{\pi}(8\sin\theta)^2\, d\theta = 32\int_0^{\pi}\sin^2\theta\, d\theta = 32\int_0^{\pi}\dfrac{1-\cos 2\theta}{2}\, d\theta = 32\left(\dfrac{1}{2}\theta - \dfrac{\sin\theta\cos\theta}{2}\right)\Big|_0^{\pi} = 16\pi.$

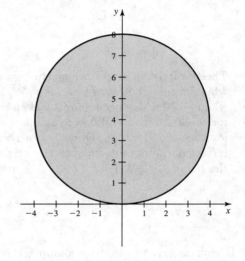

12.3.37 Using symmetry, we have $\dfrac{1}{2}\cdot 2\displaystyle\int_0^{\pi}(2+\cos\theta)^2\, d\theta = \int_0^{\pi}(4 + 4\cos\theta + \cos^2\theta)\, d\theta = \left(4\theta + 4\sin\theta + \dfrac{1}{2}\theta + \dfrac{\sin\theta\cos\theta}{2}\right)\Big|_0^{\pi} = \dfrac{9\pi}{2}.$

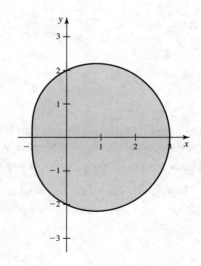

12.3.39
$$2 \cdot \frac{1}{2} \int_0^{\pi/6} \cos^2 3\theta \, d\theta = \int_0^{\pi/6} \frac{1 + \cos 6\theta}{2} \, d\theta =$$
$$\left(\theta/2 + \frac{\sin 6\theta}{6} \right) \Big|_0^{\pi/6} = \frac{\pi}{12}.$$

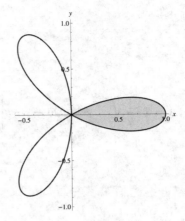

12.3.41

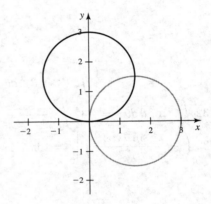

a. These curves intersect when $\sin \theta = \cos \theta$, which occurs at $\theta = \pi/4$ and $\theta = 5\pi/4$, and when $r = 0$ which occurs for $\theta = 0$ and $\theta = \pi$ for the first curve and $\theta = \pi/2$ and $\theta = 3\pi/2$ for the second curve. Only two of these intersection points are unique: the origin and the point $(3\sqrt{2}/2, \pi/4) = (-3\sqrt{2}/2, 5\pi/4)$.

b. By symmetry, we need to compute the area inside $r = 3\sin\theta$ between 0 and $\pi/4$ and then double that result. We have $2 \cdot \frac{1}{2} \int_0^{\pi/4} 9\sin^2\theta \, d\theta = \frac{9}{2} \int_0^{\pi/4} (1 - \cos(2\theta)) \, d\theta = \frac{9}{2} \left(\theta - (1/2)\sin(2\theta) \right) \Big|_0^{\pi/4} = \frac{9}{2} \left(\frac{\pi}{4} - \frac{1}{2} \right) = \frac{9}{8}(\pi - 2)$.

12.3.43

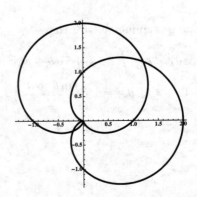

a. The curves intersect when $\sin\theta = \cos\theta$, which occurs for $\theta = \pi/4$ and $\theta = 5\pi/4$. The corresponding points are $\left(\frac{2+\sqrt{2}}{2}, \frac{\pi}{4} \right)$ and $\left(\frac{2-\sqrt{2}}{2}, \frac{5\pi}{4} \right)$. They also intersect at the pole: the first curve is at the pole at $(0, \pi)$ and the other at $(0, 3\pi/2)$.

12.3. Calculus in Polar Coordinates

b. By symmetry, we can compute the area between $\pi/4$ and $5\pi/4$ inside $r = 1 + \cos\theta$ and then double it. This will include both the bigger and smaller enclosed regions. We have $2 \cdot \dfrac{1}{2}\displaystyle\int_{\pi/4}^{5\pi/4} (1 + \cos\theta)^2\,d\theta = \displaystyle\int_{\pi/4}^{5\pi/4}(1 + 2\cos\theta + (1/2)(1+\cos(2\theta)))\,d\theta = \displaystyle\int_{\pi/4}^{5\pi/4}((3/2) + 2\cos\theta + (1/2)(\cos 2\theta))\,d\theta = (3\theta/2 + 2\sin\theta + (1/4)\sin 2\theta)\Big|_{\pi/4}^{5\pi/4}$

$= \left(\dfrac{15\pi}{8} - \sqrt{2} + \dfrac{1}{4}\right) - \left(\dfrac{3\pi}{8} + \sqrt{2} + \dfrac{1}{4}\right) = \dfrac{3\pi}{2} - 2\sqrt{2}.$

12.3.45 The area is given by $2 \cdot \dfrac{1}{2}\displaystyle\int_0^{\pi/3}(\cos^2\theta - (1/2)^2)\,d\theta = \displaystyle\int_0^{\pi/3}(\cos(2\theta)/2 + \dfrac{1}{4})\,d\theta = (\sin(2\theta)/4 + \theta/4)\Big|_0^{\pi/3} = \dfrac{\sqrt{3}}{8} + \dfrac{\pi}{12}.$

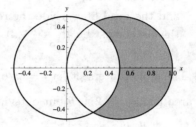

12.3.47 The region inside the circle between 0 and $\pi/3$ is $1/6$ the area of a circle of radius $1/\sqrt{2}$ so it has area $(1/6)\pi(1/2) = \pi/12$. The rest of the area is represented by $\dfrac{1}{2}\displaystyle\int_{\pi/3}^{\pi}\cos\theta\,d\theta = \dfrac{1}{2}(\sin\theta)\Big|_{\pi/3}^{\pi} = \dfrac{1}{2}\left(1 - \sqrt{3}/2\right).$ The total area is therefore $\dfrac{\pi}{12} + \dfrac{1}{2} - \dfrac{\sqrt{3}}{4}.$

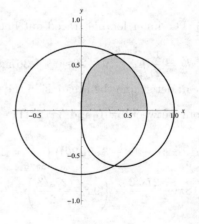

12.3.49 Using symmetry, we compute the area of $1/2$ of one leaf, and then double it. We have $A = \dfrac{1}{2}\displaystyle\int_0^{\pi/10}\cos^2(5\theta)\,d\theta = \dfrac{1}{10}\displaystyle\int_0^{\pi/2}\cos^2 u\,du = \dfrac{1}{10}\left(\dfrac{1}{2}u + \dfrac{\cos u\sin u}{2}\right)\Big|_0^{\pi/2} = \dfrac{\pi}{40}.$ So the area of one leaf is $2 \cdot \dfrac{\pi}{40} = \dfrac{\pi}{20}.$

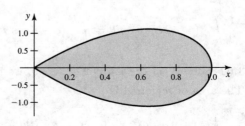

12.3.51 Note that the area inside one leaf of the rose but outside the circle is given by $\frac{1}{2}\int_{\pi/12}^{5\pi/12} (16\sin^2(2\theta) - 4)\,d\theta = (2\theta - \sin(4\theta))\Big|_{\pi/12}^{5\pi/2} = \sqrt{3} + \frac{2\pi}{3}$.
Also, the area inside one leaf of the rose is $\frac{1}{2}\int_0^{\pi/2} 16\sin^2(2\theta)\,d\theta = (4\theta - \sin(4\theta))\Big|_0^{\pi/2} = 2\pi$. Thus the area inside one leaf of the rose and inside the circle must be $2\pi - (\sqrt{3} + \frac{2\pi}{3}) = \frac{4\pi}{3} - \sqrt{3}$, and the total area inside the rose and inside the circle must be $4(\frac{4\pi}{3} - \sqrt{3}) = \frac{16\pi}{3} - 4\sqrt{3}$.

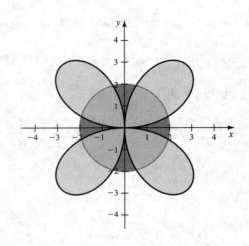

12.3.53 The circles intersect for $\theta = \pi/6$ and $\theta = 5\pi/6$.
The area inside $r = 2\sin\theta$ but outside of $r = 1$ would be given by $\frac{1}{2}\int_{\pi/6}^{5\pi/6} (4\sin^2\theta - 1)\,d\theta = \frac{1}{2}(x - \sin(2x))\Big|_{\pi/6}^{5\pi/6} = \frac{\pi}{3} + \frac{\sqrt{3}}{2}$. The total area of $r = 2\sin\theta$ is π. Thus, the area inside both circles is $\pi - (\frac{\pi}{3} + \frac{\sqrt{3}}{2}) = \frac{2\pi}{3} - \frac{\sqrt{3}}{2}$.

12.3.55 The inner loop is traced out between $\theta = \pi/6$ and $\theta = 5\pi/6$, so its area is given by $\frac{1}{2}\int_{\pi/6}^{5\pi/6}(3 - 6\sin\theta)^2\,d\theta = \frac{1}{2}\int_{\pi/6}^{5\pi/6}(9 - 36\sin\theta + 36\sin^2\theta)\,d\theta = \frac{3}{2}(3\theta + 12\cos\theta + 6\theta - 3\sin(2\theta))\Big|_{\pi/6}^{5\pi/6} = 9\pi - \frac{27\sqrt{3}}{2}$.
We can determine the area inside the outer loop by using symmetry and doubling the area of the region traced out between $5\pi/6$ and $3\pi/2$. Thus the area inside the outer region is $2 \cdot \frac{1}{2}\int_{5\pi/6}^{3\pi/2}(3 - 6\sin\theta)^2\,d\theta = 3(3\theta + 12\cos\theta + 6\theta - 3\sin(2\theta))\Big|_{5\pi/6}^{3\pi/2} = 18\pi + \frac{27\sqrt{3}}{2}$. So the area outside the inner loop and inside the outer loop is $18\pi + \frac{27\sqrt{3}}{2} - (9\pi - \frac{27\sqrt{3}}{2}) = 9\pi + 27\sqrt{3}$.

12.3.57 One half of the area is given by $\frac{1}{2}\int_0^{\pi/2} 6\sin 2\theta\,d\theta = -\frac{3}{2}\cos 2\theta\Big|_0^{\pi/2} = 3$, so the total area is 6.

12.3.59 The area is given by
$\frac{1}{2}\int_0^{2\pi}(4 - 2\cos\theta)^2\,d\theta = \int_0^{2\pi}(8 - 8\cos\theta + 2\cos^2\theta)\,d\theta = (8\theta - 8\sin\theta + \theta + \frac{1}{2}\sin(2\theta))\Big|_0^{2\pi} = 18\pi$.

12.3.61 Solving $4 - 2\cos\theta = 2 + 2\cos\theta$, we have $\cos\theta = \frac{1}{2}$, so $\theta = \pm\frac{\pi}{3}$.
The area of region A is
$\int_0^{\pi/3} \frac{1}{2}((2 + 2\cos\theta)^2 - (4 - 2\cos\theta)^2)\,d\theta = \frac{1}{2}\int_0^{\pi/3}(24\cos t - 12)\,dt = \frac{1}{2}(24\sin t - 12t)\Big|_0^{\pi/3} = 6\sqrt{3} - 2\pi$.

12.3. Calculus in Polar Coordinates

The area of region B is

$$\int_0^{\pi/3} \frac{1}{2}(4-2\cos\theta)^2 \, d\theta + \int_{\pi/3}^{\pi} \frac{1}{2}(2+2\cos\theta)^2 \, d\theta$$

$$= \int_0^{\pi/3} (8 - 8\cos\theta + 2\cos^2\theta) \, d\theta + \int_{\pi/3}^{\pi} (2 + 4\cos\theta + 2\cos^2\theta) \, d\theta$$

$$= \int_0^{\pi/3} (9 - 8\cos\theta + \cos 2\theta) \, d\theta + \int_{\pi/3}^{\pi} (3 + 4\cos\theta + \cos 2\theta) \, d\theta$$

$$= \left(9\theta - 8\sin\theta + \frac{1}{2}\sin 2\theta\right)\Big|_0^{\pi/3} + \left(3\theta + 4\sin\theta + \frac{1}{2}\sin 2\theta\right)\Big|_{\pi/3}^{\pi}$$

$$= 3\pi - \frac{15\sqrt{3}}{4} + 2\pi - \frac{9\sqrt{3}}{4}$$

$$= 5\pi - 6\sqrt{3}.$$

The area of region C is

$$\int_{\pi/3}^{\pi} \frac{1}{2}\left((4-2\cos\theta)^2 - (2+2\cos\theta)^2\right) d\theta = \int_{\pi/3}^{\pi} (6 - 12\cos\theta) \, d\theta = (6\theta - 12\sin\theta)\Big|_{\pi/3}^{\pi} = 4\pi + 6\sqrt{3}.$$

12.3.63 Note that the diameter of the circle is a, and that the complete circle is traversed for $0 \le t \le \pi$.
$$L = \int_0^{\pi} \sqrt{(a\sin\theta)^2 + (a\cos\theta)^2} \, d\theta = \int_0^{\pi} a \, d\theta = \pi a.$$

12.3.65 $L = \int_0^{2\pi} \sqrt{\theta^4 + 4\theta^2} \, d\theta = \int_0^{2\pi} \theta\sqrt{\theta^2 + 4} \, d\theta$. Let $u = \theta^2 + 4$, so that $du = 2\theta \, d\theta$. Substituting gives
$$\frac{1}{2}\int_4^{4\pi^2+4} \sqrt{u} \, du = \frac{1}{2} \cdot \frac{2}{3}\left(u^{3/2}\right)\Big|_4^{4\pi^2+4} = \frac{1}{3}\left(8(\pi^2+1)^{3/2} - 8\right) = \frac{8}{3}\left((\pi^2+1)^{3/2} - 1\right).$$

12.3.67

$$L = 2\int_{-\pi/2}^{\pi/2} \sqrt{(4+4\sin\theta)^2 + (4\cos\theta)^2} \, d\theta = 8\int_{-\pi/2}^{\pi/2} \sqrt{2+2\sin\theta} \, d\theta = 8\sqrt{2}\int_{-\pi/2}^{\pi/2} \sqrt{1+\sin\theta} \cdot \frac{\sqrt{1-\sin\theta}}{\sqrt{1-\sin\theta}} \, d\theta$$

$$= 8\sqrt{2}\int_{-\pi/2}^{\pi/2} \frac{\cos\theta}{\sqrt{1-\sin\theta}} \, d\theta.$$

Let $u = 1 - \sin\theta$, so that $du = -\cos\theta \, d\theta$. Then we have

$$L = 8\sqrt{2}\int_0^2 u^{-1/2} \, du = 16\sqrt{2}\left(\sqrt{2} - 0\right) = 32.$$

12.3.69 $L = \int_0^{\ln 8} \sqrt{4e^{4\theta} + 16e^{4\theta}} \, d\theta = 2\sqrt{5}\int_0^{\ln 8} e^{2\theta} \, d\theta = \sqrt{5} \cdot e^{2\theta}\Big|_0^{\ln 8} = \sqrt{5}(64-1) = 63\sqrt{5}.$

12.3.71 $L = \int_0^{\pi/2} \sqrt{\sin^6(\theta/3) + \sin^4(\theta/3)\cos^2(\theta/3)} \, d\theta = \int_0^{\pi/2} \sin^2(\theta/3) \, d\theta = \frac{1}{2}\int_0^{\pi/2} 1 - \cos(2\theta/3) \, d\theta =$
$\frac{1}{2}\left(\theta - \frac{3}{2}\sin(2\theta/3)\right)\Big|_0^{\pi/2} = \frac{1}{2}\left(\frac{\pi}{2} - \frac{3\sqrt{3}}{4}\right) = \frac{2\pi - 3\sqrt{3}}{8}.$

12.3.73 $L = \int_0^{2\pi} \sqrt{(4-2\cos\theta)^2 + 4\sin^2\theta} \, d\theta = \int_0^{2\pi} \sqrt{20 - 16\cos\theta} \, d\theta = 26.73.$

12.3.75

a. False. The area is given by $\dfrac{1}{2}\int_\alpha^\beta f(\theta)^2\,d\theta$.

b. False. The slope is given by $\dfrac{dy}{dx}$, which can be computed using the formula
$$\frac{dy}{dx} = \frac{dy/d\theta}{dx/d\theta} = \frac{f'(\theta)\sin\theta + f(\theta)\cos\theta}{f'(\theta)\cos\theta - f(\theta)\sin\theta}.$$

c. True. For example, see number 23 above.

12.3.77 The first horizontal tangent line is at the origin. The next is at approximately $(4.0576, 2.0288)$, and the third at approximately $(9.8262, 4.9131)$. The first vertical tangent line is at approximately $(1.7206, 0.8603)$, the next is at about $(6.8512, 3.4256)$, and the next at approximately $(12.8746, 6.4373)$.

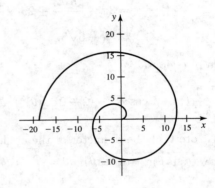

12.3.79 $r'(\theta) = -ae^{-a\theta}$. Thus,
$$L = \int_0^\infty \sqrt{e^{-2a\theta} + a^2 e^{-2a\theta}}\,d\theta = \sqrt{1+a^2}\int_0^\infty e^{-a\theta}\,d\theta$$
$$= \lim_{b\to\infty}\sqrt{1+a^2}\int_0^b e^{-a\theta}\,d\theta = \frac{\sqrt{1+a^2}}{-a}\lim_{b\to\infty} e^{-a\theta}\Big|_0^b$$
$$= \frac{\sqrt{1+a^2}}{-a}(0-1) = \frac{\sqrt{1+a^2}}{a}.$$

12.3.81

a. $A_n = \dfrac{1}{2}\int_{(2n-2)\pi}^{(2n-1)\pi} e^{-2\theta}\,d\theta - \dfrac{1}{2}\int_{2n\pi}^{(2n+1)\pi} e^{-2\theta}\,d\theta = -\dfrac{1}{4}e^{-(4n-2)\pi} + \dfrac{1}{4}e^{-(4n-4)\pi} + \dfrac{1}{4}e^{-(4n+2)\pi} - \dfrac{1}{4}e^{-4n\pi}.$

b. Each term tends to 0 as $n\to\infty$ so $\lim\limits_{n\to\infty} A_n = 0$.

c. $\dfrac{A_{n+1}}{A_n} = \dfrac{e^{-(4n+2)\pi} + e^{-(4n)\pi} + e^{-(4n+6)\pi} - e^{-(4n+4)\pi}}{e^{-(4n-2)\pi} + e^{-(4n-4)\pi} + e^{-(4n+2)\pi} - e^{-4n\pi}} = e^{-4\pi}$, so $\lim\limits_{n\to\infty}\dfrac{A_{n+1}}{A_n} = e^{-4\pi}$.

12.3.83

a. If $\cot\varphi = \dfrac{f'(\theta)}{f(\theta)}$ is constant for all θ, then $\varphi = \cot^{-1}\left(\dfrac{f'(\theta)}{f(\theta)}\right)$ is constant. Then $\dfrac{d}{d\theta}\ln(f(\theta)) = \dfrac{1}{f(\theta)}\cdot f'(\theta) = \cot\varphi$ is constant.

b. If $f(\theta) = Ce^{k\theta}$, then $\cot\varphi = \dfrac{f'(\theta)}{f(\theta)} = \dfrac{kCe^{k\theta}}{Ce^{k\theta}} = k$.

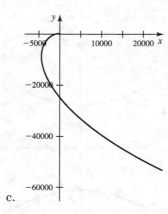
c.

12.3.85 Suppose that the goat is tethered at the origin, and that the center of the corral is $(1, \pi)$. The circle that the goat can graze is $r = a$, and the corral is given by $r = -2\cos\theta$. The intersection occurs for $\theta = \cos^{-1}(-a/2)$.

The area grazed by the goat is twice the area of the sector of the circle $r = a$ between $\cos^{-1}(-a/2)$ and π, plus twice the area of the circle $r = -2\cos\theta$ between $\pi/2$ and $\cos^{-1}(-a/2)$. Thus we need to compute
$$A = \int_{\cos^{-1}(-a/2)}^{\pi} a^2\, d\theta + \int_{\pi/2}^{\cos^{-1}(-a/2)} 4\cos^2\theta\, d\theta = a^2\pi - a^2\cos^{-1}(-a/2) + (2\cos\theta\sin\theta + 2\theta)\Big|_{\pi/2}^{\cos^{-1}(-a/2)} =$$
$a^2(\pi - \cos^{-1}(-a/2)) - \pi - \frac{1}{2}a\sqrt{4 - a^2} + 2\cos^{-1}(-a/2)$. Note that $\pi - \cos^{-1}(-a/2) = \cos^{-1}(a/2)$, so this can be written as $(a^2 - 2)\cos^{-1}(a/2) + \pi - \frac{1}{2}a\sqrt{4 - a^2}$. Note that for $a = 0$ this is 0, and for $a = 2$, this is π, as desired.

12.3.87 Again, suppose that the goat is tethered at the origin, and that the center of the corral is $(1, \pi)$. The equation of the corral fence is given by $r = -2\cos\theta$. Note that to the right of the vertical line $\theta = \pi/2$, the goat can graze a half-circle of area $\pi a^2/2$. Also, there is a region in the 2nd quadrant and one in the 3rd quadrant of equal size that can also be grazed. Let this region have area A, so that the total area grazed will then be $\dfrac{\pi a^2}{2} + 2A$.

Imagine that the goat is walking "west" from the polar point $(a, \pi/2)$, and is keeping the rope taut until his whole rope is along the fence in the third quadrant. Let φ be the central angle angle from the origin to the polar point $(1, \pi)$ to the point on the fence that the goat's rope is touching as he makes this walk. When the goat is at $(a, \pi/2)$, we have $\varphi = 0$. When the goat is all the way to the fence, we have $\varphi = a$. Then length of the rope not along the fence is $a - \varphi$. Thus, the value of A is $\dfrac{1}{2}\int_0^a (a-\varphi)^2 d\varphi = \dfrac{1}{2}\left(a^2\varphi - a\varphi^2 + \dfrac{\varphi^3}{3}\right)\Big|_0^a = \dfrac{a^3}{6}$.

Thus, the goat can graze a region of area $\dfrac{\pi a^2}{2} + \dfrac{a^3}{3}$.

12.4 Conic Sections

12.4.1 A parabola is the set of points in the plane which are equidistant from a given fixed point and a given fixed line.

12.4.3 A hyperbola is the set of points in the plane with the property that the difference of the distances from the point to two given fixed points is a given constant.

12.4.5

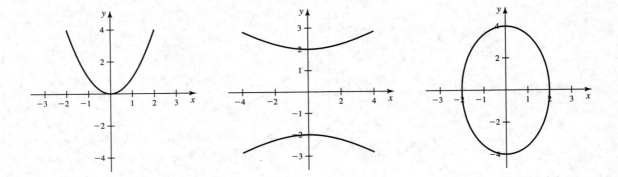

12.4.7 $\left(\dfrac{x}{a}\right)^2 + \dfrac{y^2}{a^2 - c^2} = 1.$

12.4.9 The foci for both are $(\pm ae, 0)$.

12.4.11 The asymptotes are $y = -\dfrac{b}{a} \cdot x$ and $y = \dfrac{b}{a} \cdot x$.

12.4.13 This is a parabola. Directrix: $y = -3$. Focus: $(0, 3)$.

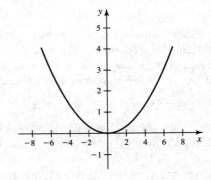

12.4.15 This is an ellipse. Vertices are $(\pm 2, 0)$, and the foci are $(\pm\sqrt{3}, 0)$. The major axis has length 4 and the minor axis has length 2.

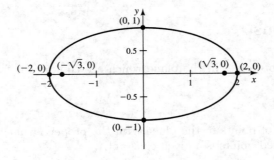

12.4.17 This is a parabola. Directrix: $x = 4$. Focus: $(-4, 0)$.

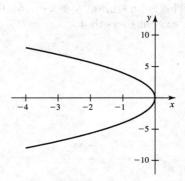

12.4.19 This is a hyperbola. The vertices are $(\pm 2, 0)$, and the foci are $(\pm\sqrt{5}, 0)$. The asymptotes are $y = \dfrac{\pm 1}{2} \cdot x$.

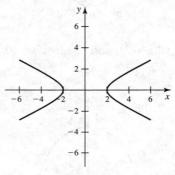

12.4.21 This is a hyperbola. The vertices are $(\pm 2, 0)$, and the foci are $(\pm 2\sqrt{5}, 0)$. The asymptotes are $y = \pm 2x$.

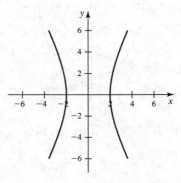

12.4.23 This is a parabola. Directrix: $y = \dfrac{2}{3}$. Focus: $\left(0, -\dfrac{2}{3}\right)$.

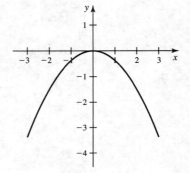

12.4.25 This is an ellipse. Vertices are $(0, \pm 4)$, and the foci are $(0, \pm 2\sqrt{3})$. The major axis has length 8 and the minor axis has length 4.

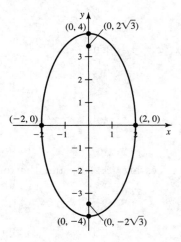

12.4.27 This is an ellipse. Vertices are $(0, \pm\sqrt{7})$, and the foci are $(0, \pm\sqrt{2})$. The major axis has length $2\sqrt{7}$ and the minor axis has length $2\sqrt{5}$.

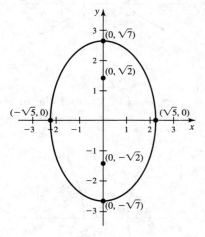

12.4.29 This is a hyperbola. The vertices are $(\pm\sqrt{3}, 0)$, and the foci are $(\pm 2\sqrt{2}, 0)$. The asymptotes are $y = \pm\sqrt{\dfrac{5}{3}} \cdot x$.

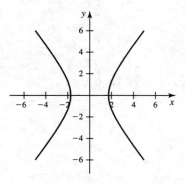

12.4. Conic Sections

12.4.31

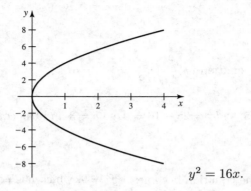
$y^2 = 16x$.

12.4.33

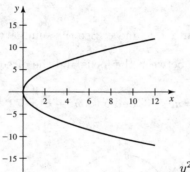
$y^2 = 12x$.

12.4.35 $x^2 = 4py$ and $4 = 4p(-6)$, so $p = -\frac{1}{6}$ and $x^2 = -\frac{2}{3} \cdot y$.

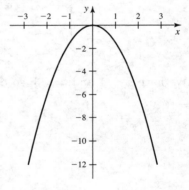

12.4.37 Because the vertex is $(-1, 0)$ and the parabola is symmetric about the x-axis, we have $y^2 = 4p(x+1)$ and because the directrix is one unit left of the vertex, we obtain $p = 1$ and $y^2 = 4(x+1)$.

12.4.39 $a = 4$, and $b = 3$, so the equation is $\dfrac{x^2}{16} + \dfrac{y^2}{9} = 1$.

12.4.41 We have $a = 4$ and $c = 6$, so $b^2 = c^2 - a^2 = 20$, so the equation is $\dfrac{x^2}{16} - \dfrac{y^2}{20} = 1$.

12.4.43 $a = 5$, and the equation is of the form $\dfrac{x^2}{25} + \dfrac{y^2}{b^2} = 1$. Because $\left(4, \dfrac{3}{5}\right)$ is on the curve, we have $\dfrac{16}{25} + \dfrac{9}{25b^2} = 1$, so $b = 1$. The equation is $\dfrac{x^2}{25} + y^2 = 1$.

Copyright © 2019 Pearson Education, Inc.

12.4.45 We have $a=2$, and because the asymptotes are $y = \dfrac{\pm bx}{a}$, we have that $b=3$, so the equation is $\dfrac{x^2}{4} - \dfrac{y^2}{9} = 1$.

12.4.47 $a=3$ and $b=2$, so the equation is $\dfrac{x^2}{4} + \dfrac{y^2}{9} = 1$.

12.4.49 We have $a=4$ and $c=5$, so $b^2 = 25 - 16 = 9$, so $b=3$ and the equation is $\dfrac{x^2}{16} - \dfrac{y^2}{9} = 1$.

12.4.51

a. True. Note that if $x=0$, the equation becomes $-y^2 = 9$, which has no solution.

b. True. The slopes of the tangent lines range continuously from $-\infty$ to 0 to ∞ and then back through 0 to $-\infty$ again.

c. True. Given c and d, one can compute a, b, and e. See the summary after Theorem 10.3.

d. True. The vertex is exactly halfway between the focus and the directrix.

12.4.53 We have $a=9$ and $e=\dfrac{1}{3}$, so $c=ae=3$, and $b^2 = a^2 - c^2 = 72$, so the equation is $\dfrac{x^2}{81} + \dfrac{y^2}{72} = 1$.

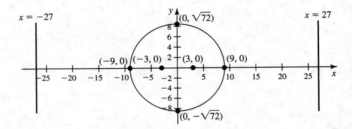

12.4.55 We have $a=1$ and $e=3$, so $c=ae=3$ and $b^2 = c^2 - a^2 = 9 - 1 = 8$. Thus, the equation is $x^2 - \dfrac{y^2}{8} = 1$.

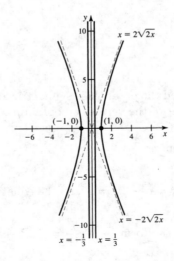

12.4.57 The vertex is $(2,0)$. The focus is $(0,0)$, and the directrix is the line $x=4$.

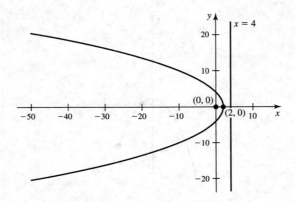

12.4.59 The vertices are $(1,0)$ and $(-1/3, 0)$. The center is $(1/3, 0)$. The directrices are $x = -1$ and $x = 5/3$.

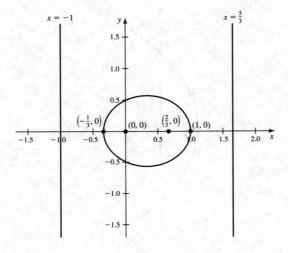

12.4.61 The vertex is $(0, -1/4)$, and the focus is $(0, 0)$. The directrix is the line $y = -\dfrac{1}{2}$.

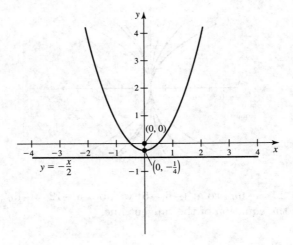

12.4.63 The parabola starts at $(1,0)$ and goes through quadrants I, II, and III for $\theta \in [0, 3\pi/2]$. It then approaches $(1,0)$ by traveling through quadrant IV for $\theta \in (3\pi/2, 2\pi)$.

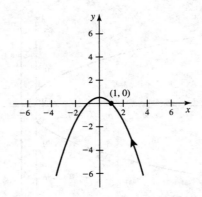

12.4.65 The parabola begins in the first quadrant and passes through the points $(0,3)$ and then $(-3/2, 0)$ and $(0-3)$ as θ ranges from 0 to 2π.

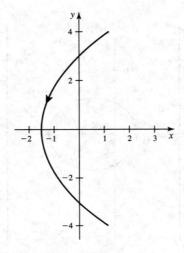

12.4.67 For negative p, the parabola opens to the left and for positive p it opens to the right. As p increases to 0, the parabola opens wider and as p decreases (for $p > 0$), it gets narrower.

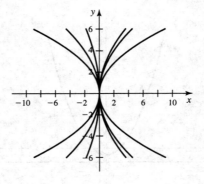

12.4.69 Differentiating gives $2x = -6y'$, so at $(-6, -6)$ we obtain $-12 = -6y'$, so $y' = 2$. Thus $y - (-6) = 2(x - (-6))$, or $y = 2x + 6$ is the equation of the tangent line.

12.4.71 Differentiating implicitly, we have $2yy' - \dfrac{x}{32} = 0$, so at $(6, -5/4)$ we have $-\dfrac{5}{2}y' - \dfrac{3}{16} = 0$, so $y' = -\dfrac{3}{40}$. The equation of the tangent line is $y + \dfrac{5}{4} = -\dfrac{3}{40}(x - 6)$, or $y = -\dfrac{3x}{40} - \dfrac{4}{5}$.

12.4. Conic Sections

12.4.73 Using implicit differentiation, we have $\frac{2x}{a^2} + \frac{2yy'}{b^2} = 0$, or $y' = -\frac{b^2x}{a^2y}$. If (x_0, y_0) is the point of tangency, then $-\frac{b^2 x_0}{a^2 y_0} = \frac{y - y_0}{x - x_0}$, so $\frac{x_0(x - x_0)}{a^2} = -\frac{y_0(y - y_0)}{b^2}$, so $\frac{x_0 x}{a^2} + \frac{y_0 y}{b^2} = \frac{x_0^2}{a^2} + \frac{y_0^2}{b^2} = 1$.

12.4.75 We have a hyperbola with focal point at the origin and directrix $y = -2$. Furthermore $P = (0, -4/3)$ is a vertex. Thus, $e = \frac{|PF|}{|PL|} = \frac{4/3}{2/3} = 2$, and $r(\theta) = \frac{2(2)}{1 - 2\sin\theta} = \frac{4}{1 - 2\sin\theta}$.

12.4.77 The points on the intersection of the two circles are a distance of $2a + r$ from F_1 and a distance of r from F_2. So for P an intersection point, we have $|PF_1| - |PF_2| = 2a$ for all r, and the set of all such points form a hyperbola with foci F_1 and F_2.

12.4.79 $V_x = \pi \int_{-a}^{a} \left(b^2 - \frac{b^2 x^2}{a^2}\right) dx = \pi b^2 \int_{-a}^{a} \left(1 - \frac{x^2}{a^2}\right) dx = \pi b^2 \left(x - \frac{x^3}{3a^2}\right)\bigg|_{-a}^{a} = \frac{4\pi b^2 a}{3}$.

$V_y = \pi \int_{-b}^{b} \left(a^2 - \frac{a^2 y^2}{b^2}\right) dy = \pi a^2 \int_{-b}^{b} \left(1 - \frac{y^2}{b^2}\right) dy = \pi a^2 \left(y - \frac{y^3}{3b^2}\right)\bigg|_{-b}^{b} = \frac{4\pi a^2 b}{3}$.

These are different if $a \neq b$. In the case $a = b$, both volumes give $\frac{4\pi a^3}{3}$, the volume of a sphere.

12.4.81

a.
$$V_x = \pi \int_a^c \left(\sqrt{\frac{b^2 x^2}{a^2} - b^2}\right)^2 dx = \pi \int_a^c \left(\frac{b^2 x^2}{a^2} - b^2\right) dx = \pi b^2 \left(\frac{x^3}{3a^2} - x\right)\bigg|_a^c$$
$$= \pi b^2 \left(\frac{c^3}{3a^2} - c - \frac{a}{3} + a\right) = \frac{\pi b^2}{3a^2}(c^3 - 3ca^2 + 2a^3) = \frac{\pi b^2}{3a^2}(a - c)^2(2a + c).$$

b. $V_y = 2 \cdot 2\pi \int_a^c a^2 b \sqrt{\frac{x^2}{a^2} - 1}\, dx = 2\pi \int_0^{b^2/a^2} a^2 b \sqrt{u}\, du = 2\pi a^2 b \left(\frac{2}{3} u^{3/2}\right)\bigg|_0^{b^2/a^2} = 2\pi a^2 b \frac{2b^3}{3a^3} = \frac{4\pi b^4}{3a}$.

12.4.83

a. The slope of a line making an angle θ with the horizontal is $\tan\theta$. The slope of the tangent line at (x_0, y_0) is $y' = \frac{x}{2p}$, so $y' = \frac{x_0}{2p}$, so $\tan\theta = \frac{x_0}{2p}$.

b. The distance from $(0, y_0)$ to $(0, p)$ is $p - y_0$, and $\tan\varphi = \frac{\text{opposite}}{\text{adjacent}} = \frac{p - y_0}{x_0}$.

c. Because l is perpendicular to $y = y_0$, we have $\alpha + \theta = \pi/2$, or $\alpha = \frac{\pi}{2} - \theta$, so $\tan\alpha = \cot\theta = \frac{2p}{x_0}$.

d. $\tan\beta = \tan(\theta + \varphi) = \dfrac{\dfrac{x_0}{wp} + \dfrac{p - y_0}{x_0}}{1 - \dfrac{p - y_0}{2p}} = \dfrac{x_0^2 + 2p^2 - 2py_0}{x_0(p + y_0)}$. Now because $x_0^2 = 4py_0$, we obtain $\tan\beta =$

$\dfrac{4py_0 + 2p^2 - 2py_0}{x_0(p + y_0)} = \dfrac{2p(p + y_0)}{x_0(p + y_0)} = \dfrac{2p}{x_0}$.

e. Because α and β are acute, we have that $\tan\alpha = \tan\beta$, so $\alpha = \beta$.

12.4.85 Assume the two fixed points are at $(c, 0)$ and $(-c, 0)$. Let P be the point $(0, b)$, and note that P is equidistant from the two given points, so we must have $b^2 + c^2 = a^2$ by the Pythagorean theorem. Now let $Q = (u, 0)$ be on the ellipse for $u > c$. Then $u - c + (c + u) = 2a$, so $u = a$. Now let $R = (x, y)$ be an arbitrary point on the ellipse (assume $x > 0$ and $y > 0$ – the other cases are similar.) Using the triangles formed between the foci, R, and the projection of R onto the x-axis, we have $\sqrt{(x + c)^2 + y^2} = 2a - \sqrt{(c - x)^2 + y^2}$. Squaring both sides gives $(x + c)^2 + y^2 = 4a^2 - 4a\sqrt{(c - x)^2 + y^2} + (c - x)^2 + y^2$. Isolating the root gives $\sqrt{(c - x)^2 + y^2} = \frac{1}{4a}\left((c - x)^2 + y^2 - (c + x)^2 - y^2 + 4a^2\right)$, so $\sqrt{(c - x)^2 + y^2} = a - \frac{c}{a}x$. Squaring again yields $(c - x)^2 + y^2 = a^2 - 2xc + \frac{c^2}{a^2}x^2$, so $c^2 - 2cx + x^2 + y^2 = a^2 - 2cx + \frac{c^2}{a^2}x^2$, or $x^2\left(1 - \frac{c^2}{a^2}\right) + y^2 = a^2 - c^2$. Thus $\frac{x^2}{a^2} + \frac{y^2}{a^2 - c^2} = 1$, which can be written $\frac{x^2}{a^2} + \frac{y^2}{b^2} = 1$, because $b^2 = a^2 - c^2$.

12.4.87 Let the parabola be symmetric about the y-axis with vertex at the origin. Let the circle have radius r and be centered at $(r + a, 0)$, and let the line be $y = -a$. The distance form the point $P(x, y)$ to the line is $u = y + a$. The distance from the point P to the circle is $v = \sqrt{x^2 + (r + a - y)^2} - r$. Setting $u = v$ yields $y + a = \sqrt{x^2 + (r + a - y)^2} - r$, so $y + r + a = \sqrt{x^2 + (r + a - y)^2}$, and squaring gives $y^2 + 2(r + a)y + (r + a)^2 = x^2 + (r + a - y)^2$, so $y^2 + 2(r + a)y + (r + a)^2 = x^2 + (r + a)^2 - 2(r + a)y + y^2$, and thus $4(r + a)y = x^2$, so $y = \frac{1}{4(r + a)}x^2$, the equation of a parabola.

12.4.89 Let the hyperbolas be centered at the origin with equations $\frac{x^2}{a^2} - \frac{y^2}{b^2} = 1$ and $\frac{y^2}{B^2} - \frac{x^2}{A^2} = 1$ and eccentricities $e = \frac{c}{a}$ and $E = \frac{C}{B}$, respectively. Because the hyperbolas share a set of asymptotes $A = ra$ and $B = rb$ fro some $r > 0$, and

$$C^2 = A^2 + B^2 = (ra)^2 + (rb)^2$$
$$= r^2(a^2 + b^2) = r^2 c^2.$$

Then we have

$$e^{-2} + E^{-2} = \left(\frac{c}{a}\right)^{-2} + \left(\frac{C}{B}\right)^{-2} = \frac{a^2}{c^2} + \frac{B^2}{C^2}$$
$$= \frac{a^2}{c^2} + \frac{r^2 b^2}{r^2 c^2} = \frac{a^2 + b^2}{c^2} = \frac{c^2}{c^2} = 1.$$

12.4.91 The latus rectum L intersects the parabola at $x = p$, $y = \pm 2p$. The distance between any point $P(x, y)$ on the parabola to the left of L and L is $p - x$. The distance from F to P is $\sqrt{(x - p)^2 + y^2} = \sqrt{x^2 - 2px + p^2 + 4px} = \sqrt{x^2 + 2px + p^2} = x + p$ (because both x and p are positive.) Thus $D + |FP| = p - x + x + p = 2p$.

12.4.93 Let P be a point on the intersection of the latus rectum and the ellipse. The length of the latus rectum is twice the distance from P to the focus. Let l be the length from P to the focus, and let L be the distance from P to the other focal point. Then $l + L = 2a$, so $L^2 = 4c^2 + l^2$, and thus $(2a - l)^2 = 4c^2 + l^2$, and solving for l yields $l = a - \frac{c^2}{a}$. Because $c^2 = a^2 - b^2$, this can be written as $l = a - \frac{a^2 - b^2}{a} = a - (a - \frac{b^2}{a}) = \frac{b^2}{a}$. The length of the latus rectum is therefore $\frac{2b^2}{a}$. Now because $e = \frac{c}{a}$, we have $\sqrt{1 - e^2} = \sqrt{1 - \frac{a^2 - b^2}{a^2}} = \sqrt{\frac{b^2}{a^2}} = \frac{b}{a}$. The length of the latus rectum can thus also be written as $2b \cdot \frac{b}{a} = 2b\sqrt{1 - e^2}$.

12.4.95 Let the equation of the ellipse be $\frac{x^2}{a^2} + \frac{y^2}{a^2 - c^2} = 1$ and let the equation of the hyperbola be $\frac{x^2}{r^2} - \frac{y^2}{c^2 - r^2} = 1$. Let (x_0, y_0) be a point of intersection. By evaluating both equations at the point of

12.4. Conic Sections

intersection and subtracting, we obtain the result

$$\frac{x_0^2}{a^2} - \frac{x_0^2}{r^2} + \frac{y_0^2}{a^2 - c^2} + \frac{y_0^2}{c^2 - r^2} = 0,$$

which can be written

$$\frac{r_0^2 x_0^2 - a^2 x_0^2}{a^2 r^2} + \frac{(c^2 - r^2) y_0^2 + (a^2 - c^2) y_0^2}{(a^2 - c^2)(c^2 - r^2)} = 0.$$

This equation can be rewritten in the form $\dfrac{x_0^2}{y_0^2} = \dfrac{a^2 r^2}{(a^2 - c^2)(c^2 - r^2)}$, which we will use later.

Now implicitly differentiating the equation for the ellipse yields $\dfrac{2x}{a^2} + \dfrac{2yy'}{a^2 - c^2} = 0$, and thus the slope of the tangent line to the ellipse at (x_0, y_0) is $y'_e = -\dfrac{x_0}{y_0} \cdot \dfrac{a^2 - c^2}{a^2}$. Differentiating the equation of the hyperbola gives $\dfrac{2x}{r^2} - \dfrac{2yy'}{c^2 - r^2} = 0$, so the slope of the tangent line to the hyperbola at the point of intersection is $y'_h = \dfrac{x_0}{y_0} \cdot \dfrac{c^2 - r^2}{r^2}$.

Now consider the product

$$-1 \cdot y'_e \cdot y'_h = \frac{x_0^2}{y_0^2} \cdot \frac{(a^2 - c^2)(c^2 - r^2)}{a^2 r^2}.$$

By the result of the first paragraph, this is equal to 1, and thus the two curves are perpendicular at the point of intersection.

12.4.97

a. The curve and the line intersect when $x^2 - m^2(x^2 - 4x + 4) - 1 = 0$, which occurs for $\dfrac{2m^2 \pm \sqrt{1 + 3m^2}}{m^2 - 1}$, assuming $m \neq \pm 1$. So there are two solutions in this case – but if $-1 < m < 1$, one of the solutions is negative (the intersection lies on the other branch of the hyperbola.) If $m^2 = 1$, then the equation becomes $4x - 5 = 0$, and there is only the solution $x = \dfrac{5}{4}$. So there are two intersection points on the right branch exactly for $|m| > 1$. We have $v(m) = \dfrac{2m^2 + \sqrt{1 + 3m^2}}{m^2 - 1}$ and $u(m) = \dfrac{2m^2 - \sqrt{1 + 3m^2}}{m^2 - 1}$.

b. $\displaystyle\lim_{m \to 1^+} u(m) = \lim_{m \to 1^+} u(m) \cdot \frac{2m^2 + \sqrt{1 + 3m^2}}{2m^2 + \sqrt{1 + 3m^2}} = \lim_{m \to 1^+} \frac{4m^4 - 3m^2 - 1}{(m^2 - 1)(2m^2 + \sqrt{1 + 3m^2})} =$
$\displaystyle\lim_{m \to 1^+} \frac{(m^2 - 1)(4m^2 + 1)}{(m^2 - 1)(2m^2 + \sqrt{1 + 3m^2})} = \frac{5}{4}.$

$\displaystyle\lim_{m \to 1^+} v(m) = \lim_{m \to 1^+} v(m) \cdot \frac{2m^2 - \sqrt{1 + 3m^2}}{2m^2 - \sqrt{1 + 3m^2}} = \lim_{m \to 1^+} \frac{4m^4 - 3m^2 - 1}{(m^2 - 1)(2m^2 - \sqrt{1 + 3m^2})} =$
$\displaystyle\lim_{m \to 1^+} \frac{(m^2 - 1)(4m^2 + 1)}{(m^2 - 1)(2m^2 - \sqrt{1 + 3m^2})} = \lim_{m \to 1^+} \frac{(4m^2 + 1)}{(2m^2 - \sqrt{1 + 3m^2})} = \infty.$

c. $\displaystyle\lim_{m \to \infty} u(m) = \lim_{m \to \infty} \frac{2 - \sqrt{\dfrac{1}{m^4} + \dfrac{3}{m^2}}}{1 - \dfrac{1}{m^2}} = 2.$

$\displaystyle\lim_{m \to \infty} v(m) = \lim_{m \to \infty} \frac{2 + \sqrt{\dfrac{1}{m^4} + \dfrac{3}{m^2}}}{1 - \dfrac{1}{m^2}} = 2.$

d. The expression $\displaystyle\lim_{m \to \infty} A(m)$ represents the area of the region bounded by the hyperbola and the line $x = 2$. It is given by $\displaystyle 2 \int_1^2 \sqrt{x^2 - 1}\, dx = 2 \left(\dfrac{x}{2} \sqrt{x^2 - 1} - \dfrac{1}{2} \ln(x + \sqrt{x^2 - 1}) \right) \bigg|_1^2 = 2\sqrt{3} - \ln(2 + \sqrt{3}).$

Chapter Twelve Review

1

a. False. For example, $x = r\cos t$, $y = r\sin t$ for $0 \le t \le 2\pi$ and $x = r\sin t$, $y = r\cos t$ for $0 \le t \le 2\pi$ generate the same circle.

b. False. Because $e^t > 0$ for all t, this only describes the portion of that line where $x > 0$.

c. True. They both describe the point whose cartesian coordinates are $(3\cos(-3\pi/4), 3\sin(-3\pi/4)) = (-3\cos(\pi/4), -3\sin(\pi/4)) = (-3/\sqrt{2}, -3/\sqrt{2})$.

d. False. The given integral counts the inner loop twice.

e. True. This follows because the equation $0 - x^2/4 = 1$ has no real solutions.

f. True. Note that the given equation can be written as $(x-1)^2 + 4y^2 = 4$, or $\dfrac{(x-1)^2}{4} + y^2 = 1$.

3 Note that $\left(\dfrac{x}{4}\right)^2 + \left(\dfrac{y}{3}\right)^2 = 1$. This represents an ellipse generated counterclockwise.

5 $x = y^2$, so $y = \sqrt{x}$. This is a segment of a parabola opening to the right, starting at $(4, 2)$ and ending at $(9, 3)$.

7

a. $(x-1)^2 + (y-2)^2 = 64\cos^2 t + 64\sin^2 t = 64$. This is the circle centered at $(1, 2)$ with radius 8.

b. $\left.\dfrac{dy}{dx}\right|_{t=\pi/3} = -\left.\dfrac{\cos t}{\sin t}\right|_{t=\pi/3} = -\dfrac{1/2}{\sqrt{3}/2} = -\dfrac{1}{\sqrt{3}}$.

c. Note that the point on the curve corresponding to $t = \pi/3$ has coordinates $(5, 2 + 4\sqrt{3}) \approx (5, 8.9)$.

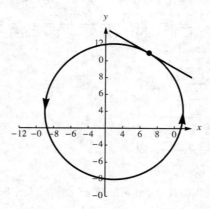

9 Let $y = t$, then $x = 5(t-1)(t-2)\sin t$.

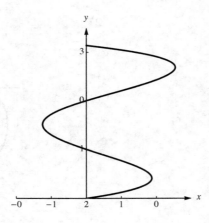

11

a. $x^2 + (y+1)^2 = 9\cos^2(-t) + 9\sin^2(-t) = 9$, so $x^2 + (y+1)^2 = 9$.

b. This is the lower half of a circle of radius 3 centered at $(0,-1)$ which starts at $(3,-1)$ and ends at $(-3,-1)$.

c. The slope at $(0,-4)$ is $\left.\dfrac{dy}{dx}\right|_{t=\pi/2} = -\left.\dfrac{3\cos(-t)}{3\sin(-t)}\right|_{t=\pi/2} = 0$.

13 $\dfrac{dy}{dx} = \dfrac{dy/dt}{dx/dt} = \dfrac{\sin t}{1-\cos t}$. At $t = \pi/6$, the slope of the tangent line is $\dfrac{1}{2-\sqrt{3}} = 2+\sqrt{3}$. So the equation of the tangent line is $y - (1-\sqrt{3}/2) = (2+\sqrt{3})(x - (\pi/6 - 1/2))$, or $y = (2+\sqrt{3})x + \left(2 - \dfrac{\pi}{3} - \dfrac{\pi\sqrt{3}}{6}\right)$.

At $t = 2\pi/3$, the slope of the tangent line is $\dfrac{\sqrt{3}}{3}$, so the equation of the tangent line is $y - \dfrac{3}{2} = \dfrac{\sqrt{3}}{3}\left(x - \left(\dfrac{2\pi}{3} - \dfrac{\sqrt{3}}{2}\right)\right)$, or $y = \dfrac{x}{\sqrt{3}} + 2 - \dfrac{2\pi}{3\sqrt{3}}$.

15 From P to Q, we use $(x(t), y(t)) = tQ + (1-t)P = (t,t) + (t-1, 0) = (2t-1, t)$. So $x(t) = 2t-1$, $y(t) = t$, $0 \le t \le 1$.

From Q to P, we use $(x(t), y(t)) = tP + (1-t)Q = (-t, 0) + (1-t, 1-t) = (1-2t, 1-t)$, for $0 \le t \le 1$. Thus $x(t) = 1-2t$, $y(t) = 1-t$, $0 \le t \le 1$.

17 $x = 3\sin t$, $y = 3\cos t$, $0 \le t \le 2\pi$.

19 The area is $\displaystyle\int_0^1 (t-t^3)(2t)\,dt = \int_0^1 (2t^2 - 2t^4)\,dt = \left.\left(\dfrac{2t^3}{3} - \dfrac{2t^5}{5}\right)\right|_0^1 = \dfrac{2}{3} - \dfrac{2}{5} = \dfrac{4}{15}$.

t

21 The volume is $2\displaystyle\int_0^{\pi/4} 2\pi \cos t\sqrt{\cos^2 t + 4\sin^2 2t}\,dt \approx 9.1$.

23 The arc length is

$$\int_0^{\pi/4} \sqrt{(-2\sin 2t)^2 + (2-2\cos 2t)^2}\,dt = 2\sqrt{2}\int_0^{\pi/4} \sqrt{1-\cos 2t}\,dt$$

$$= 2\sqrt{2}\int_0^{\pi/4} \sqrt{1-(1-2\sin^2 t)}\,dt = 4\int_0^{\pi/4}\sin t\,dt = \left.-4\cos t\right|_0^{\pi/4} = 4 - 2\sqrt{2}.$$

25

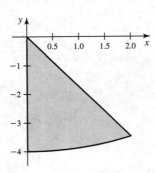

27

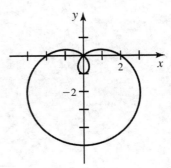

29

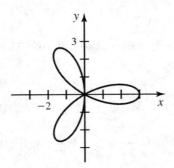

31

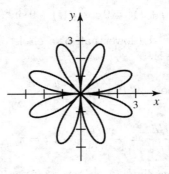

33 Liz should choose the cardioid, which is $r = 1 - \sin\theta$.

35 Letting $x = r\cos\theta$, $y = r\sin\theta$, and $r^2 = x^2 + y^2$, we have $x^2 + y^2 + 2y - 6x = 0$, which can be written as $x^2 - 6x + 9 + y^2 + 2y + 1 = 10$, or $(x-3)^2 + (y+1)^2 = 10$, so this is a circle of radius $\sqrt{10}$ centered at $(3,-1)$.

37 If we let $r = 1 + \cos t$, then $x = r\cos t$ and $y = r\sin t$. The curve $r = 1 + \cos t$ is a cardioid.

39 If $x = r\cos\theta$ and $y = r\sin\theta$, then $(r\cos\theta - 4)^2 + r^2\sin^2\theta = 16$, so $r^2\cos^2\theta - 8r\cos\theta + 16 + r^2\sin^2\theta = 16$, so $r^2 = 8r\cos\theta$, and thus $r = 8\cos\theta$. The complete circle can be described by $-\pi/2 \le \theta \le \pi/2$.

41

a. The curve intersects the origin at $\theta = \pi/2$, and there is a vertical tangent at $(0, \pi/2)$.

$$\frac{dy}{dx} = \frac{-\cos\theta\sin\theta + (1-\sin\theta)\cos\theta}{-\cos^2\theta - (1-\sin\theta)\sin\theta} = \frac{\cos\theta(1 - 2\sin\theta)}{\sin^2\theta - \cos^2\theta - \sin\theta} = \frac{\cos\theta(1 - 2\sin\theta)}{2\sin^2\theta - \sin\theta - 1}.$$

There are horizontal tangents when $\sin\theta = 1/2$, which occurs for $\theta = \pi/6, 5\pi/6$, and when $\cos\theta = 0$ (but not $\sin\theta = 1$) which occurs at $\theta = 3\pi/2$. So the horizontal tangents are at $(1/2, \pi/6)$, $((1/2, 5\pi/6)$, and $(2, 3\pi/2)$. There are vertical tangents when $2\sin^2\theta - \sin\theta - 1 = (2\sin\theta + 1)(\sin\theta - 1) = 0$, or $\theta = 7\pi/6$ and $\theta = 11\pi/6$. The vertical tangents are thus at $(3/2, 7\pi/6)$, $(3/2, 11\pi/6)$, and $(0, \pi/2)$, as well as the aforementioned $(0, \pi/2)$.

b. This is undefined.

c.

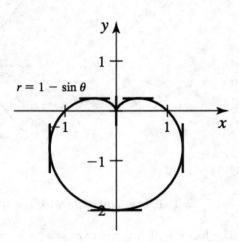

43 $\sqrt{\cos 3t} = \sqrt{\sin 3t}$, so for $t \neq 0$, $\tan 3t = 1$, so $3t = \frac{\pi}{4}, \frac{9\pi}{4}, \frac{17\pi}{4}$. So $t = \frac{\pi}{12}, \frac{9\pi}{12}, \frac{17\pi}{12}$. The curves intersect at $\left(\frac{\pi}{12}, \frac{1}{\sqrt[4]{2}}\right)$, $\left(\frac{3\pi}{4}, \frac{1}{\sqrt[4]{2}}\right)$, $\left(\frac{17\pi}{12}, \frac{1}{\sqrt[4]{2}}\right)$. Also, they intersect at $(0,0)$.

45 The area is

$$2\int_{7\pi/6}^{3\pi/2} \frac{1}{2}(1 + 2\sin t)^2 \, dt = \int_{7\pi/6}^{3\pi/2} (1 + 4\sin t + 4\sin^2 t) \, dt$$

$$= \int_{7\pi/6}^{3\pi/2} (1 + 4\sin t + 2 - 2\cos 2t) \, dt = (3t - 4\cos t - \sin 2t)\bigg|_{7\pi/4}^{3\pi/2}$$

$$= \left(\frac{9\pi}{2} - 0 - 0\right) - \left(\frac{7\pi}{2} + 2\sqrt{3} - \frac{\sqrt{3}}{2}\right) = \pi - \frac{3\sqrt{3}}{2}.$$

47 $2 = 4\cos 2\theta$, so $\cos 2\theta = \frac{1}{2}$, so $\theta = \pm\frac{\pi}{6}$. The area is $4\int_0^{\pi/6} \frac{1}{2}(4\cos 2\theta - 2) \, d\theta = 4\int_0^{\pi/6} (2\cos 2\theta - 1) \, d\theta =$
$4(\sin 2\theta - \theta)\bigg|_0^{\pi/6} = 4\left(\frac{\sqrt{3}}{2} - \frac{\pi}{6}\right) = 2\sqrt{3} - \frac{2\pi}{3}.$

49 The area is given by $2\int_0^{\pi/2} \frac{1}{2}\left((1 + \cos\theta)^2 - (1 - \cos\theta)^2\right) d\theta = 4\int_0^{\pi/2} \cos\theta \, d\theta = 4\sin\theta\bigg|_0^{\pi/2} = 4..$

51 $L = \int_0^{2\pi} \sqrt{(3 - 6\cos\theta)^2 + (6\sin\theta)^2} \, d\theta = \int_0^{2\pi} 3\sqrt{5 - 4\cos(\theta)} \, d\theta \approx 40.09.$

53

a. This represents a hyperbola with $a = 1$ and $b = \sqrt{2}$.

b. The vertices are $(\pm 1, 0)$, the foci are $(\pm c, 0)$ where $c^2 = a^2 + b^2 = 3$, so they are $(\pm\sqrt{3}, 0)$. The directrices are $x = \frac{\pm a^2}{c} = \frac{\pm 1}{\sqrt{3}}.$

c. The eccentricity is $e = \dfrac{c}{a} = \sqrt{3}$.

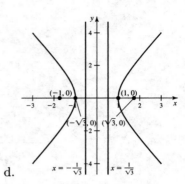

d.

55

a. This can be written as $\dfrac{y^2}{16} - \dfrac{x^2}{4} = 1$. It is a hyperbola with $a = 4$ and $b = 2$.

b. The vertices are $(0, \pm 4)$. The foci are $(0, \pm c)$ where $c^2 = a^2 + b^2 = 16 + 4 = 20$, so they are $(0, \pm\sqrt{20})$. The directrices are $y = \dfrac{\pm a^2}{c} = \dfrac{\pm 16}{\sqrt{20}} = \dfrac{\pm 8}{\sqrt{5}}$.

c. The eccentricity is $e = \dfrac{c}{a} = \dfrac{\sqrt{20}}{4} = \dfrac{\sqrt{5}}{2}$.

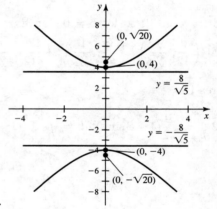

d.

57

a. This can be written as $\dfrac{x^2}{4} + \dfrac{y^2}{2} = 1$, so it is an ellipse with $a = 2$ and $b = \sqrt{2}$.

b. The vertices are $(\pm 2, 0)$. The foci are $(\pm c, 0)$ where $c^2 = a^2 - b^2 = 4 - 2 = 2$, so they are $(\pm\sqrt{2}, 0)$. The directrices are $x = \dfrac{\pm a^2}{c} = \dfrac{\pm 4}{\sqrt{2}} = \pm 2\sqrt{2}$.

c. The eccentricity is $e = \dfrac{c}{a} = \dfrac{\sqrt{2}}{2}$.

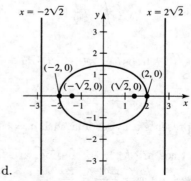

d.

59 $2y\dfrac{dy}{dx} = -12$, so at the point in question, $\dfrac{dy}{dx} = 3/2$. So the equation of the tangent line is $y + 4 = \dfrac{3}{2}\left(x + \dfrac{4}{3}\right)$, or $y = \dfrac{3}{2}x - 2$.

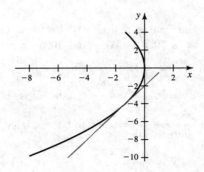

61 The eccentricity is 1, and the directrix is $y = 2$. The vertex is $(0, 1)$ and the focus is $(0, 0)$.

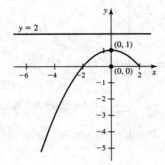

63 The eccentricity is $\dfrac{1}{2}$, and the directrices are $x = 4$ and $x = -\dfrac{20}{3}$. The vertices are $\left(\dfrac{4}{3}, 0\right)$ and $(-4, 0)$ and the foci are $(0, 0)$ and $\left(-\dfrac{8}{3}, 0\right)$.

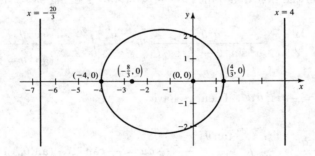

65 Because the center is halfway between the vertices, it is $(0, 0)$. We must have $a = 2$ and because $\dfrac{a^2}{c} = d = 1$, we have $c = 4$. So $b^2 = 16 - 4 = 12$. The hyperbola has equation $\dfrac{y^2}{4} - \dfrac{x^2}{12} = 1$. The eccentricity is $\dfrac{c}{a} = \dfrac{4}{2} = 2$.

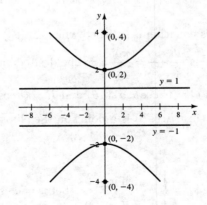

67 Because the center is halfway between the vertices, it is $(0,0)$. We must have $a = 4$ and because $\dfrac{a^2}{c} = d = 10$, we have $c = \dfrac{8}{5}$. So $b^2 = a^2 - c^2 = 16 - \dfrac{64}{25} = \dfrac{336}{25}$. The ellipse has equation $\dfrac{25x^2}{336} + \dfrac{y^2}{16} = 1$. The eccentricity is $\dfrac{c}{a} = \dfrac{8/5}{4} = \dfrac{2}{5}$.

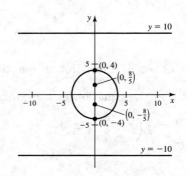

69 We have $a = 6$, $c = 4$ and $e = \dfrac{c}{a} = \dfrac{4}{6} = \dfrac{2}{3}$. Also, $b^2 = a^2 - c^2 = 36 - 16 = 20$, and the equation is $\dfrac{y^2}{36} + \dfrac{x^2}{20} = 1$. The vertices are $(\pm 2\sqrt{5}, 0)$. The directrices are $y = \pm \dfrac{a^2}{c} = \pm 9$.

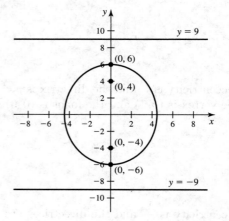

71 The area of the ellipse in the first quadrant is $\dfrac{\pi ab}{4}$, so we are seeking θ_0 so that

$$\dfrac{\pi ab}{8} = \dfrac{1}{2} \int_0^{\theta_0} \dfrac{a^2 b^2}{a^2 \sin^2 \theta + b^2 \cos^2 \theta} \, d\theta = \dfrac{a^2}{2} \int_0^{\theta_0} \dfrac{\sec^2 \theta}{\dfrac{a^2}{b^2} \tan^2 \theta + 1} \, d\theta.$$

Let $u = \dfrac{a}{b} \tan \theta$ so that $du = \dfrac{a}{b} \sec^2 \theta \, d\theta$. Then we have

$$\dfrac{\pi ab}{8} = \dfrac{ab}{2} \int_0^{\frac{a}{b} \tan \theta_0} \dfrac{1}{1 + u^2} \, du = \dfrac{ab}{2} \tan^{-1}\left(\dfrac{a}{b} \tan(\theta_0)\right).$$

Note that this equation is satisfied when $\tan(\theta_0) = \dfrac{b}{a}$, because then the expression on the right-hand side of that equation is $\dfrac{ab}{2} \cdot \dfrac{\pi}{4} = \dfrac{\pi ab}{8}$. So the desired value of m is $\tan(\theta_0) = \dfrac{b}{a}$.

73 Note that $Q = (a \cos \theta, a \sin \theta)$ and $R = (b \cos \theta, b \sin \theta)$, where θ is the angle formed by l and the x-axis. Then $P = (a \sin \theta, b \cos \theta)$ is a point on the ellipse $\dfrac{x^2}{a^2} + \dfrac{y^2}{b^2} = 1$, because it satisfies that equation.

75

a. Let $\operatorname{sgn}(x) = \begin{cases} 1 & \text{if } x \geq 0 \\ -1 & \text{if } x < 0. \end{cases}$. Let $x = a \cdot \operatorname{sgn}(\cos t) |\cos(t)|^{2/n}$ and $y = b \cdot \operatorname{sgn}(\sin(t)) |\sin(t)|^{2/n}$.

b.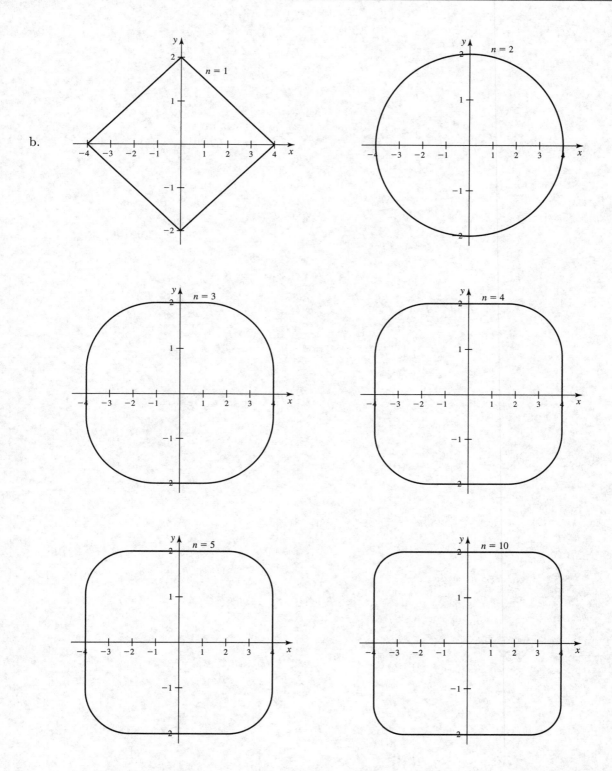

c. As n increases from near 0 to near 1, the curves change from star-shaped to a rectangular shape with corners at $(\pm a, 0)$ and $(0, \pm b)$. As n increases from 1 on, the curves become more rectangular with corners at $(\pm a, \pm b)$.